**STO**

**ACPL ITEM
DISCARDED**

# Safety in the agri-food chain

edited by:
P.A. Luning
F. Devlieghere
R. Verhé

Wageningen Academic
Publishers

**Subject headings:**
**Quality systems**
**Food hazards**
**Risk assessment**

**ISBN 9076998779**

**First published, 2006**

**Wageningen Academic Publishers**
**The Netherlands, 2006**

All rights reserved.
Nothing from this publication may be reproduced, stored in a computerised system or published in any form or in any manner, including electronic, mechanical, reprographic or photographic, without prior written permission from the publisher, Wageningen Academic Publishers,
P.O. Box 220, NL-6700 AE Wageningen, The Netherlands.

The individual contributions in this publication and any liabilities arising from them remain the responsibility of the authors.

The publisher is not responsible for possible damages, which could be a result of content derived from this publication.

# Acknowledgements

The publication of the book "Safety in the agri-food chain" is the result of activities by an extensive network of European institutions in the framework of the Socrates Programme.

During the period 1999-2004 European institutions of higher education in Vienna (Austria), Ghent (Belgium), Sofia, Plovdiv (Bulgaria), Copenhagen (Denmark), Helsinki (Finland), Rennes (France), Athens (Greece), Riga (Latvia), Kaunas (Lithouania), Wageningen (Netherlands), Olsztyn, Warsaw (Poland), Cluj-Napoca (Romania), Maribor (Slovenia), Burgos, Murcia, Valencia (Spain), and Uppsala (Sweden), have been involved in Socrates programmes dealing with curriculum development: "Total Food Quality Management" 27910-IC-3-98-1-BE-Erasmus-CDA7 (1999-2003) and "Safety in the Agri-Food Chain" 27910-IC-6-BE-Erasmus-EPS-1 (2001-2004) coordinated by Prof. Roland Verhé, Ghent University, Faculty of Bioscience Engineering.

Special thanks to Andreja Rajkovic (University of Ghent) for his support in lay-out editing of the chapters and Mieke Uyttendaele (University of Ghent) for specific editing of Chapter 11 (analyses of microbial hazards). Furthermore, detailed descriptions of microbial analyses, linked to Chapter 11 will be separately published in a monograph.

In the framework of these programmes, teaching programmes and curricula for MSc degrees in food quality management and safety in the agri-food chain have been developed and organised at various locations in Europe.

Teaching material was prepared by various experts and used as syllabi in the partner institutions. This teaching material consists of books, course syllabi, CD-roms and long distance educational material.

The book "Safety in the agri-food chain" involves an overview of all modules developed during these curricula activities and establishes a European dimension in the field of safety and quality of food products.

The coordinator and all the participants are extremely thankful for the extensive financial contribution of the EU Socrates Programme.

# Preface

Food safety has become a major issue in the agri-food chain in the last decade. The increased interest of agribusiness and the food industry in safety has been enhanced by the number of serious food incidences in various chains in Western Europe. Examples are Bovine Spongiform Encephalopathy (BSE) and classical swine fever (CSF) in 1997, the dioxin affair in 1999, foot and mouth disease (FMD) in 2001, the nitrophen and medroxyprogesterone acetate (MPA) incidents in 2002, and Avian Influenza in 2003. Moreover, the occurrence of foodborne diseases is still increasing. WHO estimated that, worldwide, thousands of millions of cases of foodborne disease occur every year resulting in human suffering and economic losses running into billions of US dollars.

Agri-food production chains are subject to a dynamic environment. Changes in food supply systems (such as increased complexity of supply chains, booming food service establishments), in demographic situations (e.g. more elderly people), in social situations, consumption behaviour and lifestyle (such as, increased outdoor consumption, raised consumption of imported exotic products, changes in food preparation habits), in environmental conditions (such as increased pollution, changes in climate and ecological systems), and changes in food production systems (such as mild processing techniques) complicate the production of safe foods. Moreover, more critical consumer attitudes towards food production, new European food legislation, and ethical discussions on wholesome production, put additional demands on the agri-food chain.

In anticipation of this increased concern for food safety, a large number of quality and safety systems, guidelines, and standards (such as Good Practice codes, hygiene codes, HACCP, BRC, EUREP-GAP, ISO 9000 family, ISO 22000, and SQF), decision-supporting systems, tools and techniques (e.g. predictive modelling, risk assessment, tracking and tracing), and policies (such as TQM) have been developed to assess, monitor, and/or assure safety levels of agri-food products.

To be able to manage and analyse the production of safe food requires a multidisciplinary approach using theories and concepts from areas such as food microbiology, chemistry, quality management, law, ethics, mathematics, consumer and management sciences. Detailed and accurate knowledge of the nature, level, and behaviour of food hazards, and how they can be monitored, are the basis for formulating strategies to reduce foodborne risks. Moreover, insight is required into how the different systems and tools (e.g. Q-systems, tracking and tracing systems, risk management systems, modelling techniques, etc) contribute to the realisation of food safety. Increased consumer awareness and governmental concerns about food safety require insight into consumer perception, legislative matters, and ethical issues on food production.

The objective of this book is to explain and discuss a wide range of issues that are relevant to food safety from different disciplinary viewpoints. The disciplinary backgrounds of the contributors range from food microbiology, chemistry, quality management, law, ethics, consumer science, and management studies.

The book consists of four distinct sections. The first section aims at obtaining insight into which factors in the agri-food production chain, and which mechanisms, contribute to the occurence of food-related hazards. Typical systems that support the realisation of safe food are discussed in section II, whereas in section III new and traditional analytical methods and measurements to assess and monitor food hazards are described in detail. The last section deals with the concerns of various stakeholders in the agri-food chain, such as consumers, government and society in general.

Section I comprises five chapters: the first provides a concise overview of the agri-food chain with typical factors influencing safety and common systems to manage it. Chapters 2 and 3, respectively, provide detailed information about the nature and behaviour of microbial and chemical hazards that cause foodborne-related risks. Chapters 4 and 5 give insight into which characteristic physical and other miscellaneous hazards may constitute a risk for human health.

Section II starts with a concise description of major quality systems and discusses the importance of recognising both the typical food technological issues and the managerial factors that influence the behaviour of people involved in food production (Chapter 6). In this chapter the contribution of the common QA systems to food safety assurance is also evaluated. Chapter 7 provides a detailed description of the development and application of an HACCP system. The principles of risk assessment and essential tools to manage food safety and assess their risks are described in detail in Chapter 8. A major technique to support risk assessment, i.e. modelling, is further elaborated on in the next chapter, and an overview is given of different modelling principles and applications (Chapter 9). The last chapter concerns tracking and tracing, that has become an important system for agribusiness and food companies in the management of safe food. Principles and supporting systems and tools in tracking and tracing are explained in Chapter 10.

Section III deals with the methods of analysis used to identify and monitor hazards in the agri-food chain that negatively contribute to human health. Both traditional and modern analytical and measurement techniques are described in detail. Chapters 11 and 12 concern the analysis of microbial and chemical hazards, respectively. In each chapter the assessment of typical microbial and chemical hazards are worked out in two case studies as illustration.

Finally, section IV considers the concerns of various stakeholders. Nowadays, government is very much concerned about regaining and maintaining consumer trust, and new food legislation has been developed in anticipation of raised food safety concerns. Chapter 13 gives an overview of European food safety law from the perspective of European Union (EU) law. The rationale for taking consumer behavioural issues into account in agri-food safety debates is provided in Chapter 14. Basic principles of consumer behaviour and a selection of topical case studies are described in this chapter. In conclusion, Chapter15 deals with ethical concerns about food safety in our society. This chapter will offer the readers the opportunity to gain some knowledge in an important field of applied ethics, and to define their position, their ethical implications and obligations in the food production and consumption process. It will allow the readers to reflect on specific topics and to form their own opinions on food ethics, since, in contrast to pure science, there are in many cases no straight answers.

The editors hope that the readers of the book will get a broad overview of the relevant principles, methods, issues and concerns in food safety.

*Pieternel Luning, Frank Devlieghere and Roland Verhé*

# Table of contents

**Acknowledgements** 7

**Preface** 9
*Pieternel Luning, Frank Devlieghere and Roland Verhé*

**1. Agri-food production chain** 19
*Susanne Knura, Stefanie Gymnich, Ewa Rembialkowska and Brigitte Petersen*
    1.1 Introduction 19
    1.2 Quality attributes of food in the production chain 22
    1.3 Overview of the Food Production Chain and factors affecting quality attributes 30
    1.4 Quality management activities with respect to safe food production chain 48
    1.5 Organic food production and safety 54
    References 61

**2. Biological hazards** 67
*Jordi Rovira, Avrelija Cencic, Eva Santos and Mogens Jakobsen*
    2.1 Study objectives and structure of chapter 67
    2.2 Introduction 68
    2.3 Types of biological hazards 76
    2.4 Principal means of food contamination (How microorganisms reach food) 120
    2.5 Microbial growth in food (how microorganisms survive and grow in food stuffs) 127
    References 136

**3. Chemical hazards** 145
*Bruno De Meulenaer*
    3.1 Introduction: food intoxications and food sensitivities 145
    3.2 Basic toxicological considerations 146
    3.3 Food sensitivities 149
    3.4 Food additives 156
    3.5 Residues 163
    3.6 Environmental contaminants 184
    3.7 Process contaminants 190
    3.8 Microbiological contaminants 198
    3.9 Endogenous toxicants 204
    3.10 Legislative information and further reading 207
    References 208

**4. Physical hazards in the agri-food chain** 209
*Anna Aladjadjiyan*
    4.1 Introduction 209
    4.2 Physical Contaminants: definition and classification 209

| | |
|---|---|
| 4.3 Non-radioactive physical contamination | 210 |
| 4.4 Radioactive contamination | 217 |
| References | 222 |

## 5. Miscellaneous hazards 223
*Avrelija Cenčič and Krzysztof Krygier (Illustrations by Anton Cenčič)*

| | |
|---|---|
| 5.1 Objectives | 223 |
| 5.2 Background | 223 |
| 5.3 Genetically modified organisms (GMOs) and associated foods | 224 |
| 5.4 Risk to human health | 231 |
| 5.5 Risk assessment | 238 |
| 5.6 Legislative measures | 239 |
| 5.7 Novel and functional foods | 241 |
| 5.8 Potential risks | 245 |
| 5.9 Legislative measures | 246 |
| References | 248 |

## 6. Quality assurance systems and food safety 249
*Pieternel Luning, Willem Marcelis and Marjolein van der Spiegel*

| | |
|---|---|
| 6.1 Introduction | 249 |
| 6.2 Technological and managerial aspects in quality assurance | 253 |
| 6.3 Principles of Quality Assurance systems | 268 |
| 6.4 Quality assurance systems and assuring food safety | 277 |
| 6.5 Total Quality Management | 289 |
| 6.6 Conclusion | 294 |
| References | 295 |
| Appendix I. Global overview of contents of GMP topics of parts I-III. | 298 |

## 7. Design and implementation of an HACCP system 303
*Isabel Escriche, Eva Doménech and Katleen Baert*

| | |
|---|---|
| 7.1 Introduction | 303 |
| 7.2 Preparation for management in the food industry | 309 |
| 7.3 Requirements before applying HACCP | 311 |
| 7.4 Development of an HACCP plan | 330 |
| 7.5 Implementation | 345 |
| 7.6 Verification and maintenance | 347 |
| 7.7 Future proposal | 350 |
| Acknowledgements | 350 |
| References | 351 |
| Definition of terms used | 352 |

## 8. Steps in the risk management process 355
*Adriane Mack, Thomas Schmitz, Gereon Schulze Althoff, Frank Devlieghere and Brigitte Petersen*

| | |
|---|---|
| 8.1 Introduction | 355 |

|  |  |
|---|---|
| 8.2 Risk assessment | 365 |
| 8.3 Risk management and food safety issues | 369 |
| 8.4 Risk communication and food safety issues | 373 |
| 8.5 Principles of enterprise risk management | 386 |
| References | 393 |

## 9. Modelling food safety — 397
*Frank Devlieghere, Kjell Francois, Bruno De Meulenaer and Katleen Baert*

|  |  |
|---|---|
| 9.1 Introduction | 397 |
| 9.2 Modelling microbial behaviour: predictive microbiology | 397 |
| 9.3 Modelling of migration from packaging materials | 417 |
| 9.4 Risk assessment: a quantitative approach | 428 |
| References | 435 |

## 10. Traceability in food supply chains — 439
*Jacques Trienekens and Jack van der Vorst*

|  |  |
|---|---|
| 10.1 Introduction | 439 |
| 10.2 Relationship between quality systems and traceability in food chains | 443 |
| 10.3 Traceability systems | 447 |
| 10.4 International benchmark study on traceability systems | 457 |
| 10.5 Technology for traceability systems | 463 |
| 10.6 Major research themes and questions for discussion | 467 |
| References | 469 |

## 11. Microbial analysis of food — 471
*Anna Maraz, Fulgencio Marin and Rita Cava (cases by Andreja Rajkovic)*

|  |  |
|---|---|
| 11.1 Introduction | 471 |
| 11.2 Objectives | 471 |
| 11.3 Detection, identification and typing of foodborne pathogens | 472 |
| 11.4 Case studies in detection and quantification/enumeration of pathogens and their toxins | 518 |
| References | 522 |

## 12. Analysis of chemical food safety — 525
*Carmen Socaciu*

|  |  |
|---|---|
| 12.1 Introduction | 525 |
| 12.2 Chemical risk evaluation via food analysis | 529 |
| 12.3 Some examples of chemical analysis protocols for the identification and quantification of chemical risk factors | 546 |
| Abbreviations | 556 |
| References | 556 |

## 13. Modern european food safety law — 559
*Bernd van der Meulen and Menno van der Velde*

| | |
|---|---:|
| 13.1 Introduction | 559 |
| 13.2 Food law: development, crisis and transition | 564 |
| 13.3 The General Food Law: general provisions of food law | 573 |
| 13.4 The General Food Law: institutional aspects | 584 |
| 13.5 The composition of food | 590 |
| 13.6 Food handling | 600 |
| 13.7 Informed choice: presentation of food products | 605 |
| 13.8 Globalisation | 611 |
| 13.9 Conclusion | 616 |

## 14. Consumer perception of safety in the agri-food chain — 619
*Wim Verbeke, Joachim Scholderer and Lynn Frewer*

| | |
|---|---:|
| 14.1 Introduction | 619 |
| 14.2 Consumer motivations for food choice | 620 |
| 14.3 Consumer decision-making and consumer behaviour | 620 |
| 14.4 Food safety facts versus consumer perception | 625 |
| 14.5 Food risk perception | 626 |
| 14.6 Selected cases in food safety perception | 629 |
| 14.7 Conclusions | 639 |
| References | 641 |

## 15. Ethics in food safety — 647
*Kriton Grigorakis*

| | |
|---|---:|
| 15.1 General introduction to ethics | 647 |
| 15.2 Principles of food ethics and safety | 652 |
| 15.3 Special issues in food ethics and safety | 666 |
| 15.4 Conclusive remarks | 674 |
| References | 675 |

## Concluding remarks — 679
*Frank Devlieghere, Pieternel Luning and Roland Verhé*

## Authors — 681

# 1. Agri-food production chain

*Susanne Knura, Stefanie Gymnich, Ewa Rembialkowska and Brigitte Petersen*

## 1.1 Introduction

The aim of this chapter is to provide an overview of major safety-related issues in the food production chain and to put forward different points of view on realising the quality and safety of food. The process-oriented instead of the former product-oriented strategy is discussed with respect to consumer orientation.

Agricultural and food markets have changed drastically in the last few decades. Perhaps the most fundamental change is the shift from production to market orientation. In order to establish a strong competitive position in their market, agribusiness and the food industry have to produce products which comply with the wants and needs of consumers. In this context consumer orientation, competitive strength and marketing efficiency are key words in agribusiness and the food industry, whereby effective control of both production and logistical processes are essential for the maintenance of product quality. Quality assurance and food safety are of paramount importance to all companies and organisations involved in the food production chain. Modern trading conditions and legislation require food businesses to demonstrate their commitment to food quality (Jongen and Meulenberg, 2005).

In this chapter changes in food production concepts are put in perspective. First, an overview of the historical development of chain concepts is given. Subsequently, factors within the agri-food chain are described, which influence food safety as well as other quality attributes, and strategies to manage them are described. Finally, there is a discussion of how these factors are currently monitored and controlled and how they should be monitored and controlled. Finally, the opportunities for organic food production in relation to food safety are presented.

### 1.1.1 Importance of the chain perspective of food production

The food industry is challenged with increasingly competitive and global markets in which process efficiency, product and process quality, consumers` trust, elimination of food hazards, tracking and tracing of food products throughout the production chain, animal welfare, and environmental protection (environmental management or eco-management) have become crucial elements for competitiveness and market success.

*Improvement of business processes*
To meet these challenges, the traditional focus on organisation and production management is not sufficient. This has spurred initiatives in industry and research to identify and utilise possibilities for management improvements, which could support the market position and profitability of enterprises in today's increasingly complex and difficult market environment (Schiefer, 1997).

*Susanne Knura, Stefanie Gymnich, Ewa Rembialkowska and Brigitte Petersen*

One major initiative, that has received much attention in industry, attempts to reach the objectives through a renewed focus on improvement of business processes as regards their organisation and control. Research suggested that this approach could have a major impact on business performance (Brocka and Brocka, 1992). It involves efforts to improve market orientation and efficiency of processes and to coordinate process improvements throughout production chains.

Food production, manufacturing, distribution and retailing are becoming a highly complex business. For example, raw materials are nowadays obtained from sources worldwide, an ever-increasing number of processing technologies are utilised, and a vaster array of products are available to consumers. Such complexity necessitates the development of comprehensive control procedures to ensure the production of safe and high quality food. In addition, consumer expectations are changing, with a desire for convenience, less-processed and fresher foods with more natural characteristics. Against this background the total food chain has to ensure that the highest standards of quality and safety are maintained. At all stages of the food chain, from acquisition of raw materials through manufacture, distribution and sale, whether it retail or catering outlets, consideration must be given to quality issues associated with specific products, processes and methods of handling. There are a number of reasons why, especially in agri-food business, the implementation of quality assurance systems is an issue of the greatest importance:

- Agricultural products are often perishable and subject to rapid decay due to physiological processes and microbiological contamination.
- Most agricultural products are harvested seasonally.
- Products are often heterogeneous with respect to desired quality parameters such as the content of important components, size and colour. This kind of variation is dependent on cultivar differences and variables, which cannot be controlled.
- Primary production of agricultural products is performed by a large number of farms operating on a small scale (Hoogland *et al.*, 1998).

*Food safety policy*
The guiding principle of EU-food safety policy is based on a comprehensive, integrated approach. This means throughout the food chain ('farm to table') across all food sectors to ensure a high level of human health and consumer protection. In this way, the 'farm to table' policy covering all sectors of the food chain, including feed production, primary production, food processing, storage, transport and retail sale, will be implemented systematically and in a consistent manner. A successful food policy demands the traceability of feed and food and their ingredients. Adequate procedures to facilitate such traceability must be introduced. This includes the obligation of feed and food businesses to ensure that adequate procedures are in place to withdraw feed and food from the market where there's a risk to consumer's health. Operators should also keep adequate records of suppliers of raw materials and ingredients so that the source of a problem can be identified. However, it must be emphasised that unambiguous tracing of feed and food and their ingredients is a complex issue and should take into account the specificity of different sectors and commodities.

## Agri-food production chain

The General Food Law (GFL) concerns (EC) No 178/2002, which aims at harmonising food safety related laws for the European Union. The focus of GFL is on all actors of the food production chain including animal feed processing units. It introduces the obligation for all players in the food sector to introduce traceability systems.

Roughly speaking, the food sector consists of fresh chains and industrial chains. The first are characterised by trading a fresh, unprocessed, agricultural product. The latter are characterised by food processing operations on an industrial scale. Without underestimating the complexity of traceability in fresh chains, the industrial chains pose the biggest challenge for implementing traceability, due to sequences of diverging and converging operations (Vernède, 2003).

These comprehensive, integrated approaches will lead to a more coherent, effective and dynamic food policy. The policy needs to address the shortcomings which result from the current rigid approach of the sector and which has limited its ability to deal rapidly and flexibly with risks to human health.

The policy needs to be kept under constant review and, where necessary, be adapted to respond to shortcomings, to deal with emerging risks and to recognise new developments in the production chain. At the same time, the development of this approach needs to be transparent, involving all the stakeholders and allowing them to make effective contributions to new developments (White paper on food safety, 2000).

### 1.1.2 Historical development of the chain concept

Besides the chain perspective there has also been an historical development of the chain concept. Because of the Dioxin- and BSE crisis in the 90's the food safety legislation and the resulting control system needed to be reorganized in a more coordinated and integrated way at EU level. Essential elements of this new concept are the integration of the entire food production chain including feed legislation and primary production (from stable to table), the traceability of feed and food and their ingredients, the precautionary principle, and risk analysis as a foundation of food safety policy. An independent scientific support will be guaranteed by establishing a European Food Safety Authority (Hartig and Untermann, 2003).

Traditionally, production chains are characterised by two distinct features:
1. The one-way communication from producers of raw materials (most often farmers) to the users of the end products (consumers).
2. No common understanding of the concept of quality.

Quality was, and in a number of cases still is, predominantly based on production costs and productivity (i.e. at the level of primary production: kg per hectare, and at industrial level: kg product per kg raw material). Nowadays, many actors in the production chain still have these two aspects of quality in common. However, each actor may also employ a number of additional quality determinants, such as homogeneity and storability of raw materials at industrial level and ease of handling at retail level. Such quality determinants are sometimes incompatible, whereby the quality determinants formulated by one actor in the chain may conflict with those used by another actor or even with those of consumers.

The traditional approach observed in the operation of production chains was also found in the organisation and implementation of scientific research. Research was organised by discipline and could not deal with the new challenges that arose from rapidly changing consumer preferences. In the past, agronomists focused their efforts predominantly on raising crop yield levels through improved cultivation practices and efficient use of inputs such as fertilisers and pesticides. Plant breeders were concerned with creating new high-yield varieties that were less susceptible to pests and disease. The combined efforts of breeders and agronomists resulted in products that were delivered to the industry for further processing. Industries assessed the quality of the raw material they received and consequently tried to make the most of it.

This traditional approach to research and production leaves the door wide open for many improvements in product quality, and, as such, is inadequate to respond to the changes that present market developments dictate. First of all, integration of the complete production chain is necessary, and consumers' demands must act as the driving force for the production chain (Jongen *et al.*, 1996).

## 1.2 Quality attributes of food in the production chain

Food production systems are developing continuously. Increasing demands are being put on both the products and the production process. Consumer expectations for food safety, animal welfare, and the environment as well as progress in gene technology and biotechnology are the main driving forces behind the change (website 1).

In this section we will firstly give a short overview of intrinsic and extrinsic factors that influence quality perception. Subsequently, emphasis will be put on specific concepts of food safety as a basic part of food quality.

Food quality is not a precise term. There are several definitions of the term quality. According to the definition in international standards, quality is a total sum of features, characteristics and properties of a product, which bear on its ability to satisfy stated or implied needs. Such a definition is clear and understandable, but it does not help very much in an individual situation or in the attitude of an individual person. Quality is a flexible term and is composed of many parameters or properties having different significance for the overall quality of a food product (Paulus, 1993).

Moreover, to control and to assure quality the critical factors and parameters which affect quality attributes in the agri-food production chain have to be known. They can be divided into intrinsic and extrinsic quality attributes. The sum of intrinsic and extrinsic factors as a combination of attributes that affect final quality perception determines the attractiveness of products for consumers. The combinations may differ for different products under different circumstances of use and/or consumption.

## 1.2.1 Intrinsic quality attributes

However, different classifications with respect to intrinsic and extrinsic quality attributes are to be found in the literature. The intrinsic factors refer to physical product characteristics such as taste, texture and shelf life. These intrinsic factors can be measured in an objective manner. In addition there are several other attributes like taste, nutritional value, freshness, safety, appearance and health. The combination of all these attributes determines the intrinsic end product quality.

Luning and co-authors (2002) considered as intrinsic attributes:
- safety and health aspects of a product;
- shelf life and sensory properties;
- convenience and product reliability.

They described the attributes in more detail as shown below:

*Product safety and health*
Health aspects refer to food composition and diet. For example, nutritional imbalance can have negative consequences on human health. Nowadays, the food industry anticipates these nutritional needs by the development of functional foods like low-fat and low-cholesterol products, but also vitamin or mineral enriched foods. These products are assumed to contribute positively to human health.

Product safety and food safety, respectively, refers to the requirement that products must be 'free' of hazards with an acceptable risk. Whereas hazard can be defined as a potential source of danger, risk can be described as a measure of the probability and severity of harm to human health. According to Shapiro and Mercier (1994) a food product can be considered safe if its risks are judged to be acceptable.

Food safety can be affected by different sources. As an example, four main sources are described below according to Luning and co-authors (2002):
- growth of pathogen micro-organisms;
- presence of toxic compounds;
- presence of foreign objects;
- and occurrence of calamities.

Pathogenic micro-organisms include bacteria as well as moulds. With respect to the two types of pathogen micro-organisms a distinction have been made between food infection and food poisoning. While food infection is caused by the presence of living pathogens (e.g. *Salmonella, Campylobacter jejuni* or *Listeria monocytogenes*) in food, food poisoning is caused by toxic compounds produced by pathogenic bacteria (enterotoxins) and moulds (mycotoxins) in the food.

Toxic compounds can originate from different sources in the agri-food production chain. Toxins can occur as natural compounds in raw materials (i.e. natural toxins), but they can also be formed during storage (i.e. aflatoxin) and processing (heterocyclic amines). Other sources of toxic compounds are contaminants from the environment (polychlorinated biphenyls -PCB's- and nitrate), residues from pesticides, veterinary drugs and disinfectants.

Foreign objects include physical objects like pieces of glass, wood, stones but also pests and insects. Calamities can be defined as all other (unexpected) phenomena that negatively affect food safety. A typical example is the disaster at the nuclear plant in Chernobyl which was unexpected, but had major consequences for food safety.

*Sensory properties and shelf life*

The sensory perception of food is determined by the overall sensation of taste, odour, colour, appearance, texture and sound (e.g. the sound of crispy chips). The physical features and chemical composition of a product determine these sensory properties. In general, agri-food products are perishable by nature. After the harvesting of fresh produce or the processing of foods, the deterioration processes starts, which negatively affects the sensory properties.

Processing and/or packaging are aimed at delaying, inhibiting or reducing the deterioration processes in order to extend the shelf life. For example, freshly harvested peas are spoiled within 12 hours, whereas canned peas can be kept for 2 years at room temperature.

The shelf life of a product can be defined as the time between harvesting or processing and packaging of the product and the point at which it becomes unacceptable for consumption. The unacceptability is usually reflected in decreased sensory properties, for example, formation of rotten odour or sour taste by bacteria spoilage. The actual shelf life of a product depends on the rate of the deterioration processes. Often one type of deterioration process is limiting for the shelf life; for example, cured ham can turn grey very quickly upon exposure to oxygen. Although the product is still safe, because bacteria did not spoil it, it will become unacceptable because of the grey colour (i.e. the shelf life limiting reaction). Below (Table 1.1) are listed some major shelf life limiting deteriorating processes typical for food products, such as microbiological, and/or (bio)

*Table 1.1. Overview of major shelf life limiting processes, typical sources and typical effects (short summary of Luning et al., 2002).*

| Process character | Typical sources | Typical effects |
| --- | --- | --- |
| Microbiological reactions | Food spoilage | Loss of texture, off-flavour, off-colour, pathogens |
| Chemical reactions | Processing and storage | Non-enzymatic: browning and oxidative reactions |
| Biochemical reactions | Cutting of fresh vegetables | Enzymatic: browning by phenolase and formation of off-flavours by lipoxygenase |
| Physical changes | Mishandling during harvesting, processing and distribution | Desiccation, swelling, phase changes |
| Physiological reactions | Post-harvest storage | Production of ethylene |

chemical and/or physiological and/or physical processes (Fennema and Tannenbaum, 1985; Singh, 1994; Jay, 1996). However, although the shelf life of a product is often limited by one major reaction, sometimes a typical quality defect may be due to different mechanisms (Ellis, 1994).

*Product reliability and convenience*
Product reliability refers to the compliance of actual product composition with product description. For example, the weight of the product must be correct within specified tolerances. But also claims, such as enriched with vitamin C, must be in agreement with actual concentration in the product after processing, packaging and storage. Deliberate modification of the product composition will cause damage to the product reliability, i.e. product falsification. One example is when (cheaper) alternative raw materials are used and not mentioned on the label. Product reliability is generally an implicit expectation; consumers just expect that a product is in compliance with the information mentioned on the packaging.

In addition to product reliability, time value is increasingly of interest (Leitzmann, 1993). Convenience relates to the ease of use or consumption of the product for the consumer and thus contributes to product quality. Product convenience can be accomplished by preparation, composition and packaging aspects. Convenience food has been defined as food offered to the consumer in such a manner that purchase, preparation and consumption of a meal costs less physical and mental effort and/or money than when the original and/or separate components are used.

The increased consumption of convenience food is assumed to be due to decreasing volume of households, decreasing appreciation for housekeeping, increasing labour participation of women and increasing welfare (Van Dam *et al.*, 1994). Convenience foods range from sliced and washed vegetables to complete ready-to-eat meals that only have to be warmed in the microwave or oven. Much attention is paid in the food industry to developing "ready-to-eat meals" that can be easily and quickly prepared while still having good sensory and nutritional properties. Furthermore, packaging concepts are designed more and more to fulfil the consumer's need for convenience; typical examples are easy-to-open and close, good pouring properties and extra light packaging (Luning *et al.*, 2002).

## 1.2.2 Extrinsic quality attributes

Extrinsic quality attributes are related to the way in which the food was produced such as the use of pesticides, the type of packaging material, a specific processing technology or the use of genetically modified organisms during the production of ingredients. Extrinsic quality attributes do not necessarily have a direct influence on physical product properties, but can influence consumers' quality perception.

Luning and co-authors (2002) considered as extrinsic attributes:
- production system characteristics;
- environmental aspects;
- marketing (communication).

Typical extrinsic attributes have been described by Luning *et al.* (2002) in more detail as shown below:

*Production system characteristics*
Production system characteristics refer to the way a food product is manufactured. It includes factors such as the use of pesticides while growing fruit and vegetables, animal welfare during breeding, use of genetic engineering to modify product properties or use of specific food preservation techniques. The influence of production systems characteristics on product acceptance is very complex. For example, there has been much concern about public acceptance of new genetically modified food products and genetic engineering in general. Nevertheless, Frewer and co-authors (1997) indicated in their study on the consumer's attitude towards different food technologies used in cheese, that product benefits (such as product quality, animal welfare, environment) were a more important factor in their decision than process considerations. However, they emphasised that the results could not be simply translated to other applications of genetic engineering in food production; each application should be considered on a case-by-case basis.

*Environmental aspects*
Environmental implications of agri-food products refer mainly to the use of packaging and food waste management. Wandel and Bugge (1997) proposed that intrinsic quality properties, such as taste or nutritional value, are related to personal interests, whereas environmental properties of food may be related to wider community-oriented interest. Consumers may express an interest in buying foods from environmentally sound production, either because of concern for their own health or because of concern for the external environment. With respect to the environmental consequences of packaging waste, European directives have been implemented to reduce this environmental impact. Since 1997, the food packaging industry is legally liable for improving material recycling and reducing the amount of packaging materials, which means reducing packaging waste.

With respect to waste management, inefficient processing is mainly a cash problem for processors and not yet a major quality concern of consumers. For example, in fresh produce processing, recoveries vary from 47 to 52% of finished product yields; the product loss can be found on the processing floor. Hurst and Schuler (1992) suggested that more emphasis should be put on waste management, to improve waste reduction methods and thus decrease the environmental waste load.

*Marketing (communication)*
The effect of marketing on product quality is complex. According to Van Trijp and Steenkamp (1998) consumers form an impression about the product's expected fitness for use at purchase; this is the consumer's judgement of 'quality expectation'. In their quality guidance model they proposed that marketing efforts (e.g. communication via branding, pricing and labelling) determine extrinsic quality attributes, affecting quality expectation. However, marketing can also affect credence attributes (which can not be checked by consumers themselves), influencing quality experience.

## 1.2.3 Food safety concepts

Quality together with safety in the food chain is one of the most important issues facing the agricultural and food industries. While quality of food is important to consumers, the safety of food is essential (Early and Shepherd, 1997). Food safety implies absence or acceptable and safe levels of contaminants, adulterants, naturally occurring toxins or any other substance that may make food injurious to health on an acute or chronic basis (website 2).

Today's integrated production and distribution systems mean that a contaminated food product can be consumed by a large number of people in a broad geographical range in a short period of time. The desire to limit the risks and to control safety of food has lead to the development of various food safety concepts. The purpose of these concepts is to reduce the risk of unsafe food products and to assure both processors and consumers that products supplied are safe and of high quality. One of the biggest factors encouraging the food industry to adopt food safety concepts is the shift in public attitude and awareness about their food. Consumer awareness and expectations of safety have increased along with the ability to detect and link food safety problems to a particular processor, farmer or activity (website 3).

Food safety concepts and programs are designed to limit exposure to foodborne risks. They will educate processors as well as consumers about the importance of safe food handling and how to reduce the risks associated with foodborne illness. Figure 1.1 shows legally binding guidelines and concepts as well as standards under private law that assure food safety (explanation below).

*Basic requirements*
The basic requirement of safe food production is the consideration of generally accepted principles and procedures. Without these the production of defect-free products with consistent quality is nearly impossible. On the level of processing these requirements are described as "Good

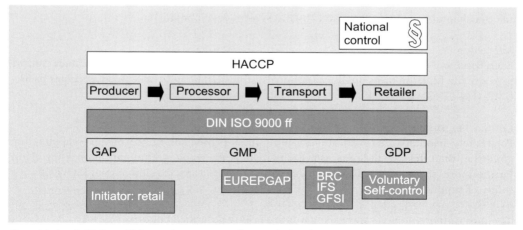

*Figure 1.1. Legally binding guidelines and concepts as well as standards under private law to assure food safety.*

*Susanne Knura, Stefanie Gymnich, Ewa Rembialkowska and Brigitte Petersen*

Manufacturing Practice" (GMP). Less well established are "Good Agricultural Practice" (GAP) and "Good Distribution Practice" (GDP) that are similar to GHP (Good Hygienic Practice) in processing.

**Good Manufacturing Practice (GMP) -** GMP describes the basic requirements in food processing. It includes regulations concerning cleaning, personal hygiene, infrastructure and traceability. GMP is that part of quality assurance which ensures that products are consistently produced and controlled to the quality standards appropriate for their intended use and as required by the marketing authorisation or product specification. GMP is concerned with both production and quality control (for details see Chapter 6 on QA systems).

**Good Agricultural Practices (GAP) -** Good Agricultural Practices (or GAPs) are a collection of principles applied to on-farm production and post-production processes, resulting in safe and healthy food and non-food agricultural products, while taking into account economical, social and environmental sustainability.

**Good Distribution Practices (GDP) -** GDP is that part of quality assurance which ensures that products are consistently stored, transported and handled under suitable conditions as required by the marketing authorisation or product specification.

*HACCP*
A central instrument of food safety is the Hazard Analysis and Critical Control Point concept (HACCP). HACCP is an effective and rational means of assuring food safety from harvest to consumption (see Chapter 7). Preventing problems from occurring is the paramount goal underlying any HACCP system. Under such systems, if a deviation occurs indicating that control has been lost, the deviation is detected and appropriate steps are taken to re-establish control in a timely manner to ensure that potentially hazardous products do not reach the consumer. In order to assure food safety, properly designed HACCP systems must also consider chemical and physical hazards in addition to other biological hazards (website 6). For details see the chapters on QA systems and HACCP system development.

*Self-control*
Each company can use a system of self-control (e.g. audits and random end product control) to verify the function of its quality assurance systems (GMP and HACCP) on a regular basis. GMP, HACCP as well as self-control are legally binding.

*Certification systems*
Food Safety Standards are certification systems under private law. Concerning traceability, they go further than the law: the processor has to include the whole preliminary processing chain. Furthermore, these standards also include other quality aspects (elements of ISO 9000), e.g. safety of supply.

The most important food safety standards are briefly described below:

**EUro-REtailer Produce Working Group (EUREP-GAP) -** EUREPGAP started in 1997 as an initiative of retailers belonging to the Euro-Retailer Produce Working Group (EUREP). It has subsequently evolved into an equal partnership of agricultural producers and their retail customers. Its mission is to develop widely accepted standards and procedures for the global certification of Good Agricultural Practices (GAP).
http://www.eurep.org/about.html

**British Retail Consortium (BRC) -** In 1998 the British Retail Consortium (BRC) developed and introduced the BRC Technical Standard and Protocol for Companies Supplying Retailer Branded Food Products (the BRC Food Technical Standard). Although originally developed primarily for the supply of retailer branded products, in recent years the BRC Food Technical Standard has been widely used across a number of other sectors of the food industry such as food service and ingredients manufacturing.

The Standard requires:
- the adoption and implementation of an HACCP system;
- a documented and effective quality management system;
- control of factory environment standards, product, process and personnel.

http://www.brc.org.uk/standards/index.htm

**European Food Safety Inspection Service (EFSIS) -** EFSIS is the premier third party independent inspection service, providing retailers, manufacturers and caterers, throughout the world, with expert inspection of their operations and suppliers to ensure only the highest standards are maintained. These range from accredited inspection of food premises to ISO 9000 system audits to HACCP certification and beyond. http://www.efsis.com

**International Food Standard (IFS) -** Food retailers (and wholesalers) regularly make audits to check the food safety aspects of their suppliers of own branded food products. These audits are made by independent auditors from qualified bodies. In 2002, German food retailers from the HDE (Hauptverband des Deutschen Einzelhandels) developed a common audit standard called International Food Standard or IFS, in order to create a common food safety standard. The IFS was officially launched by the Global Food Safety Initiative (GFSI) in January 2003.
http://www.food-care.info/

**Global Food Safety Initiative GFSI -** GFSI is an initiative of more than 200 retailers from more than 50 countries. Their aim is a uniform international standard of food safety.
www.globalfoodsafety.com

Based on these standards and guidelines numerous methods have been developed and are used to reduce risks posed by pathogenic micro-organisms including pasteurisation, cooking, addition of preservatives and proper storage conditions. With regard to the danger of food contamination by disease-causing micro-organisms during production, storage, transport and preparation of food, four basic food safety steps are recommended (website 4, 5):

**Clean -** According to food safety experts, bacteria can spread throughout the kitchen and get on to cutting boards, knives, sponges and counter tops. Thus, hands and surfaces should be washed often.

**Separate -** Separate raw meat, poultry, and egg products from cooked foods to avoid cross-contamination. Cross-contamination is how bacteria spread from one food product to another. This is especially true for raw meat, poultry and seafood. Experts caution on keeping these foods and their juices away from ready-to-eat foods.

**Cook -** Food safety experts agree that foods are properly cooked when they are heated for a long enough time and at a high enough temperature to kill the harmful bacteria that cause foodborne illness. Raw meat, poultry, and egg products need to be cooked thoroughly. The use of a food thermometer helps to ensure that foods have reached a high enough temperature to kill any harmful bacteria that may be present.

**Chill -** Food safety experts advise consumers to refrigerate foods quickly because cold temperatures keep most harmful bacteria from growing and multiplying.

## 1.3 Overview of the Food Production Chain and factors affecting quality attributes

This section deals with the food production chain and specifies factors affecting quality attributes and more specific food safety. First, the special assembly of the production chains in the agri-food industry is described. Accordingly, each step of this chain - from primary production to the end user - will be considered from the viewpoint of how it might affect food safety.

Agri-chains differ fundamentally from industrial production or assembly chains in a number of ways (Van Beek, 1990; Bloemhof-Ruwaard *et al.*, 1995):
- Restricted quality life of primary, intermediate and final products. This implies that storage technology and conditioning play a prominent role.
- Large variation in quality, quantity and availability of primary products, due to regional and seasonal conditions, which require storage and or transportation (fruits and vegetables, plants and flowers).
- Unintended and or unwanted by-products, remainders and refuse (especially from decomposition processes such as slaughtering, cheese-making or potato processing).
- High turnover of volume products.
- Many suppliers of primary products; centralised marketing for only a few product groups.
- Substantial environmental impact of production, processing, distribution and consumption (packaging material, surplus product, used product).
- Large public interest of organisations in all links of the chain (health and safety aspects of food, animal friendly production and environmentally friendly processing and distribution).

Irrespective of size and complexity, all food businesses must have an appropriate food quality assurance programme. Production and trade of food products involve a number of regulations and laws intended to protect consumer health and to prevent unfair competition. Both elements

## Agri-food production chain

are of primary importance, and uniformity of regulations is necessary in order to promote the trade at a worldwide level.

There is an increasing need for effective and efficient chains with regard to international competition, implying a great variation of (new) products to satisfy market demand, which must be high quality on delivery, cost effective, and produced under increasingly restrictive conditions.

Value chains may consider different scopes ranging from suppliers of ingredients (e.g. primary producers) and packaging for products, manufacturers, wholesalers, distributors/transportation companies, to retailers and consumers. Actors within chains depend upon information and ICT applications for organising, staffing, executing, managing, coordinating and controlling business activities.

*General chain model*
Even more than in other branches agriculture is characterised by a segregated structure. In contrast to the sequential functioning of the industry in agri-food chains, there is normally no specialisation in different sections of one company but there is a specialisation of companies relating to the special segment of the production.

A general chain model for the kind of specialisation and segregation in agricultural production chains is shown in Figure 1.2. Circles describe raw material and products (e.g. feed, fertiliser, meat, straw, etc.) and the quadrangle stands for services and modes of operation.

As indicated in Figure 1.2 multiple connections between customers and suppliers exist in the agri-food chain. Each element of the chain (breeder, producer, processor and consumer) is mostly both the customer and the supplier. They need resources and by-products to produce, and by-products arise from production and processing. Furthermore, several services are offered by consulting and administrative organisations.

Specialisation in the agri-food sector has led to an immense growth of productivity. However, it has also led to an increase in complex customer-supplier relationships in agri-food chains (Figure 1.3).

As a result of the big food scandals, legislative actors have introduced new framework conditions on agricultural production. The newly established EU General food law ((EC) No 178/2002) demands transparency and a "stable to table" approach throughout agricultural production while governmental food safety inspection is currently undergoing reorganisation towards a „control of control" (Schulze Althoff, 2004).

The "stable to table" approach to food safety is an holistic approach embracing all elements, which may have an impact on the safety of food, at every level of the food chain from "stable to table" and from "farm to fork". The phrase is used to encompass the production of all foods of animal origin and can be applied not only to meat but also to milk, eggs, fish and other products from aquaculture, as well as fruits and vegetables (FVE, 2004).

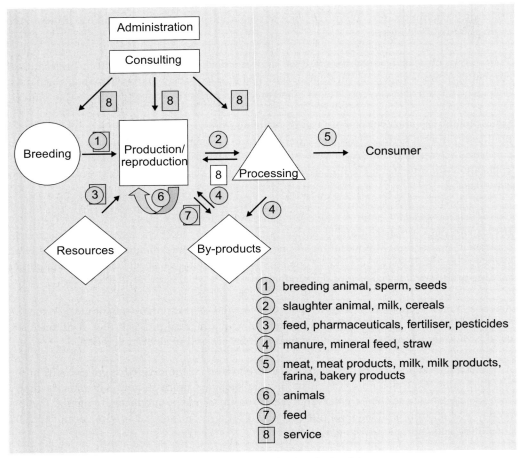

Figure 1.2. Schematic representation of the specialisation in agricultural production chains.

By integrating the feed manufacturing, production, transport, processing and distribution stages, the "stable to table" approach to food safety aims to increase the quality and safety of food in order to achieve the highest possible level of health protection.

## 1.3.1 Conventional primary production

The potential for food to become contaminated with chemical substances or micro-organisms starts from the moment it is harvested and continues right through until the moment it is eaten. Subsequently in this section, animal as well as vegetable production will be described. Emphasis will be put on the different production factors and how they may influence food safety.

*Agri-food production chain*

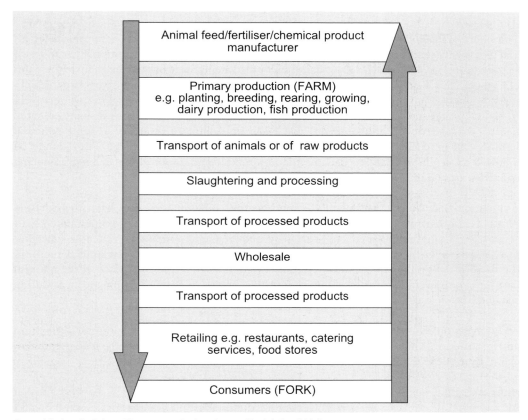

Figure 1.3. Agri-food chain: The stages in food production (Food Safety, 2004).

*Animal production*
Wholesome safe food of animal origin can only be produced from healthy animals kept in hygienic conditions and in husbandry systems that cause them minimal stress, combined with the responsible use of veterinary medicinal products. The stable to table approach to food safety can therefore only be successful if the health and welfare of animals are fully integrated into the approach (FVE, 2004).

In this section there will be an emphasis on the following production factors:
- animal feed;
- animal identification;
- animal health and welfare;
- animal waste and;
- animal movement.

*Animal feed*
The safety of food of animal origin begins with safe animal feed. Certain chemical substances and biological agents incorporated into feed (either intentionally or unintentionally) can result in hazards in food of animal origin and may enter feed at any stage of production up to the point of feeding. Thus, feed manufacturers, farmers and food operators have the primary responsibility for food safety. But the hazards to human health associated with animal feeding are relatively minor in comparison to foodborne hazard from other sources. However, there are many potential hazards associated with animal feed. Table 1.2 is an exemplary summary of some potential hazards and practices in controlling them. Good Manufacturing Practice (GMP) should be followed at all times. GMP has become the minimum standard for all major feed producing countries around the world (website 7).

As mentioned in Table 1.2 an example of how bacterial contamination does relate to foodborne illness is the harmful bacterium *Salmonella*. Animal feed is frequently contaminated with *Salmonella*. When contaminated feed is eaten by food-producing animals such as cattle, chickens, pigs, and turkeys, the *Salmonella* may take up residence in the animal's intestine. Sometimes the bacteria cause illness but most often they reveal no sign of their presence. If the harmful bacteria continue to live in the animal's intestine, they can then contaminate meat during slaughter or cross-contaminate other food products.

Additionally animal faeces containing *Salmonella* can contaminate water used for agriculture and for drinking. Thus, harmful bacteria in animal feed can travel through the food production chain, from "farm to fork", and cause human foodborne illness (website 8).

*Animal identification*
Animal identification is a basic requirement for the establishment of food safety and quality management systems. Without it traceability of food from animal origin is impossible (for detail see the chapter on tracking and tracing).

*Table 1.2. Potential hazards and practices of help in controlling them.*

| Potential hazard | Practices aiding in controlling feedborne hazards |
|---|---|
| Mycotoxins | Feeds contaminated with mycotoxins in excess of acceptable levels should not be fed to animals producing milk, eggs or other tissues used for human consumption |
| Infectious agents such as Salmonella | Salmonella is sensitive to heat and readily killed if the manufacture of feed involves a process with sufficient heating |
| Veterinary drugs such as antibiotics | Only veterinary drugs licensed for administration to food producing animals should be used and withholding time should be observed before animals are sent for slaughter |
| Environmental and industrial contaminants such as pesticides, herbicides fungicides and polychlorinated biphenyl's and heavy metals | Agricultural chemical levels in feed need to be sufficiently low so that their concentrations in food are consistently lower than maximum residue levels (MRL) |

A foolproof system of animal identification is required and must be harmonised across the EU. Public health would benefit from the development of such a system because this would permit accountability for and prevention of food safety hazards, including residues and harmful pathogens. In addition, the public would benefit because data collection and long-term research studies are currently hampered by the lack of animal identification. Understanding of the ecology of food-borne pathogens in the production and handling period before slaughter needs to be improved. Animal identification will permit packers and consumers to reward producers for using food safety- related production practices (Vitiello and Thaler, 2001).

*Animal health and welfare*
The slogan "healthy animals - healthy humans" is often used to demonstrate that a clear relationship exists between the health status of animals and that of human beings. Prevention of the transmission of zoonotic agents has become an even bigger challenge because - due to an increase in the human population - there has also been a contaminant increase in production animals (Notermans and Beumer, 2002).

Public health, animal health and animal welfare are interrelated. As an example of this, stressed animals are more likely to develop diseases which require veterinary treatment. This in turn may increase the presence of residues of veterinary medicinal products in animal produce, thereby potentially affecting public health. As a key factor of animal health and food safety, animal welfare must therefore be a component of herd health surveillance programmes.

Herd health surveillance schemes must be introduced at farm level. These schemes are intended to improve the health of the herd by focusing on husbandry practices and farm management. They must also include routine animal health visits carried out by a vet.

In addition, since many pathogens can be transferred from animals to man by direct contact or through vectors such as food, an effective interaction between herd health and epidemiological surveillance schemes is essential.

*Animal waste*
Transport, handling and disposal of animal waste such as faeces or offal are critical elements of the stable to table approach, which must be thoroughly controlled.

For example, because of the BSE (Bovine Spongiform Encephalopathy) scandal in the 80's and with the detection of a possible epidemiological connection between Animal and Human TSE (Transmissible Spongiform Encephalopathy), the principle outstanding issue was to protect public health from the risk of being exposed to the BSE agent in food via infected tissues of clinically healthy incubating cattle, killed for human consumption. Therefore, six tissues (brain, spinal cord, thymus, tonsils, spleen and intestines from duodenum to rectum inclusive) were specified for removal from all cattle over six month of age and for disposal without risk (Bradley, 2002). The disposal ensures that this offal will not enter the food chain or animal feed.

Moreover, animal faeces are a known source of pathogens that can cause foodborne illness (compare to section on animal feed).

Farming practices that emphasise the use of animal manure, manure slurries and animal manure-based compost play an important role in the recycling of organic nutrients and in developing a rich soil structure. The recycling of bacterial pathogens and protozoan parasites from animals to humans through water, soil and crops has created a serious challenge for producers, processors, and consumers of fresh products (website 9). Table 1.3 summarises the key bacteria concerning fruit and vegetables mainly originating from faeces.

*Animal movement*
Movement of animals causes stress and stressed animals are more likely to cause contamination of the meat. In relation to transport, disease can result from:
- tissue damage and malfunction in transported animals;
- pathogens already present in transported animals;
- pathogens transmitted from one transported animal to another;
- pathogens transmitted from transported animals to other animals directly or indirectly, e.g. via equipment.

Transport and associated handling can affect the susceptibility of individuals to pathogens, the infectiousness of individuals carrying pathogens or the extent of contact, which might result

Table 1.3. Microbial pathogens occurring in fruits and vegetables and their major carriers and primary sources (modified website 9).

| Microbial pathogen | Type of organism | Major carriers | Primary source |
|---|---|---|---|
| Salmonella | Bacterium | Production animals | Faeces |
| | | Domestic pets | carcasses |
| | | Wild animals | soil (persistent) |
| | | Cross-contamination from meat, poultry, egg | |
| Escherichia coli O157:H7 | Bacterium | Production animals | Faeces |
| | | Domestic pets | water-borne |
| | | Wild animals | soil (persistent |
| | | Humans | unclear) |
| | | Cross-contamination from meat, poultry | |
| Cryptosporidium | Protozoan | Production animals | Faeces |
| | | Domestic pets | water-borne |
| | | Wild animals | |
| | | Humans | |
| Toxoplasma | Protozoan | Production animals | Faeces |
| | | Cats | |
| Cyclospora | Coccoid parasite | Humans | Faeces |
| | | | water-borne |

## Agri-food production chain

in transmission of pathogens (Report of the Scientific Committee on Animal Health and Animal Welfare, 2002).

Furthermore, transportation of animals from the farm to the abbatoir, or between farms, necessitates farm animal transport vehicles, taking into account appropriate loading density, transport time, climate, possible mixing of animals or groups of animals from different origins.

After the time spent in transit, animals need enough time for rest. Obviously, resting time is proportional to transport duration. Lairage facilities should have enough space for appropriate animal density and animals must have free access to water and feed.

The reduction of stress situation results in decreased incidence of lower meat quality, well known as PSE (Pale Soft Exudative) and DFD (Dark Firm Dry) (Murray, 1995).

*Vegetable production*
From planting to consumption, there are many opportunities for bacteria, viruses, and parasites to contaminate produce. On the farm, soil, manure, water, animals, equipment, and workers may spread harmful organisms. Produce may be harvested on a farm, processed in one plant, repackaged in another, then stored, displayed, or served by an institution or in the home. Each of these steps is an opportunity for harmful micro-organisms to enter the food supply. Table 1.4 gives a short overview of potential food safety hazards in vegetable production.

Many factors can have an impact on vegetable crop quality throughout the whole production system, from the field to the consumer. All these circumstances are important and in order to provide the best quality parameters of vegetable production, all of them should be included in everyday practical activities.

The most important factors are given below:
- climate and weather (sun, rainfall, soil humidity, temperature, wind, etc.);
- cultivation conditions;
  - soil type and structure;

*Table 1.4. Potential food safety hazards in vegetable production.*

| Potential hazard | Results |
|---|---|
| Environmental contamination of soil and water (industry, transport, communal wastes, intensive agriculture) | Contaminants in plants detrimental to human health (heavy metals, aromatic hydrocarbons, PCB, dioxins, etc.) |
| Improper mineral and organic fertilization (over-fertilization) | Harmful levels of nitrates in plants |
| Improper use of pesticides | Harmful level of pesticide residues in plants |
| Use of fresh animal and human faeces as fertilizers | Microbial contamination, parasite afflictions |
| Improper storage conditions (too warm and too dry) | Increased nitrites level, bad nutritive quality (decomposition of vitamins) |

- soil abundance and fertility;
- sowing conditions;
- system of cultivation (greenhouse, plastic tunnel, open field);
- harvesting conditions (time of day, weather conditions);
- Species and cultivar impact;
- Storage, transport and trade conditions.

*Climate and weather*
General climate conditions are dictated by nature and are not alterable by human activity. That also applies to weather conditions, except for the water content in soil which can be regulated by proper irrigation, but this is only available for richer farmers or farm cooperatives.

Scientists have proved that weather conditions during cultivation have a strong impact on the crops' composition. High average temperature and drought in the growing season decreases the yield and increases the content of nitrates and heavy metals in crops, and the opposite - moderate average temperature and sufficient water content - increases the yield and decreases the concentration of the afore-mentioned substances in crops (Rembialkowska 2000). Therefore, an adequate irrigation system can help to maintain a good yield but also proper quality of the produced vegetables (Sady and Cebulak 2000).

As a consequence of the above, climate and weather conditions have a strong influence not only on the quality but also the safety of vegetable products. For example, it is recognised that very warm and dry weather during the growing season increases the concentration of nitrates and heavy metals in vegetables. High levels of nitrates in vegetables are dangerous, because nitrates can easily convert to nitrites in the human digestive tract, and nitrites can combine with amines (protein derivatives) and can produce potentially carcinogenic nitrosamines within the body. Heavy metals consumed with vegetables over a long period can cause serious disorders and cancers in human organism.

*Cultivation conditions*
Among cultivation conditions soil type and structure are the most important factors. For example, the intake of cadmium by plants has a very negative correlation with the soil reaction (pH): lower pH, higher cadmium intake (Szteke, 1992). Soil acidification increases the mobility and availability of cadmium for plants and leads to a humus degradation. On the other hand, low humus content in soil means poor soil structure (exchange sorption complex) and it leads to higher heavy metals intake (Sady et al., 1999).

Soil structure is strictly connected to the abundance of nutrients and fertility. The main factors influencing soil fertility are the natural soil quality class and the fertilisation system. Of course, the better the soil fertility, the higher the general quality of the produced vegetables. However, in heavy soils vegetables accumulate more nitrates and diminishing nitrate accumulation can be observed as follows: organic soils > clayey soils > sandy soils (Sady et al., 1999).

Sowing conditions are also important; for example the level of nitrogen fertilisation at sowing time will determine the level of vitamin C and ß carotene in carrots (Brandt and Mølgaard, 2001). Early varieties of potato and vegetables accumulate more nitrates -it's especially important in the case of early vegetables with a short vegetation period like lettuce and spinach (Sady *et al.*, 1999).

The system of cultivation (greenhouse, plastic tunnel, open field) is significant for the crop quality. The same cultivar of plant will have different features when cultivated in the open field or in a greenhouse, because the vegetation conditions are different (light, wind, sun, irrigation, etc).

Harvesting conditions are extremely important for vegetable quality; even the time of day that harvesting takes place. Leafy vegetables harvested in the afternoon or on sunny days contain fewer nitrates than those harvested in the morning or on cloudy days (Sady *et al.*, 1999).

*Species and cultivar impact*
The genetic background of a particular species is responsible for its quality parameters. It is well known that the ability to accumulate contaminants is dependent on the plant species. For example, some plants accumulate more nitrates (leafy vegetables, like spinach, lettuce, dill, and some root vegetables like radish, garden beet, turnip, kohlrabi), and other plants less nitrates in the same cultivation conditions (parsley, garlic, cucumber, cauliflower, leek, celery, carrot, tomato, potato, onion, paprika, rhubarb, bean, broccoli). Therefore, the highest permissible content of nitrates in vegetables and potatoes is different for different vegetables (e.g. according to the Polish Directive of Ministry of Health and Social Care from 27 DEC 2000). The ability to accumulate other contaminants like heavy metals (cadmium, lead) is also diverse; for example, grown in the same soil potatoes will contain less cadmium than carrots and garden beets will contain more cadmium than carrots (Rembialkowska, 2000). The importance of the cultivar is also big. In the same species different cultivars have a different ability to accumulate contaminants. For example, in the experiment with head lettuce Vogtmann (1985) found that some cultivars accumulated significantly less nitrates than others. Also, Rembialkowska (2000) proved that some potato and carrot cultivars cumulated less nitrates than others. Similarly, the content of desirable compounds like vitamins, sugars, amino acids as a rule differs very much between cultivars of the same species.

*Storage, transport and trade conditions*
The storage system after harvesting is very important for vegetable quality. There are three main systems used nowadays: professional storage, clamping and cellar storage.

Professional storage means the use of specially built cooling chambers with regulation of temperature, humidity and air circulation. This is, of course, the best but the most expensive system, and not all producers can afford it.

Clamping is very popular among small farmers; they construct clamps covered with soil and straw for the wintertime. This system is quite efficient if the clamp is prepared properly, but is normally not effective beyond March. When the spring warmth comes, even the best clamp is not sufficient to keep the temperature low for potatoes and vegetables.

Cellar storage is used mostly for own consumption in rural families. If there is no heating system in the cellar, potatoes and vegetables can be kept in such conditions for most of the wintertime.

In all countries there are special standards regulating the conditions of potato and vegetable storage. For example, Polish standards (PN - 77/R - 74457, PN - 91/R - 75411) prescribe that optimal storage conditions for potatoes are as follows: storage in pallet boxes and bags from loosely woven artificial materials, temperature 4-6°C, humidity 85-90 %, active ventilation usage. Similarly, for vegetables the conditions prescribed are, optimal temperature 1-5°C, humidity 90-95 %, storage in boxes or pallet boxes, active ventilation usage.

The same or very similar rules are compulsory for transport and trade of vegetables. Unfortunately only large trading companies can afford special transport cars with cooling systems and a professional repository at the trading place (supermarket, store). Many small stores don't have a proper place to keep vegetables. A survey of several small stores with organic food in Poland (Rembialkowska, 2000) showed improper storage of those vegetables expected to sell. As a consequence, the potatoes and vegetables contained too many nitrites compared to the health standards, because at higher temperatures nitrates convert easily to nitrites (Lecerf, 1995). As a result all the efforts of the organic farmer to produce high quality vegetables are unfortunately wasted.

### 1.3.2 Food processing and packaging

During the journey from farm to consumer, food commodities are likely to be exposed to a multitude of hazards that may lead to contamination by dust, dirt, weeds, mechanical injury, physicochemical changes accelerated by heat, light, metal ions, contamination or spoilage due to micro-organisms, insects and rodents, or biochemical changes brought about by enzymes that may be endogenous or contributed by the invading biological agents. Thus, the wholesomeness of the product often depends on the product history (Singhal *et al.*, 1997).

In this section different types of food processing and factors affecting food safety are described. Emphasis is put on food preservation through processing. Additionally, future prospects concerning minimal processing are mentioned. Subsequently, the importance of packaging as well as new packaging technologies for food safety is highlighted.

*Food processing*
Food preservation through processing is an extremely broad area in food science. There are textbooks on every method of processing. Below are some of the most important processes, and examples of their contribution to food safety:

**Refrigeration and freezing -** Refrigeration and freezing are common and cheap methods of preserving perishable foods. Ice, for example, is still used in addition to mechanical refrigeration to extend the shelf life of fish, seafood, poultry and some vegetables. If done properly, freezing preserves foods without causing major changes in shape, size, colour, texture and flavour.

Low temperatures necessarily result in higher stability and provide an intrinsic advantage, if properly exploited, in bringing nutritionally high value foods to the consumer. The freezing of vegetables after harvest, for example, will guarantee the consumer a higher vitamin C content than could be attained by any other form of preservation and distribution. Furthermore, frozen foods have always enjoyed a safety record second to none since, if properly handled before freezing and during distribution, there is no possibility of the growth of microbial contaminants between freezing and thawing (Kennedy, 2000).

**Irradiation of foods -** Commercial irradiation equipment uses gamma rays, electron beams, or X-rays to expose food products to ionising radiation that causes changes at the molecular level, damaging or destroying living cells. Depending on the type of food and radiation dosage, irradiation can be used to sterilise food for storage at room temperature, control pathogenic micro-organisms, delay spoilage of fresh foods, control insect infestations, delay ripening of certain fruits, or inhibit sprouting of certain vegetables. Extensive scientific research reviewed by the World Health Organization (WHO) and the U.S. Food and Drug Administration has indicated that irradiated food is safe to eat.

**Thermal processing -** Most foods are submitted to thermal processes such as cooking, baking, roasting, extrusion cooking, pasteurisation or sterilisation. The first processes serve mostly to obtain particular sensory or texture features, the last two to ensure microbiological safety and to eliminate some enzymatic activities that reduce food preservation.

Heating food destroys a large proportion of the micro-organisms and natural enzymes that reduce shelf life. Heating will not completely sterilise a food product, so even if the food is protected from recontamination, spoilage may eventually occur. Milk, for example, is heated to improve its shelf life and to kill disease-causing micro-organisms and viruses (Arnoldi, 2001).

**Transformation process -** Raw materials are transformed into foods by means of a variety of processes, such as fermentation (using micro-organisms or enzymes), extrusion (e.g. snacks from starch containing raw materials), hydrogenation of fats, emulsification, and extraction (e.g. fruit juices). Frequently, the resulting food product does not resemble the raw material in appearance or properties. For instance, cheese is completely different from milk, bread is different from wheat, orange juice different from oranges (Luning *et al.*, 2002).

Fermentation, for example, is a very old method of preserving food in which more flavourful foods are produced from original products. Reasons to ferment food products are:
- improved shelf life of a product;
- improved nutritional value and digestibility;
- removal of anti-nutritional factors from some plant derived base materials.

Some famous examples of fermented food products are bread, beer and wine. These products are obtained by alcoholic fermentation by yeasts. Yogurt, buttermilk, cheese, sauerkraut, olives and salami are mostly obtained by lactic acid fermentation by *Lactobacillae* (website 10).

**Minimal processing -** The growing demand for convenience and fresh-like, healthy foods is driving the industrialised markets for chilled prepared foods. Traditional means to control food spoilage and microbiological safety hazards, such as sterilisation, curing or freezing, are not compatible with market demands for fresh-like convenience food. Therefore, food manufacturing industries seek to satisfy these consumer demands through the application of new and mild preservation techniques such as refrigeration, mild heating, modified atmosphere packaging and the use of natural antimicrobial systems. Food types that fall into the category of so-called "minimally processed foods" are, for example, fresh-cut vegetables, prepared sandwiches, ready-to-eat meals and chilled prepared foods (Zeuthen, 2002).

The idea of minimal processing is to preserve food materials with minimum damage but still retain as much freshness in taste and other sensory properties as possible. Figure 1.4 summarises the main aims of minimal processing in consideration of food safety.

The concept of minimal processing has developed much during recent years. Nowadays minimal processing is regarded as the mildest possible preservation technique tailored to a particular

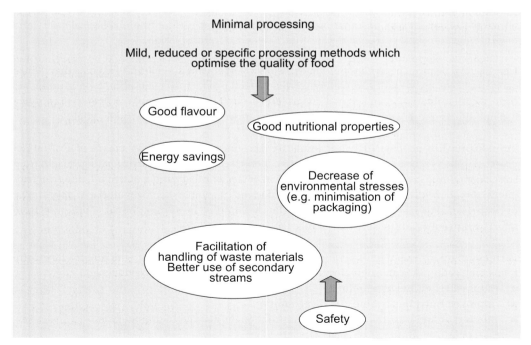

*Figure 1.4. The main aims of minimal processing without compromising the safety of foodstuffs (Ahvenainen et al., 2000).*

raw material and food. The aim is to maintain the natural properties of a foodstuff by non-obtrusive processing using a minimum of preservatives.

Ahvenainen (2002) basically classified preservation methods in the minimal processing of foods into three different categories:
1. Optimised traditional preservation methods (e.g. canning, blanching, freezing, drying and fermentation) in order to improve sensory quality and to save energy, not forgetting microbiological safety.
2. Novel mild preservation techniques, such as high-pressure technology, electric field pulses (so-called non-thermal processing), various mild heat treatments, sous-vide cooking, post-harvest technologies, protective microbiological treatment, etc.
3. Combinations of various methods and techniques (hurdle effect), whereby a synergistic effect is obtained.

*Packaging*
A very important aspect of all preservation technologies is packaging. It forms the barrier between the food and its environment. It can protect the food from recontamination and other undesirable influences from the environment. Many cooking and preservation processes have been and are still largely based on proper packaging. For example, packaging is an integral part of canning, aseptic, sous-vide and baking processes.

Additionally, the benefits of several preservation methods (e.g. drying, freezing) would be lost without protective packaging after processing (Ahvenainen, 2002).

Packaging development has also changed the preservation methods used for food products: Ten to fifteen years ago, all poultry products and industrially prepared raw minced meat were sold as frozen. Nowadays, thanks to modified atmosphere packaging (map) based on protective gas and novel gas impermeable packaging materials, they are mainly sold chilled. Even industrial preparation of fresh-cut fruit and vegetables for retail sale is possible today, due to respiring packaging films (Ahvenainen, 2002).

The use of modified or controlled atmospheres is a relatively new concept in food processing, whereby the atmospheric condition around the product is different from normal air. The gas composition of headspaces in packed foods together with barrier properties of packaging materials greatly influences shelf life and food safety. Especially lower oxygen concentrations are applied to delay oxidative reactions *(e.g.* lipid oxidation, discolouration), to inhibit growth of aerobic micro-organisms and to decrease respiration rates, thus extending the shelf life of food products (Phillips, 1996). Several packaging concepts such as vacuum, modified atmosphere and active oxygen scavenging are used to obtain these conditions. Since low oxygen levels may favour the growth of same anaerobic foodborne bacteria, preventive measures must be taken, such as correct heat treatment, and/or low pH, and/or low $a_w$, and/or hygienic handling. A new development with respect to gas composition is the use of extreme high oxygen conditions to extend shelf life (Luning *et al.*, 2002).

*Susanne Knura, Stefanie Gymnich, Ewa Rembialkowska and Brigitte Petersen*

Packaging has developed during recent years mainly because of increased demands in the areas of product safety, shelf life extension, cost efficiency, environmental issues and consumer convenience. In order to improve the performance of packaging, innovative active and smart packaging concepts are being developed and tested in laboratories and companies around the world. Active and smart packaging concepts have great commercial potential to ensure the quality and safety of food without (or at least with fewer) additives and preservatives, thus reducing food wastage, food poisoning and allergies (Ahvenainen, 2002).

*Combination of factors - hurdle technology*
Nowadays there is a shift from individual preservation factors to a combination of factors to guarantee food safety, while maintaining sensory properties and nutritional value - also called hurdle technology (Gould, 1996). Some typical combination treatments include:
- enhancement of the effectiveness of an anti-microbial acid by lowering the pH;
- anti-microbial efficacy of carbon dioxide in modified atmosphere is greatly enhanced at reduced temperature;
- foods with low water activity and/or reduced pH require a milder heating treatment because of the synergy between $a_w$, pH and temperature;
- the combination of mild heating of vacuum-packaged food with well-controlled chill storage also appears to be successful, i.e. "sous-vide" production (Luning *et al.*, 2002).

### 1.3.3 Distribution and retail

The likelihood of adversely affecting the quality and safety of food should be taken into account during food storage and final distribution to the consumer. Optimal storage conditions will depend on the food, local climatic conditions and whether it is intended for final consumption or further processing (FAO, 2004).

To consider the effects of storage and distribution on physical product features, a distinction must be made between fresh produce and manufactured foods. Fresh produce, such as fruit and vegetables, are characterised by their respiratory activity after harvesting. Most fleshy fruits show an increased respiration rate accompanied by ripening (i.e. colour, flavour and texture changes). Temperature and composition of the storage atmosphere markedly influence the respiration process in plant tissues. Normally the rate of respiration decreases as the temperature decreases, within the physiological temperature range for a particular commodity. Exposure outside the recommended range for more than short periods causes injury and a decrease in quality and shelf life, i.e. chilling injury. Visible effects of chilling injury include tissue necrosis (e.g. apples), failure to ripen (e.g. bananas), black spots (e.g. bell peppers, aubergines) and wooliness in texture (e.g. peaches) (Haard, 1985). Furthermore, manipulation of the oxygen and carbon dioxide concentration of the storage atmosphere can considerably delay ripening and subsequent decay process. Another important factor for fresh produce is the relative humidity (RH). A proper RH regulation is required to prevent both desiccation (RH too low) and mould growth (RH too high). Furthermore, chemicals can be applied during storage to prevent, for example, sprouting or damage by insects (Luning *et al.*, 2002).

Suboptimal storage and distribution conditions may lead to poisoning of the food. Thus, storage facilities must be appropriate to the food, for example, large warehouses for the storage of raw commodities or chilled refrigerated units for perishable products, etc. Food distribution is the final step in the chain prior to the final consumer. It may include different forms of transportation, and a range of retail outlets, such as open-air markets, shops, supermarkets, etc. (FAO, 2004).

In common with all points in the food chain, conditions during storage and distribution should ensure the quality and safety of the food product, through the application of Good Hygienic Practices (GHPs) as set out in the Codex General Principles of Food Hygiene (FAO, 2004).

**1.3.4 End users**

The food chain does not end with distribution and retail; even end users (catering & consumer) have to be included in a comprehensive chain management. In this section emphasis will be put on how end users can influence food safety. Even if a product reaches the end user in a wholesome and safe manner it can also be contaminated by consumers.

*Catering*
Catering is about providing meals or snacks either as a business or service, and can be roughly divided into trade catering, in which the main purpose is to make a profit, and service catering, in which the main purpose is to provide a service in meals. Examples of trade catering are takeaways and restaurants. Examples of service catering are work's canteens, school meals, meals on wheels and hospital catering. All catering operations share the need for basic food hygiene such as cleanliness and temperature control, as described in earlier chapters. However, there are special problems linked with certain types of catering (Berg, 1999). Figure 1.5 gives an overview of different catering systems concerning warm meals.

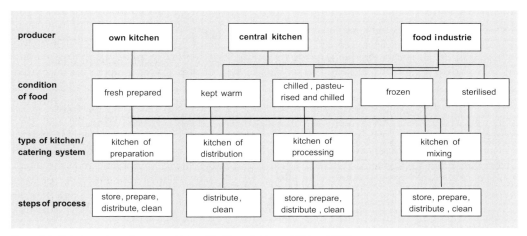

*Figure 1.5. Catering systems (warm meals) (modified from Reiche and Kleiner, 2002).*

In all forms of catering the main consideration must be the safety of the food. All catering operations should be carefully planned. The planning should include hazard analysis and/or risk assessment. The procedures should be carefully monitored to ensure that they are correctly carried out. All staff involved in handling food should be trained to an appropriate level in food hygiene.

Unfortunately service catering is often responsible for many cases of food poisoning, e.g. by *Clostridium perfringens* as well as *Salmonella* and *Staphylococcus aureus*. Critical factors with respect to food safety associated with service catering are (Berg 1999):
- badly planned kitchens;
- poor handling of cooked food;
- inadequate cooking systems;
- poor staff time management;
- excessive concentration on special diets.

Good temperature control is the best weapon any food business has in the prevention of food poisoning. All high-risk food must be stored below 8°C (although 5°C is recommended) or above 63°C (Figure 1.6).

**Cook & chill/Cook & freeze -** Cook & chill and Cook & freeze are normally used in service catering, especially for school meals and hospital food. Some airlines utilise Cook & freeze for in-flight catering. There are economic advantages for large-scale organisations to utilise either Cook & freeze or Cook & chill, as these systems lead to a more efficient use of equipment and staff. These methods can also provide a wider choice of meal types to consumers (Berg, 1999).

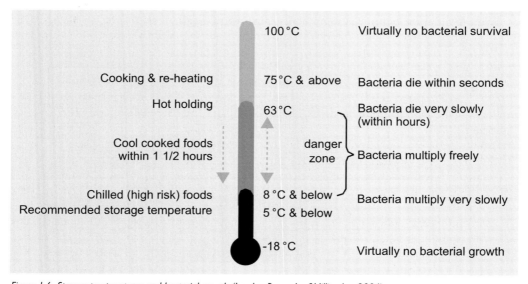

*Figure 1.6. Storage temperatures and bacterial growth (London Borough of Hillingdon, 2004).*

In both systems food is prepared, then cooked in a central kitchen, then chilled or frozen, before being transported cold to a feeder kitchen, where the food is reheated before being served. Assurance is realised by strict temperature control and hygienic handling, therefore these systems are safer than conventional service catering (Reiche and Kleiner, 2002). As staff can work at a steady rate there is no mad rush to meet meal times. Also careful planning can eliminate many of the risks of cross-contamination. Moreover, since staff are under less pressure it is easier to integrate cleaning into the system (Berg, 1999).

*Consumer*

The consumer is the final element of the food chain. Food that is perfectly safe at the point of purchase needs to be handled carefully to avoid contamination at home. During purchase & transport, storage as well as food preparation, several measures can be taken to ensure that food is not spoiled by the risks of falling ill (Eufic, 2004).

**Purchase & transport -** Once food has been purchased it becomes the consumer's responsibility to ensure that the food remains safe until it is eaten. The first problem the consumer faces is transporting the food home. Consumers should always check the "Use-By" date or "Best before" date marking on packaged food and make sure that the packaging of food is not damaged.

They should not purchase products marked as "Keep refrigerated", "Keep chilled" or "Keep frozen" that have not been stored under adequate refrigeration. Furthermore, food that needs refrigeration should be taken home quickly and placed in the refrigerator or freezer promptly. If frozen products have been thawed, they must not be re-frozen (Eufic, 2004).

**Storage -** Most food that does not need refrigeration, including dried foods and most canned food, are best stored in cool dry conditions away from direct sunlight. Many types of dried food, including dried fruit, flour and sugar, will grab moisture from the air if they have a chance. This often leads to spoilage by moulds. Direct sunlight can cause chemical changes in food, especially in food with high fat contents. All food should be stored wrapped or covered to protect it from insects and physical contamination (Berg, 1999).

Contact between raw and cooked food should be avoided. This reduces the risk of cross-contamination (bacteria passing from one food to another) (Eufic, 2004). To prevent cross-contamination in fridges, care has to be taken where the food is placed. Dairy products should be put on the top shelf, cooked food should be placed on the middle shelf and raw meat or fish on the bottom shelf. All food in the fridge should be wrapped or covered to prevent cross-contamination and odour migration. Hot food should not be placed in the refrigerator, as it will cause the temperature to rise, but kept outside until it has cooled to room temperature (Berg, 1999).

**Food Preparation -** Raw food contains a large number of microbes, which will spread to any surface or equipment that comes into contact with the raw food. Hence all knives, boards, and other equipment used in preparing raw food should be washed immediately after use. Also hands, which carry a large number and variety of micro-organisms should be washed before and after handling raw food (Berg, 1999).

*Susanne Knura, Stefanie Gymnich, Ewa Rembialkowska and Brigitte Petersen*

To reduce the risk of cross-contamination separate cutting boards and utensils should be used for raw and cooked foods. All kitchen surfaces have to be cleaned by washing with hot soapy water and disinfectant to prevent cross-contamination. Frozen food should be thawed in the refrigerator and cooked immediately it has thawed. Cooked food has to be cooled as quickly as possible to slow down the growth of bacteria and should be reheated thoroughly to kill any bacteria, which may have developed during storage (Eufic, 2004).

## 1.4 Quality management activities with respect to safe food production chain

Changes in markets are forcing the agri-food sector to focus much more than before on meeting the increasing requirements of consumers on quality of products as well as on the quantity of the production processes throughout the production chains, ranging from the farm to the processing industry and the trade. Therefore, quality management is of paramount importance to all companies and organisations involved in the production, processing, sale and handling of food.

Modern trading conditions and legislation require food businesses to demonstrate their commitment to food quality and establish an appropriate quality management programme. Such programmes can usefully be applied to both organisational and technological issues. The practical success of any quality assurance programme will depend on the proper use of appropriate methods and tools.

Today in the European Union (EU), the United States, Canada and other countries the implementation of HACCP is mandatory. The EU established new legislation and rules on food safety, the "Directive on the Hygiene of Foodstuffs", where general rules and procedures are laid down to increase the consumer's confidence in the safety of foodstuffs for human consumption (website 11).

The following sections briefly characterise existing customer-supplier relationships and give an overview of common contract types. Furthermore, basic requirements for quality management activities with respect to a safe food production chain are described. It will be emphasised that food safety management is an integral part of overall quality management and a key component of the longer-term managerial strategy to enhance the safety and quality of products (Jouve *et al.*, 1998).

### 1.4.1 Customer-supplier relationship

The traditional price-based relationship among customers and suppliers is changing. Long-term relationships are being built up on total cost, trust, innovation, quality, and flexibility. Developing mutually beneficial relationships with the suppliers is critical to any company. Today, we look at the entire supply chain to deliver high-value products and services to our customers. The key to building excellent customer-supplier relationships is communication. Communication cannot be one-way; there must be a 360 degree exchange of information. This helps to improve the quality of products and services you receive from your supplier and therefore reduce the total cost (La Londe and Raddatz, 2000).

Suppliers are all those whose input is essential for their customers to do their job effectively. Obvious suppliers are those outside an organisation that deliver raw materials for production processes, stock for re-sale, capital equipment or equipment consumables. Furthermore, they provide other organisations with services or with information essential for them to do their job. Customers are the sole reason for continuously improving processes. It is they who decide whether a product or service offers the best value for money when they make their purchase decisions. While the importance of this relationship is obvious, there are a number of factors which complicate it. Many companies do not deal directly with the ultimate users of their products or services; they deal through intermediaries. These intermediaries (such as wholesalers and retailers) have needs, which are as important to satisfy as those of the ultimate customers (HCI, 2004). Figure 1.7 clarifies internal and external customer-supplier relationships.

To establish a legal relationship between customers and suppliers contractual endorsements should be built up. A contract could be a tool that regulates the relationship between two or more partners (Warberg, 1997). The general purpose of the contract is to manifest and to bind the contracting parties to perform as agreed. The best contractual outcome would be that both parties could experience the same situation, i.e. a win-win situation. Contracts free the companies to focus on their core competence and to outsource all other production and service activities needed in business (Warsta, 2001). Table 1.5 gives a short overview concerning common contract types.

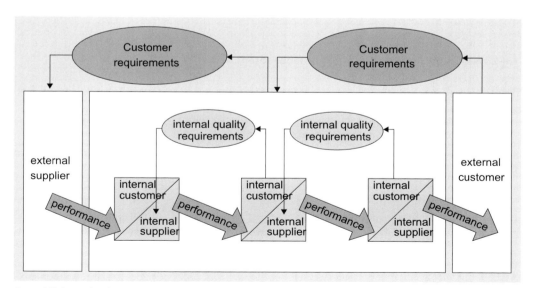

Figure 1.7. Internal and external customer-supplier relationships (Kwaliteg, 2004).

Table 1.5. Common contract types (Warberg 1997).

| Contract type | Characterisation |
|---|---|
| Spot | Used in procuring generic products. Buying does not demand further cooperation and the contracts are typified by standard clauses. |
| Traditional contracts | Specified usually for a limited time frame. All conditions regulating exchange are written in the contract, according to the customer's needs. |
| Simple framework contract | Agreement is characterised by future exchange of specific product in own contract. The selection of the supplier has been done after normal competition. |
| Binding framework contract | In this agreement the customer commits to buy a certain amount of the product during specific time frame. The selection of the supplier has been done after competition. |
| Complex framework contract | This agreement form denotes an often close relationship between the customer and supplier. Synergy effects generate the joint profits. |
| Cooperation contract | Cooperation contracts differ from the complex framework contracts with inbuilt change and development aspects. The agreements are not specified in detail when it is signed. Profound negotiations are required before and after signing. |
| Integrated cooperation contract | This contract form expects the partners to work together in a joint project organisation. |
| Joint venture | This cooperation form requires the establishment of a joint company to carry out the assignment. |
| Full integration | The partnering companies are fully integrated either using the joint venture as described above or merging two companies. This type of integration is more usual between two suppliers, but not between customer and supplier where the customer is an end-user. |

### 1.4.2 Basic requirements for quality management activities

Since the food crisis in the nineties, very much attention has been focused on quality and food safety issues. To obtain a good quality end product, quality is managed more and more along the whole food chain from the supplier of raw materials to consumption. Customer understanding of food quality and the ultimate concern for health and food safety forces the actors in agri-business and the food industry to use quality management as a strategic issue in innovation and production.

The current situation in the food markets and agri-food production chain is very turbulent due to changing consumer requirements, increased competition, environmental issues and governmental interests. It requires continuous improvement in quality management activities, in which knowledge of modern technologies and management methods play a crucial role (website 12).

*Agri-food production chain*

Management of food quality is a rather complicated process. It involves the complex characteristics of food and their raw materials, such as variability, restricted shelf life, potential safety hazards, and the large range of (bio)chemical, physical, and microbial processes. Therefore, food products can be considered as dynamic systems that have a variable behaviour in time. Moreover, human handling plays a crucial role in quality management. Human handling is rather unpredictable and changeable. As a consequence, the result for agri-business and the food industry, as the combined action of individuals striving for quality, is much more uncertain than is often assumed (website 12).

The International Organisation for Standardisation (ISO) developed the 9000-series, which provided a framework for quality management and quality assurance (see also Chapter 6 on QA systems). The primary objectives of the quality management system requirements of ISO 9001:2000 are to achieve customer satisfaction by meeting customer requirements through application of the system, continuous improvement of the system, and prevention of nonconformity (website 12).

At the heart of the ISO 9001-2000 concept is the continual improvement model (Figure 1.8). This represents a system that is continually in cycle. Management has complete responsibility for knowing customer requirements, providing the necessary resources, planning and managing the process under control then measuring, analysing and acting on opportunities to improve future outcomes. Customer satisfaction will be measured as well.

The four blocks in the circle represent the core elements of the standard. This is a structure that will in fact improve any business of any type (Clause, 2001).

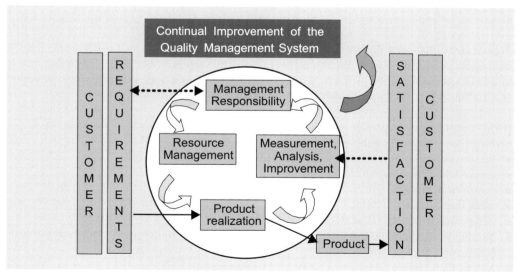

*Figure 1.8. Continual improvement model (Clause, 2001).*

*Susanne Knura, Stefanie Gymnich, Ewa Rembialkowska and Brigitte Petersen*

To realise such chain-wide management system activities, two key dimensions need to be taken into consideration (for more details see the chapter on risk management):
1. Organisational dimension: To set up organisational structures, such as "Trusted third parties", which support communication and cooperation between actors in the chain,
2. Technical dimension: To develop chain-wide IT systems, which integrate existing solutions, standard systems and available data sources to reduce implementation time and cost.

### 1.4.3 Food safety tools: an integrated approach

As already described, food safety is an essential element of quality. Thus, food safety tools have to be implemented intimately into quality management programmes. The practical success of a food safety programme will be dependent on the proper use of appropriate methods and tools. This comprises the use of elements of Good Manufacturing Practice (GMP) and Good Hygiene Practice (GHP), respectively (see also section 1.1.3).

The product safety requirements for the manufacture of a specific foodstuff, and the design and operation of the associated plant such that production of a safe food is assured, can be established by applying HACCP (Jouve *et al.*, 1998).

Other tools of more general application are quality assurance methods and systems such as the ISO 9000 series of standards (ISO 1994) and the Total Quality Management (TQM) approach (Jouve *et al.*, 1998). Note that TQM is not a system but a philosophy (see also Chapter on QA systems and food safety).

This section does not contain detailed information on the characteristics of these tools. For details, see the chapter on QA systems and HACCP implementation respectively. However, it describes the application and integration of these tools in a food safety program (Figure 1.9).

Figure 1.10 illustrates the intimate relationship between HACCP and the overall quality management system. Within the framework of a quality system, HACCP provides an approach to a food safety assurance plan which sets out the specific practices, resources and sequence of activities relevant to a particular product's safety (Jouve *et al.*, 1998).

Other tools of a more specialised nature may have application in areas of food safety management. These include:
- hazard analysis and operability studies (HAZOP);
- cause-and-effect diagram (fishbone or Ishikawa diagram);
- event tree analysis;
- fault tree analysis;
- failure mode and effect analysis (FMEA);
- predictive mathematical modelling (process modelling, microbial growth, death or survival, etc.); and
- probabilistic safety assessment (PSA).

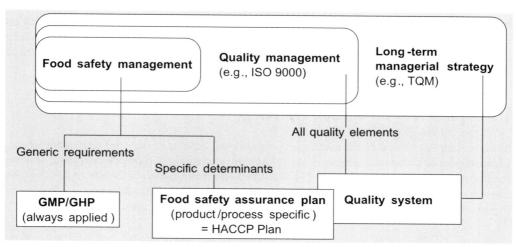

Figure 1.9. Food safety tools: an integrated approach (Jouve et al., 1998).

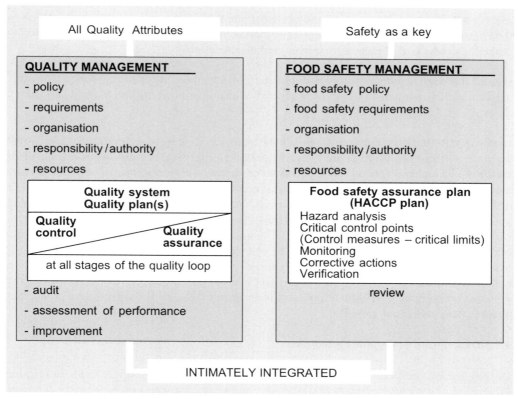

Figure 1.10. Food safety and quality management (Jouve et al., 1998).

Expert guidance should be sought on the use of these techniques, particularly those based on statistical concepts. Increasing numbers of software-based food safety packages are commercially available (Jouve *et al.*, 1998). Further information on these tools can be found in Mortimore and Wallace (2000), Montgomery (1996) and Whiting (1995).

## 1.5 Organic food production and safety

The previous sections focused mainly on conventional aspects of the food production chain, because most of European agriculture is still based on conventional methods, including mineral fertilisers and synthetic pesticides. This kind of agriculture has to be better controlled and consequently improved in respect of final product safety. Moreover, the previous sections of this chapter considered the food safety aspects in a predominantly primary food chain.

Intensive agriculture has been strongly criticised for the last several decades because of its negative influence on the environment, wildlife, quality of foodstuffs and overproduction. As a result, a new Common Agricultural Policy (CAP) was initiated in 1992, changing the main goals of agriculture in the European Union. The general direction is to make farming more extensive and friendlier to the environment, and to change the focus from quantity to quality aspects in food production. One of the most important goals is to improve food quality and safety and to make the production system more transparent to consumers.

One of the main sectors developed within the new Common Agricultural Policy is organic agriculture, using only natural methods of production and protection, both in plant and animal areas.

In this chapter the main factors influencing the quality and safety of organic food products will be presented and analysed. Then, the principal role of the legal regulations in organic plant production, animal production and organic food processing will be described, because these rules have a basic impact on food safety in the organic sector. In the last part of the chapter an overview of differences between organic and conventional food production is presented, as a result of all measures undertaken in the whole organic food production chain. This overview is based on scientific research conducted over the last 20 years, mostly in Europe. It shows the actual state-of the-art principles and practices within the wide topic of "organic food quality and safety". It is important to emphasise that organic agriculture is a coherent system comprising both plant and animal production; therefore the construction of this subchapter is different from the previous subsections. The holistic approach is necessary if the idea of organic food production chain is to be presented properly.

### 1.5.1 Description of organic agriculture

Organic agriculture is defined as determined in Council Regulation no 2092/91 on Organic Production of Agriculture Products and Indications referring thereto on Agriculture Products and Foodstuffs. Organic crops are obtained in controlled conditions according to the guidelines of the mentioned Regulation. Organic plant crops are produced without chemical pesticides and easy soluble mineral fertilisers, and with the application of natural animal manure and composts, green manure and diversified turnover. Communal waste and compost are forbidden. Animal production is conducted according to the animals' needs and on-the-farm produced fodder is used as the main food for animals. Certification in organic farming means that a control unit testifies the product as produced according to the accepted rules and the production system is obligatorily controlled.

Organic farming is becoming more and more popular in Europe and other parts of the world. At an overall constant growth rate in the EU countries of around 25 % per year, for the last 10 years, organic agriculture is doubtless one of the fastest growing sectors of agricultural production (Kouba, 2003). The main reason is the growing demand of consumers seeking more safe and controlled foodstuffs, and consumer concern about healthy environment and life. Organic methods in agriculture are considered friendly to the environment, mostly because of the basic principles of harmonious cooperation with nature and due to the lack of extensive use of chemicals. There is already much evidence that the condition of environment, soil and groundwater improves as a result of organic farming applications (Haas *et al.*, 2000). Organic farming is also often regarded as a system of improving crop quality. It is important to emphasise that food safety problems are more widely considered in organic farming than in conventional farming. This comprises not only the quality of the food product as such, but also the quality of the agri-food system: its ecological safety, transparency of production, control and information system, proximity to the consumer and, as a consequence, increased confidence of the consumers (Hansen *et al.*, 2002).

### 1.5.2 Factors influencing the quality of organic food products

In Figure 1.11 the main factors determining the final quality and safety of organic foods are presented. Environmental conditions are the most important factors in this respect. Without clean soil, water and air it is impossible to produce high quality crops. Next, the principal role of legal regulations in organic plant and animal production should be noted, because these rules have a basic impact on food safety in the organic sector. Equally significant is regulation of food processing in the organic sector - many methods and compounds commonly used in conventional food processing cannot be used in the organic sector and this creates essential differences in food safety.

Efficiency of the control and certification system is also a fundamental factor of the final product value. Even if legal rules are transparent and strong as in organic farming, farmers and processors have to follow these rules precisely. Without good control systems it is not possible to guarantee that the whole food production chain is in accordance with the regulations. "Confidence is good but control is better" - this motto should be followed by all stakeholders in the food chain, but especially in the organic sector, where consumers pay a higher price for the highest quality and safety of the products.

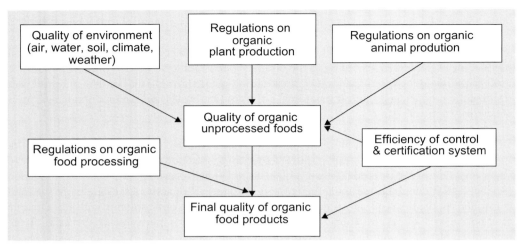

Figure 1.11. Factors influencing the quality of organic food products.

## 1.5.3 Environmental conditions and methods of farming as the basic factors of crop quality and safety

It is obvious that plant and animal food materials produced in a contaminated environment can contain harmful substances and - consequently - cannot be healthy. Therefore, it is very important to keep soil, air and water in the agricultural areas uncontaminated. When considering the contamination of food products, the main problem is that the same contaminant can be introduced into the food chain from different sources. Therefore, it is very often impossible to assess whether the particular substance is present in food because of a particular human activity. As a result of this complex situation it is better to present the potential contamination threat and include all the different sources (industry, transport, agriculture, communal sources), as in Table 1.6.

Industrial sources introduce a variety of dangerous substances into the food chain, such as heavy metals, chlorinated biphenyls (PCB) being the precursors of dioxins, and aromatic hydrocarbons. Transport sources (cars, buses, ships, airplanes, trains) also emit many contaminants into the environment. Communal sources (liquid sewage and solid waste) create more and more serious problems for the quality of the environment, especially in countries where improper utilisation of waste is common practice (countries in transition and developing). Other sources (medicine, weapon production) can have a big impact locally though generally their importance is not so great.

In Table 1.6 the **bold letters** indicate the food contaminants coming from conventional methods in agriculture, animal production and food processing. It appears that the extensive use of chemicals can have considerable effects in the agri-food chain.

Table 1.6. Food contamination, its sources and negative health impacts (Philp 1995; Rejmer 1997, own survey study).

| Food contamination | Sources | Negative health impacts |
|---|---|---|
| Heavy metals | | |
| ■ Cadmium (Cd) | Industry (non-ferrous metallurgy) **Agriculture (phosphoric fertilisers)** | Damage to kidneys, liver, osseous system Prostate cancer |
| ■ Lead (Pb) | Industry Transport (lead petrol) Pots, canned food | Disturbances of protein synthesis Anaemia Neurological and cerebral changes |
| ■ Mercury (Hg) | Chemical, electromechanical and dyeing industry **Agriculture (seed dressing)** | Paralysis of nervous system Mutagenic and teratogenic impact |
| ■ Arsenic (As) | Metal-forming industry Petroleum refinery **Agriculture (pesticides)** | Carcinogenic impact Metabolic disturbances |
| Nitrates, Nitrites | **Agriculture (nitric mineral fertilizers)** **Food processing** | Methemoglobinaemia |
| Nitrosoamines | **Agriculture (nitric mineral fertilizers)** **Food processing** | Carcinogenic |
| Pesticides, Seed dressings | **Agriculture (insecticides, herbicides etc.)** | Chronic intoxication Damage to nervous and digestive system Carcinogenic |
| Mycotoxins | **Improper grain storage** **Intensive farming → soil fungi start to produce mycotoxins** | Strongly toxic Carcinogenic |
| Chlorinated biphenyls (PCB) | Plastic packages Greases, paints, insulators **Insecticides** | Toxic impact on the whole organism |
| Aromatic hydrocarbons e.g. benzo (a) pyren | **Food processing (drying of cereals, smoking)** Contamination from industry, transport and communal sources | Carcinogenic |
| Plant growth stimulators, e.g. choline chloride | **Agriculture** | Toxic impact |
| Antibiotics, hormones | **Animal breeding** | Metabolic disturbances Reduction of resistance Asthma, anaemia, allergy |
| Radioactive isotopes | Radiation accident Trials with nuclear weapons Medicine | Leukaemia, cancer Radiation sickness |
| Plastic substances (monomers) | Packing from plastic (polypropylene, polystyrene) | Toxic impact |

*Susanne Knura, Stefanie Gymnich, Ewa Rembialkowska and Brigitte Petersen*

Most of the agricultural products are still produced and processed with conventional methods. This means that a lot of chemicals are used in the production system. They are introduced into the food chain via soil, plants, animals, and humans, and food contamination can sometimes be high. As a consequence, it is very difficult to control the food chain system in conventional agriculture, even if the level of applied chemicals is considered as moderate.

For example, cadmium can be introduced into the food chain from synthetic phosphoric fertilisers. Sowing grain had been treated with mercury compounds for many years (though it has been forbidden since 1988), and arsenic is still a component of some pesticides. Aromatic hydrocarbons, e.g. benzo (a) pyren can be found in improperly dried food grain and smoked meat preserves. Residues of the nitrates, nitrites, nitrosamines, pesticides, seed dressings, plant growth stimulators, antibiotics and hormones can be found in food products mainly because of the conventional methods used in agriculture, animal breeding and food processing.

It's important to emphasise that contaminants listed in Table 1.6 should be regarded as „critical points" to consider during all stages of the agricultural production.

### 1.5.4 Influence of regulations in the organic sector on food quality and safety

*Impact of regulation on organic food plant production*
The regulations for organic plant production are versatile and detailed. The EU regulation published in 1991 (2092 / 91) contains sections which relate directly to the composition of organic plant products (Council Regulation, 1991). The most important aspects of this regulation include (Hansen *et al.*, 2002):
- a ban on genetic engineering and GMOs;
- lower nitrogen levels: maximum limits for manure application of 170 kg N / ha / yr;
- a ban on synthetic pesticides;
- a ban on synthetic mineral fertilisers;
- a ban on growth promoters.

Organic farmers have to follow the above regulations if they want to pass the annual inspection procedure and receive a certification document. If all requirements are complied with, several qualitative results can be expected. The most important conclusions from scientific comparisons are given below.

*Impact of regulation on organic animal production*
The regulation of organic animal production is many-sided and comprises several aspects of feeding, housing, demarcation, care, medical treatment and slaughter (Council Regulation EC 1804 / 99). An EU regulation on organic animal husbandry was published in 1999 (1804 / 99), and several aspects of it are supposed to have a direct influence on the quality of organic animal products. This regulation obliges (Hansen *et al.*, 2002):
- extended access to outdoor areas with a lower stocking density;
- restriction on animal feeds:
  - compulsory use of roughage feeds;

- ban on antibiotics, growth promoters and additives;
- ban on GMOs;
- ban on meat and bone meal.
- double retention time after medicinal treatment.

Following the above instructions should have a clear impact on the composition of the animal products from the organic husbandry.

*Impact of regulation in organic food processing on food quality and safety*
The processing of organic food aims to maintain nutritional value and limit the number and quantity of additives and processing aids in food products (IFOAM, 2001). The regulations for organic processing in relation to food safety prohibit (Hansen *et al.*, 2002):

- the use of more than 5% non-organic constituents;
- irradiation, colouring agents, sweeteners;
- synthetic additives;
- the flavouring of animal products and artificial flavouring in vegetable food products;
- GMOs;
- artificial (trans) fatty acids.

These regulations have a direct influence on the composition and nutritional value of food products, so they are strictly connected with food safety (Hansen *et al.*, 2002). The presumption is that the impact of organic processed foods on human health ought to be more positive than that of conventional foods, but till now there has been no scientific evidence to support this hypothesis.

### 1.5.5 Quality of organic food products

Recent food crises (BSE and foot and mouth disease, food contamination by dioxins, toxic fungi, *Salmonella* and *Campylobacter* bacteria) have pushed consumers to look for more authentic and safer food. Organic food production is widely recognised as more friendly to the environment, more controlled and better for animal welfare. On the other hand many data indicate that a lot of the food contaminants have their source in the conventional methods of agriculture, animal production and food processing. The negative role of the extensive use of chemicals in agriculture can be significant as previously explained. Therefore, safer methods of agricultural production, mostly organic methods, are very important.

Studies conducted in several countries (Brandt and Mølgaard, 2001; Lund and Algers, 2003; Lundegårdh and Mårtensson, 2003; Rembialkowska, 2000; Worthington, 2001) have indicated several positive aspects of organic food quality but also a few negative aspects and some uncertainties (Table 1.7).

It is necessary to emphasise that present knowledge about the quality and safety of organically produced foods is incomplete and further research is necessary. Nevertheless, based on the above overview it is possible to assume that several important features of organic foods are better

Table 1.7. Overview of positive and negative aspects of organic food quality as well as uncertainties.

**Positive aspects of organic food quality**
- Organic crops contain fewer nitrates and nitrites and fewer residues of pesticides than conventional ones. There is no clear difference in the content of heavy metals between organic and conventional crops;
- Organic plant products contain as a rule more dry matter, more vitamin C and B-group vitamins, more phenolic compounds, more exogenous indispensable amino acids and more total sugars; however, the level of $\beta$ carotene is often higher in conventional plant products; organic plant products contain statistically more iron, magnesium and phosphorus; they also tend to contain more chromium, iodine, molybdenum, selenium, calcium, boron, manganese, copper, potassium, sodium, vanadium and zinc;
- Organic plant products usually have a better sensory quality - they have a more distinct smell and taste and they are sweeter and more compact because of higher dry matter content; preference for organic products is typical not only for humans but also for animals like rats, rabbits and hens. This phenomenon was also observed in cases in which - according to chemical analyses - both organic and conventional fodder satisfied the physiological needs of tested animals;
- Vegetables, potatoes and fruits from organic production show better storage quality properties during winter keeping - clearly fewer mass decrements, caused by transpiration, decay, and decomposition processes. This is possibly connected to a higher content of dry matter, minerals and total sugars. It brings not only nutritive but also economic profits; in the conventional system high yields are produced but big storage decrements undermine the economic sense of this production;
- Farm animals from organic herds show fewer metabolic diseases like ketosis, lipidosis, arthritis, mastitis and milk fever. Milk and meat from organically reared animals has a more profitable fatty acids profile and regularly contains more CLA (conjugated linoleic acid), which is thought to have an immunomodulating and anticarcinogenic effect on human health;
- Animals fed organically grown feed show better health and fertility parameters: fewer incidents of prenatal deaths, larger litters, and higher weight of young at birth and after 90 days, and better immunity to illness. It is especially interesting that the fertility and condition of animals fed organically improves over several generations. In the light of diminishing human fertility in civilised countries this fact might be of great importance;
- Organically processed foods contain much fewer synthetic additives (colouring and flavouring agents, sweeteners, artificial trans fatty acids, etc.);
- There is only a little evidence that human health could be improved as a result of eating organically. Some reports from the past show that an organic diet can improve general health and immunity, and some recent evidence, coming from the alternative medicine sector, indicates that a completely organic diet is essential for successful recovery after cancer. Further research in this respect is necessary;
- Organic food should be recommended for everyone but especially for young babies, pregnant and breast-feeding women, elderly or chronically ill people and vegetarians. The last group consumes a lot of vegetables which can contain too high levels of carcinogenic substances when produced conventionally;
- The lower content of nitrates and higher content of phenolic compounds and vitamin C in organic crops are of special importance for health. Nitrates are easily converted in our digestive tract into poisonous nitrites, which are the precursors of carcinogenic nitrosamines. This process is hampered by vitamin C, and carcinogenesis is retarded by phenolic compounds. Therefore, organic vegetables can play an important role in cancer prevention.

*Table 1.7. Continued.*

**Negative aspects of organic food quality**
- Plants cultivated in the organic system have as a rule a significantly lower yield; on average 20 % lower than conventionally produced crops. The same problem concerns animal production, where the yield of milk and meat is significantly lower. This means a lower profit for the organic producer, an increase in prices, and a barrier for many consumers to buy organic food;
- More frequent parasitic afflictions in organically reared animals are typical. This can create some problems for consumers though proper technological and culinary treatment can prevent health risks.

**Uncertainties**
- Environmental contamination (heavy metals, PCB, dioxins, aromatic hydrocarbons) can be similar in organic and conventional crops because the impact of industrial, transport and communal sources is similar on organic and conventional farms located in the same area. However, the level of cadmium sometimes appears to be higher in conventional crops, probably due to relatively high cadmium content in the phosphoric mineral fertilisers and communal sludge used in the conventional system;
- Bacterial contamination, mainly Salmonella and Campylobacter, can sometimes be higher in organic produce, but the scientific evidence is so far not clear;
- Mycotoxins can contaminate both organic and conventional foods. Scientific data is, however, contradictory, so more research into the problem is needed.

compared to conventionally produced foods. Consequently, the health and fertility of some organically fed experimental animals is also better. This statement enables us to make a powerful assumption: that perhaps the impact of organic foods on human health is positive. However, in order to prove it, serious long-term studies are vital.

# References

Ahvenainen R., K. Autio, I. Helnader, K. Honkapää, R. Kervinen, A. Kinnunen, T. Luoma, T. Lyijynen, L. Lähteenmäki, T. Mattila-Sandholm, M. Mokkila and E. Skyttä, 2000. VTT Research programme on minimal processing. Final report. VTT Research Notes 2052, VTT Technical Research Centre of Finland, Espoo.

Ahvenainen, R., 2002. Minimal processing in the future: integration across the supply chain, In: T. Ohlsson and N. Bengtsson (eds.), Minimal processing technologies in the food industry, Woodhead Publishing Limited.

Arnoldi, A., 2001. Thermal processing and food quality: analysis and control, In: P. Richardson (ed.), Thermal technologies in food processing, Woodhead Publishing Limited.

Berg, R.T.L., 1999. Food Hygiene in Catering,
http://www.berg1a.freeserve.co.uk/Java%20Out%20Home%20Page/Chapt11.HTM (01.06.2004)

Bloemhof-Ruwaard, J.M., P. van Beek, L. Hordijk and L.N. van Wassenhove, 1995. Interaction between Operational Research and Environmental Management, European Journal of Operational Research and Environmental Management, European Journal of Operational Research 85, 229-243.

Bradley, R., 2002. An overview of current research on animal transmissible spongiform encephalopathy (TSE), In: J.M. Smulders and J.D. Collins (eds.), Food safety assurance in the pre-harvest phase Wageningen Academic Publishers, The Netherlands.

Brandt, K. and J.P. Mølgaard, 2001. Organic agriculture: does it enhance or reduce the nutritional value of plant foods? J. Sci. Food Agric. 81: 924 - 931.

Brocka, B. and M.S. Brocka, 1992. Quality Management: Implementing the Best Ideas of the Masters, Irwin.

Clause, R., 2001. ISO 9001-2000 - a brief discussion, AgDM newsletter article, July 2001, http://www.extension.iastate.edu/agdm/articles/others/ClaJuly01.htm (13.09.2004).

Council Regulation EC 1804 / 99 of 19 July 1999 supplementing Regulation EC no. 2092 / 91 on organic production of agricultural products and indications referring thereto on agricultural products and foodstuffs to include livestock production, 1999.

Council Regulation, 1991. no. 2092 / 91 of 24 June 1991 on organic production of agricultural products and indications referring thereto on agricultural products and foodstuffs.

Early, R. and D. Shepherd, 1997. A Holistic Approach to Quality with safety in the Food Chain, In: G. Schiefer and R. Helbig (eds.), Quality Management and Process Improvement for Competitive Advantage in Agriculture and Food, Volume 1, Proceedings of the 49th Seminar of the EAAE, February 19-21, 1997, Bonn, Germany, ISBN: 3-928332-89-9.

EC No 178/2002, 2002. Regulation of the European Parliament and of the Council of 28 January 2002; laying down the general principles and requirements of food law, establishing the European Food Safety Authority and laying down procedures in matters of food safety.

Ellis, M.J.,1994. The methodology of shelf life determination, In: C.M.D. Man and A.A. Jones (eds.), Shelf life evaluation of foods. Blackie Academic & Professional, Chapman & Hall, Glasgow, 27-39.

Eufic, 2004. European Food Information Council, The Basics: Backgrounder on food safety, http://www.eufic.org/en/quickfacts/food_safety.htm#14 (01.06.2004).

FAO (Food and Agriculture Organization of the United Nations), 2004.
http://www.fao.org/es/ESN/food/foodquality_storage_en.stm (14.09.2004).

Fennema, O.R. and S.R. Tannenbaum,1985. Introduction to Food Chemistry, In: O.R. Fennema (ed.), Food Chemstry, Marcel Dekker, Inc., 1-23, New York.

Food Safety Authority of Ireland, 2004. Traceability - From Farm to Fork,
http://www.fsai.ie/publications/leaflets/farm_industry/farm_traceability.pdf (25.05.2004).

Frenzen, P.D., A. Majchrowicz, J.C. Buzby and B. Imhoff, 2000. Consumer Acceptance of Irradiated Meat and Poultry Products, Food Safety Economics, Agriculture Information Bulletin No. 757.

Frewer, L.J., C. Howard, D. Hedderley and R. Shepherd, 1997. Consumer attitude towards different food-processing technologies used in cheese production, The influence of consumer benefit, Food Quality and Preferences, 8, 271-280.

FVE, 2004. Federation of Veterinarians of Europe, Food safety, The stable to table approach, http://www.fve.org/papers/pdf/fhph/position_papers/stabletotable.pdf (26.05.2004).

Gould, G.H., 1996. Methods for preservation and extension of shelf life, International Journal of Food Microbiology, 33, 51-64.

Haard, N.F., 1985. Characteristics of edible plants tissues, In: O.R. Fennema (ed.), Food Chemistry, Marcel Dekker, Inc., 857-911, New York.

Haas, G., F. Wetterich and U. Köpke, 2000. Life cycle assessment of intensive, extensified and organic grassland farms in southern Germany, In: T. Alföldi, W. Lockeretz and U. Niggli (eds.), Proceedings 13th International IFOAM Scientific Conference, 28-31 Sept. 2000, Basel, Switzerland, s. 157.

Hansen, B., H.F. Alrøe, E.S. Kristensen and M. Wier, 2002. Assessment of food safety in organic farming, DARCOF Working Papers no. 52.
Hartig, M. and F. Untermann, 2003. The new European Food Law - 1. White paper - The concept for food safety, Fleischwirtschaft 83, 9, 143-145.
HCI, 2004. http://www.hci.com.au/hcisite2/toolkit/customer.htm (07.06.2004)
Hoogland, J.P., A. Jellema and W.M.F. Jongen, 1998. Quality Assurance Systems, In: W.M.F. Jongen and M.T.G. Meulenberg (eds.), Innovation of food production systems, Wageningen Pers, the Netherlands.
Hurst, W.C. and G.A. Schuler, 1992. Fresh produce processing- An Industry perspective, Journal of Food Protection, 55, 824-827.
IFOAM, 2001, Basic standards for organic farming and processing, 2nd draft 2001.
Jay, J.M., 1996. Modern Food Microbiology, 5th edition, (D.R. Heldmann, ed.) Chapman and Hall, 661 pp., New York.
Jongen, W.M.F., A.R. Linnemann and M. Decker, 1996. Productkwaliteit uitgangspunt bij aansturen product(ie)technologie vanuit keten, Voedingsmiddelentechnologie 29, 26, 11-15.
Jongen, W.M.F. and M.T.G. Meulenberg, 2005. Innovation in agri-food systems: Product quality and consumer acceptance. Wageningen Academic Publishers, Wageningen, The Netherlands, 398 pp.
Jouve, J.L., M.F. Stringer and A.C. Baird-Parker, 1998. Report on Food Safety Management Tools, ILSI Europe Risk Analysis in Microbiology Task Force, 83 Avenue E. Mounier, B-1200 Brussels, Belgium.
Kennedy, C.J. (ed.), 2000. Managing frozen foods, Woodhead Publishing Limited, Cambridge England.
Kouba, M., 2003. Quality of organic animal products. Livestock Production Science 80: 33-40.
Kwaliteg, 2004. http://www.kwaliteg.co.za/tqm/customer%20supplier%20relationship.htm (07.06.2004).
La Londe, P.C. and J.R. Raddatz, 2000. Tool time for customer-supplier relationship improvement, Customer-Supplier Division Conference, http://www.asqcsd.org/TapinCSD/ToolTime-SupplierImprovement.PDF (03.06.2004).
Lecerf, J.M., 1995. L'agriculture biologique. Interet en nutrition humaine? Cah. Nutr. Diet. 30, 6:349-357.
Leitzmann, C., 1993. Food Quality - Definition and a Holistic View, In: H. Sommer, B. Petersen and P. v. Wittke (eds.), Safeguarding Food Quality, Springer Verlag Berlin Heidelberg, ISBN 3-540-56368-7, New York.
London Borough of Hillingdon, 2004. A Step by Step Guide For Food Businesses - Food Safety Handbook, http://www.hillingdon.gov.uk/environment/food_safety/food_safety_booklet.pdf (01.06.04).
Lund, V. and B. Algers, 2003. Research on animal health and welfare in organic farming - a literature review. Livestock Production Science 80: 55 - 68.
Lundegårdh, B. and A. Mårtensson, 2003. Organically produced plant foods - evidence of health benefits. Acta Agric. Scand., Sect. B, Soil and Plant Sci. 53: 3 - 15.
Luning, P.A., W.J. Marcelis and W.M.F. Jongen, 2002. Food quality management - a techno-managerial approach, Wageningen Pers, ISBN 9074134815.
Montgomery, D.C., 1996. Introduction to Statistical Quality Control, 3rd ed. New York: John Wiley & Sons.
Mortimore, S. and C. Wallace, 1994. HACCP: A Practical Approach. Practical Approaches to Food, Control and Food Quality Series No.1. London: Chapman and Hall.
Murray, A.C., 1995. The evaluation of muscle quality, In: S.D.M. Jones (ed.), Quality and grading of carcasses of meat animals, CRC Press, Inc., 83-108, Florida, USA.
Notermans, S. and H. Beumer, 2002. Microbiological concerns associated with animal feed production, In: J.M. Smulders and J.D. Collins (eds.), Food safety assurance in the pre-harvest phase, Wageningen Academic Publishers, the Netherlands.

Paulus, K., 1993. Quality Assurance: The Strategy for the Production of Safe Food Products with High Quality, In: H. Sommer, B. Petersen and P. v. Wittke (eds.), Safeguarding Food Quality, Springer-Verlag, Berlin, Heidelberg.

Phillips, C.A., 1996. Review: Modified atmosphere packaging and its effects in the microbiological quality and safety of produce, International Journal of Food Science and Technology, 31, 463-479.

Philp, R.B., 1995. Environmental hazards & human health, CRC Press, Inc., Lewis Publishers, 306 ss, PN - 77/R - 74457: Indications of potato storage in the actively ventilated compartments are the standard's subject (seed-potatoes, edible potatoes provided for direct consumption and processing).

PN - 91/R - 75411: Indications regarding long-lasting storage of root vegetables provided for fresh consumption are the standard's subject (garden beet, carrot, parsley, celery, and Spanish salsify)

Polish Directive of Ministry of Health and Social Care from 27 DEC 2000, The highest permissible residues of the nitrates, heavy metals and pesticides in food products.

Reiche, Th. and U. Kleiner, 2002. Cook & Chill in Theorie und Praxis, Planung-Umsetzung-Kosten, Behrs Verlag, ISBN 3-89947-022-2, Hamburg.

Rejmer, P., 1997. Principles of eco-toxicology, Wydawnictwo Ekoinzynieria, Lublin, 208 ss.

Rembialkowska, E., 2000. Wholesomeness and sensory quality of potatoes and selected vegetables from the organic farms, Fundacja Rozwój SGGW, Warszawa.

Report of the Scientific Committee on Animal Health and Animal Welfare, 2002. The welfare of animals during transport (details for horses, pigs, sheep and cattle), Adopted on 11 March 2002 http://www.dyrenes-beskyttelse.dk/db/db.nsf/aa2d63c6d52d81afc125698b00753221/-a8557a407a258503c1256ba6002abd15/$FILE/Eutransp.pdf (14.09.2004).

Sady, W., R. Grys and S. Rozek, 1999. Changes of nitrate and cadmium content in carrots as related to soil and climatic factors, Folia Horticulturae 11/2: 105-114.

Sady, W. and T. Cebulak, 2000. The effect of irrigation and cultivation methods on some mineral compounds in storage roots of carrot, Folia Horticulturae 12/2: 35-41.

Schiefer, G., 1997. Quality Management and Process Improvement - The Challenge, In: G. Schiefer and R. Helbig (eds.), Quality Management and Process Improvement for Competitive Advantage in Agriculture and Food (Volume I), Dept. of Agricultural Economics, University of Bonn, Germany.

Schulze Althoff, G., 2004. Chain quality information system: development of a reference information model to improve transparency and quality management in pork netchains along the Dutch German border. In: H.J. Bremmers, S.W.F. Omta, J.H. Trienekens and E.F.M. Wubben (eds.), Dynamics in Chains and Networks, Wageningen Academic Publishers, Wageningen, The Netherlands.

Shapiro, A. and C. Mercier, 1994. Safe food Manufacturing, The Science of Total Environment, 143, 75-92.

Singh, R.P., 1994. Scientific principles of shelf life evaluation, In: C.M.D. Man and A.A. Jones (eds.), Shelf life evaluation of foods, Blackie Academic & Professional, Chapman & Hall, Glasgow, 3-26.

Singhal, R.S., P.R. Kulkarni and D.V. Rege (eds.), 1997. Handbook of indices of food quality and authenticity, Woodhead Publishing Limited, Cambridge, England.

Szteke, B., 1992. Impact of the chemical contaminants on the wholesomeness of the raw and processed fruits and vegetables, Przem. Ferm. Owoc. Warz. 8: 13-17.

Van Beek P., 1990. Het bestaansrecht van agrologistiek, interview, Tijdschrift voor Inkoop en Logistiek, 6, June, 23-26.

Van Dam Y.K., C. de Hoog and J.A.C. van Ophen, 1994. Reflecties op gemak bij voeding, In: Y.K. van Dam, C. de Hoog and J.A.C. van Ophen (eds.), Eten in de jaren negentig, Serie Economie van Landbouw en Milieu, Eburon Delft, 1-7 (in Dutch).

Van Trijp J.C.M. and J.E.B.M. Steenkamp, 1998. Consumer-oriented new product development: principles and practice, In: W.M.F. Jongen and M.T.G. Meulenberg (eds.), Innovation of food production systems, Wageningen Pers, Wageningen, The Netherlands, 37-66.

Vernède, P., F. Verdenius and J. Broeze, 2003. Traceability in Food Processing Chains - State of the art and future developments, Klict Position paper, Version 1.0, 20 October 2003, Agrotechnology & Food Innovations bv; Wageningen, 2003.

Vitiello D.J. and A.M. Thaler, 2001. Animal identification: links to food safety, Rev. sci. tech. Off. int. Epiz., 20 (2), 598-604.

Vogtmann, H., 1985. Ökologischer Landbau - Landwirtschaft mit Zukunft, Pro Natur Verlag, Stuttgart.

Wandel, M. and A. Bugge, 1997. Environmental concern in consumer evaluation of food quality, Food Quality and Preference, 8, 19-26.

Warberg, E., 1997. Mulige kontraktsstrategier for bedre og/eller rimeligere anskaffelser (In Norwegian). Kjeller, Forsvarets Forskningsinstitutt, FFI: 50.

Warsta, J., 2001. Contracting in software business - Analysis of evolving contract processes and relationships, ISBN 951-42-6599-8, ISSN 0355-3191, URL: http://herkules.oulu.fi/issn03553191/, Oulu University Press, Oulu, Finland.

website 1 http://agriculture.de/acms1/conf6/ws3sum.htm (07.09.2004)
website 2 http://www.fao.org/trade/docs/LDC-foodqual_en.htm (05.09.2005)
website 3 http://www1.agric.gov.ab.ca/$department/deptdocs.nsf/all/afs4361?opendocument
website 4 http://www.fightbac.org (07.09.2004)
website 5 http://www.lsuagcenter.com/Communications/LouisianaAgriculture/agmag/43_2_articles/-overview.asp (08.09.2004)
website 6 http://www1.agric.gov.ab.ca/$department/deptdocs.nsf/all/afs4338?opendocument (08.09.2004)
website 7 http://www.afma.co.za/AFMA_Template/sept_therole.htm (09.09.2004)
website 8 http://www.cdc.gov/ (09.09.2004)
website 9 http://vric.ucdavis.edu/veginfo/foodsafety/foodsafety.htm (09.09.2004)
website 10 http://www.foodmicrobiology.wur.nl/content/research.html#Fermentation_and_Fungi (14.09.2004)
website 11 http://www.dnv.com/binaries/Food_Safety_brochure_tcm4-10831.pdf (14.09.2004)
website 12 http://www.ftns.wau.nl/pdq/education/thesis/msc_fqm/topics_mscfqm.htm (13.09.2004)

White Paper on food safety, 2000. Brussels, 12 January 2000, COM, 719 final.

Whiting, R.C., 1995. Microbial modelling in foods. Critical Reviews in Food Science and Nutrition 35(6): 467-494.

Worthington, V., 2001. Nutritional Quality of Organic Versus Conventional Fruits, Vegetables, and Grains. The Journal of Alternative and Complementary Medicine 7/2: 161 - 173.

Zeuthen, P., 2002. Safety criteria for minimally processed foods, In: T. Ohlsson and N. Bengtsson (eds.), Minimal processing technologies in the food industry, Woodhead Publishing Limited.

# 2. Biological hazards

*Jordi Rovira, Avrelija Cencic, Eva Santos and Mogens Jakobsen*

## 2.1 Study objectives and structure of chapter

In this chapter biological hazards are discussed, which together with chemical and physical hazards (described in Chapters 3 and 4, respectively) present a potential danger in the production of "safe food". An understanding of these hazards forms the basis for the development of a good hazard analysis in an HACCP system (Chapter 7) and the performance of good Risk Management (Chapter 8).

In the introduction some basic concepts of food microbiology and related food safety are discussed, highlighting the importance of biological hazards with regard to food safety. For instance, the fact that, although most organisms associated with food are useful, some of them can exhibit pathogenic characteristics, causing severe adverse health effects under certain circumstances. It is important to clearly understand what an "emerging food pathogen" is and the reasons for its emergence, as this may facilitate easier control of such organisms. However, it also has to be noted that most of the factors contributing to the emergence of these pathogens are beyond human control.

In the second section, all types of biological hazards that can appear along the food chain will be described. The aim is to collect the most relevant information for the identification of hazards and design of appropriate corrective actions. It is important for the student to identify the different sources and routes of contamination in order to prevent or combat, where feasible, entrance of pathogens into the food chain.

In the third section, students will be shown how the different biological risks can reach and affect foodstuffs.

In the fourth section, we will try to answer the question: how do microorganisms survive and grow in food? A variety of intrinsic and extrinsic food factors which affect the survival and growth of microorganisms in the food matrix will be analysed.

To sum up, this chapter will provide the student with an overview of the main characteristics of the different types of biological hazards associated with the microbial contamination of foods, how different microorganisms can enter the food chain, and how they can survive or multiply in food. Control strategies are explained in Chapters 11 and 12.

*Jordi Rovira, Avrelija Cencic, Eva Santos and Mogens Jakobsen*

## 2.2 Introduction

### 2.2.1 Food safety and biological risks

Consumers in developed countries have never before enjoyed such a wide variety of foodstuffs with their balanced composition, broad sensory characteristics, long shelf lives, and packaging presentations, as are available today. Life expectancy and survival rates of children are also higher now than in any other period of human history. All these conquests have been achieved in part by a great revolution in food production, technology and nutrition. Moreover, food safety and control have been recognised as important issues in many countries for decades. However, this idyllic situation presents two paradigms. The first one is that almost two thirds of the world's population do not enjoy this idyllic situation and struggle to eat adequately every day, resulting in the death of a large number of people due to starvation. This imbalance cannot be ignored and work must still be done in developing food science and technology to try and address these differences. The other paradigm is related to food safety, as despite the considerable advances made in food science and technology, the safety of our food supply is still a matter of concern today (Käferstein, 2003). Recent outbreaks of food-borne diseases and food-related scandals have shaken consumer confidence in the food safety system. Terms such as BSE, dioxins, *E. coli* O157:H7, *Listeria*, *Salmonella*, virtually unknown only 10 years ago, are now household names (Schlundt, 2002). In 1983 the World Health Organization (WHO) Expert Committee on food safety concluded that "illness due to contaminated food was perhaps the most widespread health problem in the contemporary world and an important cause of reduced economic productivity (WHO, 1984).

Foods can be contaminated by physical, chemical and biological hazards. In food safety terminology, risk is the probability that hazard occurs. Therefore, a biological risk can be defined as the probability of a biological agent (hazard) contaminating food during any step of the food production, which if ingested by the consumers can cause a health disorder. Every living organism able to colonise foods and in some cases grow or survive in the food matrix, or any structure that needs the participation of a living being to reproduce itself, or produce toxic metabolites, can be considered as a biological agent.

There are some important differences between biological, chemical, and physical hazards. Biological hazards involve mainly living organisms, implying strong growth/no growth/death depending on the intrinsic and extrinsic factors of the food matrix. Intrinsic factors are mainly related to the structure and composition of the foodstuff, while extrinsic factors are related to some environmental parameters that affect the food. Examples of extrinsic factors are temperature, oxygen availability, the presence of $CO_2$, pressure, type of packaging, and method of processing.

Biological hazards seem to have a more notorious impact on public opinion than physical or chemical risks, probably because biological risks are more frequently reported, and normally affect to a certain extent a large number of consumers. In addition, biological hazards generally induce acute symptoms, which make them noticeable by the consumer. Several data show that the incidence of foodborne microbiological diseases has increased over the last few decades. In that sense, data collected by WHO of *Salmonella* serotype distribution from human cases shows a dramatic relative increase in *S. enteritidis* in the recent past. Reports of campylobacteriosis cases have also been continuously increasing in many parts of the world since 1985 (Schlundt, 2002). In 1990, research carried out in 11 European countries showed a rate of 120 cases of foodborne illnesses per 100,000 people; however, more recent studies indicate that in some European countries, at least 30,000 people out of every 100,000 suffer from acute gastroenteritis every year. Most of these cases are thought to be foodborne in nature (Notermans and van der Giessen, 1993). In the USA, an average of 76 million cases per year of foodborne illnesses has been reported, resulting in 325,000 hospitalisations and 5,000 deaths (Mead *et al.*, 1999). According to Käferstein (2003), sentinel studies in industrialised countries showed an unexpectedly high annual prevalence of foodborne diseases in around 10-15 % of the population. Recent data from the USA shows a much larger figure of 30% (Mead *et al.*, 1999), and it is easy to imagine that this figure is even higher in developing countries, with more severe consequences. It is accepted that reported cases of foodborne diseases are just the "tip of the iceberg". Wheeler *et al.* (1999) in a recent sentinel study in England estimated that for every case detected by laboratory surveillance, there were 136 in the community. The degree of under-reporting varies according to the geographical area, type of microorganism, and the severity of the foodborne illness that it causes. For instance, in England, the ratio is 3.2:1 for *Salmonella*, 7.6:1 for *Campylobacter*, 35.1:1 for rotavirus, and 1562:1 for Noroviruses (Norwalk-like viruses). Under-reporting of foodborne illnesses is due to several reasons: many foodborne pathogens cause mild symptoms and the victim may not seek medical help, in many countries only a few foodborne diseases appear on the list of notifiable diseases and in many cases the lists from different countries report different diseases. In several cases only data from outbreaks and not from sporadic cases is collected, and in most countries the surveillance infrastructure is either weak or nonexistent (Forsythe, 2000; Schlundt, 2002).

Apart from the health costs, foodborne illness implies an important economic cost as well. According to the EU Commissioner for Public Health D. Bryne, salmonellosis affects up to 160,000 people and causes 200 deaths per year in Europe and the cost of the illness is up to €2.8 billion. In the same way, the cost of campylobacteriosis in the USA has been estimated at between US$1.5 and US$8 billion (Forsythe, 2000).

Biological hazards can affect the food chain at any step, and they may be macro- and microbiological hazards. Macrobiological hazards are those which are seen without the aid of a microscope, like insects or small mammals. They hardly ever pose a real hazard for food safety, because although unpleasant it is easy for the consumer to remove them from the food. Of course, these animals could indirectly contaminate food with some pathogenic bacteria, like *Salmonella* spp., if they are portable. Although some parasites are considerable in size, they are traditionally considered as microbiological hazards (Mortimore and Wallace, 1994).

*Jordi Rovira, Avrelija Cencic, Eva Santos and Mogens Jakobsen*

## 2.2.2 Microorganisms and food

For many centuries human beings have been witnesses of the special relationship between microorganisms and food: at times utilising this relationship to our advantage, and at other times suffering infections or health disorders caused by the presence of microorganisms in foodstuffs. In fact, it is possible to find three main kinds of microorganisms in foods, which reminds us of the old western film entitled "The Good, The Ugly and The Bad". The "Good" microorganisms are those which produce a positive change in the raw food matrix, to obtain a different product, whose sensory properties have been changed and whose shelf life has been increased as with fermented foods. For many centuries fermentation has been used by humans to preserve food; although we still consume fermented foods, fermentation is applied now more for the desired alteration of a product's sensorial properties than for preservation. Among fermented products it is easy to find some of the most popular foods and beverages, like cheese, dry fermented sausages, wine, and beer. The "Ugly" are spoilage microorganisms that reduce the shelf life of foods by changing their sensory properties, producing an awful, disgusting aspect to foods, which prevents consumers from eating them. Some of these spoilage microorganisms are the same as or are related to the "Good" ones, which are utilised in fermentations. However, in this case they grow in other products or under different circumstances resulting in spoilage. This is the case with some lactic acid bacteria that contaminate beer or spoil vacuum packaged meat. The "Bad" are those microorganisms that cause a health disorder when they are ingested together with a contaminated foodstuff. A wide spectrum of these "Bad" microorganisms can contaminate human food and water supplies, and can cause illness after being consumed. These microorganisms are referred to as foodborne pathogens, and the diseases they produce are known as foodborne illnesses. Foodborne pathogens include a variety of enteric pathogens and parasites, as well as marine dinoflagellates, bacteria that produce biotoxins in fish and shellfish, moulds that produce mycotoxins, and the self-inducing prions of the transmissible encephalopathies. Whereas some of these pathogens require human hosts as part of their life-cycle, others infect humans accidentally. Some are professional foodborne pathogens that are only transmitted by food, while others can follow different infection routes (Tauxe, 2002).

In general, foodborne infections occur when pathogenic microorganisms are ingested, and sometimes they invade the mucosa or other tissues. It seems, however, that some pathogens can cause infections by other alternative routes such as wound infection. An example of which are *Streptococcus iniae* and *Vibrio vulnificus* biotype 3, both newly recognised as human pathogens, that cause severe wound infections in handlers who get pricked with a fish bone in the process of manipulating live tilapia fish (Bisharat *et al.*, 1999).

It is possible to distinguish different modes of action of foodborne pathogens. Foodborne toxico-infections arise when a microorganism ingested in contaminated food grows in the gastrointestinal tract and elaborates a toxin(s) that damages tissues or interferes with normal tissue/organ functions. Foodborne microbial intoxications occur by ingestion of a food containing harmful toxins or chemicals produced by the microorganisms usually during their growth in the food (IFT, 2002). The symptoms of foodborne illnesses vary depending on the causative microorganism, its mode of action, and the status of the host's health. Most infectious illnesses are characterised by acute symptoms such as vomiting and diarrhoea affecting the gastrointestinal tract, these being typical symptoms of infections caused by *Bacillus cereus*, *Clostridium perfringens* and Noroviruses. Other pathogens such as *Campylobacter jejuni* or the protozoa *Cryptosporidium parvum* also affect the gastrointestinal tract, but with greater severity in immunocompromised people. As mentioned above, not all foodborne diseases are limited to the gastrointestinal tract; some of them, such as Non-typhoidal *Salmonella* and *E. coli* O157:H7 can invade deeper tissues or produce toxins that produce systemic symptoms, including fever, headache, kidney failure, anaemia, and even death. Other pathogens, such as *Listeria monocytogenes* or *Clostridium botulinum* exert their pathogenicity outside the gastrointestinal tract, affecting foetuses and the central nervous system. Apart from acute illness, which is most noticeable to the public, pathogenic microorganisms also cause chronic diseases within sensitive target groups of the population. Among these chronic diseases are toxoplasmal encephalitis which results in dementia, inflammatory bowel disease, and the Guillain-Barré Syndrome, a disorder of the peripheral nervous system and a common cause of neuromuscular paralysis. This syndrome has also been correlated with previous infections by the individual *C. jejuni*. And finally biotoxins, which are substances produced by some foodborne microorganisms and that can cause food intoxication. Biotoxins are also responsible for both acute and chronic illness manifestations that affect different organs. Although the microorganisms producing biotoxins can disappear from the foodstuff, the biotoxins themselves remain active. Biotoxins include toxins produced by bacteria (i.e. botulism toxin), by moulds (i.e. aflatoxins, patulin), by dinoflagellates (ciguatoxin), and by plants (phytotoxins).

The pathogenic spectrum can change substantially over the course of time, and some of them disappear or reduce their presence whereas others may emerge. At the beginning of the last century, typhoid fever, brucellosis, and trichinosis, among others, were recognised as foodborne illnesses. The introduction of some new processes and methods as a result of technological advances, like milk pasteurisation, good manufacturing practices, and not feeding pigs with uncooked garbage, made these foodborne illnesses almost disappear (Tauxe, 2002). It is now known that more than 200 diseases are transmitted through food, and more than half of all recognised foodborne disease outbreaks have an unknown origin. The implications of this are that the real number of disease-causing agents is greater than 200, and despite advances made during the 20th century there are still many foodborne pathogens to be identified (IFT, 2002). Table 2.1 shows a list of the 27 principal foodborne pathogens in the USA that account for only 19 % of the total estimated number of cases attributable to foodborne illnesses and 36 % of deaths during 1997. In this list there are 13 microorganisms that have been recognised as foodborne pathogens in the last 25 years, e.g. *E. coli* O157:H7, *Cryptosporidium parvum* and the noroviruses. These 13 pathogens account for 82 % of all estimated cases. In contrast, 5 pathogens that were major causes of foodborne illnesses in the 19th century, such as *Brucella, C. botulinum, Salmonella Typhi, Trichinella*, and toxigenic *V. cholerae*, now represent only 0.01% of the cases (Tauxe, 2002). The situation is quite similar in Europe, meaning that new emerging foodborne pathogens are now the leading diseases transmitted through food all around the world. As a result of this and in a broad context, emergence can be used to describe a recent significant change. Using this idea, a foodborne pathogen could be described as emerging when it is first linked to a human

*Jordi Rovira, Avrelija Cencic, Eva Santos and Mogens Jakobsen*

Table 2.1. Principal foodborne infections as estimated for 1997, ranked by estimated number of cases caused by foodborne transmission each year in the United States (Mead et al., 1999). (Values over 1000 are rounded to the nearest 1000).

| | |
|---|---|
| Norwalk-like viruses* | 9,200,000 |
| Campylobacter* | 1,963,000 |
| Salmonella (nontyphoid) | 1,342,000 |
| Clostridium perfringens | 249,000 |
| Giardia lamblia | 200,000 |
| Staphylococcus food poisoning | 185,000 |
| Toxoplasma gondii | 112,000 |
| Escherichia coli O157:H7 and other Shiga-toxin producing E. coli* | 92,000 |
| Shigella | 90,000 |
| Yersinia enterocolitica* | 87,000 |
| Enterotoxigenic E. coli* | 56,000 |
| Streptococci | 51,000 |
| Astrovirus* | 39,000 |
| Rotavirus* | 39,000 |
| Cryptosporidium parvum* | 30,000 |
| Bacillus cereus | 27,000 |
| Other Escherichia coli | 23,000 |
| Cyclospora cayetanensis* | 14,000 |
| Vibrio (noncholera)* | 5000 |
| Hepatitis A | 4000 |
| Listeria monocytogenes* | 2000 |
| Brucella | 777 |
| Salmonella typhi (typhoid fever) | 659 |
| Botulism | 56 |
| Trichinella | 52 |
| Vibrio cholerae, toxigenic* | 49 |
| Vibrio vulnificus* | 47 |
| Prions* | 0 |

Those that have emerged in the last 30 years are indicated by an asterisk

disease, when the illness it causes suddenly increases in frequency or severity, or even when a previously recognised pathogen suddenly "reappears" (IFT, 2002).

It would be interesting to know which factors make a microorganism "emerge" as an important foodborne pathogen and cause illness, in order to improve safety throughout the food chain. A foodborne illness appears as a result of three main factors: the pathogen, the host, and the environment in which they exist and interact (Figure 2.1). Current strategies to control foodborne illnesses are based on reducing the contribution of one of these three factors, but probably a more effective way to succeed would be the reduction of several of these factors to obtain a

# Biological hazards

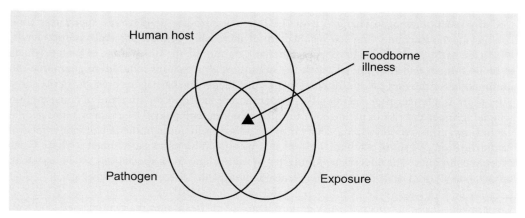

Figure 2.1. The main factors determining the occurrence of a foodborne illness.

combined effect. When one or more of these three factors change there is an opportunity for new foodborne pathogens to "emerge".

## 2.2.3 Causes of foodborne "emerging" pathogens

There are different causes that make pathogenic microorganisms "emerge", all of which could be explained by change that affects the host, the pathogen, and the environment. Analysing the causes behind these changes could be of interest in developing ways to diminish or avoid new foodborne illness incidences.

*Host factors*
Demographic changes have affected the number of cases of foodborne illness, and as the world population continues to increase so do our food safety problems. However, it is the uneven distribution of people that causes more serious problems than the continuing increase in the number of people. In developed countries these demographic changes imply a change in the structure of the population pyramids: as a consequence of the drop in the number of births and the increase in life expectancy levels, the population is getting older. Population ageing has a big influence on the prevalence of foodborne diseases, because it creates a large sector of people with a less effective immune system who are more susceptible to such diseases. There is a continuous increase in the number of people who suffer from chronic diseases like diabetes, those who are infected with HIV, or those receiving chemotherapy or drugs to combat the rejection of transplanted organs. The immune systems of these people are weaker and consequently less resistant to foodborne pathogens.

International travelling, especially air travel, causes people to acquire foodborne infections in one part of the world, develop clinical symptoms and spread the causative agent in another part of the world (Käferstein, 2003).

Cooking habits have also changed over the last few decades; people now spend less time cooking in the kitchen. This has resulted in an increasing demand for convenience foods, which are easier and quicker to prepare at home. Commonly, the preparation of such products at home is such that they do not reach the minimum temperature required to guarantee the elimination (or prevent the multiplication) of some foodborne pathogens present. In the same way, there is a longer time between the preparation of food and the moment of consumption which increases the risk of pathogens multiplying in the food matrix. Nowadays, more people eat in restaurants, where certain factors may influence an increase in the risk of food infection or intoxication. We must take into account that most food outbreaks are linked to these types of establishments. The incessant popularity of eating raw food specialities, such as "steak tartar", "sushi" or "sashimi" and rare beef, also increases the risk of food poisoning.

The growing desire of consumers to eat more "natural" or "fresh" minimally processed food with a long shelf life also constitutes a high risk for food safety. The danger is further increased when these kinds of foods undergo temperature abuse and/or insufficient heating prior to consumption. That is why it is very important to test, from the point of view of food safety, all changes in food formulations or food processing.

Finally, a lack of education in the basic rules of the hygienic preparation of foods also plays an important role in the rise of foodborne illness. More attention should be paid to this step of the food chain because it is the definitive step: the moment at which food pathogens come into contact with the host (consumer).

*Pathogen factors*
Among pathogenic factors, evolution and natural selection play an important role. This natural mutability of microorganisms is part of their nature and happens without human intervention. In addition to this unstimulated hypermutability, food production systems and the processing environment can increase the rate of changes in foodborne pathogens (IFT, 2002). In this sense, environmental stress can provoke a series of survival mechanisms in microorganisms enabling them to adapt to their new environment. These survival mechanisms include the activation of virulence determinants or the adoption of a viable but nonculturable state (VBNC) as a response to changes to extrinsic factors such as the temperature, osmotic stress or nutrient levels. The VBNC state is currently still under discussion (Forsythe, 2000). Among the stress situations that accelerate the rate of bacterial evolution are starvation, pH, oxidation, heat, cold, osmotic imbalance, and high pressure. It has been proved that genetic resistance to one stress may protect the microbe from a different stress. This phenomenon is known as cross protection, an example of which is the observation that some *E. coli* O157:H7 strains that are resistant to mild heat treatments, acid, peroxides or osmotic stress, are also more resistant to high pressure treatments (Benito *et al.*, 1999).

The use of antimicrobial agents in livestock production is a cause of great concern in food safety. It has been demonstrated that the abuse of these substances may select resistant strains to those and related antimicrobials, which in turn can reduce the effectiveness of human treatments. As an example, the use of fluoroquinolone in a chicken flock can change the prevalence of *C.*

*jejuni* strains from 100% susceptibility to 100% resistance within days (McDermott *et al.*, 2001). Among Campylobacter strains isolated from human infections in the USA, 14% were resistant to fluoroquinolone and were found to be linked to poultry consumption (CDC, 2002).

The transfer of genetic information between separate lineages of pathogenic bacteria seems to be more frequent than expected. This mechanism of changing genetic information is also one way by which a pathogen can "emerge" relatively quickly. Such a mechanism might have been used in the evolution of the highly virulent O157:H7 strain of *E. coli*.

In addition to the factors mentioned above, it is important to emphasise the improvement of analytical techniques, which enable us to identify more accurately some pathogens that were impossible to identify some years ago.

*Environmental factors*
There are also some environmental factors that can affect the emergence of foodborne pathogens. Some of them are related to the factors that have already been pointed out. Globalisation or international trade can move, directly or indirectly, pathogens and toxins from one country or region to another. For example, the first outbreak of the protozoa *Cyclospora cayetanensis* in the USA and Canada, which affected 1465 people, was caused by the importation of contaminated raspberries from Guatemala (IFT, 2002). In an indirect way, the new serotype O3:K6 of *Vibrio parahaemolyticus* was introduced to the USA from Southeast Asia as a result of the common practice of using harbour water as ballast in empty tankers. This practice involves the movement of huge volumes of water from the harbour of origin to the final destination and may play an important role in spreading pathogenic microorganisms all around the world (Tauxe, 2002).

As a consequence of the rising world population, food production has also increased by means of intensified agricultural and animal production. This situation is favourable for the spreading of zoonotic microorganisms that can introduce new foodborne illnesses. In order to achieve the maximum growth rate in intensive farming, some antimicrobials are used to promote growth. A good example is the appearance of the multi-resistant DT104 strain of *Salmonella Typhimurium* (penta-resistant), which has a selective advantage where ampicillin-like agents, florfenicol, streptomycin, sulphonamides, or tetracyclines are used. To counteract this, recent studies in Denmark show that it is possible to grow chickens in an intensive manner without using antimicrobials or animal antibiotics in the farm. There is also much concern over the use of manure to fertilise agricultural fields, which occurs in organic farming, mainly if the manure is not properly managed.

Global warming causes the warming of the oceans. A small difference in sea water temperature can affect the ecology of some toxic dinoflagellates, which are expected to increase in number when this happens, which in turn enhances the risk that certain fish and shellfish become poisonous (Käferstein, 2003).

*Jordi Rovira, Avrelija Cencic, Eva Santos and Mogens Jakobsen*

## 2.3 Types of biological hazards

### 2.3.1 Bacteria

Bacteria are the most well known biological hazards. With the increase in awareness, many people today know what *Salmonella* or *Listeria* are and the tremendous impact that these bacteria have on public health. Bacteria are prokaryotic microorganisms, which mean they do not have a true nucleus inside their protoplasm. According to the different behaviours they show in response to the Gram stain, it is possible to differentiate between Gram-positive and Gram-negative bacteria. Gram-positive bacteria show a purple colour with Gram stain, whereas Gram-negative bacteria show a pink colour. These chromatic differences are related to the different cell wall structures that these two different types of bacteria have. Gram-positive bacteria have a more or less developed cell wall, called the peptidoglycan above the cell membrane. In contrast, Gram-negative bacteria have a small peptidoglycan layer and a second cell membrane over the peptidoglycan.

In this section, the most prominent Gram negative and Gram-positive foodborne pathogens will be described in detail.

**Gram-negative foodborne pathogens**

*Salmonella*
**General description -** *Salmonella* spp. are probably the most well-known bacterial foodborne pathogens. Most people have probably been in contact with them directly or indirectly. They are rod-shaped (0.3-1 µm x 1.0-6.0 µm) Gram negative, non-sporeforming, motile bacteria, with the exception of *S. pullarum* and *S. gallinarum* which belong to the family *Enterobacteriaceae*. They are also facultative aero-anaerobes, and have an oxidative and fermentative metabolism. They produce acid and frequently gas during fermentation of D-glucose or other sugars and the G + C content ranges from 50 to 53 %.
There are different serotypes of *Salmonella* according to the antigenic structure of the strains.
- O antigens or somatic antigens (in German "ohne hauch": these are antigens from the cell wall related to the lipopolysaccharide. These antigens are heat stable and alcohol resistant.
- K antigens or capsule antigens.
- H antigens or flagellar antigens (in German "hauch"): related to the protein flagellin, its amino acid composition is constant for each antigenic type.

For a long time the antigenic structures or serotypes, of which there are more than 2000, were used to name the different Salmonellas. Now, most *Salmonella* species are classified as *Salmonella enterica* and then further identified by a serovar, e.g. *Salmonella enteridis* becomes *S. enterica* serovar Enteridis (see Figure 2.2; IFT, 2002).

**Physiological data -** *Salmonella* spp. are capable of growing between 5 and 45/47 °C with an optimal growth temperature range of 35-37 °C. Below 10 °C they usually grow very slowly. Like most bacteria belonging to the family *Enterobacteriaceae*, *Salmonella* also show a high sensitivity

## Biological hazards

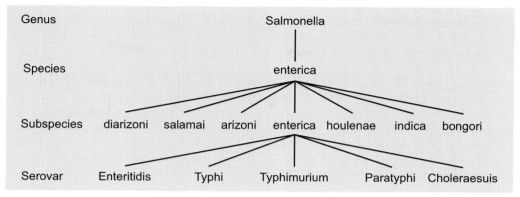

*Figure 2.2. Nomenclature of Salmonella.*

to heat, pasteurisation at 72 °C for 15 seconds being enough to destroy them in milk. However, different strains show different behaviours in different types of food.

Although *Salmonella* can grow at cold temperatures, an important decrease in population occurs after freezing. Despite this fact, one should by no means assume elimination of present Salmonellas by freezing.

*Salmonella* can grow in a pH range between 4.5 and 9.0, with optimum growth at pH values between 6.5 and 7.5. Resistance to pH depends on the type of acid that is determining the pH of the growth medium. An example of this being that certain strains can grow till pH 4.5 with citric acid in the medium, but can persist in acidic mayonnaise at pH 3.2 (Gledel, 1988).

With regards to their resistance to $a_w$, *Salmonella* can grow easily between 0.945 and 0.999, and can still multiple at 0.930. At lower $a_w$ values, i.e. 0.200 which are typical for dehydrated foods, they can persist for a long period of time. They are quite sensitive to high NaCl concentrations (5.8% maximum), but are rather resistant to nitrites (curing agent). Lactic acid bacteria are better competitors than *Salmonella* in foodstuffs and normally inhibit their growth.

Finally, *Salmonella* has a very low resistance to radiation and is destroyed by radiation doses less than 10KGy.

**Source** - *Salmonella* can be isolated from the intestines of a broad range of wild and domestic animals, especially those of poultry and swine. *Salmonella* can infect and cause clinical disorders in some domestic animals such as cattle, horses, and sheep. However, in other animals, such as poultry, swine, and dogs, they do not produce any sintomatology (Gledel, 1988). The ubiquitous nature of *Salmonella* may facilitate a cyclic lifestyle consisting of passage through a host to the environment and back into a new host. The long-term survival of *Salmonella* in the secondary habitat (non-host environments) ensures its passage to the next host (Winfield and Groisman, 2003). Secondary habitats include water, soil, insects, factory and kitchen surfaces (biofilms), animal faeces (manure), and raw meat, poultry and seafood. *Salmonella* can live in non-host environments for a long time. For example, it can survive for more than 21 days in waste slurry from infected pigs that is spread on farmland as an agricultural fertiliser (Baloda, *et al.*, 2001). It can also survive for up to 4 weeks in flies that are able to transmit it (Mian *et al.*,

2002), or for longer than 50 days on toilet bowls by forming part of the biofilms there (Barker and Bloomfield, 2000). *Salmonella* can also live for a long time in water and soil, where it can survive and multiply for at least one year (Winfield and Groisman, 2003).

**Associated foods -** The majority of the 1.3 billion annual cases of *Salmonella* that cause human gastroenteritis result from the ingestion of contaminated food products, such as raw or undercooked meat, poultry, eggs, milk and dairy products, fish, shrimp and seafood, sauces and salad dressings, cream-filled desserts, cakes, and chocolate (IFT, 2002; Pang *et al.*, 1995). Several *Salmonella* species were isolated from eggshells, but recently *S. enteridis* has been isolated from the egg yolk. This suggests vertical transmission, and the existence of the microorganism inside the egg, before deposition on the shell (IFT, 2002).

*Salmonella* infections are also associated with the consumption of fresh fruits or vegetables that have been contaminated by infected fertiliser (Tauxe, 1997). Strangely, on the other hand, *Salmonella agona* has been isolated from dry breakfast cereal, a ready-to-eat food normally considered to be devoid of microbial contamination (Zottola, 2001).

As mentioned above, flies can cause cross-contamination, and poor handling or handling of food by an infected person can also result in the contamination of foods.

**Mode of action -** The infective dose capable of causing disease is relatively low at 15-20 cells. The dose depends upon the age of the host (affecting mainly children and elderly people), and the health state of both the host and the strain. AIDS patients suffer from recurrent episodes of infection, which are estimated to be 20-fold more than the general population (IFT, 2002). The cause of the disease is the penetration of the *Salmonella* cells from the lumen of the gut into the epithelium of the small intestines, giving rise to an inflammation. Some authors have found evidence of enterotoxin production within the enterocytes in determinate circumstances (Sadruddin *et al.*, 1983; IFT, 2002). It also occurs quite frequently that *Salmonella* harbours some plasmids encoding antibiotic resistance (Gledel, 1988).

**Disease -** The disease caused by *Salmonella* is called salmonellosis, and affects 160,000 people annually in the EU, causing 200 deaths, and costing €2.8 million per year. The first signs of *Salmonella* infection appear between 6-48 hours of ingestion of contaminated food. There are two main manifestations of the disease; the first one produces typhoid or typhoid-like fever in humans, whereas the second one produces gastroenteritis.

Typhoid and typhoid-like fever are caused mainly by *S. typhi* and *S. paratyphyi* and various organs may be infected leading to lesions. The fatality of typhoid fever is 10% compared to less than 1% for most forms of salmonellosis. These strains also produce septicaemia, which has been associated with the subsequent infection of virtually every organ system (IFT, 2002).

Other *Salmonella* species are responsible for acute symptoms such as nausea, vomiting, abdominal cramps, diarrhoea, fever, and headaches. These symptoms may last for 1 to 2 days or more, depending on strain, ingested dose, and host related factors. It is rare but possible that salmonellosis could be fatal, the chance of fatality being higher in immuno-depressed patients, in newborns and elderly people. Chronic consequences, such as arthritic symptoms, may follow 3-4 weeks after the acute phase. Reactive arthritis may occur with a frequency of about

2% of culture-proven cases (IFT, 2002). The most common *Salmonella* serovars involved in gastro-intestinal disease are Enteridis and Typhimurium.
Although there are target groups more susceptible to *Salmonella* infection, such as the elderly, infants and the infirm, all age groups can be affected.

*Shigella spp.*
**General description -** *Shigella* are Gram-negative, non-motile, non-sporeforming rod-shaped bacteria. The four species of the genus are sometimes designated a letter related to the serotype pattern: Serotype A: *S. dysenteriae*, serotype B: *S. flexneri*, serotype C: *S. boydii* and serotype D: *S. sonnei*. The most usual Shigella are *S. dysenteriae* and *S. sonnei*, the later being the causative agent of most *Shigella*-related diarrhoea.

**Physiological data-** *Shigella* and *Escherichia* are closely related, which has led some authors to believe that they are actually metabolically inactive biogroups of *E. coli* (Rowe and Gross, 1984). They are facultative anaerobes, having both a respiratory and fermentative type of metabolism. They do not produce gas from carbohydrates and are also lactose negative.

**Source -** *Shigella* are highly infectious bacteria that are often found in water polluted with human faeces, in foods contaminated by infected handlers with poor hygiene standards, and in vegetables irrigated with sewage. Infected flies can also contaminate foods. Drinking or swimming in contaminated water, due to contact between the water and sewage or with someone with shigellosis, can also cause an infection. The person to person infection route has also been reported.

**Associated foods -** As mentioned above, raw vegetables and salads (potato, tuna, shrimp, macaroni and chicken), milk and dairy products, and poultry are among the foods associated with *Shigella* infections. Contamination of these foods is usually via the faecal-oral route.

**Mode of action -** The infective dose of *Shigella* is as few as 10 cells, depending on the strain and the target population (age and health status of the host). *Shigella* produces infection by invasion of epithelial cells of the intestine, where they multiply inside the cells and spread to contiguous epithelial cells resulting in tissue destruction. Some strains also produce enterotoxin and Shiga toxin (CFSAN, 2003).

**Disease -** The name of the disease is shigellosis or bacillary dysentery. Shigellosis is endemic throughout the world. Worldwide there are approximately 164.7 million cases, of which 163.2 million are in developing countries and 1.5 million in industrialised countries. Each year 1.1 million people are estimated to die from a *Shigella* infection and 580,000 cases of shigellosis are reported among travellers from industrialised countries (WHO, 2003).
The onset of shigellosis is between 12 to 50 hours after ingestion of the contaminated food or contact with the infectious agent. It produces acute symptoms characterised by abdominal pain, cramps, diarrhoea, fever, vomiting, and bloody faeces, sometimes accompanied by pus and mucus. Shigellosis usually resolves itself in 5 to 7 days. For the elderly and infants the diarrhoea

can be so severe that hospitalisation is necessary. A total of 69% of all episodes and 61% of all deaths attributable to shigellosis involve children less than 5 years of age.
Some infections are associated with mucosal ulceration, rectal bleeding, and drastic dehydration. Fatality may be as high as 10-15% with some strains (CFSAN, 2003).
Although all groups of the population are susceptible to infection, the most problematic groups are, as in most cases, infants, the elderly, and people with impaired immunological systems.

*E. coli.*
**General description -** *E. coli* are Gram-negative, motile or non-motile, non-sporeforming rod-shaped bacteria that belong to the *Enterobacteriaceae*. Most of the strains of *E. coli* are commensal, forming part of the natural anaerobic microflora that lives in the gastrointestinal tract of humans and warm-blooded animals.
Different *E. coli* serotypes can be distinguished according to the 3 surface antigens: O (somatic), H (flagellar) and K (capsule). Currently 174 O antigens, 56 H antigens and 80 K antigens have been identified (Doyle *et al.*, 1997).
Although most *E. coli* strains are harmless, some are pathogenic and produce gastroenteritis. The pathogenic strains are classified into four groups according to their different virulence characteristics, pathogenicity mechanisms, clinical symptoms, and O and H antigens:
- EPEC: Enteropathogenic *E. coli*;
- ETEC: Enterotoxigenic *E. coli*;
- EIEC: Enteroinvasive *E. coli*;
- EHEC: Enterohaemorrhagic *E. coli*.

The principal characteristics of these different enterovirulent groups of *E. coli* are shown in Table 2.2. As part of these main pathogenic groups, two more groups have recently been identified: DAEC (Diffusely Adhering *E. coli*) and EAggEC (Enteroaggregative *E. coli*). DAEC use adhesins to adhere to the epithelial lining of the intestine, and produce a kind of bloody diarrhoea in children and the presence of leucocytes in faeces. EAggEC is also related to persistent diarrhoea in newborns and infants.
Of all the *E. coli* strains, *E. coli* O157:H7 probably most typically represents what might be expected in terms of an emerging pathogen (IFT, 2002). That is why it will be described in detail.

*E. coli O157:H7*
This pathogenic microorganism was first identified as a human pathogen related to food consumption in 1982, when it was isolated in two outbreaks of haemorrhagic colitis. Therefore, *E. coli* O157:H7 belongs to the EHEC group (Karmali, *et al.*, 1983; Riley *et al.*, 1983). The strain O157:H7 (EHEC 1) is the most predominant EHEC in North America, Japan, and the U.K. Strains O26:H11 and O111:NM (EHEC 2) dominate in other areas of the world, such as Central Europe and Australia. These two EHEC groups clearly comprise two distinct genotypic lineages (IFT, 2002).

**Physiological data -** *E. coli* O157:H7 is extremely acid tolerant and can survive in mayonnaise of pH 3.6-3.9 at 5°C for 5-7 weeks or for 10 to 31 days at 8°C in apple cider of pH 3.6-4.0. Pasteurisation of milk (72°C for 16.2 sec) produces a decrease of $10^4$ cfu/ml of *E. coli* O157:H7. Cooking temperatures of up to 68.3°C are necessary to ensure the inactivation of *E. coli* O157:H7 in all foods of animal origin (Doyle *et al.*, 1997).

Table 2.2. The most common types of enterovirulence E. coli.

| | EPEC Enteropathogenic E. coli | ETEC Enterotoxigenic E. coli | EIEC Enteroinvasive E. coli | EHEC Enterohaemorrhagic E. coli |
|---|---|---|---|---|
| Mode of action | Not well known, but unrelated to the excretion of typical enterotoxin. Some of them are invasive | Toxin which induces fluid secretion in the intestine. | Invasion and destruction of intestine epithelial cells. Closely related to *Shigella spp.* | Potent toxins that cause severe damage to the lining of the intestine. Verotoxin (VT) Shiga-like toxin |
| Infective dose | Very low in children Adults more than $10^6$ organisms | High $10^8$ to $10^{10}$ organisms | Low, less than 10 organisms | Unknown but probably similar to *Shigella spp.*, as few as 10 organisms. |
| Disease | Infantile diarrhoea | Gastroenteritis Travellers diarrhoea Usually self-limiting | Bloody diarrhoea with mucus Bacillary dysentery | Hemorrhagic colitis, watery diarrhoea with severe abdominal pain Lasts for an average of 8 days Self-limiting |
| Target groups | Most often affects infants, especially bottle fed | Infants and visitors of less developed countries | All people | All people, but with more severe symptoms in infants and elderly people |
| Associated foods | Raw beef, chicken and any food exposed to faecal contamination | Foods contaminated by polluted water, human sewage, or by infected handlers | Unknown. Any food contaminated with infected human faeces. | Undercooked and raw ground beef, alfalfa sprouts, unpasteurised fruit juices, dry-cured salami, lettuce, raw milk |
| Outbreaks | Sporadic | Sporadic in developed countries, more frequent in underdeveloped countries | | |
| Serotypes | O55; O119; O126; O127; O142 | O6; O8; O15; O148; O149; O167 | O29; O112; O124; O136; O143; O144; O152; O164; | O157:H7; O26:H11; O103; O104; O111; O157:H⁻ |

**Source -** Cattle and other ruminants have been established as the major natural reservoirs for *E. coli* O157:H7 and play a significant role in the epidemiology of human infections (Rassmussen *et al.*, 1993; Omisakin *et al.*, 2003). *E. coli* O157:H7 lives commensally in these animals and does not cause any illness in them. Cross-contamination usually occurs between different foods and contaminated beef, cattle, and human faeces. Transmission of *E. coli* O157:H7 also occurs from person to person or via contaminated water.

**Associated foods -** Outbreaks of *E. coli* O157:H7 have mostly been associated with the consumption of undercooked beef and less frequently with pasteurised milk. Recent studies have established a direct link between contaminated cattle on farms and contaminated beef and ground beef (Aslam *et al.*, 2003; Omisakin *et al.*, 2003). It does not seem likely that other domestic animals (except sheep) could act as reservoirs. The existence of *E. coli* O157:H7 in other animals looks like a problem of cross-contamination during cutting in retail shops (Doyle *et al.*, 1997). Nevertheless, other foods including vegetables, cider, melons, sauce dressings, mayonnaise, and dry cured salami have been also involved in several outbreaks (IFT, 2002).

**Mode of action -** The infective dose of *E. coli* O157:H7 (as few as 10 cells) is very low and similar to that of *Shigella* spp. Recently, scientists have made significant advances in understanding the virulence mechanisms of EHEC strains. Two major virulence pathways contributing to disease have been identified:
a) the attaching and effacing genes (AE) within the LEE pathogenicity island that enables the bacteria to adhere to the surfaces of epithelial cells, upon which the bacteria reside. These attaching and effacing genes are also found in enteropathogenic *E. coli* (EPEC).
b) the possession of Shiga-like toxins (related to the toxin1 of *Shigella dysenteriae*). These toxins have names typical of verotoxins. The Shiga toxins (Stx) are comprised of two components, namely the A and B subunits. The B subunit is tissue-specific, enabling the toxin to adhere to a specific glycolipid receptor, globotriaosylceramide, (Gb3), on cell surfaces. Subunit A is the active part of the toxin and is delivered into the host cell, where it inhibits protein synthesis consequently leading to the death of the host cell (IFT, 2002).
It is believed that EHEC is a recent evolution of EPEC due to the fact that the Shiga toxin is encoded within a mobile genetic element (a bacteriophage) that enables it to move to different strains of bacteria (Doyle *et al.*, 1997; IFT, 2002).

**Disease -** *E. coli* O157:H7 produces haemorrhagic colitis, which is characterised by severe cramping (abdominal pain) and watery diarrhoea within 24-48 hours, and followed in more than 90% of cases by bloody diarrhoea. Vomiting occurs occasionally. A low-grade fever can occur. The symptoms are usually self-limiting and last for an average of 8 days (CFSAN, 2003).
Very young people can develop the haemolytic uremic syndrome (HUS), characterised by kidney failure and haemolytic anaemia. The disease can produce the permanent loss of renal function. HUS can develop in 15% of haemorrhagic colitis victims (Doyle *et al.*, 1997; CFSAN, 2003).
In the elderly HUS, in addition to fever and neurological symptoms, constitutes thrombotic thrombocytopenic purpura (TTP), which has a mortality rate of almost 50% in elderly victims. Although all people are susceptible to haemorrhagic colitis caused by an *E. coli* toxico-infection, young children and elderly people suffer more serious symptoms more frequently.

## Biological hazards

*Campylobacter spp.*
**General description** - *Campylobacter* are slender, spirally-curved, relatively small, non-sporeforming Gram-negative rods. They have a characteristic corkscrew-like motility due to polar flagellum in one or both cell ends. The family *Campylobacteriaceae* comprises 20 species and subspecies within the genus *Campylobacter* and 4 species within *Arcobacter*, previously classified as aerotolerant campylobacters (Vandamme et al., 1991). *C. jejuni* and *C. coli* are responsible for more than 95% of the infections caused by campylobacters. They have genomes of approximately 1.7 Mb in size, which is about one-third the size of the *E coli* genome (Nachamkin, 1997).

**Physiological data** - *Campylobacters* have very fastidious growth due to their high number of requirements; this results in them being relatively fragile and sensitive to environmental stress. They are microaerophilic organisms that require between 3-5% oxygen and 2-10% of carbon dioxide for optimal growth. They are very sensitive to osmotic stress and do not grow in NaCl concentrations of 2 or more percent. At the same time, *Campylobacters* have a very low sensitivity to pH, being unable to grow and actually dying at pH values below 4.9 (Park, 2002). Pasteurisation temperatures kill them, their decimal reduction time at 55°C being about 1 min and their Z value being 5°C (Nachamkin, 1997). They are also very sensitive to drying processes and more sensitive to radiation (1 kGy) than other foodborne pathogens such as *Salmonella* and *Listeria monocytogenes*.

As they are microaerophilic and unable to grow in the presence of air and are also only able to grow within a narrow temperature range (30-42°C), they are restricted in their ability to multiply outside of an animal host. Consequently, unlike most other bacterial foodborne pathogens, these bacteria are not normally capable of multiplying in food during processing and storage (Park, 2002). Freezing significantly reduces their survival.

**Source** - The gastrointestinal tract of certain warm-blooded animals is thought to be the natural habitat and environmental reservoir of *Campylobacteriaceae*. Many healthy chickens carry these bacteria in their intestinal tracts. The bacteria are also often carried by healthy cattle and by flies on farms (CFSAN, 2003).

The ecological niches of campylobacters in poultry are the intestinal mucus layer in the crypts of the intestinal epithelium, as an optimal temperature of 42°C and microaerophilic conditions occur in this area. The special corkscrew-like motility is the feature that enables them to remain motile in mucus, a highly viscous environment that rapidly paralyses other motile rod-shaped bacteria. In addition, they have mechanisms that allow them to compete for essential nutrients with the other bacteria in the gastrointestinal tract. One example is the possibility to express an iron transport system, which enables them to scavenge siderophores, such as enterochelin, that have been generated by other gastrointestinal bacteria (Park, 2002).

**Associated foods** - *Campylobacter*-caused infection results primarily due to the ingestion of contaminated food of animal origin. Raw chicken is frequently contaminated with campylobacters. Surveys show that 20 to 100% of retail chickens are contaminated, and the contamination level may vary between $10^2$-$10^5$ cfu/carcass. Contamination of poultry occurs mostly in the abbattoir during evisceration. Consequently, the consumption of undercooked poultry is the main infective route of *Campylobacter* in industrialised nations.

Raw milk is also a source of infection, as well as some pets, and surface waters contaminated by waste slurry from farms. Cross-contamination may also present a significant source of infection in the kitchen, following the preparation of raw chicken.

**Mode of action** - The mechanisms by which campylobacters are pathogenic and induce illness are not clearly understood, and are still being studied. Two mechanisms have been proposed based on experimental data: intestinal adherence followed by toxin production, and bacterial invasion and proliferation with the intestinal mucosa. *C. jejuni* strains possess at least one cytotoxin, the cytolethal distending toxin (CDT) (Pickett, 2000). However, as certain CDT-negative mutant strains of *C. jejuni* retain some toxigenic activity, it is possible that additional toxigenic activities are present in these strains (Purdy *et al.*, 2000).

The infective dose of *C. jejuni* is relatively low, and human feeding studies suggest that about 400-800 bacteria could be enough to cause illness in some individuals. Host susceptibility also plays an important role and dictates the infectious dose to some degree (CFSAN, 2003).

**Disease** - The illness caused by campylobacters is called campylobacteriosis, campylobacter enteritis or gastroenteritis. Although there are 20 species of campylobacters, almost 95% of campylobacter infections are produced by *C. jejuni* and *C. coli*. Surveys have shown that in some developed countries such as the USA or UK, *C. jejuni* is the leading cause of bacterial diarrhoeal disease, being responsible for more cases than *Shigella spp.* and *Salmonella spp.* combined (Park, 2002; CFSAN, 2003).

Campylobacteriosis causes diarrhoea which may be watery or sticky and can contain blood and faecal leukocytes. It also produces fever, cramps (abdominal pain), nausea, headaches, and muscle pain. The first symptoms of illness can appear 2-5 days after ingestion of contaminated food or water. The disease lasts for 7 to 10 days and relapses can occur in 25% of the cases. Most infections are self-limiting and antibiotic treatment is not necessary.

Although not frequent, some complications may occur as reactive arthritis, haemolytic uremic syndrome (HUS), and septicaemia. Campylobacteriosis can have some rare sequels such as Guillain-Barré syndrome (neuromuscular paralysis) or Reiter syndrome. Fatalities are rare in healthy individuals and usually occur in cancer patients. One death per 1000 cases has been reported for infections by *C. jejuni* (CFSAN, 2003).

Although all people are susceptible to infection by *Campylobacter*, children under 5 years and young adults of 15 to 29 years are usually more frequently infected. Outbreaks are usually small. What is really surprising and paradigmatic is that such an apparently sensitive microorganism is able to persist in the food chain and in doing so remains the most common cause of bacterial foodborne disease (Park, 2002).

*Arcobacter spp.*

This genus was proposed to describe organisms previously classified as "aerotolerant campylobacters" (Vandamme *et al.*, 1991). There are four species - *A. butzleri*, *A. cryaerophilus*, *A. skirrowii* and *A. nitrofrigilis*. Morphologically they are similar to *Campylobacter*, the key distinguishing features being: ability to grow at 15°C but not at 42°C, and optimal growth under aerobic conditions at 30°C and G+C 27-30 mol%. Human *Arcobacter* infections are mostly related to poultry, mainly chicken and turkey, including mechanically separated meat (MSM) from these species. They have also been found in pork and water. The common symptoms of *Arcobacter* infection are persistent diarrhoea accompanied by abdominal pain and stomach cramps (Lerner *et al.*, 1994; Phillips, 2001).

*Yersinia spp.*
**General description -** *Yersinia* spp. are Gram negative, small rod-shaped, facultative anaerobes that belong to the *Enterobacteriaceae* family. They include three species pathogenic for humans. *Yersinia pestis* is a causative agent of the plague, an infectious disease that has caused thousands of deaths over many years of human history. This microorganism infects humans by routes other than food. The other two, *Yersinia paratuberculosis* and *Yersinia enterocolitica,* are clearly related to food infections, especially *Yersinia enterocolitica* (Robins-Browne, 1997; CFSAN, 2003).

**Physiological data -** *Yersinia enterocolitica* are psychotrophic bacteria that can grow between 0 and 44°C, growing optimally at 28-30°C. They are very resistant to freeze-thawing processes, but are very heat sensitive, being destroyed by heating at 71.8°C for 18 sec. *Y. enterocolitica* can grow over quite a broad pH range of 4 to 10, the optimum pH for growth being around 7.6. This tolerance to acidic mediums is due to their urease activity, which enables them to convert urea to ammonia thereby increasing the intracellular pH. They are also very sensitive to radiation, UV light, and to nitrates and nitrites added to foodstuffs. However, they are quite resistant to osmotic pressure and can survive in solutions with up to 5% NaCl.

**Source -** This microorganism has been isolated from different environments, and in the gastrointestinal tracts of different mammals, birds and frogs. Although, it is possible to find *Yersinia* in soil and in different fresh water ecosystems, almost all strains lack virulence markers and are not relevant for human health.

**Associated foods -** Although *Yersinia* strains are quite ubiquitous, the foods mostly associated with it are undercooked pork and beef, lamb, poultry, and milk and dairy products. They can live and multiply in raw and cooked foods, surviving better at room and refrigerated temperatures than at intermediate temperatures. They can also multiply in cooked pork and beef to more than $10^6$ cfu/g during 24 hours at 25°C. *Yersinia* can grow in a broad range of heat-treated or cooked foods, and vacuum-packaged meat stored at refrigeration temperatures, and in pasteurised liquid eggs (Robins-Browne, 1997).
As *Yersinia* can grow at quite low temperatures, it is a more common foodborne pathogen in temperate zones than in the tropics. Unlike campylobacters, *Yersinia* are ubiquitous and can grow in a broad range of foods and environments. Outbreaks are, however, very rare.

**Mode of action -** *Yersinia* are invasive microorganisms that produce an inflammatory response in the involved tissues. They have a special tropism for lymphoid tissue, tissue invasion exclusively occurring through the epithelial M cells that cover the intestinal lymphoid tissue (Preyer's plaques).

**Disease -** The symptoms of yersiniosis are gastroenteritis with diarrhoea and vomiting. However, the more characteristic symptoms are fever and abdominal pain in the lower right quadrant, which makes yersiniosis infections mimic appendicitis and mesenteric lymphadenitis. The bacteria may also cause infections in other sites such as wounds, joints, and the urinary tract. The first symptoms appear between 24 and 48 hours after ingestion of contaminated foods. Complications like reactive arthritis and bacteremia appear sometimes. Yersiniosis is relatively common in Northern Europe and Scandinavia (CFSAN, 2003).

The most susceptible population groups are children, the elderly, and in general all people undergoing therapy with immunosuppressive effects.

*Vibrio spp.*
**General description -** The genus *Vibrio* includes Gram-negative, rod or curved rod shaped, facultative anaerobes. To date 20 species of *Vibrio* have been identified, of which 12 are pathogenic for human beings and 8 are related to foodborne infections. The most studied species is *Vibrio cholerae*. It is transmitted mainly by contaminated water and produces a heat-sensitive enterotoxin that causes the characteristic cholera symptoms. Apart from *V. cholerae* and *V. mimicus*, the other species of *Vibrio* are halophilic and therefore need NaCl in the media to grow. *Vibrio parahaemolyticus* and *Vibrio vulnificus* are the most prevalent in food intoxications (West, 1989).

**Physiological data -** Vibrios are quite sensitive to low temperatures, although contaminated foods can protect vibrios from the effect of the temperature. Vacuum packaging can help in limiting the resistance of *Vibrio vulnificus* in refrigerated foods.

The minimum and maximum temperature for growth are 12.8 and 43°C respectively, with optimum growth occurring at 37°C. They are sensitive to heat treatment and are inactivated with treatments of 15-30 min at 60°C or 5 min at 100°C. They survive in quite a broad range of pH (4.5 to 11), and above a water activity ($a_w$) of 0.94.

**Source and associated foods -** Normally halophilic vibrios are found in warm marine and estuarine environments and have been isolated from many species of fish, shellfish, and crustaceans. Infections with these organisms have been associated with the consumption of raw, improperly cooked, or cooked and re-contaminated fish or shellfish. Raw oysters are the major source of foodborne disease caused by *V. vulnificus* (CFSAN, 2003). It seems there is a correlation between the probability of infection and the warmer months of the year (Elliot *et al.*, 1998).

**Mode of action -** It seems that the mechanisms used by vibrios comprise of adherence to the small intestinal epithelial lining, and the production and excretion of a toxin. The nature of the toxin and whether there are one or more toxins involved[0] is unclear at the moment. For *V. Parahaemolyticus* at least four haemolytic components have been identified, and among them is the Kanagawa hemolysin (TDH) (Oliver and Kaper, 1997).

The infective dose seems to be different for *V. vulnificus* and *V. parahaemolyticus*. For *V. vulnificus* the dose for healthy individuals is unknown. The dose for predisposed people is very low, less than 100 cells. The infective dose for *V. parahaemolyticus* is around $10^6$ cells, although this dose can be lowered by the coincidental consumption of antacids.

**Disease** - *V. parahaemolyticus* produces an acute mild or moderate self-limiting gastroenteritis, characterised by diarrhoea, abdominal cramps, nausea, vomiting, headache, fever and chills. The first symptoms appear between 4 to 96 hours after the ingestion of contaminated foods and last for an average of 2.5 days (CFSAN, 2003).

*V. vulnificus* causes a more complicated disease. Among healthy individuals ingestion of contaminated seafood can cause vomiting, diarrhoea, and abdominal pain. In susceptible people, particularly those with chronic liver disease, it can infect the bloodstream and cause a syndrome known as primary septicaemia. Primary septicaemia is fatal about 50% of the time. *V. vulnificus* can also produce wound infections with skin breakdown and ulceration, due to contact with contaminated warm seawater. This infection can have potentially fatal complications in immunocompromised persons. Infection is treated with antibiotics, i.e. doxycycline or ceftazidine, and those who recover should not encounter any long-term consequences (CDC, 2000).

*Aeromonas spp. and Plesiomonas spp.*

*Aeromonas* and *Plesiomonas* are Gram-negative, rod-shaped, facultative anaerobes. In general they are motile by means of polar flagella, one in *Aeromonas spp* and 2 to 7 in *Plesiomonas spp*. They were previously located in the family *Vibrionaceae* as they are found mainly in water environments. However, recent molecular studies have shown that these species are not closely related to *Vibrio*. *Aeromonas* is now considered as a new family *Aeromonadaceae*, while *Plesiomonas* is more closely to related *Enterobacteriaceae* (Kirov, 1997). Although there is some evidence of the enteropathogenicity of these microorganisms, they are still considered as putative foodborne pathogens as there is no definitive evidence of their pathogenicity in human volunteer trails or in animal models. The main pathogenic species of both microorganisms are *Aeromonas hydrophila* and *Plesiomonas shigelloides*, their principal characteristics are summarised in Table 2.3.

## Gram-positive foodborne pathogens

*Clostridium spp.*

The members of the genus *Clostridium* are Gram-positive, anaerobic, sporeforming, rod-shaped bacteria. The spores are heat-resistant and can survive in foods that are incorrectly or minimally processed. There are two very important species that constitute an important risk with regards to food safety: *C. botulinum* and *C. perfringens*.

*C. botulinum*

**General description** - Apart from the general morphology described above, *C. botulinum* produces a protein with characteristic neurotoxicity. Antigenic types of *C. botulinum* are identified by complete neutralisation of their toxins by the homologous antitoxin. There are seven recognised antigenic types, which are named with capital letters from A to G. Only types A, B, E and F cause foodborne diseases in humans. Aside from toxin type, *C. botulinum* can be differentiated into general groups on the basis of cultural, biochemical, and physiological characteristics. All cultures that produce type A toxin and some that produce B and F toxins are proteolytic, and all type E and the remaining B and F producing strains are non-proteolytic (Dodds and Austin, 1997; CFSAN, 2003).

Table 2.3. *Main characteristics and features of* Aeromonas hydrophila *and* Plesiomonas shigelloides.

| | Aeromonas hydrophila | Plesiomonas shigelloides |
|---|---|---|
| Source | All freshwater environments and in brackish water | Freshwater, freshwater fish and shellfish, and different animals such as swine, cattle and goats |
| Associated foods | Fish, shellfish, red meat, and poultry | Contaminated water and raw shellfish |
| Disease symptoms | Two different gastrointestinal disorders: 1. cholera-like illness (watery diarrhoea) 2. dysenteric illness (loose stools with blood and mucus) | Gastroenteritis Suspected of being toxigenic and invasive |
| Infective dose | Not known, but presumably high | High (> 106 organisms) |
| Target population | All people, but more frequently in young children, and more severe in susceptible population | All people. Complications in infants, children and chronically ill people |
| Physiological features ■ Growth temperature | Maximum 42°C Optimum 28°C Minimum 1-4°C | Maximum 44°C Optimum 37°C Minimum 8°C |
| ■ pH | Maximum 10 Minimum 4 | Maximum 7.7 Minimum 5 |

Based on: CFSAN, 2003 and Mortimore and Wallace, 1994.

**Physiological data** - All relevant data concerning *C. botulinum* can be found in Table 2.4.

**Source** - This organism and its spores are widely distributed in nature, and it can be found in different kinds of soils, in sediments of diverse water ecosystems, and in the intestinal tracts of fish, shellfish and mammals (Dodds and Austin, 1997; CFSAN, 2003).

**Associated foods** - A broad range of foods enable the organism to grow and produce toxins, including processed foods in which spores survive which are not heated before consumption. Also foods that are not very acidic (pH above 4.6) can support growth and toxin production by *C. botulinum*. There is also increasing concern about certain food products that do not use any preservative and are packaged with low oxygen concentrations, as used in vacuum packaging or modified atmosphere packaging. One example being fish products that should be kept at refrigerated temperatures to prevent growth and toxin production by *C. botulinum* (Dodds and Austin, 1997).

The presence of this organism has been demonstrated in a great variety of foods like canned vegetables (peppers, green beans, asparagus, mushrooms and olives), chicken and chicken by-products, luncheon meats, ham, sausages, tuna, smoked and salted fish, and lobster. Apart from inadequately processed industrial foods, it is also remarkable that botulinum toxin can be found in home-canned foods.

## Biological hazards

Table 2.4. Principal physiological characteristics of C. botulinum and C. perfringens.

| Characteristics | C. botulinum | | C. perfringens |
|---|---|---|---|
| | proteolytic strains | non proteolytic strains | |
| Minimal growing temperature | 10°C | 3.3°C | 12°C |
| Maximal growing temperature | 48°C | 40-45°C | 50°C |
| Optimal growing temperature | 35-40°C | 18-25°C | 43-45°C |
| Thermal inactivation of spores at 121°C ($D_{121°C}$) | 0.2 min | < 0.001 min | could be very resistant |
| Toxin inactivation | 1 min at 85°C  1.5 h at 65°C | | 5 min at 60°C |
| Minimal growing pH | 4.6 | 5.0 | 5.0 |
| Maximal growing pH | 9.0 | | 8.3 |
| Optimal growing pH | 6.5-7.0 | 6.5-7.0 | 6.0-7.0 |
| Minimal growing $a_w$ | 0.94 | 0.97 | 0.93-0.97 |
| Inhibitory concentration of NaCl | 10% | 5% | 6-8% |

**Mode of action -** As mentioned above, *C. botulinum* produces 7 different neurotoxins (A-G). These neurotoxins are quite similar in structure and mode of action. They are polypeptides of high molecular weight (150 KDa) formed by two protein subunits, the H and L subunits of 100 KDa and 50 KDa respectively. Botulinum toxin is considered to be one of the most powerful known toxins ($LD_{50}$ mice < 0.1 ng/kg) (Dodds and Austin, 1997). These neurotoxins block the release of the neurotransmitter acetylcholine in the peripheral terminals of the motor nerves by the selective hydrolysis of the proteins involved in the fusion of the synaptic vesicles with the presynaptic plasmatic membrane (Dodds and Austin, 1997,). The infective dose is very low and a few nanograms of toxin can cause illness.

**Disease -** Four types of botulism are recognised: foodborne, infant, wound, and an undetermined form of botulism (a sort of infant botulism that affects adult people). Foodborne botulism is a typical example of a food intoxication caused by the ingestion of foods containing neurotoxins produced by *C. botulinum*. Infant and undetermined botulisms can be acquired by eating contaminated foods, among other ways, but in these cases botulism is produced by the ingestion of botulinum spores which colonise and produce toxin in the intestinal tract of the individual. Honey is a typical reservoir for infant botulism (CFSAN, 2003).

The most common symptoms of foodborne botulism usually appear after 18 to 36 hours of ingestion of contaminated foods, consisting of double vision, drooping eyelids, progressive difficulty in speaking and swallowing, dry mouth, and muscle weakness. If untreated, the symptoms may progress leading to paralysis of most of the muscles including those of the respiratory system. Respiratory failure and paralysis may require the use of an assisted breathing machine for weeks, plus intensive medical and nursing care. Patients that survive an episode of botulism have sequels including fatigue and shortness of breath for years. Respiratory failure produced by botulism can result in death. However, the mortality percentage has fallen during the last 50 years, from 50% to 8% (CDC, 2002). With an early diagnosis botulism can be treated with an antitoxin that blocks the action of toxin circulating in the blood.

*C. perfringens*
**General description -** This microorganism shares more or less the same morphological features of most *Clostridium* species (see above). *C. perfringens* is a very efficient and aggressive foodborne pathogen. It has the ability to produce active toxins in the gastrointestinal tract, the vegetative cells being able to duplicate in 10 minutes (Labbe, 1989). In some conditions it produces very thermo-resistant spores, and although it is anaerobic it is able to grow and multiply in atmospheres with 5% oxygen (Poumeyrol, 1988).
To date five types of *C. perfringens* (A-E) have been identified according to their toxin profile production. In general, there are 4 types of toxins: alpha ($\alpha$), beta ($\beta$), epsilon ($\epsilon$) and jota ($\varphi$).

**Physiological data -** See Table 2.4.

**Source -** *C. perfringens* is ubiquitous, widely distributed in the environment and usually found in the gastrointestinal tract of domestic and wild animals as well in humans, where it can reach populations of $10^3$ to $10^6$ cfu/g in faeces (McClane, 1997). However, recent data shows that only 5% of *C. perfringens* isolates encode the gene *cpe*, which is necessary to produce the food-borne intoxication.

**Associated foods -** Most of the outbreaks of *C. perfringens* intoxications are related to temperature abuse of prepared meals. Small numbers of organisms or spores can survive cooking processes and multiply to food poisoning levels during cooling down and storage of prepared meals. Meat and meat products are the foods most frequently implicated (CFSAN, 2003).

**Mode of action -** The ingestion of large numbers of vegetative cells ($> 10^8$ cfu/g) as a consequence of incorrect cooling of heat-treated foods results in some of these organisms surviving the acidity of the stomach, and reaching the small intestine where they multiply and form spores. During sporulation toxins are expressed and released into the intestinal lumen. Afterwards, they fix to the surface of epithelial cells producing morphological damage to the cells lining the intestines (McClane, 1997).

**Disease -** The disease is a food infection produced by ingestion of *C. perfringens* type A and is referred to as perfringens food poisoning. A more serious but rare disease is produced by the consumption of foods contaminated by type C strains. This illness is known as necrotic enteritis or pig-bel disease.
The first symptoms of perfringens poisoning appear 8 to 22 hours after the ingestion of foods containing large numbers of toxin-producing strains. The main disorders are an intense abdominal pain and diarrhoea. The illness is usually self-limiting and lasts 24 hours, although in some individuals less severe symptoms can persist for 1 or 2 weeks more. Although perfringens poisoning can affect everybody, the young and elderly are the most frequent victims.

Perfringens poisoning is one of the most commonly reported foodborne illnesses and in many cases is related to institutional feeding, where large quantities of food are prepared several hours before serving (CFSAN, 2003).

Necrotic enteritis is often fatal as consequence of infection and necrosis of the intestines followed by septicaemia.

*Staphylococcus aureus*

**General description** - *S. aureus* is a Gram-positive, spherical, facultative anaerobe which on microscopic examination can appear in pairs, short chains or grape-like clusters. It belongs to the genus *Staphylococcus* which comprises 23 species and subspecies. Although some of them can grow in foods and produce enterotoxins, *S. aureus* is responsible for almost all foodborne illness due to staphylococci.

**Physiological data** - *S. aureus* grows optimally at 37°C, and has minimum and maximum growth temperatures of 10 and 48°C, respectively. In general, it is heat sensitive and most cooking and industrial heat treatments can eliminate it from foods. It can grow over a broad pH range of 4.0 and 9.9, growth being optimal between pH 6.0 and 7.0. The development of competitive bacterial populations in different media can inhibit their growth. *S. aureus* is quite halophilic and can grow at high concentrations of NaCl (3.5 M), and at really low $a_w$ values (0.86) with a generation time of 300 min. It can also survive for a long time in dehydrated and frozen foods, although most of the typical food preservation methods can eliminate it (Buyser, 1988; Jablonski and Bohach, 1997). Physiological data related to the *S. aureus* enterotoxin are shown in Table 2.5.

**Source** - Although *S. aureus* is ubiquitous and exists in air, dust, sewage, soil, water, and on food processing equipment, the primary and main reservoirs are humans and animals where they are on most occasions saprophytic microflora. Staphylococci are present in the nasal passages, throats, hair, and skin of 50 percent or more of healthy individuals (CFSAN, 2003).

**Associated foods** - There are a lot of foods that can be contaminated by staphylococci, including meat and meat products, poultry and egg products, some pasta salads, bakery products

Table 2.5. Physiological data about growth and enterotoxin production of S. aureus.

| Factor | Growing | Enterotoxin production |
|---|---|---|
| Range of temperature | 6-46°C | 10-45°C |
| Optimal temperature | 37°C | 40°C |
| Range of pH | 4-9.8 | 5-8 |
| Optimal pH | 6-7 | 6.5-7 |
| NaCl | 0-20% | 0-10% |
| Minimal $a_w$ | 0.83 | 0.86 |

Adapted from Buyser, 1988.

such as cream-filled pastries and cream pies, chocolate, milk and dairy products. Generally, *S. aureus* can develop in all kinds of foods which need a lot of handling and manipulation during their preparation and are kept at slightly elevated temperatures before consumption (Jablonski and Bohach, 1997; CFSAN, 2003).

**Mode of action** - *S. aureus* causes typical food intoxication by means of enterotoxins (SEs), which are small single-strain globular proteins (MW around 30,000). Seven enterotoxins have so far been identified according to their antigenic pattern: SEA, SEB, SEC, SED, SEE, and more recently SEG and SHE. Enterotoxin A (SEA) is most usually associated with foodborne illness, being responsible for about 80% of total intoxications due to *S. aureus* (Buyser, 1988). Although the staphylococci enterotoxin was identified a long time ago (in 1914), its mode of action is still quite confusing. It seems that the emetic effect (causing vomiting) is produced by the stimulation of neural receptors in the abdomen connected to the medullar emetic centre, which provokes vomiting (Sugiyama and Hayama, 1965). Currently there is no evidence about the nature of the target cells in the intestinal epithelium for SEs. It seems clear that these enterotoxins produce an inflammatory reaction in the mucosa of the gastrointestinal tract, mainly in the stomach and upper part of the small intestine (Jablonski and Bohach, 1997).
1 to 10 µg of toxin are enough to cause staphylococcal intoxication in humans, and this amount is reached when the population of *S. aureus* is higher than $10^5$ cfu/g of food.

**Disease** - The disease caused by *S. aureus* is known as staphylococcal food poisoning or staphyloenterotoxicosis. It appears 1 to 8 hours after ingestion of contaminated food. The symptoms appear abruptly and the most common are nausea, vomiting, abdominal cramping, frequently with diarrhoea and prostration. Fever is rare, although slight hyperthermia can occur (38°C). Although the intoxication lasts for only about 24 hours, fatigue can occur for several days more. This disease is self-limiting. In some sensitive people (children or the elderly) complications can appear mainly in the form of headaches, muscle cramping, and transient changes in blood pressure that may sometimes require hospitalisation. Fatality is very rare. In summary, staphyloenterotoxicosis is a short but very intense foodborne disease (Buyser, 1988; CFSAN, 2003).

*Listeria monocytogenes*
**General description** - *Listeria* is a Gram-positive, non-sporeforming, microaerophilic bacteria with an ovoid (coccobacillar) form. It is quite similar to *Staphylococcus*, *Streptococcus*, *Lactobacillus* and *Brocothrix*, from a phylogenetic point of view, due to its low content of G+C (36-42%). Although 6 different species of *Listeria* have been identified, only *L. monocytogenes* and *L. ivanovii* are considered virulent. Of these two, only *L. monocytogenes* is considered of great concern to public health. There are 13 recognised serotypes of *L. monocytogenes* that can cause illness, but only 3 of them (1/2a, 1/2b, and 4b) represent 95% of all human isolates. Strains of serotype 4b are involved in 33 to 50% of human sporadic cases around the world.

**Physiological data** - *L. monocytogenes* is a very stress resistant bacteria and it can grow in hard environmental conditions. The optimal growth temperature is around 25-30 °C, the minimum and maximum temperatures for growth being 0 and 45°C respectively. Therefore, *L.*

*monocytogenes* is psychrotrophic and it can multiply at refrigeration temperatures. It is also able to survive after freezing at -18 °C. It grows optimally at pH 7-7.5 and can survive at pH 4.4 at 30 °C or between pH 5.0 and 9 at 4 °C. It can also live in foods with an $a_w$ value of 0.92 and with 10% of NaCl in the growth medium. Due to its phylogenetical proximity to lactic acid bacteria (LAB), *Listeria monocytogenes* is sensitive to some bacteriocins produced by LAB, i.e. nisin, pediocin, and bavaricin A, and some problems with resistant mutants have been reported (Larsen and Norrung, 1993; Davies and Adams, 1994).

**Source -** This organism is ubiquitous and can be found in soil and water, from where it can contaminate plants and vegetables including silage, and thereafter animals. Humans can be infected by both vegetables and animals. A small percentage of healthy asymptomatic persons (1-10%) who can be carriers and transmit the infection have been reported (CFSAN, 2003).

**Associated foods -** Traditionally, *L. monocytogenes* infection is associated with foods such as raw milk, pasteurised milk, soft cheeses, dairy products with a high fat content (cream, butter, ice creams), raw vegetables, raw and cooked meat and poultry, fermented raw sausages, and raw and smoked fish and shellfish. Recently, the Food and Drug Administration (FDA) of the USA published a risk assessment for *L. monocytogenes*. One of the conclusions of this work is that soft cheeses and related dairy products should be considered as risky foods instead of high risk foods, due to the efforts of the dairy industry to reduce and control the prevalence of this organism (FDA, 2003).

**Mode of action -** It is possible to define *L. monocytogenes* as a highly invasive intracellular pathogen, which is able to invade different kinds of phagocytic or non-phagocytic cells, i.e. macrophages, fibroblasts (connective tissue cells), hepatic cells and epithelial cells. The infective dose is unknown and varies with the strain and host susceptibility, but is believed to be less than 1000 microorganisms. It is not clear exactly were *L. monocytogenes* enters the intestinal lining. However, the invasive process of *Listeria* seems quite clear and always follows the same scheme (Figure 2.3). The bacteria enters the cytoplasm of the host cell by phagocytosis (directly or induced in the non-phagocytic cells), and becomes surrounded by the cell membrane in a

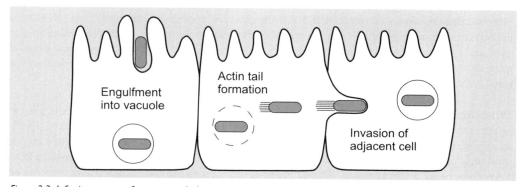

*Figure 2.3. Infection pattern of gut mucosa by* Listeria monocytogenes.

vacuole. In less than 30 minutes the membrane of the vacuole is lysed and bacteria remain free in the cytoplasm of the host cell, where they start to multiply with an estimated duplication time of around 1 hour. At this stage the bacteria is surrounded by cellular actin filaments, which concentrate asymmetrically in one pole of the cell producing a comet tail in the opposite direction to the movement of the cell. When bacterial cells come into contact with the host's cell membrane, they produce a long arm or extension to the cytoplasm of the neighbouring cell. This continues until the bacterium is surrounded by two membranes, one from the first infected cell and another from the new infected cell. After this happens the infective cycle can start again. The whole infective cycle lasts around 5 hours, and this is the way in which the infectious focus spreads. It seems that actin polymerisation plays an important role in the propagation of the infection. This way of spreading the infection from cell to cell enables *L. monocytogenes* to avoid contact with host antibodies. When *Listeria* invades macrophages it can be transported through the blood stream to different tissues of the body.

**Disease -** Listeriosis is the name of the illness caused by infection with *L. monocytogenes*. The symptoms of listeriosis appear within a few days to three weeks after ingestion of the contaminated food. Typical symptoms are septicaemia, meningitis or meningoencephalitis, encephalitis, and intrauterine or cervical infections in pregnant women which may result in spontaneous abortion, mainly in the third trimester, or stillbirth. Prior to abortion pregnant women suffer influenza-like symptoms, including persistent fever. It has been reported that gastrointestinal symptoms such as nausea, vomiting, and diarrhoea may precede more serious forms of listeriosis or may be the only symptoms expressed (CFSAN, 2003).
The target population is quite well defined in the case of listeriosis and includes pregnant women, who are about 20 times more likely than other healthy adults to get listeriosis, and comprise one third of all listeriosis cases. Newborns suffer serious effects from infection during pregnancy. People with immune systems compromised for different reasons, i.e. cancer, diabetes, kidney failure, and AIDS, have a 300 times greater probability of getting listeriosis than those with normal immune systems. People who take glucocorticosteriod medications and the elderly are also highly susceptible to listeriosis. Healthy people and children rarely become seriously ill when they are infected by *L. monocytogenes*.

*Bacillus cereus*
**General description -** *Bacillus cereus* is a Gram-positive, sporeforming, motile, big rod-shaped bacterium. Although aerobic it can also grow in anaerobic conditions. *Bacillus cereus* belongs to the bacilli group of *Bacillus subtilis* and is very closely related to other bacilli species such as *B. thuningensis*, *B. anthracis* and *B. mycoides*. In fact, they could be considered a subspecies of *B. cereus* (Ash et al., 1991).

**Physiological data -** This organism can grow over a broad temperature range of 5 to 55°C, growth being optimal at 30-37°C. Some strains can grow in a slightly narrower range of 15-50°C, but in these cases the germination of spores takes place, strangely enough, between 5-8°C. It is able to grow in a pH range of 4.5-9.3, in foods with $a_w$ up to 0.95, and with concentrations of NaCl of 7.5%. In general, *B. cereus* grows well in most food products; especially in rice and more so when this is combined with animal proteins. Nisin and sorbic acid (0.2%) are able to inhibit its growth.

**Source** - *B. cereus* is widespread in nature and frequently associated with soil, which is why it is very common to find them on growing plants and vegetable products. From these they can cross-contaminate animal foods such as meat and meat products, and milk and dairy products.

**Mode of action** - Two different kinds of food illnesses, the diarrhoeal and the emetic type, are caused by two different metabolites produced by *B. cereus* (Granum and Lund 1997). The main characteristics of the distinct toxins involved, the main symptoms of the illnesses, and associated foods are shown in Table 2.6. Other species of *Bacillus* have been isolated from foodstuffs incriminated in food poisoning episodes, especially *B. subtilis* and *B. licheniformis*. The reservoir of these species is also soil and vegetables, from where they can contaminate a lot of different foods, i.e. meat and meat products, vegetables, rice, custards and bakery products. They also produce thermo-resistant spores and highly heat-stable toxins which may be similar to the emetic toxin produced by *B. cereus*.

*Biogenic amines*
It is possible that some non-pathogenic bacteria including some raw material contaminants, spoilage bacteria and fermentative bacteria can produce toxic substances such as biogenic amines in certain conditions.

**General description** - Biogenic amines are organic molecules of low molecular weight, i.e. aliphatic mono-, di- and polyamines, aromatic amines, and related compounds. They are

Table 2.6. Characteristics of the different toxins involved in food illness caused by B. cereus.

| Characteristics | Diarrhoeal syndrome | Emetic syndrome |
|---|---|---|
| Infective dose | $10^5$-$10^7$ cells | $10^5$-$10^8$ cfu/g |
| Site of toxin production | Hosts small intestine | In foodstuffs |
| Nature of toxin | Large molecular weight protein (1 or more?) | Low molecular cyclic peptide |
| Related toxins | *C. perfringens* toxin, although is 100 times more potent | *S. aureus* toxin |
| Thermal resistance of toxins | 5 min at 56°C | 90 min at 126°C |
| Incubation time (hours) | 6-15, sometimes more than 24 | 0.5-6 |
| Symptoms | Nausea, watery diarrhoea and abdominal cramps | Nausea, vomiting, occasionally abdominal cramps and diarrhoea |
| Duration of the illness (hours) | 12-24, occasionally some days | 6-24 |
| Target group | All people | All people |
| Associated foods | Meat products, milk and dairy products, soups, vegetables, sauces | Rice, pasta, bakery products |
| Geographical distribution | More frequent in Europe and North America | More frequent in Japan |

Adapted from Granum, 1997

biologically active and are normally formed by bacterial decarboxylation of their precursor amino acids (Table 2.7). Biologically active amines are generally either psychoactive or vasoactive. Psychoactive amines act on neural transmitters in the central nervous system, while vasoactive amines act directly or indirectly on the vascular system. Although biogenic amines are normal metabolites in plants, microorganisms, and animals, the ingestion of large amounts of these substances or the failure of the normal detoxification mechanisms by inhibition or genetic deficiency, can result in a hazard for the health of affected individuals.

There are three main factors that govern the formation of amines in foods: the presence of free amino acids in the medium by autolytic or microbial proteolysis (including fermentation processes), the existence of decarboxylating microorganisms, and favourable environmental conditions for the growth of microorganisms and for enzyme production. Among these favourable conditions are temperatures between 20-37 °C and even as low as 0-5 °C in some cases. Optimal growth occurs over a pH range of 5-6.5, which is a normal pH in many foods. Decarboxylation is a part of protein degradation and may also be a protective mechanism used by bacteria to survive in an acidic environment.

**Microorganisms involved** - There are a lot of non-pathogen foodborne bacteria that are able to produce biogenic amines. Some of them are contaminants of raw material, mainly enterobacteria such as *Proteus morganii*, *P. mirabilis*, *Klebsiella pneumoniae*, *Hafnia alvei*, *E. coli*, and *Enterobacter aerogenes* or other bacteria, such as some species of *Pseudomonas*, *C. perfringens* or *Clostridium spp.*, and lactic acid bacteria such as *Enterococcus faecalis*, some species of *Lactobacillus*, and *Pediococcus*. In fact, any bacteria capable of decarboxylation in the presence of free amino acids and in the conditions described above are able to produce these compounds.

**Associated foods** - Many different kinds of foods can contain biogenic amines, they are, however, mainly found in those foods where protein degradation can occur, such as fish (especially from the histidine rich scombroids), cheese, dry fermented sausages, and even in some alcoholic beverages and fruits (bananas).

**Disease** - Ingestion of these substances produces episodes of acute illness some of whose symptoms resemble those of microbial intoxications. Secondary amines such as dimethylamine

Table 2.7. Principal biogenic amines found in foods.

| Chemical type | Name | Precursor amino acid |
|---|---|---|
| Aliphatic | putrescine | arginine |
| | cadaverine | lysine |
| | spermidine | arginine |
| | spermine | arginine |
| Aromatic | histamine | histidine |
| | tryptamine | tryptophan |
| | tyramine | tyrosine |
| | phenylethylamine | Phenylalanine |

can contribute to the formation of carcinogenic nitrosamines in products where nitrites are used as chemical preservative, i.e. meat products. They frequently cause skin reactions, and neurological disorders; gastrointestinal symptoms are less frequent.

Histamine poisoning is the most frequent intoxication from biogenic amines, and fish mainly from the scombroid family are most often implicated (scombroid poisoning). The intoxication is a mild self-limiting illness with a short incubation period (< 1 hr to a few hours) and lasting just 6-9 hours. It appears when the detoxification system existing in the intestines fails, is inhibited or when there is a large amount of histamine present. The most significant symptoms are flushing, nausea, cramps, and an oral burning sensation. Diarrhoea, headache, vomiting, urticaria, and/or palpitations can also occur. The toxigenic dose of histamine is between 10 to 100 mg/kg.

Tyramine poisoning is characterised by high blood pressure, headache, fever and sometimes vomiting. It can also cause a hypertensive crisis in patients treated with monoamine oxidase inhibitor drugs (MAOI), which block the pathway for catabolism of this amine. Therefore, such patients need very clear dietary specifications to avoid the consumption of foods that can provoke tyramine intoxication. The ingestion of 10 to 25 mg/K of tyramine can cause a sudden increase in blood pressure in these patients, whereas for other people a much higher toxigenic dose of 100-800 mg/K is required for intoxication (ten Brink *et al.*, 1990). Tyramine can be found in cheese and in dry fermented meat products (González-Fernández *et al.*, 2003).

Dietary migraine is also another illness caused by biogenic amines such as phenethylamine and tyramine in quite low concentrations of 3 mg/kg and 125 mg/kg respectively. Chocolate, cheese, and citrus fruits are associated with this type of migraine.

Diamines, putrescine, and cadaverine increase the toxigenicity of other biogenic amines such as histamine and tyramine, and are considered as precursors of nitrosamines (ten Brink *et al.*, 1990).

### 2.3.2 Moulds

Moulds perform a double role in foodstuffs. Some of them are necessary to obtain different foods by fermentation, i.e. soy sauces, or by ripening as with Camembert cheese or all different blue cheeses. On other occasions moulds act as spoilage microorganisms. The moulds themselves do not represent a hazard for consumers, but some secondary metabolites that they produce during their growth (and referred to as mycotoxins) could be dangerous for human health. Mycotoxins can be defined as a group of structurally diverse, naturally occurring chemical substances produced by moulds that have a toxigenic effect on animals and humans. The term mycotoxin is restricted to describing the metabolites of microfungi, as opposed to the toxic principles of certain macrofungi like mushrooms (Cole and Dorner, 1993).

The toxic effect of mycotoxins on humans depends on the dose ingested, and may include (1) acute toxicity and death as a consequence of exposure to high amounts of a mycotoxin, (2) suppression of immune functions and reduced resistance to infections from chronic exposure to low levels of toxins, and (3) tumour formation and cancer from prolonged exposure to very low levels of a toxin. In addition, mycotoxins may be mutagenic and teratogenic and may sometimes produce neurological disorders (Cole and Dorner, 1993).

*Jordi Rovira, Avrelija Cencic, Eva Santos and Mogens Jakobsen*

There are a lot of mould species that can produce mycotoxins in foods (Table 2.8), but most of the well-known mycotoxins are produced by three main genera; *Aspergillus, Penicillium,* and *Fusarium*. Members of the genera *Aspergillus* and *Penicillium* tend to be saprophytic and attack foods during storage, whereas members of the genus *Fusarium* attack growing plants and are also saprophytic. Toxin production is affected by a lot of environmental factors, such as moisture content (water activity), pH, relative humidity, temperature, the presence and nature of different substrates, and competition with other microorganisms. Not all species of these genera produce mycotoxins, and within one species some strains can produce them whereas others cannot. Some mycotoxins can also be produced by different strains of the same genus and strains of another genus. Mycotoxins can be ingested by humans in two different ways: (1) directly via

*Table 2.8. Most relevant mycotoxin-producing moulds and associated foods and diseases.*

| Mycotoxin | Mould | Associated foods | Disease |
|---|---|---|---|
| Aflatoxins | *Aspergillus flavus* *Aspergillus parasiticus* *Aspergillus nomius* | Peanuts, corn, wheat, rice, nuts, milk, eggs and cheese | Hepatotoxicity, Intestinal tract haemorrhage, kidney failure, liver carcinogenesis |
| Ochratoxin A | *Aspergillus ochraceus* *Penicillium viridicatum* | Cereal grains, dry beans, mouldy peanuts, cheese, wine | Nephropathy, teratogenic effects, enteritis |
| Ergot alkaloids | *Claviceps purpurea* *Claviceps paspalli* *Claviceps fusiformis* | Cereal grains, i.e. barley and rye | St Anthony's fire, Gangrene, Convulsions |
| Cyclopiazonic acid | *Penicillium cyclopium* *Aspergillus flavus* *Aspergillus versicolor* *Aspergillus tamarii* other *Penicillium* and *Aspergillus* | Corn, peanuts, cheese | Intestinal haemorrhage, muscle necrosis, oral lesions |
| Patulin | *Penicillium expansum* *Byssochlamys nivea* *Penicillium spp.* *Aspergillus spp.* | Mouldy feed, rotten apples, and apple juice | Oedemas, Haemorrhage, paralysis of motor nerves, convulsions, carcinogenesis |
| Penitrem A | *Penicillium crustosum* *Penicillium cyclopium* | Mouldy cream cheese, walnuts, and beer | Tremors, lack of coordination, bloody diarrhoea, death |
| Sterigmatocystin | *Aspergillus versicolor* | Green coffee, mouldy wheat, Dutch cheeses | Carcinogenesis, hepatotoxin |
| Trichothecenes | *Fusarium spp.* | Corn, wheat, commercial cattle feed, mixed feed | Digestive disorders, dermatitis, blood disorders |
| Zearalenone | *Fusarium graminearum* | Corn, mouldy hay, pelleted commercial feed | Estrogenic effects. Atrophy of testicles and ovaries, abortion |

Based on Cole and Dorner, 1993; Hocking, 1997; Pitt, 1997; Bullerman, 1997.

the consumption of contaminated vegetables, and (2) indirectly via the consumption of animal products, such as milk. Some of the most relevant mycotoxins are shown in Table 2.8.

Of all mycotoxins, aflatoxins are of greatest concern because they are highly toxic and potentially carcinogenic. Chemically aflatoxins are derivates of difuranocumarin (Hocking, 1997), and there are four main aflatoxins produced by moulds of the genus *Aspergillus*: B1, B2 G1, and G2. Two more aflatoxins, M1 and M2, first identified in milk also occur and they are produced from the original aflatoxins B1 and B2, by hydroxylation in lactating animals. Although most aflatoxin producing strains produce all the different types of toxins, aflatoxin B1 is usually predominant and is the most toxic (CFSAN, 2003).

The ingestion of foods contaminated with aflatoxins can result in a disease called aflatoxicosis. Depending on the dose and frequency of aflatoxin ingestion, acute or chronic disorders may appear. These disorders include hepatotoxicity, carcinogenesis, teratogenesis, and an immunosupressive effect that can increase sensitivity to infectious diseases (Cole and Dorner, 1993). The chance of human exposure to acute levels of aflatoxin is remote in well-developed countries. In 1974, one of the most significant outbreaks of aflatoxins intoxication occurred in northwest India; 397 people were affected of whom 108 died. The cause of this outbreak was determined as the consumption of corn contaminated with aflatoxins at concentrations of 0.25 to 15 mg/kg (CFSAN, 2003). Several studies have revealed a significant correlation between hepatic cancer and the ingestion of aflatoxin-contaminated foods in Central Africa and in some countries of southeast Asia (Hocking, 1997).

Aflatoxin B1 is metabolised in the liver by the enzymatic oxidase system into a 2, 3 epoxid of aflatoxins. The epoxid derivative is a very reactive intermediate metabolite that is able to bind with the DNA strand, causing the interruption of transcription which consequently gives rise to mutagenesis or carcinogenesis. In addition, aflatoxins may also inhibit some enzymes of the electron transporting chain causing a reduction in TPA levels.

The production of aflatoxins by different strains of *Aspergillus* is influenced by the growth conditions. *A. flavus* and *A. parasiticus* grow between 10-12°C to 42-43°C, the optimal temperature for growth being 32-33°C. Aflatoxins are produced almost across the entire temperature range for growth. Moulds are quite xerobiotic and can grow well at low water activity values. Although *Aspergillus* strains can grow at $a_w$ values down to 0.80, aflatoxin production is stopped below 0.85 and optimal production takes place at 0.98-0.99. Both *Aspergillus* species can grow over a broad pH range, with *A. parasiticus* being able to grow between 2.0 and 10.5, and *A. flavus* between 2 and 11.2. However, aflatoxin production can only occur between pH 3.0 and 8.0. Reduction of available oxygen as in vacuum and modified atmosphere packaging (MAP) inhibits the production of aflatoxins.

Aflatoxins are controlled by legislation in most of the well-developed countries and the maximum limit has been established at around 5-20 μg/kg of food.

*Jordi Rovira, Avrelija Cencic, Eva Santos and Mogens Jakobsen*

Other mycotoxins, i.e. ochratoxin A, patulin, ergotoxins, and tricothecenes should be taken into consideration in hazard analysis mainly dealing with raw material evaluation (Table 2.8).

## 2.3.3 Parasites

Parasites are a group of biological hazards that are normally underestimated or are not taken into account when an HACCP study is developed. The reason for this underestimation is probably the common assumption that parasites are only hazards in particular geographical areas of the world, mainly the underdeveloped countries. On the contrary, cases of food parasitosis have increased dramatically over the last decade. In 1990, about 13 species of parasites were of concern to food scientists in the USA, but today this figure has multiplied by more than a factor 8 (Orlandi *et al.*, 2002). This spectacular increase in concern is due to different socio-economic factors, which were mentioned in the introduction to this chapter. In short, this trend can also be explained by the globalisation of the food trade, preference for raw and undercooked dishes, ease of international travel (tourism), and the increasing numbers of immunocompromised people. This is further complicated by the emergence of parasites not previously associated with pathogenesis in humans (Orlandi *et al.*, 2002). Currently more than 107 species of parasites that can be transmitted by food have been identified; some of them are shown in Table 2.9. According to a CDC study, 7% of the annual food and beverage-borne disease incidences are caused by parasites, 13% by bacteria, and 80% by viruses (Mead *et al.*, 1999). Two different big groups of parasites can be distinguished: parasitic protozoa and parasitic metazoa (which comprise mainly helminths or worms).

Contamination of food products by parasites may occur at several steps along the food chain, from growing and harvesting food at the farm or catching fish, through to consumption by consumers. Water is one of the most important sources for transmission of foodborne parasites, directly and indirectly due its use in irrigation, its role as an ingredient, or for cleaning support. Soils as well as contact with infected animals or infected handlers are also different ways by which a parasite can be introduced into the food chain.

**Control measures -** Different measures can be taken to prevent food contamination by parasites including proper cleaning. Also, most parasites are quite sensitive to heat treatment, an example being the elimination of the infective capability of some helminths by heating at 56-60°C for several minutes (Fayer *et al.*, 2001) and the elimination of some protozoa such as *Cryptosporidium parvum* by 1 min at 72°C (Steiner, *et al.*, 1997). In all cases it should be ensured that the proper temperature is attained in the deepest part of the foods. Application of cold temperatures is also useful in inactivating most parasites. Freezing of *C. parvum* at -20°C for 24 hours was found to render oocysts non-infectious and non-viable (Fayer and Nerad, 1996). The same happens with helminths such as *Anisakis simplex*, when fish is frozen. Irradiation and high-pressure treatments are also effective to some extent (FAD, 1985; Slifko *et al.*, 2000a). Some traditional preservation methods, i.e. fermentation in brine, hot smoking or drying, are also effective for the elimination of some helminths. However, it is remarkable that chlorine does not inactivate protozoan cysts and oocysts, an example of which is the resistance of *Giardia lamblia* (*intestinalis*) to levels of chlorine with anti-bacterial effects (Venczel, *et al.*, 1997).

Table 2.9. Main characteristics of some foodborne parasites.

| Parasite | Infective form | Associated foods | Disease | Control measures |
|---|---|---|---|---|
| **Protozoa** | | | | |
| *Entamoeba histolytica* | Cyst | Drinking water or food contaminated by faeces and raw vegetables | Amebiasis: gastrointestinal distress, dysentery | |
| *Giardia lamblia* | Cyst | Drinking water or food contaminated by faeces and raw vegetables | Diarrhoea | |
| *Toxoplasma gondii* | Oocyst, Bradizoites | Contaminated water, meat | Severe illness in foetuses | T>61°C for 3.6 min, and freezing at -13°C to eliminate intracellular cysts |
| *Cyclospora cayetanensis* | Oocyst | Contaminated water or foods | Cyclosporidiosis: watery diarrhoea, fatigue, nausea, vomiting, mialgia, weight loss | Similar to *Cryptosporidium* |
| **Helminths** | | | | |
| *Fasciola hepatica* | Metacercaria (Trematode) | Vegetables, contaminated water | Hepatic tissue destruction | |
| *Taenia solium* | Cysticercoid/Egg (Cestode) | Raw or undercooked pork meat | Slight abdominal pain, anorexia, nervous disorders | Heat treatment up to 56°C, or freezing at -10°C for 14 days |
| *Diphyllobothrium latum* | Plerocercoid (Cestode) | Fish | Abdominal distension, cramping, diarrhoea, pernicious anaemia | |
| *Trichinella spiralis* | Larva (Nematode) | Raw or undercooked meat | Gastrointestinal disorders, diarrhoea and fever. Muscular dysfunction. | |

*Protozoa*

Most of the parasitic protozoa belong to the phyla *Apicomplexa*, *Ciliophora*, *Sarcomastigophora* and *Microspora*. Members of the phylum *Apicomplexa* are known as coccidian; they have very complex biological cycles which alternate between asexual and sexual reproduction. To this group belong *Cryptosporidium parvum*, *Cyclospora cayetanensis* (which are considered as emergent parasites), *Toxoplasma gondii,* and *Isospora belli*. The species of *Cliophora* and *Sarcomastigophora* have a very simple biological cycle that normally occurs in the same host (monoxenous). Among *Sarcomastigophora* the most relevant members are *Giardia lamblia* (*Giardia intestinalis*) and *Entamoeba histoytica* (Speer, 1997).

*Cryptosporidium parvum*
**General description -** It is a ubiquitous obligate intracellular parasite that infects the microvillus border of the epithelium in the gastrointestinal tract and the respiratory system of humans and various animal hosts (Clark, 1999). The infective stage of the organism is the oocyst, an environmentally resistant form, which is very small in size (3-6 µm). Once this oocyst is ingested, excysts in the intestine release four sporozoites which infect the epithelial cells. They undergo asexual and sexual development in the epithelial cells, finally producing new infective oocysts that can produce an auto-infection or be released back into the environment (Speer, 1997).

**Disease -** Cryptosporidiosis is the illness caused by *Cryptosporidium parvum*, which happens to be the most recent and significant microbiological pathogen to emerge and cause concern within the food processing sector. The disease is transmitted through contaminated water and food. The organism is capable of causing a high degree of morbidity in healthy communities and mortality in immunocompromised populations, as there is no effective anti-microbial treatment to eliminate it from the gastrointestinal tract (Millar, *et al.*, 2002).
This is a self-limiting disease in most healthy individuals, persisting for around 2-4 days. Infection mainly occurs in the gastrointestinal tract, and most relevant symptoms are severe watery diarrhoea, dehydration, abdominal cramps, vomiting, weight loss, and electrolyte imbalance. Sometimes this organism may parasite the respiratory tract (Millar, *et al.*, 2002, CFSAN, 2003). The infective dose is less than 10 organisms and, presumably, only one of them is needed to initiate an infection. To date, there is no known effective drug for the treatment of cryptosporidiosis.

**Target group -** All people can be infected by *C. parvum*, but children at day-care centres, and immunocompromised people (especially AIDS patients) are much more susceptible to having the disease for life, with the severe watery diarrhoea contributing to death (CFSAN, 2003). It is now estimated that 32% of AIDS patients have suffered at some stage from cryptosporidiosis (Hunter and Nichols, 2002).

**Source and associated foods -** As this organism is an intracellular parasite of the gastrointestinal tract of warm-blooded animals, this is considered the primary source of contamination. The faeces of animals can contaminate meat during slaughter operations. Water is the most important vehicle for transmission of the illness, and it may become contaminated from human or animal faeces either by a direct (agricultural runoff) or indirect (accidental contamination from human sewage) route (Millar, *et al.*, 2002). The largest waterborne outbreak of cryptosporidiosis occurred in April 1993 in Milwaukee (USA) and affected 403,000 people (MacKenzie *et al.*, 1994; CFSAN, 2003).
Oocysts of *C. parvum* have been isolated from several foods such as fruits, vegetables and shellfish (they can survive for long periods in seawater). The presence of oocysts in this sort of produce is alarming from a public health point of view, as they can be consumed raw without any thermal processing that would inactivate the oocysts (Millar *et al.*, 2002). In fact, any foodstuff could be contaminated by infected food handlers.

## Biological hazards

**Control measures** - The prevention and control of oocysts in foodstuffs may be achieved through an integrated HACCP approach, tailored specifically for the elimination of viable oocysts in food processing. Lateral transfer of other HACCP plans for other enteric bacterial pathogens should be discouraged, because the survival kinetics for this parasite differs significantly from those of viral, bacterial, and fungal pathogens (Millar *et al.*, 2002). Water quality control by combining filtration with disinfection is necessary as chlorination alone has not been successful in eliminating *Cryptosporidium* oocysts.

In summary, *C. parvum* can enter the food chain in different ways and at different stages of food processing, i.e. through contaminated raw ingredients, use of contaminated water in the foodstuff formula and for cleaning industrial equipments, through pest infections, and from contaminated food handlers. All the different possibilities for contamination should be taken into account when designing a realistic and effective HACCP system.

Other protozoan parasites, such as *Cyclospora cayetanensis, Toxoplasma gondii, Giardia lamblia* (*intestinalis*) and *Entamoeba histolitica,* are summarised in Table 2.9. More information can be obtained from specialised literature.

*Helminths*
The helminths are generally referred to as tapeworms and roundworms and include members of the classes *Trematoda, Cestoda, Acanthocephala* and *Nematoda*. These worms can infect humans via foods of animal origin (i.e. meat or fish and shellfish), foods of vegetable origin, by the ingestion of some infected invertebrates that can contaminate some foods or drinking water, or by faecal contamination (Table 2.9). Although, food-borne helminths are taxonomically very different, all of them need more than one host to complete their biological cycle. Normally, these parasites can be eliminated by suppressing the intermediate host or by heating the contaminated food.

*Anisakis simplex* and related worms
*Anisakis* is one of the foodborne parasites that nowadays causes most concern among consumers. They are nematodes (roundworms) of the genera *Anisakis, Pseudoterranova,* and *Contracaecum* and are associated with the consumption of raw or undercooked seafood. The illness produced by these kinds of worms is called anisakiasis. There are two forms of anisakiasis: (1) the non-invasive form, that is in general asymptomatic and produces a tingling or tickling sensation in the throat caused by the movement of the worms migrating through the oesophagus, and (2) the invasive form, where the worms penetrate the mucosa and submucosa of the stomach or small intestine causing the infected individuals to suffer an acute abdominal pain, similar to appendicitis and accompanied by a nauseous feeling and sometimes vomiting. These symptoms may start as soon as 1 hour and up to as long as 2 weeks after the ingestion of raw or undercooked seafood. Only good heat treatment or a prolonged freezing period can destroy the parasite. Neither cold smoking nor brine preservation methods can eliminate *Anisakis*.

Relevant information about other foodborne helminths is shown in Table 2.9, and more information could be obtained in specific literature.

*Algae*
There are some planktonic algae, generally belonging to the family of dinoflagellates, that can indirectly cause food poisoning when they are ingested by several marine animals, which are in turn consumed by humans. Dinoflagellates produce different kinds of toxins that are concentrated in the flesh of the marine animals that consume them. It is possible to distinguish between two different groups of poisoning according to the intermediate or reservoir animal: (1) ciguatera fish poisoning which occurs when toxins are accumulated in the flesh of tropical and subtropical fin fishes (reef-associated fishes), such as barracudas, snappers, jacks, mackerel, tiger fish, and many other species of warm-water fish, and (2) shellfish poisoning which actually represents several diseases produced by a number of toxins produced in shellfish by different species of dinoflagellates. The more important shellfish poisonings are paralytic shellfish poisoning (PSP), diarrheic shellfish poisoning (DSP), neurotoxic shellfish poisoning (NSP), and amnesic shellfish poisoning (ASP). The severity of the symptoms and the length of the disease is very much dependent on the type of toxin(s) present in the contaminated food, their concentration, and the amount of contaminated fish or shellfish consumed. The main characteristics of these different diseases are shown in Table 2.10.

Diagnosis of these kinds of food poisonings is entirely based on observed symptomatology and recent dietary history. There are no good statistical data on the occurrence of these kinds of poisoning, but it is believed that their incidence is largely under-reported probably due to their non-fatal nature and the short duration of these diseases (CFSAN, 2003).

## 2.3.4 Virus

Foodborne and waterborne viral infections are increasingly recognised as causes of illness in humans. This increase is partly explained by changes in food processing and consumption patterns, e.g. the increased consumption of ecologically-produced food. The European directive has established no specific microbiological criteria concerning the presence of viruses even though it has been shown that viruses have been detected in foods that have met bacteriological standards. The lack of microbiological criteria for viruses is mainly attributed to the fact that quantitative methods for their detection are not yet available.

The globalisation of infectious diseases due to the rapidly increasing international food trade and travel will certainly raise the incidence of these viral infections in the years to come. It may therefore happen that a single contamination of food by a single food-handler at a single source will cause vast outbreaks of foodborne and waterborne viral infections. Of concern with regards to possible outbreaks is the ageing population, which together with very young children and people with weak immune systems are considered the most vulnerable groups (Koopmans *et al.*, 2003).

Viruses are microorganisms, which cannot be seen by the naked eye or by a light microscope, whose size ranges from 15 to 400 nanometers. Unwanted and uninvited, they are totally dependent on their hosts (bacteria, animals, and plants) for survival, making them obligate intracellular parasites (McKane and Kandell, 1986).

Table 2.10. Main characteristic features of the most important food poisoning produced by dinoflagellates toxins.

| | Microorganism involved | Toxins | Associated foods | Disease symptoms | Target population |
|---|---|---|---|---|---|
| Ciguatera fish poisoning | Ciguatera Gambierdiscus toxicus Ostroepsis lenticularis | Ciguatoxin Maitotoxin Scaritoxin | Tropical and sub-tropical marine finfish | Appear six hours after consumption of contaminated food and are combination of gastrointestinal, neurological and cardiovascular disorders | All humans |
| Paralytic shellfish poisoning (PSP) | Alexandrium spp. Pyrodinium babanense Gymnodinium catenatum | More than 20 toxins involved, all derivatives of saxitoxin i.e. neosaxitoxin, gonyautoxins etc. | Mussels, clams, cockles, and scallops | First symptoms appear 0.5 to 2 hours after ingestion. Tingling and burning sensation, numbness, incoherent speech, and respiratory paralysis | All humans |
| Diarrheic shellfish poisoning (DSP) | Dinophysis spp. Prorocentrum spp. | High molecular weight polyethers, i.e. Okadaic acid Dinophysis toxin Pectenotoxin Yessotoxin | Mussels, oysters and scallops | Onset of disease between 30 minutes and 2-3 hours after ingestion. Mild gastrointestinal disorders, i.e. nausea. vomiting, diarrhoea and abdominal pain. | All humans |
| Neurotoxic shellfish poisoning (NSP) | Gymnodinium breve | Group of polyethers referred to as brevetoxins | Oysters, mussels, clams and scallops | It causes gastrointestinal and neurological symptoms (tingling and numbness of lips, muscular aches, reversal of the sensation of hot and cold) | All humans |
| Amnesic shellfish poisoning (ASP) | Pseudonitzschia pungens | Unusual amino acid: domoic acid | Mussels | Gastrointestinal disorders and neurological problems (Alzheimer-like symptoms) | All humans, especially elderly people |

*Jordi Rovira, Avrelija Cencic, Eva Santos and Mogens Jakobsen*

They are incapable of any independent metabolism, often wearing just a thin protein coat wrapped around a small cluster of genes (Figure 2.4).

Viruses possess only a single type of nucleic acid, either deoxyribonucleic acid (DNA) or ribonucleic acid (RNA), but never both. In addition, the nucleic acid of some viruses is single-stranded or double stranded. Four possible configurations exist:
- double stranded DNA (e.g. Herpes virus group);
- single stranded DNA (e.g. Parvovirus group);
- double-stranded RNA (e.g. Reovirus group);
- single-stranded RNA (e.g. Retrovirus group).

*Foodborne viruses*
The most relevant in foodborne infections are those viruses that infect the cells lining the intestinal tract. They are transmitted essentially via the faecal-oral route as humans are the reservoirs for food-borne viruses (Figure 2.5). Food can be contaminated directly by man or indirectly by contaminated water. Among viral groups for whom foods can serve as vectors, are mainly poliovirus, hepatovirus (hepatitis A and E), various gastroenteritis viruses like rotavirus, astrovirus, and small round structured viruses (SRSVs), like the Norwalk and other caliciviruses now referred to as Noroviruses. The main characteristics of those viruses are given in Table 2.11 (Varman and Evans, 1996; Untermann, 1998; Koopmans *et al.*, 2002). It should be mentioned that with the new lifestyle trends (i.e. more "natural" food) potential exists for a higher incidence of foodborne infections of Tickborne encephalitis (TBEV), as it has been reported that up to 10% of infections are due to the consumption of raw, non-pasteurised milk (Jurceviciene et al., 2002).

Although the current opinion is that there is no known risk of transmission of the human immunodeficiency virus (HIV) via foods, given that basic sanitary precautions are observed, its should also be mentioned in such discussions.

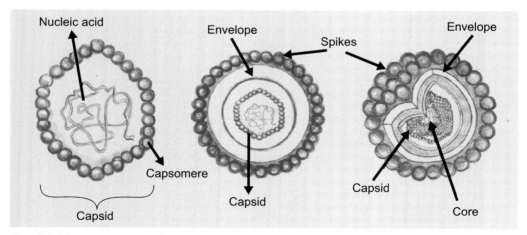

*Figure 2.4. Schematic presentation of viral structure a) nonenveloped virus, b) and c) enveloped viruses.*

## Biological hazards

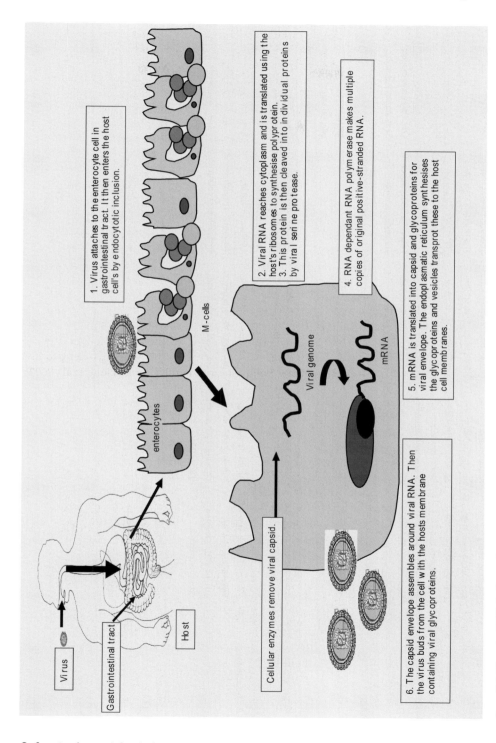

Figure 2.5. Life cycle of RNA foodborne viruses.

Table 2.11. The main characteristics of foodborne viruses.

|  | Noroviruses | Rotaviruses | Hepatitis A | Hepatitis E | TBEV |
|---|---|---|---|---|---|
| Mode of action | Infection and multiplication in the gastrointestinal tract | Infection and multiplication in the gastrointestinal tract | Infection of enterocytes and multiplication in hepatocytes | Infection of enterocytes and multiplication in hepatocytes | Not known |
| Infective dose | Not known, but presumably very low | Not known, but presumably very low | Presumably 10-100 virus particles | Not known | Not known |
| Disease | Gastroenteritis with nausea, vomiting, diarrhoea and abdominal pain | Gastroenteritis with vomiting and diarrhoea | Fever, malaise, nausea, anorexia, jaundice | Fever, malaise, nausea, anorexia, jaundice | Occurs in two stages: First stage (1-8 days): fever, headache, myalgia, leuko and trombocytopenia, second stage: meningitis or meningoencephalitis. |
| Target groups | Babies and young children | Infants and children | Elderly people, people with other forms of hepatitis | Young to middle age adults (15 - 40 years old) pregnant women | All |
| Associated foods | Water, shellfish and salad ingredients, raw or insufficiently steamed clams or oysters | All foods that require handling and no further cooking, such as salads, fruits (faecal-oral route) | Cold-cuts and sandwiches, milk and dairy products, vegetables, salads and iced drinks (faecal-oral route) | Untreated water zoonotic through swine in developed countries | Raw milk |
| Outbreaks | Sporadic in developed countries, more frequent in underdeveloped countries | Endemic | Sporadic | Sporadic in developed countries, more frequent in underdeveloped countries | Sporadic |

## Biological hazards

According to the type of illness they produce, foodborne viruses can be classified into three main groups:
- viruses that cause gastroenteritis (e.g. astrovirus, rotavirus, the enteric adenoviruses, and two genera of enteric caliciviruses: "Norwalk-like viruses" (NLV) and typical caliciviruses or "Sapporo-like viruses" (SLV));
- faecal-orally transmitted hepatitis viruses: hepatitis A virus (HAV), hepatitis E virus (HEV);
- group of viruses that replicate in the human intestine but cause illness after they migrate to other organs such as the central nervous system (CNS), i.e. the poliovirus or tick-borne encephalitis virus (TBEV).

*Gastroenteritis causing viruses*

Recent studies have shown that the most common cause of gastroenteritis in people of all ages are enteric caliciviruses (SRSVs) and rotaviruses (Figure 2.6).

*Small round structured viruses (SRSVs)*

**General description -** The structure of SRSVs is amorphous lacking surface geometry (Figure 2.6A). The virus contains a positive strand RNA genome of 7.5kb and a single structural protein of Mw 60.000.

The Norwalk virus is the prototype of the family of unclassified small round structured viruses (SRSVs), which may be related to the caliciviruses. It was described as the cause of "winter vomiting disease" in 1968. The family consists of several serologically distinct groups of viruses that have been named after the places where the outbreaks occurred. In the U.S., the Norwalk and Montgomery County agents are serologically related but distinct from the Hawaii and Snow Mountain agents. The Taunton, Moorcroft, Barnett, and Amulree agents were identified in the U.K., and the Sapporo and Otofuke agents in Japan.

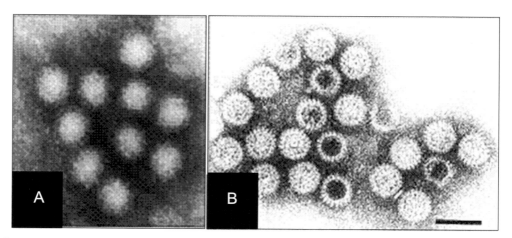

Figure 2.6. Viruses that cause gastroenteritis A) Norwalk virus - caliciviridae and B) Rotavirus - reoviridae (www.cdc.gov/ncidod/eid/vol4no4/parasharG.htm).

**Source and associated foods** - Caliciviruses were originally observed in specimens from sporadic cases of gastroenteritis in babies and later from outbreaks among babies and young children.

In the environment, water is the most common source of outbreaks and may include water from municipal supplies, wells, recreational lakes, swimming pools, and water stored aboard cruise ships. Transmission of caliciviruses is essential via the faecal-oral route, contaminated water and foods. Secondary person-to-person transmission has also been documented. Shellfish and salad ingredients are the foods most often implicated in Norwalk outbreaks. Ingestion of raw or insufficiently steamed clams and oysters poses a high risk for infection with Norwalk virus. Foods other than shellfish that are contaminated by ill food handlers can also be a source of infection. There have been several outbreaks published. Up-to-date data can be followed at web page: http://www.cdc.gov/mmwr/index.html.

**Nature of acute disease** - Common names of the illness caused by the Norwalk and Norwalk-like viruses are:
- viral gastroenteritis;
- acute non-bacterial gastroenteritis;
- food poisoning;
- and food infection.

The infectious dose is unknown but presumed to be low. The disease is self-limiting usually lasting 24-60 hours. Symptoms are mild and characterised by nausea, vomiting, diarrhoea, and abdominal pain. Headache and low-grade fever may occur. The onset time is between 10 to 50 days, dependent on the individual.

*Rotaviruses*

**General description** - Rotaviruses are classified within the Reoviridae family. They have a genome consisting of 11 double-stranded RNA segments surrounded by a distinctive two-layered protein capsid (Figure 2.6B). The particles are 70 nm in diameter.

Six serological groups have been identified, three of which (groups A, B, and C) infect humans:

*Group A*
Group A rotavirus is endemic worldwide. It is the leading cause of severe diarrhoea among infants and children, and accounts for about half of the cases requiring hospitalisation. The number of cases attributable to food contamination is unknown.

*Group B*
Group B rotavirus is also known as the adult diarrhoea rotavirus or ADRV. Several large outbreaks of group B rotavirus involving millions of people as a result of contamination of drinking water supplies by sewage have occurred in China since 1982.

*Group C*
Group C rotavirus is associated with rare and sporadic cases of diarrhoea in children in many countries. The newly recognised group C rotavirus has been implicated in rare and isolated cases of gastroenteritis. However, it was associated with three outbreaks among school children, of which one was in Japan in 1989 and two were in England in 1990.

## Biological hazards

**Source and associated foods -** Rotaviruses are quite stable in the environment and have been found in estuary samples at levels as high as 1-5 infectious particles/L. Sanitary measures adequate for bacteria and parasites seem to be ineffective in the endemic control of rotavirus, as a similar level of incidence of rotavirus infection is observed in countries with either a high or low standard of health.

Transmission of rotaviruses is essentially by the faecal-oral route. Person to person transmission through contaminated hands is probably the most important route by which rotaviruses are transmitted. Infected food handlers may contaminate foods that require handling and no further cooking, such as salads, fruits, and hors d'oeuvres (Untermann, 1998).

**Nature of acute disease -** Rotaviruses cause acute gastroenteritis. Infantile diarrhoea, winter diarrhoea, acute non-bacterial infectious gastroenteritis, and acute viral gastroenteritis are names applied to the infection caused by the most common and widespread group A rotavirus. The incubation period ranges from 1-3 days. Symptoms often start with vomiting followed by 4-8 days of diarrhoea. Temporary lactose intolerance may occur.

Rotavirus gastroenteritis is a self-limiting, mild to severe disease characterised by vomiting, watery diarrhoea, and low-grade fever. Although recovery is usually complete, diarrhoea without fluid and electrolyte replacement may result in severe diarrhoea and death.

As a person with rotavirus diarrhoea often excretes large numbers of the virus, $10^8$-$10^{10}$ infectious particles/ml of faeces, the infective doses can be readily acquired through contaminated hands, objects, or utensils. Asymptomatic rotavirus excretion has been well documented and may play a role in perpetuating endemic disease.

*Faecal-orally transmitted hepatitis viruses: hepatitis A virus (HAV), hepatitis E virus (HEV)*

*Hepatitis A (HAV)*

**General description -** In practice most reported incidents of viral foodborne illnesses are due to gastroenteritis viruses and hepatitis A virus (HAV). Hepatitis A virus (HAV) is classified with the enterovirus group of the Picornaviridae family (Figure 2.7). HAV has a single molecule of RNA surrounded by a small protein capsid, 27nm in diameter. Many other picornaviruses can

*Figure 2.7. Electron micrograph of hepatitis A viruses (http://www.cdc.gov/).*

cause human disease including polioviruses, coxsackieviruses, echoviruses, and rhinoviruses (cold viruses).

**Source and associated foods -** Although such viruses are usually transmitted from person to person, it is increasingly recognised that food may have an important role in their transmission (Figure 2.8). It is likely that foodborne transmission of viral infections is greatly under-reported, and the actual extent of the problem is not known.

HAV is excreted in the faeces of infected people and can produce clinical disease when susceptible individuals consume contaminated water or foods. Cold cuts, sandwiches, fruits and fruit juices, milk and milk products, vegetables, salads, shellfish, and iced drinks are commonly implicated in outbreaks. Water, shellfish, and salads are most frequently associated with HAV. Contamination of foods by infected workers in food processing plants and restaurants is common. Since the fatality rate of HAV infections increases with age, the risk of more serious illness is higher for older people who were not exposed to the virus at an earlier age. Persons with other forms of hepatitis, i.e. hepatitis B, are at higher risk of hepatitis following super-infection with HAV (Koopmans *et al.*, 2003; WHO, 2000).

**Nature of acute disease -** The term hepatitis A (HA) or type A viral hepatitis has replaced all previous designations such as infectious hepatitis, epidemic hepatitis, epidemic jaundice, catarrhal jaundice, infectious icterus, Botkins disease, and MS-1 hepatitis.

The incubation period for hepatitis A, which varies from 10 to 50 days (mean 30 days), is dependent on the number of infectious particles consumed. Infection with very few particles results in longer incubation periods. The period of communicability extends from early in the

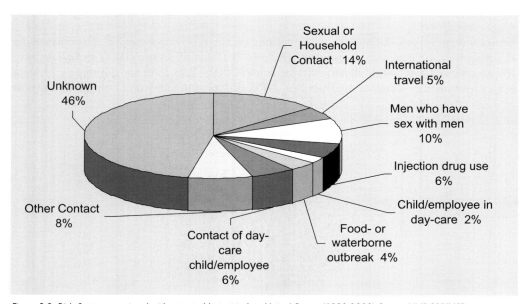

Figure 2.8. Risk factors associated with reported hepatitis A, in United Status (1999-2000) Source NNDSS/VHSP.

## Biological hazards

incubation period to about a week after the development of jaundice. The greatest danger of spreading the disease to others occurs during the middle of the incubation period, well before the first presentation of symptoms. Many infections with HAV do not result in clinical disease, especially in children. Hepatitis A is usually a mild illness characterised by the sudden onset of fever, malaise, nausea, anorexia, abdominal discomfort, and followed after several days by jaundice. The infectious dose is unknown but is presumed to be 10-100 virus particles. The symptoms are occasionally severe and convalescence can take several months.

*Hepatitis E (HEV)*
**General description -** Hepatitis E (HEV) is an unenveloped single-stranded RNA virus (Figure 2.9). HEV has a single-stranded polyadenylated RNA genome of approximately 8 kb and is extremely labile. Based on its physicochemical properties it is presumed to be a calici-like virus.

**Source and associated foods -** It is prevalent in the areas where hygiene standards are poor. In developed countries it is shown to be zoonotic through the swine industry. HEV is epidemic and usually transmitted by contaminated food or water. Person to person transmission may also occur (Koopmans *et al.*, 2003; WHO, 2000).

HEV is expected in non-endemic countries as well. Although HEV is mainly transmitted via the faecal-oral route, a high potential exists for foodborne transmission. Cases of acute hepatitis of novel HEV variants have been reported in humans in Europe and Japan, showing that HEV is not limited or geographically distributed, but is endemic. 43,5% of sewage samples were positive in Barcelona (Spain), 20% in Washington (USA), and 25% in Nancy (France). These results show that HEV may be more prevalent than previously considered. There have been an increasing number of reports from Europe and the USA of sporadic hepatitis attributable to HEV, leading to suggestions that HEV may be endemic at low levels in developed countries (Figure 2.10) (Clemente-Casares *et al.*, 2003; Widdowson *et al.*, 2003). Although HEV is often a self-limiting

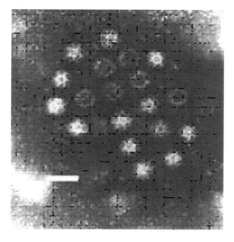

*Figure 2.9. Electron micrograph of hepatitis E (HEV) virus (www.epa.gov).*

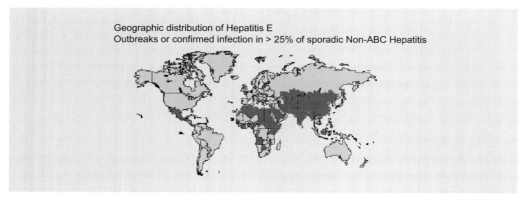

*Figure 2.10. The map of HEV infection that generalises available data and patterns (http://www.cdc.gov/ncidod/diseases/hepatitis/).*

disease, agricultural workers who use untreated wastewater for irrigation are at high risk and the mortality of pregnant women (25%) is considerably high (Ceylan *et al.*, 2003; Chibber *et al.*, 2003). It is also of importance that HEV-RNA was found to be present in the colostrum of breast-feeding mothers that were HEV infected. Animal reservoirs, like the occurrence of HEV in swine native to industrialised countries, should be considered a potential public health hazard and relevant food related studies initiated (Widowson et al., 2003).

**Nature of acute disease -** The disease is most often seen in young to middle aged adults (15-40 years old). Pregnant women appear to be exceptionally susceptible to severe disease, and excessive mortality has been reported in this group. The disease caused by HEV is called hepatitis E, or enterically transmitted non-A non-B hepatitis (ET-NANBH). Other names include faecal-oral non-A non-B hepatitis, and A-like non-A non-B hepatitis.

Hepatitis caused by HEV is clinically indistinguishable from the hepatitis A disease. Symptoms include malaise, anorexia, abdominal pain, arthralgia, and fever. The infective dose is not known (Krawczynski *et al.*, 2001).

*Tick-borne encephalitis virus (TBEV)*

**General description -** TBE is a major European arbovirosis, endemic in a number of areas, mostly in Central and Eastern Europe. It is normally spread by the bite of ticks of the genus *Ixodes*. In addition, it can be spread by raw contaminated milk. Related viruses within the same group, Louping ill virus (LIV), Langat virus (LGTV) and Powassan virus (POWV), also cause human encephalitis but rarely on an epidemic scale. Three other viruses within the same group, Omsk haemorrhagic fever virus (OHFV), Kyasanur Forest disease virus (KFDV) and Alkhurma virus (ALKV), are closely related to the TBEV complex viruses and tend to cause fatal haemorrhagic fevers rather than encephalitis.

Tick-borne encephalitis virus (TBEV) has spherical enveloped virions, uniform in shape and ranging from 40 to 60nm in diameter. Surface projections of the envelope are small (surface appears rough), or distinct and obvious fringes in negative stains. Nucleocapsids are isometric and 25-30 nm in diameter. Symmetry is polyhedral (McKane et Kandell, 1986).

**Source and associated foods -** TBEV is believed to cause at least 11,000 human cases of encephalitis in Russia and about 3000 cases in the rest of Europe annually. Very few food-related studies have been carried out for TBEV. Although TBEV is an arbovirus, it is also an emerging pathogen as it was found that approximately 10% of its infections were associated with the consumption of goat, sheep and cow raw milk and raw milk products. TBEV is currently mainly a problem in the Southern and Eastern parts of Europe. Generally, the best protection against viral diseases is vaccination, however, the prophylactic use of probiotics may be a valuable complement to vaccination in endemic TBEV areas, where the consumption of raw milk from goats and sheep is common. Probiotics may also be used in prophylactics in western European countries to prevent entry through the gastrointestinal tract, especially in those countries (i.e. France) where raw goat milk is used in dairy production. With the new lifestyle trends in Europe (i.e. towards more ecological or organic food production, processing, and consumption) the potential for much higher incidences of foodborne infections by TBEV should not be neglected (Jurceviciene et al., 2002).

**Nature of acute disease -** Tick-borne encephalitis (TBE) is typically a biphasic disease. The initial stage (1-8 days) includes non-characteristic symptoms such as fever, headache, and myalgia. Leuko- and trombocytopenia are common. After an asymptotic interval of a few days to three weeks, about one fourth of the patients develop a second stage of the disease afflicting the central nervous system (CNS) in the form of meningitis or meningoencephalitis with or without myelitis. In the acute stage of the disease paresis is seen in 3.0-23.5% of the patients and the reported mortality in Europe is 0-3.9%. These viruses can infect via the alimentary tract and also intranasally when experimental animals are inoculated. Concentrated aerosols or high virus concentrations delivered as a powder contaminating food, are presumed to be infectious. TBEV are excreted in the urine and faeces of experimentally infected animals but it is unlikely that this form of the virus would provide an efficient route of infection for humans.

*Spread and persistence of foodborne viruses from farm to fork*
Foods at highest risk are those that require intensive manual handling, those that are manually handled under poor hygienic conditions, those that are 'close to fork', and end products that are manually handled. This problem arises especially with ecologically-produced food where no additives are introduced into the food chain. Most of the outbreaks can be traced to infected food handlers at the end of the food chain. For example, carriers of hepatitis A typically shed high quantities of the virus 10 to 14 days after infection.

Another factor determining the risk for contamination of foods is the stability of some of the foodborne viruses (e.g. rotaviruses) in the environment. Viruses can persist for up to 60 days on several types of materials (i.e. paper and cotton) in domestic and institutional environments. In mineral water, viruses may survive for over 1 year at 4°C. As viruses are strict intracellular parasites, infectivity never increases during processing, transport, or storage (Koopmans *et al.*, 2002; Untermann, 1998).

**Prevention -** Although currently the data available is insufficient to determine which steps are critical for all foods in an HACCP system, the following points should be addressed:

*Jordi Rovira, Avrelija Cencic, Eva Santos and Mogens Jakobsen*

- Water of drinking quality should be used in the culturing or preparation of food.
- Guidelines specifically aimed at the reduction of viral contamination are needed as it has become clear that current indicators for water and shellfish quality are insufficient as predictors of viral contamination.
- Raw milk should not be consumed.
- Harmonising and enhancing epidemiological surveillance should be initiated through studies of molecular detection techniques.
- Web access to a central database to track viruses, which provides the foundation for an early warning system of foodborne and other common-source viral outbreaks, should be established.

**Summary -** Viruses are obligate intracellular microorganisms invisible to the naked eye or light microscope, with a size range from 15 to 400 nanometers, containing either DNA or RNA.
Foodborne viruses are viruses that cause infections of humans and animals through their gastro- intestinal tract and are transmitted essentially via the faecal-oral route or through contaminated water. The main groups of foodborne viruses include the poliovirus, hepatoviruses (hepatitis A and E), and gastroenteritis viruses like rotavirus, astrovirus and norovirus.
Transmission and persistence of foodborne viruses from 'farm to fork' occurs most often by infected food handlers at the end of the food chain. The infectivity will never increase during the processing, transport, and storage.
Water used in combination with the culturing or preparation of food should be of drinking quality.

Guidelines specifically aimed at the reduction of viral contamination are needed as it has become clear that current indicators for water and shellfish quality are insufficient as predictors of viral contamination. Raw milk should not be consumed. Harmonising and enhancing epidemiological surveillance should be initiated through studies of molecular detection techniques.

### 2.3.5 Prions

Transmissible spongiform encephalopathies (TSE) are a group of rare, progressively non-inflammatory neurodegenerative diseases affecting humans and animals. They are characterised by long incubation periods with no antibodies detectable in the blood. TSEs can occur as genetic, sporadic, or transmissible forms and are fatal.

The abnormal prion protein PrP$^{sc}$ is the causative agent of several fatal neurogenerative diseases (transmissible spongiform encephalopathies (TSE)) that effect animals and humans. The common characteristic of all diseases caused by PrP$^{sc}$, are deposits of protein aggregates in the brain that can be diagnosed post-mortem as sponge-like brains (Prusiner, 1998a).

TSEs are not new diseases, scrapie (a TSE occurring in sheep and goat) was first identified in the 17$^{th}$ century as a natural and rather common disease affecting about 0,5 - 1,0% of these animals each year. Scrapie can be transmitted horizontally and vertically and although severe measures have been taken to eradicate the disease, further incidences in areas with households having sick sheep have occurred. Despite this, the transmission of scrapie to humans has never been documented. Spongiform encephalopathy has occurred as infrequent outbreaks in commercially bred mink and as sporadic cases in wild or semi-captive ungulates, as well as, in wild and domestic cats that are fed infected meat. In humans, a disease of cannibals in New Guinea (Kuru) was found to be caused by the ability of ingested prion protein to resist degradation by intestinal proteases. The infection may be inherited, or acquired. In this respect the acquired protein was "infectious" agent transffering the diseae from a a person to a person.

Another human spongiform encephalopathy occurring sporadically due to prion protein transformation is Creutzfeld-Jacob disease (CJD). CJD was first described in 1920 and is a rare sporadic progressive, neurogenerative disorder causing death to approximately one person per million worldwide (Prusiner, 1998b).

In the last two decades, scientific and public awareness as well as high anxiety has surfaced after officials in the United Kingdom (UK) published in 1996 a possible association with bovine spongiform encephalopathy (BSE), better known as mad cow disease. Evidence suggested that the epidemic in the UK principally resulted from feeding cattle with scrapie and BSE contaminated recycled sheep carcases and cattle meat in the form of meat and bone meal (MBM) protein nutritional supplements. The epidemic has mostly involved cattle that were born in dairy herds where calves are fed concentrated feeds from a very early age. The age-specific incidences of BSE reflect a minimum incubation period of 2 years and a median of 4 to 6 years. Clinical signs of BSE in cattle involve behavioural disorders (sensitivity to light, noise and touch) and postural abnormalities that result in death after a period of 1 to 6 months.

The appearance of a new variant of the Creutzfeld-Jacob disease (vCJD) began about 10 years after BSE first appeared. The recent cases mostly affect young people, appearing with the same symptoms as sporadic CJD and are fatal. The molecular and histopathological analysis of vCJD and BSE revealed that the infection crosses species lines. Thus, on ingesting prion-containing meat humans can acquire "mad cow disease".

The emergence of vCJD following the BSE epidemic in the UK has united the European Union on issues including food safety risk assessment, regulation and communication, agricultural practices, consumer confidence in beef and beef products, and the food market in general.

**General description -** Prions are defined as proteinaceous infectious particles that lack nucleic acid. The normal form of prion protein ($PrP^c$) exists in every eukaryotic cell and its function is still not known. Under certain conditions that are not yet known, it can be transformed into a pathological isoform ($PrP^{sc}$). The conversion of normal cellular membrane protein into an abnormal disease-causing isoform involves a conformational change from the alpha-helical protein secondary structure to an increased amount of beta-sheet conformation. Conformational changes result in profound changes in the function of a protein such as its ability to resist proteinases.

In addition to proteinase resistance, PrP$^{sc}$'s are also resistant to heat, ultra violet light, ionising radiation, and disinfectants such as formaldehyde and glutaraldehyde (Cohen and Prusiner, 1998).

**Source and associated food -** It has recently been proposed that BSE may have originated via exposure of bovine embryos to various specific high-dose lipophilic formulations of organophosphates. It was suggested that the metabolites of organophosphates could penetrate the foetus and covalently phosphorylate various active sites on foetal prion protein. It was also suggested that the timing, distribution and dynamics of usage of these specific organophosphates correlates with the epidemiology of BSE. The main associated food in prion infections is without doubt meat and meat products from infected cattle. Cattle are consumed not only as beef, but also in a wide range of beef products such as mechanically recovered meat (MRM), gelatine, tallow, and glycerol. During the 1990's the British public was continually reassured that BSE did not constitute a risk to humans and that British beef and beef products were safe. Such reassurances now sound hollow, and indeed they were much criticised by the official BSE Inquiry held in the UK. Most people were simply unaware that TSEs could be transmitted by very small amounts of the most infectious material, and that the infectivity was not destroyed by cooking. Despite recent epidemiological studies concluding that farms with cows giving the highest amount of milk have the greatest probability of BSE developing in their herds. It is important to note that milk from BSE infected animals has not shown any BSE infectivity and there is evidence from other animal and human spongiform encephalopathies to suggest that milk will not transmit these diseases. Milk and milk products are therefore considered safe even in countries with a high incidence of BSE. In fact, it has proven to be very difficult to transmit BSE or other TSEs in experiments involving feeding with milk from an infected animal. Although this is reassuring and the risks are likely to be small, it is impossible to say categorically that milk is safe. Furthermore, milk and dairy products are important sources of calcium and vitamins in many countries and therefore especially important for growing children. A healthy diet free of dairy products is of course possible, but requires a lot of care and attention. To conclude, risks from milk cannot be said to be zero, but the evidence suggests that milk is unlikely to be a problem.

Another product causing some doubt is gelatine. According to published data the food safety risk from all foods made from gelatine is considered low as its preparation involves a chemical extraction process that destroys BSE infectivity (Dormont, 2002).

**Nature of acute disease -** The long incubation period (presumed to be more than 2 years and most commonly 5 years) means that the number of cases has appeared to decrease according to MAFF statistics since 1994. There is an argument about the validity of this data. Clinically, the cow appears alert but agitated, anxious, and apprehensive. However, as the disease progresses the animal starts to take a wide base stance, the abdomen is drawn up, and the gait becomes abnormal and exaggerated giving rise to tumbling and skin wounds. Fine muscle contractions are seen involving small muscle groups over the surface of the neck and body with occasional larger muscle jerks. The animal loses weight and is taken to frenzied movements including aimless head butting.

## Biological hazards

The animal is asymptomatic for a long period before the disease becomes clinically apparent. During this time, many of the tissues of the body are infectious but at a relatively low titre compared to that in the nervous system during the symptomatic period. This titre is however adequately high enough to permit infection of other animals by intracerebral inoculation and possibly by parenteral or oral routes.

TSE infectivity is present in most tissues tested, e.g. liver, kidney, muscle, brain, thymus, and spleen, the distribution varying among species. The finding of infectivity in the buffy coat of the blood has led to fears that CJD may be transferred by blood transfusions, but to date there have been no reports of this actually happening. The discovery that the scrapie agent was present in peripheral and central nervous tissue and in lymphoid tissue has meant little surprise upon finding TSE agent in the muscles of goats, hamsters, mink and possibly humans.

Research has shown that the agent will pass along peripheral nerves and travel in this way from the site of absorption to the brain. Other studies have shown that it is present in the buffy coat (probably the macrophages or lymphocytes) of the blood. The exact mode of transmission of TSE inside the body is unclear.

**Prevention measures** - The transmission of prions to humans is proven. The consumption of infectious meat and bone meal (MBM) is considered one of the major sources of BSE. The EU's Steering Committee decided that it is essential to have a clear understanding of the consumption and source of MBM in a given geographical area in order to assess the probability that BSE might occur in its animal populations. As the infectious agent is very resistant, the recent epidemiological situation does not leave room for the possibility of establishing suitable control measures in the context of HACCP (i.e. in a meat processing company). Therefore, the suppression of specified bovine offal and specified risk material (SRM) like cattle brains, spinal cords, spleens, intestine, eyes, lungs, skull vertebral column testes, and tonsils, from entering the food and feed chain would contribute to a decreased risk of BSE or vCJD infection. Strict regulations to govern animal feeding, testing, slaughter, the age of cattle slaughtered for human consumption, and removal of SRM are in place. Although still of concern, the incidence of BSE in cattle in the UK has fallen significantly over recent years and cases in mainland Europe remain low. The risk of contracting vCJD from food is now believed to be very low. Although muscle tissue is considered to be low in infectious prion protein, scientific evidence clearly shows their presence in sheep and in the urine of infected cattle and humans.

Following the epidemic in the UK, a protocol for a control case study examining risk factors for CJD is being implemented linked to active surveillance in all EU countries. These measurements fall into three groups:
- to avoid BSE-contaminated animals and meat products entering the food chain or to be used as medicinal products;
- to provide monitoring and surveillance of BSE and CJD;
- to identify research needs and support their implementation.

Food safety is only ensured by the shared responsibility of everybody involved with food from the professional to the consumer. All along the food chain, various procedures and control mechanisms are implemented to assure that the food that reaches the consumer's table is fit for consumption and that the risks of contamination are minimised, so that the population as a whole is healthier from the benefits of safe quality food. However, zero-risk food does not exist and we must also be aware that even the best legislation and control systems cannot fully protect us against those with criminal intentions.

The best way to practice food safety remains to be well informed about the basic principles of food production and safe food handling at home (Fishbein, 1998).

**Terminology** - BSE (bovine spongiform encephalopathy) and CJD (Creutzfeldt-Jakob disease) are incurable and fatal diseases marked by the appearance of spongy areas of damage ("spongiform lesions") to the brain. BSE is a disease of cattle, and CJD is a similar disease in humans.

Since diseases of the brain are called "encephalopathies", they are collectively called "spongiform encephalopathies".

Diseases such as BSE may also be passed from one animal to another (cow to cow in the case of BSE) in some generally unexplained manner in the field, or by deliberate transmission in laboratory experiments (feeding infected material or injection of infected material). Because of this characteristic, the word "transmissible" is often added, hence the use of the term "transmissible spongiform encephalopathies" or TSEs.

Many other TSEs are found in the animal kingdom, including scrapie in sheep, and chronic wasting disease (CWD) in deer.

## 2.4 Principal means of food contamination (How microorganisms reach food)

Microorganisms are naturally found in the environment surrounding food, i.e. in the air, water, and soil. Thus, their presence in food is very common. Contaminated food may withstand sensory alterations of a low or high level, and the contamination may be of such a quality and degree that it can cause intoxication or a toxic infection. Today, it is estimated that there are approximately 4 billion cases of diarrhoeal disease annually in the world. Food, including drinking water, is the major vehicle for transmission of these diseases and is associated with up to 70% of the cases. In industrialised countries, up to 10% of the population is estimated to suffer from a foodborne disease annually (Motarjemi and Käferstein, 1999).

Food contamination may be caused by using contaminated raw materials which have been in contact with water, air, or soil. Furthermore, there may also be additional food contamination during the processing and alteration of raw material, as well as during transportation and delivery of food. Besides that, we have to take into account food contamination during its preparation at home.

### 2.4.1 Biological contamination of raw materials

*Contamination by water*
Water is an inevitable element of a food chain. It is needed to grow crops, vegetables, and rear livestock. Water is also used as a raw material in several foods and plays a very important role in food processing (washing, refreshing, cleaning, cooling, etc.), all these being reasons for water

to be considered as a very important agent with regards to food contamination. Due to close contact between water and soil, microorganisms frequently isolated from water are the same as the ones located in soil, and belong to the following bacterial genuses: *Micrococcus, Alcaligenes, Corynebacterium, Pseudomonas, Acinetobacter, Aeromonas, Chrombacterium, Moraxella,* etc. (Bourgeois *et al.*, 1988). Water may also be contaminated by moulds such as *Aspergillus, Penicillium, Rhizopus,* and *Botrytis*. The presence of yeast in water is less common.

Water can also be contaminated by protozoa, including *Giardia* and *Cryptosporidium* (two of the most well-known trouble-causing microorganisms), and by viruses such as Hepatitis A and Polio. Bivalves for example, act as a transport host by concentrating viable *Cryptosporidium* oocysts and *Giardia* cysts from the environment (faecally contaminated fresh, estuarine, and marine waters) and have been suggested as reservoirs for zoonotic transmission (Slifko *et al.*, 2000b). Hepatitis A is a virus that can occur in undercooked shellfish, oysters and water polluted with human waste. Bacteria from the genus *Vibrio* are usually associated with fresh water, estuarine and marine environments and related outbreaks have been linked to the consumption of raw or undercooked shellfish and fish (Labbé and García, 2001).

Microorganisms isolated from faecal matter, such as enterobacteria and enterococci, are commonly found in water supplies. The faecal material can enter the water from:
1. Sewage discharged into the water via cross-contamination of sewage and water lines.
2. Sewage directly discharged from small sewage plants into lakes or streams.
3. Animals depositing their faecal material directly into the water.
4. Rainfall moving raw or improperly treated animal manure from the soil to a body of water.

Included in this group are some pathogens responsible for foodborne illnesses like *E. coli, Salmonella, Shigella,* and *Campylobacter*. The contamination of foods by these microorganisms constitutes a sanitary risk, so that coliforms (although not all are pathogens) are systematically investigated in foods as a measure for controlling hygiene.

The role played by organic agriculture in the contamination of water needs, in particular, to be addressed. According to the FAO (Food and Agricultural Organisation of the United Nations) organic agriculture is a system that relies on ecosystem management rather than external agricultural inputs. It is a system that takes into consideration potential environmental and social impacts and eliminates the use of synthetic inputs such as synthetic fertiliser, pesticides, veterinary drugs, genetically modified seeds and breeds, preservatives, additives, and irradiation. Avoiding the use of chemical pesticides and composted manure as fertiliser contributes to the elimination or reduction of water pollution.

However, there are claims that although the use of manure as fertiliser in organic crops reduces contamination, it increases the exposure of organic foods to microbial foodborne pathogens such as *Salmonella, Campylobacter and Listeria* originating from animal manure. When raw or improperly composted manure is deposited on the soil, irrigating water or rain water spreads the microorganisms into the soil, increasing the contamination of the crops and pasture for livestock. In order to avoid microbial contamination the farmer has to be aware of the critical times and temperatures needed to compost and make manure microbiologically safe.

In general, water supplies in developed countries are tested and treated regularly, so no microbial risk is expected when this water is used in the food chain as raw material or in different operations. However, waterborne diseases are common in many parts of the world where water is not tested and treated.

*Contamination by air*
Air contains microorganisms, especially bacteria and moulds. Contamination by these kinds of microorganisms is more problematic in highly-exposed products like vegetables, dairy, and meat products (Bourgeois *et al.*, 1988). In order to avoid contamination from the air, filters are commonly installed in the food industry to purify the air by retaining atmospheric dust and microorganisms. However, some studies indicate that atmospheric dust deposited on air filters may serve as a nutrient for moulds when there is sufficient humidity and no air flowing through the filters, leading to the filters actually becoming a source of microorganisms (Maus *et al.*, 2001).

*Contamination by soil microorganisms*
Soil is considered a favourable habitat for a wide range of microorganisms, including bacteria, fungi, algae, viruses and protozoa. Microorganisms are found in large numbers in soil, bacteria and fungi being the most prevalent. These microorganisms are to a large extent responsible for the fertility of soils as a result of their ability to decompose organic matter, which releases nutrients for use by plants for growth. Also, different microorganisms can be found in the soil which are antagonistic to some pathogens and can prevent the infection of crop plants. These microorganisms are generally called biopesticides, and they are considered an emerging alternative to the use of chemical pesticides for the protection of crops against certain pathogens and pests (EU IMPACT Project, 1993-1996).

In contrast to these beneficial soil microorganisms, other soil microorganisms are pathogenic to plants and may cause considerable damage to crops. Taking into account the relationship between water and soil, microorganisms from the soil are the same as those from water. Spores of *Clostridium* or *Bacillus* species as well as *Listeria* spp., enterobacteria, and pseudomonads among others, are commonly found in soil. Bacteria from soil can reach vegetables when rain or irrigation water hits the soil, splashing some of the soil onto the plants. Therefore, vegetables may harbour pathogenic microorganisms that remain viable through subsequent handling up to the point of consumption, unless effective sanitising procedures are administered.

*Microorganisms that grow on the surface of raw materials*
According to the above, vegetables may carry microorganisms on their surface which come from the soil, water, and the environment. In general, coliforms, micrococci and pseudomonad's, and moulds are the microorganisms deposited on vegetables. Pathogens can also be detected on the surface if they are present in the irrigation water or soil.

## Biological hazards

The surface of mammary glands, including those of healthy animals, contains microorganisms, mainly bacteria, which are able to invade the milk during milking, even when this is done under the cleanest conditions. Usually micrococcus prevails, *Staphylococcus* or *Micrococcus*, but streptococci are also present. The kind and number of microorganisms, though, may depend on the presence of infections and sickness in the animals, and on the milking conditions (ICMSF, 1980a). There may also be bacteria in the nipples, such as *Bacillus, Clostridium*, and coliforms, coming from the animals' environment (soil, dung, etc.), that are able to enter the milk during extraction.

Eggshells can be contaminated by microorganisms coming from the bowels of birds, the nest, dust, food products, packing and storage, people, and other animals. Generally, the flora located in the eggshell is composed of gram-positive cocci (mainly *Micrococcus* spp.), gram-positive bacilli (*Bacillus and Arthrobacter*), and gram-negative bacilli, including strains from the genus's *Pseudomonas, Escherichia,* and *Enterobacter,* etc. Poultry is recognised as a major reservoir of various *Salmonella* serotypes, therefore eggs may also be contaminated by *Salmonella* through the intestinal tract (ICMSF, 1980b).

Healthy cattle are reservoirs for the major foodborne pathogens like *E. coli* O157, *Salmonella* spp., *Listeria* spp. and *Campylobacter* spp.. These microorganisms may be transferred onto meat during slaughter and dressing of carcasses (Vanderlinde *et al.*, 1998). Apart from pathogenic bacteria, isolated spoilage microorganisms include lactic acid bacteria, *Pseudomonas* spp., *Acinetobacter* spp. and *Moraxella* spp. (Kraft, 1992). Microorganisms are in the intestinal tract of the animals and on their hides, hair and hooves (Gill and Newton, 1978). Generally, the internal surfaces of carcasses are sterile, but during slaughter errors in dressing and skinning can lead to contact of the carcasses with dust, dirt, and faeces facilitating surface contamination.

Equipment and utensils used in slaughtering plants play an important role in the cross-contamination of carcasses. An example in the case of swine and cattle being that, although bacteria on the skin are largely destroyed by scalding, the skin is recontaminated by spoilage and pathogenic bacteria during passage of the carcass through the dehairing equipment (Gill and Bryant, 1993). Workers could also be another microbial contamination source.

Different sanitising methods or strategies have been developed in order to reduce the microbial load in carcasses, especially pathogens. Some of these methods consist of steam pasteurisation in specially designed cabinets, the use of trisodium phosphate (TSP), and spraying with one or more several organic acids (Jay, 1996). For example, the combination of 1% acetic acid and 3% hydrogen peroxide administered by spray washing provides reductions higher than 3 Log cfu $cm^2$ for *E. coli* on beef carcass tissue (Bell *et al.*, 1997).

Also, safety programs have been adopted in slaughtering plants in order to improve meat safety. Good manufacturing practices (GMP) emphasise sanitary effectiveness and hygienic practices during the processing of foods. When effective GMP programs are implemented in the plants, the level of pathogenic and spoilage microorganisms is reduced (Silliker, 1980). Hazard Analysis Critical Control Point (HACCP) is another food safety program mainly directed at the prevention of occurrence of foodborne pathogens in the final food product. In the USA every slaughter plant that operates under federal inspection is committed to establish and carry out an HACCP program as well as to apply sanitation standard operating procedures (SSOP) (USDA, 1996). In the European Union the EU Commission Decision (2001/471/EC) requires validated HACCP systems in slaughter plants and regular checks on general hygiene.

### 2.4.2 Biological contamination during food processing

It is known then that the flora found in food is due to the environment. In the case of fruit and vegetables the flora comes from soil and water, and in the case of meat the flora comes from the skin of animals and their enteric contents in addition to contamination during slaughter. The particular flora of milk is due to the nature of the microorganisms found on the nipple's surface and also depends on the levels of hygiene during extraction.

In addition to the amount of microorganisms living in raw materials, food contamination may increase during processing due to cross-contamination, contact with contaminated surfaces, equipment, tools, and workers. Furthermore, there will be physical and chemical changes during the process which will select for specific microbial genuses and species (Bourgeois *et al.*, 1988).

There are technological operations during the process which reduce the number of microbes. For instance, operations based on high temperatures such as boiling, sterilisation, dehydration, etc, may at least modify the flora suppressing pathogenic microorganisms. When pasteurisation occurs in the production process, all vegetative bacteria that contaminate the product are eliminated, and when sterilisation is applied all microorganisms are eliminated from the product. The drying process involves the suppression of *Campylobacter* spp. and *Vibrio* spp. from the contaminated product (van Gerwen *et al.*, 1997). Application of curing salts favours the growth of lactic acid bacteria, inhibiting pathogens like *Clostridium botulinum*. Modified atmosphere packaging favours the growth of pseudomonads when oxygen is available or lactic acid bacteria and *Brochothrix thermosphacta* when oxygen is not present (Brody, 1989).

Other operations in the food industry may increase pathogenic flora and affect raw materials if the correct hygienic procedures are not respected. This frequently happens during chopping, especially with meat. When meat is chopped there is an homogenisation of the mass and an increase in the contact surface for microbial contamination, which increases the microbial risk to these products. Due to several foodborne outbreaks reported concerning the consumption of hamburgers and related to the presence of *E. coli* O157:H7, the USA has a program for *E. coli* O157:H7 in hamburgers where this pathogen has been identified as an adulterant by regulation, and no product that is contaminated can be offered for sale (USDA, 1999).

Surfaces and utensils play an important role in the contamination of foods. It is really important that the equipment which comes in contact with food has even surfaces, and does not have any angles, cracks, or ruptures. They must have good curves to avoid the accumulation of food and to facilitate easy cleaning. Equipment, including utensils and surfaces, can be insufficiently cleaned between uses and be responsible for indirect cross-contamination via "equipment to food".

With this in mind, special attention has been focussed over the last few years on the role of biofilms as a source of microbial contamination, a frequent occurrence in heat exchangers. Biofilm is the term used to describe microbial communities attached to surfaces (Forsythe, 2000). Microorganisms can be embedded in a polymeric matrix of mainly exopolysaccharides (they produce) and food residues, which increases resistance of the build-up to anti-microbial agents (Druggan et al., 1993). Therefore, food in contact with biofilms during processing can be contaminated giving rise to serious hygiene problems and economical losses. Generally, an effective cleaning and sanitation program is enough to avoid biofilm formation. However, when biofilms are present mechanical treatment as well as chemical breakage of the polysaccharide matrix are necessary because of the resistance of the matrix to the penetration by disinfectants (Forsythe, 2000).

Food contamination may also result from contact of the final product with raw materials. This leads to direct 'food-to-food' cross-contamination (Panisello et al., 2000).

Water can also be considered as a source of contamination during food processing as it is used in washing operations, as an additional ingredient, to elaborate pickles, etc. As a general principle, only potable water free from harmful microorganisms should be used in food handling. Finally, considering in particular the significance of the faecal-oral transmission route, good personnel hygiene constitutes a significant preventive measure against contamination during the food processing.

As mentioned in the previous section, different safety programs such as Good Manufacturing Practices (GMP) and Hazard Analysis Critical Control Point (HACCP) have been adopted in the food industry.

### 2.4.3 Biological contamination during distribution of foods

When storage and transportation of foods take place under improper conditions, the growth of pathogenic and spoilage microorganisms can occur if those organisms are present in the product or have access through the environment or as a result of inadequate manipulation. Although packaging of foods can prevent the entry of microorganisms and prevent changes in the relative humidity, maintenance of the cold chain during the distribution of foods is one of the most important factors to control in chilled and frozen foods, like meat and meat products, dairy products, and ready-to-eat foods. Time-temperature integrators, mainly colour dispositives, have recently been included in some chilled packed food to inform consumers when a product has been subject to temperature abuse.

Possible contamination during distribution is a significant risk in the case of catering where food is prepared well in advance and is frequently distributed at warm temperatures. When refrigeration or reheating temperatures are not respected and the service personnel are not properly educated about the risks involved, microorganisms like *Staphylococcus aureus*, *Clostridium perfringens* or *Salmonella* can develop resulting in a food-borne outbreak.

## 2.4.4 Biological contamination at home

Worldwide, analyses of foodborne disease outbreaks has shown that the great majority of foodborne disease outbreaks result from malpractice during food preparation in small food businesses, canteens, homes, and other places where food is prepared for consumption. Considering that most outbreaks resulting from the preparation of food at homes or in small food service establishments are not investigated and reported, it can be concluded that the number of outbreaks related to industrially produced foods is below 5% (Motarjemi and Käferstein, 1999).

Modern lifestyles have imposed new food habits. Consumer demand for convenience has resulted in the increased consumption of ready-to-eat or minimally processed products. When this handling is followed by time/temperature combinations that permit pathogens to survive and grow, the risk of foodborne diseases increases (de Roever, 1999). Other sources of microbial contamination in the domestic environment are people, pets, and pests. The problem of food contamination at home is very important for vulnerable groups such as infants, children, the elderly, pregnant women, and immunocompromised persons where the consequences of foodborne diseases may be more serious. These observations highlight the importance of consumer education and information on hazardous food.

The following basic hygiene measures are recommended for preparation of food at home:
- Clean the surfaces where food is prepared (cutting boards, dishes, utensils) after preparing each food item and before going on to the next food item, preferably with hot soapy water. Wash hands with hot soapy water before and after preparing food, and especially after using the toilet and playing with pets. Also, wash kitchen cloths and sponges with hot soapy water. According to Beumer and Kusumaningrum (2003) faecal coliforms, *Staphylococcus aureus*, *Bacillus cereus* and *Listeria monocytogenes* were some of the pathogens found in a previous study carried out on 15 sites of 250 domestic kitchens (including dishcloths, cutting boards and utensils), which provided evidence that places in the kitchen or kitchen objects are reservoirs and disseminators of microorganisms.
- Separate foods in order to avoid cross-contamination. Keep raw materials away from ready-to-eat foods. Never place cooked food on a plate that previously held raw meat, poultry, eggs or seafood unless the plate has been thoroughly cleaned between uses.
- Cook or reheat foods until they reach an internal temperature of 160°F. In the case of leftovers reheat to at least 165°F.
- Refrigerate or freeze perishables, ready-to-eat foods and leftovers within two hours of purchasing or preparation. Make sure the refrigerator is set no higher than 4°C and the freezer is set at -20°C.

## 2.5 Microbial growth in food (how microorganisms survive and grow in food stuffs)

Food is considered to be a chemically complex matrix, and predicting whether or how fast microorganisms will grow in any given food is difficult. Most foods contain sufficient nutrients to support microbial growth. Several factors encourage, prevent, or limit the growth of microorganisms in food and some of the most important are $a_w$, pH, and temperature. Some of the parameters are intrinsic to the product like pH, $a_w$, redox potential, presence of natural antimicrobials, food composition or structure, while others are related to the environment around the food, i.e. temperature, relative humidity and gas composition, and are called extrinsic parameters.

### 2.5.1 Food structure and composition

*Physical factors that prevent or facilitate the microbial contamination of foods*
Food can often be protected by a structure that constitutes an excellent physical barrier to entrance or penetration of microorganisms. Examples of this structure are the teguments in fruits, husk in grains, skin in animals, and shells in molluscs or eggs.

However, after the harvesting of vegetables or the slaughtering of animals this protective structure can be damaged by decomposition, cuts or lesions, or removed as a step in the elaboration process (peeling, grinding, crushing, or pressing). When this happens the food is exposed to the action of the microorganisms. In the case of eggs, the shell and the membrane prevent the entrance of the majority of microorganisms when they are in optimal condition.

*Food composition*
Microorganisms as well as living organisms need water, an energy and carbon source, nitrogen, and minerals and vitamins to survive and develop, and foods are an excellent source of the majority of those cited compounds. However, the proportion of these nutrients will determine the kind of microbial flora present in the food.

With regards to their energy source, microorganisms can be classified as phototrophs when they use radiant energy (light), chemotrophs when they use the oxidation of an organic form of carbon as their primary energy source, or lithotrophs when they oxidise inorganic compounds. Carbon is the structural backbone of the organic compounds that make up a living cell. According to the source of carbon, bacteria can be divided into autotrophs or heteretrophs. Autotrophs are organisms that use $CO_2$ as the sole source of carbon, while microorganisms that use organic carbon are called heterotrophs (website 1).

According to the source of energy and carbon, all organisms in nature can be placed in one of four groups described below and in Table 2.12.

Photoautotrophs transform carbon dioxide and water into carbohydrates and oxygen through photosynthesis. Cyanobacteria, algae, and green plants use hydrogen atoms from water to

*Table 2.12. Classification of organisms according to energy and carbon sources.*

| Nutritional type | Energy source | Carbon source | Examples |
|---|---|---|---|
| Photoautotrophs | Light | $CO_2$ | Photosynthetic bacteria (green and purple sulphur bacteria, cyanobacteria), algae and green plants |
| Photoheterotrophs | Light | Organic compounds | Green and purple nonsulphur bacteria. |
| Lithoautotrophs (chemolithoautotrophs) | Inorganic compounds: $H_2S$, $S_2$, $NH_3$, $H_2$, Fe, $NO_2$ | $CO_2$ | A few bacteria and many *Archaea* |
| Chemoheterotrophs (chemoorganoheterotrophs) | Organic compounds | Organic compounds | Most bacteria, protozoans, fungi and animals |

reduce carbon dioxide to form carbohydrates, and during this process oxygen is given off (an oxygenic process). Other photosynthetic bacteria (the green sulphur bacteria and purple sulphur bacteria) carry out an anoxygenic process, using sulphur, sulphur compounds or hydrogen gas to reduce carbon dioxide and form organic compounds (website 2).

According to Table 2.12 most bacteria, all protozoans, fungi, and animals are chemoheterotrophs. Generally, microorganisms use carbohydrates as their primary energy and carbon sources before they utilise fatty acids and nitrogenous compounds. Simple sugars are preferred although some microorganisms are capable of degrading complex carbohydrates like starch or cellulose.

Apart from a carbon source, both autotrophic and heterotrophic microorganisms may require small amounts of certain essential organic compounds for growth that they are unable to synthesise from the available nutrients. Such compounds are called growth factors (website 1). Growth factors are required in small amounts by cells because they fulfil specific roles in biosynthesis. The need for a growth factor results from either a blocked or missing metabolic pathway in the cells. Growth factors are organised into three categories:
- purines and pyrimidines which are required for synthesis of nucleic acids (DNA and RNA);
- amino acids which are required for the synthesis of proteins;
- vitamins which are needed as co-enzymes and functional groups of certain enzymes.

Although amino acids are the main source of nitrogen used by microorganisms, some microbial groups degrade complex proteins due to the presence of specific enzymes, thereby producing simple substrates like peptides and amino acids for the growth of other microbial species.

Although vitamins are essential for microbial growth, they are required in very small quantities and foods possess generous amounts of them. Some of the vitamins required by microorganisms include vitamins from group B like pyridoxine (B6), riboflavin (B2), thiamine (B1),

## Biological hazards

cyanocobalamin (B12), folic acid (B9), biotin, nicotinic acid (B3), pantothenic acid (B5) and those from other groups, i.e. lipoic acid and vitamin K.

Some bacteria, e.g. *E. coli,* do not require any growth factors, as they are capable of synthesising all essential purines, pyrimidines, amino acids, and vitamins as part of their own intermediary metabolism. Certain other bacteria, e.g. *Lactobacillus,* require purines, pyrimidines, vitamins and several amino acids in order to grow. These compounds must be added in advance to culture media that are used to grow these bacteria. The growth factors are not metabolised directly as sources of carbon or energy; they are instead assimilated by cells to fulfil their specific role in metabolism (website 1).

With regard to minerals, sulphur is needed to synthesise sulphur containing amino acids and certain vitamins. Depending on the microorganism sulphates, hydrogen sulphide, or sulphur-containing amino acids may be used as a source of sulphur. Phosphorous is also needed to synthesise phospholipids, DNA, RNA and ATP. Potassium, magnesium, iron, and calcium are required for certain enzymes to function. Other minerals which function as co-factors in enzyme reactions, like sodium, zinc, copper, molybdenum, manganese, and cobalt, are required in very small amounts and are present as "contaminants" of the water or food components (website 2).

### 2.5.2 Factors affecting microbial survival or growth rate in foods

*pH*
pH is defined as the negative logarithm of the hydrogen ion concentration measured on a scale of 1 to 14 (7 is neutral, < 7 is acid, and > 7 is basic). The formula is:

$$pH = -\text{Log}\,[H^+]$$

pH has an influence on chemical and biochemical reactions and therefore on the growth of microorganisms. The pH range of a microorganism is defined by a minimum value (at the acidic end of the scale) and a maximum value (at the basic end of the scale). There is a pH optimum for each microorganism where its growth is maximal. Moving away from the pH optimum in either direction slows down the microbial growth.

Most microorganisms grow better at pH's near neutrality (5.5-8.0); they are called neutrophiles and most human pathogens are included in this category. Acidophiles are capable of growing at pH's between 0 and 5.5, e.g. the *Thiobacillus* species. Alkalophiles proliferate at pH levels of 8.0 to 11.5 while extreme alkalophiles can grow in environments with a pH of 10 or higher. *Vibrio cholerae* is considered an alkalophile.

Both moulds and yeasts can grow over a wide pH range, growth being possible at extreme values of pH 1-2 and pH 9-11. Limits of growth are described for some microorganisms in Table 2.13.

Table 2.13. pH limits of growth for some microorganisms (Jay, 1992, Bourgeois et al., 1988).

|  | Minimum | Optimal | Maximum |
|---|---|---|---|
| Moulds | 1.5-3.5 | 4.5-6.8 | 8-11 |
| Yeasts | 1.5-3.5 | 5-6.5 | 8-8.5 |
| Bacillus | 4.5-6 | 6.8-7.5 | 9.4-10 |
| Lactic acid bacteria | 3.2 | 5.5-6.5 | 10.5 |
| Lactobacillus acidophilus | 4.0-4.6 | 5.8-6.6 | 7.0 |
| Thiobacillus thiooxidans | 0.5 | 2.0-2.8 | 4.0-6.0 |
| Enterobacteria | 5.6 | 6.5-7.5 | 9.0 |
| Salmonella | 3.6 | 6.5-7.5 | 9.5 |
| E. coli | 4.4 | 6-7 | 9.0 |
| Clostridium botulinum | 4.8 | - | 8.2 |
| Clostridium perfringens | 5.5 | 6.0-7.6 | 8.5 |
| Clostridium sporogenes | 5.0-5.8 | 6.0-7.6 | 8.5-9.0 |
| Staphylococcus aureus | 4.2 | 7.0-7.5 | 9.3 |
| Pseudomonas aeruginosa | 4.4-4.5 | 6.6-7.0 | 8.0-9.0 |
| Streptococcus pneumoniae | 6.5 | 7.8 | 8.3 |

However, these values cannot be considered in isolation, as they also depend on other parameters which affect the growth of organisms. For example, a food with a pH of 5.0 (within the range for growth of *C. botulinum*) and $a_w$ of 0.935 (above the minimum for growth of *C. botulinum*) may not support the growth of this bacterium due to the interplay of other factors (website 3). *E. coli*, *Salmonella* and *Staphylococcus* are more resistant to pH changes than other foodborne pathogens (Bourgeois *et al.*, 1988). The type of acid used to decrease the pH must also be considered as it has been observed that acetic acid seems to be more inhibitory to microbial growth than lactic or citric acid (ICMSF, 1980a).

The pH found in foods can be a result of the intrinsic characteristics of the food or can be due to the activity of microorganisms, as in fermented sausages and many dairy products. In these products the food may start with a pH which precludes bacterial growth, but as a result of the metabolism of certain microorganisms, such as the lactic acid bacteria, the pH can change and allow or inhibit bacterial growth (biological acidity). Some foods show resistance to pH changes and are said to have a buffering capacity.

The pH of a food product will determine the kind of microorganisms that can grow in it. An example being the low pH of fruits (compared to other foods) that inhibits the growth of bacteria leaving moulds and fungi, which are able to grow at pH's lower than 3.5, as the organisms responsible for their spoilage. In the case of meat, the pH of muscle is close to neutrality when the animal is alive, but after slaughter and development of rigor mortis the pH decreases to values around 5.4. The exact extent of decrease in pH will depend on the species, type of muscle, rest or stress status of the carcass, etc. The pH of muscle from a rested animal may differ from that of a fatigued animal resulting in meat with a pH near neutrality which facilitates the growth of microorganisms. Pseudomonads and a few other Gram-negative psychrotrophic organisms dominate proteinaceous foods such as meat stored aerobically at chill temperatures, while LAB, *Enterobacteriaceae* and *Brochothrix thermosphacta* are more abundant when meat is stored under vacuum (Dainty and Mackey, 1992). Meat that is cured by salt and low in pH is usually shelf-stable and is not spoiled by microorganisms (ICMSF, 1998).

Fish usually present a pH higher than that found in meat, which rises during storage as a result of the release of $NH_3$ and amines during protein degradation and by the action of microorganisms that grow better at a higher pH (Bourgeois *et al.*, 1988). Increasing the shelf life by decreasing the pH to below 5, or by increasing the NaCl concentration to greater than 6%, or by adding sorbate and/or benzoate, eliminates the Gram-negative microflora and favours the LAB and yeasts in semi-preserved fish products (Gram *et al.*, 2002).

Despite the habitat external to the cell, the internal pH is usually close to neutral. The main effects of acids on microorganisms are an increase in the energy required to maintain a cell's internal pH. When changes in pH of the environment are introduced adjustments are often made by the organism to maintain the cell's internal pH in order to survive in the inhospitable medium. This can be done through the production of metabolic waste products, but this process cannot be maintained for a long time. Finally, membrane permeability and enzyme activity are affected as a result of the denaturation of proteins, DNA, and other compounds.

*Water activity*
Water is the solvent in which microbial reactions take place, and the availability of water is therefore a critical factor that affects the growth of all organisms. The availability of water for a cell depends upon its presence in the atmosphere (relative humidity) or its presence in foods in two different forms: bound water and available water. Bound water has a physical bond to other compounds in the growth environment and is not available for use in cellular functions. Available water is free and available for microorganisms to use. The degree of water availability for chemical activity and growth is called the water activity ($a_w$). Water activity is defined as the ratio of the vapour pressure of the growth medium to the vapour pressure of pure water. The values range from 0 to 1. The activity of pure water is equal to one. The addition of solute decreases the $a_w$ to values less than 1.00 as is shown in Table 2.14.

Osmotic pressure and relative humidity are concepts closely related to water activity in the sense that water activity is inversely proportional to osmotic pressure and directly proportional to relative humidity. Increasing the solute concentration, lowering the relative humidity, or removing water results in a reduction in water activity. In general, $a_w$ varies very little with temperature over the temperature range that supports microbial growth.

An $a_w$ value generally stated for a microorganism is the minimum $a_w$ which allows its growth. At the minimum $a_w$, growth is usually minimal, and increases with $a_w$. A decrease of $a_w$ is associated with an increase in the lag phase and a diminution of the growth rate. Although a proportion of the microorganism population dies at $a_w$ values below that minimal for growth, the rest remain dormant.

Table 2.14. Relationship between $a_w$ and NaCl concentration (website 3).

| % NaCl (w/v) | Moles | Water activity ($a_w$) |
|---|---|---|
| 0.9 | 0.15 | 0.995 |
| 1.7 | 0.30 | 0.99 |
| 3.5 | 0.61 | 0.98 |
| 7.0 | 1.20 | 0.96 |
| 10.0 | 1.77 | 0.94 |
| 13.0 | 2.31 | 0.92 |
| 16.0 | 2.83 | 0.90 |
| 22.0 | 3.81 | 0.86 |

Most bacteria cannot grow in environments with water activities lower than 0.9 and pathogens usually require $a_w$ values greater than 0.95. Fungi are generally more tolerant of low $a_w$ values than bacteria. Microorganisms that can live in dry environments are called xerophiles whereas those that can withstand low $a_w$ are called xerotolerant.

With respect to NaCl concentration microorganisms can be classified according to their growth response to this salt, which is commonly used to decrease $a_w$. Microorganisms that require some NaCl for growth are called halophiles. Mild halophiles require 1-6% salt, moderate halophiles require 6-15% salt, and extreme halophiles require 15-30% NaCl for growth. Bacteria that are able to grow at moderate salt concentrations, even though they grow best in the absence of NaCl, are called halotolerant (website 1). The terms osmophilic and osmotolerant are reserved for microorganisms that are able to live in environments high in sugar. Therefore, the term halophile is usually applied to bacteria, osmophile to yeasts, and xerophile to moulds.

Microorganisms differ in their ability to grow in various water activities as is shown in Table 2.15. In general, microorganisms protect themselves against low $a_w$ and osmotic pressure through intracellular accumulation of compatible solutes, usually $K^+$ ions, glutamate, glutamine, proline, γ-aminobutirate, saccharose, trealose and glycosyl-glycerol. Xerotolerant moulds tend to produce polyhydric alcohols like glycerol, eritriol and arabitol (Jay, 1992).

Table 2.15. Minimum Water Activity required for the growth of different Microbial Groups (Jay, 1992).

| Group | Minimum $a_w$ |
|---|---|
| Bacteria | 0.90 |
| Yeast | 0.87-0.88 |
| Moulds | 0.80 |
| Halophilic bacteria | 0.75 |
| Xerophilic moulds | 0.65 |
| Osmophilic yeasts | 0.62 |

## Biological hazards

The minimal $a_w$ for growth will depend on the interplay with other factors like pH, growth temperature, redox potential, and nature of the solute used to reduce the $a_w$. Most fresh foods, vegetables, fruits, meat, fish, milk and dairy products usually have $a_w$ values higher than 0.96 which makes these products vulnerable to microbial spoilage, especially by bacteria. The exception being for fruits where the low pH favours the growth of moulds and yeasts as discussed above.

The concept of lowering water activity in order to prevent bacterial growth is the basis for preservation of foods by drying (under the sun or by evaporation) or by the addition of high concentrations of salt or sugar. As an example, decreasing $a_w$ eliminates bacterial growth, and only estremophiles (such as halophiles, osmophiles or xerophiles) are capable of developing on dried, salted fish, meat and fruit products (Pitt and Hocking, 1997).

*Redox potential*
Oxidation is defined as the loss of electrons whereas reduction is the gain of electrons. When a substance is oxidised it releases electrons and when a compound is reduced it gains electrons. The redox potential (Eh) measures the ease with which electrons are lost or gained and reflects the availability of free oxygen. A medium that loses electrons is said to be reducing and has a negative Eh, whereas one which gains electrons is an oxidising medium and has a positive Eh. The presence of reducing agents in foods like SH groups, ascorbic acid, and simple sugars, contributes to reduction of the Eh, while the presence of oxygen makes the product have a positive Eh.

According to their relationship with oxygen, microorganisms can be classified as follows:
- Obligate aerobes require $O_2$ for growth; they use $O_2$ as a final electron acceptor in aerobic respiration. These microorganisms cannot ferment and they can only live in the presence of oxygen (they require positive Eh in the medium). Examples of this kind of microorganism are *Pseudomonas* spp., *Bacillus subtilis*, *Legionella* spp. and moulds.
- Obligate anaerobes or strict anaerobes do not need or use $O_2$ as a final electron acceptor (in order to produce ATP). They require environments with a negative Eh to survive. $O_2$ is toxic to them and either kills them or inhibits their growth. Obligate anaerobic prokaryotes may live by fermentation, anaerobic respiration, or bacterial photosynthesis, etc. *Clostridium botulinum* is a good example.
- Facultative anaerobes (or facultative aerobes) are organisms that can switch between aerobic and anaerobic types of metabolism. Then the microorganism can utilise fermentation when molecular oxygen is absent, and can utilise aerobic cellular respiration when $O_2$ is present. The Enterobacteria are included in this group.
- Aerotolerant *anaerobes* are bacteria with an exclusively anaerobic (fermentative) type of metabolism but are insensitive to the presence of $O_2$. They live by fermentation alone whether or not $O_2$ is present in their environment; oxygen is not required and not utilised. Certain species from the genus *Clostridium* are slightly aerotolerant.
- Microaerophiles are microorganisms which are unable to grow when oxygen concentrations reach those found in air (20%), but nevertheless their growth requires the presence of some oxygen (e.g. 2 to 10%) or else slightly reducing conditions are required. Microaerophilic

conditions may be found in some environments particularly at aerobic/anaerobic interfaces such as those found in soil, water, or part of the bodies of animals and some plants. An example of microaerophilic microorganisms are strains from the genus *Lactobacillus spp.*

The redox potential is affected by microbial growth as well as pH. This occurs especially when aerobic microorganisms grow in a product and consume the oxygen, resulting in its enrichment in reducing substances like $SH_2$ which decrease the Eh (Bourgeois et al., 1988). Also, a food can present different Eh's. For example in the case of fermented cheese, although microorganisms are distributed throughout the curd, substantially higher populations are located on the outer surface because of the availability of oxygen. Thus, oxidative microorganisms (brevibacteria, micrococci, *D. hansenii*, *Y. Lipolytica*, and moulds) are more prevalent on the curd surface, while fermentative (lactic acid bacteria) are more predominant within the curd (Fleet, 1999).

In relation to the pH, the Eh of a substance is generally determined at pH 7.0 and it tends to be more negative when the alkalinity of the environment increases.

*Temperature*
Higher temperatures can speed up biological reactions. Generally, organisms will grow faster with rising temperatures until proteins and nucleic acids become denatured, and the cell dies. Organisms have temperatures at which they grow best (optimum), minimum temperatures below which they will not grow (but are not usually killed), and maximum temperatures above which they will die (forming the basis of many sterilisation methods).

Organisms can be classified according to their temperature optima:
- Psychrophiles: these are microorganisms that can grow at temperatures between 0 and 20°C with an optimum growth temperature of 15°C or lower. Many psychrophiles are found in deep oceans, Arctic and Antarctic habitats. The cellular functions of these organisms are adapted to withstand cold environments. For instance, the amount of unsaturated fatty acids in their cellular membranes is high enabling the cell to remain semi-fluid at low temperatures.
- Psychrotrophs: these are microorganisms than can grow at 0°C, but their optimum growth temperatures range from 20 to 30°C. The maximum temperature they can grow at is 35°C. They are much more common than psychrophiles. These organisms are a factor in the spoilage of refrigerated foods since they are invariably brought into the refrigerator from mesophilic habitats and continue to grow in the refrigerated environment where they spoil the food, though at a slow rate. Psychrotrophic strains and species can be found in the genus *Alcaligenes*, *Shewanella*, *Brochothrix*, *Corynebacterium*, *Flavobacterium*, *Lactobacillus*, *Micrococcus*, *Pseudomonas*, *Psychrobacter*, and *Enterococcus*.
- Mesophiles: these are microorganisms which grow optimally between 20°C and 45°C. Most microorganisms and importantly most pathogens are mesophiles.
- Thermophiles: these are microorganisms which grow optimally at temperatures greater than 50°C. Thermophiles are mostly bacteria but also include some algae and fungi. Typical habitats are composts, soils, and hot springs. The enzymes of these organisms are also heat stable. Like the psychrophiles their membranes are adapted to the environment and contain

## Biological hazards

more saturated fatty acids with higher melting points that enable them to remain more stable at high temperatures. Most thermophilic bacteria of special interest to the food industry belong to genus *Bacillus* and *Clostridium*, these being very important in the canned industry.
- Extreme thermophiles or hyperthermophiles: these are microorganisms that grow poorly at 55°C and have optimum growth temperatures between 80 and 100°C.

As with pH most moulds are able to grow between higher intervals of temperature than bacteria. Prokaryotic cells can grow at much higher temperatures than eukaryotes, and archaebacteria can generally grow at higher temperas than eubacteria.

### Natural antimicrobials and food additives

Certain foods of vegetable or animal origin contain substances which can act as antimicrobial agents. In fact, there is considerable interest in these kinds of compounds for possible use as natural preservatives in deference to the use of chemical preservatives. Their effect could either be by inhibiting the growth of pathogen microorganisms or delaying the spoilage in foods.

In the case of plants many naturally occurring compounds such as phenols (phenolic acid, polyphenols, tannins) and organic acids (acetic, lactic, citric, succinic, malic, tartaric) have been considered in this context (Nychas, 1995). In the case of organic acids: cell walls, cell membranes, metabolic enzymes, protein synthesis system and genetic material are the main targets of their action in a wide range of microorganisms (Eklund, 1989).

Also, many spice and herb extracts possess antimicrobial activity mainly due to the essential oil fraction (Deans and Ritchie, 1987, Rajkovic *et al.*, 2005). Among the compounds having antimicrobial properties are eugenol and cinnamic aldehyde from cloves and cinnamon, carvacrol and thymol from oregano and thyme, allicin (allyl isothiocyanate) from garlic, onions, mustard, and horseradish, etc. In general, these compounds are active against Gram-positive and Gram-negative bacteria, although it seems that Gram-positive bacteria are more sensitive to the antimicrobial compounds in spices (Nychas *et al.*, 1995). There are several studies about the effect of different extracts from spices or plants on foodborne bacterial pathogens like *S. aureus*, *Listeria monocytogenes*, *Aeromonas hydrophila*, *Clostridium botulinum*, and *Salmonella*, *B. cereus*, etc (Shelef *et al.*, 1980; Hall and Maurer, 1986; Deans and Ritchie, 1987; Paster *et al.*, 1990; Ting and Deibel, 1992, Rajkovic *et al.*, 2005).

With regard to antimicrobial compounds from animals, the lactoperoxidase system is one of the most well-known antimicrobial mechanisms present in milk. The lactoperoxidase system causes oxidation of thiol groups on enzymes and affects the cytoplasmic membrane of sensitive microorganisms such that ions, amino acids and even polypeptides are lost from the cell contents. This system is more efficient against gram-negative bacteria than it is against gram-positive bacteria (Björck, 1978). Despite this the lactoperoxidase system inhibits gram-positive and -negative pathogens like *S. aureus*, *L. monocytogenes* and *Campylobacter jejuni* (Beumer *et al.*, 1985; Siragusa and Johnson, 1989; Kamau *et al.*, 1990). Milk also contains lactoferrin and transferrin (iron chelating glycoproteins) which contribute to the antimicrobial defence systems of the animal through chelation of iron; it seems however that this is not the only mechanism of inhibition (Arnold et al., 1982). Lysozyme (1,4-β-N-acetylmuramidase) is another antimicrobial agent present in milk and other secretions. This enzyme cleaves the glycosidic bond between C-1 of N-acetylmuramic acid and the C-4 of N-acetyl glucosamine compounds of the polymer in the peptidoglycan in the cell walls. Lysozyme activity is greatest against gram-positive bacteria where the peptidoglycan of the wall is more exposed (Davidson, 1997).

Lysozyme and ovotransferrin or conalbumin (also an iron chelating glycoprotein) are the antimicrobial compounds found in egg whites. Avidin is a glycoprotein present in the albumen of chicken eggs, which combines with the vitamin biotin and inhibits the growth of some bacteria and yeasts that need biotin.

Finally, bacteria produce many inhibitory compounds which can inhibit the growth of potential spoilage or pathogenic microorganisms. These include fermentation end products such as organic acids, hydrogen peroxide, and diacetyl, in addition to bacteriocins (Hill, 1995). The production of bacteriocins and their use to control spoilage and pathogenic bacteria have been extensively studied in recent years. Bacteriocins are protein compounds which act in a detrimental nature towards closely related species. They act by disrupting cellular processes or by creating holes in the membrane causing cellular leakage. Gram-positive bacteriocins have a greater spectrum of activity than those produced by Gram-negative bacteria. In fact, the food industry is more concerned with those bacteriocins produced by food grade organisms such as members of the lactic acid bacteria (Hill, 1995). Nisin, a bacteriocin produced by a *Streptococcus lactis*, is the only one of the LAB bacteriocins that is being commercially produced and used (De Vuyst and Vandamme, 1994, Delves-Broughton et al., 1996, Rajkovic et al., 2005). Despite the research on natural antimicrobial substances, the use of chemical additives to preserve foods is still common practice in the food industry. In this group, preservatives worth mentioning are organic acids and esters, especially lactic and acetic acids which have important antimicrobial activity. Sorbic acid and sorbate inhibit fungi and certain bacteria including *Salmonella, S. aureus*, and *E. coli* (Sofos and Busta, 1993). The principal use of nitrites as antimicrobial compounds is based on their ability to inhibit the growth and toxin production of *Clostridium botulinum;* their effect on other microorganisms is variable (Davidson, 1997). NaCl, $SO_2$, and their salts, or phosphates are other preservatives commonly used in the food industry. $SO_2$ is mainly used to control the undesirable growth of spoilage microorganisms in fruits, juices, wines, sausages and fresh shrimps (Davidson, 1997). Gram-positive bacteria are more sensitive to phosphates than gram-negatives (Post et al., 1963).

## References

Arnold, R.R., J.E. Russell, W. J. Champion, M. Brewer and J. .J. Gauthier, 1982. Bactericidal activity of human lactoferrin: differentiation from the stasis of iron deprivation. Infect. Immun., 35, 792-797.

Ash, C., J.A.E. Farrow, S. Wallbanks and M.D. Collins, 1991. Comprative analysis of *Bacillus anthracis, Bacillus cereus,* and related species on the basis of reverse transcriptase sequencing of 16S rRNA. Int J Systematic Bacteriol 41: 343-346.

Aslam, M., F. Nattress, G. Greer, Ch. Yost, C. Gill and L. McMullen, 2003. Origin of contamination and genetic diversity of *Escherichia coli* in beef cattle. Appl Environ Microbiol 69: 2794-2799.

Baloda, S.B., L. Christensen and S. Trajcevska, 2001. Persistence of a *Salmonella enterica* serovar Typhimurium DT12 clone in a piggery and in agricultural soil amended with *Salmonella*-contaminated slurry. Appl. Environ. Microbiol. 68: 4758-4763.

Barker, J. and S. F. Bloomfield, 2000. Survival of *Salmonella* in bathrooms and toilets in domestic homes following salmonellosis. J.Appl. Microbiol. 89: 137-144.

Bell, K.Y., Cutter, C. N. and Summer, S. S. 1997. Reduction of foodborne micro-organisms on beef carcass tissue using acetic acid, sodium bicarbonate, and hydrogen peroxide spray washes, Food Microbiol., 14, 439-448.

Benito, A. G. Ventoura, M. Casadei, T. Robinson and B. MacKey, 1999. Variation in resistance of natural isolates of Escherichia coli O157:H7 to high hydrostatic pressure, mild heat and other stresses. Appl Environ Microbiol 65: 1564-1569.

Beumer R.R. and H. Kusumaningrum, 2003. Kitchen hygiene in daily life. Int. Biodeterioration & Biodegradation, 51, 299-302.

Beumer, R.R., A. Noomen, J.A. Marijs and E.H. Kampelmacher, 1985. Antibacterial action of the lactoperoxidase system on Campylobacter jejuni in cow's milk. Neth. Milk Dairy J., 39, 107-114.

Bisharat, N., V. Agmon, R. Finkelstein, R. Raz, G. Ben-Dror, L. Lerner, S. Soboh, R. Colodner, D.N. Cameron, D.L. Wykstra, D.L. Swerdlow and J.J. Farmer 3rd, 1999. Clinical, epidemiological, and microbiological features of *Vibrio vulnificus* biogroup 3 causing outbreaks of wound infection and bacteraemia in Israel. Lancet 354: 1421-1424.

Björk, L, 1978. Antibacterial effect of the lactoperoxidase system on psychrotrophic bacteria in milk. J. Dairy Res., 45, 109-118.

Bourgeois, C.M., J.F. Mescle and J. Zucca, 1988. In: Microbiologie Alimentaire 1. Aspect microbiologique de la sécurité et de la qualité alimentaires, Technique et Documentation, Lavoisier.

Brody, A.L. (ed), 1989. In: Controlled-Modified Atmosphere/Vacuum Packaging of Foods, Food and Nutrition Press, Trumbull.

Bullerman, LL.B., 1997. *Fusarium* and other toxigenic moulds different from *Aspergillus* and *Penicillium*. In: Doyle, M.P., L. Beuchat and T.J. Montville (Eds.), Food microbiology. Fundamentals and frontiers. ASM Press, Washington, D.C., pp. 441-456.

Buyser, M.L., 1988. *Staphylococcus*. In: Aspect microbiologique de la securité et de la qualité alimentaires. Technique et Documentation (Lavoisier), Paris. France, pp. 67-79.

CDC (Center for Disease Control and Prevention), 2000. Disease information of Division of Bacterial and Mycotic Disease. www.cdc.gov.

CDC (Center for Disease Control and Prevention), 2002. National Antimicrobial Resistance Monitoring System for Enteric Bacteria. Annual Report for 2000, CDC. 2002. www.cdc.gov.

Ceylan, A, M. Ertem, E. Ilcin and T. Ozekinci, 2003. A special risk group for hepatitis E infection: Turkish agricultural workers who use untreated waste water fro irrigation. Epidemiol Infect. 131:753-756.

CFSAN (U.S Food & Drug Administration. Center for Food Safety & Applied Nutrition), 2003. Foodborne pathogenic microorganisms and natural toxins handbook. www.cfsan.fda.org.

Chibber, R.M., M.A. Usmani, M.H. Al-Sibai and H.E.V. Should, 2003. infected mothers breast feed? Arch Gynecol Obstet.

Clark, D.P., 1999. New insights into human Cryptosporidiosis. Clinical Microbiol Reviews 12: 554-563.

Clemente-Casares, P., S. Pina, M. Buti, R. Jardi, M. Martin and R. Bofill-MasGirones, 2003. Hepatitis E virus epidemiology in industrialized countries. Emerg Infect Dis. (4):448-54.

Cohen, F.E., Prusiner, S.B. 1998. Pathologic conformations of prion proteins. Ann Rev Biochem 67: 793-819.

Cole, R.J. and J.W. Dorner, 1993. Mycotoxins. In: Macrae, R., Robinson, R. K. and Sadler, M. J. (Eds.), Encyclopaedia of Food Science, Food Technology and Nutrition. Academic Press, London, pp. 3196-3214.

Dainty, R.H. and B.M. Mackey, 1992. The relationship between the phenotypic properties of bacteria from chill-stored meat and spoilage processes, J. Appl. Bacteriol. Symp. Suppl., 73, 103S-144S.

Davidson, P.M., 1997. In: Food Microbiology. Fundamentals and Frontiers, (M.P. Doyle, L.R. Beuchat, T.J. Montville, eds), American Society for Microbiology, Washington.

Davies, E.A. and M.R. Adams, 1994. Resistance of *Listeria monocytogenes* to the bacteriocin nisin. Int J Food Microbiol 21: 341-347.

Delves-Broughton, J., P. Blackburn, R.J. Evans and J. Hugenholtz, 1996. Applications of bacteriocin, nisin. Antonie van Leeuwenhoek 69, 193-202.

De Roever, C., 1999. Microbiological safety evaluations and recommendations on fresh produce, Food Control 10, 117-143.

De Vuyst, L. and E.J. Vandamme, 1994. In: Bacteriocins of lactic acid bacteria (L. De Vuyst and E.J. Vandamme, eds), Blackie Academic and Professional, Glasgow.

Deans, S.G. and G. Ritchie, 1987. Antibacterial properties of plant essential oils, Int. J. Food Microbiol., 5, 165-180.

Dodds, K.L. and J.W. Austin, 1997. In: M.P. Doyle, L. Beuchat and T.J. Montville (eds.), Food microbiology. Fundamentals and frontiers. ASM Press, Washington, D.C., pp. 301-318.

Dormont, D., 2002. Les risques alimentaires liés aux agents transmissibles non conventionnels (ATNC), Revue Française des Laboratoires, 45-51.

Doyle, M.P., T. Zhao, J. Meng and S. Zhao, 1997. Escherichia coli O157:H7. In: Doyle, M. P., Beuchat, L., Montville, T. J. (Eds.), Food microbiology. Fundamentals and frontiers. ASM Press, Washington, D.C., pp. 177-198.

Druggan, P., S.J. Forsythe and P. Silley, 1993. Indirect impedance for microbial screening in the food and beverage industries. In: New Techniques in Food and Beverage Microbiology. R.G. Kroll and A. Gilmour (eds.), Society for Applied Bacteriology, Technical Series No 31. Blackwell Science, Oxford.

Eklund, T., 1989. In: Mechanisms of action of food preservation procedures, G.W. Gould (ed.), Elsevier, London.

Elliot, E.L., Ch.A. Kaysner, L. Jackson N.L. Tamplin, 1998. *Vibrio cholerae, V. parahaemolyticus, V. vulnificus* and other *Vibrio spp.* In: Bacteriological Analytical Manual. 8$^{th}$ Edition. www.cfsan.fda.gov.

EU IMPACT project "Interactions between microbial inoculants and resident populations in the rhizosphere of agronomically important crops in typical soils" European Community Biotechnology Programme BIO2 CT930053, (1993-1996).

Fayer, R. and T. Nerad. 1996. Effects of low temperatures on viability of *Cryptosporidium parvum* oocysts. Appl Environ Microbiol 62: 1431-1433.

Fayer, R., H.R. Gamble, J.R. Lichtenfels and J.W. Bier, 2001. Waterborne and foodborne parasites. In: Downes, F.P. and Ito, K. (Eds), Compendium of methods for the microbiological examination of foods. 4th ed. Am Publ Health Assn., Washington, D.C., pp. 429-438.

FDA, 1985. Irradiation in the production, processing, and handling of food. Food and Drug Admin., Fed. Reg. 50: 29658-29659.

FDA, 2003. Risk Assessment Reinforces That Keeping Ready-To-Eat Foods Cold May be the Key to Reducing Listeriosis. www.fda.gov.
Fishbein, L., 1998. Transmissible spongiform encephalopathies, hypotheses and food safety: An overview. The Science of the total environment, 217: 71- 82.
Fleet, G.H., 1999. Microorganisms in food ecosystems, Int. J. Food Microbiol., 50, 101-117.
Forsythe, S.J., 2000. "The microbiology of safe food". Blackwell Science Ltd. Ed. Oxford.
Gledel, J., 1988. *Salmonella*. In: Aspect microbiologique de la securité et de la qualité alimentaires. Technique et Documentation (Lavoisier), Paris. France, pp. 53-66.
Gill, C.O. and J. Bryant, 1993. The presence of Escherichia coli, Salmonella and Campylobacter in pig carcass dehairing equipment, Food Microbiol., 10, 337-344.
Gill, C.O. and K.G. Newton, 1978. The ecology of bacterial spoilage of fresh meat at chill temperature, Meat Sci. 2, 207-217.
González-Fernández, M.C., E.M. Santos, I. Jaime and J. Rovira, 2003. Influence of starter cultures and sugar concentrations on biogenic amine contents in chorizo dry sausage. Food Microbiol 20: 275-284.
Gram, L., L. Ravn, M. Rasch, J.B. Bruhn, A.B. Christensen and M. Gibskov, 2002. Food spoilage-interactions between food spoilage bacteria, Int. J. Food Microbiol., 78, 79-97.
Granum, P.E. and T. Lund, 1997. *Bacillus cereus* and its food poisoning toxins. FEMS Microbiology Letters 157: 325-333.
Hall, M.A. and A.J. Maurer, 1986. Spice extracts and propylene Glycols as inhibitors of *Clostridium botulinum* in turkey frankfurter slurries. Poult. Sci., 65, 1167-1171.
Hill, C., 1995. In: G.W. Gould (ed), New methods of food preservation, Blackie Academic and Professional, London.
Hocking, A.D., 1997. Toxigenic species of *Aspergillus*. In: Doyle, M. P., Beuchat, L., Montville, T. J. (Eds.), Food microbiology. Fundamentals and frontiers. ASM Press, Washington, D.C., pp. 413-425.
Hunter, P.R. and G. Nichols, 2002. Epidemiology and clinical features of Cryptosporidium infection in immunocompromised patients. Clinical Microbiol Reviews 15: 145-154.
ICMSF, International Commission for the Microbiological Specifications for Foods, 1980a. in: Microbial Ecology of Foods, vol I. Factors affecting life and death of microorganisms, Academic Press Inc., New York.
ICMSF, International Commission for the Microbiological Specifications for Foods, 1980b. in: Microbial Ecology of Foods, vol II. Food commodities, Academic Press Inc., New York.
ICMSF, International Commission for the Microbiological Specifications for Foods, 1998. In: Microorganisms in Foods 6. Microbial Ecology of Food Commodities, Blackie Academic and Professional, London.
IFT (Institute of Food Technologists), 2002. Emerging Microbiological Food Safety Issues: Implications for Control in the 21$^{st}$ Century. Report in: www.ift.org.
Jablonski, L.M. and G.A. Bohach, 1997. *Staphylococcus aureus*. In: Doyle, M. P., Beuchat, L., Montville, T. J. (Eds.), Food microbiology. Fundamentals and frontiers. ASM Press, Washington, D.C., pp. 371-393.
Jay, J.M., 1992. In: Modern Food Microbiology 4$^{th}$ ed, Van Nostrand Reinhold, New York.
Jay, J.M., 1996. Microorganisms in fresh ground meats: the relative safety of products with low versus high numbers, Meat Sci, 43, S59-S66.
Käferstein, F.K., 2003. Actions to reverse the upward curve of foodborne illness. Food Control 14: 101-109.
Kamau, D.N., S. Doores and K.M. Pruitt, 1990. Antibacterial activity of the lactoperoxidase system against *Listeria monocytogenes* and Staphylococcus aureus in milk, J. Food Prot., 53, 1010-1014.

Karmali, M.A., B.T. Steele, M. Petric and C. Lim, 1983. Sporadic cases of haemolytic uremic syndrome associated with fecal cytotoxin and cytotoxin-producing *Escherichia coli*. Lancet I: 619-620.

Kirov, S., 1997. *Aeromonas* and *Pleisomonas*. In: M.P. Doyle, L. Beuchat and T.J. Montville (eds.), Food microbiology. Fundamentals and frontiers. ASM Press, Washington, D.C., pp. 277-300.

Koopmans, M., C.H. von Bonsdorff, J. Vinje, D. de Medici and S. Monroe, 2002. Foodborne viruses. FEMS Microbiology Reviews 26: 187-205.

Koopmans, M., H. Vennema, H. Heersma, E. van Strien, Y. van Duynhoven, D. Brown, M. Reacher and B. Lopman, 2003. Early Identification of Common-Source Foodborne Virus Outbreaks in Europe. Em.Infect.Dis. 9:1136-1142.

Kraft, A.A., 1992. Psychrotrophic bacteria in foods: diseases and spoilage, Boca Raton, FL:CRC Press.

Krawczynski, K., S. Kamili and R. Aggarwal, 2001. Global epidemiology and medical aspects of hepatitis E. Forum(Genova) 11:166-79.

Labbé, R., G. 1989. *Clostridium perfringens*. In: M.P. Doyle (ed). Foodborne Bacterial Pathogens. Marcel Dekker, New York., pp. 192-234.

Labbé, R.G. and S. García, 2001. In: Guide to foodborne pathogens, John Wiley and Sons, Inc. Publication.

Larsen, A.G. and B. Norrung, 1993. Inhibition of *Listeria monocytogenes* by bavaricin A, a bacteriocin produced by *Lactobacillus bavaricus* MI401. L Appl Microbiol 17: 132-134.

Lerner, J., J. Breynaert and V. Preac-Mursic, 1994. Severe diarrhoea associated with *A. butzleri*. European J Clinical Microbiol and Infectious Disease 13: 660-662.

MacKenzie, W.R., N.J. Hoxie, M.E. Proctor, M.S. Gradus, K.A. Blair, D.E. Peterson, J.J. Kazmierczak, D.G. Addiss, K.R. Fox, J.B. Rose and J.P. Davis, 1994. A massive outbreak in Milwaukee of Cryptosporidium infection transmitted through the public water supply. New England J of Med 331: 161-167.

Maus, R., A. Goppelsröder and H. Umhauer, 2001. Survival of bacterial and mould spores in air filter media, Atmospheric environment, 35, 105-113.

McClane, B.A., 1997. *Clostridium perfringens*. In: Doyle, M.P., Beuchat, L., Montville, T.J. (Eds.), Food microbiology. Fundamentals and frontiers. ASM Press, Washington, D.C., pp. 319-342.

McDermott, P.E., S.M. Bodeis, L.L. English, D.G. White and D.D. Wagner, 2001. High level ciprofloxacin MICs develop rapidly in Campylobacter jejuni following treatments of chickens with sarafloxacin. American Society for Microbiology, 101st Annual Meeting, Orlando, Florida. ASM Press, Washington, D.C., p. 742.

McKane, L. and J. Kandell, 1986. Microbiology, essentials and applications, McGraw-Hill Book Company.

Mead, P.S., L. Slutsker, V. Dietz, L.F. McCraig, J.S. Bresee, C. Shapiro, P.M. Griffin and R.V. Tauxe, 1999. Food-related illness and death in the United States. Emerg Infect Dis 5: 607-625.

Mian, L.S., H. Maag and J.V. Tacal, 2002. Isolation of *Salmonella* from muscoid flies at commercial animal establishments in San Bernardino county, California. J. Vector Ecol. 27: 82-85.

Millar, B.C., M. Finn, L. Xiao, C.J. Lowery, J.S.G. Dooley and J.E. Moore, 2002. Cryptosporidium in foodstuffs-an emerging aetiological route of human foodborne illness. Trends Food Scie Tech 13: 168-187.

Mortimore, S. and C. Wallace, 1994. HACCP. A practical approach. 1st Edition. Chapman & Hall Ed., London.

Motarjemi Y. and F. Käferstein, 1999. Food safety, Hazard Analysis and Critical Control Point and the increase in foodborne diseases: a paradox?, Food Control, 10, 325-333.

Nachamkin, I., 1997. *Campylobacter jejuni*. In: Doyle, M. P., Beuchat, L., Montville, T. J. (Eds.), Food microbiology. Fundamentals and frontiers. ASM Press, Washington, D.C., pp. 165-176.

Notermans, S. and A. van der Giessen, 1993. Foodborne diseases in the 1980's and 1990's: the Dutch experience. Food Contam 4: 122-124.

Nychas, G.J.E., 1995. In: G.W. Gould (ed), New methods of food preservation, Blackie Academic and Professional, London.

Oliver, J.D. and J.B. Kaper, 1997. Vibrio species. In: Doyle, M.P., Beuchat, L., Montville, T.J. (Eds.), Food microbiology. Fundamentals and frontiers. ASM Press, Washington, D.C., pp. 239-276.

Omisakin, F., M. MacRae, I.D. Ogden and N.J.C. Strachan, 2003. Concentration and prevalence of *Escherichia coli* O157 in cattle faeces at slaughter. Appl Environ Microbiol 69: 2444-2447.

Orlandi, P.A., D.M.T. Chu, J.W. Bier and G.J. Jackson, 2002. Parasites and the food supply. Food Tech 56: 72-81.

Pang, T., Z.A. Bhutta, B.B. Finlay and M. Altwegg, 1995. Typhoid fever and other salmonellosis: a continuing challenge. Trends Microbiol. 3: 253-255.

Panisello, P.J., R. Rooney, P.C. Quantick and R. Stanwell-Smith, 2000. Application of foodborne disease outbreak data in the development and maintenance of HACCP systems, Int. J. Food Microbiol., 59, 221-234.

Park, S.F., 2002. The physiology of *Campylobacter* species and its relevance to their role as foodborne pathogens. Int J Food Microbiol 74: 177-188.

Paster, N., B.J. Juven, E. Shaaya, M. Menasherov, R. Nitzan, H. Weisslowicz and U. Ravid, 1990. Inhibitory effect of oregano and thyme essential oils on moulds and foodborne bacteria. Ltr Appl. Microbiol., 11, 33-37.

Phillips, C.A., 2001. *Arcobacter* spp in food: isolation, identification and control. Trends Food Scie Tech 12: 263-275.

Pickett, C.L., 2000. *Campylobacter* toxins and their role in pathogenesis. In: Nachamkin, I., Blaser, M.J. (Eds.), *Campylobacter*, 2$^{nd}$ edn., ASM Press, Washington, D.C., pp. 179-190.

Pitt J.L. and A.D. Hocking, 1997. In: Fungi and Food Spoilage, 2$^{nd}$ ed. Blackie Academic and Professional, London.

Pitt, J.I., 1997. Toxigenic species of Penicillium. In: Doyle, M. P., Beuchat, L., Montville, T. J. (Eds.), Food microbiology. Fundamentals and frontiers. ASM Press, Washington, D.C., pp. 427-439.

Post, F.J., G.B. Krishnamurty and M.D. Flanagan, 1963. Influence of sodium hexametaphosphate on selected bacteria, Appl. Microbiol., 11, 430-435.

Poumeyrol, M., 1988. *Clostridium perfringens*. In: Aspect microbiologique de la securité et de la qualité alimentaires. Technique et Documentation (Lavoisier), Paris. France, pp. 93-106.

Prusiner, S.B., 1998a. Prions, Proc.Natrl Acad Sci USA. 10: 13363-83.

Prusiner, S.B., 1998b. The prion diseases. Brain Pathol. 8: 499-513.

Purdy, D., C.M. Buswell, A.E. Hodgson, K. McAlpine, I. Henderson and S.A. Leach, 2000. Characterisation of cytolethal distending toxin (CDT) mutants of *Campylobacter jejuni*. J Med Microbiol 49: 473-479.

Rajkovic, A., M. Uyttendaele, T. Courtens and J. Debevere, 2005. Antimicrobial effect of nisin and carvacrol and competition between *Bacillus cereus* and *Bacillus circulans* in vacuum packed potato puree. Food Microbiol. 22: 189-197.

Rasmussen, M.A., W.C. Cray, T.A. Casey and S.C. Whipp, 1993. Rumen contents as a reservoir of enterohemorrhagic *Escherichia coli*. FEMS Microbiol Lett 114: 79-84.

Riley, L.W., R.S. Remis, S.D. Helgerson, H.B. McGee, J.G. Wells, B.R. Davis, R.J. Hebert, E.S. Olcott, L.M. Johnson, N.T. Hargett, P.A. Blake and M.L. Cohen, 1983. Hemorrhagic colitis associated with a rare *Escherichia coli* serotype. N Engl J Med 308: 681-685.

Robins-Browne, R.M., 1997. Yersinia enterocolitica. In: Doyle, M. P., Beuchat, L., Montville, T.J. (Eds.), Food microbiology. Fundamentals and frontiers. ASM Press, Washington, D.C., pp. 199-237.

Rowe, B. and B.J. Gross, 1984. Shigella. In: Bergey's manual of systematic bacteriology. Vol. 1. Editors: Krieg, N.R. and Holt, J.G. Williams & Wilkins. Baltimore. USA.

Sadruddin, F.H. and I. Mansaon, 1983. Hemaglutinating and hydrophobic surface properties of Salmonella producing enterotoxin neutralized by cholera-antitoxin. Vet. Microbiol. 8: 443-458.

Schlundt, J., 2002. New directions in foodborne disease prevention. Int J Food Microbiol 78: 3-17.

Shelef, L.A., O.A. Naglik and D.W. Bogen, 1980. Sensitivity of some common food-borne bacteria to the spices sage, rosemary and allspice. J. Food Sci., 45, 1042-1044.

Silliker, J.H. (ed.), 1980. Microbial ecology of Foods, Vol. 2, New York: Academic Press.

Siragusa, G.R. and M.G. Johnson, 1989. Inhibition of *Listeria monocytogenes* growth by the lacteroxidase-thiocyanate-$H_2O_2$ antimicrobial system. Appl. Environ. Microbiol., 55, 2802-2805.

Slifko, T.R., E. Raghubeer and J.B. Rose, 2000a. Effect of high hydrostatic pressure on *Cryptosporidium parvum* infectivity. J Food Protect 63: 1262-1267.

Slifko, T.R., H.V. Smith and J.B. Rose, 2000b. Emerging parasite zoonoses associated with water and food. Int. J. Parasitol., 30, 1379-1393.

Sofos, J.N. and F.F. Busta, 1993. In: P.M. Davidson and A.L. Branen (eds), Antimicrobials in foods $2^{nd}$ ed., Marcel Dekker, Inc, New York.

Speer, C.A., 1997. Parasitic protozoa transmitted by food and water. In: Doyle, M.P., Beuchat, L., Montville, T.J. (Eds.), Food microbiology. Fundamentals and frontiers. ASM Press, Washington, D.C., pp. 501-516.

Steiner, T.S., N.M. Thielman and R.L. Guerrant, 1997. Protozoal agents: What are the dangers for public water supply?. Ann Rev Med 48: 329-340.

Sugiyama, H. and T. Hayama, 1965. Abdominal viscera as site of metic actino for staphylococcal enterotoxin in the monkey. J. Infect Dis 115: 330-336.

Tauxe, R.V., 1997. Emerging foodborne diseases: an evolving public health challenge. Emerg. Infect. Dis. 3: 425-434.

Tauxe, R.V., 2002. Emerging foodborne pathogens. Int J Food Microbiol 78: 31-41.

Ten Brink, B., C. Damink, H. Joosten and V. in 't Huis, 1990. Occurrence and formation of biologically active amines in foods. Int. J. Food Microbiol. 11: 73-84.

Ting, E.W.T. and K.E. Deibel, 1992. Sensitivity of *Listeria monocytogenes* to spices at two temperatures. J. Food Safety, 12, 129-137.

Untermann, F., 1998. Microbial hazards in food. Food Control 9: 119-126.

USDA, 1999. Ground beef processing guidance material, Fed. Regist., 64(11):2872-2873.

USDA, 1996. Pathogen reduction: Hazard Analysis and Critical Control Point (HACCP) Systems; Final Rule. Fed. Regul., 61, 38805-38855.

Van Gerwen S.J.C., J.C. de Wit, S. Notermans and M.H. Zwietering, 1997. An identification procedure for foodborne microbial hazards, Int. J. Food Microbiol., 38, 1-15.

Vandamme, P., E. Falsen, R. Rossau, P. Segers, R. Tygat and J. Delay, 1991. Revision of *Campylobacter*, *Helicobacter* and *Woinella* taxonomy: emendation of genetic descriptions and proposal for *Arcobacter* gen. nov. Int J Systematic Bacteriol 41:88-103.

Vanderlinde, P.B., B. Shay and J. Murray, 1998. Microbiological quality of Australian beef carcass meat and frozen bulk packed beef, J. Food Protect., 61, 437-443.

Varman, A.H. and M.G. Evans, 1996. Foodborne pathogens, Manson Publishing

Venczel, L.V., M. Arrowood, M. Hurd and M.D. Sobsey, 1997. Inactivation of *Cryptosporidium parvum* and *Clostridium perfringens* spores by mixed-oxidant disinfectant and by free chlorine. App Environ Microbiol 63: 1598-1601.

West, P.A., 1989. The human pathogenic vibrios -a public health update with environmental perspectives. Epidemiol Infect 103: 1-34.

Wheeler, J.G., D. Sethi, J.M. Cowden, P.G. Wall, L.C. Rodrigues, D.S. Tompkins, M.J. Hudson and P.J. Roderick, 1999. Study of infectious intestinal disease in England: rates in the community, presenting to general practice, and reported to national surveillance. Br Med J 318: 1046-1050.

WHO (World Health Organization), 1984. The role of food safety in health and development. Report of a Joint FAO/WHO Expert Committee on Food Safety. Technical Report Series No. 705.

WHO (World Health Organization), 2000. Foodborne diseases: focus on health education.. Geneva, Switzerland.

WHO (World Health Organization), 2003. Infectious diseases: Shigella. www.who.int.

Widdowson, M.C., W.J.M. Jaspers, W.H.M. van der Poel, F. Verschoor, A.M. de Roda Husman, H.L.J. Winter, H.L. Zaaijer and M. Koopmans, 2003. Cluster of cases of acute hepatitis associated with hepatitis E virus infection acquired in the Netherlands. Clin Infect Diseas, 36:29-33.

Winfield, M.D. and E.A. Groisman, 2003. Role of nonhost environments in the lifestyles of *Salmonella* and *Escherichia coli*. Appl. Environ. Microbiol. 69:3687-3694.

Zottola, E.A., 2001. Reflections on *Salmonella* and other "wee beasties" in foods. Food Technology 55: 60-67.

# 3. Chemical hazards

*Bruno De Meulenaer*

## 3.1 Introduction: food intoxications and food sensitivities

Food safety can be affected by microbiological, physical and chemical causes. The former two causes are discussed elsewhere in this textbook. This chapter and Chapter 12 consider the chemical causes affecting food safety. In this particular chapter, some general aspects on chemical food safety are discussed, together with basic principles about toxicology. Furthermore, this chapter gives a non-exhaustive overview of chemicals present in our food, which may affect our health. In Chapter 12, more details about the analysis of these compounds are compiled.

Adverse reactions to chemicals present in food can be classified as food intoxications or food sensitivities. Typically every individual in a population is vulnerable to food intoxications, as long as the exposure is high enough. Food sensitivities, however, are individual adverse reactions towards particular food components.

The most well-known examples of food sensitivities are food allergies. These are typically induced due to an abnormal reaction of the immune system. Apart from the immunological mediated sensitivities, non-immunological food intolerances and secondary food sensitivities can be distinguished, as shown in Figure 3.1. The latter occur with or after the effects of other conditions. A typical example of these secondary food sensitivities includes drug-induced sensitivities such as the increased sensitivity to histamine among patients on monoamine oxidase-inhibiting drugs (anti-depressive agents).

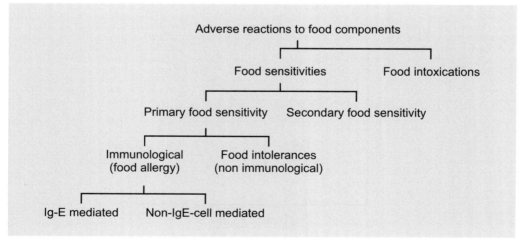

*Figure 3.1. Overview of the adverse reactions to food components.*

*Bruno De Meulenaer*

Chemicals provoking food intoxications can be classified in various manners. The classification used in this textbook should therefore not be considered as absolute. Moreover, in the classification presented, it will be obvious that some toxicants can be classified in several classes. Nevertheless, the following types will be considered:
- food additives;
- residues;
- contaminants;
- endogenous substances.

Some representative examples of each of these classes will be presented.

## 3.2 Basic toxicological considerations

As already stated more then 500 years ago by Parcellus, every chemical may induce adverse reactions in our body. A crucial factor, however, determining the severity of the adverse reaction is the dose. This is the intake of a compound expressed per unit of body weight. Dose-response curves, as shown in Figure 3.2, represent the relationship between the dose (logarithmic scale) and the cumulative fraction of the population exhibiting an adverse effect. These curves are typically obtained in model systems, such as animals or *in vitro* systems as cells.

The toxic effect studied may be acute or chronic. Although in chemical food safety the chronic toxicological effects are considered, there is growing concern about the acute toxicological effects, such as those provoked by food allergens or pesticides.

Dose response curves can also be obtained for essential nutrients, such as vitamins and minerals, as shown in Figure 3.3. In contrast to the curve shown in Figure 3.2, however, it follows from Figure 3.3 that if the dose of an essential nutrient is too low, adverse effects are induced due to

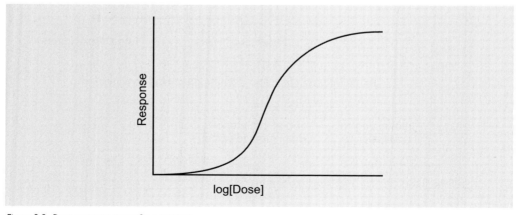

*Figure 3.2. Dose-response curve for a toxicant.*

## Chemical hazards

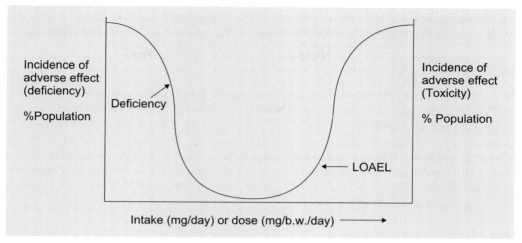

Figure 3.3. Dose-response curve for an essential nutrient.

a deficiency. At intermediate doses, no adverse effects are observed until the dose is too high. The second part of the curve at higher exposure levels is similar to the dose response relationship as shown in Figure 3.2, indicating that essential nutrients may also exhibit some toxicity.

A crucial observation in Figure 3.2 and Figure 3.3 is the presence of a so-called threshold dose, below which no effects can be observed. The existence of such a threshold level can be explained by detoxification and repair mechanisms in a living organism.

The highest daily dose, at which no adverse effects are observed in the most susceptible animal species or model system during chronic exposure, is identified as the No Observable Adverse Effect Level (NOAEL). The NOAEL value is used as the basis for setting human safety standards for chemicals present in our diet such as the TDI or the ADI. TDI or Tolerable Daily Intake is the daily intake of a particular chemical, expressed per unit of bodyweight, which during a lifetime span will not induce any adverse effects in the individual. The ADI or Acceptable Daily Intake is similar to the TDI, but is typically used for food additives and residues, while the TDI is used for contaminants. These safety standards are calculated from the NOAEL value, by considering two supplementary uncertainty factors, $UF_1$ and $UF_2$, as shown in the following formula:

$$\text{TDI or ADI} = \frac{\text{NOAEL}}{UF_1 \times UF_2}$$

The first uncertainty factor accounts for the extrapolations from the animal or model systems to humans, while the second factor considers the inter-individual variability in the human species. Both factors are typically equal to 10.

For chemicals which tend to bio-accumulate in the human body, such as dioxins, a Tolerably Weekly Intake (TWI) is usually considered instead of a TDI, which is calculated as follows.

$$TWI = \frac{NOAEL}{UF_1 \times UF_2} \times 7$$

In some cases, acute food intoxications may occur due to, for example, consumption of an apple that accidentally received too high a dose of a particular pesticide. In such a case, chronic toxicity studies are often not relevant to consider and, consequently, the threshold levels obtained in such chronic exposure studies are not relevant either. In acute toxicity studies, however, threshold levels are observed analogously as for chronic toxicity. A so-called Acute Reference Dose (ARfD) is defined at which during a single exposure no acute toxic effects can be observed in an individual. Although this concept is currently not applied yet for food allergens, it follows from Figure 3.4 that for food allergies too, such a threshold dose is likely to be defined.

Some chemicals are believed to have no threshold below which toxic effects are observed. The most common groups of chemicals in this respect are genotoxic carcinogens. These compounds induce cancer as a result of their interactions with DNA. However, it is obvious that we are all subjected to a continuous exposure to low doses of carcinogens present in our food and in our diet. Therefore, some protective mechanisms probably neutralise small doses of genotoxic carcinogens in our body. However, these mechanisms are currently poorly understood, and therefore no reliable threshold level could be determined either.

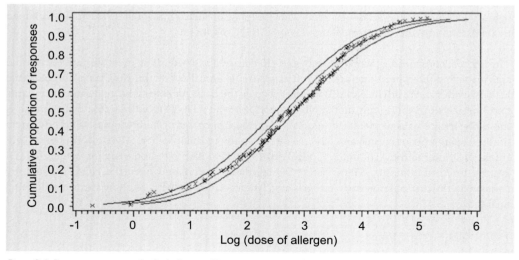

Figure 3.4. Dose-response curve for food allergens (the x-axis represents the log of the amount of allergen in mg, and the y-axis the cumulative proportion of responders) (reproduced from Bindslev et al., 2002).

## 3.3 Food sensitivities

### 3.3.1 Terminology

Food sensitivities can be defined as individualistic adverse reactions to food. Many types are involved in these individual reactions, as already shown in Figure 3.1.

True food allergies are abnormal responses of the immune system to naturally occurring proteins in the foods. They can be divided into two categories, based upon the nature of the immune response. If the response of the immune system involves the formation of allergen-specific immunoglobulin E antibodies (IgE), an immediate hypersensitivity reaction begins to develop within minutes or an hour after ingestion of the offending food.

In delayed hypersensitivity reactions, symptoms begin to appear up to 24 hours or longer after ingestion of the offending food. The most well-known example is celiac disease, also known as celica sprue or gluten-sensitive enteropathy. It involves an abnormal response of the cellular immune system, through a T-cell mediated mechanism, resulting in an inflammatory process. Celiac disease is associated with the naturally-occurring gliadin or related protein fractions of wheat, rye, barley and oats. After consumption of these offending grains, the absorptive epithelial cells in the small intestine are damaged by an inflammatory reaction. As a result, the absorption of nutrients through the epithelium is compromised, giving rise to a severe malabsorption syndrome characterised by diarrhoea, bloating, weight loss, anaemia, fatigue, weakness, growth retardation, etc. Its treatment involves the total avoidance of the above-mentioned cereals. If a gluten-free diet is followed, the symptoms will resolve and the above-mentioned symptoms disappear.

Certain foods may elicit adverse reactions that resemble food allergies. In contrast to food allergies, however, the adverse reactions are elicited by the presence of elevated levels of histamine, which is one of the principal endogenous mediators of allergic reactions in our body. The elevated levels of histamine are provoked by microbiologically-induced decarboxylation of the amino acid histidine, which occurs typically in scombroid fish (tuna, mackerel) or some cheeses. Unlike food allergies however, all individuals are susceptible to histamine poisoning, and consequently, it should not be considered as a food allergy but as food intoxication.

Food intolerances do not involve the immune system. The following types of food intolerances are known:
- metabolic food disorders;
- anaphylactoid reactions;
- idiosyncratic reactions.

Some typical examples will be discussed in the following sections.

### 3.3.2 IgE mediated food allergies

*Mechanism*
Immunoglobulin E antibodies (IgE) are one of the five classes of antibodies present in the human body. Although IgE antibodies are particularly involved in fighting off parasitic infections, allergic individuals produce IgE antibodies that are specific for certain environmental antigens. These antigens are typically proteins that can be found in pollen, mould spores, bee venom, dust, mites, animal danders and in foods. The initial step of the IgE-mediated allergy is the sensitisation phase in which the antigen stimulates the production of specific IgE antibodies (Figure 3.5). These antibodies bind subsequently to mast cells in various tissues and to the basophils in the blood. In such a way, the surface of these sensitised cells is covered with allergen-specific IgE antibodies. Sensitisation most commonly occurs in young infants. On subsequent exposure to the allergen, this substance cross-links two IgE antibodies on the surface of the mast cells or the basophils. As a further consequence the granules contained in these cells, and which are filled with physiologically active chemicals that mediate the allergic response, are stimulated to release their content into tissues and blood. Although various mediators are described, histamine is probably one of the primary mediators responsible for many of the immediate symptoms associated with IgE-mediated allergic reactions.

*Symptoms and diagnosis*
A variety of clinical manifestations of food allergy are described as summarised in Table 3.1. As can be observed, these symptoms vary from mild and annoying to severe and life-threatening.

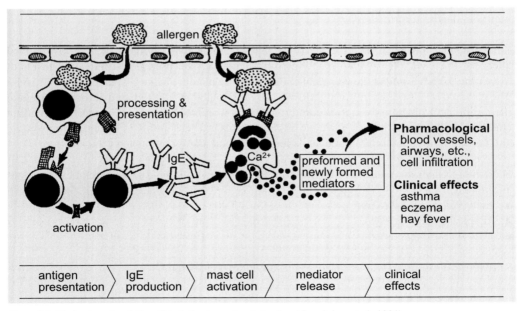

*Figure 3.5. Mechanism of an IgE-mediated allergic reaction (reproduced from Lehrer et al., 1996).*

*Table 3.1. Some clinical manifestations of food allergy.*

| Some clinical manifestations of food allergy |
| --- |
| Anaphylaxis |
| Oral allergy symptoms |
| Cutaneous symptoms |
| Urticaria |
| Eczema or atomic dermatitis |
| Angiodema |
| Gastrointestinal reactions |
| Nausea |
| Vomiting |
| Diarrhoea |
|      abdominal pain |
|      colic |
| Respiratory reactions |
| Rhinitis |
| Asthma |

The most frightening symptom is probably the anaphylactic shock. Anaphylaxis is considered as a systemic reaction involving gastrointestinal and cutaneous symptoms combined with a reaction of the respiratory and cardiovascular system. Due to a cardiovascular and/or respiratory collapse, these shocks may be fatal and numerous deaths have been attributed to inadvertent exposure to the offending allergen.

The nature and severity of the clinical manifestations are dependent upon every individual and may also vary with the dose of the offending food ingested, the degree of sensitisation to the offending food and probably some other factors.

As can be observed as well, the symptoms mentioned in Table 3.1 may also be provoked by a variety of other illnesses, making it sometimes difficult to associate the observed medical problems with a food allergy. Using food diaries or elimination diets often facilitates this quest for association. Once the association between the ingestion of particular foods is associated with the adverse reactions, proof of the IgE-mediated mechanism should be considered. This can be achieved using *in vitro* tests in which the presence of allergen-specific IgE in the patient's serum is assessed. The skin prick test is a frequently used diagnosis of allergies in general, including food allergies, despite the fact that false positive results are frequently obtained. The golden standard, however, remains the Double Blind Placebo Controlled Food Challenge (DBPCFC). In such a test, a patient is challenged with varying amounts of the offending allergen in such a way that neither the patient nor the physician is aware whether the allergen is present in the administered food or not. Such tests, however, need to be performed in a clinical environment, because of the risks associated with an allergic reaction.

*Bruno De Meulenaer*

*Allergenic foods and prevalence*
Eight types of food groups are thought to account for more then 90 % of food allergies worldwide (Table 3.2). These are known as the "big eight". Tree nuts include almonds, walnuts, pecans, cashews, Brazil nuts, pistachios, hazelnuts, pine nuts, macadamia nuts, chestnuts and hickory nuts. Crustaceans include shrimp, crab, lobster and crayfish.

In addition to the "big eight", however, more then 160 other foods have been documented as causing food allergies, but on a less frequent basis. Remarkably some of the "big eight", and in addition some other foods, are known to cause food allergies frequently in particular geographic regions. Peanut allergy, for instance, is more common in North America than in other parts of the world. Similarly buckwheat and celery are important allergens in respectively Southeast Asia and Europe, respectively. Presumably, the prevalence of a typical food allergy in a population is linked to the presence of the food allergen in the population's local diet.

As mentioned before, specific proteins in these foods are responsible for the food allergy. Various allergic food proteins have been identified and characterised as mentioned in Table 3.3. As can be observed, a typical nomenclature is used to identify food allergens using the accepted taxonomic name of their source. The first three letters of the genus are used, followed by the first letter of the species and then an Arabic number, which is assigned according to the order of their identification. Some proteins in these allergic foods are considered as major allergens,

*Table 3.2. Most common foods inducing food allergies (the "big eight").*

| | |
|---|---|
| Cereals containing gluten | Crustaceans and products |
| Soybeans and products | Fish and fish products |
| Peanuts and products | Milk and milk products |
| Tree nuts and nut products | Eggs and egg products |

*Table 3.3. Some major food allergens and their sources.*

| **Allergen source** | **Allergens (systematic and original name)** |
|---|---|
| *Arachis hypogea* (peanut) | Ara h 1 |
| *Bertholletia excelsa* (Brazil nut) | Ber e 1; 2S albumin |
| *Gadus callarias* (Cod) | Gad c 1 |
| *Gallus domesticus* (chicken eggs) | Gal d1; *ovomucoid* |
| | Gal d2; *ovalbumin* |
| | Gal d3; *ovotranferin* |
| | Gal d4; *lysozym* |
| *Glycine max* (soybean) | Gly 1; *minor protein of 7S fraction* |
| *Penaeus aztecus* (brown shrimp) | Pen a 1; *tropomyosin* |
| *Metapenaeus enis* (greasyback shrimp) | Met e 1; *tropomyosin* |

## Chemical hazards

since more then 50 % of the allergic patients have specific IgE antibodies. The other allergic proteins are considered as minor allergens.

An important observation in allergic patients is the cross-reactivity which occurs between proteins of different sources. A well-documented example is the case of tree pollen allergic patients that also frequently suffer allergy to tree nuts, fruits and vegetables. Apples and hazelnuts are the most common cross-reacting foods. This cross-reactivity can be explained by the similarity between the epitopes of the various allergic proteins involved. An epitope is the recognition site for the allergen-specific IgE antibody. Thus, IgE binding similarities have been shown between the 18kDa hazelnut allergen and the major birch pollen allergen *Bet v1*. Similar cross-reactions are reported between various crustaceans and eggs of various avian species or cows' and goats' milk, respectively.

It is very difficult to reveal the common physical and chemical properties of food allergens. Despite this fact, some common features are thought to be of importance with regard to the allergic potential of a particular protein. These features include IgE-binding properties that are resistant to heat treatments, digestion, proteolysis or hydrolysis. Assessment of the allergenic potential of a food protein is an important issue in the approval of novel foods or novel food proteins.

Concerning the prevalence of food allergies, it is obvious that infants and children are more affected by food allergies then other groups. Among infants, the prevalence appears to be in the range of 5 to 8%, while for the total population between 1 and 2 % of individuals are estimated to be allergic. Some allergies are known to diminish with age, such as cow's milk allergy. Peanut allergy, on the other hand, is not out grown.

### 3.3.3 Food intolerances

As mentioned before, typically three major classes of non-immunological food sensitivities can be distinguished: anaphylactoid reactions, metabolic disorders and idiosyncratic reactions. In contrast to food allergies, food intolerances are caused by other substances then proteins, which moreover may be naturally occurring in the food or additive substances.

Anaphylactoid reactions result in symptoms that are very similar to the clinical manifestations of a food allergy. In fact, the physiological reaction is indeed the same as in a food allergy, since the anaphylactoid reaction is also provoked by a release of histamine and other mediators from mast cells. In contrast to a food allergy, however, this release is not mediated via IgE antibodies. Despite the fact that no histamine-releasing substances have been isolated from a food, this mechanism is well established for particular drugs. The best example of an anaphylactoid reaction is probably strawberry sensitivity. Although intolerance towards strawberries is well-documented, there is little evidence that IgE-mediated mechanisms are responsible. Strawberries contain little protein and no strawberry allergens or strawberry specific IgE antibodies were identified. Therefore, an unidentified substance probably causes a histamine release from the mast cells, giving rise to allergy-like symptoms.

*Bruno De Meulenaer*

Metabolic food disorders involve genetically-determined altered metabolic patterns resulting in deficient metabolism or increased sensitivity. A typical example of a deficient metabolic food disorder is lactose intolerance. It is caused by a deficiency of β-galactosidase and results in an inability to digest lactose. Since the lactose cannot be digested in the small intestine, it is fermented in the large intestine resulting in abdominal cramps, flatulence and diarrhoea.

Favism can be considered as a metabolic food disorder increasing the patient's sensitivity to a particular food, in this case fava beans. Favism is caused by a genetic deficiency of glucose-6-phosphate dehydrogenase in the red blood cells causing an increased sensitivity to several haemolytic factors in fava beans. This particular enzyme prevents oxidative damage of erythrocytes since it maintains the levels of reduced glutathione and nicotine amide dinucleotide phosphate. Fava beans contain several naturally occurring oxidants, such as vicine and convicine as shown in Figure 3.6. These substances may cause oxidative damage to the erythrocyte membrane if insufficient protection is available. Consequently, symptoms typical for haemolytic diseases are developed: pallor, fatigue, nausea, fever, chills, etc.

Many adverse reactions that affect certain individuals have an unknown mechanism. Many different mechanisms may be involved in these idiosyncratic reactions. Similarly, a variety of symptoms is associated with these reactions, as illustrated in Table 3.4. As can be observed, three

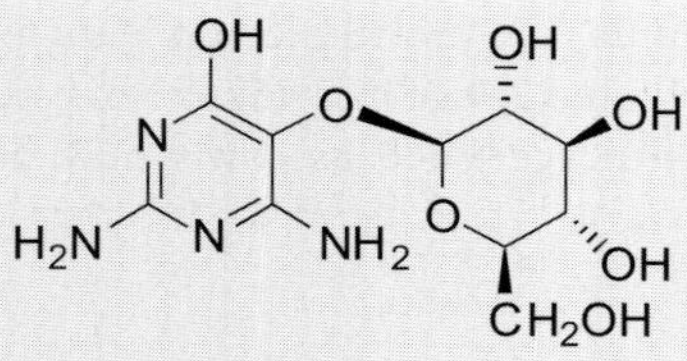

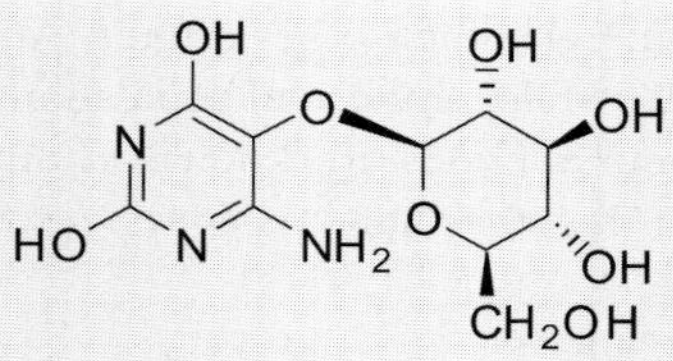

*Figure 3.6. Structure formulae of vicine 1 and convicine 2.*

*Table 3.4. Some food-associated idiosyncratic reactions.*

| Category | Implicated food or ingredient | Reaction |
|---|---|---|
| Proven | Sulphites | Asthma |
|  | Aspartame | Urticaria |
| Unproven | BHA, BHT, benzoates | Chronic urticaria |
|  | Tartrazine | Asthma, urticaria |
|  | Aspartame, chocolate | Migraine headache |
|  | Sugar | Aggressive behaviour |
|  | Monosodium glutamate | MSG symptom complex, formerly known as the Chinese restaurant syndrome; asthma |
| Disproven | Food colouring agents | Hyperkinesis |

## Chemical hazards

categories can be distinguished as well: reactions for which the association with a specific food ingredient is well-documented: reactions for which the association is not well-established or controversial; and reactions associated with particular food ingredients, despite the fact that considerable evidence is available to the contrary.

In the vast majority of food idiosyncrasies, the role of the specific food or food ingredient remains unclear. A major reason for this is the lack of challenge tests, similar to the DBPCFC, used for food allergens.

### 3.3.4 Consequences for the food industry and legislative aspects

In the food industry, major concerns exist about immunological food sensitivities. Although food intolerances may have serious consequences as well, the presence of the offending food or ingredient is in most cases quite clear to the consumer. The presence of food additives, for example, needs to be mentioned in the ingredient list.

For food allergens and gluten, however, the problem is far more complicated. As revealed above, particular proteins are responsible for the adverse immunological reactions. It is frequently unclear to the allergic patient whether the offending protein is present in the food, unless the use of the offending protein or food is mentioned on the ingredient list. The food industry, however, is confronted with the problem of hidden allergens. Allergic proteins may be carried over from one food formulation to another. Plain chocolate and milk chocolate production may serve as a typical example. If the same line is used for the production of these two products, a shift in the packaging from milk chocolate to plain chocolate is typically performed if the plain chocolate's colour is considered to be dark enough, instead of considering the case in content of the end product.

Some companies try to circumvent the allergy problem by preventive labelling. These labels warn the consumer that particular allergens may be present in the food (e.g. "may contain peanuts"). Frequently, however, these labels are not used because the offending food or ingredient *is* present, and the company wants to safeguard against possible claims by allergic patients. In the future, however, the possible presence of the most important food allergens will probably have to be included in the ingredient list, and food industry will need to adopt HACCP programs in order to avoid the presence of undeclared allergens.

In order to avoid an allergic attack, it is obvious that a patient should avoid ingestion of food containing the offending proteins. Due to the ever-increasing complexity of food production and food formulation, however, it is very difficult for the consumer to decide whether a food may contain the offending allergen. In accordance with the Codex Alimentarius Commission's recommendations, the European Commission issued a proposal to amend the European Food Labelling Directive 2001/13/EC. The proposal abolishes the '25% rule' which specifies that for some foods it is not obligatory to label the ingredients that make up less than 25 % of the final food product. Consequently, all ingredients intentionally added will have to be included on the label's ingredients list. Particular attention is paid to the labelling of the 8 allergens mentioned

*Bruno De Meulenaer*

above, in addition to some other relevant European allergic foods such as celery, sesame, mustard and sulphite. Because of these changes, the food industry is confronted with the need for reliable analytical methods that are able to detect and quantify particular allergens in food.

## 3.4 Food additives

### 3.4.1 Regulatory aspects in the EU

According to the European Union's legislation (Directive 89/107/EEC), a food additive should be considered as "any substance not normally consumed as a food in itself and not normally used as a characteristic ingredient of food whether or not it has nutritive value, the intentional addition of which to food for a technological purpose in manufacture, processing, preparation, treatment, packaging, transport or storage of such food, results or may be reasonably expected to result, in its or its by-product becoming directly or indirectly a component of such foods". Although this definition does not include processing aids, in the cited directive, these substances are defined as substances intentionally used for the processing of raw materials, foods or ingredients to fulfil a certain technological purpose during treatment or processing. The use of a processing aid may result in an unintentional but unavoidable residue in the final product, which however must not present any health risk.

This framework directive also lists in an annex the various categories of food additives used. These categories, together with their definitions and their technical function, are listed in Table 3.5. In a second annex, general criteria concerning technological need, safety and use of food additives are specified.

In addition to the framework directive, three 'comprehensive' or so-called 'specific' directives are issued concerning the use of sweeteners (94/35/EC), colours (94/36/EC) and other classes of additives (95/2/EC) and amendments. These directives are rather technical but are of key interest for food producers. These directives contain the so-called positive lists that specify the food in which these additives may be used and, in addition for many additives, specify maximum levels of use. If the use of the food additive is considered to be safe under envisaged conditions of use, it is not considered necessary to prescribe a maximum quantity. In such cases, the term *quantum satis* is used, which means that the additive may be used according to good manufacturing practices.

Another important element with regard to the food additive legislation is the labelling duty. Food additives used in the manufacture of food products must be identified as prescribed in Directive 2000/13/EC. The category name as included in Table 3.5, followed by the specific name of the additive or the E number should be mentioned in the correct position in order by weight (greatest first).

## Chemical hazards

Table 3.5. Categories of food additives listed in annex I of Directive 89/107/EEC, their definition and technical function.

| Functional classes | Definition |
|---|---|
| 1. Acid | Increases the acidity and/or imparts a sour taste to a food |
| 2. Acidity regulator | Alters or controls the acidity or alkalinity of a food |
| 3. Anti-caking agent | Reduces the tendency of particles of food to adhere to one another |
| 4. Anti-foaming agent | Prevents or reduces foaming |
| 5. Antioxidant | Prolongs the shelf-life of foods by protecting against deterioration caused by oxidation, such as fat rancidity and colour changes |
| 6. Bulking agents | A substance, other than air or water, which contributes to the bulk of a food without contributing significantly to its available energy value |
| 7. Colour | Adds or restores colour in a food |
| 8. Colour retention agent | Stabilizes, retains or intensifies the colour of a food |
| 9. Emulsifier | Forms or maintains a uniform mixture of two or more immiscible phases such as oil and water in a food |
| 10. Emulsifying salt | Rearranges cheese proteins in the manufacture of processed cheese, in order to prevent fat separation |
| 11. Firming agent | Makes or keeps tissues of fruit or vegetables firm and crisp, or interacts with gelling agents to produce or strengthen a gel |
| 12. Flavour enhancer | Enhances the existing taste and/or odour of a food |
| 13. Flour treatment agent | A substance added to flour to improve its baking quality or colour |
| 14. Foaming agent | Makes it possible to form or maintain a uniform dispersion of a gaseous phase in a liquid or solid food |
| 15. Gelling agent | Gives a food texture through formation of a gel |
| 16. Glazing agent | A substance which, when applied to the external surface of a food, imparts a shiny appearance or provides a protective coating |
| 17. Humectant | Prevents food from drying out by counteracting the effect of a wetting agent atmosphere having a low degree of humidity |
| 18. Preservative | Prolongs the shelf-life of a food by protecting against deterioration caused by microorganisms |
| 19. Propellant | A gas, other than air, which expels a food from a container |
| 20. Raising agent | A substance or combination of substances which liberate gas and thereby increase the volume of a dough |
| 21. Stabilizer | Makes it possible to maintain a uniform dispersion of two or more immiscible substances in a food |
| 22. Sweetener | A non-sugar substance which imparts a sweet taste to a food |
| 23. Thickener | A substance which enhances the consistency of the food |

### 3.4.2 Exposure assessment

An exposure assessment is a crucial element in the risk analysis of food additives. The goal is to find out if any individual has potential intakes that might exceed the ADI for a particular additive and if so, by how much. In order to do so, two crucial elements should be investigated:
- usage data, which represent the concentration of the food additive in the food;
- food consumption data, which represent the amount of affected foods eaten, eventually specified for various subgroups in the population.

From these data, the Estimated Daily Intake (EDI) can be calculated as follows, taking into account the bodyweight:

$$\text{EDI (mg/kg bw/day)} = \frac{\text{usage (mg/kg) x consumption (kg/day)}}{\text{bodyweight (kg)}} \qquad 3.1$$

Since bodyweight is taken into consideration, direct comparison with the ADI is possible.

It is evident that the reliability of such exposure studies largely depends upon the quality of the data used. Moreover, such calculations should not be made using population average figures, since this would result in serious underestimation of consumers at the upper end of the range of possible intake levels. Consequently, advanced probabilistic modelling methods are required to combine these two databases in order to obtain an exposure distribution that can be used as a tool to carry out the total risk analysis study.

In order to avoid the use of excessive resources for analysing the risks provoked due to the intake of food additives, a so-called tiered approach is frequently used. In the first step or tier of this approach, a simple yet conservative intake estimate is made in order to identify those additives for which no further action is required, since intake will always be lower then the ADI. For the usage data, the maximum permitted concentration level of the food additive in a particular food is considered. Food consumption is estimated using a so-called budget method, which is based on the fact that there is a physiological upper limit to the amount of food and drink that can be consumed on any given day. Commonly used figures in this respect are included in Table 3.6, considering as well the fact that only a proportion of the diet is likely to contain additives.

Consequently the potential intake can be calculated as follows:

$$\text{potential intake} = \frac{\text{maximum level in drinks (mg/kg)}}{40} + \frac{\text{maximum level in food (mg/kg)}}{160} \qquad 3.2$$

Table 3.6. Figures most commonly used in the standard budget method.

|  | Upper limit of daily consumption | Proportion likely to contain additives |
|---|---|---|
| Beverages | 100 ml/kg bw | 25 % |
| Food | 25 g/kg bw | 25 % |

## Chemical hazards

Using such an approach, 58 additives have been examined on an EU level. For 36 of these the estimated intakes for adults were higher then the ADI. For children, 48 of the examined additives exceeded the ADI. These additives included several colorants, some preservatives (sorbic acid, benzoic acid, sulphite, nitrite, fumaric acid) and an antioxidant (gallates). It was recommended that all these additives were considered for further examination. This could be done using the second step in the tiered approach as shown in Figure 3.7.

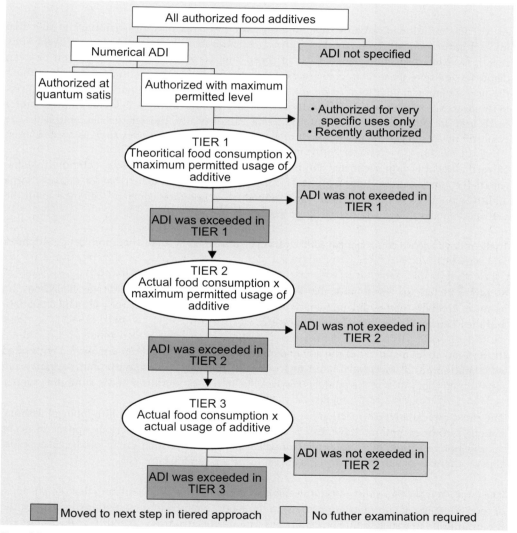

Figure 3.7. Tiered approach to estimate intake of food additives (reproduced from 'Report of the Commission on dietary food additive intake in the European Union. http://europa.eu.int/comm/food/fs/sfp/addit_flavor/flav15_en.pdf).

In the second tier, the actual concentration of the food additive in the particular food products is considered. Of course, in this tier, considerable analytical resources should be available. If the ADI is still exceeded, despite the fact that more realistic exposure data are obtained, an even more reliable estimation should be performed in the third tier. In this final tier, both the actual food additive concentration together with the actual food consumption data should be considered as explained above.

### 3.4.3 Adverse reaction to particular food additives

As already mentioned, adverse reactions to food additives may be attributed to so-called idiosyncratic food intolerances. Frequently, considerable doubt about the reality of such reactions exists especially since most of these adverse reactions are not proven or even disproven. Despite these facts, consumers still prefer foods that are "natural" and made without the use of chemical additives. In recent surveys, consumers still attribute a considerable risk to the presence of additives in foods, despite the fact that the same individuals claim to prefer foods that are nutritious, convenient and fresh. Paradoxically, these latter characteristics are typically the results of the use of nutritional additives and preservatives such as antioxidants.

Nevertheless, it is quite clear that some additives remain controversial (e.g. colorants, olestra, aspartame, monosodium glutamate), and from exposure surveys it seems that the intake of some additives might exceed the ADI for some individuals. Therefore, constant vigilance about the use and toxicity of food additives is appropriate.

In the following, two controversial additives will be discussed in somewhat more detail: sulphites and nitrites.

Sulphites include six compounds that are used as multifunctional food additives. They may act as antimicrobials; they inhibit enzymatic and non-enzymatic browning; they prevent oxidation; and they can be used as bleaching agents, etc.

In the USA, about a quarter of the adverse reactions from sulphites were associated with salad bars, and about 20 % were related to fresh fruit and vegetables. Wine and seafood each represented about 10 % of the products responsible for complaints attributed to sulphiting agents.

The most typical adverse reaction, as mentioned in Table 3.4, is a sulphite-induced asthma, although other symptoms have also been reported. The adverse reactions appear to occur mainly among a small percentage of asthmatics, but other individuals may also be susceptible. The mechanism of sulphite-induced asthma, however, remains unclear.

The joint FAO/JECFA expert committee advised obligatory labelling similar to that the presence of food allergens, if sulphites are present in the food. In contrast to regulations in the EU, the use of sulphite in wines should be labelled as well if concentrations exceed 10 ppm, despite the fact that sulphite is considered as a food additive as well in the EU.

## Chemical hazards

Sulphite continues to be a problem and even in recent years, there have been several recalls in the USA because foods were found to contain undeclared sulphites.

The case of nitrite is even more complicated then the presence of sulphite. Sodium nitrite and sodium nitrate have a long history of use as curing agents in various meat products. They render desirable properties such as colour, flavour, texture and preservation (specifically inhibiting *Clostridium botulinum*). The use of nitrite and the presence of a high nitrate concentration in particular foods such as leafy vegetables and water due to fertilisation, became however quite controversial. This is because nitrite and also nitrate are involved in the formation of N-nitroso compounds that have been shown to be carcinogenic. The formation of N-nitrosoamines is considered to be of particular importance, because of their stability compared to the related N-nitrosoamides. The general nitrosation reactions resulting in N-nitroso compounds are summarised in Figure 3.8. As can be observed, respectively secondary amines and amides are involved together with a nitrosation reagent. The classical N-nitrosation reaction occurs between nitric acid and an unprotonated secondary amine, as shown in Figure 3.9.

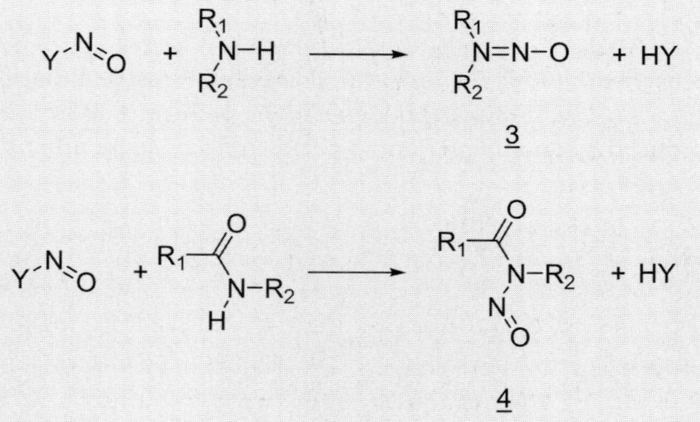

Figure 3.8. Formation of N-nitroso amines 3 and N-nitroso amides 4 due to the nitrosation of amines and amides (Y=NO2, NO3, H2O+, SCN-, halides).

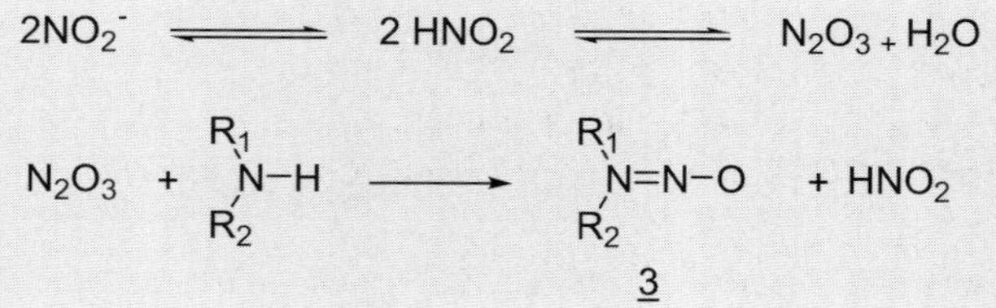

Figure 3.9. Acidic nitrosation of amines.

From Figure 3.9, it follows that the nitrosation reaction will proceed at some optimum pH because of the opposing requirements of the unprotonated amine and nitrous acid formation. For dialkylamines, the optimal pH is about 2.5 to 3.5, which is near to the pH of the mammalian stomach. The reaction and equilibria formed by nitrosating agents and their precursors are quite complex as shown in Figure 3.10.

As can be observed, ascorbate may inhibit nitrosation, as long as excess is present, since in aerobic conditions, NO may be further oxidised to dinitrogentetraoxide, which is a nitrosating agent itself, together with dinitrogentrioxide. Nitrosation agents are lost by NO and $NO_2$ gas or via formation of nitrate. Since precursors of N-nitroso compounds occur widely in the environment and in biological systems, it is not surprising that N-nitrosamines have been widely found in low concentrations in many foods and other environments.

Therefore exposure to exogenous N-nitrosoamines may be due to several sources, of which tobacco is by far the most important. Food and drink represent the broadest exposure to N-nitrosoamines, albeit at much lower levels then tobacco products. Other consumer products such as cosmetics, rubber, drugs, pesticides have been shown to contain carcinogenic nitrosamines as well.

Apart from the exogenous exposure to nitrite, however, it should be realised that nitrite itself can also be formed through a variety of endogenous metabolic pathways. Moreover, it should be realised that nitrate is reduced by oral micro flora to nitrite and, consequently, is an indirect

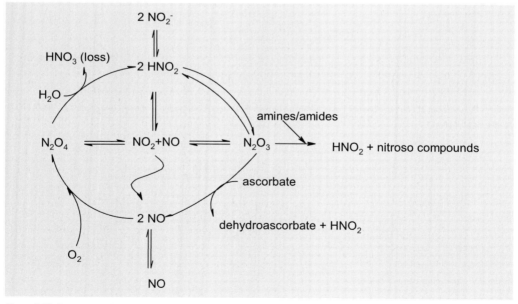

Figure 3.10. Reactions and equilibria of nitrogen oxides.

*Chemical hazards*

source of an endogenous nitrosating agent. Some studies indicate that this indirect route of nitrite exposure is far more important than the direct intake of nitrite via cured foods. The question arises, however, as to whether endogenously-generated nitrite might result in endogenously formed N-nitroso compounds. Some *in vitro* studies indicate the potential for endogenous nitrosation in the stomach, although they were not conclusive. Probably the presence of pre-formed N-nitroso compounds in cured food products is a greater health risk than the formation of N-nitroso compounds in the digestive system.

Although epidemiological evidence is weak, it was recently suggested that N-nitroso compounds increase risk for cancer. Moreover, the evidence that cured food products would increase the risk of stomach or pancreatic cancer was considered to be insufficient.

## 3.5 Residues

The presence of residues in food may result from the application of a particular chemical during agricultural production, without the intention that the involved compound or its degradation products remain present in the agricultural commodity when it is transformed into food. Consequently, pesticides and veterinary drug residues are involved. Residues may originate from cleaning and disinfection solutions used during agricultural or food processing practice. Finally, residues may also originate from the material brought into contact with food. These are the so-called migration residues. In the following section the pesticide and veterinary drug residues will be highlighted, followed by the migration residues.

### 3.5.1 Pesticides

*Introduction*
According to the Codex Alimentarius of the FAO/WHO, pesticides are considered as chemical agents that prevent, suppress, destroy, and repel particular pests during the production, storage, transport, or processing of food, feed and agricultural commodities. In addition, these chemical substances may also be used as growth regulators for plants, as defoliants, as withering agents or to inhibit sprouting. In conclusion, pesticides should not be considered only as biocides since they may also regulate the growth of a plant or act as a repellent. In the first instance, pesticides are used to ensure potential production in agriculture. Nevertheless, it is evident that due to the use of pesticides, the yield in agricultural production increases.

Pesticides are classified according to the target organism which is causing the pest:
- insecticides (against insects), acaricides (against mites), nematicides (against nematodes), rodenticides (against rodents), molluscicides (against molluscs) and avicides (against birds);
- fungicides (against moulds), bactericides (against bacteria) and soil disinfectants;
- herbicides (against weeds) and plant growth regulators.

In the following section some insecticides, fungicides and herbicides will be highlighted.

*Insecticides*

Insecticides can be classified according to various criteria such as the place of application (soil or leaves), according to their systemic character in the plant, or according to their uptake by the insect. In this overview, however, a classification according to their chemical structure or origin will be used. Thus, the following five categories can be distinguished:
- organo chlorine insecticides;
- organic phosphorus insecticides;
- carbamate insecticides;
- botanical insecticides;
- synthetic pyrethroids.

Of course, other classes of insecticides have been developed as well, but these will not be discussed in the framework of this contribution.

The organo chlorine insecticides should be considered as the start of the development of synthetic pesticides. These substances are characterised by their high chemical and biochemical stability. Combined with their hydrophobic character, this resulted in a bioaccumulation of these substances in humans and animals. The presence of organo chlorine insecticides in mother milk can be used as an indicator of the exposure of a population towards these substances. Because of their persistent character the use of several organo chlorine insecticides is prohibited in many industrialised countries. Despite its disadvantages, DDT has proven to be very efficient in the prevention of malaria and, therefore, it is still used in some tropical countries.

Organo chlorine insecticides interfere with the central nervous system of insects and exhibit a low selectivity. They are relatively toxic for mammals (Table 3.7) and very toxic for fish. Because of their excessive use throughout the years, several cases of resistance have been reported.

Three classes of organo chlorine insecticides can be distinguished:
- DDT group, including the DDT isomers and metabolites such as DDE, which is the typical persistent metabolite found in humans and animals;
- the Hexachlorine Cyclo Hexane group, including lindane and its isomers;
- the cyclodiene group including aldrin, dieldrin, etc.

*Table 3.7. Acute toxicity data of some organo chlorine pesticides.*

| Substance | $LD_{50}$ (rats) in mg/kg |
|---|---|
| DDT | 200-500 |
| γ-lindane | 150-230 |
| aldrin | 10-70 |
| dieldrin | 40-100 |
| endosulfan | 100 |

## Chemical hazards

The chemical structures of some of these substances are given in Figure 3.11.

Four different chemical classes of organo phosphorus insecticides can be distinguished as indicated in Figure 3.12. The nature of the residual R groups can be very diverse, resulting in a high amount of possible active substances, which moreover exhibit a large spectrum of physicochemical and biological properties.

The working mechanism of organo phosphorus insecticides is based on the inhibition of a crucial enzyme in the central nervous system, acetylcholine esterase. Due to this inhibition, the degradation of the neurotransmitter acetylcholine is inhibited resulting in an overload of the nervous signal transmission system. Since this neurotransmitter and this enzyme are also endogenous to mammals, organo phosphorus insecticides exhibit in general a higher acute toxicity in these organisms then organo chlorine compounds (Table 3.8). Due to the high

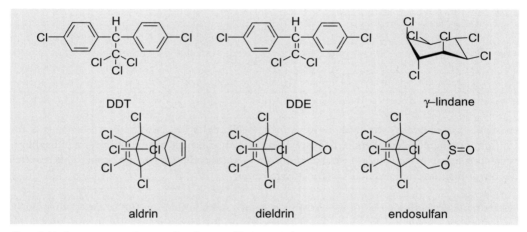

Figure 3.11. Some representative examples of organo chlorine pesticides.

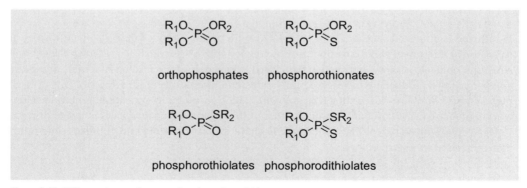

Figure 3.12. Different classes of organo phosphorus insecticides.

Table 3.8. Chemical structure of some selected phosphorus insecticides and their acute toxicity.

| Name | Structure | $LD_{50}$ (rats) in mg/kg |
|---|---|---|
| mevinfos | | 3-7 |
| dichlorvos | | 25-60 |
| parathion | | 6-13 |
| malathion | | 1200-2800 |
| demeton-S-methyl | | 2-60 |

variability in chemical structure, however, some selectivity can be introduced because of the different metabolic degradation pathways. Due to their chemical structure, organo phosphorus insecticides are moreover less persistent and do not accumulate in the fat tissue. This resulted in fewer residues in both food and the environment.

Carbamate insecticides exhibit the same working mechanism as the organo phosphorus insecticides. Nevertheless, their acute and chronic toxicity is generally lower (except for aldicarb). Moreover, they exhibit a higher stability resulting in sustained activity. Due to their systemic character, the use of carbamate insecticides has increased. As indicated in Figure 3.13, three different classes can be distinguished.

Pyrethrine is a natural insecticide of botanical origin produced by *Chrysanthemum* flowers. It consists of a mixture of various compounds such as pyrethrin I (Figure 3.14) of which the general structure consists of a cyclopropane structure. Because of its restricted stability, synthetic analogues were produced which were more stable and often even more effective. Due to their high activity, only low doses should be applied, which results in low residual amounts. Their working mechanism is probably similar to that of the organo chlorine insecticides, resulting as well in reported resistance. The structure and acute toxicity data of some of these compounds is summarised in Figure 3.14.

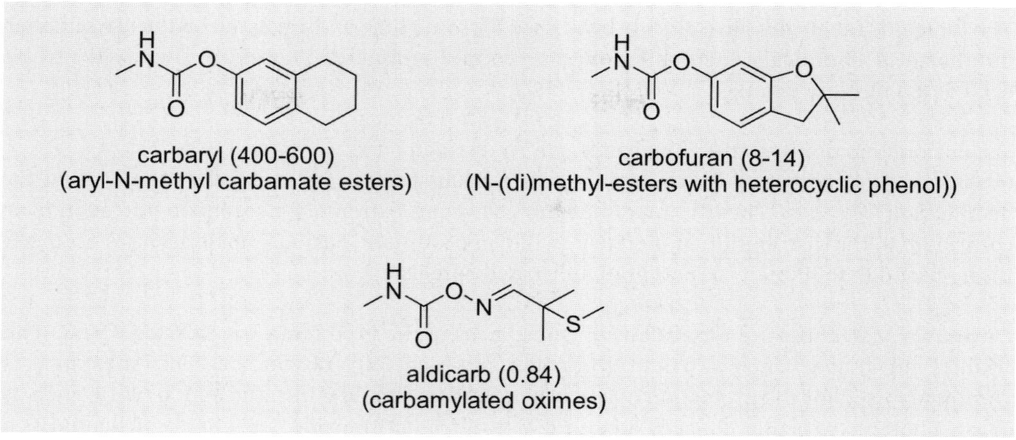

*Chemical hazards*

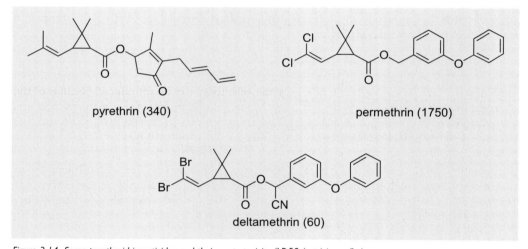

Figure 3.13. *Some carbamate insecticides, their class and their acute toxicity (LD50 (rats) in mg/kg).*

Figure 3.14. *Some pyrethroid insecticides and their acute toxicity (LD50 (rats) in mg/kg).*

*Fungicides*
Fungicides are applied in order to prevent plant diseases that are caused by moulds. Again, different classification systems are applied, but as for the insecticides, the restricted overview given here will be based on the chemical structure. Thus, the following classes can be distinguished:
- inorganic fungicides;
- organic non-systemic fungicides or contact fungicides;
- organic systemic fungicides.

The inorganic fungicides have already been used for some time and are based on the use of rather non-harmful chemicals such as sulphur or copper sulphate. Therefore, they will not be considered in more detail.

Various organic non-systemic fungicides can be distinguished:
*Organo mercury* compounds are broad active substances that are currently mainly used for disinfection of seeds. These compounds are, however, extremely neurotoxic and moreover contribute to the distribution of mercury in the environment. Serious incidents have occurred in the past due to human consumption of treated cereals.

*Organo tin* substances are tetravalent tin substances of the type $RSnX_3$, $R_2SnX_2$ or $R_3SnX$. The trialkyl substances, such as tributyltinoxide, are especially phytotoxic. This substance is commonly used for the protection of constructions that should withstand a lot of wear, such as ships. There is particular concern about the use of tributyltinoxide, since it is considered as a xeno-oestrogenic compound. These are compounds that mimic the female sex hormone oestrogen and are associated with the increased incidence of testes cancer and decreased male fertility. It has been shown that xeno-oestrogenic substances may also affect the marine and aqueous fauna. In this respect the hormone-like toxicity is observed especially if the compound is present in very low concentrations.

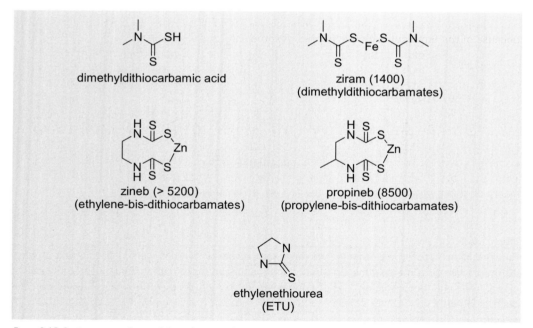

Figure 3.15. *Basic structure of some dithiocarbamates, the parent molecule and the important degradation product ETU, together with some acute toxicity data (LD50 (rats) in mg/kg).*

## Chemical hazards

*Dithiocarbamates* are probably the most commonly used fungicides in the world. Three different classes can be distinguished which are all derived from dimethyldithiocarbamic acid, as indicated in Figure 3.15. These substances are quite persistent on plants due to their low solubility in water and, therefore, long waiting periods should be respected especially if these substances are applied on leafy vegetables. The working mechanism is based on the binding of the electrophilic thiogroups with various substances or the binding of essential oligominerals. Serious toxicological concern has arisen about the presence of ethylenethiourea (ETU) in some formulations as a side product. This substance is a suspected carcinogen. Moreover, ETU may be produced from ethylene-bis-dithiocarbamates during metabolisation or degradation during storage or food preparation.

*Aromatic hydrocarbons* such as biphenyl, which is applied on citrus fruits, have a long tradition of use. Again, however, the use of chlorinated aromatic hydrocarbons such as pentachloronitrobenzene and pentachlorophenol has resulted in problems with regard to their persistency as already addressed. Pentachlorophenol is typically used as a wood protection agent. Due to the use of contaminated wood shavings in poultry houses, residues of metabolites were found in chickens, causing a typical musty taste deterioration of the meat. Moreover, during the production of these chlorinated aromatic hydrocarbons, dioxins can be produced, as will be discussed later in more detail.

The *phthalic imides and sulphamides*, such as captan (Figure 3.16), are released during their metabolisation by the mould thiophosgene, which is a highly reactive compound, resulting in a broad-spectrum agent to which no resistance has been found. Their use is however restricted because of the teratogenic characteristics of some similar compounds.

*Dicarboxyimides* such as iprodion (Figure 3.16) are especially used in vegetable and plant production. Due to its intensive use in grape production, residues may cause production problems. The use of vinclozolin has been banned in the UK, since it may act as an androgen-receptor blocker, and therefore might interfere with sexual development in boys. Its use was prohibited because of the risks for workers applying this particular compound in greenhouses.

Organic systemic fungicides have been developed since the '60s. Several classes can be distinguished, which cannot be discussed in full detail in this textbook.

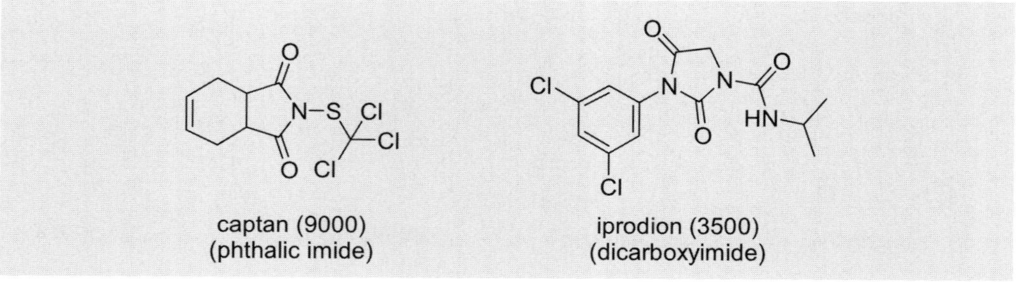

captan (9000)
(phthalic imide)

iprodion (3500)
(dicarboxyimide)

*Figure 3.16. Chemical structure and acute toxicity of a phthalic imide and dicarboxyimide fungicide (LD50 (rats) in mg/kg).*

The *benzimidazoles*, with carbendazim as the most important example, are typical systemic leaf fungicides. The working principle is based on the inhibition of cell replication and DNA synthesis. Due to this working principle, some toxicological concern about carbendazim residues has arisen. Supplementary attention to the benzimidazoles has been given because of their persistent character, both in the plant as well as in the soil.

Other systemic fungicides include several classes of ergosterol biosynthesis inhibitors, phenylamides, organo phosphorus fungicides, etc.

*Herbicides*
Herbicides are typically classified according to the type of target plant (mono- or dicotyls), and in addition according to their systemic character and place of application (soil or leaves). This classification, however, is not so straightforward in the framework of this restricted overview. Therefore, a number of selected herbicides will be discussed in somewhat more detail, without any further classification. Chemical structures of the treated examples are given in Figure 3.17.

*Dinitrophenols* are used as leaf contact herbicides for the elimination of dicoltyls. These substances are quite toxic for humans ($LD_{50}$ 25-65 mg/kg.day, depending upon the compound considered). Moreover, the use of dinoseb in particular, has been questioned because of its proven teratogenic effects.

*Phenylcarbamates* are used as herbicides and growth controlling agents. The best-known example is probably chlorpropham, which is used to inhibit sprouting in potatoes. Its use is also under pressure because of presumed genotoxic effects.

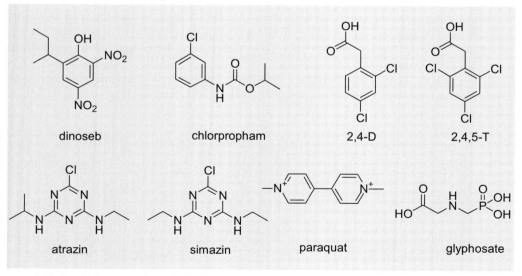

Figure 3.17. Chemical structure of some herbicides.

## Chemical hazards

*Phenoxyalkyl carbonic acids* have a working principle that is comparable to the plant growth hormone, indolacetic acid. They are typically used for the elimination of dicotyls. The famous 2,4D and 2,4,5 T were used as well, however, on a large scale during the Vietnam War as a defoliant (Agent Orange). These compounds are not persistent and have a moderate toxicity. It seemed, however, that quite high amounts of the toxic 2,3,7,8-tetrachlorodibenzo-p-dioxine and other related compounds were present as impurities (up tot 30 ppm). Nowadays, chemical synthesis is better controlled in order to avoid such high dioxin concentrations.

*Phenylurea* substances, such as diuron, are broad-spectrum herbicides which are typically quite persistent in soils due to absorption into humous substances.

*Triazines* such as atrazin and simazin are N-heterocyclic compounds. Maize is quite resistant to atrazin, resulting in extensive use of this particular herbicide in this crop. Because of its high water solubility however, groundwater and surface water have become contaminated with atrazin residues. Simazin on the other hand has low water solubility, resulting in a quite persistent character.

*Quaternary ammonium compounds* are broad-spectrum herbicides as well, which carry a positive charge. Therefore, persistency in clay soils may be extremely high. Paraquat is toxic for humans ($LD_{50}$ 150 mg/kg) and death due to accidental intake of high amounts is almost inevitable. Glyphosate is an *organic phosphorus herbicide* which is better known as Roundup. It is particularly well known because of the development of genetically modified soy and maize, which are glyphosate-resistant.

*Legal aspects*
The use of pesticides is typically regulated on a national basis, also within the EU. This legislation includes the specification of the use of particular products on particular crops. The principle of the positive list is also applicable, as in the case of food additives. In order to bring a new product on the market, a full toxicological evaluation must be presented to the appropriate authorities. A very important element in the legislation includes the specification of maximal residue limits, or so-called MRL values. Attempts are in progress to harmonise these MRL levels on an EU basis. In addition, on an even more international level, the FAO/WHO Codex Alimentarius Commission is trying to issue MRL's for pesticides.

An MRL value is calculated on the basis of the ADI level, taking into account the average bodyweight (60 kg) and maximal daily consumption of a particular crop (400 g). Thus, the following tolerance level is calculated

$$\text{Tolerance level (mg/kg)} = \frac{\text{ADI (mg/kg)} \times 60}{0.4} \qquad 3.3$$

As can be seen in Table 3.9, however, it is obvious that there is no straightforward relationship between the ADI and the MRL levels. This is because residual amounts of pesticides should be kept as low as good agricultural practice allows. This implies that the use of a particular pesticide should be kept as low as is necessary to protect the crop from pests. Therefore, the

Table 3.9. ADI and MRL levels of various pesticides in Belgium.

| Pesticide | ADI (mg/kg.day) | MRL (mg/kg) |
|---|---|---|
| endosulfan | 0.008 | 0.5 |
| dichlorovos | 0.004 | 0.1 |
| DDT | 0.005 | 0.1-1 |
| permethrin | 0.05 | 1 |
| carbaryl | 0.01 | 1.2-2.5 |
| thiram | 0.005 | 2 |
| zineb | 0.05 | 2 |
| captan | 0.010 | 15 |

tolerance levels are typically corrected on the basis of field trials in which the minimal necessary dose is determined experimentally. Therefore, it is quite important to realise that if the MRL is exceeded, this does not imply that there is an immediate risk for the consumer.

Finally in the legislation, some withdrawal periods can be specified after the application of a particular pesticide.

*Residues of pesticides: a real threat?*
If consumer exposure to pesticide residues is to be considered, it is obvious that only the edible parts of the agricultural commodity should be taken into account. This is particularly important, since some parts of the crop are more contaminated then others, and in some cases these parts are removed before consumption. In addition, it should be realised that the residual load may be influenced by food preparation as well, as illustrated in Table 3.10.

Results of surveys conducted in Belgium (1997) are summarised in Table 3.11.

It is obvious that most problems occur in leafy vegetables. The use of bromide substances as a soil disinfectant in greenhouses should be taken into particular consideration in this respect. Problems also occur with pesticides whose use is not allowed on a particular crop. In general, however, it can be concluded that the number of samples having residues higher then the MRL is relatively low. Similar results are obtained in other countries.

Table 3.10. Pesticide residues (mg/kg) in various vegetables as a function of processing in spinach.

| Pesticide | untreated | washing | blanching | sterilisation |
|---|---|---|---|---|
| captan | 4.71 | 1.84 | not detectable | not detectable |
| methoxychlor | 13.03 | 8.21 | 9.44 | 11.03 |
| parathion | 0.38 | 0.33 | 0.22 | 0.41 |
| malathion | 0.23 | 0.14 | 0.06 | 0.07 |
| zineb | 2.70 | 1.12 | 1.99 | 1.28 |

## Chemical hazards

Table 3.11. Results of residue survey in Belgium.

| Samples | National survey | | EU survey | |
|---|---|---|---|---|
| | number | % | number | % |
| All samples | | | | |
| ■ total samples | 1244 | | 263 | |
| ■ without residues | 587 | 47.2 | 198 | 75.3 |
| ■ residues < MRL | 530 | 42.6 | 64 | 24.3 |
| ■ residues > MRL | 127 | 10.2 | 1 | 0.4 |
| ■ residues > EC MRL | 43 | 3.5 | 0 | 0 |
| Without leafy vegetables | 722 | | | |
| ■ residues > MRL | 45 | 6.2 | | |
| Leafy vegetables | 522 | | | |
| ■ residues > MRL | 82 | 15.7 | | |

Residues in animal products are typically due to persistent organo chlorine residues, of which the main ones have been discussed previously.

Generally, it can be concluded that the perceived risk of the presence of pesticides by the consumer is far bigger then the actual risk. It has been suggested that the environmental risk may be more important compared to the risk for human health. Nevertheless, surveillance and monitoring programmes are still necessary in order to make sure that the legislation is still implemented so that food safety is assured.

### 3.5.2 Veterinary drugs

*Introduction*
Due to more intensive animal production, the importance of the administration of veterinary drugs has grown over the years. Animals can be treated with drugs for four reasons. Therapeutic agents are used to control infectious diseases caused by pathogenic microorganisms, parasites or fungi in farm and domestic animals. Typically, individual animals are treated or, if a larger group of animals is involved, administration can be accomplished via the feed.

Due to the increasing intensification of animal production, however, the prophylactic use of drugs as a precaution against disease has been introduced as well. As a matter of fact, the development of intensive animal production systems would probably not have been possible without the prophylactic use of drugs. These agents are usually administered via the feed.

Two kinds of growth-promoting substances can be distinguished. Some antibiotics, mixed with the feed at sub-therapeutic concentrations, alter the microbiological flora of the gut, resulting in increased feed conversion ratios. Anabolic growth promoters and β-agonists exert their effects via the animal's metabolism. Administration can be accomplished via an implant in the animal's ear (in countries were the use is allowed) or via injections (in countries where the use is prohibited).

Finally, drugs are also used in order to facilitate herd and flock management. Hormonal drugs, for instance, are also used to control reproduction on animal farms. Tranquillisers are used on a large scale as well, in order to reduce stress and aggressive behaviour.

These four goals can be accomplished by the use of one or more active compounds, which can be classified as follows:
- antimicrobial and antibiotic agents;
- anabolic agents;
- anthelmintic agents;
- coccidostats;
- tranquillisers and β-agonists;
- non-hormonal growth promoters.

Some of these groups will be discussed in the following section.

*Antibiotics, anthelmintics and coccidostats*
Antimicrobial and antibiotic agents are generally used as therapeutic, prophylactic or growth-promoting agents. For the treatment of tapeworms, roundworms and flukes, anthelmintics are typically used in cattle, sheep and pigs. Coccidia are intracellular protazoal parasites that are found in intestinal epithelial cells or other tissues such as the liver. Typically, poultry is susceptible to this kind of parasite. In the following section some important classes, and legislative and safety aspects will be considered.

As can be observed from Figures 3.18 and 3.19, various classes of antibiotic or related compounds are used. These are usually classified according to their chemical structure.

Sulphonamides were developed for human medicine, but are currently extensively used for the treatment of systemic bacterial diseases in animals. Several hundred individual compounds have been synthesised (Figure 3.18). The use is particularly widespread in pig farming for the control of pneumonia. Sulphonamides have also been shown to have coccidostatic activity.

β-Lactams and β-lactamase inhibitors such as clavulinic acid are also broad-spectrum antibiotic substances. The chemical structure of the penicillins and cephalosporins is based on the core β-lactam ring to which either a five-membered thiazolidine (penicillin) or a six-membered thiazine ring is bound (Figure 3.18). These substances are typically used for the treatment of mastitis in dairy cattle.

Tetracyclins (Figure 3.18) were developed after sulphonamides and penicillins. These broad-spectrum antibiotics are used therapeutically in humans, animals and fish, and at subtherapeutic levels for growth promotion. Their intensive use in the farming of salmon and trout has been questioned because of the large doses required to control the spread of infections.

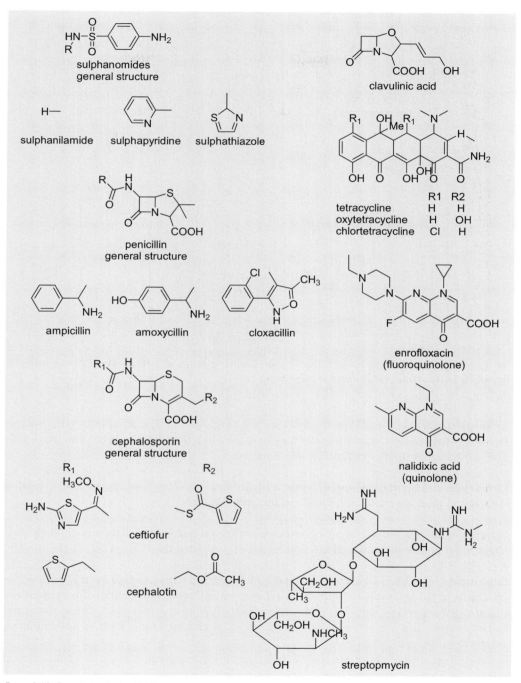

Figure 3.18. Some important antibiotics.

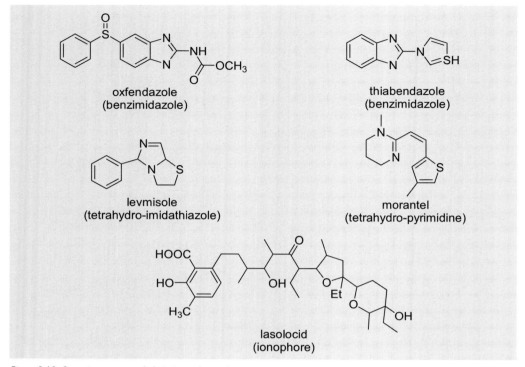

Figure 3.19. Some important anthelmintics and coccidiostats.

Aminoglycosides such as streptomycin (Figure 3.18) exhibit antibacterial properties towards Gram-negative bacteria. Again, initially they were used in human medicine, but nowadays wide use in the treatment of animal diseases is reported.

Quinolones and fluoroquinolones (Figure 3.18) are broad-spectrum antibiotics widely used to control and treat infections in farmed species, including fish.

Nitrofurans and chloramphinecol are broad-spectrum antibiotics as well, but due to their toxic character, their use is prohibited in some countries, including the EU.

Benzimidazoles (Figure 3.19) are very important substances with regard to the control of worm infections in cattle, sheep and horses. Tetrahydro-imidathiazoles and tetrahydro-pyrimidines (Figure 3.19) are widely used for the control of stomach, intestinal and long worm in sheep and cattle. Avermectins are active against both ecto- and endoparasites, such as anthelminths, nematodes, ticks and mites.

Apart from the sulphonamides already mentioned, several other classes of compounds act as coccidiostats. The ionophores (Figure 3.19) are probably the largest group. In addition, these compounds are used as growth promoters in beef cattle.

Within the EU, harmonised legislation with regard to the application of veterinary drugs is being constructed. Within this legislation, compounds are listed in four annexes:
- annex 1 for pharmacologically active substances for which an MRL has been fixed;
- annex 2 for substances not subjected to an MRL;
- annex 3 for pharmacologically active substances used in veterinary medicine for which a provisional MRL has been issued;
- annex 4 for pharmacologically active substances for which no MRL can be fixed.

Substances in annex 4 exhibit too strong a toxicity in humans and are therefore prohibited (e.g. chloramphinecol). In the other lists, the application of particular substances is specified in a particular animal. In addition, they should be considered as positive lists (except annex 4). Similar as to the pesticide legislation, MRL levels are specified based on toxicological data and good agricultural practice. In addition, withdrawal periods are included.

Typically, three kinds of risks are associated with the use of antibiotics or similar compounds. First of all, there is the toxicological risk for the human being itself. Some drugs are considered to be too toxic and are prohibited from use in particular countries. Sometimes, imported animal products seem to contain these prohibited substances and in accordance with the legislation within the EU these shipments are refused.

Another potential risk arises from the allergenicity to particular antibiotics. Allergic reactions to penicillins deserve a special mention.

A serious problem related to the large-scale use of antibiotics, as in fish farms, is the risk of the development of resistant strains of bacteria. This resistance problem is becoming a major problem in human medicine. Due to the presence of antibiotic residues in food, a supplementary change in the bacterial flora of the human gut may be induced as well.

Finally, the presence of antibiotic residues in agricultural commodities that undergo fermentation (cheese, yoghurt, sausage) may create serious technological problems.

*Anabolic agents and β-agonists*
These substances are used because of their growth-promoting characteristics. As regards anabolic agents, both steroidal and non-steroidal anabolics are used. As for the steroidal compounds, both the naturally-occurring sex hormones (testosterone, progesterone) are administered as their synthetic analogues. Some examples are shown in Figure 3.20. As regards the non-steroidal compounds, the stilbenes have an especially bad reputation because of their toxicity.

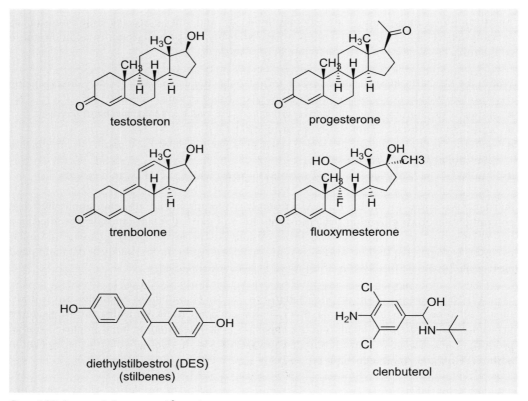

Figure 3.20. Some anabolic agents and β-agonists.

For the β-agonists, clenbuterol (Figure 3.20) is very well known. These substances will induce a redistribution of fat to muscle tissue, if they are used over a prolonged period.

Apart from the substances already mentioned, anti-thyroidic substances should be discussed. Despite the fact that they increase the weight of the muscle tissue, the resulting meat is of an extremely bad quality (low water-holding capacity), which explains the bad reputation of the use of 'hormones' in meat production.

The use of growth promoters within the EU is forbidden because of fears about the health effects of residues. However, the hormone concentrations in animals treated with naturally-occurring steroids are within the same range as for non-treated animals. For the synthetic analogues, MRL levels have been specified based on toxicological data, in countries where the use of these substances is allowed. As mentioned before, the use of DES and other stilbenes is prohibited because of the clear toxicity of DES in humans. For β-agonists, several human intoxications have been described.

*Chemical hazards*

Typically, the administration of growth promoters is accomplished via an implant to the base of the animal's ear. At slaughter, the ear is discarded to prevent contamination of the food by residual amounts of the drug. In countries where the use of these substances is prohibited, administration is accomplished via injection. It is obvious that the highest concentrations of the active compound can be found at the injection side, but since the administration is illegal, the injected muscle enters the food chain as such. For this and other reasons, it seems that the consumer is arguably at greater risk from hormone residues than when some of these products were licensed.

Since the use of these substances is forbidden in the EU, a stringent surveillance system has been introduced. It should be realised, however, that the number of positive samples is quite low (<1 %) and is still decreasing.

### 3.5.3 Migrants from food contact materials

*Food contact materials*
During the handling of agricultural raw materials, during their processing and transformation into foods and during the transport of these products from the producers to the consumers, contact with other materials frequently occurs. The most common example for the end user of the food is probably the packaging material. However, apart from a variety of packaging materials, a lot of other contact materials should be considered as well, e.g. stainless steel processing, transport or storage equipment, tubing for food transport, sealing materials in piping equipment, protection foils or lacquers used in storage facilities, etc. These contact materials can be of a varying chemical nature, as indicated in Table 3.12.

Of course, packaging materials represent probably the most important group within the food contact materials listed in Table 3.12. And within the range of packaging materials, polymeric packaging materials are used in particular, e.g. in plastics, or in combination with other materials, e.g. the coating on a tin can or a component in a multilayer cardboard material.

*Table 3.12. Groups of contact materials requiring legislation within the EU.*

| | |
|---|---|
| 1. | Plastics including varnish and coatings |
| 2. | Regenerated cellulose |
| 3. | Elastomers and rubber |
| 4. | Paper and board |
| 5. | Ceramics |
| 6. | Glass |
| 7. | Metals and alloys |
| 8. | Wood, including cork |
| 9. | Textile products |
| 10. | Paraffin and micro-crystalline waxes |

Therefore, in the following section the migration from polymeric food contact materials will be considered, although it should be noted that other materials may also exhibit migration.

Plastics can be regarded as macromolecular organic compounds which can be produced synthetically or by modification of naturally-occurring products, for example regenerated cellulose. Various kinds of plastics may be used, as indicated in Figure 3.21.

Apart from the fundamental component -the polymer- plastics contain other chemical components as well. These so-called additives alter the properties of the polymer in a desired way. Table 3.13 shows the main additives used in plastic manufacture together with their function. Generally, these additives are applied in relatively small concentrations, although fillers (e.g. silica) and plasticisers (e.g. phthalates) are used at high concentrations as well. Apart from the additives, processing aids, necessary to the polymerisation process, may also be present.

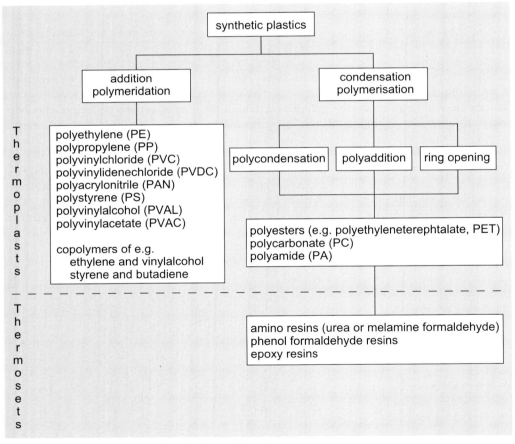

Figure 3.21. Overview of the main synthetic plastic materials used.

## Chemical hazards

*Table 3.13. Additives for plastics and their function.*

| Additive | Function |
|---|---|
| 1. Nucleating agents | Induce regular crystallisation |
| 2. Lubricants | Improve processing above glass transition temperature, alters rheology |
| 3. Anti-static agents | Reduce the chargeability of the plastic |
| 4. Blowing agents | Generate inert gases to produce expanded plastics |
| 5. Plasticisers | Gel the polymer, improve flexibility and processibility |
| 6. Anti-fogging agents | Avoid water droplets on films used to pack moisture rich foods |
| 7. Dyes and pigments | Impart colour of the plastic |
| 8. Fillers and reinforcing agents | Increase bulk and improve physical properties |
| 9. Stabilisers | |
| 9.1. Antioxidants | Avoid polymer oxidation by trapping free radicals or by inducing decomposition of peroxides |
| 9.2. UV absorbers | Reduce harmful effect of UV radiation |
| 9.3. Heat stabilisers | Prevent dehydrochlorination during processing of PVC |
| 9.4. Anti-acids | Neutralise acids arising from catalysts or PVC thermo degradation |

Typical examples include: surfactants, initiators, cross linkers, catalysts, etc. In addition to the low molecular weight compounds, residual monomers and oligomers can be present within the plastic material. Finally, impurities, degradation or reaction products of plastic ingredients were found to be present in polymeric contact materials as well. Because all these low molecular weight compounds are not covalently bound to the polymer chain, they are able to diffuse throughout the polymer matrix. This diffusion is one of the basic processes of migration from plastic food contact materials.

*Migration from food contact materials*
In a food packaging system, three phases should be considered: the food, the package and the environment (Figure 3.22). In between these phases interactions may occur, resulting in an energy or a mass transfer. The mass transfer can be macroscopic as in the chipping of a glass container or microscopic as in the contamination of food by microorganisms. The sub-microscopic mass transfer, however, involves the diffusion of individual molecules in one phase and their sorption by the other.

If the mass transfer involves the three phases of the packaging system, volatiles are transported from the environment via the contact material to the food or vice versa. This phenomenon is known as permeation. No net uptake or removal of chemical substances from the food contact material takes place. The permeation process may significantly affect the quality of the food. Mild preservation techniques such as modified atmosphere packaging, which are used successfully to prolong the shelf life of minimally processed foods, are based on the selective permeation of particular gases through the packaging material. Chemical contamination due to permeation of organic volatiles (e.g. solvents) through the packaging material has been reported as well.

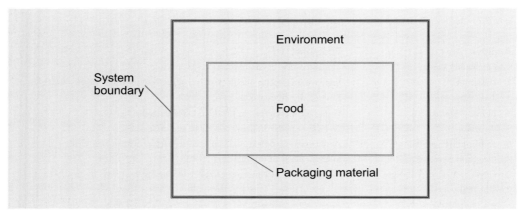

*Figure 3.22. The three phases of a packaging system.*

If mass transfer is restricted only to the food and the packaging material, the phenomenon is also known as migration. Migration can take place from the contact material to the food and vice versa. The latter case is also known as negative migration, while the former is simply identified as migration. A typical example of negative migration is the flavour scalping in fruit juices due to the partial absorption of flavour compounds by the plastic contact material. Due to this phenomenon the fruit juice aroma deteriorates. Another example resulting, however, in an improvement of food quality, is the use of oxygen scavenging materials in the packaging of foodstuffs sensitive to oxidation.

The mass transfer from the packaging material to the food can have both deteriorating and improving consequences for the food. Migration of toxic packaging compounds to the food is a serious risk to food safety. Similarly, migration of particular substances could induce sensorial deterioration of the food (e.g. styrene migration). On the other hand, migration of particular food additives such as antioxidants and anti-microbial agents, could improve the shelf life of the product and at the same time minimise the direct use of these additives in food manufacture.

More fundamental aspects with regard to the migration from plastics to foods are discussed in the chapter on modelling of risks. Since the fundamental mechanisms of migration from polymeric materials are fairly well understood, mathematical models can be formulated in order to predict the migration from these materials.

*Legislative aspects on the migration from plastics*
Given the principles of migration from plastics to foods, it is obvious that this phenomenon can be a food safety issue. Plastics may contain compounds which should be considered as carcinogens (e.g. vinylchloride, acrylonitrile) or which exhibit another type of toxicity. In order to ensure consumer protection, legislation has been developed over the years in various countries with regard to food contact materials in general, and plastic food contact materials in particular.

## Chemical hazards

Within the European Union, a process of harmonisation, started in 1972, tries to bring all existing legislation in the various member states in line or introduces new directives. Especially issues concerning plastic food contact materials got a lot of attention and were elaborated in this harmonization process. In addition to the European initiatives, legislation was developed in the United States. In this chapter, however, only the basic principles about the EU legislation on plastics will be presented.

Basically, the established legislation deals with the following aspects:
1. a list of authorised substances;
2. a restricted amount of migration;
3. a system of checking migration.

The list of authorised substances consists of monomers, other starting substances and most types of additives (except colorants and catalysts) which can be used for the production of plastic food contact materials. This list rules out the use of unlisted materials, so they should be considered as so-called positive lists.

Apart from the fact that compositional restrictions are imposed on plastic materials, the migrated amount of substances to the food should also be limited. Two general restrictions are applied:
- an overall migration limit;
- a specific migration limit.

The overall migration limit, set at 60 $mg.kg^{-1}$ of food or 10 $mg.dm^{-2}$ contact material is the total amount of substances which can migrate out of a plastic material to the food. This limit is set to ensure the inert character of the packaging material. In addition, it avoids specifying migration limit for every listed compound.

The specific migration limit refers to the restricted migration of particular substances of toxicological relevance. The specific migration limit (SML, [$mg.kg^{-1}$]) is calculated on the basis of the acceptable daily intake (ADI) or tolerable daily intake (TDI) laid down by the SCF for the particular substance. Supposing that 1 kg of the food containing the migrating substance is consumed daily by a 60 kg adult, the specific migration limit is given as:

SML = 60 x ADI
or
SML = 60 x TDI

The specific migration limit for a particular substance is specified in the positive lists issued by the European legislator.

Finally, the EU legislation specifies in which practical manner the migration from a plastic material can be estimated from laboratory experiments. Therefore, food simulants are defined, depending for example on the lipophilic or hydrophilic character of the food making contact with the material. Furthermore, specifications with regard to contact time and contact temperature are made. More details with regard to the practical methodology of estimating the potential migration from a plastic food contact material fall out of the scope of this textbook.

## 3.6 Environmental contaminants

### 3.6.1 Introduction

Industrial processes emit several thousands of inorganic and organic chemicals. Due to their emission, agricultural commodities, and thus also our food, may become contaminated with these compounds. It would not be feasible to monitor all the possible environmental contaminants in our food. Therefore, a simple and pragmatic scheme was developed in order to select those chemicals that are considered to be of prime importance. These criteria include the following:
- production volume, since the amount of chemicals emitted is probably related to the total amount produced;
- pattern of usage or emission, since for example highly diffuse usage or emission would affect a higher amount of commodities;
- possible fate in the environment, since some contaminants may accumulate in the water, the soil or other environmental compartments;
- likelihood of entering the food chain;
- mechanism of entry into the food chain;
- persistence in the food chain, which is a key issue because of the problem of bioaccumulation;
- toxicity.

Consequently, a restricted overview of some important environmental contaminants will be presented. Both organic and inorganic (heavy metals, nitrate) will be considered. Some of them (nitrate, polyaromatic hydrocarbons) are discussed in other sections, since they may also be present in the food for other reasons than those mentioned here.

### 3.6.2 Aromatic hydrocarbons

Aromatic hydrocarbons include benzene and alkylated benzenes such as toluene, ethyl benzene, xylenes and naphthalene (Figure 3.23). Due to the introduction of lead-free fuel, motor vehicle exhaust gases are probably the main contributor to the release of toluene and, to a lesser degree, of the other aromatic hydrocarbons. The other aromatic hydrocarbons are in majority released by industrial solvents used in paints and adhesives.

Once released into the atmosphere, aromatic hydrocarbons may enter the food chain via various routes including direct absorption from the atmosphere by fatty foods. It has also been suggested that food on sale at petrol shops could contain higher concentrations of these compounds compared to similar foods on sale at other shops.

## Chemical hazards

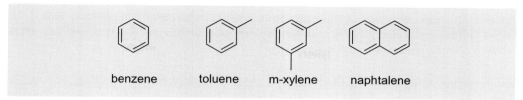

Figure 3.23. Some important aromatic hydrocarbons found in foods.

Apart from dietary exposure, however, exposure to these chemicals is to a large extent due to the presence of these chemicals in urban air. Benzene is considered as a genotoxic carcinogen, while other substances are considered to be suspect as well, although less toxicological data are available.

### 3.6.3 Dioxins and dioxin-like compounds

*Structure, physicochemical characteristics and sources*
Dioxins is a group name for two kinds of compounds, namely the polychlorinated dibenzo-*p*-dioxins (PCDD's) and the polychlorinated dibenzofurans (PCDF's). These compounds are planar, tricyclic compounds composed of two benzene rings, interconnected by respectively two and one oxygen atom, with at least one chlorine atom. (Figure 3.24).

Obviously, a high number of different PCDD and PCDF's exist, depending upon the degree of chlorination and the place of chlorination. Since these compounds all have similar physicochemical properties and a similar chemical structure, they are considered as so-called congeners. Consequently, theoretically 75 PCDD congeners and 135 PCDF congeners exist. All congeners are attributed with an Arabic number, according to an IUPAC protocol. Similarly, 209 polychlorinated biphenyl (PCB) congeners can be produced.

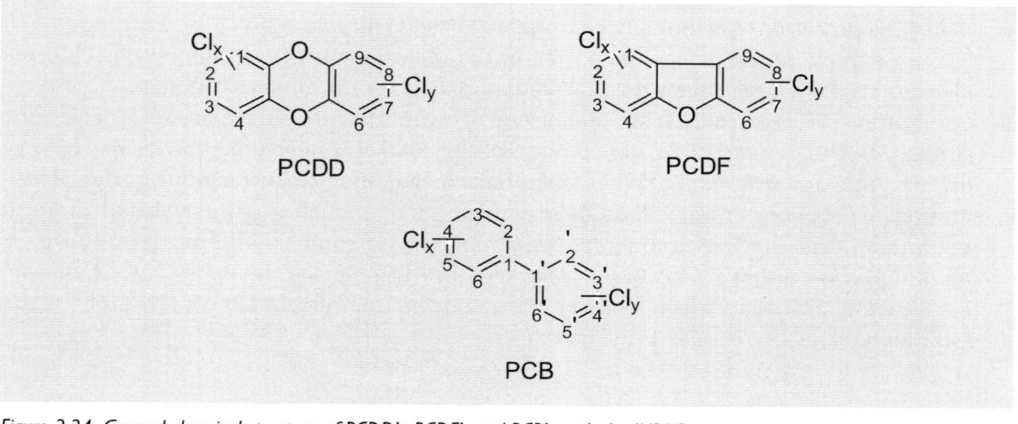

Figure 3.24. General chemical structure of PCDD's, PCDF's and PCB's, with the IUPAC numbering system.

Apart from dioxins and PCB's, similar compounds may also be present in the food chain, such as polybrominated biphenyls (PBB's) and polychlorinated napthalenes. Again various congeners of each of these classes of compounds exist.

All the compounds mentioned exhibit similar physicochemical properties, including a highly lipophilic character that is increased with the degree of chlorination. Like the organo chlorine pesticides, they are chemically and biochemically resistant towards degradation. As a result they tend to bioaccumulate and biomagnify in the food chain. The half-life of dioxins in the human body, for instance, is estimated to be about 10 years.

PCB's, PPB's and polychlorinated napthalenes were or are still produced industrially. The polychlorinated napthalenes were in fact the first group of dioxin-like compounds produced on a large scale. They were produced about 20 years before the introduction of PCB's, which subsequently became a main substitute for them. These oil-like fluids were used as heat transfer fluids in industrial installations such as transformers, as hydraulic fluids, plasticisers, lubricant inks, paint additives, etc. Because of their persistent and toxic character, industrial production was banned in 1979 in industrial countries, while in developing countries production should finish by 2006. Despite the fact that production has drastically decreased, older industrial installations may still contain large amounts of PCB oils in particular. Upon renewal of these installations, the oil should be selectively collected and incinerated in specially designed furnaces in order to avoid their emission into the environment.

Polybrominated biphenyls are also industrially produced and are typically used as flame-retardants. Dioxins on the other hand are not intentionally produced. Basically, three sources of PCDD's and PCDF's are known:

- Chemical manufacture; as already mentioned before, during the production of organo chlorine compounds, such as pesticides, dioxins may be produced as side products. Due to better control of chemical processes, however, new environmental contamination from this source is quite unlikely, but due to the persistent character of these compounds, an environmental load is still present.
- Bleaching processes; traditionally chlorine was used to bleach wood pulp, resulting as well in the production of dioxins. Again, due to the replacement of chlorine by non-chlorine bleaching agents, new environmental contamination via this route is unlikely.
- Combustion processes; during the combustion of organic material in the presence of chlorine substances, PCDD's and PCDF's are produced. The kind of combustion process may be very diverse, from cigarette smoke, to emissions from incineration or steel manufacturing plants, to volcano eruptions or forest fires. Stringent control on industrial plants helps to diminish the human causes of environmental contamination via combustion processes. However, natural processes that are beyond human control cause part of the emissions. Obviously, combustion processes are nowadays the major source of PCDD's and PCDF's in the environment.

*Toxicity*

Although all dioxins and PCB's exhibit similar physicochemical properties, not all of these compounds are considered as toxicants. Apparently, the degree of chlorination and even more importantly, the position at which the molecules are chlorinated, is very significant. Those PCDD's and PCDF's chlorinated at positions 2,3,7 and 8 exhibit the greatest toxicity. Consequently, out of the 75 PCDD congeners, only 7 congeners are considered to be toxicologically relevant. Similarly for the PCDF's, only 10 congeners are toxic (Table 3.14). For the PCB's, those molecules which exhibit a planar structure, such as dioxins, are toxicologically relevant. The non-ortho PCB substituted on both para and at least two meta positions, together with the PCB's chlorinated at only one ortho-position (mono-ortho PCB's) are known as dioxin-like PCB's (Table 3.15). Consequently, out of the 209 congeners, only 12 exhibit dioxin-like toxicity.

Dioxin and dioxin-like compounds may induce cloracne (typically during incidental high exposure), endocrine disruption, immunotoxicity, neurological alterations and teratogenic effects. The 2,3,7,8-TCDD is classified as a known human carcinogen.

The mechanism of toxicity is probably related to the interactions these molecules show with the aryl hydrocarbon receptor in our cells. This receptor protein is a ligand-dependent transcription factor that induces transcription of various genes, resulting in toxicity.

Table 3.14. TEF values of PCDD's and PCDF's.

| PCDDs and PCDFs | Toxic Equivalency Factor (TEF) | |
| --- | --- | --- |
| | I-TEF (NATO/CCMS, 1988) | WHO-TEF (van den Berg et al., 1998) |
| 2,3,7,8-TCDD | 1 | 1 |
| 1,2,3,7,8-PnCDD | 0.5 | 1 |
| 1,2,3,4,7,8-HxCDD | 0.1 | 0.1 |
| 1,2,3,6,7,8-HxCDD | 0.1 | 0.1 |
| 1,2,3,7,8,9-HxCDD | 0.1 | 0.1 |
| 1,2,3,4,6,7,8-HPCDD | 0.01 | 0.01 |
| OCDD | 0.001 | 0.0001 |
| 2,3,7,8-TCDF | 0.1 | 0.1 |
| 1,2,3,7,8-PnCDF | 0.05 | 0.05 |
| 2,3,4,7,8-PnCDF | 0.5 | 0.5 |
| 1,2,3,4,7,8-HxCDF | 0.1 | 0.1 |
| 1,2,3,6,7,8-HxCDF | 0.1 | 0.1 |
| 1,2,3,7,8,9-HxCDF | 0.1 | 0.1 |
| 2,3,4,6,7,8-HxCDF | 0.1 | 0.1 |
| 1,2,3,4,6,7,8-HPCDF | 0.01 | 0.01 |
| 1,2,3,4,7,8,9-HPCDF | 0.01 | 0.01 |
| OCDF | 0.001 | 0.0001 |

Table 3.15. TEF values of PCDD's and PCDF's.

| PCBs (IUPAC number) | Toxic Equivalency Factor (TEF) | |
|---|---|---|
| | PCB-TEF (Ahlborg et al., 1994) | WHO-TEF (van den Berg et al., 1998) |
| Non-*ortho* PCBs | | |
| 3,3'4,4'-TCB (77) | 0.0005 | 0.0001 |
| 3,4,4',5-TCB (81) | - | 0.0001 |
| 3,3'4,4',5-PnCB (126) | 0.1 | 0.1 |
| 3,3'4,4',5,5'-HxCB (169) | 0.01 | 0.01 |
| Mono-*ortho* PCBs | | |
| 2,3,3'4,4'-PnCB (105) | 0.0001 | 0.0001 |
| 2,3,4,4',5-PnCB (114) | 0.0005 | 0.0005 |
| 2,3',4,4',5-PnCB (118) | 0.0001 | 0.0001 |
| 2,3,4,4'5-PnCB (123) | 0.0001 | 0.0001 |
| 2,3,3'4,4',5-HxCB (156) | 0.0005 | 0.0005 |
| 2,3,3'4,4',5'-HxCB (157) | 0.0005 | 0.0005 |
| 2,3',4,4',5,5'-HxCB (167) | 0.00001 | 0.00001 |
| 2,3,3',4,4',5,5'-HpCB (189) | 0.0001 | 0.0001 |
| Di-*ortho* PCBs | | |
| 2,2',3,3'4,4',5-HpCB (170) | 0.0001 | - |
| 2,2',3,4,4',5,5'-HpCB (180) | 0.00001 | - |

Because of the complex nature of dioxin and PCB mixtures, risk evaluation is very difficult. Therefore, the concept of Toxic Equivalent Factors (TEF value) was introduced. TEF indicates an estimate of the toxic potency of a compound related to the reference and most toxic compound 2,3,7,8-TCDD. Consequently, the latter compound is attributed a TEF value of 1. These TEF values are shown in Tables 3.14 and 3.15. It should be noted that in some cases there is as yet no scientific agreement with regard to the attribution of a TEF value to some compounds. As can be seen, PCB's are far less toxic compared to most of the PCDD's and PCDF's. Combining the results of chemical analysis with the respective TEF values for each toxic compound, the Toxic EQuivalent (TEQ value) can be calculated, which represents the total 2,3,7,8-TCDD equivalent toxic potency present in a particular sample:

$$TEQ = \Sigma_{i=1-7} (PCDD_i \times TEF_i) + \Sigma_{i=1-10} (PCDF_i \times TEF_i) + \Sigma_{i=1-12} (PCB_i \times TEF_i)$$

Since these compounds tend to bioaccumulate in the body, a tolerable weekly intake has been proposed by the WHO in the range of 7-20 pg/kg bodyweight for 2,3,7,8-TCDD. So, a maximal daily intake of about 1-3 pg TEQ/kg bodyweight can be tolerated.

*Exposure to PCDD's, PCDF's and PCB's*
Dietary exposure to these compounds accounts for more then 90 % of the total exposure in humans. In a European study, the mean daily exposure expressed in pg 2,3,7,8-TCDD/kg

bodyweight of dioxins amounted to 0.4-1.5, while for the dioxin-like PCB's an estimate of 1-3 pg TEQ/kg bodyweight was made. The $95^{th}$ percentile's exposure amounted to 2-3 times the mean exposure. This indicates that a significant proportion of the population is exposed to too high levels of dioxins and dioxin-like compounds, taking into account the above-mentioned TWI.

In this study the contribution of various foods to the total exposure was also estimated, as indicated in Table 3.16.

As can be observed, seafood contributes in a large part to the total dietary exposure. This is in agreement with the quite high contamination levels found in fish, which range for PCB's only from 10-75 pg dioxin-like PCB TEQ/g of fat. It should be realised in this respect that since the TEF values for PCB's are considerably lower then those for the dioxins, the real PCB contamination of fish oil is several orders of magnitude higher.

Due to all the measures already taken to reduce dioxin emission it can be observed, however, that as a function of time, exposure is decreasing. It should also be mentioned in this regard that, within the EU, MRL levels and action levels for food and feed have recently been specified.

Table 3.16. Contribution of various foodstuffs in the total dietary dioxin exposure.

| Food | Percentage |
|---|---|
| Seafood | 11-63 |
| Dairy products | 16-39 |
| Meat products | 6-32 |
| Vegetable products | 6-26 |

### 3.6.4 Inorganic environmental contaminants

Inorganic environmental contaminants are typically heavy metals, such as lead, mercury and cadmium. However, nitrate can also be considered as an inorganic contaminant due to its presence in surface and drinking water as a result of its extensive use in agriculture as a fertiliser. The nitrate and nitrite problem, however, has already been dealt with.

Lead exposure was typically caused by the use of lead pipes in water supply systems in the past. Due to the extensive used of tetraethyl lead in car fuel, lead emission in the environment increased intensively. Nowadays, non-leaded fuels are used instead. Other possible contamination routes are the use of lead in the soldering of some tin cans, the use of lead-containing crystal, and the environmental deposition of lead-contaminated dust on agricultural commodities. Only 10 % of ingested lead is taken up in the digestive system. Lead is stored in the bones of our body, but is in partial equilibrium with the lead present in the blood system. Chronic exposure to lead may result in anaemia, which is rarely observed in the case of food intake. More worrying effects on young children are the various neuropsychological indicators that show definite negative correlations with serum lead levels.

Mercury may give rise to incidental and serious poisoning caused by the consumption of treated grains, as already mentioned before. Environmental contamination however is typically caused by industrial pollution of water. In addition, natural processes such as volcano eruptions are thought to be important in the total emission of mercury to the environment. Mercury tends to bioaccumulate as methyl mercury into the food chain. Consequently, it is not surprising that seafood seems to be more highly contaminated than other foods. In fact, cases of mercury poisoning due to the consumption of fish caught in highly contaminated bays have been described. Toxicological effects are particularly situated in the central nervous system. Foetus exposure to lead may cause serious brain damage in the unborn child. Therefore, consumption of predatory fish such as tuna by pregnant women should be limited, because these fish are much more contaminated with methyl mercury other fish.

Cadmium toxicity was discovered in Japan, where rice paddies were irrigated with contaminated water due to mining operations. Due to the chronic exposure, an extremely painful demineralisation of the skeleton occurred, especially amongst post-menopausal women. Other industrial activities may also result in the emission of cadmium into the environment. In contrast to lead, cadmium is absorbed by plants and may accumulate in them. Sewage sludge and sludge collected from water treatment plants may be contaminated with cadmium as well. Other sources of cadmium include the use of cadmium-containing materials, which come into contact with food, although they are not intended for this purpose.

Within the EU, legal limits are only available for the heavy metals discussed here (EC 466/2001).

## 3.7 Process contaminants

During food manufacture, various chemical reactions occur giving rise to the production of desirable and undesirable compounds. Some of these undesirable compounds seem to exhibit a relevant toxicity. The number of compounds produced is yet unknown but a large amount of them have already been identified: polyaromatic hydrocarbons, heterocyclic amines, nitrosoamines, oxidised sterols, oxidised triacylglycerols, lysinsoalanine, etc. Recently, acrylamide was identified as a yet unknown process contaminant. Due to this discovery, it became clear that other toxicologically relevant, and as yet unknown substances, are probably produced during food processing.

In the following section, only three compounds or classes of compounds will be considered in somewhat more detail:
- polyaromatic hydrocarbons;
- heterocyclic amines;
- acrylamide.

## 3.7.1 Polyaromatic hydrocarbons

Polyaromatic hydrocarbons (PAH) are a large and complex group of substances with the common structural feature of two or more fused benzene rings, as shown in Figure 3.25. A kind of classification can be made according to the number of benzene rings present in the molecule. If the number of benzene rings exceeds four, the compounds are classified in the heavy fraction, otherwise in the light fraction.

The compounds are produced as the result of a pyrolysis of organic matter. Consequently, many sources of contamination can be identified, including industrial and geochemical activities. Therefore, PAH's are also considered as typical environmental contaminants. In this regard it should be noted that PAH's tend to bioaccumulate as well, especially those compounds in the heavy fraction. Apart from the environmental contamination, however, they can also be produced in food themselves, during heating (e.g. grilling), smoking or drying processes, for example. Possible chemical routes of their origin are shown in Figure 3.26.

Polyaromatic hydrocarbons are associated with carcinogenesis. However, not all compounds exhibit the same degree of toxicity. Therefore, a similar system for assessing the risk of PAH exposure as used for dioxins and dioxin-like compounds has been introduced. Using the most toxicologically potent PAH as a reference, i.e. benzo(a)pyrene, TEF values are attributed to various PAH's as indicated in Table 3.17. Again it should be noted that for some substances no agreement on the TEF value exists.

No TDI levels have been specified yet, but despite this fact, some MRL levels are used by particular industries. These range from 25 to 50 µg/kg for the total amount of PAH, and from 5 to 10 µg/kg for the heavy fraction only.

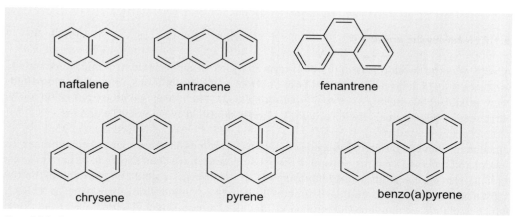

Figure 3.25. Chemical structure of some polyaromatic hydrocarbons.

*Bruno De Meulenaer*

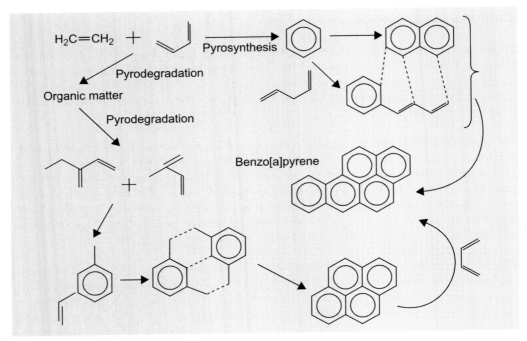

*Figure 3.26. Possible chemical routes for PAH formation.*

Apart from grilled and smoked foods, the major contribution to human exposure originates from oils, vegetables and fruits. In vegetable oils, deodorisation processes can remove the light fraction. If the heavy fraction also needs to be removed, active carbon treatments should be applied.

### 3.7.2 Heterocyclic amines

Shortly after the development of the short-term assay for the determination of mutagenic activity by Ames in 1975, it became clear that under normal cooking conditions, mutagenic compounds were produced in a variety of animal protein rich foods. The mutagenic properties could not be attributed to polyaromatic hydrocarbons, but were identified as heterocyclic amines.

Since then, about 20 different mutagenic heterocyclic amines have been identified in numerous model and real food systems. Several of these compounds have been shown to be carcinogenic in long-term animal studies and genotoxic in DNA repair tests as well. Two compounds are the most abundant present: 2-amino-3,8-dimethylimidazo[4,5-*f*]quinoxaline (abbreviated as MeIQx), which is an imidazoquinoline (IQ compound), and the 2-amino-1-methyl-6-imidazo[4,5-*b*]pyridine (abbreviated as PhIP), which is an imidazoquinoxaline (Figure 3.27). Generally, PhIP is produced to a higher extent then MeIQx, typically at a maximum level of 480 ng/g and 50 ng/g, respectively.

## Chemical hazards

Table 3.17. TEF values for different polyaromatic hydrocarbons.

| Component | TEF (Nisbet and LaGoy, 1992) | TEF (U.S. EPA, 1993) |
|---|---|---|
| dibenz[a,h]anthracene | 5 | 1 |
| benzo[a]pyrene | 1 | 1 |
| dibenzo[a,h]pyrene | - | 1 |
| dibenzo[a,i]pyrene | - | 1 |
| dibenzo[a,l]pyrene | - | 1 |
| benzo[b]fluoranthene | 0,1 | 1 |
| dibenzo[a,e]pyrene | - | 0,1 |
| benzo[j]fluoranthene | - | 0,1 |
| benzo[k]fluoranthene | 0,1 | 0,1 |
| benz[a]anthracene | 0,1 | 0,1 |
| indeno[1,2,3-c,d]pyrene | 0,1 | 0,1 |
| anthracene | 0,01 | 0,01 |
| benzo[g,h,i]perylene | 0,01 | 0,01 |
| chrysene | 0,01 | 0,01 |
| acenaphthene | 0,001 | - |
| acenaphthylene | 0,001 | 0,01 |
| fluoranthene | 0,001 | 0,01 |
| fluorene | 0,001 | 0 |
| 2-methylnaphthalene | 0,001 | - |
| naphthalene | 0,001 | - |
| phenanthrene | 0,001 | 0 |
| pyrene | 0,001 | 0 |

A complex combination of several chemical reactions may induce the production of these heterocyclic amines in cooked foods. The Maillard reaction, which is the condensation reaction between a reducing sugar and an amino compound and which results in brown coloration and aroma production, is thought to be of prime importance with regard to the production of imidazoquinolines or IQ compounds. In addition, however, typical reaction products of a side reaction of the Maillard reaction, known as the Strecker degradation, such as pyrazines and pyridines, are thought to react with carbonyl and amino compounds as well to give rise to

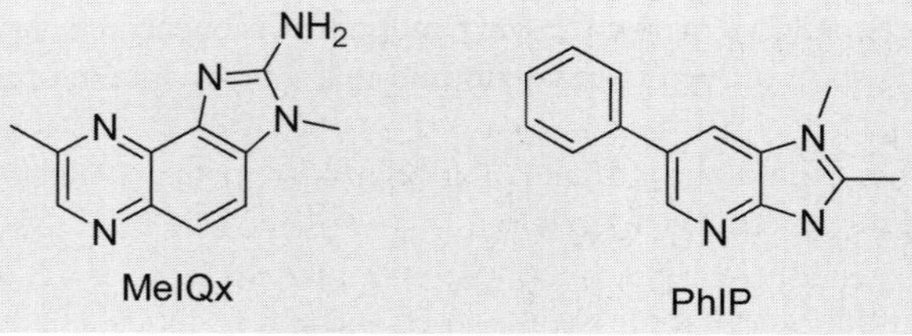

Figure 3.27. Chemical structures and common names of two heterocyclic compounds typically found in cooked foods.

heterocyclic amines. Typically, creatine and creatinine are supposed to play a central role in the production of IQ compounds (Figure 3.28). Phenylalanine and creatine are supposed to be important precursors for PhIP. Apart from phenylalanine, however, other amino acids may also be involved. Reducing sugars such as glucose were found to have a considerable influence on the formation of these compounds as well, although for the production of PhIP they are not considered to be essential. In the absence of carbonyl compounds, heterocyclic compounds seem to originate from a pyrolysis of amino acids.

Cooked meat and fish are important sources of exposure to heterocyclic amines, as are pan residues in countries where they are used to make gravy. They are primarily found in the crust of cooked meat and fish. Meats are generally more sensitive then fish. Prepared meat products generally contain lower amounts compared to pure meat, probably because of the presence of various additives, such as sulphite, nitrite or citric acid, which inhibit their production. Commercially cooked food products generally contain very low or undetectable amounts of heterocyclic amines, with a few exceptions. This implies that the cooking practice at home or in restaurants is especially important in order to reduce the intake of these mutagenic compounds. Individual daily intakes range, depending upon the country, from 0-2 up to 13.8 µg.

Figure 3.28. Structure of two important precursors in heterocyclic amine formation: creatine and creatinine.

## 3.7.3 Acrylamide

A compound very much related to the heterocyclic amines, although its formation in foods was only recently discovered, is acrylamide. Previously, human exposure to acrylamide was supposed to be very low, since acrylamide is essentially used as the monomer for the production of polyacrylamide. This polymer is used for technical applications, but also sometimes as a food contact material. From this food contact material, a migration of the unreacted acrylamide to the food could arise, as discussed before. Apart from this source, acrylamide exposure was considered to be mainly occupational (e.g. smoking). However, Swedish researchers discovered that apart from the occupational exposure, every individual is exposed to a particular background contamination, which soon proved to be attributed to a dietary intake of acrylamide.

Particular concern about the dietary intake of acrylamide exists because of its carcinogenicity, its neurotoxicity and its reproductive toxicity.

Like heterocyclic amines, acrylamide seems to be formed in foods as a result of the Maillard reaction. In contrast to heterocyclic amines, however, the amide is produced in carbohydrate rich foods, such as potatoes or cereal products, as shown in Table 3.18. From this table, however, it follows that the manner in which the food is prepared has a big influence on the acrylamide levels (e.g. potato, cooked and fried). Therefore, the mechanism of acrylamide formation should be explained.

As with the heterocyclic amines, the Maillard reaction seems to play an important role in the formation of acrylamide in foods, however creatine is not involved, although asparagine and especially free asparagine seem to be a detrimental factor in its formation. As shown in Figure 3.29, illustrating two of the possible reaction mechanisms, acrylamide formation should be considered as a side-reaction of the classical Maillard reaction. Since other compounds, which do not participate in the Maillard reaction, also react with asparagine to form acrylamide, other pathways may also be operative. Apart from reducing sugars, other carbonyl compounds may give rise to acrylamide formation upon reaction with asparagine.

The amount of free asparagine is considered to be important with regard to the acrylamide formation. This is reflected in the kind of foods susceptible to acrylamide formation and their asparagine content. Unfortunately, important staple foods seem to be susceptible to acrylamide

Table 3.18. Acrylamide levels in processed foods and asparagine content of some foods (fresh).

| Food | acrylamide (µg/kg) | asparagine (mg/kg) |
|---|---|---|
| Almonds, roasted | 260 | 980-6410 |
| Asparagus, roasted | 143 | 11000-94000 |
| Baked products: bagels, breads, cakes, cookies, pretzels | 70-430 | |
| Beer, malt and whey drinks | 30-70 | |
| Biscuits, crackers | 30-3200 | 1500 (wheat) |
| Breakfast cereals | 30-1346 | 1500 (wheat) |
| Chocolate powder | 15-90 | 300 (cocoa) |
| Coffee powder | 170-351 | |
| Corn chips, crisps | 34-416 | |
| Crisp bread | 800-1200 | |
| Fish products | 30-39 | |
| Gingerbread | 90-1660 | |
| Meat and poultry products | 30-64 | 0.4-11 |
| Peanuts | 140 | |
| Potato, boiled | 48 | 1700-3500 |
| Potato chips, crisps | 170-3700 | |
| Potato, French fried | 200-12000 | |
| Potato puffs, deep-fried | 1270 | |
| Snack, other then potato | 30-1915 | |
| Soybeans, roasted | 25 | |
| Sunflower seeds, roasted | 66 | |
| Taco shells, cooked | 559 | |

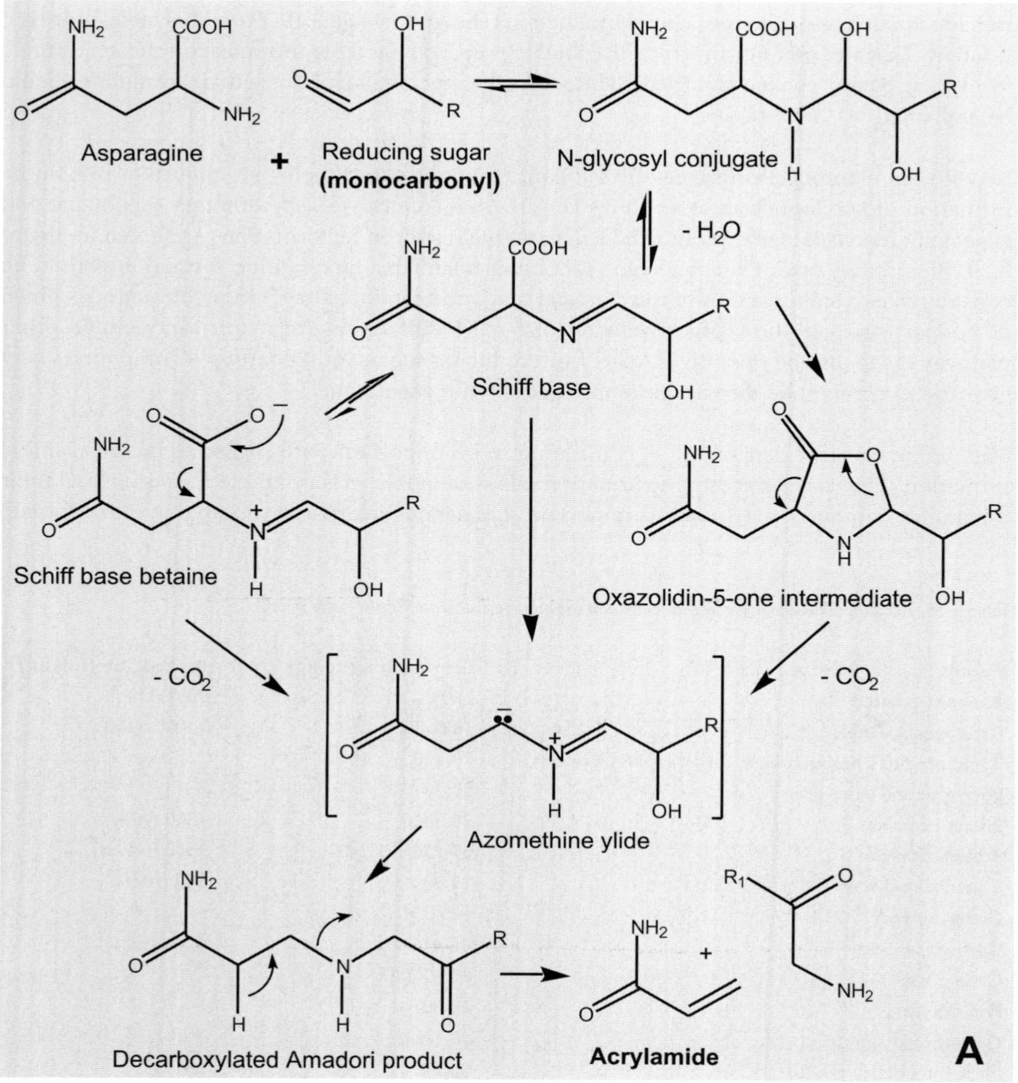

Figure 3.29A. Possible pathways of acrylamide formation in foods.

formation. In some of these products, such as potatoes, the content of free asparagine and reducing sugars, respectively, may vary during storage and from variety to variety.

Apart from endogenous factors, it is quite clear that process parameters will also influence acrylamide formation. During boiling, no relevant acrylamide production is reported. Unless temperatures are considerably lower in a cooking process compared to a frying process, it is

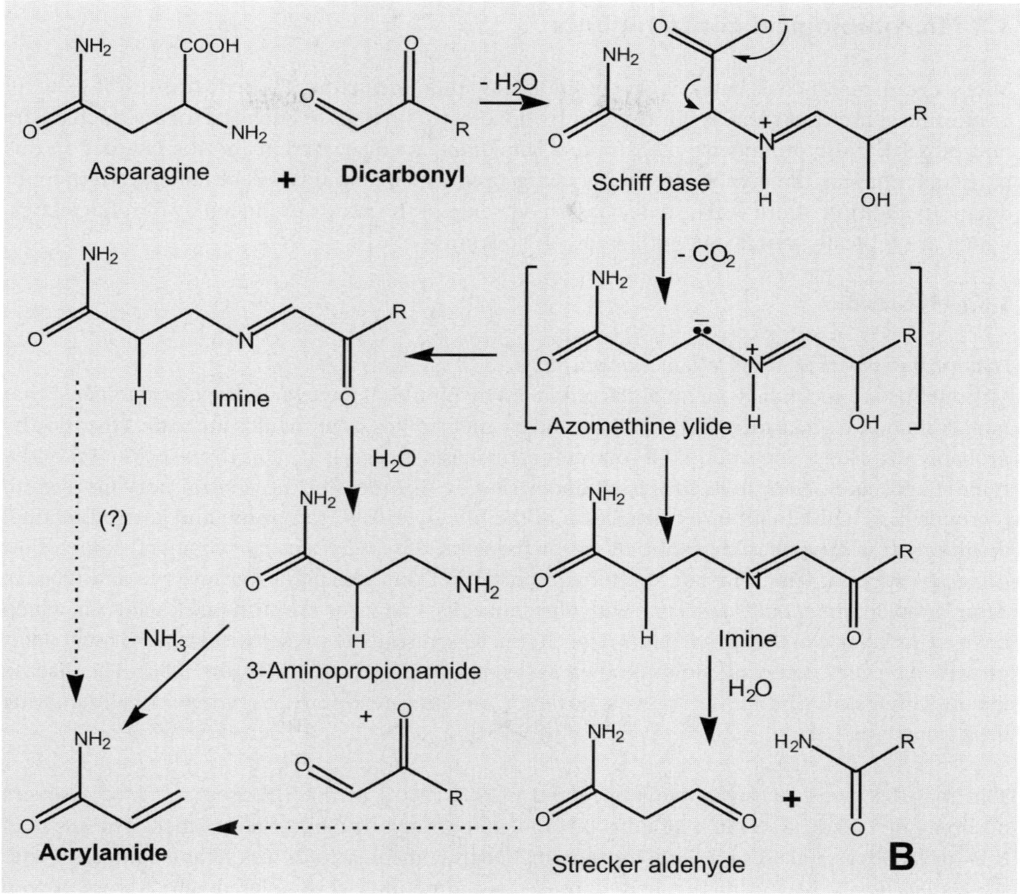

*Figure 3.29B. Possible pathways of acrylamide formation in foods.*

the excessive amount of water that really inhibits acrylamide production. Heating processes that induce a surface dehydration of the food, such as a frying or oven baking, give rise on the other hand to appreciable acrylamide production. Temperature is also important, but in practice remains a parameter difficult to alter, since at lower temperatures desirable processes such as drying, colour and aroma production are slowed down as well. Despite the fact that acrylamide is apparently degraded at higher temperatures (> 200°C), higher process temperatures are frequently not an option either, since other undesirable reactions will be favoured.

## 3.8 Microbiological contaminants

Microorganisms, such as bacteria and moulds, may produce toxins. These toxins result in acute or chronic symptoms such as vomiting or liver cancer. The risks of bacterial toxins, such as the neurotoxin botulin produced by *Clostridium botulinum*, are discussed in another chapter. In this particular chapter, the mycotoxins or the toxins produced by moulds will be considered in more detail. In addition, there will be a discussion on some of the toxins produced by dinoflagellates, which are typically associated with shellfish poisoning.

### 3.8.1 Mycotoxins

*Historical perspective: discovery of mycotoxins*
Mycotoxins are secondary metabolites produced by moulds. However, it is well established that not all moulds are toxigenic and not all secondary metabolites from moulds are toxic. Historically, probably the oldest documented human mycotoxicosis is ergotism. This disease, known in the middle ages as St Anthony's fire, is characterised by disorders of the central nervous system (convulsions, hallucinations), contraction of the blood vessels (gangrene) and gastrointestinal disorders. It is caused by ingestion of grains parasitised by *Claviceps purpurea* and some other *Claviceps* species, which invade the female portion of the host plant (barley, rye and wheat). After invasion, the moulds associate with dormant cells, known as purplish-black sclerotia, which have a similar size to the cereal grain. The chemicals causing the mycotoxicosis were much later proven to be very potent alkaloids, known as lysergic acid derivatives (Figure 3.30). The disease is now rather rare. Similar derivatives, however, are also present in a very potent hallucinating drug known as LSD.

The real discovery of mycotoxins occurred in the 1960's in the UK as a result of a severe outbreak of Turkey 'X', which killed about 100, 000 turkeys and other farm stock. The cause of this disease was traced to a feed component, peanut meal, which was heavily infested with *Aspergillus flavus*. Analysing the feed, fluorescent compounds were isolated, which were proven to be responsible for the outbreak, and were later called aflatoxins.

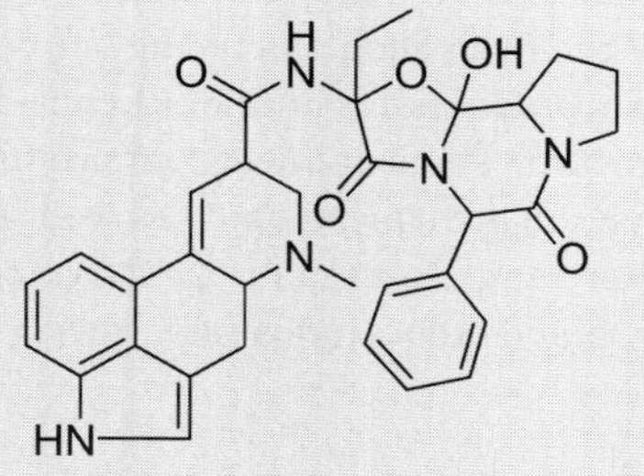

*Figure 3.30. Chemical structure of ergotamine, the most abundant ergot alkaloid.*

## Chemical hazards

From then on, it became clear that many other human diseases were caused by a variety of mycotoxins. In Table 3.19, the most probable routes for mycotoxin contamination of foods and feeds are summarised.

*Important mycotoxins*
Depending upon the kind of agricultural commodities, some important mycotoxins or classes of mycotoxins can be distinguished. Generally, moulds can produce mycotoxins on living plants, on decaying plant material and on stored plant material.

As regards stored plant material, cereals and cereal products are extremely important. For these products, moulds in the genera *Aspergillus*, *Penicillium*, *Fusarium* and *Alternaria* are the most important. For living plants, especially the *Fusarium*, *Aspergillus* and *Claviceps* genera are important to consider. Apart from the *Aspergillus*, these genera are also found on decaying plant material.

The *Aspergillus* species produces a class of closely related difurancoumarin derivatives, known as aflatoxins (Figure 3.31). Presently, 18 different aflatoxins have been identified, but the most potent aflatoxin $B_1$ and $G_1$ occur most frequently. Apart from their most prominent toxic characteristic, their carcinogenicity, aflatoxins have proven to be immunotoxic and haematotoxic. A lot of agricultural commodities are susceptible to aflatoxin contamination, including peanuts and their derived products, corn, pistachio nuts, Brazil nuts, pumpkin seeds, oilseeds, etc. In

Table 3.19. Most probable routes for mycotoxin contamination of foods and feeds.

**Mould damaged foods**
- agricultural products
    - cereals
    - oilseeds
    - fruits
    - vegetables
- consumer foods (secondary infections)
- compounded animal feeds (secondary infections)

**Residues in animal tissues and animal products**
- milk (animal and human)
- dairy products
- meats (organs such as liver and kidney in particular)

**Mould ripened foods**
- cheeses
- fermented meat products
- oriental fermentations

**Fermented derived products**
- microbial proteins
- food additives

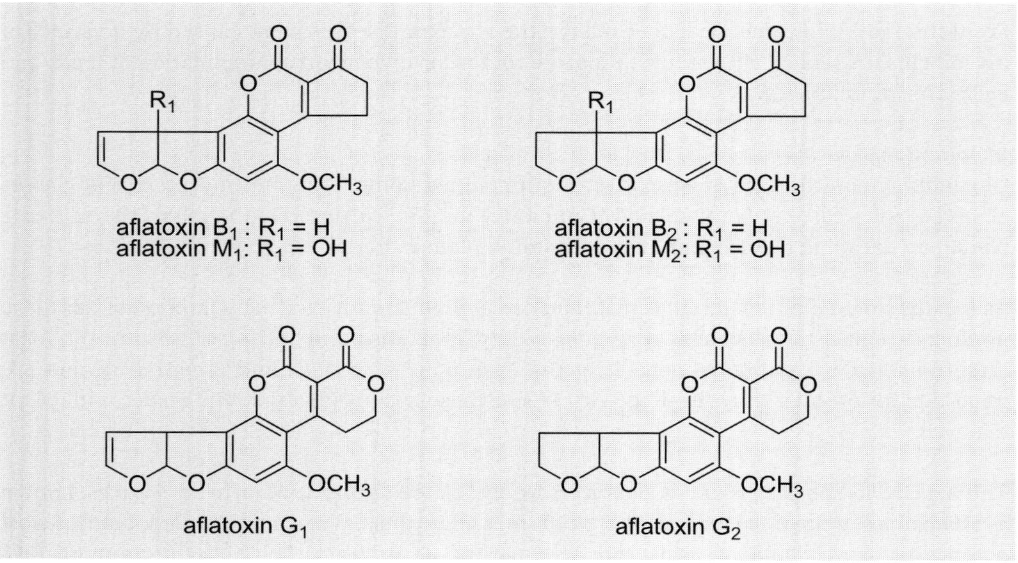

Figure 3.31. Chemical structure of various aflatoxins.

addition, aflatoxins may be metabolised and occur therefore in animal products as well, such as aflatoxin $M_1$ in milk (aflatoxin $B_1$ metabolite).

*Aspergillus ochraceus*, some other *Aspergillus* species and *Penicillium* verrucosum produce ochratoxin A (OTA) (Figure 3.32). The latter mould is typically associated with stored cereals in northern countries. In addition, grapes and consequently wine are susceptible to *Aspergillus* infection. This may induce nephropathy and chronic intestinal nephritis. It is also considered as a possible human carcinogen, as immunotoxic, cyto-, geno- and neurotoxic.

*Fusarium* species are another important group of toxigenic moulds. They can produce a variety of mycotoxins as summarised in Table 3.20. Their chemical structures are given in Figure 3.33.

T-2 toxin is the most potent of the trichothecene toxins and its first outbreak was associated with an epidemic of aleukia in Russia in 1913 due to the consumption of contaminated grain. This

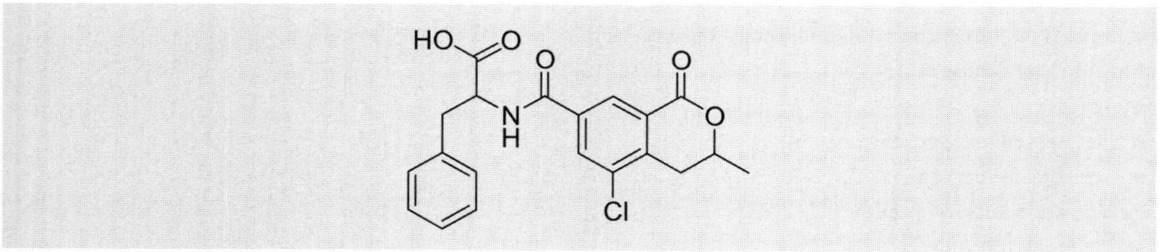

Figure 3.32. Chemical structure of ochratoxin A.

Table 3.20. Overview of toxins and their properties produced by various Fusarium moulds.

| Mycotoxin | Toxicity | Prevalence |
|---|---|---|
| trichothecenes | | |
| ■ deoxynivalenol (DON, vomitoxin) | decreased growth, reduced food consumption, vomiting (higher doses), immunotoxic, neurotoxic | cereals |
| ■ T-2 toxin | alimentary toxic aleukia, pellagra, growth effects, immunotoxic, haematotoxic | corn, wheat |
| ■ HT-2 toxin | similar as T-2 toxin | corn, wheat |
| zearalenone | oestrogenic substance, carcinogenicity, reproductive toxicity | corn, liver of contaminated animals |
| fumonisins | | |
| ■ fumonisin $B_1$ | sphingolipid synthesis inhibition, carcinogenicity, growth effects | corn |

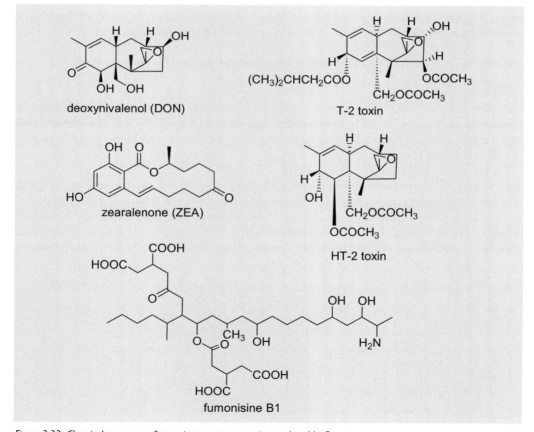

Figure 3.33. Chemical structures of some important mycotoxins produced by Fusaria.

disease causes leukopania, agranulocytosis, bleeding from the nose and throat, exhaustion of the bone marrow and fever. Fumonisins were only recently discovered (1988) and a high incidence of contaminated corn is typically associated with an increased incidence of oesophageal cancer in men as reported in the South African Transkei region.

Apart from the presence of mycotoxins in relatively dry products such as cereals and nuts, fresh vegetables and fruits may also be susceptible to mycotoxin contamination. Two typical examples can be given in this regard: patulin and alternariol.

Patulin (Figure 3.34) can be produced by a variety of genera including *Penicillium expansum* and the *Aspergillus* species. The former mould is the principal cause of apple rot, and therefore it is not surprising that apples, pears, peaches, tomatoes and other fruits and their derived products, such as purees and juices, are very susceptible to patulin contamination. Patulin intoxication is associated with mutagenicity, teratogenicity and carcinogenicity.

Alternariol (Figure 3.34) is produced by the *Alternaria* species and its presence is typically associated with post harvest decay of many fruits and vegetables. In addition to this secondary metabolite, a variety of other toxic metabolites are produced by these genera as well.

In Table 3.21, toxicological data of some of the discussed mycotoxins are summarised.

*Legislative aspects*
Since the discovery of aflatoxins in the 1960's, regulations have been established in order to protect the consumer from the harmful effects of mycotoxins. Therefore, tolerance levels in food and feed are specified for a variety of mycotoxins. Sometimes, tolerance levels are provided for specific commodities.

Typically, the industrialised countries have fairly detailed legislation in place, although there is still a lack of harmonisation and completeness. Aflatoxin tolerance levels are most widespread in their application. In African and Asian countries, however, serious shortcomings in the food legislative framework still exists. Some of these countries are still frequently confronted with food shortages, and therefore destruction of contaminated batches is probably not a realistic scenario.

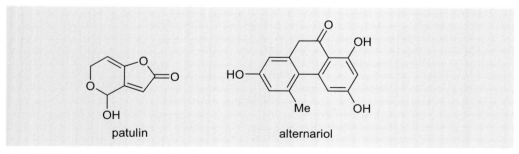

Figure 3.34. Chemical structures of patulin and alternariol.

Table 3.21. TDI values of some important mycotoxins.

| Mycotoxin | TDI (ng/kg bodyweight/day) |
|---|---|
| Aflatoxin ($B_1$, $B_2$, $G_1$, $G_2$ en $M_1$) | 0.016 ($AFB_1$) |
| Ochratoxin A | 5 |
| Fumonisin $B_1$ | 2000 |
| Deoxynivalenol (DON) | 1000 |
| T-2 and HT-2 toxin | 60 |
| Nivalenol | 70 |
| Zearalenone (ZEA) | 200 |
| Patulin | 400 |

## 3.8.2 Shellfish poisoning

There are three known classes of toxins, produced by dinoflagellates: paralytic, diarrheal and amnesic toxins. Typically these intoxications are associated with the consumption of edible shellfish species, such as mussels, cockles, clams and scallops. Problems occur both in warm and cold water. Saxitoxin and okadaic acid are typical paralytic and diarrheal algae toxins, respectively (Figure 3.35).

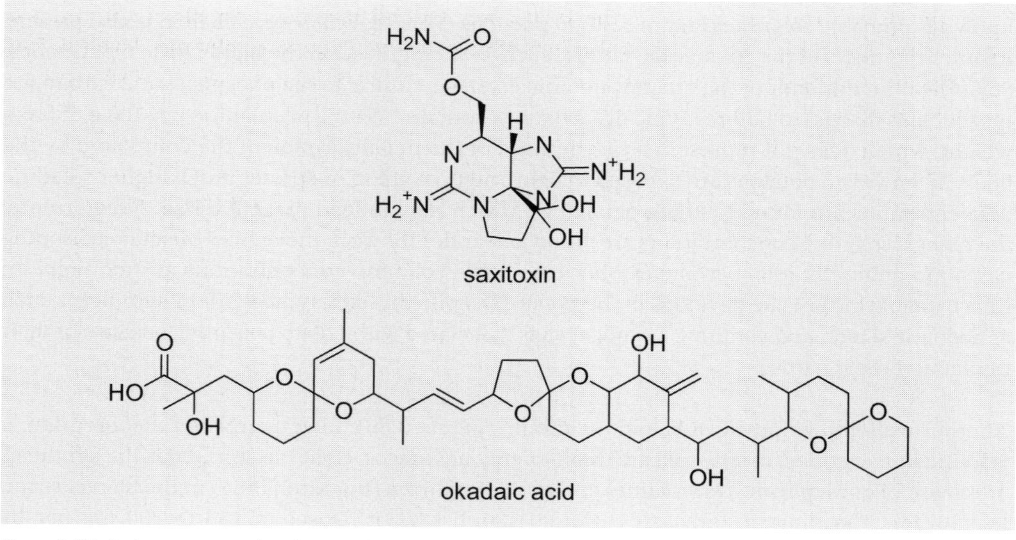

Figure 3.35. An important paralytic (saxitoxin) and diarrheal (okadaic acid) algae toxin.

*Bruno De Meulenaer*

## 3.9 Endogenous toxicants

Currently, society is seized with a kind of chemophobia. This phobia is characterised by distrust in any (synthetic) chemical that may come in contact with human beings. This distrust probably explains the success of organic foods nowadays. People are convinced that organic foods are healthier compared to regular foods, because, for example, no pesticides have been used in their production. In line with this conviction is the great confidence people have in naturally-occurring compounds, which they believe are not toxic. This, however, is not true, since numerous compounds occurring in nature and in our food may indeed be harmful to human beings.

Most of these endogenous toxic compounds are of plant origin. The most well-known exception to this is tetrodotoxin, occurring in various organs of the puffer fish, which is a popular delicacy in Japan. Great skill is required in order to separate the toxic parts (liver, ovaries, skin and intestines) from the edible parts (muscle and testes). Since the toxin blocks the movement of sodium ions across the membranes in the nerve fibres, ingestion of the toxin leads to total paralysis, respiratory failure and finally death.

With regard to the endogenous toxins in plants, a restricted overview will be presented within the framework of this textbook.

Plants contain numerous alkaloid compounds with various physiological activities. The best known is probably solanine (Figure 3.36) in potatoes. This steroidal glycoalkaloid is also present in other members of the Solanaceae family, such as aubergines and the highly toxic nightshades. Solanine is an inhibitor of the enzyme choline esterase, similar to the phosphate and carbamate insecticides discussed before. Typically, potatoes contain 2-15 mg of solanine per 100 g of fresh weight, which does not represent a risk because of the malabsorption of the compound by the body. If, however, potatoes are exposed to light and turn green or sprout, much higher solanine concentrations are found (100 mg per 100 g), which may indeed present a risk. Furthermore, the highest solanine concentrations are found just under the skin. Incidences of potato poisoning cases are generally low, particularly because if the solanine concentrations are too high, an unacceptable bitter taste develops in the potato. Despite this fact, typical minor symptoms such as abdominal pain and vomiting are not always associated with potato poisoning because of their uncharacteristic nature.

Another well-known purine alkaloid is caffeine (Figure 3.36). Like the related theobromine, it is found in tea, coffee, cocoa, cola and high-energy beverages. Caffeine stimulates the synthesis or release of epinephrine (adrenaline) and norepinephrine (noradrenaline) in the bloodstream. In addition, the diuretic effect of caffeine is well known. Doses of 150 to 200 mg per kg bodyweight are documented to be fatal for humans. Taking into account that a cup of coffee contains 50 to 125 mg of caffeine, it is obvious that this sort of dosage cannot be achieved even by drinking excessive amounts of coffee. Nevertheless, other symptoms such as interference with the heart rhythm and sleeplessness are well known.

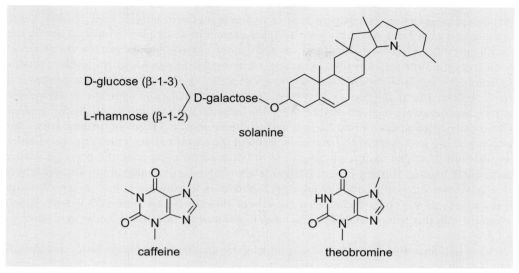

Figure 3.36. Chemical structure of some toxic alkaloids present in food.

Chocolate is also well known to induce headaches in susceptible individuals. This is due to its elevated concentrations of phenylethylamine (10 mg/100g), which is a biogenic amine. Biogenic amines are vasoconstrictors that occur widely in foods. Despite the fact that they can also originate from microbiological protein putrification reactions, they may also be naturally present in, for example, banana (concentrations ranging from 5-70 mg/kg, depending upon the kind of amine), plantain and tomato. Some typical examples are shown in Figure 3.37.

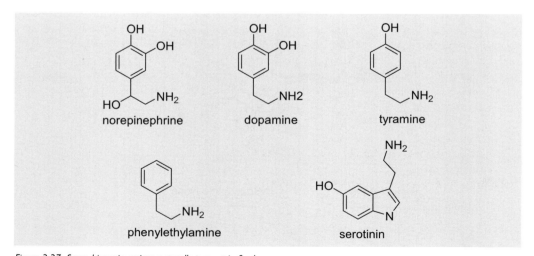

Figure 3.37. Some biogenic amines naturally present in foods.

*Bruno De Meulenaer*

Another group of naturally occurring toxic compounds are the cyanogenetic glycosides. These compounds, such as amygdalin (Figure 3.38) are present in many plant families. They do not usually present a problem because they are there in the non-edible parts, such as the stones of some fruit (e.g. almonds). However, in cassava, a major staple food in Africa, the levels of various cyanogenetic glycosides are quite high, giving rise to 50 mg of hydrogen cyanide per 100 g of fresh product. Due to the activity of two enzymes, β-glycosidase and hydroxynitrilelyase, the active compounds are released (Figure 3.38). Enzyme activation occurs essentially due to plant tissue damage occurring during harvest and food preparation. Therefore, in West Africa for instance, it is a well-established tradition to ferment the cassava prior to consumption in order to allow the liberated hydrogen cyanide to be released as a gas from the porridge during subsequent heating. Chronic exposure to low levels of hydrogen cyanide induces ataxia, which is a degenerative neurological condition, and blindness. Furthermore, due to detoxification mechanisms in the body, higher serum levels of thiocyanate are observed as well, which interfere with the iodine metabolism. As a result of this latter interaction, goitre may develop.

Protease inhibitors such as the Kunitz trypsin inhibitor and the Bowman Birk factor in soybeans typically inhibit the digestive trypsin and chymotrypsine. These proteins bind to these proteases and inactivate them. As a result, proteins present in foods are only partially digested. During chronic exposure pancreas size may increase since this organ is responsible for the synthesis of these proteases. Protease inhibitors are typically present in legumes, although also in egg white. They can be (partially) inactivated due to heat treatment, but this may also affect the nutritive quality of the proteins.

As a final example of a group of endogenous food toxicants, phenols can be considered. Many different phenols are present in plants. Despite the fact that many phenols in edible plants or edible plant parts are associated with protective properties against oxidation reactions, various plants contain numerous toxic phenols as well. Only one example will be considered here,

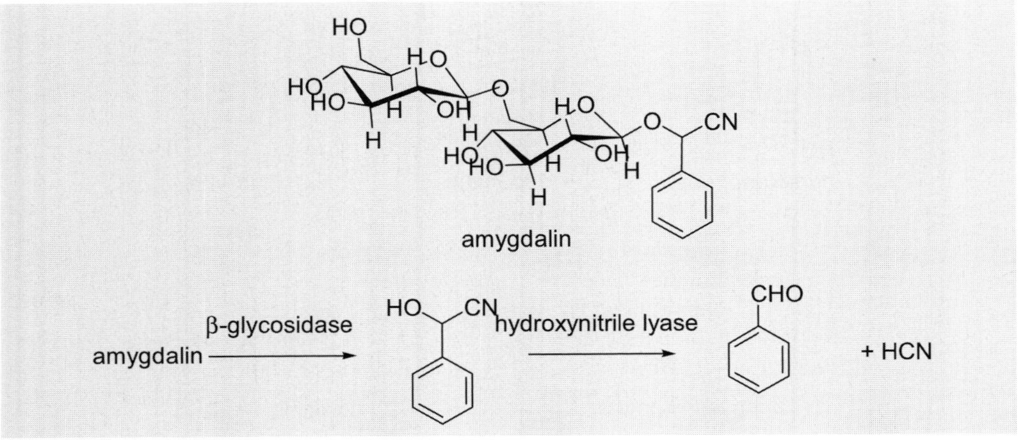

*Figure 3.38. Chemical structure of amygdalin and the mechanism of hydrogen cyanide release.*

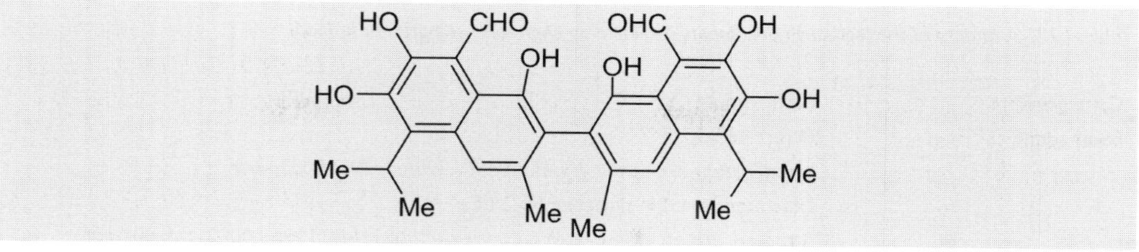

*Figure 3.39. Chemical structure of gossypol*

gossypol (Figure 3.39). Gossypol is a yellow pigment typically presenting in cotton. Its toxic properties gained importance because of the use of cottonseed meal in animal and also human diets. Common symptoms due to chronic exposure are loss of appetite, weight loss, diarrhoea, lowering of red blood cell count and serum protein. In addition, oedematous fluids may occur in body cavities such as the lungs and heart. Moreover, serious liver, spleen, stomach and small intestine damage may be induced.

## 3.10 Legislative information and further reading

Despite the fact that an overview of legislative information with regard to chemical hazards is inevitably a snapshot at a particular time, and is therefore prone to incompleteness, an overview is given at the end of this chapter of the harmonised legislation in place within the European Union with regard to this particular topic. Sometimes reference is made to a website instead of the specific legislative document. This is summarised in Table 3.22.

In addition to these legislative documents, some more extensive textbooks can be consulted for a more in-depth and complete overview with regard to the aspects discussed in this chapter. Again, a non-exhaustive list is given:

Belitz, H.D. and W. Grosch, 1999. Food Chemistry, Springer, Berlin, Germany, 992 p.
Coultate, T.P., 2002. Food, The chemistry of its components, Royal Society of Chemistry, Cambridge, UK, 432p.
Macholz, R. and H.J. Lewerenz, 1989. Lebensmitteltoxikologie, Springer Verlag, Berlin, Germany, 664 p.
Piringer, O.G. and A. Baner, (Eds), 2000. Plastic Packaging Materials for Food, Wiley-VCH, Weinheim, Germany, 576 p.
Taylor, S.L. and R.A. Scanlan (Eds.), 1989. Food Toxicology. A perspective on the relative risks, Marcel Dekker, New York, USA, 466 p.
Watson, D. (Ed), 2002. Food Chemical Safety. Volume 1. Contaminants, CRC Woodhead Publishing Ltd. Cambridge, UK, 322 p.
Watson, D. (Ed), 2002. Food Chemical Safety. Volume 2. Additives, CRC Woodhead Publishing Ltd. Cambridge, UK, 308 p.
Weidenbörner, M., 2001. Encyclopedia of Food Mycotoxins. Springer, Berlin, Germany 294 p.

Table 3.22. Summary of the legislative documents available on chemical risks present in food.

| Component | References |
|---|---|
| Food additives | Directive 95/2/EC: |
| | http://europa.eu.int/eur-lex/en/lif/dat/1995/en_395L0002.html |
| | Lists mentioned in directive 95/2/EC: |
| | http://europa.eu.int/eur-lex/en/consleg/pdf/1995/en_1995 L0002_do_001.pdf |
| | Guideline 96/77/EG |
| | Guideline 95/45/EG |
| | Guideline 95/31/EG |
| Lead, cadmium, mercury | Regulation 466/2001: |
| | http://europa.eu.int/eur-lex/pri/en/oj/dat/2001/l_077/l_07720010316en00010013.pdf |
| Radio-isotopes | Regulation 2218/89 |
| Nitrates in vegetables | Regulation 466/2001: |
| | http://europa.eu.int/eur-lex/pri/en/oj/dat/2001/l_077/l_07720010316en00010013.pdf |
| Pesticides | Guideline 76/895/EEG |
| | Guideline 86/362/EEG |
| | Guideline 86/363/EEG |
| | Guideline 90/642/EEG |
| Veterinary drugs | |
| ■ Antibiotics, disinfectants) | Regulation 2377/90 |
| ■ Growth stimulators | Guideline 88/146/EEG, 96/22/EEG |
| PCB's, PCDF's, dioxins | Regulation 2375/2001 |
| Residues of extraction solvents | Guideline 88/344/EEG |
| Process contaminants | |
| ■ Monochloro-propanediol | Regulation 466/2001: |
| | http://europa.eu.int/eur-lex/pri/en/oj/dat/2001/l_077/l_07720010316en00010013.pdf |
| Migration from food contact materials | |
| ■ Plastics | See http://cpf.jrc.it/webpack/legislat.htm |
| Mycotoxines | |
| ■ Aflatoxinen | Regulation 466/2001: |
| | http://europa.eu.int/eur-lex/pri/en/oj/dat/2001/l_077/l_07720010316en00010013.pdf |
| ■ Patuline | |
| ■ Ochratoxine A | |

# References

Bindslev-Jensen, C., D. Briggs and M. Osterballe, 2002.Can we determine a threshold level for allergenic foods by statistical analysis of published data in the literature?, Allergy 8, 741-746.

Lehrer, S.B., W.E. Horner and G. Reese, 1996. Why are some proteins allergenic? Implications for biotechnology, Critical Reviews in Food Science and Nutrition, 36, 553-564.

# 4. Physical hazards in the agri-food chain

*Anna Aladjadjiyan*

## 4.1 Introduction

Quality management has recently played an important role in food production. Quality defines the place of food products on the world market. Substantial parts of quality management are food-quality assessment and food-quality control.

Food-quality assessment must be considered from different perspectives and its improvement requires the united efforts of multidisciplinary research. The integrated view of producers, consumers and researchers on food quality depends on the different steps in food preparation and needs thorough examination. An important feature of this examination process is research and prevention of different hazards appearing in the agri-food chain. Consumers' perception of risk associated with potential food hazards can strongly affect their quality concept.

The objective of this chapter is to present and describe some of the current physical hazards in the agri-food chain, to reveal their nature, their influence on food quality and safety, and discuss possible ways of preventing food contamination.

Acquaintance with physical hazards, and methods of their control and avoidance, is an important part of the management of safety in the agri-food chain.

After studying this chapter, students will have a better idea of the sources of physical contamination in the agri-food chain. They need to acquire knowledge about the classification of physical food contaminants, some methods of control, and ways to avoid physical risks in the agri-food chain. After defining physical contaminants, the different types of physical contaminants are discussed in more depth.

## 4.2 Physical Contaminants: definition and classification

The presence of foreign bodies in food is of major concern to the producer. 25 % of consumer complaints are related to the physical contaminants in food. Physical contamination can occur at any stage of the food chain and therefore all reasonable precautions must be taken to prevent this type of contamination. There is a requirement in EU legislation that food business operators must have a quality system to ensure the safety of food (Council Directive, 1989; 1994)

Definition: Physical contaminants are additional matter or alien objects normally not existing in food that could cause injury, disease or psychological trauma to the organism.

In a review of the literature on physical hazards in food products (Olsen, 1998) it is estimated that 1-5% of the foreign objects ingested (swallowed) by people result in minor to serious injury.

According to the same review, most ingested foreign objects (80-90%) pass through the gastrointestinal tract spontaneously with the remainder requiring removal either by endoscopy or, less frequently, surgery. Morbidity estimates are higher for certain kinds of sharp objects. For example, perforation injury is estimated to occur in 15-35% of the patients who are examined by physicians for ingestion of slender, pointed objects. Ingested objects that are slender and sharp or pointed are widely recognised as presenting greater risk of injury from perforation of tissues of the gastrointestinal tract and greater need for surgical removal than other kinds of ingested foreign objects. The types of injury resulting from eating hard or sharp objects also include laceration of mouth or throat tissues and breaking or chipping of teeth.

In the framework of this book, radioactive contaminants are also considered as physical contaminants. They are invisible and therefore have no direct significant effect on the quality perception of the consumer. They can only be detected by special instruments. Radioactive contamination can be dangerous, causing serious diseases.

## 4.3 Non-radioactive physical contamination

### 4.3.1 Classification

Non-radioactive physical contaminants are foreign bodies found in food that can cause injuries or morbidity to the consumer when ingested. Almost any hard or sharp object could be a physical hazard if found in a food product, because it causes mouth or throat wounding - cuts, bleeding, choking, laceration. Non-radioactive physical contamination also includes any filth that can produce nausea, vomiting, or at least a bad appearance of the food product. An overview is given in Table 4.1.

Different methods have been developed for avoiding non-radioactive physical contaminants. These methods are usually the objects of HACCP preparation for any specific action. The Food & Drug Administration has actually established maximum levels of natural or unavoidable defects in foods for substances that present no major human health hazard (from tiny insect parts to dirt). They are called "food defect action levels" and will be discussed further.

According to their nature, mechanical (non-radioactive) contaminants can be classified in three groups:
- mineral (soil, stones, dust, metals, glass, fibre, paint flakes, etc.);
- plant (weeds, leaves, stems, wheat-ears);
- animal (mites, insects, rodents, fowl).

## Physical hazards in the agri-food chain

Table 4.1. Major materials of concern as physical hazards and most popular sources of the major non-radioactive physical hazards of concern are presented along with possible resulting injuries.

| Materials | Potential Injuries | Sources |
|---|---|---|
| Glass | Cuts, bleeding, surgery may be needed for its location and extraction | Bottles, jars, lamps, utensils, gauge covers |
| Wood | Cuttings, infection, choking, surgery may be needed | Fields, land, chests, boxes, buildings |
| Stones | Choking, tooth breakage | Fields, buildings |
| Metal: nail, key, coin, machinery part | Cuttings, infection, surgery may be needed | Machinery, fields, wires, workers |
| Insulation materials | Choking; if it is asbestos wool choking might be serious | Construction materials |
| Bones | Choking, trauma, injury | Fields, improper industrial technology |
| Insects and other pests | Disease, trauma, choking | Fields, factory production area |
| Plastic materials | Choking, cuttings, infection, surgery may be needed | Fields, industrial packing materials, packages, workers |
| Personal effects: jewellery, button | Choking, cuts, tooth breakage, surgery may be needed | Workers |

Contaminants from the three listed groups can appear during the harvest of raw materials, during their storage as well as during food processing. Included in the food, these contaminants can affect quality perception of the consumers and moreover, can cause injury.

Mineral contaminants may often cause tooth breakage, cuts and bleeding in the mouth, or similar damages later in the oesophagus. They can cause perforation of tissues of the gastrointestinal tract and a subsequent need for surgical removal.

Animal contamination may cause disease and even stronger sequences - secondary infections associated with this type of injury. Allergies and poisoning are also possible.

Plant contaminants may cause both effects - cutting and diseases related to allergies and poisoning. Wound location and heeling of possible infections must be performed by surgeon or other medical specialists.

According to the sources of the mechanical contamination, they can be divided into:
- raw materials;
- water;
- floor covers and construction and building materials;
- staff.

Water can be used in food production both for cleaning and rinsing instruments and raw materials as well as an ingredient.

The most significant way of contaminating food with foreign materials is by people participating in different stages of food preparation and distribution. These people are: production staff, maintenance staff, cleaning staff, delivery staff, and visitors. Personal effects include: fingernails, hair, sweet papers, cigarettes, jewellery, plasters, clothing.

Another important source of mechanical contamination can be the packaging materials. Packaging is an important step in food preparation. It shapes the final look of the food and plays an important role in quality perception. As a source of contamination, packaging can include cardboard, plastic, wood, metal, and staples. Most hazardous to consumers are pieces of glass packaging.

Food contamination can also be due to the presence of structural elements, such as bulbs, paint, fragments of plaster, grease, nuts, and bolts. It can occur during the storage of raw materials and final production.

This kind of contamination can occur at any stage of food preparation.

Contamination by pests should also be mentioned. Food contaminants of pest origin can include droppings, fur feathers, dead bodies, eggs, and larvae. This kind of contamination occurs mainly during the storage of raw materials.

### 4.3.2 Measures for preventing non-radioactive physical contamination

Physical hazards find their way into food at the food production level, as well as at the restaurant level. Their prevention or elimination is often a difficult task. At the processing level, there are many methods to remove stones, metal pieces, insects, dirt, wood, plastic, and glass, etc, from raw materials or from finished products. At the restaurant level, the physical hazard may even be a natural part of the food itself, such as stems in produce and bones in poultry products, fish, or meats (especially ground or chopped meats). The following preventive measures can be taken to avoid physical contamination:

- thorough cleaning and washing of raw materials before use;
- inspecting raw materials for some inclusions;
- filtering liquids before use;
- sieving powders;
- protecting filling hoppers, elevators and belts conveying open food from overhead contamination;
- selecting machinery with guards that are easy to remove and clean;
- installing scanners;
- installing metal detection equipment;
- avoiding temporary repairs;
- during maintenance and repairs, special precautions are recommended, such as:

- covering or removing all food;
- checking that all equipment used during maintenance is removed;
- cleaning machinery and halls prior to restarting food production;
- checking machinery before restarting.

## 4.3.3 Control measures

In Europe, no particular legislation or limits exist for non-radioactive physical contamination. Food defect action levels established by the Food and Drug Administration represent a part of food-quality control in United States regulation. These limits define the acceptable dimensions of foreign body inclusions that are not considered as a defect. Their definition is based on the investigation of frequencies for types of injuries reported to the FDA through the agency's consumer complaint system. The most frequent injuries are injuries of the mouth and throat; in second position is minor dental damage, followed by gastrointestinal distress. The topic of a number of records is glass fragments in food (Olsen, 1998).

Almost any hard or sharp object could be a physical hazard if found in a food product. Hard or sharp natural components of a food (e.g. bones in seafood, shell in nut products) are unlikely to cause injury because of awareness on the part of the consumer that the component is a natural and intrinsic component of a particular product. This is not the case when the label on the food claims that the hard or sharp component has been removed from the food, e.g. pitted olives. The presence of the naturally occurring hard or sharp object in those situations (e.g. pit fragments in pitted olives) is unexpected and may cause injury. In an HACCP plan, control measures must be calibrated to detect contaminants of the sizes considered hazardous. The FDA has established Defect Action Levels for many of these types of unavoidable defects in other Compliance Policy Guides (FDA, 1999).

The FDA's Health Hazard Evaluation Board (HHEB) considered hard or sharp foreign objects to represent a potential hazard in one or more of the following areas:
- laceration of the mouth or throat (most often cited by the Board);
- laceration/perforation of the intestine (possible secondary infection);
- damage to the teeth or gums that would normally be treated by a dentist.

HHEB has conducted health hazard evaluations based on consumer complaints. The definitions of the terms used by the Board to classify the general nature of a hazard are as follows:
- severe: when there is reasonable probability of significant disability, but deaths are rare;
- moderate: in the case of a reasonable possibility of transient but significant disability or of permanent minor disability;
- limited: when there is a reasonable possibility of minor transient disability and/or annoying physical complaints.

In addition, the Board classified the clinical nature of a hazard as follows:
- acute: maximum general effect attained in minutes/hours/one day;
- sub acute: maximum general effect attained in days/one week;

- chronic: maximum general effect attained in weeks/ months/years.

The definitions for the above general and clinical natures of a hazard are established in the FDA regulations governing the classification of voluntary product recalls (Food and Drug Administration, 1997).

Glass fragments as well as metal ones can cause injury to the consumer. The FDA's Health Hazard Evaluation Board has supported regulatory action against products with metal fragments of 0.3" (7 mm) to 1.0" (25mm) in length and also against products with glass fragments of the same measurements (see FDA Compliance Policy Guide #555.425).

The following conclusions are based on the literature and the HHEB evaluations:
1. Hard or sharp objects that are 7 mm or longer, maximum dimension, represent a potential physical hazard in food.
2. Hard or sharp objects less than 7 mm, maximum dimension, represent a possible physical hazard in food, especially if a special-risk group is among the intended consumers of the product.
3. Very large objects, such as those that meet or exceed the Consumer Product Safety Commission (CPSC) safety standard for small parts in toys, are easily and readily detectable by a consumer prior to consuming a food product and do not normally represent a health hazard.
4. Intended use, processing steps that eliminate the hazard, guidance and requirements concerning unavoidable natural defects, and other factors are important considerations when determining the potential hazard from hard or sharp foreign objects in food.

As a result of the above listed HHEB evaluations, the following requirements concerning inclusion of hard or sharp objects have been formulated:

Foods are considered adulterated if:
- the product contains a hard or sharp foreign object that measures 7 mm to 25 mm in length;
- the product is ready-to-eat, or according to instructions or other guidance or requirements, it requires only minimal preparation steps, e.g., heating, that would not eliminate, invalidate, or neutralize the hazard prior to consumption (FDA, 1999).

Concerning metal inclusion:
- no metal fragments in finished product. (Note: FDA's Health Hazard Evaluation Board has supported regulatory action against products with metal fragments of 0.3" [7 mm] to 1.0" [25mm] in length. See also FDA Compliance Policy Guide #555.425.), or
- no broken or missing metal parts from equipment at the CCPs for "metal inclusion" (FDA, 2001a).

### Physical hazards in the agri-food chain

Concerning glass inclusion:
- no glass fragments in finished product. (Note: FDA's Health Hazard Evaluation Board has supported regulatory action against products with glass fragments of 0.3" [7 mm] to 1.0" [25 mm] in length. See also FDA Compliance Policy Guide #555.425.);
- no broken glass at the Critical Control Points for "glass inclusion" (FDA, 2001b).

#### 4.3.4 Detection of non-radioactive physical contaminants

Non-radioactive physical contaminants usually represent foreign bodies present in the mass of the main food product or raw material. Foreign body detection assumes an important place in the food industry, with a view to avoiding food defect prosecution. Many different techniques have been developed in order to detect foreign body defects in food production. Foreign body detection techniques can be categorised in three main groups:
- Mechanical techniques - find foreign objects by means of size or weight differences between the foreign body and the food product.
- Optical inspection techniques - find foreign objects by means of shape and/or colour analysis.
- Techniques based on the interaction between the foreign body and some part of the electromagnetic spectrum penetrating the food product.

Mechanical techniques include simple techniques like sieving, filtering and centrifuging. They are applicable for loose food or fluids.

Optical techniques are based on the interaction of light (the visible part of the electromagnetic spectrum) with food products and foreign objects. The reflectance of light is most frequently measured. Reflectance can be measured at one wavelength or at two wavelengths. The second technique allows avoiding the influence of the sample and apparatus. Both detect only surface defects. Measuring light transmission reveals subsurface defects, but is not too spread out.

A lamp is often used as a light source; the received image can be recorded by a camera and later inspected by a computer. Some systems use laser as a light source. Laser systems generally give better contaminants detection, but are more expensive. Optical inspection techniques are applicable, for example, in the detection of stones and soil clods in vegetables or cereals.

Electromagnetic interaction includes a large number of techniques that are based on the use of different parts of the electromagnetic radiation spectrum.

X-ray inspection uses Roentgen (X-) rays with a wavelength shorter than 10 $nm$ ($10^{-9}m$). Their high energy enables them to penetrate biological tissues and other objects that are opaque to visible radiation. The absorption of X-rays leads to dissociation or ionization of atoms or molecules. The absorption depends on the thickness of the sample and the density of material. Transmitted X-rays are detected by X-ray sensors. There are three types of X-ray sensors:

- phosphor screens that absorb directly energy from X-rays and convert it into light, followed by the use of an optical imaging system - so-called "open screen" system;
- phosphor screens and imaging system built into one unit under high vacuum;
- solid-state X-ray sensitive elements arranged in linear series, oriented perpendicular to the food product flow.

X-ray inspection techniques are applicable for the detection of metal, glass, stone, bone, rubber and some plastic.

Microwave inspection uses microwaves with frequencies of $10^{10}$ - $10^{12}$ $Hz$. Water molecules absorb the energy of microwave radiation very effectively. The transmission of microwave radiation can be used to determine the presence of a foreign body containing water. It is applicable for detecting pits in cherries, and can be sensibly combined with some optical techniques.

Near-infrared inspection uses radiation with a wavelength of 700-2500 $nm$. This technique is based on the characteristic of some molecular bonds to absorb infrared-radiation energy at different well-defined wavelengths. These types of measurements concern the food surface and are applicable in moisture detection and protein analysis.

Ultraviolet inspection uses the interaction of the matter with ultraviolet radiation with a wavelength of 100 - 400 $nm$. Some organic and inorganic compounds re-radiate after absorbing ultraviolet radiation (or another high-energy radiation). This phenomenon is known as photoluminescence. The photoluminescence spectrum is characteristic of the emitting substrate. This can be used to detect the presence of fat, sinews and bones in meat. Another possible application is the detection of aflatoxins, produced by fungal attack in corn, dried dates and figs.

Nuclear magnetic resonance uses a strong magnetic field (1 - 10 $T$) in which the nuclei take up a certain orientation. Combination with broadband microwave pulses adds more information. The magnetic field used can be uniform (nuclear magnetic resonance - NMR) or non-uniform (magnetic resonance imaging). It is a powerful method for determining maturity, damage, ripeness and decay, but these magnetic field values can only be achieved with expensive helium-cooled semiconductors.

Magnetic field system uses the phenomenon of electromagnetic induction: if a coil is placed in a magnetic field and the magnetic field is changed, it induces electric voltage in the coil. A food flow passes through the apparatus. If it contains magnetic metal, this will change the value of the magnetic field and thus generate voltage in the coil.

Electrostatic techniques are based on the fact that the capacity of a parallel-plate capacitor increases when the air space between two plate electrodes is replaced by another dielectric material. When a food product flow is passing between the plates of the capacitor, it acts as dielectric material and changes the capacity. The electrostatic method allows for control of the homogeneity of food.

Ultrasonic techniques use ultrasound with frequencies in the range of 16-200 $kHz$. Ultrasound has the ability to propagate through biological materials. At an interface between two media with different densities, ultrasound will be partly reflected and partly transmitted. This special technique offers the ability to build up a three-dimensional image, but it is a very expensive method and therefore not yet applied in the food industry.

Table 4.2 presents a review of techniques used for detecting non-radioactive contaminants.

*Table 4.2. Summary of techniques used for the detection of non-radioactive physical food contaminants (Graves et al., 1998).*

| Technique | Wavelength | Food product | Foreign bodies | Availability |
|---|---|---|---|---|
| Magnetic | Not applicable | Loose and packaged food | Metals | Widely commercially available |
| Capacitance | Not applicable | Products < 5mm thick | | Research |
| Microwave | 1-100 $mm$ | Fruits, possibly others containing water | Fruit pits | Research |
| Nuclear magnetic resonance | 1-10 $mm$ + magnetic field | Fruits and vegetables | Fruit pits and stones | Research |
| Infrared | 700 $nm$ -1 $mm$ | Nuts, fruits, vegetables | Nut shells, stones, pits | commercially available |
| Optical | 400-700 $nm$ | Any loose product, fruits, vegetables | Stones, stalks | Widely commercially available |
| Ultraviolet | 1 - 400 $nm$ | Meat, vegetables, fruits | Fat, sinews, stones, pits | commercially available |
| X-rays | <1 $nm$ | All loose and packaged goods | Stone, metal, rubber, glass, plastic, bone | Widely commercially available |
| Ultrasonic | Not applicable | Potatoes in water | Stones | Research |

## 4.4 Radioactive contamination

### 4.4.1 Basics of radioactivity

*Definition*
Radioactivity is the ability of some heavy nuclides to decay into light nuclides by emitting radioactive particles - $\alpha$, $\beta$ and $\gamma$.

*Sources of $\alpha$, $\beta$ and $\gamma$ radioactive contamination*
The nature of $\alpha$, $\beta$ and $\gamma$ radiation was revealed by the English physicist Rutherford at the beginning of the 20$^{th}$ century. He has found that $a$ - particles are helium (He) nuclides, $\beta$ - particles are high-velocity electrons and $\gamma$ - particles are electromagnetic radiation with wavelengths less than $10^{-11}$ $m$.(Table 4.3).

Table 4.3. Properties of radioactive particles.

| Type of Radiation | Alpha particle | Beta particle | Gamma ray |
|---|---|---|---|
| Symbol | α | β | γ |
| Mass (atomic mass units) | 4 | 1/2000 | 0 |
| Charge | +2 | -1 | 0 |
| Speed | slow | fast | very fast (speed of light) |
| Ionising ability | high | medium | 0 |
| Penetrating power | low | medium | high |
| Stopped by: | paper | aluminium | lead |

Sources of α, β and γ radioactive contamination are nuclides with radioactive decay.
α - decay is the process whereby the parent nuclide $X$ is transformed in the product nuclide $Y$ according to the reaction

$$^{A}_{Z}X \rightarrow\ ^{A-4}_{Z-2}Y + ^{4}_{2}He,$$

where A is the mass number of nuclide, z - its atomic number.

β - decay is the process whereby the parent nuclide $X$ is transformed into the product nuclide $Y$ according to one of the following reactions:

$$^{A}_{Z}X \rightarrow\ ^{A}_{Z+1}Y + ^{0}_{-1}e$$

or

$$^{A}_{Z}X \rightarrow\ ^{A}_{Z-1}Y + ^{0}_{1}e.$$

The first reaction is accompanied by the rising of an electron $^{0}_{-1}e$ and the second by a positron $^{0}_{1}e$.

### 4.4.2 Half-life of radioactive nuclides

*Definition*
The half-life of radioactive nuclides is the time interval necessary for the decay of half of the primary quantity of the radioactive nuclides. If the number of heavy nuclides in the beginning is $N_0$ and their number is decreasing according to the law

$$N = N_0 e^{-\lambda t},$$

where $\lambda$ is the constant of radioactive decay, $t$ is the time and $e$ is the base of Napierian (natural) logarithm, the half-life $T_{1/2}$ satisfies the condition

$$\frac{N_0}{2} = N_0 e^{-\lambda T_{1/2}}$$

Therefore

$$T_{1/2} = \frac{\ln 2}{\lambda}$$

Half-lives of different nuclides can vary from parts of a second to millions of years.

*Physical Dose Measurements*
Injury of the human organism by radioactive contamination is due to the absorption of high-energy irradiation by body organs and tissues. For quantitative assessment of the injury it is necessary to measure the absorbed irradiation.

Physical measurements of irradiation include quantities of the "absorbed dose" and the "exposure." Exposure refers to the radiation incident upon the object, while the absorbed dose refers to the actual interaction of that radiation within the object.

The absorbed dose can be measured in terms of:
- the energy delivered per volume;
- the energy delivered per mass;
- ion-pairs created per volume;
- ion-pairs created per mass.

The system (SI - Systeme International) unit of absorbed dose is **Gray (Gy)**. The dose $D$ is 1 $Gy$ when the total energy $E$ (J) absorbed by the object with mass $m$ (kg), in any medium from any type of ionising radiation, is:

$$D = \frac{E}{M} \; ; \; 1 Gy = 1 J/kg.$$

Previously, *rad*, was used as a unit for the absorbed dose and can be translated into SI-units by the formula:

$1 \, rad = 10^{-2} \, Gy.$

For externally originating radiation, it is commonly appropriate to measure the **exposure (X)**, i.e. the radiation incident on the sample, in terms of the ionisation produced in air when that radiation passes through it. The exposure is measured by the total electric charge Q (measured in C - Coulombs) of ions produced by X- or gamma-radiation in a given volume of air, divided by its mass m:

$$X = \frac{Q}{m}$$

Previously Roentgen(R) was used as a unit for X.

$1 \, R = 2{,}58 * 10^{-4} C/kg.$

### 4.4.3 Natural and anthropogenic radionuclides

Generally, two types of radionuclides can be found in the environment - those of natural origin and those of anthropogenic origin. The natural atmospheric radioactivity, cosmic radiation and radioactive gases, released from the ground surface, are the sources of the natural radioactivity. Natural radionuclides are formed in the upper atmospheric layers under the influence of different cosmic radiation. They comprise $^{14}$C, $^{3}$H, $^{10}$Be, $^{7}$Be, and $^{26}$Al. Natural radioactivity also comprises some radionuclides with a long half-life, formed during primary synthesis, such as $^{238}$U, $^{235}$U and $^{232}$Th, and the products of their decay.

$^{238}$U → $^{206}$Pb
$^{235}$U → $^{208}$Pb
$^{232}$Th → $^{207}$Pb

Radionuclides of anthropogenic origin are related to human activities in research and the exploitation of radioactive materials. Over the last 25 years, new artificial radioactive environmental contaminants have appeared. These have resulted from the use of nuclear power for peaceful and military purposes, nuclear electric stations and satellites. Accidents in nuclear electric stations as well as in operating nuclear stations release radioisotopes in spite of the high-degree protection of the systems. They are a source of radioactive contamination.

The Chernobyl accident completely changed the public perception of nuclear risk. During the Chernobyl accident (in 1986), large areas of half-natural ecosystems were damaged by radioisotope deposition. Pastures, fruit and vegetable fields and fertile lands were polluted. Radioactive contamination spread later to animals and directly or indirectly into food products. The pathways of food contamination with radioactive nuclides are presented Figure 4.1. The most important radionuclides influencing food contamination according to Vosniakos, 2001, are presented in Table 4.4.

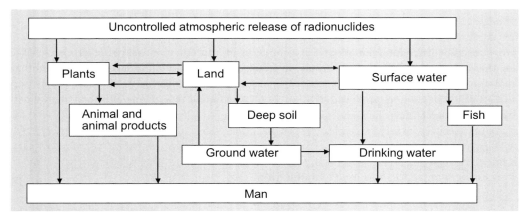

Figure 4.1. The pathways of food contamination.

Table 4.4. Most important radionuclides having biggest influence on food contamination (Vosniakos, 2001).

| | |
|---|---|
| Air | $^{131}I$, $^{134}Cs$, $^{137}Cs$ |
| Water | $^{3}H$, $^{89}Sr$, $^{90}Sr$, $^{131}I$, $^{134}Cs$, $^{137}Cs$ |
| Milk | $^{89}Sr$, $^{90}Sr$, $^{131}I$, $^{134}Cs$, $^{137}Cs$ |
| Meat | $^{134}Cs$, $^{137}Cs$ |
| Other food products | $^{89}Sr$, $^{90}Sr$, $^{134}Cs$, $^{137}Cs$ |
| Vegetation | $^{89}Sr$, $^{90}Sr$, $^{95}Zr$, $^{95}Nb$, $^{103}Ru$, $^{106}Ru$, $^{131}I$, $^{34}Cs$, $^{137}Cs$, $^{141}Cs$, $^{144}Cs$ |
| Soil | $^{90}Sr$, $^{134}Cs$, $^{137}Cs$, $^{238}Pu$, $^{239+240}Pu$, $^{241}Am$, $^{242}Cm$ |

## 4.4.4 Detection of radioactive contamination

Radioactivity is invisible, has no smell, and makes no sound - in fact it cannot be detected by any of our senses. However, because radioactivity affects the atoms that it passes through, we can easily monitor it using a variety of methods.

Geiger Counters, or a Geiger-Mueller (GM) tube with some form of counter attached, usually give the number of particles detected per minute ("counts per minute"). GM tubes use the ionising effect of radioactivity. This means that they are best at detecting alpha particles, because α-particles ionise very well. Different models of GM tubes are available for detecting α, β and γ radiation. The tube is filled with Argon gas, and voltages around 400 V are applied to a thin wire in the middle. When a particle enters the tube, it pulls an electron from an Argon atom and converts it to an ion. The electron is attracted to the central wire, and as it rushes towards the wire, the electron will knock other electrons from other Argon atoms, causing an "avalanche". Thus, one single incoming particle will cause many electrons to arrive at the wire, creating a pulse which can be amplified and counted resulting in a sensitive detector, although modern instruments are much more sensitive.

"Scintillation Detectors" works by the radiation striking a suitable material (such as Sodium Iodide), and producing a tiny flash of light. This is amplified by a "photomultiplier tube" which results in a burst of electrons, large enough to be detected. Scintillation detectors form the basis of the hand-held instruments used to monitor contamination in nuclear power stations. They can recognise the difference between α, β and γ radiation, and make different noises (such as bleeps or clicks) accordingly.

"Solid-State Detectors" are the most up-to-date instruments. They are used in particle-accelerator laboratories to show the results of high-energy collisions, with banks of them clustered around the collision site, feeding data into super-computers. The way they work is related to their characteristic of changing their physical parameters under the influence of radiation.

## References

Council Directive 1989, 89/109/EEC. 28 Nov 1989. Council Directive on the Approximation of the laws of member states relating to materials and articles intended to come into contact with foodstuff. Off J Eur Communities L347:37-44

Council Directive 1993, 93/43/EEC. 19 Jul 1993. Council Directive on the hygiene of foodstuff. Off J Eur Communities L347:37-44

Food and Drug Administration, 1997. Definitions. Title 21 Code of Federal Regulations, Section 7.3, 71.

Food and Drug Administration, 1999. Foods - Adulteration Involving Hard or Sharp Foreign Objects. FDA/ORA Compliance Policy Guide, Chapter 5, Sub Chapter, 555, Section 555.425 (Issued: 3/23/1999). Department of Health and Human Services, Public Health Service, Food and Drug Administration, Washington, DC.

Food and Drug Administration, 2001a. Metal Inclusion. Ch. 20. In Fish and Fishery Products Hazards and Controls Guide. 3rd ed., p. 249-258. Food and Drug Administration, Center for Food Safety and Applied Nutrition, Office of Seafood, Washington, DC.

Food and Drug Administration, 2001b. Glass Inclusion. Ch. 21. In Fish and Fishery Products Hazards and Controls Guide. 3rd ed., p. 259-268. Food and Drug Administration, Center for Food Safety and Applied Nutrition, Office of Seafood, Washington, DC.

Graves, M., A. Smith and B. Batchelor, 1998. Approaches to Foreign Body Detection in Foods. Trends in Food Science &Technology, 9, 21-27

Olsen, A., 1998. Regulatory Action Criteria for Filth and Other Extraneous Materials. Regulatory Toxicology and Pharmacology, 28, 181-189

Vosniakos, F.K., 2001. Studies on the radioactive transfer in air, soil, plants, foods. PhD Thesis. Sofia

# 5. Miscellaneous hazards

*Avrelija Cencič and Krzysztof Krygier (Illustrations by Anton Cencič)*

## 5.1 Objectives

The objectives of this chapter are:
- to become familiar with the terms GMOs, novel and functional food;
- to become familiar with the potential risks of GMOs, novel and functional food;
- to become acquainted with food safety and legislative measures regarding GMOs, novel and functional food in Europe.

## 5.2 Background

These days, we come across more and more different food types and individual products in our daily life, for example, genetically modified, convenience, regional, ethnic (e.g. kosher, halal), functional, medical or pharmafood, minimally processed, zero and low- fat, -cholesterol, -sugar, or -salt, light, organic, novel, healthy and many others. Generally, safety problems associated with such foods are well known and are the same as those for normal, typical foods. However, in some cases we come across foods of abnormal composition, of a new, unknown technology or from new sources. Such products have to be carefully examined from the safety point of view. In Europe such products generally come under the regulation on "novel foods and novel food ingredients". The most important groups of these novel foods are genetically modified (GM) foods and functional foods. In this chapter safety problems of the whole novel foods group will be discussed first, and then individual problems with GM foods and functional foods.

GMOs, and some novel and functional food are terms closely connected to the term Biotechnology. Biotechnology, and its pros and cons, is an issue that can be discussed endlessly, from many different angles. The consumer population is left either in fear or in restless convenience. There is almost nobody who feels completely comfortable with the fact that biotechnology occupies bigger and bigger parts of our everyday life (e.g. foods, clothes, and drugs). Biotechnology is based on biology and technology developing together with human knowledge.

What does the term biotechnology mean? Many would answer that biotechnology means genetically modified organisms (GMOs), cloning, genes, production organisms, tools and so on. It is obvious that the general perception about biotechnology does not readily cover the wider aspects as defined in the dictionary: "biotechnology is the industrial use of living cells".

We should not confuse genetic engineering with molecular biotechnology. Humans have been using genetic engineering for a long time ago already, for example, to breed two varieties of plants in order to obtain a better one. Since the discovery of the DNA structure by Watson and Crick in 1953 and the isolation of the first restriction enzymes in 1970, important progress has been made in biotechnology, especially in upstream processing.

Food production is a good and concrete example of a biotechnological process. It can be involved from the very beginning (improvement of raw material; plant breeding, animal breeding) or in upstream processing like strain development. Bacteria, yeast and moulds play the principal role in fermentation and biotransformation. Finally, in order to get a pure product, the purification is carried out during the downstream processing. Mankind suffers from a lack of food (or its poor distribution) and a lack of drugs that can cure diseases. Therefore, it is of extreme importance that judgements on biotechnology are not simply black and white. The term "sustainable development", used on many occasions, could be easily translated into "conditions of existence". No other method can guarantee a long and prosperous future for mankind. Poverty and environmental degradation go hand in hand. Tackling these problems is closely related to the policies that will be followed in transforming agriculture in developing countries. In spite of the great success of the " Green revolution" that brought to the fore organic or ecological farming, there are still some questions about whether environmentally sustainable as well as yield increasing methods can help food needs over the next two decades. Biotechnology - one of many tools of agricultural research and development - could contribute to food security by helping to promote sustainable agriculture centred on smallholder farmers, especially in developing countries.

Whatever the arguments on this sensitive issue may prove or disprove, it is evident that biotechnology has evolved from a pure science into a real life application. As long as the know-how is based on the policy of giving, getting and sharing, there should be no misinterpretation of the benefits of this symbiosis.

It is hard to believe that pure humanity is behind the arguments in favour of biotechnology, but there is great potential in this field; however, there are also many unknown parameters and the relatively short period of use has so far not yielded convincing or reliable statistical interpretations. Therefore, it is necessary to find out whether a product is really genetically modified, how it was modified and whether there are negative impacts to our health, before judging that all GMOs are harmful to us.

## 5.3 Genetically modified organisms (GMOs) and associated foods

### 5.3.1 Genetically modified organisms (GMOs)

Genetically modified organisms - these words are usually associated with a kind of »Frankenstein« creature that frightens the average consumer in Europe. The term GMO does not refer only to genetically modified plants that will be used in a food chain and in food products. After processing most such products do not contain traces of genetic engineering materials used in the initial steps of genetic engineering.

*General description*
Genetically modified organisms (GMOs) and genetically modified microorganisms (GMM's) can be defined as organisms (and micro-organisms) in which the genetic material (DNA) has been altered in a way that does not occur naturally by mating or natural recombination. The technology is often called "modern biotechnology" or "gene technology" and even "recombinant DNA technology" or "genetic engineering". It allows selected individual genes to be transferred from one organism into another, also between non-related species. Genetically modified organisms contain a gene or genes that have been artificially inserted. The inserted gene sequence (known as the transgene) may come from another unrelated organism, or from a completely different species: transgenic Bt corn, for example, which produces its own insecticide, contains a gene from a bacterium. As GMM's are not of specific issue in this chapter, the emphasis will be given to the genetically modified plants and animals. Plants containing transgenes are often called genetically modified or GM crops, although in reality all crops have been genetically modified from their original wild state by domestication, selection and controlled breeding over long periods of time. Identifying and locating genes for agriculturally important traits is currently the most limiting step in the transgenic process. We still know relatively little about the specific genes required to enhance yield potential, improve stress tolerance, modify chemical properties of the harvested product, or otherwise affect plant characters. Usually, identifying a single gene involved with a trait is not sufficient; scientists must understand how the gene is regulated, what other effects it might have on the plant, and how it interacts with other genes active in the same biochemical pathway. Public and private research programs are investing heavily in new technologies to rapidly sequence and determine functions of genes of the most important crop species. These efforts should result in identification of a large number of genes potentially useful for producing transgenic varieties. Once a gene has been isolated and cloned (amplified in a bacterial vector), it must undergo several modifications before it can be effectively inserted into a host organism (Figure 5.1). A constructed transgene must contain the necessary components for successful integration and expression:

1. A promoter sequence must be added for the gene to be correctly expressed (i.e. translated into a protein product). The promoter is the on/off switch that controls when and where the gene will be expressed in the organism. To date, most promoters in transgenic crop varieties have been "constitutive", i.e., causing gene expression throughout the life cycle of the plant in most tissues. The most commonly used constitutive promoter in plants is CaMV35S, from the cauliflower mosaic virus, which generally results in a high degree of expression in plants. Other promoters are more specific and respond to cues in the plant's internal or external environment. An example of a light-inducible promoter is the promoter from the cab gene, encoding the major chlorophyll a/b binding protein.
2. Sometimes, the cloned gene is modified to achieve greater expression in a plant. For example, the Bt gene for insect resistance is of bacterial origin and has a higher percentage of A-T nucleotide pairs compared to plants, which prefer G-C nucleotide pairs. In a clever modification, researchers substituted A-T nucleotides with G-C nucleotides in the Bt gene without significantly changing the amino acid sequence. The result was enhanced production of the gene product in plant cells.
3. The termination sequence signals to the cellular machinery that the end of the gene sequence has been reached.

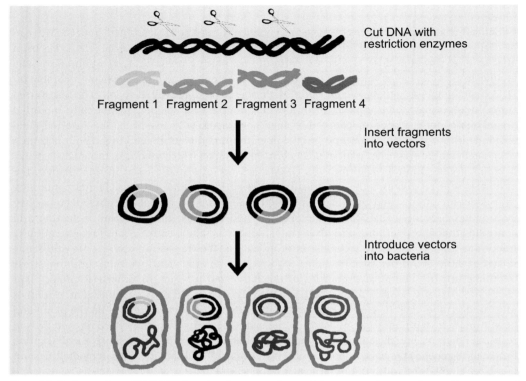

*Figure 5.1. To clone a piece of DNA, DNA is cut into fragments using restriction enzymes that recognize specific sequences of bases in DNA. The fragments are pasted into vectors that have been cut by the same restriction enzyme. Vectors (e.g., plasmids or viruses) are needed to transfer and maintain DNA in a host cell. Collections of clones are called libraries (www.fhcrc.org/.../basic/proaches/cloning.html).*

4. A selectable marker gene is added to the gene "construct" in order to identify plant cells or tissues that have successfully integrated the transgene. This is necessary because achieving incorporation and expression of transgenes in plant cells is a rare event, occurring in just a few percent of the targeted tissues or cells. Selectable marker genes encode proteins that provide resistance to agents that are normally toxic to plants, such as antibiotics or herbicides. As explained below, only plant cells that have integrated the selectable marker gene will survive when grown on a medium containing the appropriate antibiotic or herbicide. As for other inserted genes, marker genes also require promoter and termination sequences for proper functioning.

Cloning hosts are multiplied to amplify a recombinant gene. End products of such clones can be gene libraries, proteins, recombinant plants or animals (Figure 5.2 and Figure 5.3).

*Miscellaneous hazards*

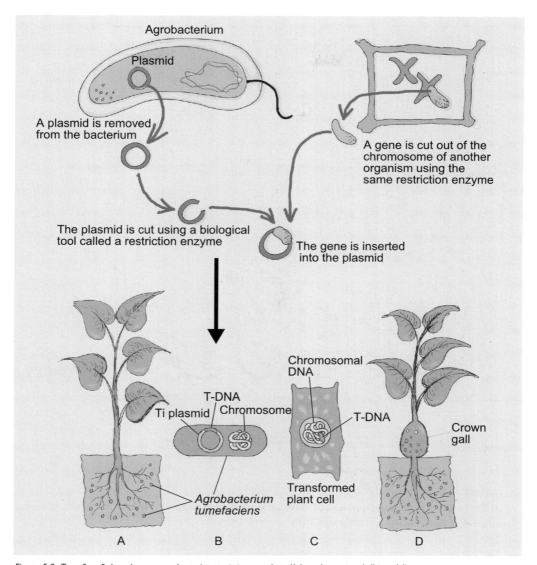

Figure 5.2. Transfer of cloned genes to plants (strategis.ic.gc.ca; http://elmo.shore.ctc.edu/biotech/).

After the gene has been efficiently transformed to a plant, usually with a plant pathogen bacteria *Agrobacterium tumefaciens* and after selection of recombinants, a transgenic plant undergoes an extensive evaluation process to verify whether the inserted gene has been stably incorporated without detrimental effects to other plant functions, product quality, or the intended agro ecosystem. Initial evaluation includes attention to:

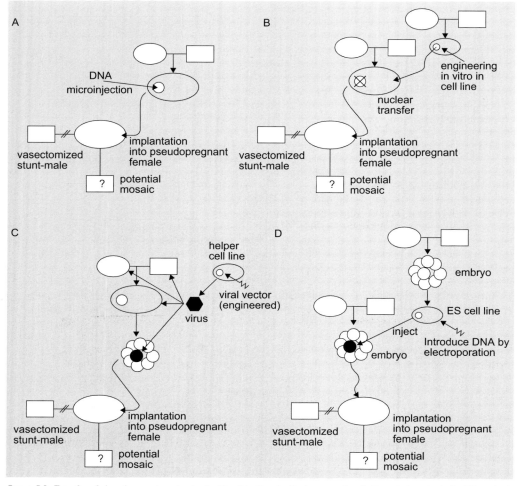

Figure 5.3. Transfer of cloned genes to animals. A. Microinjection; B. Nuclear transfer; C. Retroviral methods; D. ES methods (www.csun.edu/.../biotechnology/ ecl4/lecl4.html).

- activity of the introduced gene;
- stable inheritance of the gene;
- unintended effects on plant growth, yield, and quality.

If a plant passes these tests, it will probably not be used directly for crop production, but will instead be crossed with improved varieties of the crop. This is because only a few varieties of a given crop can be efficiently transformed, and these generally do not possess all the producer and consumer qualities required of modern cultivars. The initial cross to the improved variety must be followed by several cycles of repeated crosses to the improved parent, a process known

## Miscellaneous hazards

as backcrossing. The goal is to recover as much of the improved parent's genome as possible, with the addition of the transgene from the transformed parent.

The next step in the process is multi-location and multi-year evaluation trials in greenhouse and field environments to test the effects of the transgene and overall performance. This phase also includes evaluation of environmental effects and food safety.

Most of the transgenic crop varieties currently grown by farmers are either herbicide tolerant or insect and pest resistant. In addition to the crops listed in Table 5.1, minor acreages are planted with transgenic potato, squash, and papaya.

Transgenic crops are regulated at every stage in their development, from research planning through field-testing, food and environmental safety evaluations, and international marketing.

In spite of the fact that genetically modified animals are of main importance to the pharmaceutical industry, breeding of genetically modified animals for food is on the increase.

Table 5.1. Transgenic crop production area by country.

| Country | Area planted in 2000 (millions of acres) | Crops grown |
|---|---|---|
| USA | 74.8 | soybean, corn, cotton, canola |
| Argentina | 24.7 | soybean, corn, cotton |
| Canada | 7.4 | soybean, corn, canola |
| China | 1.2 | cotton |
| South Africa | 0.5 | corn, cotton |
| Australia | 0.4 | cotton |
| Mexico | minor | cotton |
| Bulgaria | minor | corn |
| Romania | minor | soybean, potato |
| Spain | minor | corn |
| Germany | minor | corn |
| France | minor | corn |
| Uruguay | minor | soybean |

### 5.3.2 Associated foods

Over the past decade, the development and adoption of transgenic technology has progressed more rapidly in the United States than in Europe. The majority (68%) of global GM crops acreage is now planted in the USA. The major reason for such a situation was the European regulatory climate, where a moratorium on GM crops only ended in May 2004, and a lack of public acceptance of GM plant material in the food chain. In spite of that fact, genetically engineered crops and additives are already used in the preparation of over 75% of processed foods. Millions of people are now unknowingly consuming novel genetically engineered foods in every meal. Some estimate that 70% to 80 % of processed foods contain ingredients made from genetically engineered corn, soybeans or cottonseed oil. These foods are sold without correct labelling, yet several studies show that they could contain toxins and may cause cancers, allergies and illness like conventional foods. Most supermarket processed food items now "test positive" for the presence of GMO ingredients. A substantial amount of GM soy has been detected in organic products. In addition, several dozen more GE crops are in the final stages of development and will soon be released into the environment and sold in the marketplace. According to the biotechnology industry almost 100% of US food and fibre will be genetically engineered within 5-10 years. There are approximately 40 foods that have been approved for commercial sale. Examples include: Canola oil, Chicory, Corn, Cotton, Papaya, Potato, Rice, Soybean, Squash, Sugar beet, Tomato and Dairy products.

Various groups have also done random testing on products to determine if the food contains genetically modified ingredients, for example: Frito-Lay Corn Chips, Kellogg's Corn Flakes, General Mills Total Corn Flakes Cereal, Post Blueberry Morning Cereal, Heinz 2 Baby Food, Enfamil ProSobee Soy Formula, Similac Isomil Soy Formula, Nestle Carnation Alsoy Infant Formula, Quaker Chewy Granola Bars, Nabisco Snackwell's Granola Bars, Ball Park Franks, Duncan Hines Cake Mix, Ultra Slim Fast, Quaker Yellow Corn Meal, Aunt Jemima Pancake Mix, Alpo Dry Pet Food, Gardenburger, Boca Burger Chef Max's Favorite, Morning Star Farms Better'n Burgers, Ovaltine Malt Powdered Beverage Mix, Betty Crocker Bac-O's Bacon Flavor Bits, Old El Paso Taco Shells, Jiffy Corn Muffin Mix. A list of GMO and associated food products approved under directive 90/220/EEC can be found at http://europa.eu.int/rapid/start/cgi and http://www.ucsusa.org/food_and_environment/biotechnology/page.cfm?pageID=337.

Animals used for xenotransplantation are not considered safe for human consumption and are excluded from the food chain by current regulations. Animals that are genetically modified to produce non-food products in their milk, eggs or blood might be introduced into the food chain - at least half of the born male population that produces no protein. Therefore, the safety of food products from such animals that were culled from transgenic lines might present concerns.

Meat and food products from non-genetically modified - cloned animals are considered to have a low level of food safety concern, although no real scientific analytical data have been published. The products from the offspring of cloned animals were regarded as posing no food safety concern because the animals are the result of natural mating.

Meat and food products of animals cloned from somatic cells are considered low risk because, currently, there is no evidence that the food products would pose a hazard. But there is also no published comparative analytical data assessing the composition of meat and milk products.

## 5.4 Risk to human health

We do not know what the real dangers are for humans or animals consuming genetically modified (GM) food. We can consider GM food to be as safe as conventional food. In the main, the associated risk is considered to be the unexpected consequences of inserting foreign DNA into the genome of a plant or animal. Genetic engineering, either in an animal or a plant, always has unpredictable outcomes and they are frequently greater than the intended change. This is because it is wrong to consider genes as independent units of information that can be accurately slotted into the genetic code of any organism. Genes have evolved within a given organism to work in combinations in the context of a complex genetic, biochemical and ecological environment. In GMOs, genes are taken from animals, plants, insects, bacteria, and viruses and then forcibly inserted into the DNA of food crops, bypassing natural safety barriers and creating pathways for diseases and genetic weaknesses to cross over from completely unrelated species (Figure 5.2 and Figure 5.3). Human health risks vary on a case-by-case basis. The level of risk is determined by the nature of the gene inserted and the stability and location of the inserted gene. Some proteins are allergens - they promote an allergic reaction in some consumers. If the gene for this protein were incorporated into a GM crop then consumers would probably be allergic to the GM crop too. Another risk to human health occurs if the inserted gene blocks the function of another gene. If the blocked gene is responsible for removing or altering a toxic substance in the crop, then a higher level of this toxin may be present in the GM variety. As a consequence of the nature of the gene inserted, the stability and location of the inserted gene as well as newly synthesised proteins and possible blocking of another gene's function, the main harmful effects of GMOs can be classified in the following categories:
1. allergenicity;
2. increased toxicity;
3. increased cancer risks;
4. horizontal transfer and antibiotic resistance;
5. changed nutrient levels.

*Allergenicity*
Allergic reactions in humans occur when a normally harmless protein enters the body and stimulates an immune response. If the novel protein in a GM food comes from a source that is known to cause allergies in humans or a source that has never been consumed as human food, the concern that the protein could elicit an immune response in humans increases. The possibility that we might see an increase in the number of allergic reactions to food as a result of genetic engineering has a powerful emotional appeal because many of us experienced this problem before the advent of transgenic crops, or know of someone who did.

However, there is no evidence so far that genetically engineered foods are more likely to cause allergic reactions than are conventional foods. The genetic engineering process itself does not create allergens. The nature of the genes that are chosen for transfer will determine whether allergens are introduced into the engineered host plant. The Food and Drug Administration (FDA) in USA, reviews proposed transgenic foods and compares possible allergens to a checklist of characteristics that have been found to be associated with allergenicity. In several years of testing dozens of proposed transgenic crops, only two potential problems have been uncovered: a soybean that was withdrawn from development and the now-famous StarLink corn. Although no allergic reactions to GM food by consumers have been confirmed, in vitro evidence suggesting that some GM products could cause an allergic reaction has motivated biotechnology companies to discontinue their development. StarLink, developed by the company Aventis, was originally intended to be sold as an all-purpose corn, but concern that it might be allergenic led to its approval only for animal feed. Although the issue of allergenicity remains unresolved, the panel noted that the government's estimate in early 2001 of about 0.4% StarLink in the nation's human food supply of corn was probably an overestimate and that aggressive actions by the government and by Aventis had reduced the amount to less than 0.125% by the summer of 2001.

The genetic engineering of animals intended for use as food will involve the expression of new proteins in animals since the safety, including the potential allergenicity of the newly introduced proteins, will have to be assessed. While most known allergens are proteins, only a few of the innumerable proteins found in foods are allergenic under typical circumstances of exposure. For example, consumption of food (especially milk) containing bioactive proteins or peptides could result in the transfer of such molecules into the bloodstream of newborn infants as the digestive epithelium of newborns permits the transient absorption of the whole proteins or large protein fragments until closure of the gut epithelium occurs.

Many genetically engineered fish and shellfish express and introduce a growth hormone (GH) gene - in order to promote rapid growth. Hence, it is particularly important to make sure that such a transgene product has no biologic activity in humans or animals that consume fish or shellfish expressing such a transgene.

Therefore, an adequate allergenicity assessment will require an understanding of several factors, including the source of the transferred protein, its level of expression, the physical and chemical properties of the protein and structural similarities to known allergens.

*Nature of acute disease*
If the proteins in transgenic and non-transgenic material produce an allergic reaction, the site of toxicity is the immune system. The protein is the antigen that is seen as foreign to the individual, and therefore, the production of antibodies is stimulated. Since proteins are larger and more complex molecules, they are more antigenic than more simple molecules such as polysaccharides. The antigens (also called allergens) usually cause an immediate hypersensitivity that can produce allergic rhinitis (chronic runny or stuffy nose); conjunctivitis (red eyes); allergic asthma; atopic dermatitis (uticaria or hives); as well as other symptoms. These symptoms occur due to the production of IgE antibodies, which attach to mast cells and basophiles, stimulating the secretion of histamine. The histamine as well as other chemicals produces the symptoms of the allergic reactions. Other food allergy symptoms include diarrhoea and colic, which are mediated by prostaglandins. If the immune response is severe enough, an anaphylactic reaction will occur in which life-threatening symptoms present, including restriction of the person's airway and profound hypotension. This type of allergic reaction requires immediate medical intervention.

*Increased toxicity*
Most plants produce substances that are toxic to humans. Most of the plants that humans consume produce toxins at levels low enough that they do not produce any adverse health effects. There is concern that inserting an exotic gene into a plant could cause it to produce toxins at higher levels that could be dangerous to humans. This could happen through the process of inserting the gene into the plant. If other genes in the plant become damaged during the insertion process it could cause the plant to alter its production of toxins. Alternatively, the new gene could interfere with a metabolic pathway causing a stressed plant to produce more toxins in response. Although these effects have not been observed in GM plants, they have been observed through conventional breeding methods creating a safety concern for GM plants. For example, potatoes conventionally bred for increased diseased resistance have produced higher levels of glycoalkaloids. In 1989 there was an outbreak of a new disease in the USA, contracted by over 5000 people caused by a genetically engineered brand of L-tryptophan, a common dietary supplement. Although contained less than 0,1 % of a toxic compound, it killed 37 Americans and permanently disabled or afflicted more than 5,000 others with a potentially fatal and painful blood disorder, eosinophilia myalgia syndrome (EMS), before it was recalled by the Food and Drug Administration. The manufacturer, Showa Denko, Japan's third largest chemical company, has already paid out over $2 billion in damages to EMS victims.

In 1999, front-page headline stories in the British press revealed Rowett Institute scientist Dr. Arpad Pusztai's explosive research findings that GE potatoes, spliced with DNA from the snowdrop plant and a commonly used viral promoter, the Cauliflower Mosaic Virus (CaMv), are poisonous to mammals. GE-snowdrop potatoes, found to be significantly different in chemical composition from regular potatoes, damaged the vital organs and immune systems of lab rats fed the GE potatoes.

*Increased cancer risks*
In 1994, the FDA approved the sale of Monsanto's controversial GE recombinant Bovine Growth Hormone (rBGH) - injected into dairy cows to force them to produce more milk - even though scientists warned that significantly higher levels (400-500% or more) of a potent chemical hormone, Insulin-Like Growth Factor (IGF-1), in the milk and dairy products of injected cows, could pose serious hazards for human breast, prostate, and colon cancer.

A number of studies have shown that humans with elevated levels of IGF-1 in their bodies are much more likely to get cancer. In addition the US Congressional watchdog agency, the GAO, told the FDA not to approve rBGH, arguing that increased antibiotic residues in the milk of rBGH-injected cows (resulting from higher rates of udder infections requiring antibiotic treatment) posed an unacceptable risk for public health.

In 1998, Monsanto/FDA documents were released by government scientists in Canada, showing damage to laboratory rats fed dosages of rBGH. Significant infiltration of rBGH into the prostate of the rats as well as thyroid cysts indicated potential cancer hazards from the drug. Subsequently the government of Canada banned rBGH in early 1999. The European Union has had a ban in place since 1994. Although rBGH continues to be injected into 4-5% of all US dairy cows, no other industrialised country has legalised its use. Even the GATT Codex Alimentarius, a United Nations food standards body, has refused to certify that rBGH is safe.

Another increased cancer risk from GMOs is considered to be a cauliflower mosaic virus promoter. When scientists use transgenic technology to put a new gene into a plant, they put in additional pieces of DNA to direct the activity of that gene. Each gene needs a "promoter" to turn it on under specified conditions. The most widely used promoter is the cauliflower mosaic virus 35S promoter, often abbreviated as the CaMV promoter or the 35S promoter. This promoter was obtained from the virus that causes cauliflower mosaic disease in several vegetables, such as cauliflower, broccoli, cabbage, and canola.

A multi-step chain of events would have to occur for the CaMV promoter to escape the normal digestive breakdown process, penetrate a cell of the body, and insert itself into a human chromosome. There is some evidence that fragments of DNA sneak into the blood stream and travel to some internal organs, and that these fragments sometimes become closely associated with host DNA. There is some evidence that the CaMV promoter poses little threat to human health. People have been eating it in small quantities for hundreds of years when we eat vegetables that are infected with the disease. Although vegetables heavily infected with CaMV are unappetising, there have been no documented negative effects on health from eating the virus or its promoter. In spite of that, some scientists say that the cauliflower mosaic virus promoter could increase the risk of stomach and colon cancers. The water and food contaminated by this GM material could hasten the growth of malignant tumours, therefore the health of people who live near the farm of such GMOs should be monitored.

*Horizontal transfer and antibiotic resistance*
The use of antibiotic resistance markers in the development of transgenic crops has raised concerns about whether transgenic foods will play a part in our loss of ability to treat illnesses with antibiotic drugs. The public is sensitised to this danger because of reports that some of the antibiotics that we rely on have lost efficacy after years of misuse through over-prescribing by doctors and improper use by patients. Animal husbandry enterprises such as cattle feedlots and chicken farms routinely use, and often over use, antibiotics as part of their feeding or treatment regimens, leading to publicity about the consequences of widespread overuse of these defences against disease.

At several stages of the laboratory process, developers of transgenic crops use DNA that codes for resistance to certain antibiotics, and this DNA often becomes a permanent feature of the final product although it serves no purpose beyond the laboratory stage. Will transgenic foods contribute to the existing problems with antibiotic resistance?

One of the concerns is related to the risk of horizontal gene transfer, that is, transfer of DNA from one organism to another outside of the parent-to-offspring channel. Transfer of a resistance gene from transgenic food to microorganisms that normally inhabit our mouth, stomach, and intestines, or to bacteria that we ingest along with food, could help those microorganisms to survive an oral dose of antibiotic medicine.

Although horizontal transfer of DNA does occur under natural circumstances and under laboratory conditions, it is probably quite rare in the acid environment of the human stomach.

Another concern is that the enzyme product of the DNA might be produced at low levels in transgenic plant cells. While high processing temperatures would inactivate the enzyme in processed foods, ingestion of fresh or raw transgenic foods could result in the stomach containing a small amount of an enzyme that inactivates an orally administered dose of the antibiotic. This issue was raised during the approval processes for Calgene's FlavrSavr tomato and Ciba-Geigy's Bt corn 176. In both cases, tests showed that orally administered antibiotics would remain effective. While the risks from antibiotic resistance genes in transgenic plants appear to be low, steps are being taken to reduce the risk and to phase out their use.

Researchers at the University of Newcastle have reported that microorganisms in the human digestive system took up a herbicide resistance gene after the human subjects ate a meal of GM soy. The experiment was small, involving 12 people with intact digestive systems and 7 who had undergone colostomies, in which their lower intestines were removed. In people with intact digestive systems, no GM DNA was found in the stools and no microorganisms took up GM DNA. But in people who had undergone colostomies, about 4 percent of the GM DNA survived the trip through the abbreviated intestinal tract and a small number of microorganisms took up GM DNA. The GM DNA in this experiment was a herbicide resistance gene rather than an antibiotic resistance gene, but the results suggest that horizontal transfer of GM DNA can occur in the human digestive tract under some circumstances.

Transgene transfer from plants to rumen bacteria in ruminants is theoretically possible when ruminants are fed with GMOs, but probably occurs at an extremely low frequency. Again, the risks resulting from dissemination will also depend on the nature and the function of the gene, and on the selective force acting on the outcome. Accurate prediction of possible consequences of the introduction of novel genes in an open environment is thus a highly complex issue that requires substantial further research and must currently proceed on a case-by-case basis.

With respect to antibiotic resistance marker genes, antibiotic resistances have become common and widespread since the corresponding antibiotics have become widely used in medicine and agriculture. The *bla*-TEM ampicilin resistance gene used in some varieties of transgenic maize is already detectable in *E.coli* strains, and 10 to 50% of the human gut strains are already ampicillin-resistant. It is conceivable that a new route of acquisition might be opened up for bacterial pathogens by feeding animals with transgenic plant material.

The EU has engaged scientists to assess the risk of horizontal gene transfer from genetically modified foods. A balance has to be found based on a combination of meaningful experimentation and careful reasoning of the use of GMOs for foods and feed.

*Changed nutrient levels*
How do genetically engineered foods compare with conventional foods in nutritional quality? This is an important issue, and one for which there will probably be much research in the future,

as crops that are engineered specifically for improved nutritional quality are marketed. However, the potential for engineering nutritional traits has so far taken a back seat to goals such as pest resistance and herbicide tolerance. Thus, there have only been a few studies to date comparing the nutritional quality of genetically modified foods with their unmodified counterparts.

The central question for GE crops that are currently available is whether plant breeders have accidentally changed the nutritional components that we associate with conventional cultivars of a crop. The isoflavone content of soybeans is a nutritional component that has been investigated because of its potential for preventing disease. The body converts some isoflavones into phytoestrogens, which are believed to help prevent heart disease, breast cancer, and osteoporosis. Soy is added to many health foods to increase dietary isoflavones. People who eat soy products for health benefits are interested in whether RoundupReady soybeans, which have been genetically engineered to allow them to survive being sprayed with Roundup herbicide, contain the same amount of isoflavones as conventional soybeans. The definitive study has not been done, but a comparison of available results reveals only small differences. On the other hand, there are no scientific publications showing the nutrient changes between transgenic animals (especially their milk) and their non-transgenic variety.

*Absorption, distribution, metabolism*
When scientists make a transgenic plant, they insert pieces of DNA that did not originally occur in that plant. Often these pieces of DNA come from entirely different species, such as viruses and bacteria (as shown in Figure 5.4). Is there any danger from eating this "foreign" DNA?

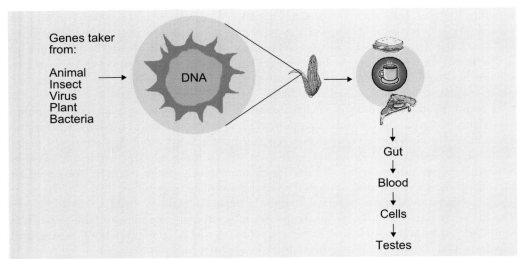

*Figure 5.4. Genetic engineering is starting to magnify health risks into actual disaster www.naturallaw.org.nz/genetics/HandBook/4.htm.*

We eat DNA every time we eat a meal. DNA is the blueprint for life and all living things contain DNA in many of their cells. What happens to this DNA? Most of it is broken down into more basic molecules when we digest a meal. A small amount is not broken down and is either absorbed into the blood stream or excreted in the faeces. We suspect that the body's normal defence system eventually destroys this DNA. In terms of the absorption, distribution, and metabolism of the specific transgenic material that is not naturally occurring in the plant, it is not a new type of material to our digestive systems. Humans consume approximately 0.1 to 1 gram of DNA in their diet each day, so it is present in very small amounts. Research has shown that dietary DNA has no known toxicity itself.

Further research in this area would help to determine exactly how humans have managed to eat DNA for thousands of years without noticing any effects from the tiny bits that sneak into the bloodstream.

When someone ingests a genetically modified food product, it goes through the same process of digestion in the body as would a non-modified food product. Therefore, the absorption, distribution, and metabolism of GMOs can be described by following the pathway for normal human digestion. During and after ingestion, the food is broken down into polymers as it passes through the mouth, oesophagus, and stomach. These larger organic molecules are digested into monomers, which are then transported across the wall of the small intestine into the blood and lymph. This transportation of the organic food molecules is the absorption process. Absorption occurs at a rapid rate because of the extensive surface area that exists within the folds of the small intestine.

In terms of distribution, different areas of the small intestine absorb and therefore distribute the different food monomers. For example, the absorption of carbohydrates and lipids occurs in the duodenum and jejunum, however the water and electrolytes are absorbed primarily in the ileum. Finally, metabolism of the broken-down GMO occurs, but varies from person to person. The body's metabolic rate is influenced by exercise, body temperature, age, sex, and rate of thyroid secretion. In general, however, individual differences in metabolic rate are mainly due to differences in physical activity.

What happens to foreign DNA that finds its way into the tissues of an organism? We suspect that the body's normal defence system eventually destroys fragments of foreign DNA. If some fragments are incorporated into the DNA of the host, they might be inactivated by mechanisms that control the activity of genes. Further research in this area would help to determine exactly how humans have managed to eat DNA for thousands of years without noticing any effects from the tiny bits that sneak into the bloodstream. So far there is no evidence that DNA from transgenic crops is more dangerous to us than DNA from the conventional crops, animals, and their attendant microorganisms that we have been eating all our lives.

It should also be pointed out that GMO DNA presents a potential risk only where the transgenes are not normally present at any significant concentration in the diet or in the commensal flora. The risks from oncogenic DNA are well known to regulatory authorities, but there is little information with which to assess the possibility of deleterious effects resulting from rare insertion of other sequences.

*Dose response*
Several published studies have attempted to show dose-response data on genetically modified foods in animals but have not yielded useful results. Some of these studies attempt to feed animals the whole food, others administer concentrated doses of the food, while others use the purified novel protein from the GM food. Nearly all of the GM foods that have been tested for toxicity have shown No Adverse Effect Levels (NOAEL's) at dosages far beyond the expected level of human consumption.

One example of an attempt to find dose-response information comes from a class of GMOs originally developed in the 1980's and now in widespread use. These are the so-called "Roundup Ready" plants that contain a protein allowing the plants to tolerate glyphosate, the active ingredient in the herbicide Roundup. In acute toxicity studies mice that were fed the GM protein by oral gavages at dosages up to 572mg/kg body weight experienced no adverse effects. There were no significant differences between treatment and control groups in body weight, cumulative body weight, and food consumption. This dosage can be said to represent a NOAEL for this particular protein. According to the authors, 572mg/kg body weight is approximately 1300 times the highest potential human consumption if the protein were expressed in soybean, corn, tomato, and potato.

Studies do exist that have shown toxic effects of GM foods in animals. The most well known involves a transgenic potato engineered to produce a natural insecticide. In this study, rats were fed GM potatoes for ten days and presented toxic effects in their stomach linings. The results were controversial because there was disagreement over whether it was the novel protein causing the effects or a component of the genetic engineering process.

## 5.5 Risk assessment

In response to the increased supply of genetically modified (GM) foods to international markets, the Ad Hoc Intergovernmental Task Force on Food Derived from Biotechnology of the Codex Alimentarius Commission (Rome) agreed in March 2003 on principles for the human health risk analysis of GM foods. These principles dictate a case-by-case pre-market assessment that includes an evaluation of both direct and unintended effects. They state that a safety assessment of GM foods needs to investigate direct health effects (toxicity), tendency to provoke allergic reactions (allergenicity), specific components thought to have nutritional or toxic properties, the stability of the inserted gene, nutritional effects associated with genetic modification and any unintended effects that could result from the gene insertion. Of particular note, the task force broadens risk assessment to encompass not only health-related effects of the food itself, but also the indirect effects of food on human health (*e.g.* potential health risks derived from out-crossing).

*Miscellaneous hazards*

In any risk assessment, however, it is important to differentiate between hypothetical and proven risks. And, to date, no food-derived health problems have been identified with the use of GM plants. However, it must be acknowledged that occasional pleiotropic, unintended safety relevant effects in organisms produced with traditional or modern biotechnology can occur and need to be addressed.

The decision by the Codex to include unintended effects (*e.g.* environmental health risks) in the risk assessment is an important new development. The link between environment and human health is formed through the exposure of humans to environmental hazards, where such hazards may take many forms, wholly natural in origin or derived from human activities and interventions. There have been several attempts to conceptualise environmental-human health interactions. Indicators for environmental health and methods for the consideration of the burden of disease from environmental risk factors are presently harmonised to support and monitor policy on environment and health for many developments. These concepts may be useful in the analyses of the effects of GM organisms for food production. Such assessments need to compare different approaches to food production, such as conventional, organic or GM technologies, and may also prove valuable in assessing regional differences (health relevant decreases or increases of pesticide use according to the local agro-ecological situation) in the impact of modern methods of food production.

## 5.6 Legislative measures

There are significant differences between the regulatory frameworks of Europe and the United States, demanding that manufacturers follow different procedures and requirements, depending on the final form in which the GM crop product will be marketed. The data on legislation can be retrieved at: http://europa.eu.int/eur-lex/en/search_lif_simple.html and http://www.fda.gov/opacom/laws/lawtoc.html and http://www.epa.gov/epahome/lawreg.html. In spring 2004, the European Commission finally called an end to the existing GM moratorium, which posed a *de facto* ban on all regulatory approvals of new genetically modified agricultural products in Europe. As a result, the European Commission has authorised the import of the GM maize variety NK 603 from Monsanto for processing and use in animal feed and the import of the insect-resistant Bt-11 maize of Swiss Syngenta AG for food uses. New rules on genetically modified food and feed are the subject of Regulation (EC) No 1830/2003 of the European Parliament and of the Council of 22 September 2003 on genetically modified food and feed (Official Journal L 268 of 18.10.2003). The Council adopted the proposal on 22 July 2003 but the directive is not yet published in the Official Journal. New rules were established: to ensure a high level of protection of human life and health, animal health and welfare, the environment and consumers' interests in relation to genetically modified food and feed, whilst maintaining the effective functioning of the internal market; to lay down Community procedures for the authorisation and supervision of genetically modified food and feed; to lay down provisions for the labelling of genetically modified food and feed. The new regulation is also concerned with environmental protection, setting out measures to be taken in the event of environmental problems, whilst ensuring that the demands made of economic operators are not too burdensome to implement. In addition, as part of the endeavour to improve and bring coherence to Community legislation "from farm to table", the Commission announced in the White Paper on Food Safety its intention to review Regulation (EC) No 258/97 on novel foods and novel food ingredients, including the introduction of new requirements at least equivalent to those in the regulatory framework for the deliberate release of GMOs under Directive 90/220/EEC (now 2001/18/EC).

### 5.6.1 GM labelling rules

EU Members States rejected all attempts to approve new products, stating the lack of strict EU rules on labelling GM food. On 18 April 2004, new labelling laws finally came into force.

GM labelling rules are the subject of Regulation (EC) No 1139/98 (Official Journal L 159 of 3.06.1998), amended by Commission Regulation (EC) No 49/2000 of 10 January 2000. Council Regulation of 26 May 1998 concerns the compulsory indication of the labelling of certain foodstuffs produced from genetically modified organisms or particulars other than those provided for in Directive 79/112/EEC. Regulation (EC) No 49/2000 laid down a *de minimis* level for the accidental content of DNA or proteins resulting from the genetic modification of food, whereby food which is accidentally contaminated with genetically modified soya or maize is not subject to the labelling requirements under Regulation (EC) No 1139/98, if the proportion is no higher than 1% of the food ingredient being considered.

Under the new regulations all foods and feed produced from GMOs have to be labelled. This includes all foods produced from GMOs, without making a distinction between those containing DNA (deoxyribonucleic acid) or protein resulting from genetic modification and those which do not. The new legislation also covers all genetically modified feed, giving it the same protection as food intended for human consumption.

All products approved in accordance with the Regulation should be subject to mandatory labelling, so the consumer will have more information concerning the labelling of genetically modified products, whether for human or animal consumption.

In addition, accidental and technically unavoidable traces of GMOs in conventional or organic foods have to be indicated on the label above a threshold of 0,5%. For unclear cases, the Commission announced it would publish labelling guidelines.

### 5.6.2 Traceability

As regulated in Regulation (EC) 1830/2003 of the European Parliament and of the Council of 22 September 2003 concerning the traceability and labelling of genetically modified organisms and the traceability of food and feed products produced from genetically modified organisms and amending Directive 2001/18/EC (Official Journal L 268 of 18.10.2003):
1. At the first stage of the placing on the market of a product consisting of or containing GMOs, including bulk quantities, operators shall ensure that the following information is transmitted in writing to the operator receiving the product:

a. that it contains or consists of GMOs;
b. the unique identifier(s) assigned to those GMOs in accordance with Article 8.
2. At all subsequent stages of the placing on the market of products referred to in paragraph 1, operators shall ensure that the information received in accordance with paragraph 1 is transmitted in writing to the operators receiving the products.
3. In the case of products consisting of or containing mixtures of GMOs to be used only and directly as food or feed or for processing, the information referred to in paragraph 1(b) may be replaced by a declaration of use by the operator, accompanied by a list of the unique identifiers for all those GMOs that have been used to constitute the mixture.
4. Without prejudice to Article 6, operators shall have in place systems and standardised procedures to allow the holding of information specified in paragraphs (1), (2) and (3) and the identification, for a period of five years from each transaction, of the operator by whom and the operator to whom the products referred to in paragraph 1 have been made available.
5. Paragraphs 1 to 4 shall be without prejudice to other specific requirements in Community legislation.

## 5.7 Novel and functional foods

Two main groups of novel foods are genetically modified and functional foods. However, every new product, new technology, new ingredient and new amount of a certain ingredient can be classed as a novel food. It is clear that many, very different cases can be expected. The procedure of acceptance of novel foods is very complicated, time consuming and very expensive. The main reason for this is assurance of the consumers safety.

### 5.7.1 Novel foods

Novel food is a relatively new term connected with unusual food products. In order to ensure the highest level of protection of human health, novel foods must undergo a safety assessment before being placed on the EU market. Only those products considered to be safe for human consumption are authorised for marketing. From the practical point of view there are many difficulties connected with commercialisation of novel foods or novel food ingredients. First of all because definitions are not clear and on the other hand cannot be clear, for example: "significant changes in the composition or structure of the food or food ingredient which affect their nutritional value, metabolism or level of undesirable substances" or "substantially equivalent to existing foods or food ingredients".

Companies that want to place a novel food on the EU market need to submit their application in accordance with Commission Recommendation 97/618/EC concerning the scientific information and the safety assessment report required. Novel foods or novel food ingredients may follow a simplified procedure, only requiring notifications from the company, at which point they are considered by a national food assessment body as "substantially equivalent" to existing foods or food ingredients (as regards their composition, nutritional value, metabolism, intended use and the level of undesirable substances contained therein).

A total of 37 applications were made to the European Commission between May 1997 and March 2002. By April 2002, 6 novel foods were approved for trading in the EU; 2 products were refused. The following were approved: 'phospholopids from egg yolk', 'yellow fat spread with added phytocholesterol esters', 'dextran preparation produced by *Leuconostoc mesenteroides*', 'pasteurised fruit-based preparations produced using high-pressure pasteurisation', 'trehalose as a novel food or novel food ingredient', 'coagulated potato proteins and hydrolysates thereof as novel food ingredients'. The two refused were: 'Stevia rebaudiana Bertoni' and 'Nangai nuts *Canarium indicum* L.

### 5.7.2 Functional foods

Although functional foods are located in the European regulation under Novel Foods, the safety of functional foods is of specific importance. The term "functional food" is not widely known among ordinary consumers and has never been granted an official definition. That's why many different definitions exist, but probably the most widely used definition in Europe is the following:

"A food can be regarded as functional if it is satisfactorily demonstrated to affect beneficially one or more target functions in the body, beyond adequate nutritional effects, in a way that is relevant to either an improved state of health and well-being and/or reduction of risk of disease. A functional food must remain food and it must demonstrate its effects in amounts that can normally be expected to be consumed in the diet" (Diplock, 1999; FUFOSE, 1995). A more simple definition of functional foods is: foods which are developed, manufactured or modified in such a way that they have obtained scientifically proven (authors' emphasis) specific health-promoting or disease-preventing properties. "Scientifically proven" is underlined because it is the most important factor of functional foods. Genuine functional foods need to have official scientific or medical confirmation of their healthy function.

In general, there are four basic groups of these foods:
1. natural food;
2. food to which a component is added;
3. food from which a component is removed;
4. food in which the bioavailability of a component is increased.

Although functional foods are not specifically regulated, the European Commission has published a draft proposal for regulations on nutrition, functional and health claims of food (Working Document Sanco/1832/2002/). The proposed regulation specifies that the use of nutrition, functional and health claims in the labelling, presentation or advertising of foods must not be false or misleading or give rise to doubt about the safety and / or the nutritional adequacy of other foods. The use of claims will only be permitted if the food or ingredient for which the claim is made has been proven to have a beneficial nutritional or physiological effect. The substance must also be present in the final product in a quantity sufficient to produce the intended effect.

*Miscellaneous hazards*

The scientific evidence for functional foods and their physiologically active component can be categorised in 4 distinct areas (Collective work 1999):
1. clinical trials;
2. animal studies;
3. experimental *in vitro* laboratory studies;
4. epidemiological studies.

Much of the current evidence for - at least potential - functional foods is not backed up by well-designed clinical trials, however the evidence provided through other types of scientific investigation is substantial for several of the functional foods and their health-promoting components. It is also evident that not only nutritional value but primarily the safety of functional foods has been examined.

Before starting a discussion on functional foods safety, it is necessary to know in as much depth as possible what kind of products come under the term functional foods. Probably the longest list of different functional foods was prepared in USA by the American Dietetic Association (presented in Tables 5.2, 5.3 and 5.4) and is very valuable, because it includes healthy

Table 5.2. Natural functional foods, key components and potential health benefits.

| Product | Key component | Health benefit |
|---|---|---|
| Vegetables and fruits | Vitamins, phytochemicals, fibre | Reduces cancer risk, reduces heart disease risk |
| Oatmeal/oat bran/whole oat products | Beta glucan soluble fibre | Reduces blood cholesterol |
| Whole-grain bread/high fibre cereals | Fibre | Reduces risk of certain cancers. Reduces risk of heart disease |
| Psyllium-containing products (e.g. Pasta, bread, snack foods) | Psyllium fibre | Reduces risk of coronary heart disease |
| Carrots | Beta carotene | Reduces risk of cancer |
| Broccoli | Sulphoraphane | Reduces risk of cancer |
| Tomato products | Lycopene | Reduces risk of prostate cancer. Reduces risk of myocardial infarction. |
| Tea, green or black | Catechins (e.g. EGCG) | Reduces risk of coronary heart disease. Reduces risk of gastric, oesophageal, skin cancers. |
| Fish | n-3 fatty acids | Reduces risk of coronary heart disease |
| Beef, dairy, lamb | Conjugated linoleic acid (CLA) | Reduces risk of mammary tumours |
| Soy | Soy protein, isoflavones | Reduces risk of coronary heart disease |
| Garlic | Organosulphur compounds | Reduces risk of cancer |
| Chicory root, bananas, garlic | Fructooligosaccharides | Supports normal, healthy intestinal microflora |
| Fermented dairy products | Probiotics | Reduces blood cholesterol. Reduces risk of cancer. They act in an immunomodulatory way and control enteric pathogens |

Table 5.3. Select functional foods with key component added.

| Product | Added key component | Potential health benefit |
|---|---|---|
| Juice, pasta, rice, snack bars and other foods with calcium | Calcium | Reduces risk of osteoporosis |
| Cereal with folic acid | Folic acid | Reduces risk of neural tube defect |
| Beverages, candies and other products with antioxidants | Vitamins C and E, beta carotene, phytochemicals | Supports heart health. Supports overall health. |
| Beverages with herbal additives | Variety: echinacea, ginkgo, kava, ginseng | Variety of health benefits |
| Modified margarine products | Plant sterols/stanols esters | Supports normal, healthy cholesterol levels |
| Eggs with ω-3 fatty acids | ω-3 fatty acids | Reduces blood cholesterol |
| Soups with herbal additives | Echinacea, St John's wort | Improves immune function. Reduces risk of depression |
| Medical food bar with arginine | L-arginine | Improves vascular health |

components and potential health benefits (original American table consists also of scientific evidence and regulatory classification in USA, ADA 1999).

Functional foods with healthy ingredients added are definitely the biggest group because there are a huge number of such ingredients. Potentially, every food product with an added functional ingredient can be a functional food. And the list of functional ingredients is very long: vitamins, minerals, healthy fatty acids, a lot of so-called phytochemicals including herbs, and many others. Examples of foods with these additives are presented in Table 5.3. In this case, possible dangerous overdosing can be taken into consideration.

The next group of functional foods, with a removed unhealthy component, is relatively diverse. On the one hand this is a big group of no/low fat, sugar or cholesterol products, on the other hand it includes individual products with removed allergic compounds, e.g. rice or peaches. A few examples are presented in Table 5.4. Products with removed unhealthy ingredients are generally safe, although there can be an additional safety problem when another ingredient replaces the removed compound. For example, when sugar is replaced by sugar alcohols, laxative problems can be observed.

Table 5.4. Select functional foods with removed unhealthy components.

| Product | Key component | Potential health benefits |
|---|---|---|
| Milk - low fat | Calcium | Reduces risk of osteoporosis |
| Low fats foods as part of low fat diet (e.g. cheese, snack foods, meat, fish, dairy) | Low in total fat or saturated fat | Reduces risk of cancer. Reduces risk of coronary heart disease |
| Foods containing sugar alcohols in place of sugar (gum, candies, beverages, snack foods) | Sugar alcohols | Reduces risk of tooth decay |

The smallest group of functional foods contains those foods in which the ingredients must not have any healthy effect other than to allow an increase in the availability of an already present healthy ingredient. The most popular example is inulin and calcium: inulin can increase the availability of calcium up to 30%. It means that a dangerous lack of calcium can be reduced in two ways: higher consumption of calcium or higher consumption of inulin with an increase in calcium availability.

## 5.8 Potential risks

Beyond the normal food safety problems, there are a few other specific problems connected with functional foods, which are discussed below:
Possible overdosing; this is a commonly expressed objection. Some experts say that some consumers can eat a lot of "healthy" foods which can result in overdosing. Fortunately, in most cases overdosing is not easy or even observed at all. This can be seen below in the list of vitamins (Table 5.5). On the other hand every functional ingredient should be described in two doses: recommended intake and dangerous excessive intake. Basically, two safety aspects of functional foods can be considered: possible overdosing with extremely high consumption and possible dangerous shortage of healthy ingredients with a very low consumption of such products. The potential health benefits presented in tables 5.2 to 5.4 cannot be obtained with too low a consumption of functional foods and this is a very important safety aspect of these foods.

The safety of functional foods can be connected with the type of functional foods. It is not widely known that many natural foods are in reality functional because of their high content of natural health ingredients. A list of such products is presented in Table 5.2. The biggest group is fruit and vegetables as a whole.

Natural functional foods seem to be the most safe because the composition is one hundred percent natural. The only safety problem that can occur is with extremely high consumption of such foods and a possible overdose of initially healthy ingredients.

Table 5.5. Daily intake and possible overdosing of some vitamins.

| Vitamin | Daily intake | Overdosing | Multiple daily intake |
|---|---|---|---|
| A | 1 mg | 8 mg | 8x |
| B1 | 2 mg | Not observed | Not observed |
| B2 | 2.6 mg | Not observed | Not observed |
| B6 | 2.4 mg | 1 g | 400x |
| B12 | 0.003 mg | Not observed | Not observed |
| C | 70 mg | 10 g | 140x |
| D | 0.01 mg | 0.5 mg | 50x |
| E | 10 mg | 1 g | 100x |
| PP | 23 mg | 1 g | 40x |

Functional foods can be consumed without a doctor's prescription and possible side effects cannot be eliminated. Moreover, functional foods can be eaten with drugs and possible interaction - positive or negative - can be observed. Furthermore, it is sometimes reported that functional food can be helpful for one disease but unfavourable for another.

Functional foods are consumed primarily by weak, sick and old people, thus quality and, in particular, safety should be paramount. This has to be taken into consideration when critical control points of the HACCP system are established.

Some ingredients used for functional foods production can interfere with the bioavailability of other ingredients. One very well known example is sucrose polyester spc (better known as olestra or olean), a fat substitute that is used for the production of fat-free foods. This ingredient is not absorbed and therefore its caloric value is zero. But, as a consequence, the availability of fat-soluble vitamins is abnormally low and finally a deficiency of these vitamins can be observed. Consequently, foods with sucrose polyester are enriched with fat-soluble vitamins. Another similar but very positive example, is the use of plant sterols or stanols in reducing the availability of cholesterol. Foods, primarily spreads, with plant sterols/stanols reduce the risk of cardiovascular diseases and have become very well known and popular functional foods.

## 5.9 Legislative measures

In Europe there exists a special regulation on novel and some functional foods: Regulation (EC) No 258/97 of the European Parliament and the Council of January 1997 concerning novel foods and novel foods ingredients *(Official Journal L 043, 14/02/1997 p. 0001-0007)*. The regulation concerns the placing on the market within the Community of novel foods and novel ingredients which fall under the following categories:
1. foods and food ingredients containing or consisting of genetically modified organisms with the meaning of Directive 90/220/EEC;
2. foods and food ingredients produced from, but not containing, genetically modified organisms;
3. foods and food ingredients with a new or intentionally modified primary molecular structure;
4. foods and food ingredients consisting of or isolated from micro-organisms, fungi or algae;
5. foods and food ingredients consisting of or isolated from plants or food ingredients isolated from animals, except for food and food ingredients obtained by traditional propagating and or breeding practices and having a history of safe food use;
6. foods and food ingredients produced by a production process not currently used, where the process gives rise to significant changes in the composition or structure of the food or food ingredient which affect their nutritional value, metabolism or level of undesirable substances.

In Europe, the procedure of acceptance of any novel food or novel food ingredient is as follows. First of all the application is considered and accepted in the company's country. Later acceptance by all European countries and finally the authorisation of the European Commission are needed.

## Miscellaneous hazards

In this procedure a lot of institutions and experts are involved in examining the safety of novel foods/ingredients. Of course, every country uses its own institutions and experts. For example, in Poland there are the following:

A. Research institutions:
- Institute of Food and Nutrition;
- National Institute of Hygiene;
- Institute-Centre of Children Health;
- Institute of Plants and Herbs;
- National Veterinary Institute.

B. Experts on:
- allergy;
- internal diseases;
- diabetes;
- endocrinology;
- pharmacy;
- pharmacology;
- gastroenterology;
- cardiology;
- sport medicine;
- oncology;
- paediatrics;
- gynaecology and obstetrics;
- herbalism;
- food technology;
- human nutrition;
- food toxicology;
- food microbiology;
- environmental protection.

Both lists clearly show that within Europe the safety of novel foods and novel food ingredients are treated with special attention and precision. It is also very clear for producers that a lot of different assessments - meaning a lot of time and money - are needed to be absolutely sure of safety, and to finally place the novel food or novel food ingredients on the European market (Table 5.6).

When discussing the safety problems associated with functional foods, we must always remember that not eating them poses a safety problem. A lack of healthy ingredients can result in serious ill health.

Table 5.6. Differences in regulations for food and medicinal applications of GM plants between the US and EU (Kleter et al., 2001).

| Item | European Union | United States |
|---|---|---|
| GM foods & ingredients[a] | "Novel foods" that require safety assessment (Regulation 258/97) | Not considered different from other technologies for food production (FFDCA). Focus on differences in product; should be "generally recognized as safe", except for food additives and plant pesticides (see below)[b] |
| GM food additives[a] | Requirements not different from those for other additives, except for labeling (Regulation 50/2000) | Introduced foreign gene products may be considered "food additives" (FFDCA; FDA, 1992) |
| GM plant pesticides[a] | No specific regulations, included in evaluation for environmental releases (Directive 2001/18) and food use (Regulation 258/97) | For pesticidal products from foreign genes, EPA establishes a threshold level of tolerance (FIFRA) |
| GM food supplements[a] | No harmonized EU legislation | GM food supplements are not distinguished from other supplements (DSHEA) |
| GM medicines, chemical substances | Applications for GM medicines only through centralized procedure (Regulation 2309/93) | No specific regulation for GM medicines (FFDCA) |
| GM medicines, biological substances | Applications for GM medicines only through centralized procedure (Regulation 2309/93) | "Generic" GM biological medicines still point of discussion (FFDCA) |

[a] According to a new proposed EU regulation, the GM component of any food item, including food ingredients, food additives, and food supplements, would be subject to the same evaluation before marketing
[b] Mandatory pre-market notification of GM foods was recently proposed by FDA

# References

ADA, 1999. Functional foods. J.Am.Diet.Assoc. 99, 1278.
Diplock A.T., 1999. Scientific concepts of functional foods in Europe: consensus document. Brit.J.Nutr. 81, Suppl. 1, 1-27
FUFOSE. Functional food science in Europe, FAIR-95-0572
Kleter, G.K., W.M. van der Krieken, E.J. Kok, D. Bosch, W. Jordi and L.J.W.J. Gilissen, 2001. Regulation and exploitation of genetically modified crops. Nature Biotechnology, 19, 1105 - 1110.

# 6. Quality assurance systems and food safety

*Pieternel Luning, Willem Marcelis and Marjolein van der Spiegel*

## 6.1 Introduction

In this section the aim, study objectives and structure of the chapter are described. Subsequently, the relevance of assuring food safety and historical developments in food quality management are briefly described and discussed.

### 6.1.1 Aim, study objectives and structure of chapter

The aim of this chapter is to provide a concise description of the principles and approaches of major international quality assurance systems, and to discuss them from the viewpoint of food safety. Attention is given to technological and managerial aspects that are relevant to the performance of quality assurance systems based on the paradigm of the techno-managerial approach. The importance of assuring food safety (see also Chapter 1: Agri-Food Production Chain) is briefly described in subsection 6.1.2, followed by a short description of historical developments in quality assurance systems (6.1.3). In section 6.2, the techno-managerial approach (6.2.1) is explained, a short overview is given of typical technological aspects (6.2.2) (see also Chapters 2-4: Microbial, Physical, Chemical Hazards) and managerial aspects are discussed in more detail (6.2.3). Principles and approaches of the specific QA systems are described in section 6.3, whereas in section 6.4 these systems are considered from the specific viewpoint of food safety, taking into account technological and managerial aspects. In section 6.5, the philosophy of Total Quality Management is discussed in the context of assuring food safety. The last section (6.6) contains a case study, which can be used to expand on the topics discussed in this chapter. Cross-links with other chapters of this book are indicated between brackets in the text.

For a proper understanding of the text we have given descriptions of some terms that are often used differently in both literature and practice.

The study objectives of this chapter are defined as follows:
- The student must be able to describe the general principles/approaches of the major quality assurance (QA) systems: GMP, HACCP, ISO 9001, and BRC, and the principles of Total Quality Management.
- The student must be able to recognise and discuss technological and managerial aspects that can influence the performance of QA systems with respect to assurance of food safety.

### 6.1.2 Importance of assuring food safety

In the last decade several serious food safety incidents have occurred in several agri-food chains in Western Europe. Examples are Bovine Spongiform Encephalopathy (BSE) and classical swine fever (CSF) in 1997, the dioxin affair in 1999, foot and mouth disease (FMD) in 2001, the nitrophen and medroxyprogesterone acetate (MPA) incidents in 2002, and more recently Avian Influenza in 2003. Although not all the crises had direct consequences for human health, they had a great impact on economies, politics and consumer behaviour (LNV, 1997; 2001; Crawford, 1999; USDA, 2001; Agriholland, 2003; Tacken et al., 2003).

Besides these specific affairs, the incidence of foodborne diseases is still increasing worldwide. Miles and co-authors (1999) reviewed important pathogens involved in food poisoning and infection incidents. They mentioned that in Europe and North America the rise in notified cases was in particular due to the increased incidence of micro-organisms of animal origin such as *Salmonella*, *Campylobacter* and *E. coli* (especially *E. coli* O157:H7). For example, in England and Wales the number of reported incidents rose from 70,130 in 1993 to 93,351 in 1999, with the number of food poisonings due to *Campylobacter* and *E. coli* increasing in particular (CDR, 1999). Moreover, it should be noted that the number of recorded incidents provides only a crude estimate of the real incidents of illness, as only a proportion of food poisonings are identified. Van Logtenstijn and Urlings (1995) proposed that the real incidence of food poisoning is probably 50-100 times higher than the number of reported cases.

Van der Spiegel et al. (2003) summarised the possible reasons for increased food safety problems in agri-business and the food industry including:
- increased complexity of supply chains and networks;
- increased consumption of manufactured food (such as convenience food);
- changes in way of chopping and consumption (such as more outdoor consumption);
- increased vulnerability of consumers to infectious diseases (such as an increase in the elderly population);
- changes in food production methods and techniques (such as mild processing);
- intensification of agriculture (such as intensive rearing);
- increased international trade and travel (such as the import of more new (exotic) products).

So, there are several reasons for the rise in food safety problems. Moreover, consumer awareness about food quality and safety has increased as well, which is also reflected in the media attention given to a variety of food safety affairs (Miles et al., 1999). Therefore, companies in agri-business and the food industry not only have to deal with a wide range of risk factors (Chapters 2, section 5) but also an increased risk perception (see also Chapter 14). In anticipation of these developments and trends, quality assurance systems are being developed, implemented, improved and/or combined to give guarantees of food quality and safety to consumers, purchasers, retail, government and other stakeholders. Quality assurance includes that part of the quality management system focused on fulfilment of quality requirements and providing confidence in meeting customer requirements (Table 6.1). Quality control activities are an essential part of QA systems.

*Table 6.1. Description of terms used.*

| Term | Description |
|---|---|
| Food control | Control is the ongoing process of evaluating performance (comparing test results with targets) and taking corrective action when necessary. |
| Food infection | The presence of living pathogens in food can cause food infection, i.e. the higher the number of bacteria the higher the chance of illness. The pathogens present in the food can penetrate the intestinal mucosa and colonise in the gastrointestinal tract or in other tissues. These pathogens can be reduced in food processing by heat treatment. |
| Food inspection | Inspections are often carried out to check the actual quality of the final (manufactured) product. |
| Food poisoning | Toxic compounds that are produced by pathogen bacteria (enterotoxins) and moulds (mycotoxins) cause food poisoning. Toxic compounds are released in raw materials or food products. The formation of toxic compounds should be prevented because the compounds are often heat resistant and therefore cannot be controlled by heat treatment. |
| Food safety | Refers to the requirement that products must be 'free' of hazards with an acceptable risk. |
| Hazard | Is a potential source of danger. |
| Quality assurance | Part of a quality management system focused on fulfilment of quality requirements and providing assurance in meeting customer requirements (ISO, 1998). |
| Quality management | The broad description quality management is defined as the total of activities and decisions performed in an organisation to produce and maintain a product with the desired quality level with minimal costs. Quality management is that part of the overall management activities that focuses on quality. |
| Quality performance | Performance of production systems has been commonly evaluated by measuring costs or by measuring intrinsic quality aspects (such as safety). Quality performance considered from a broader perspective includes quality dimensions of the product (product quality, availability and costs) and those of the organisation (flexibility, reliability and service), and also includes the extended quality triangle (Luning *et al.*, 2002, van der Spiegel *et al.*, 2003). |
| Quality system | A quality system is defined as the organisational structure, responsibilities, processes, procedures and resources that facilitate the achievement of quality management. |
| Quality assurance system | This term is used for the specified systems such as HACCP, BRC, ISO that cover part of a quality system. |
| Risk | A measure of the probability and severity of harm to human health. |
| System | Interrelated and interacting processes working in harmony (ISO, 1998). |

## 6.1.3 Short historical developments in food quality management

The importance of food quality has long been recognised. However, the perspective of concern about quality has changed over the years from quality inspection to total quality management. This change was supported by different philosophies developed by gurus, such as Deming, Juran,

Crosby and Feigenbaum. The shift in perspectives on food quality management is briefly described below. For details about quality philosophies, see the literature (Evans and Lindsay, 2004; Luning et al., 2002).

Concerns about food quality started with quality inspection. During the Middle Ages in Europe, skilled craftsmen served both as manufacturer and inspector. Guilds tested meat, fish and bread to protect consumers from bad quality and overly expensive products. Product characteristics were examined, measured and compared with specific requirements to assess the conformity. For example, bread was tested on grain quality, weight and amount of butter (Spapens, 1996). This means of production was lost with the industrial revolution until the middle of the eighteenth century, when interchangeable parts were used which necessitated inspection of quality. Quality inspectors took out non-conforming products at the end of the production line. Separate departments were created which were responsible for quality, whereas senior managers focused their attention on outputs like quantity and efficiency (Evans and Lindsay, 2004). So, quality products were obtained by removing defects by inspection. The focus was on quantity and not on quality and production was driven by technology and not by what customers actually wanted.

Since the beginning of the 20$^{th}$ century, many useful techniques have been developed for the control and improvement of quality problems. Several quality philosophers advocated that appropriate quality could be obtained by reduction of uncertainty and variability in the design and manufacturing process. To achieve this, control and improvement were needed. Quality control is used to eliminate systematically problems that cause defects, and to establish a better process control by application of e.g. control charts and statistical methods. Quality control involves determining what to control, establishing units of measurements so that data may be objectively evaluated, establishing standards of performance, measuring actual performance, interpreting the difference between actual performance and the standard, and taking action on the difference in order to prevent quality problems in the next batch/production. Improvement is a kind of control in the control process, where attention is paid to structural causes and solutions (Luning et al., 2002).

Control activities form the basis of several QA systems, such as HACCP (safety guarantee by using critical control points). Improvement activities are explicitly incorporated in the new ISO 9001 family and are a basic principle of Total Quality Management (TQM).

Since the last half of the 20$^{th}$ century the complexity of the agri-food supply chain has increased considerably. Raw materials are obtained from sources worldwide, an ever-increasing number of processing technologies are used, and a broad range of products is available to the consumer. In addition, consumer expectations are continuously changing, with a desire for convenience, less processed, and fresher foods with more natural characteristics. Moreover, there are a number of reasons why, especially in the agri-food business, implementation of quality assurance systems is an issue of the greatest importance:
- Agricultural products are often perishable and subject to rapid decay due to physiological processes and/or microbiological contamination.

*Quality assurance systems and food safety*

- Most agricultural products are harvested seasonally.
- Products are often heterogeneous with respect to desired quality parameters, such as the content of important components (e.g. sugars), size and colour. This kind of variation is dependent on cultivar differences and seasonal variables, which are hard to control.
- Primary production of agricultural products is undertaken by a large number of farms operating on a small scale.

Against this background, the total food supply chain has to assure and demonstrate that the highest standards of quality and safety are maintained (Hoogland *et al.*, 1998). Quality assurance encompasses all planned and systematic actions required to ensure that a product complies with the expected quality requirements. It also provides customers and consumers with the assurance that quality requirements will be met. Quality assurance focuses on system quality instead of product quality, to control what is being done. The system must be audited to ensure that it is adequate both in design and use.

So, the focus changed from how to inspect to how to produce. Food products were not only tested on their product characteristics, but also on production, packaging, handling and distribution.

Since 1987, several non-food sectors have been striving for total quality excellence (Evans and Lindsay, 2004). Total quality management (TQM) is used to manage all aspects of business (Zhang, 1997). TQM is not merely a technique, but a philosophy anchored in the belief that long-run success depends on a uniform and firm-wide commitment to quality, which includes all activities that a firm carries out. It covers a total, company-wide effort including all employees, suppliers, and customers, and it seeks continuously to improve the quality of products and processes to meet the needs and expectations of customers (Dean and Evans, 1994, Gatewood, *et al.*, 1995). There are probably as many different approaches to TQM as there are businesses. Although no program is ideal, successful implementation programs share many characteristics. The basic attributes for successful performance of TQM are customer focus, strategic planning and leadership, continuous improvement, and empowerment and teamwork (Dean and Evans, 1994). Although TQM practices are assumed to provide the best possible conditions to meet or exceed the needs and expectations of customers and so enhance business excellence, no success stories are yet known in the food industry. Some studies indicate that drivers for TQM are present in the food industry, but a lot of effort is still required to reach an excellent organisation level (Kramer and Van den Briel, 2002, Hendriks and Sonnemans, 2002).

The principles of major QA systems and TQM are described in more detail in sections 6.3 and 6.6.

## 6.2 Technological and managerial aspects in quality assurance

In this section the principle of the techno-managerial approach is explained, and relevant technological and managerial aspects regarding safety and quality control and assurance are described.

### 6.2.1 Techno-managerial approach

Management of food quality is a rather complicated process. It involves the complex characteristics of food and their raw materials, such as variability, restricted shelf life, potential safety hazards, and the large range of (bio) chemical, physical, and microbial processes. Therefore, food products can be considered as dynamic systems that have variable behaviour over time. On the other hand, human behaviour plays a crucial role in quality management. Human handling is rather unpredictable and changeable. As a consequence, the result of agribusiness and the food industry, as the combined action of individuals working with agrifood products and striving for quality, is much more uncertain than is often assumed. So, in analysing food quality management, both the use of psychology to understand human behaviour and the use of technology to understand behaviour of living materials, are needed.

Luning, et al. (2002) proposed the techno-managerial approach as a way to analyse and solve these complex quality issues. They distinguished three different approaches, i.e. the managerial, the technological and the techno-managerial approach as illustrated in Figure 6.1. The approaches differ in the extent to which they integrate managerial and technological sciences. Typical for the managerial approach is the way that technological aspects are contemplated as facts: "We can make everything we want to make, in fact there are no technological restrictions". Typical managerial measures focus on changing organisational conditions, improving the level of knowledge and skills of operators, development of procedures and systems, etc. In the traditional technological approach, management aspects are considered as boundary restrictions: " They want everything finished yesterday and never provide the appropriate budget". Typical

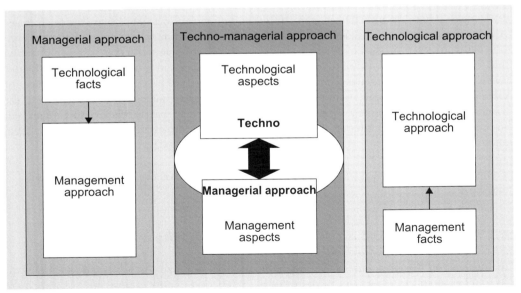

Figure 6.1. Techno-managerial approach (adapted from Luning et al., 2002).

technological measures for solving quality issues are, for example, obtaining a better understanding of the (bio) chemical mechanisms, the development of more sensitive (e.g. microbial) analyses, reducing defects by genetic modifications, etc. In contrast, the techno-managerial approach encompasses the integration of both technological and managerial aspects from a systems perspective. The core element of this approach is the contemporary use of technological and managerial theories and models in order to predict food systems behaviour and to generate adequate improvements of the system. Integration of both sciences should be developed to such an extent, that:
1. technological behaviour and changes can be judged on the effects they have on human behaviour and,
2. human behaviour and changes can be judged on the effects they have on technological behaviour.

For illustrations of applications of the techno-managerial approach in food quality management research, see the literature (Hadiprodjo, 2004; Alverado-Casanova, 2004; Rodjanatham, 2004; Van Geel, 2004). In the framework of this book, typical technological and managerial aspects are selected and described that play a role in the performance of QA systems regarding food safety. This information will support the techno-managerial way of approaching food quality and safety issues (Luning and Marcelis, *submitted*)

### 6.2.2 Technological aspects

In this section on technological aspects we focus on characteristics of food hazards and sampling requirements that are important to consider when developing or implementing a quality assurance system on a company level. Moreover, there is a schematic illustration of how food quality changes in a food production chain. It is increasingly recognised that the quality and safety of the end product depends on quality assurance in all parts of the chain and proper transfers between each actor.

*Characteristics of food hazards*
As previously mentioned, raw materials, ingredients and manufactured foods are dynamic systems, which are affected by a wide range of (process) factors along the food production chain (such as, time-temperature, product composition, hygienic conditions). These factors contribute both to desired (e.g. sensory properties, shelf life) and undesired attributes (e.g. safety risks). Safety of the final product is a result of product characteristics and process conditions (technological) on the one hand, and human behaviour and working conditions (managerial) on the other hand. According to the techno-managerial approach a good understanding of techn(olog)ical aspects is required. So as part of assuring food safety it is important to understand:
- what are the possible origins of hazards (i.e. sources of danger);
- what are their characteristics also in relation to the user;
- how can relevant hazards be controlled.

For more detailed information on biological, chemical, physical and genetic risk factors in the food chain, see Chapters 2-5.

In the framework of this chapter the hazards are considered from the viewpoint of how to control them, as shown in Table 6.2. It is clear that typical microbial hazards (especially food infections) can be controlled by proper time-temperature conditions, whereas chemical compounds should be controlled by prevention at the supply (e.g. supply control of raw materials, control of suppliers) or proper time-temperature (prevention of toxin formation). So from the perspective of quality assurance it is important to identify the relevant characteristics of food hazards and to translate these into proper control measures. Where in the production chain and how often to control, can be assessed by different systematic approaches such as HACCP (see also Chapter 7) and FMEA (Failure Mode Effect Analysis) (Luning *et al.*, 2002). The decision to control or not can be supported by a systematic risk assessment approach (see also Chapter 9). Furthermore, it is also important to be aware of the time span of the negative health effects. Some effects can be acute, such as allergic reactions or food poisoning, whereas in the long term, chronic effects can occur, such as cancer, and heart and vascular disease. Chronic effects can be due to, amongst others, unhealthy diets and long-term exposure to chemical agents.

Therefore, for the assurance of food safety a good understanding is required of the type of hazards, its characteristics and its controllability. Other attention points regarding technological food characteristics are:
- Many hazards can occur concurrently, and/or they can interact.
- Control of hazards is not restricted to one specific place. Sometimes many places in the production chain must be controlled to assure safety. For example, to maintain the shelf life and safety of ready-to-eat meals the complete chain (storage, distribution, retail, consumer) after assembly of ingredients must be controlled by a strict time-temperature regime.
- Finally, it should be mentioned that consumer demands and safety requirements are sometimes conflicting. For example, consumers prefer minimally processed food (fresh characteristics) with a considerable shelf life. However, minimal processing has consequences for the microbial stability and safety of the product, and thus for the shelf life.

*Sampling*
A major part of assuring food safety concerns control activities. For details on technological procedures of control, and microbial and chemical control, see Chapters 11 and 12. From the perspective of quality assurance, attention is paid in this subsection to the specific requirements of sampling governed by the dynamic character of (raw) food products.

In order to control a product, process or system it is important to understand the sources of variation. Variations in a production process or system can be due to different sources including people, materials, machines and tools, methods, (sampling) measurements, and environment. The combined effect of all these sources are inherent to the system and account for 80-90% of observed variation, also called common causes. Reduction of variation due to common causes can only be achieved by structural improvements (e.g. better equipment). The remaining 10-20% of variance is due to specific causes and ad hoc situations (e.g. bad batch from a material supplier) (Evans and Lindsay, 2004).

Table 6.2. Illustration of food hazards considered from their possible origin, characteristics and controllability.

| Type of hazard | Possible origins | Characteristics | Controllability (examples) |
|---|---|---|---|
| Microbial | Initial contamination of vegetable or animal (raw) materials; (Cross) contamination, e.g. from raw to processed products, unhygienic personal or working conditions, from intestinal tract, etc.; Contaminated air; Water used for e.g. cleaning, processing (if poor water quality); Inadequate time-temperature (favours growth); | Tolerance of environmental conditions increase survival chances; Microbial replication characteristics, e.g. self-replication, generation time; Virulence factors, like ability to circumvent host's immune response, ability to attach to surface, ability to synthesise various toxins; Dynamics of infection: latency (delayed onset of symptoms) and disease pattern e.g. acute, chronic. | Proper time-temperature conditions (heating for inactivation, cooling to prevent growth); Use of anti-microbial agents; Low water activity ($A_w$). |
| Chemical | Natural toxins; Toxins produced during processing (e.g. heterocyclic amines in broiled, smoked or deep-fried fish or meat) or storage; Environmental contamination (e.g. PCB's, dioxins); Allergenic compounds. | Is agent mutagenic, carcinogenic, toxic or allergic; Solubility of agents (fat-soluble, subsequent accumulation in human body); Degradability or inactivation; Actual toxicity. | Genetic engineering or plant breeding to eliminate natural toxins (prevention); Proper storage (t-T, RH) conditions to prevent formation; Rejection of batches at supply control; Sometimes t-T conditions (heating). |
| Physical | From animal materials; From (harvesting, processing) equipment; From environment. | Size; Sharpness and hardness; Decomposition rate of physical agent. | Visual inspection at e.g. supply or cleaning; Special detection equipment e.g. metal detector. |

*Pieternel Luning, Willem Marcelis and Marjolein van der Spiegel*

Sampling is a major tool in controlling safety and quality along the food production chain. Food samples can be evaluated by direct measurements (e.g. measuring of pH, visual inspection) or analysis (Luning *et al.*, 2002). The latter approach is applied when target attributes cannot be directly measured. The analysis procedure involves sampling, sample preparation, and actual measurement or analysis of the target compound(s). Interestingly, the different steps of the analysis procedure are often a major source of variance in themselves. Sampling contributes most often to the total error, whereas the actual analysis or measurement has the least effect. Ideally, the sample should be identical to the bulk and must reflect the same intrinsic properties, e.g. same textural properties, or equal concentration of toxins, or same taste. Typical causes of variation in sampling of agri-food products include irregular shape, composition changes occurring during sampling (e.g. loss of vitamins, water, volatile compounds), and the inhomogeneity of raw materials, ingredients and manufactured foods.

Sampling must be carried out according to a statistically designed sampling plan (for details, see Luning *et al.*, 2002). Which sampling plan should be chosen depends on:
1. The purpose of inspection, e.g. for acceptance sampling or process control
2. Nature of the material to be tested, e.g. homogeneity, what is the history, costs of the raw material?
3. Character of the test/lab procedure, e.g. (non) destructive, importance of test
4. Nature of lot/batch/population, i.e. sizes and number of units involved, how are sub-lots treated?
5. Level of assurance required, costs versus information obtained
6. Characteristic of sampling parameter, i.e. attribute or variable (Gould and Gould, 1993).

A critical step in the analysis of agri-food products is the sample preparation. Samples are made to minimise undesirable reactions (e.g. enzymatic and oxidising reactions), to obtain homogeneous samples, to prevent microbial spoilage of samples and/or to extract the relevant (target) compound(s). It is important to realise that during sample preparation the concentration of the target compound(s) can change, which can affect the corrective action taken upon comparison with the target value.

After sample preparation the actual analysis or measurement is performed. The reliability depends on several aspects including specificity, accuracy, precision and sensitivity of the analysis. *Specificity* is the ability to measure what actually should be measured. It can, for example, be affected by interfering substances, which react similarly to the actual compound to be measured. *Accuracy* is the degree to which a mean estimate approaches a true estimate of the measured substance. Deviation may be due to inaccuracy inherent in the method, the effects of foreign compounds, and alterations in the compound during analyses. *Precision* is the degree to which the determination of a substance yields an analytically true measurement of that substance. As a rule analyses should not be made with a precision greater than required. *Sensitivity* is the ratio between the magnitude of instrumental response and the amount of the substance (Pomeranz and Meloan, 1994).

## Quality assurance systems and food safety

Thus, sampling methods and measuring techniques must be critically evaluated for use in quality assurance systems. Although not discussed above, it should be mentioned that the use of appropriate statistics in sampling is a basis for proper control activities to assess food quality and safety. For details on specific chemical, microbial and other safety-related analyses, see Chapters 11 and 12 on microbial and chemical analyses.

*Food behaviour from a chain perspective*
From the perspective of quality and safety assurance, each actor (e.g. breeder, supplier, company, or industry) partly contributes to the overall quality assurance of the ultimate products as consumed by final users. Products from agricultural origin are living materials and keep on changing due to a complex range of (bio) chemical, microbial, physiological and physical processes at different stages along the whole production chain. Figure 6.2 shows a schematic example of the quality loss behaviour of a certain vegetable food. During breeding commonly, the concentration of desired compounds is increasing (e.g. pigments, vitamin C content, flavour components), and is hopefully retained during post-harvest storage. Processing, especially heat treatments, generally have a negative effect on natural compounds in fresh produce, while storage and distribution may give a further gradual decrease in quality. In other situations processing treatments are aimed at developing desired attributes (e.g. texture improvement, or brown colour formation of baked products), and then quality loss starts immediately after processing.

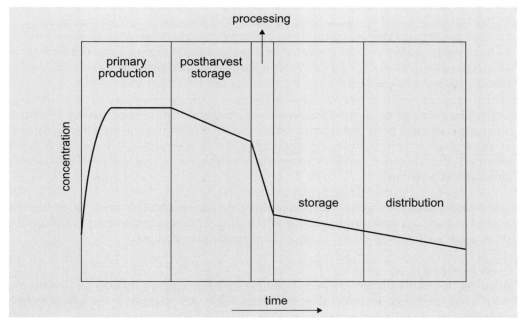

Figure 6.2. Schematic depiction of change in a quality attribute along the food processing chain (adapted from Van Boekel, 2005).

*Pieternel Luning, Willem Marcelis and Marjolein van der Spiegel*

From the viewpoint of quality assurance it is important to identify where in the chain and how (what are the critical process parameters) most loss in quality and safety can occur. Controlling critical safety and quality points along the whole food production chain will support the production of end products with a high safety and quality standard. However, common QA systems are mainly focused on actor level and only a few branch specific quality assurance systems attempt to take into account the whole production chain (Luning *et al.*, 2002).

## 6.2.3 Managerial aspects

As previously stated, human behaviour and its working environment can affect food safety as well. Where food technology involves the typical food product and production aspects, quality management is focused on human behaviour and its environment. The basic functions in (food) quality management are planning, control, leading and organising.

Planning and control with respect to decisions on quality occur at different levels, i.e. strategic, innovative and operational, which are called 'quality strategy and policy', 'quality design' and 'quality control', respectively. Additional quality management functions include 'quality improvement' and 'quality assurance', which have been developed in anticipation of, respectively, the ideas of quality philosophers and the requirements of customers and legislation concerning food safety and quality reliability. Quality assurance is the process of deciding on: the assessment of requirements of the QA system, control of performance and auditing of the QA system, and deciding on future developments as part of strategic decisions.

Leadership is focused almost exclusively on the "people" aspects of getting a job done. Leading with respect to quality includes inspiring, motivating, directing, and gaining commitment to organisational activities and quality goals. Motivation and (quality) behaviour are typical aspects that are affected by leadership.

Organising is the process of arranging people and other resources to work together in order to accomplish (quality) goals. This management function of 'organisation' involves the creation of an administrative infrastructure providing the best conditions for goal-oriented decision-making. For details about quality management functions, see Luning *et al.* (2002) and Luning and Marcelis (*submitted*).

In the framework of this book we focus on decision-making, quality behaviour and managerial aspects of changing (work) conditions (e.g. implementing systems, improvement), and, finally, some considerations with respect to chain management are mentioned.

*Decision-making*
Behaviour is reflected in the activities of farmers, growers, breeders, operators, and managers but also the consumers in the agri-food production chain. Activities can be considered a result of decision-making. From the managerial viewpoint it is important to understand the factors that affect the decision-making process. Availability of information and existence of interests are two major factors influencing both individual and group decision-making processes. When

## Quality assurance systems and food safety

decision-making takes place in groups, additional information is used, which is exchanged between group members in communication processes, besides the regular information input. With respect to interests, in decision-making in groups, people often have conflicting interests and they have the possibility of influencing each other, sometimes causing conflicts. How groups or managers deal with conflicts is called conflict management. In addition to individual decision-making, in a group situation there are differences in power. Power is the ability to drive others in the direction you prefer, and in the way that you want. The major aspects of decision-making are summarised from the following literature (Radford, 1975; Schein, 1990; Hellriegel and Slocum, 1992; Schermerhorn, 1999; Luning *et al.*, 2002) and briefly described below.

Decision-making is the process of defining problems and selecting a course of action from the generated alternatives. This process is complicated by the complexity of collected information, uncertainties in decision and the methods for implementation.

In practice, complete information is almost never available for a certain situation. One often has to deal with a particular amount of uncertainty. As a consequence, decisions are made in the context of being insufficiently informed about a problem, its alternative solutions and their respective outcomes. So, decision-making includes risks: the more information and the better its reliability, the smaller the risk. Therefore, it is important that the decision-maker has enough information to determine the probabilities associated with each alternative.

The second important factor, which reduces the amount of room for decision, is the interests of individuals or groups. Conflicting interests may result in stress situations for the decision-maker. Different types of origins of interest can be distinguished:

Interests from society ethics provide a set of rules that define right or wrong behaviour. For example, some ethical issues of relevance in food quality management are animal welfare, use of hormones, and genetic modification (see also Chapter 15).

Interests from the firm's environment, the stakeholders. Primary stakeholders, such as employees, suppliers and buyers, have formal and/or contractual relationships with the firm, whereas secondary stakeholders have a less formal connection. Secondary stakeholders are, for instance, animal rights groups, consumer groups and environmentalists.

Interest from inside the organisation refers to the organisation's culture, which can have a considerable influence on decision-making behaviour as well.

The third aspect, power, is the ability to make things happen the way one wants them to happen. Power is more than just influence; it is the possibility of getting someone to do something even when he does not agree with it. Five distinct types of power include:
1. Legitimate power, which is based on the formal position in the organisational hierarchy.
2. Reward power, which stems from the ability to reward followers.
3. Coercive power, which results from the ability to obtain compliance through fear or punishment.

*Pieternel Luning, Willem Marcelis and Marjolein van der Spiegel*

4. Referent power, which is based on followers' personal identification with the leader.
5. Expert power, which is based on having specialised knowledge and information.

The first three bases of power are related to the position in the organisation, whereas the last two are linked with personal characteristics. It is important to realise that power affects the decision-making processes, and it is also useful to understand how. Also, in chain situations power can have a big influence on decisions that are actually taken.

As decision-making plays a role at all stages in a food company (or farm or retail, etc), it can be expected to affect food safety at any time and at any place in the chain. On a management level decisions have to be made on, for example, the extent of sampling and the number of critical control points. In these situations, management has to balance the cost of measuring and control against potential costs due to safety problems, like recall or negative publicity. Management also has to decide on investment in resources such as new equipment, qualified personnel, education and training, etc. Resource decisions often contribute to structural improvements. At operational level decisions are made as well, for example, acceptance or rejection of baths, regulation of pasteurisation temperature, washing hands or not. Commonly, operators/employees can only take decisions within given administrative/working parameters (i.e. organisational structure, equipment, raw material, education level).

All these decisions, on different levels at different locations and at various moments, can have an impact on food safety and quality. Therefore, it is important to understand how decisions are made and how the decision-making process can be affected (e.g. by providing appropriate information).

*Quality behaviour*
Decision-making processes related to quality issues result in actual quality behaviour, i.e. actions related to quality and safety issues. For example, the execution of corrective actions at critical control points, decisions that are taken when accepting or rejecting a batch with certain defects, and actions that are taken when a product has to be recalled from the market.

In the framework of this book we focus on the major factors influencing behaviour with respect to quality performance. In 1990, Gerats developed a research model on quality behaviour. According to his model, two conditions should be considered when analysing quality behaviour, i.e. disposition (employee's own disposition to behave in a certain direction), and ability (objective opportunity to behave in a certain direction, i.e. the activity area), as illustrated in Figure 6.3. Gerats (1990) concluded from his research in a slaughterhouse, that the activity area (ability) for hygienic working behaviour was mainly limited by shortcomings in management with respect to hygiene, by low hygiene standards amongst workers, by low hygiene standards of first line supervisors, and by shortcomings in the hygiene facilities at the workplace. Disposition to hygienic working was mainly limited by the low knowledge level of bacteriological contamination mechanisms, by the restricted social support from colleagues, by the lack of interest in hygienic working amongst supervisors, and by the limited opportunities for hygienic

## Quality assurance systems and food safety

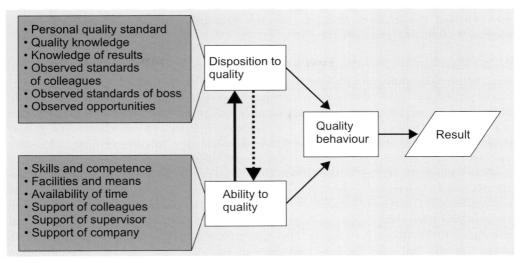

Figure 6.3. Quality Behaviour model of Gerats (1990).

working. Ivancevich and co-authors (1994) added a third factor, namely quality focus. In their view, commitment to quality includes three ingredients:
1. Quality intelligence: employees must be aware of the acceptable quality standards and how these standards can be met.
2. Quality skills: employees must have the skills and ability to achieve the quality standards set by management.
3. Quality focus: from top management to operating employees, everyone must sincerely believe that quality of all output is the accepted practice.

Commitment to quality can be obtained with motivational programs, where leadership plays a crucial role.

From a managerial point of view, leadership and organisation are the dominant factors affecting quality behaviour. In fact, leadership is one of the management functions, which is mainly focused on human behaviour aspects of getting a job done. It involves inspiring, motivating, directing, and gaining commitment to organisational activities and goals. With respect to decision-making, the effectiveness of leadership depends on personal characteristics but also on the use of power, communication capabilities and the manner of dealing with interests and conflicts. Effective leadership creates clear values that respect the capabilities and requirements of employees and other company stakeholders. It sets high expectations for performance and performance improvement. It builds loyalties and encourages teamwork based upon the values and the pursuit of shared purposes. But it also encourages and supports initiatives and risk taking amongst subordinates in the organisation, whilst avoiding chains of command that require long decision paths. The importance of leadership in implementing quality management is generally acknowledged. For details on leadership, see the literature (Ross, 1999; Schermerhorn, 1999; Robbins, 2000; Luning et al., 2002).

Besides leadership, organisational conditions also influence decision-making. For example, in small enterprises managers influence their employees directly, whereas in large companies rules, procedures, and structures are frequently needed to direct peoples' behaviour. Organisational concepts can be classified along two axes: 1) centralisation versus decentralisation of responsibility and authority, and 2) few versus many rules and procedures. The 'simple structure' concept is characterised by a flat organisation (few rules/procedures) where the decision-making authority is centralised (e.g. a powerful boss, often the owner, in a small or medium enterprise). In the 'functional structure' concept, people with similar skills and task are grouped and members of functional departments share technical expertise, interest and responsibilities. This type of organisation is characterised by many rules and procedures and a centralised responsibility and authority. Many larger food companies have a functional structure. In the 'division structure' concept, groups of people who work on the same product/process and serve the same customers are located in the same area or geographical region. This organisational structure is characterised by many rules and procedures but authority and responsibility are decentralised at division level. In the 'network structure' concept a central core is linked through networks of relationships with outside (or inside) contractors and suppliers of essential services/products. It is characterised by few rules and decentralised authority and responsibility. Network structures can cover a whole firm or parts of it, e.g. a front office in a new market having great decision authority and not hindered by procedures and rules (Luning et al., 2002). In the latter concept the quality behaviour of employees is less affected by organisational constraints and offers opportunities for flexible and efficient organisation, but it is more difficult to control and co-ordinate.

A concept that fits well in a 'less tight' organisation (fewer rules and more decentralisation of responsibility and authority) is empowerment. Empowerment is the process by which managers enable and help others to gain power and achieve influence. Effective leaders not only accept participation but they also empower others. They know that when people feel powerful, they are more willing to make decisions and take actions needed to perform their jobs. These leaders also realise that to gain power, it is not necessary for others to give it up. The success of an organisation may depend on how much power can be mobilised throughout all ranks of employees (Schermerhorn, 1999).

The concept of empowerment is associated with several developments in the applications of teams. Working in teams offer opportunities to stimulate anticipation at the workplace. Five forms of teams can be distinguished, differing in the level of empowerment (Schermerhorn, 1999; Robbins, 2000):
- Committees and task forces, i.e. small teams that work on a specific purpose outside daily job assignments (e.g. developing quality policy and quality systems) on a continuous or temporary basis.
- Cross-functional teams, i.e. teams consisting of people from different disciplines at a similar hierarchical level, who have to accomplish a task (e.g. development of a new product).

- Quality circles, i.e. a group of people that meets regularly to discuss and plan specific ways to improve work quality.
- Virtual teams, i.e. teams that use computer technology to link physically dispersed members in order to achieve common goals.
- Self-managing teams, i.e. teams that take their own responsibility, e.g. determination of work assignments, collective choice of inspection procedures, control over the pace of work, evaluation of each other's performance.

The above-mentioned examples are formal groups in the organisation. Informal groups can also exist, though these are not registered on organisation charts but rather emerge from natural and spontaneous relationships. Informal groups can have a positive impact on work performance.

With respect to the control of food safety it is important that when employees are empowered and work in teams, they are provided with enough knowledge about microbiological processes and given the means to take the right decisions so that no irresponsible risks are taken.

So, from a managerial point of view, it is important to realise that, when developing and implementing QA systems, quality behaviour aspects should be given special attention.

*Managerial aspects of implementing QA systems*
When developing and implementing a QA system, management also has to deal with resistance to change, a subject that has been well described for improvement processes. In the framework of this book there is mention of just a few major points of attention with respect to implementation of QA systems. For details, see Luning *et al.* (2002).

In studies on the implementation of advanced manufacturing technologies, it was observed that the most difficult aspect of implementing new technology and systems was changing the organisation and people (Smith and Edge, 1990). The implementation of QA systems is not just the introduction of a system with procedures and/or guidelines which have to be followed. It often requires a change in beliefs and values shared by people in the organisation (Dean and Evans, 1994).

Some of the major reasons for resistance to change include uncertainty and insecurity, reaction against the way change is presented, threats to vested interests, cynicism and lack of trust, and a lack of understanding. Moreover, resistance to change may also result in a reaction against being controlled or getting less autonomy or power. When changing a situation (e.g. implementation of a QA system) a change strategy should be developed, including elements such as education, communication, participation, facilitation, negotiation and compulsion. Different strategies require different elements depending on the specific situation (Gatewood *et al.*, 1995). Lewin (1952) developed one of the earliest 'change-models'. He distinguished three phases:
- Unfreezing i.e. disrupting forces that maintain the existing state or behaviour, e.g., by introducing new information to show discrepancies between the current and desired situation.

- Moving entails a transition period during which behaviour in the organisation or department is shifted to a new level. It involves developing new behaviours, values and attitudes through changes in structure, technology, strategy and human processes.
- Re-freezing stabilises the organisation at a new state of behavioural equilibrium.

Various change strategies can be used to implement and/or improve QA systems. Figure 6.4 summarises three common change strategies, i.e. force-coercion strategy, rational persuasion strategy and shared power strategy. Each of the strategies acts differently in the unfreezing, moving, and re-freezing phases (Schermerhorn, 1999).

In the force-coercion strategy, power is the basis of legitimacy. Rewards and punishments are the primary inducements to change. Likely outcomes of force-coercion are immediate, often temporary, compliance with targets but little commitment. The new behaviour continues only as long as the opportunity for rewards and punishments is present. For this reason, force-coercion is most useful as an unfreezing device that helps people break old patterns of behaviour and gain the initial motivation to try new ones.

The rational persuasion strategy attempts to initiate change through persuasion supported by special knowledge, empirical data, and rational arguments. The likely outcome is ultimate compliance with reasonable commitment. It is an informational strategy which assumes that facts, reason, and self-interest will guide rational people when they are deciding whether or not to

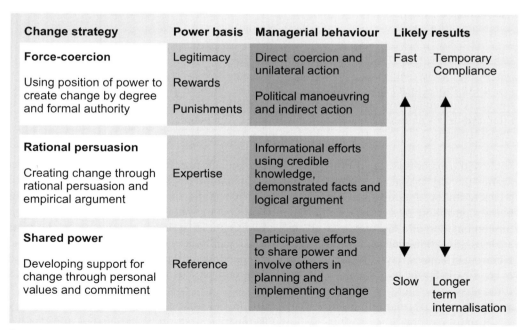

Figure 6.4. Alternative change strategies and their managerial implications (adapted from Luning et al., 2002).

support a change. This strategy largely depends on the presence of expert power. Expertise can be obtained in the form of consultants, external experts or from credible demonstration projects.

The rational persuasion strategy unfreezes and re-freezes a change situation. It is slower than force-coercion but tends to result in more sustainable change.

The shared power strategy engages people in a collaborative process of identifying values, assumptions, and goals from which support for change will naturally emerge. The process is slow, but it is likely to yield long-term commitment. This approach is based on empowerment and is highly participative in nature. It relies on involving others in examining personal needs and values, and it relies on group norms. Managers using the shared power approach to achieve change need reference power and skills to work effectively with other people in group situations. They must allow others to participate in making decisions that affect the planned change and the way it is implemented. Since it entails a high level of involvement, a normative re-education strategy is often quite time-consuming, but is likely to result in long-lasting change.

So, from a managerial point of view, it is important to choose a strategy explicitly for developing, implementing and/or improving QA systems, and to be aware of the consequences of each strategy. Improper functioning of QA systems might result in food-safety problems. For example, a study to determine the success and failure of HACCP indicated that several outbreaks of pathogens were due to improper functioning of the HACCP system; two outbreaks of *salmonellosis* appeared to be due to the improper performance of corrective actions (Motarjemi and Käferstein, 1999).

*Supply chain management*
All the handling involved in the production of perishable foods, such as cultivation, harvesting, production, supply and storage, sets high requirements for quality assurance in the whole chain. Poor handling at the beginning of the supply chain can result in serious safety and quality problems in the end product, which often cannot be eliminated by subsequent manufacturing or cooling processes. An example is the formation of aflatoxins in cereals, peanuts and corn, when the produce is not properly handled after harvesting. Too humid and too warm conditions favour the growth of *Aspergillus flavus*; this mould can produce aflatoxins which cannot be removed by any other treatment in the chain except rejection of the batch.

Safety requirements set by consumers, legislation, and new distribution techniques enabling large-scale production but requiring an availability of capital that exceeds the individual organisation, requires a supply chain approach. The supply chain is defined as all stages of the entire process from raw materials to final product. Typical motives for this approach as driven by characteristics of agri-food production are:
1. restricted shelf-life of produce and manufactured products;
2. variability of quality and quantity of supplies of farm-based inputs;
3. differences in lead time between successive stages, complementary to agricultural inputs (available in joint packages only);
4. stabilisation of consumption of many food products;

5. increased consumer awareness concerning both product and processing method;
6. fast decay of intrinsic quality after harvesting and/or processing;
7. availability of capital (Zuurbier et al., 1996).

Partnership is a crucial aspect of supply chain management. The idea behind it is that each actor in the chain adds value to the end product with minimum cost, finally obtaining a product with a high quality and safety level. Every firm is part of a value system and by means of co-operation the entire performance of the value system can be improved. Establishing value-added partnerships (VAP) is very relevant for firms operating in agribusiness and the food industry. Basically, a VAP is temporary and partial. Its structure and organisation are the result of joint activities, exchange of information, people and means (Zuurbier et al., 1996). If markets or technologies change, it might mean that partnerships are dissolved. The success of the partnership becomes apparent in aspects such as higher quality products, increased market access and more efficient processes.

However, the following aspects can complicate chain collaboration:
1. Restricted access to information for all actors/partners in the chain.
2. Opportunist behaviour of actors. Firms may take advantage of their position and provide e.g. incomplete or incorrect information, so affecting the uncertainty and complexity of the business environment.
3. The organisational structures may hinder collaboration (e.g. routine structures may impede organisational learning).
4. Balance of power between actors. Relations can vary from symmetric to asymmetric. A relationship is asymmetric if control is unbalanced.
5. Appropriation of resources. The inability to justify costs and benefits to each of the actors may hamper the development or duration of partnerships (Luning et al., 2002).
6. Mutual trust: the need for trust between partners had been identified as an essential element in creating good buyer-supplier relationships (Rousseau et al., 1998; Geyskens et al., 1998).

Although chain collaboration is very complex, with respect to the assurance of food safety, there is often a clear benefit for all participants, which enables collaboration. For more details on supply chain management, see the literature (Zuurbier et al., 1996; Hughes, 1994).

## 6.3 Principles of Quality Assurance systems

In agribusiness and the food industry QA systems like GMP, HACCP, ISO and BRC are now widely applied. However, only a part of total quality can be realised by using these specific quality systems, because they cover only part(s) of a quality system. A quality system is defined here as the organisational structure, responsibilities, processes, procedures and resources that facilitate the achievement of quality management. Whereby quality management includes the total of activities and decisions performed in an organisation to produce and maintain a product with desired quality level at minimal cost.

*Quality assurance systems and food safety*

This section provides a concise overview of the principles of established QA systems: Good Practice codes and HACCP (subsection 6.3.1), ISO 9000 family (6.3.2), and various other (combined) systems (6.3.3). The HACCP approach is only briefly mentioned because it is discussed in detail in Chapter 7.

### 6.3.1 Good Practice (GP) codes and Hazard Analysis Critical Control points (HACCP)

*Objective and guidelines of Good Practice (GP) codes*
The basic objective of GP codes is to combine procedures for manufacturing and quality control in such a way that products are manufactured consistently to a quality appropriate for their intended use (IFST, 1991). GMP consists of fundamental principles, procedures and means needed to design a suitable environment for the production of food of acceptable quality.

GP codes have been drafted, and still are being drafted, by governments, international bodies, and organised interest groups like consortia from the food industry and control authorities. For example, the Codex Alimentarius Food Standards Programme developed Codes of Hygienic Practice.

The structure of GP codes is not uniform like the HACCP guidelines. GP codes have been developed for different sectors (such as pharmacy, animal feed, food), for different types of products within a sector (such as good manufacturing practice (GMP) codes for fruits and vegetables, frozen foods), and for specific topics such as hygiene (GHP), laboratory (GLP) and agricultural (GAP) activities. Although the target (sector, group, activity) might be different, common topics are included, such as means (e.g. buildings, equipment), materials, methods (e.g. procedures, instructions) and personnel (e.g. training, job description). The food GP codes often contain additional topics, like recovery of materials, documentation, complaint and recall procedures, labelling, and infestation and pest control.

The Institute of Food Science and Technology (1991, $3^{rd}$ edition) together with interested parties has compiled a comprehensive GMP for food and drink. The GMP has two complementary interacting components, namely the manufacturing operations and the food control system. The GMP consists of four parts. Topics described in part I are: quality management, personnel and training, documentation, premises and equipment, manufacturing, recovery and reworking of materials, complaints, recall and emergency procedures, contracts (own labels) and good control laboratory practices. The second part involves supplementary guidance on specific production categories (such as heat preserved foods, frozen foods). Part III deals with specialised topics, like identification of raw materials, labelling and data processing and control systems, and infestation controls, etc. In part IV mechanisms for reviewing are provided. In order to get an idea of the content of the guidelines a summary is given of typical subtopics in part I (see Appendix I).

Quality management is responsible for the development of an appropriate and comprehensive system to ensure that specifications set for achieving the intended product quality standards are consistently met; this requires involvement and commitment of all concerned at all stages of manufacturing.

Personnel and training refers to ability, training experience, and technical qualifications of personnel and clear assignment of tasks and responsibilities. Training should also cover GMP principles and personal hygiene besides training for specific tasks.

The purpose of documentation is to define materials, operations, activities, control measures and products, to record information during complete manufacturing process and to enable tracking of non-conforming products.

Premises and equipment involve aspects of location, design and maintenance of buildings and/or equipment.

The manufacturing process must be capable of consistently yielding finished products that conform to specifications. Manufacturing procedures are necessary to ensure that all concerned know what and how it has to be done and who is responsible.

Recovery and reworking of materials must be done by appropriate and authorised methods, resulting in products that comply with relevant specifications. Processes must be accurately documented.
Complaints, recall, and emergency procedures involve the provision of appropriate procedures by those responsible. A crisis procedure and management team should be established.

Distributors' own labels (own label) when complete (part) manufacturing is carried out as own label, the obligation of the Contract Acceptor is to ensure GMP production.

Good control laboratory practice refers to appropriate premises, facilities and staff, which are organised such that they can provide an effective service for fulfilling GMP requirements.

For details, see the original GMP guidelines. In the 4$^{th}$ edition, additional guidelines were provided on novel food products and food allergens (IFST, 1991).

GMP codes in food production can function as a proper basis for the application of other specific quality systems, such as an HACCP system. According to the revised document of NACMCF (1998), an HACCP system should be built on a solid foundation of prerequisite programmes. These programmes are often accomplished through application of e.g. Good Manufacturing Practice codes or Food Hygiene codes.

*Objectives and principles of HACCP*
The objective of HACCP is to assure the production of safe food products by identifying and controlling critical production steps. It uses a systematic approach (i.e. a plan of steps) to the identification, evaluation, and control of those steps, in food manufacturing, which are critical to food safety (NACMCF, 1998).

Officially, the 'Hygiene of Foodstuffs Directive 93/43/EEC' describes five principles for ensuring food safety, but most literature sources refer to seven principles containing two additional principles. These seven principles include:
1. Hazard analysis, i.e. potential hazards along the production chain must be identified and analysed.
2. Critical control point (CCP) identification, i.e. CCP's must be identified, and then monitored to avoid or minimise occurrence of hazards.
3. Establishment of critical limits in order to control hazards at each CCP.
4. Monitoring procedures, i.e. surveillance systems for regular monitoring or observation of critical control points.
5. Corrective action procedures must be established, including measures which should be taken whenever an inadmissible deviation is recorded at critical control points.
6. Verification procedures must be established for verification of correct functioning of the HACCP system.
7. Record keeping and documentation relating to the HACCP plan must be developed for effective management.

Several procedures which facilitate the development and introduction of an HACCP plan have been described. The Codex Alimentarius described a 12-stage procedure. Leaper (1997), Early (1997), and the National Advisory Committee on Microbiological Criteria for Foods (1998) considered similar procedures. For details, see Chapter 7.

HACCP is used as a qualitative and subjective system, whereby the majority of decisions are often based on qualitative instead of quantitative data. The HACCP approach is now widely adopted. Experience reveals that the hazard analysis and subsequent critical control point establishment would benefit from more quantitative means of assessing risks. In order to achieve a more scientific basis for hazard analysis the application of quantitative risk assessment (QRA) is now recommended by the Codex Alimentarius Commission. Risk assessment consists of four parts: hazard identification, exposure assessment, hazard characterisation and risk characterisation. The QRA enables the estimation of risks of potential hazards based on severity and probability information. Modelling techniques such as predictive microbiology and Monte Carlo simulations are now being applied to get better estimations based on (often) restricted available information. For details on modelling and risk assessment, see Chapters 8 and 9 respectively.

In 2002, in anticipation of the desire to assess a company's HACCP system by an auditing system, the Food Drug and Cosmetic division developed the ASQ certification auditor - HACCP certification. The Division also published training material to support this certification. More recently, in Europe there has been a desire for food processing companies to obtain third party certification of their HACCP systems. In reaction to this need, the new work proposal was submitted to ISO Technical Committee 34 (Food products) to develop an ISO standard that defines a food safety management system (FSM). TC 34 has approved the request and is now developing ISO 22000: (Food Safety Management- Requirements). The standard has the following objectives:

- compliance with the Codex HACCP principles;
- harmonisation of the voluntary international standards on food safety;
- provision of an auditable standard that can be used for either internal audits, self-certification or third party certification;
- alignment of the structure with ISO 9001:2000 and ISO 14001:1996;
- communication of HACCP concepts on an international basis.

ISO 22000 is not intended to define a minimal food safety system but is a voluntary standard. It provides a framework for a structured food safety management system and incorporates the system into overall management activities (Surak, 2004).

### 6.3.2 International standardisation Organisation (ISO)

*Objectives of ISO*
The international organisation for standardisation has developed many standards. One of the best known is the ISO 9000 family on food quality. Recently, the structure and approach of this ISO family has been drastically reconstructed (ISO 9001: 2000 version).

ISO requires the establishment of procedures for all activities and handling, which must be followed by ensuring clear assignment of responsibilities and authority. The main objective of earlier ISO 9000 family versions (versions 1987 and 1994) was to assure customers that the products met the required specifications. In the new version (ISO 9001:2000) the focus of objectives is shifted to achievement of customer satisfaction by meeting customer requirements, continuous improvement of the system, and prevention of nonconformity (ISO, 1999).

*Principles of ISO 9000 family 2000 version*
The basis of the revised ISO 9001:2000 is the eight management principles that are derived from collective experience and knowledge of international parties who participate in ISO technical committee ISO/TC 176 (ISO, 1998). The eight management principles include:
- Customer focus: because organisations depend on their customers they should understand current and future customer needs, and they should meet and strive to exceed customer expectations. This will lead to increased market share, increased effectiveness in use of resources and improved customer loyalty leading to repeat purchases.
- Leadership: leaders should establish unity of purpose and direction of the organisation, and they should create an internal environment in which people become fully involved in achieving the organisation's objectives. This will lead to people understanding the organisation's goals and objectives and being motivated to achieve these; it will also reduce poor communication between different levels.
- Involvement of people: all people in an organisation are essential and their full involvement will contribute to the organisation's success. This will result in motivated, committed and involved people, who are accountable for their own performance and will be eager to participate and contribute to continuous improvement.

## Quality assurance systems and food safety

- Process approach: results are more efficiently achieved when activities and related resources are managed as a process. It will result in lower costs and shorter cycle times, and improved consistent and predictable results.
- System approach to management: identifying, understanding, and managing interrelated processes such as a system, contributes to effective and efficient achievement of an organisation's objectives. It will lead to integration and alignment of processes and it helps in focusing on key processes. Moreover, it will give confidence to interested parties.
- Continuous improvement: this should be a permanent objective of the organisation. It will give performance advantage through improved organisational capabilities, and will provide the flexibility needed to react quickly to opportunities.
- Factual approach to decision making: effective decisions should be taken on the basis of data and information. This will lead to informed decisions, and an increased ability to demonstrate the effectiveness of past decisions, and it will increase the ability to review, challenge and change opinions and decisions.
- Mutually beneficial supplier relationships: because organisations and suppliers are interdependent, mutual beneficial relationships will enhance the ability of both to create value. It will lead to more flexibility and speed of joint response to changing markets, and it will reduce costs and resources.

The management principles form a basis for performance improvement and organisational excellence. Many organisations will find it beneficial to set up quality management systems based on these principles (ISO, 1998).

The ISO 9001:2000 family was re-structured in 2000 into four primary standards:
- ISO 9000 includes the fundamentals and terminology, and the eight quality management principles that form the basis of ISO 9001:2000. ISO 9000:2000 has replaced the previous terminology of ISO 8402.
- ISO 9001 includes requirements for quality management systems and replaced the previous ISO 9001, 9002, and 9003 standards.
- ISO 9004 provides guidelines for the development of a comprehensive quality management system going in the direction of an excellent organisation. It provides guidelines for performance improvements and is not comparable with the previous ISO 9004 version.
- ISO 10011 contains guidelines for auditing of quality systems.

The ISO 9001:2000 version is structured and organised according to the process model. A process is considered as any activity or operation which receives inputs and converts it to outputs. Based on this basic concept the 'Quality Management Process' model was developed by ISO/TC 176 (ISO, 1999). The models covers the following aspects:
- The four major topics of the quality management system:
  1. management responsibility;
  2. resource management;
  3. product and/or service realisation;
  4. measurement, analysis and improvement.
- Continuous improvement and measurement of customer satisfaction.

- Both products and services are explicitly incorporated.
- Interaction between processes. For example, the input for product and/or service realisation is customer-driven. Subsequently, the output is measured by consumer satisfaction measurements. This information is used as feedback to evaluate and validate whether customer requirements are achieved. The management, in its turn, shall ensure that customer requirements are fully understood and met.

*Structure of ISO 9001:2000*
ISO 9001:2000 basically has a management focus. It starts with requirements for a quality management system. The organisation must define and manage processes that are necessary to ensure that products and/or services conform to customer requirements. For the implementation and demonstration of such processes, the organisation must develop, document and maintain a quality management system according to the International Standard. The contents of the four topics have been summarised and briefly described by Luning *et al.* (2002) and are shown below. For more details, see the official ISO guidelines.

In the topic management responsibility (1), requirements are established for the commitment and responsibility of (top) management. The major aspects are that (top) management must:
- demonstrate commitment with respect to realisation of customer demands by, amongst other things, creating awareness about the importance of fulfilling customer demands, by ensuring availability of resources, and by establishing quality policy, objectives and planning;
- ensure that customer needs are determined and translated into requirements for the organisation;
- establish an appropriate quality policy to ensure, amongst other things, the fulfilment of customer needs, commitment to continual improvement, and communication throughout the organisation;
- establish quality objectives for each function and level in the organisation;
- identify and plan activities and resources which are necessary to achieve quality objectives, i.e. quality planning;
- develop a quality management system that includes:
  - assignment of responsibilities and authority;
  - appointment of management representatives to ensure implementation and maintenance of the quality management system and awareness of customer requirements;
  - procedures for communication;
  - preparation of the quality manual;
  - development of procedures for controlling documents and records, on a system level;
  - establishment of a procedure for management review, i.e. management must provide a status with modifications related to the quality management system, policy and objectives, which is based on audit reports, customer complaints analysis, status of quality policy and objectives.

In the topic resource management (2), requirements are established for:
- human resources, such as the assignment of appropriate personnel, development of procedures to determine training needs and evaluate their effectiveness;

- information, such as procedures for managing information, e.g. on process, product and suppliers;
- infrastructure, such as workspace, associated facilities, equipment, but also supporting services;
- work environment, such as health and safety conditions, work methods and ethics.

The topic product and/or service realisation (3) consists of general requirements and five subtopics. The general requirements refer to determination, planning and implementation of realisation processes (e.g. criteria and methods to control processes, arrangements of measuring, monitoring and corrective actions, and maintenance of quality records).

The five subtopics are:
1. Customer-related processes, which establish requirements for identification and review of customer requirements, and customer communication (such as customer complaints, product and/or service information).
2. Development of products and/or services. The plans must include the stages of design/development processes, review, verification and validation activities, and the responsibilities and authorities of the personnel involved.
3. Purchasing puts demands on the purchasing-process to ensure that products and/or services comply with organisational requirements. The purchasing documents must contain appropriate information and purchased products and/or services must be verified. The organisation must ensure the adequacy of these documents and verification activities.
4. Product and service operations refers to the planning and control of actual operations. It establishes general requirements, for example, on the use of suitable production equipment, implementation of suitable monitoring and verification activities, and availability of clear work instructions. Moreover, it put demands on:
   - identification and traceability (where and how to record);
   - handling, packaging, storage and preservation and delivery;
   - validation of processes, i.e. to detect deficiencies, which may become apparent only after the product is in use or the service is delivered.
5. Control of measuring and monitoring devices establishes requirements for control, calibration and maintenance of measuring and monitoring devices.

In the topic measurement, analysis and improvement (4), the general requirements are planning and implementation of these measuring, analysis and improvement processes to ensure that the quality management system, processes and products and/or services comply with requirements.
- Measuring and monitoring requirements refer to the establishment of quality system performance (i.e. customer satisfaction, internal audits), measurement of processes, product and/or services.
- Control of nonconformity relates to identification, recording and reviewing of products and/or services, which do not conform to requirements.
- Analysis of data for improvement establishes requirements for data analysis for the determination of effectiveness of the quality management system, and for identification of possible improvements.

- Improvement relates to the establishment of a system procedure that describes the use of quality policy, objectives, internal audit results, analysis of data, corrective and preventive action and management review, to facilitate continual improvement. The ISO 9001:2000 can be, to a certain extent, tailored to a company by omitting requirements that do not apply to that organisation (ISO, 1999).

### 6.3.3 Structure of ISO 9004:2004

ISO 9004:2000 is the other part of the consistent pair of quality management systems, the other being ISO 9001:2000. The two standards were designed as a pair, but can be used as stand-alone documents as well.

ISO 9004 provides comprehensive guidelines for creating an excellent organisation through continuous improvement of business performance. The standard is focused on improving the internal processes of the organisation. Furthermore, the eight management principles have been integrated in this standard. These quality principles, which have been established by ISO for achieving Total Quality Management, include:
1. customer focused organisation, i.e. focus the organisation on sustained customer satisfaction while giving benefits to other interested parties as well;
2. leadership, i.e. provide visible personal leadership;
3. involvement of people, i.e. encourage employee efforts;
4. process approach, i.e. consider activities as processes with an input which is transformed into an output;
5. system approach to management, i.e. consider the total system and not sub-systems;
6. factual approach to decision-making, i.e. make decisions based on facts and not on feelings;
7. continual improvement, i.e. develop a culture of continuous improvement;
8. mutually beneficial supplier relationships, i.e. create valued relationships with suppliers.

The structure of topics in the new ISO 9004 is similar to ISO 9001, which facilitates the extension of the quality system to Total Quality Management. However, the scope of ISO 9004 is different. In ISO 9004 those elements that are necessary for performance improvement in the organisation are extensively considered, while ISO 9001 provides quality management system requirements that demonstrate its capability to fulfil customer requirements. The revised ISO 9004 version is thus not a guideline to implement the revised ISO 9001:2000.

### 6.3.4 Miscellaneous systems (BRC, SQF, EUREP-GAP)

Nowadays, (parts of) the established QA systems are often combined to assure more quality aspects. For example, combination of HACCP and ISO 9001, but also combinations with environmental quality systems, such as ISO 14000 have become more common. Furthermore, ISO is developing additional norms with requirements for food safety (HACCP) and traceability (Jonker 1997, Dobbelaar and Bergenhenegouwen 2000, de Vreeze 2001; Kolsteren and de Vreeze, 2002). Moreover, QA systems have been developed for a specific sector, such as EUREP-GAP (Euro Retailer Produce - Good Agricultural Practice), and are integrated in new systems such

## Quality assurance systems and food safety

as BRC (British Retail Consortium), and SQF (Safe Quality Food). The essence of these combined QA systems are briefly described below.

EUREP-GAP aims to maintain consumer confidence in food quality and safety, to minimise the detrimental impact on the environment and conserve nature and wildlife, to reduce the use of agrochemicals, to improve efficiency of use of natural resources, and to ensure a responsible attitude towards worker health and safety. EUREP-GAP is a worldwide standard for the production of agricultural products that are directly or indirectly delivered to supermarkets. It is a checklist that contains minimum standards for the leading retail groups in Europe with respect to food safety, welfare, environment, and health and safety at work.

The objective of BRC is to assure product quality and food safety. It is a technical standard for companies supplying retail branded food products. BRC is a checklist that combines HACCP with specific parts of GMP and parts of ISO. It is focused on both technological and managerial aspects (Smit 1999; Damman, 1999).

SQF (Safety, Quality, Food) aims to assure food quality and safety, and consists of three standards based on the systematic approach of HACCP as described in the Codex Alimentarius. SQF1000 is for the agricultural companies and low-risk processing companies, SQF2000 is for large supply and processing industry (high-risk companies), and SQF3000 is for retail. It combines ISO and HACCP and includes typical tracking and tracing aspects relevant to food supply chains (van Delst and Hendriks, 2002).

IFS (International Food Standard) was created by the federation of German distributors to enable a systematic and uniform evaluation of food product suppliers. The IFS standard concerns primarily the development of an HACCP system, the basic principles of a quality system (including plant handling issues, traceability and product recalls), good production practices and good hygienic practices in each company. An evaluation based on this standard is requested by most German and French distributors, and is accepted by Dutch and French distributors as well as by Dutch and Belgian distributors.

## 6.4 Quality assurance systems and assuring food safety

In this section the contribution of the major established QA systems (GMP, HACCP, ISO 9001:2000) in assuring food safety is considered, taking into account the typical technological and managerial aspects. Furthermore, attention is focused to the way in which QA systems can be assessed on their performance.

### 6.4.1 Role of different quality assurance systems in assuring food safety

In agribusiness and the food industry QA systems like GMP, HACCP, ISO and BRC are now widely applied. These established QA systems differ in several respects, such as perspective (e.g. technology or management), intended result (e.g. food safety, product quality, total quality), extensiveness, requirements (e.g. by legislation, by the company itself, or by retailers), quality management focus, method (e.g. plan of steps, checklists, guidelines), and suggestions for implementation. Van der Spiegel and co-authors (2003) evaluated the differences between established QA systems and concluded the following:

- GMP and HACCP focus on technology aspects, whereas ISO and TQM focus more on management aspects.
- GMP and HACCP are especially developed to assure food safety. Like HACCP, BRC deals with food safety but also with product quality, and it evaluates management aspects (ISO) and facility conditions (GMP). TQM is not a specific QA system but more a policy that aims at improving total quality by implementing four basic principles in the total organisation.
- In extensiveness GMP, HACCP and ISO are more detailed than BRC and TQM. HACCP is compulsory whereas the other systems can be applied voluntarily in the food industry, although retailers now set requirements on the implementation of BRC.
- The quality management focus also differs for the specific systems. Quality control and quality assurance are central issues in HACCP, the previous ISO 9000 (1994 version) series and BRC. Continuous improvement and customer focus are major issues in ISO 9001:2000, ISO 9004:2000 and the philosophy of total quality management.
- HACCP is the only QA system that consists of a plan of steps, in contrast to the checklists of ISO and BRC. GMP includes guidelines, and TQM uses awards or self-assessments. In fact, HACCP is the only QA system that has a normative approach (how to do), while the other systems are descriptive (what to do).

So when applying QA systems one should be aware of their possibilities and restrictions in achieving quality and/or safety objectives. We have evaluated in more detail the three common QA systems (GMP, HACCP and ISO 9001: 2001) from the viewpoint of assurance of food safety, taking into account important technological and managerial aspects. Combination systems like BRC were not considered because they (often) contain different elements of the common QA systems. TQM is also not evaluated because it is considered as a philosophy or policy and not a concrete QA system (section 6.4).

In the previous sections we discussed the relevant technological aspects, such as origin, characteristics, controllability of food hazards, and the requirements set for food properties on sampling, that can affect food quality and safety performance. With respect to the managerial aspects several topics were described, such as quality management functions, factors affecting decision-making, quality behaviour, managerial aspects of implementing QA systems, and chain management. Table 6.3 shows which elements of the three major QA systems cover relevant technological and managerial aspects that might affect food safety directly or indirectly.

*Evaluation of technological aspects of the major QA systems regarding food safety*
With respect to handling food hazards it is obvious that the GMP code focuses mainly on quality control (e.g. testing). Identification and characterisation of hazards in the food production (chain) are not explicitly incorporated in the guidelines, although some guidelines (part II) mention that physical, chemical and microbial characteristics must be considered. Guidelines

on sampling conditions are restricted to requirements for facilities, calibration and assignment of qualified persons. No specific guidelines are provided on sampling and sampling design.

HACCP aims at a systematic identification, evaluation and control of potential hazards in the food production (chain) to assure safety and prevent risk for human health. Although some attention is paid to sampling (like frequency, establishment of critical limits) no attention is paid to how samples should be taken. Also, the statistics behind sampling designs are not clearly explained.

ISO 9001:2001 is a system that is developed for a broad range of sectors (i.e. all kinds of industries but also services). As a consequence, with respect to technological aspects the guidelines are very global and give no concrete handles on food safety.

In conclusion, GMP codes provide a broad range of guidelines on many technological aspects of manufacturing and control, which are relevant for food safety. However, the codes prevent a systematic (common) approach, such as in HACCP. Moreover, the guidelines give a very general description of what to do, e.g. heated or cooked products or ingredients should be cooled as quickly as possible to below 8 °C. Thus, there is much room for interpretation and application of GMP codes. For example, how do you translate 'as quickly as possible' into operational terms? Therefore, the implementation of GMP codes may have considerably different effects on final food safety, depending on the efforts and accuracy in interpreting and translating the codes into operational procedures and actions.

Basically, ISO is not developed for assuring food safety in specific terms. It is more a management system and technological aspects of food safety are discussed only in very global terms.

With respect to sampling, it is obvious that the use of statistics for proper sampling design is not explicitly mentioned in any of the QA systems. The strictness of the sampling design (especially sample size and selection criteria), however, can have an effect on the decision to accept an improper batch or to reject a proper batch. Obviously if improper batches are accepted, there might be consequences for food safety.

So, with respect to the technological aspects of assuring food safety, GMP codes can be used as basic guidelines, whereas HACCP helps to systematically identify, evaluate and control food hazards. Both these QA systems can be used to fulfil concretely parts of the ISO 9001 system. Finally, a company can develop its own guidelines for sampling of specific products, and the use of simple statistics can help in developing proper sampling designs.

*Evaluation of managerial aspects of the major QA systems regarding food safety*
Food safety is often considered from a technological point of view. However, in our opinion managerial aspects can have a considerable influence on food safety as well. Management activities in companies aim at influencing human behaviour during food production. Human handling during production, but also during preparation and consumption of foods, can have a considerable effect on the actual safety of the end product. Although the three QA systems

Table 6.3. Global evaluation of QA systems on technological and managerial aspects that may influence food safety.

| Technological aspects | GMP code foods and drinks | HACCP (QRA)* | ISO 9001:2000 |
|---|---|---|---|
| Food hazards; Attention paid to characteristics, origin and controllability of hazards? | Guidelines (part II) on specific product categories pay some attention to typical product characteristics. Especially focus on microbial testing of raw materials, intermediates and final products. Some guidelines refer to establishment of critical control points with respect to pathogens. Guidelines often mention prevention of cross-contamination by separation of areas. Guidelines on appropriate temperature programmes in cooled food chains. | Principles of HACCP include specifically qualitative analysis (identification and evaluation) of potential safety hazards (microbial, chemical and physical), determination of critical places for control and assessment of monitoring system (critical limits, frequency, measurement) QRA aims for a quantitative approach enabling a better estimation of risks of potential hazards. | Guidelines on process management just mention that where applicable the organisation shall identify the product (or service) by suitable means throughout all processes. Control aspects are briefly mentioned in guidelines on control of nonconformity, by mentioning that the organisation shall ensure that the product (or service) which does not conform to requirements is controlled to prevent unplanned use (application or installation). |
| Sampling conditions. Attention paid to requirements set for sampling by the typical food characteristics? | Guidelines on good control laboratory practice mention facilities for testing, calibration of equipment and instruments by assigned persons, taking representative samples following established procedures. No specific guidelines on sampling and microbial testing. | With respect to development of monitoring system the following aspects are mentioned: ability to detect loss of control (i.e. specificity); information about how to adjust process before a deviation occurs; assignment of a person with relevant knowledge; frequency of sampling. Usually quick physical and chemical tests and visual inspection are used. Microbiological tests are seldom used due to the time required to obtain results. | ISO guidelines on control of measuring, inspection and test equipment mention that the organisation shall control, calibrate, maintain, handle and store applicable measuring and test equipment. Measuring, inspection and test equipment must ensure that uncertainty like accuracy and precision are known. Measuring, inspection and test equipment must be calibrated and adjusted at specified intervals and all activities must be recorded. |

Table 6.3. Continued.

| Managerial aspects | GMP code foods and drinks | HACCP (QRA) | ISO 9001:2000 |
|---|---|---|---|
| Q-management functions Attention paid to management functions: planning and control of Q-design, Q-control, Q-improvement, Q-assurance and Q-policy. Leadership and organisation | Several (aspects of) Q-management functions are implicitly mentioned. For example, quality management guidelines prescribe requirements for effective manufacturing operation and quality control: All aspects of manufacturing must be specified in advance; planning of activities. Appropriate facilities must be provided; organising appropriate conditions. Operators must be trained. Authorities and responsibilities of Production Management and Quality Control Management must be clearly defined and allocated. Quality Control Management must summarise and provide summaries of quality performance data to relevant persons; control activities. | Clear statements with respect to control and monitoring of critical control points. With respect to organisations, it is explicitly mentioned (as a prerequisite) that management must provide necessary team members for a certain period and that it must be financially supported. | ISO provides specific guidelines on management responsibility: Top management shall establish and communicate quality policy and ensure compliance with customer requirements and commitment at all levels. Organisations shall establish written quality objectives to enable implementation of quality policy. Quality planning shall determine activities to achieve Q-objectives. Top management shall be responsible for preparation of quality manual and system procedures as part of quality management system (assurance). Top management shall assign someone to ensure implementation of QMS (leadership). Organisations shall establish system procedure for controlling documents required for operation of QMS. Organisations shall establish processes for continuous improvement of QMS (improvement). |

Table 6.3. Continued.

| | | |
|---|---|---|
| Decision-making process. Attention paid to factors affecting decision-making with respect to food safety. | Decision-making processes are not explicitly mentioned. However, some aspects affecting decision-making are included. For example: Authorities' descriptions for several functions like Quality Control Manager, Production and purchase manager are given. Furthermore, it is explicitly mentioned that Quality Control Management is obliged to provide Production Management with appropriate information to minimise non-conformance and so ensure safety. | HACCP guidelines do not consider explicitly the decision-making process, but some elements of HACCP specifically deal with decision-making: HACCP provides a clear approach for deciding on where to allocate critical control points using the CCP decision-tree. An obvious place where decision-making takes place with respect to safety is at the control points where employees have to decide whether to perform corrective actions or not. HACCP guidelines require that it is clearly described who and how often samples should be taken, and what to do in case of out-of-control situations. This is often a weak point of the system. | No explicit remarks on the process of decision-making but ISO provides some guidelines with respect to information: Organisations shall define and maintain the current information necessary to achieve conformity of product/service. System procedure for managing information shall consider access and protection to ensure integrity and availability. |
| Quality behaviour. Quality disposition/intelligence. Quality skills/ability. Quality commitment. | Quality behaviour as such is not mentioned, however, several guidelines may contribute to Q-behaviour. Attention is paid to training and education with respect to hygiene, and appropriate skills in recruitment (skills/ability). Other guidelines focus on providing appropriate facilities, equipment, written procedures, etc (ability). | The following aspects are related to quality behaviour and are explicitly mentioned by the HACCP guidelines. Training of personnel in HACCP principles (skills). Commitment of management to the HACCP approach and indication of awareness of the benefits and costs (ability). HACCP clearly states that an HACCP team must be formed. Leadership affects quality behaviour and in potential team structures helps leaders to get things done by their employees. | ISO guidelines on resources mention in global terms some aspects influencing Q-behaviour, such as: Organisations shall determine and provide and evaluate effectiveness of training required to achieve conformity (skills). Organisations shall define and implement human and physical factors of the work environment needed to achieve conformity (ability). The guidelines on management responsibility require demonstration of a commitment to meet customer requirements by top management. |

Table 6.3. Continued.

| | | | |
|---|---|---|---|
| Management aspects of implementation QA systems Implementation strategy | No guidelines are provided with respect to development and implementation of the GMP codes. | A step-by-step plan is provided which indicates how to develop an HACCP system; however, no information is provided on implementation strategies | Quality objectives must be developed to enable implementation of quality policy. No information is provided on implementation strategies. |
| Chain management. Supplier relationships, factors affecting chain collaboration | Some guidelines refer to chain aspects, such as guidelines on distributors own labels and use of outside services. It deals with typical supplier-relationship aspects like clear contracts, and visits to check food safety conditions. | HACCP can be applied by all actors of the food supply chain. When safety hazards are prevented early in the chain it can reduce the number of critical control points further down in the chain. However, guidelines with respect to chain management (e.g. typical aspects like supplier relationships) are not provided. HACCP was basically developed for the company level. | Chain aspects not explicitly considered. Some guidelines are provided with respect to purchasing to ensure that purchased products conform to organisations' requirements. Organisations shall select and evaluate suppliers based on criteria. Purchasing information must be documented. |

* Quantitative Risk Assessment (QRA) is not part of HACCP but can be used to assess more scientifically and quantitatively, if potential hazards form a risk and should therefore be controlled. QRA can complement HACCP and for this reason it is included and evaluated in this table.

do not specifically consider management activities regarding food safety, we have highlighted guidelines and principles that cover management activities that can influence food safety (Table 6.3). These are briefly discussed below.

With respect to the quality management functions, the general GMP guidelines (as described in subsection 6.3.1) do not explicitly mention the tasks and responsibilities of managers. The GMP code is basically developed as advice to management on matters affecting product safety, manufacturing under hygienic conditions and associated matters. Only some (or aspects of) quality management functions are implicitly described in several guidelines, such as planning, control, leadership and organising (see Table 6.3). In fact, GMP codes are mainly focused on quality control and assurance, whereas quality design, improvement and policy and strategy are not, or only scarcely, considered at all.

HACCP is often described as a management tool for assurance of food safety. However, typical quality management functions (like planning and control of Q-design and Q-control, and leadership) are not explicitly mentioned, except for the fact that management must provide the appropriate conditions for development and implementation of an HACCP system. With respect to quality design it has been suggested that an evaluation of the hygienic design of process equipment for agri and food machines should be part of an HACCP study (Burggraaf, 1998) to obtain a higher level of food-safety assurance.

In contrast to GMP and HACCP, ISO 9001:2002 is a typical management-oriented QA system, and several specific guidelines on management responsibility with respect to quality are provided. Attention is paid to continuous improvement, quality planning and objectives, quality policy and clear assignment of responsibilities, amongst other things. If these management guidelines are tailored to food safety objectives (as part of quality objectives), they can contribute to the assurance of food safety.

Decision-making occurs at all levels and at many locations. Wrong decisions due to uncertain or incomplete information, but also conflicting interest, can have negative effects on food quality and safety. Although decision-making as a process is considered important, it is not explicitly mentioned in the three QA systems GMP, HACCP and ISO. Only in the HACCP system is there an explicit procedure for deciding which hazards are critical, i.e. the CCP decision tree. However, no explicit attention is paid to the decision-making process when executing the monitoring of critical control points. The HACCP guidelines do mention that in an HACCP plan there must be a description of which employee must take samples, how often and which corrective action should be taken in case of out-of-control situations. In practice, however, this is often a critical decision-making process where mistakes occur (Motarjemi and Käferstein, 1999). Providing the right information at the right time is an important factor influencing the decision-making process. Only some aspects of information are considered by the established QA systems. More specifically, GMP and ISO 9001:2000 provide some guidelines about to whom information must be given and how information must be documented and managed. HACCP only mentions the registration and documentation of data and information but does not consider explicitly how information should be managed and used for decision-making.

*Quality assurance systems and food safety*

Quality behaviour, which is affected by ability/skills, disposition/intelligence and commitment, can have a considerable effect on food quality and safety as well. Gerats (1990) showed in his research in slaughterhouses that disposition to hygienic working was mainly limited by the low knowledge level regarding bacteriological contamination mechanisms, by the restricted social support from colleagues, by the lack of interest in hygienic working amongst supervisors, and by the limited opportunities for hygienic working.

The GMP code pays attention to several aspects that influence quality behaviour, especially on skills and ability with respect to hygienic working. Quality commitment by top management is not mentioned.

Within the HACCP approach, the importance of training and education about HACCP principles and the need for management commitment is now recognised as important for the successful performance of an HACCP system (NACMCF, 1998, see also Chapter 7). However, the role of human behaviour in the performance of the HACCP system is still under-explored.

IOS 9001:2002 also pays attention to several aspects that influence quality behaviour, such as guidelines on training and facilities in the work environment for achieving conformity of products, and they emphasise the importance of demonstrating management commitment towards quality.

Guidelines on how to implement the QA systems are rather underexposed. Only HACCP provides a step-by-step plan. A step-by-step plan as such, however, is no guarantee of proper implementation and assurance of food safety. Information about implementation strategies and their consequences might be helpful in achieving food safety and quality.

The need for supply chain management in relation to food quality and safety is receiving more attention. Several food safety crises over the past five years were due to problems at the beginning of the food chain (e.g. animal feed production), which emphasised the need for a supply chain approach. GMP guidelines deal with some aspects of supplier relationships. The HACCP approach has now been extended to the primary sector as well, which enables a chain HACCP approach. ISO 9001:2002 also provides some guidelines on purchasing, to ensure conformity.

A consideration of the QA systems reveals that several management aspects are implicitly and sometimes explicitly mentioned. Integration of the more technology-oriented QA systems, GMP and HACCP, with the more management-oriented ISO 9001 system offers opportunities for broader quality assurance.

### 6.4.2 Measuring the effectiveness of Quality Assurance systems

Manufacturers can use various quality assurance systems or combinations to assure food quality and safety. They have to decide which QA system is most suitable to their specific situation, and how this system should be implemented. To be able to select the right QA system and implement it properly, it would be helpful to know the effectiveness of each system. So, what is the actual contribution of a QA system to food quality assurance, and which factors influence the implementation and performance of the QA system? Not much is known yet about the actual contribution of QA systems. Several authors have emphasised the need for an instrument to assess the effectiveness of food quality systems (e.g. Newall and Dale, 1992; Van der Spiegel et al., 2003). Moreover, ISO (1999) stated that a significant new effort is required to identify objective indicators and to develop appropriate procedures to monitor them as the basis for any evaluation of strategy implementation (e.g. implementation of a QA system). Although HACCP, ISO, BRC and TQM can be evaluated on the extent of implementation and on the compliance with norms and requirements, their actual contribution to assurance of food quality cannot be measured. Consequently, the effectiveness is still unknown.

Recently, Van der Spiegel and co-authors (2003) and Van der Spiegel (2004) developed a diagnostic instrument "IMAQE-Food" (i.e. Instrument for Management Assessment and Quality Effectiveness in the Food sector) for the food industry that can assess the effectiveness of food quality management. Figure 6.5 shows the conceptual model that was used to develop the diagnostic instrument. The elements of the model include quality management, quality performance, and contextual factors that affect these elements, i.e. complexity of the supply chain, complexity of the organisation, complexity of the production process, and complexity of the product assortment. The model proposes that a higher level of quality management results in higher production quality. A higher complexity of contextual factors is expected to relate to lower production

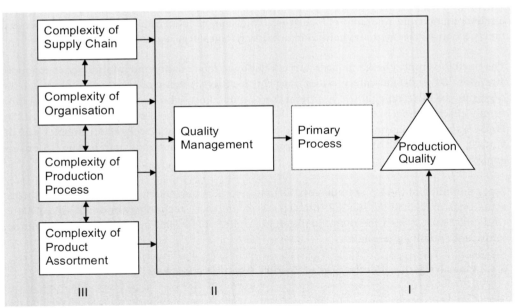

Figure 6.5. Conceptual model to measure the effectiveness of food quality systems (adapted from Van der Spiegel et al., 2000; 2003).

quality, but a higher level of quality management is assumed to reduce the influence of the contextual factors on production quality. The model was translated into performance measurement indicators to enable an assessment of the relationship between the elements. Examples of identified and validated indicators (for the bakery sector) are shown in Table 6.4.

These indicators were translated into questions in an interview, which was built up according to the following aspects: background information, contextual factors, quality management, production quality, and finishing questions about the experienced effect of implemented QA system(s). Data were collected by open questions with field coding combined with absolute answers. A scoring system was applied to measure differences within the sample. This questionnaire, with a classification system, constitutes the instrument "IMAQE-Food" that statistically analyses relationships between contextual factors, food quality management and production quality for an assessment of effectiveness.

Exploring the assumed relationships of the conceptual model in a sample of 48 bakeries (Van der Spiegel, 2004) revealed that bakeries with higher complexity implemented a higher level of quality management ($R^2 = 39\%$). However, a higher level of quality management showed no significant relationships, which means that a higher level of quality management did not necessarily result in a higher level of production quality.

For an assessment of the effectiveness of food quality management in the bakery sector, relationships were studied in detail. This study revealed six effective variables of food quality management in the bakery sector: (1) control of strategy, (2) allocation of supplying raw materials, (3) supply control, (4) control of production, (5) receiving orders, and (6) planning of distribution. These activities were effective in lowering the percentage of rejected products

Table 6.4. Examples of validated operational indicators to assess relationships between Q-management, contextual factors and production quality (Van der Spiegel et al., 2004; 2005).

| Element I: Production Quality | | | | | |
|---|---|---|---|---|---|
| Physical product quality: E.g. results of technical evaluation | | Availability: E.g. percentage of complaints about availability | | | Costs: E.g. total costs |
| **Element II: Quality management** | | | | | |
| Strategy: E.g. control of strategy | Supply control: E.g. supply control | Production control: E.g. planning of production | Distribution control: E.g. receiving orders | | Execution of production tasks: E.g. control of execution of production tasks |
| **Element III: Contextual factors *** | | | | | |
| Complexity of organisation: E.g. number of external relations E.g. number of employees | | Complexity of production process E.g. degree of automation | | | Complexity of product assortment E.g. number of product groups |

*IMAQE-Food was validated for the bakery sector. Since supply-chain complexity was not differentiating, it was omitted from the diagnostic instrument for this sector.

and percentage of complaints about availability and in improving the outcomes of technological analyses. This effectiveness was shown for bakeries with various kinds of organisation, production process and product assortment. Therefore, bakeries should look at the effective quality management activities relevant to their situation in order to obtain the required production quality.

For the improvement of food quality management, several QA systems are available for the bakery industry, i.e. a hygiene code for bakeries producing bread and confectionery, HACCP, ISO and BRC. The studied bakeries applied one or more of these QA systems and performed most effective quality management activities on the same level (see Table 6.5) (Van der Spiegel, 2004). Only supply control (i.e. purchase and selection of raw materials, selection and evaluation of suppliers, co-operation with suppliers) was performed on a lower level by bakeries that applied ISO in combination with HACCP.

Moreover, bakeries have to decide which QA system is most suitable for improving the level of production planning (i.e. development of a short-term production planning, standardisation of production methods) and control of execution of production tasks (i.e. training and education, communication, assignment of tasks, adjustments of procedures). Bakeries that applied BRC and ISO in combination with HACCP implemented production planning on a higher level. The Hygiene code and HACCP consist of no clear statements with respect to planning and control (see Table 6.3). Bakeries that applied BRC implemented control of execution of production tasks on the highest level, whereas bakeries with the Hygiene code, HACCP with or without ISO performed this activity on a lower level than bakeries with BRC and on a higher level than bakeries without a QA system. For assessment of the effectiveness of QA systems in specific terms, IMAQE-Food should be tailored to individual QA systems by the identification of specific indicators.

Table 6.5. Relationship between level of quality management activities and QA systems among 48 bakeries (Van der Spiegel, 2004).

| Quality management activities | QA systems | | | | |
|---|---|---|---|---|---|
| | No QA system | Hygiene code | HACCP | ISO + HACCP | BRC |
| Control of strategy | + | + | + | + | + |
| Allocation of supplying raw materials | + | + | + | + | + |
| Supply control | ++ | ++ | ++ | + | ++ |
| Planning of production | + | + | + | ++ | ++ |
| Control of production | + | + | + | + | + |
| Control of execution of production tasks | + | ++ | ++ | ++ | +++ |
| Receiving orders | + | + | + | + | + |
| Planning of distribution | + | + | + | + | + |

## 6.5 Total Quality Management

In the previous sections the established QA systems and their potential contribution to food safety are described and discussed. In this section the principles of Total Quality Management are explained, and the possibilities of TQM in assuring food quality and safety are considered.

### 6.5.1 Principles of Total Quality Management

Total Quality Management is not a specific QA system but a management view that strives for the creation of a customer-focused culture, which defines quality for the organisation and establishes the foundation for activities aimed at attaining quality-related goals. TQM is thus not merely a technique, but a philosophy anchored in the belief that long-term success depends on a uniform and firm-wide commitment to quality, which includes all activities, that a firm carries out. TQM can become the core of a firm's strategy to remain competitive (Gatewood *et al.*, 1995). According to Dean and Evans (1994) the term *total quality management* (TQM) covers a total, company-wide effort including all employees, suppliers, and customers and stakeholders, that seeks continuously to improve the quality of products and processes to meet the needs and expectations of customers. There are probably as many different approaches to TQM as there are businesses. Although no program is ideal, successful implementation programs share many characteristics. According to Dean and Evans (1994) the basic attributes for successful performance of TQM are customer focus, strategic planning and leadership, continuous improvement, and empowerment and teamwork, which are briefly described below. For more details, see the literature (Gatewood *et al.*, 1995; Dean and Evans, 1994; Luning *et al.*, 2002).

Customer focus. Customers finally judge the actual quality of the product. Therefore, quality systems should focus on those product and service attributes that provide value to the customers. TQM demands constant sensitivity to emerging customer and market requirements, and measurement of the factors that drive customer satisfaction. TQM also requires an awareness of developments in technology and a rapid and flexible response to customer and market needs. Such requirements extend beyond merely reducing defects and complaints or meeting specifications. Nevertheless, defect and error reduction and elimination of causes of dissatisfaction contribute significantly to the customers' views of quality and so are important parts of TQM. In addition, the company's approach to recovering from defects and errors is crucial for improving both product quality and relationships with customers.

Strategic planning and leadership. Achieving quality and market leadership requires a long-term strategy. Planning and organising improvement activities requires time, money and a major commitment from all stakeholders including customers, employees, stockholders and suppliers. Long-term strategy must also address training, employee development, supplier development, technology evolution, and other factors that support quality. A key part of the long-term commitment is the regular review and assessment of progress in long-term goals. Senior managers play an important role with respect to leadership. They must create and implement clear quality values and high expectations. The leaders must participate in development of the strategy, systems, and methods for achieving excellence. The senior managers must serve as role models through their regular personal involvement in visible activities, such as planning, reviewing company quality performance, supporting improvement teams, and recognising employees for quality achievement. They must reinforce the values and encourage leadership at all levels of management.

Continuous improvement is part of the management of all systems and processes. Achieving the highest levels of quality and competitiveness requires a well-defined and well-executed approach to continuous improvement. Such improvement needs to be part of all operations and all work unit activities in a company. Improvements may be of several types, e.g. reducing errors, enhancing values to customers or improving productivity. Improvement is not only driven by the objective to provide better quality, but also by the need to be responsive and efficient. To meet all these objectives, the process of continuous improvement must contain regular cycles of planning, execution, and evaluation. This requires a, preferably quantitative, basis for assessing progress and for deriving information for future cycles of improvement. Quality must be *measured*, and for this purpose a company should select data and indicators that represent the attributes and factors that determine customer satisfaction and operational performance.

Empowerment and teamwork. Following the TQM philosophy all functions at all levels of an organisation must focus on quality to achieve corporate goals. Everyone must participate in quality improvement efforts. For this purpose, employees must be empowered to be able to make decisions that improve quality at all stages. Empowerment often requires a radical shift in the view of senior management. Traditionally, the workforce should be "managed" to comply with existing business-systems, which is contrary to what is required in TQM.

Companies can encourage participation by recognising team and individual activities. In that way they share success stories throughout the organisation and encourage risk-taking by removing the fear of failure. Moreover, they encourage the formation of employee involvement teams, implementing suggested systems that act rapidly, provide feedback, and reward implemented suggestions. Moreover, they provide financial and technical support to employees to develop their ideas.

Employees need training in quality skills related to performing their work, and to understanding and solving quality-related problems. Training brings all employees to a common understanding of goals and objectives and the means to attain them. Training usually begins with awareness of quality management principles and is followed by specific skills in quality improvement. Training should be reinforced through on-the-job applications of learning, involvement, and empowerment.

Firms applying TQM successfully are realising a decrease in long-term costs by building a more satisfied customer base (Gatewood *et al.*, 1995). However, because short-term costs for new equipment, materials and additional labour are sometimes required to implement TQM processes, many firms are hesitating to adopt TQM. Time is another obstacle to the adoption of TQM. Many firms do not want to wait the seven to ten years typically required to obtain the first benefits of TQM.

## 6.5.2 Total Quality Management and Food Safety

As mentioned in subsection 6.1.3, quality management developments can be summarised along two lines: developments in assurance and reliability (control), and developments in attitudes to quality (total quality management). The two lines of development can be illustrated by the work of the four well-known quality guru's, Deming, Juran, Feigenbaum and Crosby. Feigenbaum is a representative of the quality assurance line. Feigenbaum proposed a system approach to quality through the definition of a quality system. It was defined as "a company and plant-wide operating work structure, documented in effective integrated technical and managerial procedures, for guiding the co-ordinated actions of work force, machines, information of the company and plant in the best and most practical ways to assure customer quality satisfaction and economical costs of quality".

Deming, Juran and Crosby should be placed in the total quality management line. The lessons they learned have been translated into the concept of prevention: doing things right the first time, and planning and designing instead of paying for failures.

Recent developments show quality assurance to play an important role in the food- and agribusiness of the future. However, it will be combined more and more with, or even made dependent on, the implementation of the policy of Total Quality Management (Luning and Marcelis, *submitted*).

For quality policy, various orientations are used depending on the organisation's culture and the individual beliefs of top management. In companies where technology plays an important role, managers tend to introduce technological solutions first for occurring problems, such as automation of processes and redesign of products and processes. In other companies managers prefer to change the culture and attitude of employees, because they strongly belief that this is the only way to accomplish improvements in future. Generally speaking, three policy orientations can be distinguished, the first one having a technological focus and the others a managerial one (Luning *et al.*, 2002):
- Product/process orientation: Technological solutions are preferred in order to make product quality, processes, and materials predictable and controllable.
- Procedure orientation: Solutions are sought within the quality system. The underlying belief is that a clear description of responsibilities accompanied by a broad description of procedures and work instructions form the basis for controlling quality performance.
- People orientation: Problems are analysed in the field of human behaviour. Solutions are sought for changing people by means of organisational development, training and empowerment.

Quality assurance systems can be perceived as originating from a procedure orientation, often combined with the product/process orientation (e.g. HACCP). QA systems do have a considerable problem in their often false underlying assumptions about people's behaviour, i.e. people accepting technological knowledge as being true and correct, and people following procedures without interpretation and discussion (Luning *et al.*, 2002).

TQM embraces all three policy orientations - product/process, procedure and people - at the same time, thus striving for the broadest approach to analysing methods and possibilities for solutions. TQM starts thinking from a people perspective and their decision-making behaviour, giving them room for manoeuvre (empowerment) for realising customer requirements (customer focus) by improving their work and working conditions (continuous improvement), stimulated and rewarded by inspiring leaders (strategic leadership).

TQM gratefully uses procedures and technical knowledge, but it makes individual employees the central elements of the quality system, thus mobilising everybody's experience, knowledge and feelings for quality.

In considering the different policy orientations in food- and agribusiness companies, it becomes clear that in food production the product-process orientation is predominant. As a matter of fact, the food industry is characterised by a technological focus and most employees do have a technical and/or technological education. Next to the product-process orientation is the procedure orientation. The latter orientation is also widely accepted in the food industry because procedures are considered as a powerful tool to provide assurance in technological processes. However, in practice the procedure-orientation often gives rise to complications because the effects of human behaviour and decision-making processes on the accurate execution of procedures are underestimated. The people-orientation approach is not widely applied in food production because it is often considered as a soft and intangible approach.

When considering quality assurance, it is not surprising that QA systems are commonly perceived as tools with a predictable outcome. This might explain why these systems (like HACCP, ISO) are rather appreciated by the food industry. Nevertheless, it is clear that these QA systems have their restrictions. In practice, employees and operators feel that the QA systems are forced on them and that the benefits are not obvious. Systems that are developed by operators and employees themselves and which are considered as useful are most successful and well accepted. This is one of the reasons why systems should be as simple and practical as possible. Overly complex QA systems will not be effective in practice. Pressing operators to comply with the enforced guidelines will be experienced as unpleasant and can result in a discrepancy between operators and the quality department or QA managers. Moreover, external certification will increase the enforcement idea.

Another typical problem for the food- and agribusiness originates from the demands of the market situation. From the viewpoint of different administrative concepts, it can be seen that the market situation has conflicting demands as reflected in Figure 6.6. On the one hand quality demands are fixed, which has a strong appeal for the bureaucratic administrative concept, i.e.:
- demands on high and advanced technological product quality, such as for functional foods;
- legislative obligations and market demands with respect to safety;
- increased requirements with respect to costs and price level.

## Quality assurance systems and food safety

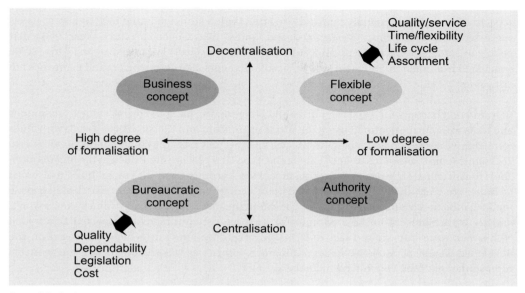

Figure 6.6. Conflicting requirements (adapted from Luning et al., 2002).

This set of demands requires a constant quality approach with a high degree of assurance. In practice, this is reflected in a high level of centralisation, and bureaucratic systems with many rules based on legislative obligations.

On the other hand, the continuously changing market and environment require a more flexible concept, i.e.:
- high demands are put on service accompanied by a customer-oriented attitude;
- high demands on flexibility due e.g. to unexpected events, seasonal influences, etc.;
- increasing reduction of life cycle of products, which leads to much focus on product development and innovation;
- increased mass customisation in the food industry leads to an enormous broadening of and large changes in product assortments.

All these aspects require a responsive quality approach that puts responsibilities low down in the organisation, and that is not hindered by detailed rules and guidelines. TQM is an approach meeting these requirements, and it can be implemented when the quality system is adjusted to the TQM policy (Luning *et al.*, 2002).

TQM can potentially improve business responsiveness in many branches of industry. However, Luning *et al.* (2002) and Kramer and Van den Briel (2002) noticed that TQM is not yet commonly adopted in the food industry. Structured QA systems, like HACCP with its use of control points and the ISO 9000 series with its use of procedures, fit better in the technological orientation. Another reason for the scarce application of TQM principles might be that the soft and intangible aspects of a people-orientation approach are often underestimated in the food industry. It was suggested by several authors (Sebastianelli and Tamini, 1998; Ghobadian and Gallear, 2001), that people or soft elements make up almost half of the barriers to TQM. Also, Lau and Idris (2001) concluded from their quantitative study that soft elements, like culture, trust and teamwork are critical for TQM.

In a study on barriers for TQM (Hendriks and Sonnemans, 2002) the terms values, techniques and tools were further defined using the terms of Hellsten and Klefsjö (2000) as a basis. Values were defined as the soft and intangible aspects that together form the culture of the organisation. Techniques were considered as decision-making processes that are embedded in the organisation. Tools must facilitate the decision-making processes. In comparison with techniques, tools must be rather concrete and well defined. Based upon a comprehensive literature search a diagnostic model has been developed to assess to what extent the three TQM drivers, 'values', 'techniques' and 'tools', are present in the food industry. In a pilot study nineteen respected Dutch food companies were interviewed face-to-face using the questionnaire, which was based on the developed diagnostic model. This set of nineteen companies was looked upon as a reasonable representation of the Dutch food industry.

The food industry was regarded as having a technological focus leaving little space for people orientation. However, the results of the pilot study showed that the TQM driver 'values' is equally present in comparison to the other TQM drivers 'techniques' and 'tools'. Since these 'soft' elements are present in the Dutch food industry, there is a basis for further development of TQM here.

While discussing the contribution of TQM to food safety, it should be made clear that the effect of TQM depends on the administrative concept upon which a company is actually built.

When the bureaucratic concept is dominant, TQM principles can help in the change towards a more flexible concept, thus creating a feeling of responsibility for food safety within a culture of more awareness for quality and a positive quality attitude. On the other hand, when the flexible concept is dominant, TQM principles can force a one-sided focus on market-responsiveness, neglecting the necessary control and assurance procedures. Whether this will be a problem or not, depends on integrating the requirements of the QA systems into the TQM policy and the changing systems and organisational structure, that will be built up as conditions for this policy.

## 6.6 Conclusion

Food safety can be supported by TQM by means of intensified attention and quality awareness, and a positive quality attitude. Therefore, TQM can overcome the most significant failing in current QA systems, i.e. the over-emphasis on control, by adopting a much broader approach, taking into account people behaviour, production processes, and procedures.

# References

AgriHolland, 2003. *Dossier klassieke vogelpest* http://www.agriholland.nl/dossiers/vogelpest/home.html.

Alverado Casanova, K., 2004. Development of strategic alliances in an international fruit supply chain generating flexibility: use of techno-managerial approach. MSc thesis Wageningen University, 65 pp.

Burggraaf, W.N.A., 1998. An approach to come to hygienic design. Voedingsmiddelentechnologie, 8, 44-45 (in Dutch).

CDR, 1999. Notifications of infectious diseases: week 52/96. Communicable Disease Report DCR Weekly, 9, 21.

Crawford, L.M., 1999. Implications of the Belgian Dioxin Crisis, Food Technology, 53, 130.

Damman, J., 1999. CBL-agreement on BRC-standard must reduce number of audits, Voedingsmiddelentechnol., 31,15-17.

Dean, J.W. and J.R. Evans, 1994. Total Quality. West Publishing Company, New York.

De Vreeze, M., 2001. ISO starts with norm for HACCP, Voedingsmiddelentechnol., 33,12-13 (in Dutch).

Dobbelaar, C. and L. Bergenhenegouwen, 2000. Due to new ISO 9000-series: better integration of quality and food safety management, Voedingsmiddelentechnol., 32,25-11 (in Dutch).

Early, R., 1997. Putting HACCP into practice. International Journal of Dairy Technology.1, 7-13.

Evans, J.R. and W.M. Lindsay, 2004. The management and control of quality. West Publishing Company, St. Paul.

Gerats, G.E.C., 1990. Working towards quality. Dissertation, University of Utrecht (in Dutch with a summary in English).

Gatewood, R.D., R.R. Taylor and O.C. Ferrell, 1995. Management: Comprehension, Analysis and Application. London, Austen Press.

Geyskens, I., J-B Steenkamp and N. Kumar, 1998. Generalization about trust in marketing channel relationships using meta-analysis. International Journal of research in marketing, vol 15.

Ghobadian, A. and D. Gallear, 2001. TQM implementation: an empirical examination and proposed generic model. The International Journal of Management Science, Vol. 29, 343-359.

Gould, W.A. and R.W. Gould, 1993. Total quality assurance for the food industries 2$^{nd}$ edition. CTI Publications, Inc. Maryland USA, 464 pp.

Hadiprodjo, I.T., 2004. Consumer-driven product development in international fruit supply chains: use of techno-managerial approach. MSc thesis, Wageningen University, The Netherlands, 69 pp.

Hellriegel, D. and J.W. Slocum, 1992. Management. Addison-Wesley, New York.

Hellsten, U. and B. Klefsjö, 2000. TQM as a management system consisting of values, techniques and tools. The TQM Magazine, vol. 12, no. 4, 238-244.

Hendriks, W. and E. Sonnemans, 2002. Drivers of TQM in the food industry. MSc thesis Wageningen University, The Netherlands.

Hoogland, J.P., A. Jellema and W.M.F. Jongen, 1998. Quality assurance. In: W.M.F. Jongen and M.T.G. Meulenberg (eds.), Innovation of food production systems, Wageningen Pers, Wageningen.

Hughes, D., 1994. Breaking with tradition: building partnerships and alliances in the European Food Industry. Wye College Press. Wye.

IFST: Institute of Food Science and Technology of the U.K., 1991. Food and Drink Good Manufacturing Practice: A guide to its responsible management, London.

ISO, 1998. 9000:2000 Quality management systems- concepts and vocabulary. ISO/TC 176/SC, 1/N 185.

ISO 1999. 9001:2000. Quality management systems- Requirements. Documents ISO/TC/SC 2/N 434.

Ivancevich, J.M., P. Lorenzi and S. Skinner, 1994. Management, quality and competitiveness. Irwin, Boston.

Jonker, J. Ed., 1997. Trends in QA systems: vision on development of quality, environmental and labor policy: Kluwer Bedrijfsinformatie (in Dutch).

Kolsteren, O. and M de Vreeze, 2002. Norms for management- and tracking and tracing systems in development. *Voedingsmiddelentechnologie*, 16/17, 36-37 (in Dutch).

Kramer, M. and S. van den Briel, 2002. Total Quality Management in the food industry. MSc thesis Wageningen University, The Netherlands, pp. 90.

Lau, H.C. and M.A. Idris, 2001. The soft foundation of the critical success factors on TQM implementation in Malaysia. The TQM Magazine, Vol.13, no.1, 51-60.

Leaper, S., 1997. HACCP: A Practical Guide. Technical Manual 38. HACCP Working Group. Campden food & drink research association. Gloucestershire, UK, 59 pp.

Lewin, K., 1952. Group decision and social change. In: G.E. Swanson e.o. (eds.), Readings in Social Psychology, Holt Rinehart, New York.

LNV, 1997. Classical swine fever file, http://www.minlnv.nl/varkenspest/dosidv03.htm (in Dutch).

LNV, 2001. BSE file http://www.minlnv.nl/infomart/dossiers/bse (in Dutch).

Luning, P.A., W.J. Marcelis and W.M.F. Jongen, 2002. Food Quality Management: a techno-managerial approach. Wageningen Press. ISBN 9074134815, Wageningen, The Netherlands. 323 pp.

Luning, P.A. and W.J. Marcelis, (submitted). A concept for Food Quality Management based on a techno-managerial approach. Trends in Food Science.

Motarjemi, Y. and F. Käferstein, 1999. Food safety, Hazard Analysis and Critical Control Point and the increase in foodborne disease: a paradox? Food Control 10, 325-333.

Miles, S., D.S. Braxton and L.J. Frewer, 1999. Public perceptions about microbiological hazards in food. British Food Journal, 10, 744-762.

NACMCF, National Advisory Committee on Microbiological Criteria of Foods, 1998. Hazard analysis and critical control point principles and application guidelines. Journal of Food Protection, 61, 762-775.

Newall, D. and B.G. Dale, 1992. Measuring quality improvement: a critical management analysis, *Sigma*, 38 (4), 13-19 (Dutch).

Pomeranz, Y. and C.E. Meloan, 1994. Food analysis. Theory and practice, 3$^{rd}$ ed. Chapman and Hall, New York, pp.778.

Radford, K.J., 1975. Managerial Decision Making. Reston Publishing Company, Inc., Reston.

Robbins, S.P., 2000. Essentials of organizational behavior. Prentice Hall, New Jersey.

Rodjanatham, T., 2004. Quality assurance in international fruit supply chains: use of techno-managerial approach. MSc thesis Wageningen University, 63 pp.

Ross, J.E., 1999. Total quality management. St. Lucie Press. Boca Raton, USA.

Rousseau, D.M., S.B. Sitkin, R.S. Burt and C. Camerer, 1998. Not so different after all: a cross discipline view of trust. Academy of management Review, vol 23, no 3.

Schein, E.H., 1990. Organizational Culture. American Psychologist, vol. 45.

Schermerhorn, J.R., 1999. Management. John Wiley, New York.

Sebastianelli, R. and N. Tamini, 1998. Barriers to Total Quality Management. Journal of Education for business, vol. 73, no 3, 158-162.

Smit, M.J., 1999. Search for a common standard. Int. Food Hygiene, 10, 5-7.

Smith, D.J. and J. Edge, 1990. Essential Quality procedures. In: Dennis Lock (ed.), Handbook of Quality Management, Gower, Aldershot.

Spapens, P., 1996. Een muis in de melk: de geschiedenis van de consumentenbescherming in Brabant. Eindhoven: Kempen Publishers.

Surak, J.G., 2004. HACCP and ISO development of a food safety management standard. http://www.saferpak.com/iso22000.htm, update may 2004.

Tacken, G.M.L., M.G.A Van Leeuwen., B. Koole, P.M.L. Van Horne, J.J. de Vlieger and C.J.A.M. de Bont, 2003. Chain consequences of the outbreak of Avian Influenza. Den Haag: LEI (Dutch report).

USDA, 2001. Foot-and-Mouth disease Q's and A's, http://www.aphis.usda.gov/oa/pubs/qafmd301.html.

Van Boekel, M.A.J.S., 2005. Kinetic Modelling of Reactions, Reader of MSc course Predictive modelling. Wageningen University, The Netherlands

Van Delst, P. and E. Hendriks, 2002. SQF links safety to quality in the chain. Voedingsmiddelentechnologie, 35, 18-19 (in Dutch).

Van der Spiegel, M., G.W. Ziggers, P.A. Luning and W.M.F. Jongen, 2000. Development of a diagnostic instrument for food quality systems: a conceptual model, In: E. Dar-El, A. Notea and A. Hari (eds.), Productivity & Quality Management Frontiers - IX. Proceedings of the 9th International Conference on Productivity and Quality Research, held at Jerusalem, Israel at June 25-27th, 2000. Bradford, MCB University Press, pp. 193-200.

Van der Spiegel, M., P.A. Luning, G.W. Ziggers and W.M.F. Jongen, 2003. Towards a conceptual model to measure effectiveness of food quality systems, Trends in Food Science, vol. 14, no.10, 424-431.

Van der Spiegel, M., P.A. Luning, G.W. Ziggers and W.M.F. Jongen, 2004. Evaluation of performance measurement instruments on their use for food quality systems. Critical Reviews in Food Science and Nutrition, 44 (7-8), 501-512.

Van der Spiegel, M., P.A. Luning, G.W. Ziggers and W.M.F. Jongen, 2005. Development of the instrument IMAQE-Food to measure effectiveness of quality management. International Journal of Quality and Reliability Management, 22 (3), 234-255.

Van der Spiegel, M., 2004. Measuring Effectiveness of Food Quality Management. PhD thesis Wageningen University, ISBN 90-8504-015-9, 181 pp.

Van Geel, A., 2004. Customer focus in management: use of the techno-managerial approach. MSc thesis Wageningen University.

Van Logtenstijn, J.G. and H.A.P. Urlings, 1995. The transmission of disease via foods of animal origin. Outlook on Agriculture, 24, 23-25.

Zhang, Z., 1997. Developing a TQM quality management method model. Groningen: University of Groningen.

Zuurbier, P.J.P., J.H. Trienekens and G.W. Ziggers, 1996. Vertical Cooperation: Concepts to Start Partnerships in Food and Agribusiness. Kluwer, Deventer (in Dutch).

# Appendix I. Global overview of contents of GMP topics of parts I-III.

| Main topic | Aspects of guidelines (not complete) |
|---|---|
| **Part I general guide** | |
| Quality management | With respect to manufacturing: |
| | Processes, equipment and activities described in advance. |
| | Necessary facilities must be provided (personnel, equipment, materials, procedures, etc). |
| | Written procedure, instructions, etc. must be provided. |
| | Records of manufacturing process must be made and documented. |
| | System for recall. |
| | With respect to effective quality control: |
| | Involvement of quality control manager in development and acceptance of specifications. |
| | Adequate facilities for sampling, testing, sensory evaluation using approved methods and established procedures under direction of quality control manager. |
| | Rapid feedback of information to enable prompt adjustment or corrective action by operators and other functions. |
| | Examination of complaints. |
| | Maintenance of contacts with relevant enforcement authorities. |
| | Clear definition of functions of quality control management and production management. |
| Personnel and training | Key personnel quality control manager and production manager: |
| | Must have sufficient authority and appropriate authority for their tasks (e.g. establishing, verifying and implementing control procedures). |
| | Assignment of responsibilities must be clear. |
| | Must have appropriate scientific/technological knowledge. |
| | Must have adequate support staff. |
| | Should be trained in principles of GMP, periodic assessment of training programmes. |
| | Aspects of food and personnel hygiene (details described in the specific topics, e.g. manufacturing) refer to: |
| | Training in hygiene. |
| | Facilities must enable personnel to comply with hygiene requirements (washing facilities, detergents, hot and cold water, etc). |
| | Personnel must wear sufficient clean and washable overclothing; wearing jewellery, etc. is not allowed. |
| | Personnel suffering from any disease must inform the manufacturer, reporting of infections must be encouraged. |
| | Adequate and clean sanitary facilities must be provided. |

| | |
|---|---|
| Documentation | Documents are divided in three classes: instructions and procedures, programmes, and records and reports.
Instructions and procedures on e.g. ingredients, packaging, bulk and finished products, quality control, plant maintenance schedules.
Programmes on production, training and quality audits.
Records and reports on e.g. receipt and examination of raw materials, release of intermediate products, process control tests, customer complaint reports, quality audit reports, quality control summaries and surveys.
A checklist is provided to design and construct records, to ensure easy record-keeping and prompt access. |
| Premises and equipment | Premises must be well maintained, appropriately constructed and provide sufficient space to carry out operations.
Working conditions (e.g. temperature, humidity, noise levels, ventilation) must have no adverse effects on operators or products.
Requirements are established for the properties of floors, ceilings, walls, pipe work, and drains.
Requirements are established for equipment with respect to microbial cleanability (hygienic designs aspects). |
| Manufacturing | Process evaluations must be carried out prior to production.
Appropriate premises, equipment, materials, trained personnel, services, information and documentation must be provided to enable production of finished products.
Clear operating instructions must be provided.
Ingredients must comply with specifications.
Raw materials must be properly stored, inspected and processed.
Records must be kept.
Packaging materials must comply with specifications; appropriate labelling, coding and information on packaging; proper storage conditions.
Process conditions must be monitored and recorded.
Finished products must be quarantined until checked and approved.
Finished products must be properly stored (hygienic, pest control). |
| Recovery or reworking of materials | Matters are classified according to:
Systematic, involving systematic use of residues from previous production.
Semi-systematic, when variable quantity of intrinsically satisfactory but extrinsically unacceptable product occurs and can be re-used (e.g. misshapen product).
Occasional, involves all other instances than those mentioned above. They must be in all cases critically assessed. Reworking should be done following established procedures. |
| Complaints, recall, and emergency procedure | With respect to complaints, written instructions must be followed, an expert (other than Quality control manager) must be responsible, the complaint must be thoroughly investigated, complaints reports must be evaluated.
With respect to recall, a responsible person must initiate and co-ordinate recall activities, written procedures must be available, followed and regularly reviewed (clear notifications with respect to required information). Recalled material must be quarantined. |

| | |
|---|---|
| Distributors' own label | Terms of the contract must be clearly stated in written specifications.
Visits should be performed e.g. to ensure that food can be safely produced, and that it agrees with product specifications to assess sampling level. |
| Good laboratory practice | Facilities must be appropriate for testing and must be well designed and equipped.
Assigned persons must calibrate equipment and instruments. Written procedures must be available; analytical steps should include verification step.
Clean and hygienic working and personnel conditions.
Reagents must be dated and prepared by competent persons, following procedures.
Samples must be representative following clear procedures, must be labelled, and contamination must be avoided.
Methods must be efficacious.
Documentation involves records of receipt of samples, test results. Analytical records should contain specified information (name, date of receipt/testing, results, reference used, etc). |
| **Part II supplementary guidance on specific product categories** | |
| Heat-preserved foods | Guidelines on aspects of heat treatment that need specific attention, such as:
Requirements are set with respect to decimal reduction of *Clostridium Botulinum*.
For efficient heat treatments, physical, chemical and microbiological characteristics of each product and critical factors must be identified.
All products must be coded for traceability.
Process equipment must comply with hygienic design criteria.
Effectiveness of sanitation programmes must be established.
Cross-contamination must be prevented by separation from preparation areas (raw materials).
Integrity and microbial contamination of packaging, etc. must be checked. |
| Chilled foods | Chilled foods may be divided into high and low risk with respect to microbial safety.
Typical guidelines refer to:
Specifications and inspection of raw materials at supply using rapid methods.
Microbial testing of raw materials, intermediate and final products should aim at monitoring standards.
A temperature regime appropriate to the product and its production chain must be determined, taking into account all kinds of variation, and customer should also be provided with adequate provision for maintenance of the cool chain.
High-risk products require special conditions with respect to production area (chill rooms), hygiene (equipment, facilities, etc), personnel (special clothes, training, etc). |

| | |
|---|---|
| Frozen foods | Guidelines focus on general aspects such as personnel, testing facilities, raw materials and processing. Attention is paid to:<br>Specifications, acceptance and storage of raw materials and microbial testing.<br>High risk materials such as poultry, meat and fish must be prepared from separate rooms from processed products.<br>Critical control points should be established at positions where microbial damage can occur.<br>Freezing conditions must be controlled and properly carried out.<br>Temperature conditions in the cold chain (including storage and distribution) must be recorded to ensure effective monitoring. |
| Dry products and materials | Guidelines focus on typical problems associated with dust, such as explosive dusty atmosphere and cross-contamination by dust particles.<br>Dry products and their ingredients can carry heavy microbial loads. When applied in manufactured and canned foods, they must be submitted to strict microbial controls. |
| Foods with functional ingredients | Guidelines focus on the critical nature of the functional ingredients, which must be emphasised in e.g. training of production supervisors, production methods and control procedures to assure the appropriate content in the final product. |
| Irradiated foods | Guidelines focus on, amongst other things, which foods may be irradiated, the approved purposes (e.g. elimination of pathogens in foods), the irradiation facilities and labelling. |
| **Part III specialised topics** | |
| Contains additional guidelines on | Identification of raw materials.<br>Correct labelling and presentation.<br>Use of outsides services (contractor and contractee).<br>Electronic data processing and control systems.<br>Warehousing, transport and distribution.<br>Foods for catering and vending operations.<br>Responsibilities of importers.<br>Export.<br>Infestation controls.<br>Foreign body controls.<br>Use of food additives and processing aids.<br>Design of products and processes: hazards should be analysed to design products and processes properly, whereby critical control points should be established and suitable controls identified.<br>Environmental, health and safety aspects not covered elsewhere. |

# 7. Design and implementation of an HACCP system

*Isabel Escriche, Eva Doménech and Katleen Baert*

## 7.1 Introduction

The aim of this chapter is to educate the students in the HACCP system and to provide them with an understanding of all the necessary requisites and stages. The goal is to supply the students with the skills and knowledge that will facilitate the design and application of their own HACCP plan in their workplace, thus enabling them to participate in the implementation and maintenance of such a system in any company in the agri-food industry for the production of safe food.

This chapter is subdivided into different sections:
- Section 7.1. This provides a general overview of the necessity and benefits of HACCP in the current agri-food sector, as an essential management system for food safety and its relationship with other Quality Assurance Systems.
- Section 7.2. This section comments on the need for adequate preparation prior to any attempt to implement a Quality Assurance System like HACCP. It involves effective communication between all the company's staff and the allocation of resources to design, implement and improve the system.
- Section 7.3. Students should be aware that even before HACCP plans are developed, it is essential that the company establishes a series of prerequisite programs or plans, to control those indirect aspects of food processing that could jeopardize food safety. In this section the most important tasks and documents needed to develop and maintain each one of the prerequisite programs are discussed.
- Section 7.4. In this section, the necessary stages for the development of an effective HACCP plan in the agri-food industry are explained. The focus is on the theoretical principles of this plan, so as to help the students understand the philosophy of the HACCP system and design the correct plan for each situation.
- Section 7.5. This provides the guidelines for the implementation of the HACCP plan, meaning the application of the plan developed for a specific process. Effective implementation is absolutely imperative to ensure food safety. Students should also understand that the best plan in the world is not worth the paper it's written on without transforming it into effective action in the operation plant.
- Section 7.6. In this section the verification and maintenance of the system is described. Students should take into account that even when the process is controlled, it is not yet over. In order to ensure that the designed plan continues to be effective it is necessary to keep it updated and verify it, as with any other management system.
- Section 7.7. This section includes a future proposal for extension of the scope.

*Isabel Escriche, Eva Doménech and Katleen Baert*

### 7.1.1 HACCP´s origin and justification

Consumers are constantly demanding safe food products. Increasingly, they are questioning food safety, especially after the latest food scandals. As a consequence, EU food regulations have become stricter and business operators require self-checking systems for their companies (CCE, 2000a; CCE 2000b; Ley 20/2002; Regulation 178/2002). Thus, the agri-food sector has to be aware that control is not the government's responsibility, but rather the business operators themselves are responsible for establishing self-checking systems to ensure safe food production.

In this respect, the HACCP (Hazard Analysis Critical Control Point) system is worldwide considered as an appropriate tool for self-checking, since it is a focused approach to identifying and subsequently controlling those points or steps of the agri-food chain essential to guarantee food safety.

The application of HACCP to food production was pioneered, during the sixties, by the Pillsbury Company with the cooperation and participation of the National Aeronautic and Space Administration (NASA), Natick Laboratories of the U.S. Army, and the U.S. Air Force Space Laboratory Project Group to ensure the safety of the food that astronauts would consume in space. NASA wanted a "zero defects" program assurance against contamination by bacterial and viral pathogens, toxins, and chemical or physical hazards that could cause illness or injury. At that time, it became clear that in order to produce safe food products it was necessary the classic monitoring systems, based on the inspection of the finished product, evolved toward a system of preventive control during production. This control of the process would control the occurrence of hazards, and thus it would ensure a high level of confidence in the finished product (Mortimore and Wallace, 1998). During the 1970's, in the chemical sector it was very common to use the well-known technique of Failure, Mode and Effect Analysis (FMEA). This technique is characterised by the study of each one of the stages of the process, by analysing what can go wrong, the possible causes of the problem and its probable effects (ICMSF, 1988).

Taking this system as a reference, another similar system was elaborated but applied to the agri-food industry. Thus, the HACCP system was established during that decade. Similar to the FMEA technique, HACCP searches for hazards related to food safety, and subsequently establishes management and control mechanisms designed to ensure the safety of the product and its safety for consumers. In summary, the HACCP is a preventive food control system whose objective is food safety.

HACCP was first described publicly in 1971 at the National Conference on Food Protection. After a public outcry following a botulism outbreak involving canned soups, the U.S. Food and Drug Administration (FDA) ordered the first use of HACCP by regulation in 1973 for all low-acid canned foods.

The United States National Academy of Science recommended in 1985 that the HACCP approach should be adopted in food processing establishments because it was the most effective and efficient means of assuring the safety of the food supply.

In 1988, the International Commission on Microbiological Specifications published a book on HACCP for Foods (ICMSF, 1988). Two years later, in 1990, the Codex Alimentarius Commission (CAC) and Codex Committee on Food Hygiene (CCFH) started to prepare draft guidelines for the application of the HACCP system. Recognising the importance of HACCP for food control, the twentieth session of the Codex Alimentarius Commission, held in Geneva, Switzerland from 28 June to 7 July 1993, adopted the *Guidelines for the application of the Hazard Analysis Critical Control Point (HACCP) system* (ALINORM 93/13A, Appendix II). The Commission was also informed that the revised General Principles of Food Hygiene draft would incorporate the HACCP approach. Also in 1993, the European Union officially recognised the HACCP methodology as a standard production method for food manufacturers to implement and maintain a production control system. The recognition was established in the Directive 93/43/EEC on the hygiene of foodstuffs. From that time, Member States had 30 months to integrate this Directive in their laws. As a consequence, all food manufacturing facilities within the EU were forced to work in accordance with 5 of the 7 principles from January 1, 1997. In 1994, the FDA announced plans to require the use of HACCP in the seafood industry beginning in 1997. In 1995, the USDA announced plans to require HACCP in all meat and poultry plants under its jurisdiction, beginning in 1998 to replace the "poke and sniff" method. The USDA program, prior to 1996 HACCP implementation, established baseline prevalence levels for the presence of microbial organisms such as *Salmonella* in meat and poultry.

In June 1997, the revised *Recommended International Code of Practice - General Principles of Food Hygiene* (CAC/RCP 1-1969, Rev 3 (1997)) was adopted by the Codex Alimentarius Commission during its twenty-second session. *The Hazard Analysis and Critical Control Point (HACCP) system and guidelines for its application* are included in this Annex.

The Codex General Principles of Food Hygiene lay a firm foundation for ensuring food hygiene. They follow the food chain from primary production to the consumer, highlighting the key hygiene controls at each stage and recommending an HACCP approach wherever possible to enhance food safety. These controls are internationally recognised as essential to ensure the safety and suitability of food for human consumption and international trade.

In 1998, the FDA announced plans for implementing HACCP for all fruit and vegetable beverages, and is now considering whether to establish HACCP as the food safety process standard throughout all segments of the food industry under its authority. Beginning in 1999, the FDA incorporated HACCP into the *Food Code*, the biennially published reference for the prevention of foodborne illness in restaurants, grocery stores, and institutions such as nursing homes and hospitals. Since the *Food Code* serves as model legislation for all states and territories, many state governments now require evidence of HACCP processes for establishments under their purview.

USDA data released in 2000 and 2001 showed significant reductions in bacterial levels across a variety of food products following HACCP implementation. Numerous academic researchers have also found evidence documenting the usefulness of HACCP in reducing the levels of foodborne pathogens in food production and food service.

*Isabel Escriche, Eva Doménech and Katleen Baert*

HACCP is here to stay, and is becoming an international requirement. The Codex Alimentarius Commission, the international food standards-setting body overseen by United Nations agencies, the Food and Agriculture Organization (FAO), and the World Health Organization (WHO) now recommend HACCP adoption worldwide, and the EEC, with the proposal of the Regulation of the European Parliament relative to the hygiene of food products, makes the fulfillment of all the principles of the HACCP establised in the Codex obligatory. Also, HACCP is now embedded in the General Agreement on Tariffs and Trade (GATT), and nations are rushing to implement the process to ensure the safety of their domestic products and to survive in fiercely competitive world food markets.

### 7.1.2 Objectives of the HACCP system

The HACCP concept is a systematic approach to hazard identification, assessment and control. From the moment something goes wrong, it is possible to act and react quickly and efficiently. This prevents large sets of products being rejected and public health being endangered. This system eliminates disadvantages inherent in control by inspectors and microbial end product control. By concentrating only on those factors, which have a direct effect on the safety of the food product, no energy and money should be spent on less important factors. This system is cheaper and more effective compared to the traditional control systems.

By focusing on the critical points, which influence the safety and quality of food products over the complete production line, producers can show now that they are controlling the production conditions and that safe products are provided. Moreover, inspectors are now able to check the effect of the measures in the long term, where earlier inspections gave a picture at a given moment.

The HACCP system is a preventive, company-specific, quality system starting at the selection and purchase of raw materials, ingredients and packaging materials, following the complete production process and ending at the final product, ready for consumption.

### 7.1.3 Benefits of the implementation of HACCP system

Nowadays HACCP is a system internationally recognised and used. The benefits of the application of HACCP to consumers, companies and the administration have been clearly demonstrated. Although its use focused initially on industry, nowadays it is applied to other sectors such as distribution chains, sales, catering, and traditional food product processing, etc. The structured approach of its application makes it useful for the products currently marketed, as well as for new products to be manufactured in the future.

Some of the most relevant benefits are:
- It increases food quality and safety, with a positive impact on the collective health and common well-being, since it is a preventive control system of food processing that maximises the possibility that these food products are safe.

## Design and implementation of an HACCP system

- It provides confidence to clients, the administration and the agri-food operators. The companies that use this system to manage the safety of their products, show that they take their responsibilities seriously and that they comply with the legislation on food safety and hygiene.
- It facilitates commercial exchanges within and outside the European Union (EU), due to the safety assurance it provides to the administrations of different countries. One of the most important obstacles to this, which was prevented for a long time by the establishment of a Single Market in the EU, was the assurance of safety of food products produced in countries outside the European Union, since the effectiveness of administrative controls was very different from country to country. The European Commission found the solution in the HACCP System, in which the responsibility for the health assurance was not under external controls, but under self-checking systems.
- It is applicable to the entire agri-food chain. Although the primary sector - production of raw materials (agriculture, livestock farming and fisheries) is exempt from the obligation to apply this system, the latest food safety scandals highlighted that this is the weakest link of the chain. Current proposals from international bodies aimed at the necessary application of HACCP in the primary sector will soon be a reality.
- It is sufficiently flexible so as to adapt the necessary changes, such as: improvements in the design of equipment, improvements in processing and technological advances associated with the product. Its application is possible for those products currently marketed, as well as for new ones that can be manufactured in the future.
- It provides documented evidence that the process is self-checking, and allows for planning of ways to avoid problems, instead of waiting for them to occur to control them. Documents and records allow, in the event of audits by public health inspectors, easier, more efficient and more precise control. It will demonstrate the company's compliance with the legislation at all times, present or past.
- The system is an important economic saving for companies as well as for the Public Administration. With regard to companies, this saving is a consequence of minimising Quality Control efforts, thanks to the preventive approach; significantly reducing the costs due to operational failures (losses due to wastes, reprocessing, production halts, loss of orders, etc.); facilitating the immediate application of corrective actions, when deviations are detected; and allowing an effective use of resources, focusing and channelling expenditure where it is really essential. With regard to the Public Administration, the saving is a consequence of: reducing medical costs caused by the alteration of consumer health; reducing productivity losses due to sick leave; and reducing costs of inspection and monitoring in companies.
- The HACCP system, in the agricultural sector, is a preliminary stage essential for the implementation of other wider quality management systems such as the ISO 9000-2000 series.

### 7.1.4 Relationship of HACCP with other quality assurance systems

The Quality Assurance System (QAS) should not be conceived as an end but as a mean for the development of a continuous improvement driving force within the company. The purpose of

QAS goes beyond meeting the buyers' requirements; its main objective is to cover the internal needs of the administration of the organisation and to cover all the activities carried out in the company that may affect, directly or indirectly, the quality of the product, and prevent faults and therefore economic losses. These activities include purchase actions, control of design, control of documentation, offers, identification of products, control of processes, inspection of products, treatment of non-compliant products, storage, personnel training, etc. In summary, QAS involves all those activities that take place in a company aimed at ensuring that its quality objectives are met. In this sense, HACCP can be considered as a QAS since it is a system that helps to achieve the objective of the hygienic-health quality of food.

Some of the different QAS models internationally applied to the whole industrial sector, and therefore to the food and agri-business industry, are: BRC (British Retail Consortium), IFS (International Food Standard), ISO 9000-2000 family, and more recently, the EFQM (European Foundation for Quality Management). The model described by the ISO 9000-2000 standard has highest international recognition and, to a large extent, is the referent for the certification of quality systems. It is mainly aimed at preventing and detecting the presence of faulty products during their production and distribution and at ensuring, by means of preventive and corrective actions, that products that do not meet the specifications will not appear again. The ISO 9000-2000 series encourages business organisations to increase client satisfaction by means of the compliance of its specifications, as well as highlighting continuous improvement.

In the agri-food industry, the ISO 9000-2000 series can be used to process food that complies with the specified requirements, as in any other industrial sector. But HACCP must also be used in this sector, as it is essential to ensure food safety,. Therefore, both systems are supplemented to provide the maximum level of confidence: ISO 9000-2000 in terms of quality and the HACCP in terms of food safety management. In the agri-food sector, the interrelationship between both quality systems is so evident that the Codex Alimentarius in its Guidelines for the application of the HACCP states (Codex, 1993): "The application of HACCP is compatible with the implementation of quality management systems, such as the ISO 9000 series, and is the system of choice in the management of food safety within such systems."

In fact, the combined application of both systems is referred to in the 2001, ISO 15161 standard, which establishes the Guidelines for the application of the ISO 9001-2000 in the food and drink products industry. The ISO 15161 standard lists the specific points of ISO 9001-2000 standard and the principles of the HACCP as follows: "ISO 9001:2000, allows an organization to integrate its quality management system with the implementation of a food safety system that is more effective than the application of either ISO 9000 or HACCP alone, leading to enhanced customer satisfaction and improved organizational effectiveness".

Figure 7.1, based on ISO 15161, presents an outline of the aspects in common between both quality systems.

# Design and implementation of an HACCP system

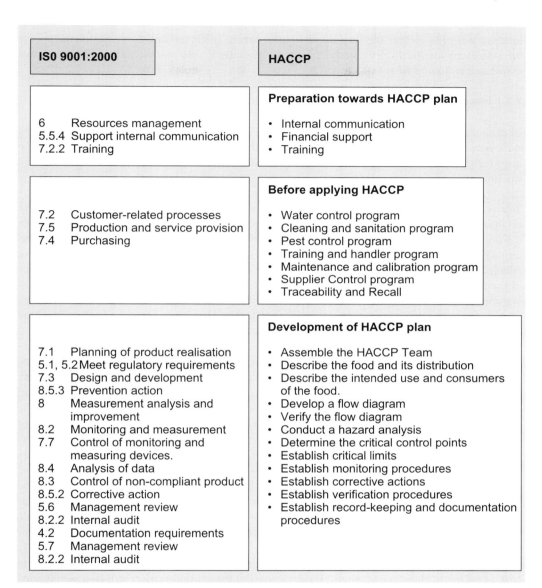

Figure 7.1. Aspects common to ISO 9001-2000 and HACCP.

## 7.2 Preparation for management in the food industry

The company's management, upon defining its quality policies and the objectives to follow regarding agri-food management, should put into practice this commitment, and start its period of preparation for management. At this point, the company's management should provide the necessary means so that the current objectives can be achieved and carry out a continuous improvement for the future. These means are translated into two necessary requirements: communication and financial support.

### 7.2.1 Internal communication to involve people at all levels

In designing and implementing HACCP systems, communication between all the company's staff is essential. At the beginning, the management should convey their commitment to improve the quality of the production process and should inform staff of the advantages and the importance of its implementation.

Top management should ensure that the appropriate communication mechanisms are established within the company. Thus, all the employees should be able to accept the demands coming from the outside and adopt a participative attitude towards all that happens in the industry, so that the problems are solved and do not recur. In this sense, it is very important that management encourages and establishes an effective communication process by means of meetings, bulletin boards, internal newsletters, audiovisuals, interviews, etc. This type of communication is called internal and it promotes the exchange of information, knowledge, results, etc., among all the workers: management, technicians in charge of the evaluation of the risk, management technicians and employees.

Within this type of communication we can distinguish two types of relationship: vertical and horizontal (Juran *et al.*, 1999). Vertical communication takes place among people that belong to different hierarchical levels and it should work in both directions, that is, upwards and downwards. It constitutes one of the important foundations that ensure business success. The downwards flow of communication usually fails when there is a lack of knowledge or trust, or simply when the management does not know what information to transmit. This situation hinders worker collaboration. In the upwards direction, the management should be extremely sensitive and receptive regarding the need to receive information and grasp the workers' opinion and point of view. They can provide a realistic take on the technical problems and the practical way to solve them. Horizontal communication takes place among people in a similar category with no difference in authority. This type of communication has several barriers due to distrust, rivalry, competition or simply due to physical or functional separation.

In summary, one of the main benefits of good communication is to be able to act quickly in case of failure in the process, since relationships and the mutual respect will be strengthened among all the personnel of the company. This will allow the promotion of trust in the safety of the food processed.

### 7.2.2 Financial support: development and maintenance

Operations related to production and, especially to management, work under certain limitations usually associated with time and money (Campanella, 2000). The company's management should ensure that the necessary resources to carry out the design of a safety plan, like HACCP,

## Design and implementation of an HACCP system

the implementation and improvement of the system, are identified beforehand and that they will be available when required. In general, costs can be classified in three different sections:

*Training and education costs*
Training should involve all the company staff, from top management down to the last employee. All workers should receive the appropriate training to carry out their duties correctly which may vary from performing tasks to monitoring them.

*Prevention costs*
Those costs that arise from the activities designed to avoid failures that result in deficient product quality. These costs are counteracted by the reduction of assessment and correction costs. Some of the prevention costs include: those due to market research activities and the consumer studies; those related to the development of new products; costs needed to ensure that the specifications of raw materials are met; and other expenses of the quality system specifically aimed at avoiding poor product quality.

*Assessment costs*
The cost of those activities related to measurement, assessment or audit of products, aimed at predicting or determining the level of quality of a product in terms of the level of compliance with the requirements established. Assessment costs include the following, among others: inspection of incoming materials from suppliers; process controls; internal and external audits.

## 7.3 Requirements before applying HACCP

### 7.3.1 Introduction

Once management has decided to start the HACCP system, it is necessary to carry out an analysis of the initial situation and of the needs for achieving the objective.

It is important to note that each HACCP study focuses on the control of the process under consideration. However, this is not enough to ensure safe food production, since in agri-food companies there are several factors, such as equipment, personal hygiene, food handlers, facilities, etc., which although they are not a stage of the process, and therefore not included in the monitoring process, their lack of control may put in jeopardy the safety of the food being processed. For this reason, the company should ensure, firstly, the control of all these factors external to the process, with the objective of consolidating a solid basis on which to put into practice the different HACCP plans later on.

In addition, the Codex Alimentarius (Alinorm 97/13, 1997) advises:
*"Prior to application of HACCP to any sector of the food chain, that sector should be operating according to:*
- *the Codex General Principles of Food Hygiene,*

- *the appropriate Codes of Practice (GMP) and*
- *appropriate food safety legislation".*

These prerequisites are the foundation and preconditions for an efficient and well-functioning HACCP system. The establishment of an HACCP plan is the ideal moment to evaluate these preconditions on their presence, completeness and implementation.

### 7.3.2 Good Manufacturing Practices (GMP)

Raw materials, which are used in the food industry, are contaminated with micro-organisms. During further processing and storage, these micro-organisms can develop, influenced by several internal, external and implicit factors, mainly resulting in spoilage. The accidental presence of pathogenic germs can lead to food infections and food intoxications. Therefore, it is necessary to maintain a low contamination level. This can be done by working towards "Good Manufacturing Practice" and by respecting the elementary rules of personal and common hygiene. Moreover, the development of the micro-organisms, which are always present, has to be inhibited by a justified choice of preservation technique.

GMP (Good Manufacturing Practices) can be defined as a package of requirements and procedures by which the work methodology takes place under controlled conditions and by which surrounding conditions are created that allow the production of hygienic and safe products.

Since the aim of the GMP regulations is to prevent or reduce the contamination to a minimum, the application of these regulations is very important. When these regulations are known and observed by everybody, hygienic working occupies a central place on every level in the company.

The most important GMP's that need to be elaborated and implemented are:
- GMP 1: cleaning and disinfection;
- GMP 2: pest control;
- GMP 3: water and air quality;
- GMP 4: temperature control;
- GMP 5: personnel (facilities, hygienic way of working, health, education);
- GMP 6: structure and infrastructure (surrounding area, building, materials, equipment);
- GMP 7: technical maintenance;
- GMP 8: waste management;
- GMP 9: control of raw material;
- GMP 10: work methodology.

### 7.3.3 Prerequisite Programs (PRP) or prerequisites

Recently, the term 'Prerequisite Program (PRP)' is more in use. PRP can be defined as every specific and documented activity or facility that is implemented corresponding to the 'Codex

## Design and implementation of an HACCP system

General Requirements of Food Hygiene', the 'Good Manufacturing Practices' and the legislation, with the purpose of creating basic requirements that are necessary for the production and processing of safe foods in all stages of the food chain.

In other words, PRP covers GHP (Good Hygienic Practices), GMP and the legislation. The term 'program' refers to the fact that PRP's are more than a working instruction, a plan or a regulation. They are general control measures that need to be verified for their effectiveness on a regular basis.

The PRP's can be divided into 14 groups:
- PRP 1: cleaning and disinfection;
- PRP 2: pest control;
- PRP 3: water and air quality;
- PRP 4: temperature control and registration;
- PRP 5: personnel (facilities, hygienic way of working, health, education);
- PRP 6: structure and infrastructure (surrounding area, building, materials, equipment);
- PRP 7: technical maintenance and calibration;
- PRP 8: waste management;
- PRP 9: control of raw material;
- PRP 10: traceability, recall, goods returned, rejections/non-conforming products;
- PRP 11: allergens;
- PRP 12: physical and chemical contamination;
- PRP 13: management of product information;
- PRP 14: work methodology.

### 7.3.4 GMP, PRP and HACCP

Although GMP, PRP and HACCP are specific elements, they have interfaces. GMPs are general hygiene measures that need to be taken and they can be sector specific or not. PRP will transform these hygiene measures into a practically manageable, effective and company specific surveillance system. Furthermore, HACCP gives a clear and systematic analysis of the hazards that are specific for the company. After the completion of a company specific hazard analysis, it will be clear, which process steps need supplementary specific control measures (CCP - Critical Control Point) in addition to the general control measures.

It can be said that HACCP is necessary to concretise the GMP and PRR measures on the one hand, and to look for additional hazards through a company specific study, and to implement additional control measures on the other hand.

However, when the PRP's are insufficiently elaborated and implemented, a high number of hazards with a quite high chance of occurrence need to be listed during the hazard analysis of HACCP. As a consequence, a lot of points will occur with high risks that are not controllable as a CCP. This can be avoided when the PRP's are well elaborated and observed. PRP's can be considered as the necessary foundation for a functional HACCP-system. The HACCP system is in its turn the ideal means to evaluate and adjust the existing PRP's on the basis of the hazard analysis.

It is essential to note that each prerequisite should be considered as an independent system, not linked to any process and applicable to the whole company. Therefore, it is a great advantage that, regardless of the processes carried out in the company and the HACCP plans associated with each one of them, the prerequisites should be unique and common to all of them. The control of factors external to the process is so important that in some EU countries like Spain, official bodies in charge of auditing product safety in the agri-food sector, require the implementation of these prerequisites, prior to HACCP. In this sense, there are many international organisations such as NACMCF, CODEX (Codex, 2001a; Codex, 2001b; NACMCF, 1998) and official bodies from different countries (GV, 2001) that have drafted guidelines to include the most common requirements and provide guidance on how to implement them in the agri-food industry.

## 7.3.5 Implementation of prerequisites

The implementation of prerequisites in the company is not immediate. As in any management system, top management will define the objectives and provide the necessary resources for their application. Also, the success of its application will not be possible without the involvement and responsibility of all the company's personnel. It will also be essential to develop a record-keeping system to record in writing all the necessary information to achieve the objectives set forth in the corresponding prerequisite, as well as all the information generated as a result of the controls carried out. Everything should be written, nothing should be left to improvisation.

This chapter attempts to structure the documentation for each requirement according to a common scheme for all of them. Thus, it will simplify their implementation and management and it will also facilitate their understanding and use by the corresponding personnel. The specific documentation for each prerequisite will constitute an independent plan that will be developed in three sections: program of activities, verification of activities carried out and control of their efficiency, and corrective actions.

*Program of activities*
Each prerequisite should contain a detailed description of all the tasks to be performed by the company to achieve the objectives for that prerequisite. To avoid omissions and to facilitate the understanding of the tasks by all staff, a simple way to present the information is to answer the following questions:
- "What"    Task to carry out.
- "How"     To carry out the task. In this case, an exhaustive explanation of how to carry it out will be provided; the steps will be described one by one; which materials or products will be used, how to apply them, etc.
- "When"    To carry out the task. At this point, the frequency and the moment when each one of the tasks will be carried out will be indicated.

- "Who"     Should carry out the task. The person in charge of carrying out the tasks will be appointed, specifying the position and even the person's name. In the case of requesting a service of external companies or laboratories it is also appropriate to specify their name.
- "Where"   The task is to be carried out. In this section, the exact place of performance will be specified.

When regulations make a given task compulsory, the company will review the corresponding legislation before answering the questions.

*Verification of activities carried out and control of their efficiency*
Verifications will provide the evidence that the tasks envisaged in the program of activities are being carried out. The control of its efficiency will be proof of whether the results obtained are as expected. If the control is appropriate and is carried out efficiently, it will facilitate the application of the corrective actions whenever necessary.

These tasks will be carried out by staff from the company or by external personnel. To present in a simple way all the information associated with control tasks, the same structure of questions outlined in the program of activities can be followed:
- "What"    Task to control. A specific control for all the tasks established in the program of activities shall be set up.
- "How"     The way to carry out that control, for example, visual with a checklist, or analytical as an analysis of micro-organisms, etc.
- "When"    The frequency and moment in which the control will be carried out will be recorded.
- "Who"     The person, laboratory or company responsible for carrying out the control will be indicated.
- "Where"   It will be specified the place where the control will be carried out.

The documents generated as a consequence of applying these controls will be properly filled in, including the date, the results obtained, signature of the person responsible in each case, etc. In addition, for a prompt reaction, it is convenient that the control record include the usual results or even the range of acceptable values. Finally, everything will be written and suitably recorded, allowing thus for a revision of results obtained later, either by the company or by the administration.

*Corrective actions*
Each plan will foresee the appropriate corrective actions in order to act quickly when, as a result of the controls, non-compliances or deviations are detected. Corrective actions, as well as the efficiency or not of their application, will be documented and recorded. The design of a model sheet, as shown in Table 7.1, can be used for all the plans and for facilitating the collection of data. In addition, this way of presenting information will allow for revision of the incidences that have taken place and action as a consequence.

Table 7.1. Report of corrective actions common to all the plans.

| Record N° sheet: | Signature of person responsible for record | Record date |
|---|---|---|
| **Description of non-compliance** | Remarks | |
| Place where the non compliance is detected | | |
| How it was detected | | |
| Description of the problem | Who detects the problem | Date |
| Analysis of possible causes | | |
| **Corrective actions** | Remarks | |
| Description of suggested corrective action(s) | Who carries out the application | Date |
| **Follow-up of corrective action** | Remarks | |
| Has the initial problem recurred? | | |
| Yes | | |
| No | Who carries out follow-up | Date |
| Partially | | |

In the following sections, the structure of the documentation is shown as an example for the following prerequisites:
- Water control program (PRP 3);
- Cleaning and sanitation program (PRP 1);
- Pest control program (PRP 2);
- Training and handler program (PRP 5);
- Maintenance and calibration program (PRP 7);
- Supplier Control program (PRP 9);
- Traceability and Recall (PRP 10).

### 7.3.6 Water quality control plan

Water in the agri-food industries plays a very important role, as it is not only a necessary key supply for carrying out the operations of cleaning and disinfection, but also on many occasions it is part of the food product as an ingredient, or it is simply part of an operation in the process.

Water quality has traditionally been considered as a supplementary part of different GMPs, but the important role that it plays during the processing operations makes it significant enough to be considered as a prerequisite.

*Water quality program of activities*
Amongst the tasks for development in this plan, the most important and widespread are:
- Activity 1: Realisation of a map of the distribution network of potable water and of the drainage of waste water.
- Activity 2: Analytical control of water.
- Activity 3: Water treatments previous to use and treatments of waste water.

## Design and implementation of an HACCP system

As previously described, the answer to a number of basic questions enables an understanding of the tasks to be carried out (Table 7.2).

*Verification of the implementation of the water quality control actions and their efficiency*
The company will perform the control of all the programmed tasks, checking its implementation with a revision of the documentation generated and reports. The maps will be verified on the field to check that they reproduce the real situation. The controls on water quality will be carried out by checking the records of the analyses carried out in an external laboratory or in the company's laboratory. In the case of carrying out water treatments previous to its use or waste water treatments, an analytical control should be carried out to verify that the treatment has been effective and will be written down in a report - specifying date, hour and place of the sampling, results obtained, person responsible, and person in the company who has registered the results.

*Table 7.2. Proposal for schedule of activities in water quality control.*

| What | How | When | Who |
|---|---|---|---|
| General map of the potable water, and waste water drainage distribution network | The map produced should describe the water distribution system showing the source (municipal network, well, spring, etc.), the pipes of hot and cold water, entrances and exits, intermediate deposits, faucets and drainages, points of intermediate treatment previous to evacuation, pipes of non potable water for authorised punctual cases, etc. | The map will be drawn up once and it will be updated when modifications take place | Quality control technician |
| Water quality self-control program | This will be carried out depending on the community legislation in force (Directive 98/83/CE) and the corresponding transpositions to the legislation of each country, for example in the case of Spain (RD 140/2003) | The legislation in force shall be observed (Directive 98/83/CE) | Laboratory performing analysis |
| Water treatment previous to its use: (Chlorination, denitrification, ozonisation, etc.) and waste water treatment | It will be indicated how to carry out this task, in compliance with legislation in force (Directive 98/83/CE and RD 140/2003 in Spain). For example, in the case of water chlorination, the process of implementation will be described, stating if chlorine should be used in the form of gas or liquid, the necessary time of contact, etc. | It will be carried out in compliance with the legislation and the needs of the company to meet the parameters required | Official |

*Corrective actions*
The company will develop a number of measures to apply when the analytical results demonstrate a deviation from the expected values. In general, these measures will focus on studying the causes, informing the corresponding person and attempting to reestablish the potability of water as soon as possible. As it happens with all the prerequisites, the detection of a problem, in this case the quality of water must be properly documented. Filling in a form like the one in Table 7.1, based on common elements of the plan, may facilitate this task. Later on, the plan should be revised to correct the cause that has motivated the deviation.

## 7.3.7 Cleaning and disinfection plan

Cross-contamination is one of the main dangers in agri-food industries, since food is in contact with food handlers, equipment and utensils. Cleaning and disinfection must be the dominant characteristic in all premises of the company and most especially in the food handling premises.

Cleaning is a process in which residues are removed, usually with the help of detergents dissolved in water. Disinfection consists of destroying most of the micro-organisms on the clean surfaces, by means of the use of chemical agents called disinfectants. Good cleaning and disinfection is based on several basic principles: the exclusive use of products authorised for use in food industries; which must be suitable for the type of dirt; and should be applied at the recommended doses following the recommended procedures correctly.

When carrying out the company's cleaning and disinfection plan, it is advisable to treat premises, equipments, tools and utensils separately, without forgetting the transport facilities and surroundings of the company.

*Cleaning and disinfection program of activities*
Among the activities for development in this plan, the most important and widespread are:
- Activity 1: Draw a map of the establishment and surroundings, detailing useful information for this specific program.
- Activity 2: Develop a detailed list of the cleaning and disinfection products used by the company.
- Activity 3: Schedule the cleaning and disinfection tasks of premises, equipments, machinery, tools, etc.

Table 7.4 shows an example of activity 3 of the program, which states how to carry out cleaning and disinfection of food processing surfaces.

*Verification of the implementation of the cleaning and sanitation control actions and their efficiency*
The company should check periodically all the activities programmed. For the cleaning and disinfection program, a revision of the establishment layout should be carried out annually, checking *in situ* and with the actual layout. Possible changes will be introduced, validating the map until the next inspection.

Table 7.3. Proposal for the scheduling of cleaning and disinfection activities.

| What | How | When | Who |
|---|---|---|---|
| Map of the establishment | The different areas of the company will be drawn on the map, indicating with colours the degree of dirt generated in them. The equipment and activities developed in each area will be signalled and the path of the product in the company will be traced with a line. | The map will be drawn once and it will be updated when modifications take place. | Quality control technician |
| Detailed list of cleaning and disinfection products | The list will include the name of the product, name of supplier, instructions, dosage, surgeon general's registration number, etc. and all those necessary data that help identify the product. | The list will be written once and it will be updated when the products used are changed. | Quality control technician |
| Schedule cleaning and disinfection tasks | For each piece of equipment, machinery, tool, etc. the procedure to follow will be described in detail: product, dose, application and rinsing instructions, ways of drying, temperature of the water used, etc. | the frequency and the moment of application regarding the productive process of the company will be recorded. | Official representative |

Table 7.4. Example of the cleaning and disinfection schedule for food processing surfaces, such as tables and chopping boards.

| What | How | When | Who |
|---|---|---|---|
| Surfaces for food processing, tables and chopping boards | ■ Remove food residues (blue cloth)<br>■ Apply "Brillo-Multi" detergent (according to label: 25mL/L) with hot water<br>■ Rinse in water<br>■ Apply "Carlim" disinfectant with a spray (according to label: 250mL/5L)<br>■ Let it act for at least 30 seconds.<br>■ Rinse with clean water (white cloth). Do not use the same cloth for rinsing as for removing the food residues.<br>■ Dry with disposable cloth | After each use | Cleaning staff |

The list of products used in the cleaning and sanitation activities should be revised once a year, checking that it includes all the products used in the company and that the identification data correspond to the real data.

The control of the cleaning procedures will be done visually. Visual observation has the advantages of being fast, simple, cheap, and if properly carried out, the information obtained allows for immediate action. However, it is a subjective monitoring procedure as it depends on the personal criteria of the person who supervises the facility. In order to provide some guidelines and to avoid omissions, it is advisable to elaborate a checklist. The checklist should include all the points to control in the premises: walls, floors, ceilings, difficult access corners, washbasins, drains, working tables, conveyor belts, switches, door handles, etc, the equipment (refrigeration chambers, freezing chambers, kneaders, etc.), and tools (can openers, knives, mixers, etc.). Table 7.5 shows an example of visual checking using a checklist.

Sanitation control should be the main objective and it should not be subject to possible interpretations; for this reason, it will be carried out using a microbiological analysis of the surfaces. The indicator micro-organisms reported in the legislation should be quantified. Sanitation plans should include all the required information: sampling procedures, when sampling should be performed, surfaces to examine, means to use, etc. In addition, to permit a rapid action in case of adverse results, it is advisable to write down the acceptable limit values in the analytical control sheet.

Table 7.5. Example of a check-list for the visual observation of daily cleaning task efficiency in a bakery company.

| Supervisor: Date: Signature: | Comments: | |
|---|---|---|
| **Where** | **What** | **Evaluation** |
| Department unit | Floor | √ |
| | Tables, ashtrays and dustbins | √ |
| Production | Floor | Flour residues |
| | Scale | √ |
| | Conveyor belt | √ |
| | Rollers | Embedded residues |
| | Tip chopper | √ |
| | Mobile trays | √ |
| | Photoelectric cells | √ |
| | Flour sprinkling brushes | √ |
| | Fermentation closets | √ |
| | Freezing tunnel | √ |
| | Kneaders | √ |
| Lavatories | Floor | √ |
| | Lavatories and restrooms | √ |
| Canteen | Tables, ashtrays and dustbins | √ |
| | Floor | √ |

*Corrective actions*
If as a consequence of the visual cleaning checking or analytical sanitation checking some non-compliances are identified, such as dirty areas, embedded food remains or contaminated surfaces, it will be necessary to proceed again with the cleaning or sanitation tasks to solve the problem. Subsequently, the possible causes of the problem will be analysed so that the required measures may be taken in order to avoid recurrence of the problem. These measures should consider changing the cleaning product or dosage, completing the description of the cleaning task, giving training courses to the operators, etc. In all cases, the non-compliance and the corrective measures taken will be indicated in the corrective action report (Table 7.1).

## 7.3.8 Training plan for food handlers

Food handlers are a key factor in the safety of food processing. In their daily work they can introduce physical, chemical or microbiological contamination in the food they are handling. In the last case, the transmission to the food may be caused by the bacteria of their own intestinal, skin, respiratory flora or by an infected injury. To minimize these risks, it is necessary to perform a health check of the personnel and to establish adequate training plans.

*Training food handlers program of activities*
In some countries, like Spain, it is a legislative requirement for food handlers to attend training courses before being hired. In addition, management must provide adequate time for thorough education and training of its personnel to guarantee food safety. Besides that, management should provide the necessary means for training and verify that the knowledge acquired in the training period is put into practice.

Once the company has analysed the situation with respect to its training needs, the company will define the program of activities (Table 7.6):
- Activity 1: Annual planning of courses and seminars for continuous training of the company's personnel. This should have the approval of the management and the supervision of the corresponding governmental department.
- Activity 2: List of food handlers.
- Activity 3: Updated record of the medical reports of the workers in those companies whose activities demand it.

Table 7.7 shows an example of how to develop workers' training programs in an agri-food company.

*Verification of the implementation of training control actions and their efficiency*
The verification of training actions for food handlers will consist of the revision of the documents generated. Once the course or seminar has finished, a record sheet will be fulfilled with all the information on the course (course title and contents, date, course leader, participant workers, etc.). If the training activity is held outside the company, the name of the external training entity must be registered and a copy of the attendance certificates should be kept. On the other hand,

Table 7.6. Planning of the training activities.

| What | How | When | Who |
|---|---|---|---|
| Courses and seminars (indicate course title) | Describe the scheduled training activities and any other relevant aspects: objectives of the course, course contents, intended audience, teaching staff, audiovisual requirements and number of hours. | The scheduled dates of the course will be stated | Quality team |
| List of food handlers | A list will be elaborated including all the food handlers, specifying their names, working area and date of entry. | The list is elaborated once and will be updated to include changes. | Quality Manager |
| Register of workers' medical reports. | A list will be elaborated with the worker's personal data and the date of issue of the medical report. The corresponding certificates will also be attached. | It will be done once and will be completed as necessary | Quality Manager |

Table 7.7. Example of training plans in an agri-food company.

| What | How | | | | When | | Who | Where |
|---|---|---|---|---|---|---|---|---|
| Course/ semi-nar | Objectives | Hours | Contents | Teaching staff | Scheduled date | Target audience | Place | |
| Course: HACCP | Recycling technicians training in Quality and Safety | 12 | – Introduction<br>– Principles<br>– Application of quantification elements and simulation models | External personnel | July 2004 | Quality control managers and technicians | The company's premises | |
| Seminar: Good Manufacture practice | Maintain and extend the concepts on good manufacture practices | 1 | – Correct behaviours<br>– Food intoxication<br>– Hazards in the production/ distribution chain<br>– Importance of the cold chain | In-house quality control technician | February 2004 | Packing and shipping personnel | The company's premises | |

to control the efficiency of the training activities and guarantee that the food handler is applying the knowledge acquired it is essential to monitor his/her behaviour in the job.

The list of food handlers and their medical reports will be controlled by verifying the authenticity of the data and by checking that all possible changes, and when these have happened, have been recorded.

*Corrective actions*
The corrective measures to be taken should be included in the plan, and consist of modifying the training courses, teaching staff, duration, etc. In addition, within the control plan for food handlers, when workers report some kind of food transmission illness, the corrective action will be to separate them from their job until the risk has disappeared. Similar to the other requirements, when corrective action is taken, it should be recorded (Table 7.1).

## 7.3.9 Preventive maintenance plan

In the agri-food industry, maintenance plans should be included in the design phase and have continuity during the operating phase. Plan development should focus on the building, facilities, equipment and utensils, as well as on the services offered by the company.

The structures and elements of the buildings and facilities (floors, walls, windows, doors, working surfaces, insulation, drains, pipes, etc.) should be built with durable, smooth, impervious materials that are easy-to-clean and disinfect. In addition, the floors should be non-slippery. Correct maintenance of the building is fundamental to achieving optimal conservation conditions. Moreover, there should be periodical control of the lighting, ventilation, fume extraction, and environmental conditions in the working place.

All the equipment, utensils and containers that will be in contact with food should be maintained such that equipment availability is guaranteed and they are certified in perfect conditions without damaged parts, cracks or fractures which could make their proper operation and cleaning of them difficult. For equipment operation to be optimal, special attention should be given to heat-cool equipment, measurement devices such as thermometers, scales, dosing devices, etc. and the cleaning instruments.

Calibration and control of tools and equipment are also considered in this plan. All the equipment and instruments with control devices or measurement parameters such as temperature, time, speed, etc. should also be included in a calibration program since some of these devices will be fundamental to control the Critical Control Points (CCP's).

*Preventive maintenance program activities*
The maintenance program covers the following activities (Table 7.8):
- Activity 1: Planning of the maintenance tasks in premises, facilities and equipment.
- Activity 2: List of devices and equipment to calibrate/verify.
- Activity 3: Calibration/Verification schedule of measurement equipment.

Table 7.8. Program of maintenance activities.

| What | How | When | Who |
|---|---|---|---|
| Maintenance of:<br>■ General structure of the building: ceilings and roof, walls, floors.<br>■ Window insulation, mosquito netting, slats, pipes, etc.<br>■ Drainage system.<br>■ Equipment (analyse each separately) | Specify step-by-step how to perform maintenance tasks, which pieces should be dismounted, how to proceed, which products to use, etc. | Frequency will depend on each case; for equipment, follow the manufacturer's indications, or as needed due to wear in the equipment, premises and installations. | Maintenance manager |
| List of devices and equipment to calibrate/verify | Make a list describing all the equipment and instruments. | The list will be drawn up once and will be modified when there are any changes. | Maintenance manager |
| Calibration/Verification schedule of measurement equipment | Describe the procedures of calibration/verification and, whenever possible, use validated measurement standards. | The manufacturer must specify frequency. In addition, indicate the scheduled date of next calibration/verification. | Maintenance manager |

*Verification of the implementation of the maintenance control actions and their efficiency*
The verification of building and installations maintenance will be done visually, by filling in a "checklist" that will help in monitoring the implementation and efficiency of the maintenance tasks. Table 7.9 shows a checklist for the premises and installations. The verification of equipment maintenance will be done by revising the documents when performing the task.

Table 7.9. Example of checklist for control of maintenance activities.

| Facilities and installations | Verification |
|---|---|
| Date        10-5-2003 | Comments: Repair the broken window pane as soon as possible |
| Ceilings and roofs | Compliant |
| Walls | Compliant |
| Floors | Compliant |
| Drains | Compliant |
| Structure | Compliant |
| Windows | Non-compliant |
| Lighting | Compliant |
| Switches | Compliant |
| Electric boards | Compliant |

## Design and implementation of an HACCP system

The efficiency control of the list of devices and equipment to calibrate/verify will be done by checking in situ that all devices are included in the list. Finally, the verification of the calibration/verification tasks will provide certificates or data records stating that the instrument calibration/verification has been performed and that the equipment is suitable for use. These documents should indicate the date of the verification/calibration and the next scheduled date for verification/calibration.

*Corrective actions*
The measures proposed in this section aim at correcting the non-compliances related to equipment maintenance and calibration/verification. In all cases, it shall be documented in the corrective action report (Table 7.1).

### 7.3.10 Pest control plan

The presence of insects, rodents and other animals constitutes a serious threat to food safety. Their control will be performed at two levels: first, by avoiding the pest outbreak in the industry, and second, by taking the required measures for their removal.

The most common animals and/or parasites in agri-food industries are birds (pigeons, sparrows, etc.) mammals (bats, rodents, etc.) and insects. The company building is one of the main barriers to pest entrance. For this reason, building design is a fundamental factor that should be associated with the adequate maintenance of the premises and facilities, and installations (piping system, siphoned drains, cracks, etc.). Another important aspect is preventing animals from getting food and shelter; to this end, an adequate control of food storage and food residues should be carried out. Waste containers must be kept clean and covered with a lid. In addition, the outside of the building should be clear and free of dense vegetation or any other material that could be a pest source.

Therefore, it is clear that the success of the pest control plan depends directly on the correct application of the action plans mentioned above: preventive maintenance and cleaning and disinfection plans

*Pest control program of activities*
The company, besides trying to avoid a pest outbreak, should be prepared to act if the pest surpasses the barriers envisaged. It is advisable that pest treatment tasks be done by an authorised external enterprise that has the required official certificate for the use of pesticides. In this way, the company will avoid the storage of toxic chemicals (pesticides). The most common activities, to be performed by the external enterprise for pest control, are the following (Table 7.10):
- Activity 1: Map with the pest control points;
- Activity 2: Programming of the activities for pest treatment and control.

Table 7.10. Pest control program of activities.

| What | How | When | Who |
|---|---|---|---|
| Map of pest control points | Indicate the traps, baits, electric traps for insects, etc. placed to control the pest | The map will be drawn up once and will be modified if there are any changes. | External enterprise |
| Schedule of the activities for pest treatment and control. | Describe each of the actions to perform for the control of each pest and of the pesticides to use. The company shall have a list of all the products, indicating: health record number, commercial name, use, manufacturer, level of hazard, handling, etc. | Frequency will be established depending on the needs and regulations (Spain: RD 3349/1983) | External enterprise |

*Verification of the implementation of the pest control actions and their efficiency*
The company hired will check that the activities programmed for pest control are being implemented and are effective. The agri-food company will require all the documents of the external enterprise stating its activities and control actions.

*Corrective actions*
If the agri-food company detects non-compliances relative to the presence of pests, it will take measures that may range from reprimanding the external enterprise hired to cancelling the contract. In any case, any incident will be recorded in the corresponding corrective action report (Table 7.1).

### 7.3.11 Supplier control plan

The supplier is the organisation or person (producer, distributor, retailer, seller, etc.), that provides the company with a product (ISO 9000:2000). The quality of the supplied products should be established through a contract between the company and the supplier, but safety aspects are not negotiable. However, the company should develop a plan that allows it to verify the quality and safety of the raw materials.

*Supplier control program of activities*
Among the possible activities that the company can develop to control its suppliers and the raw materials delivered, the most important are (Table 7.11):
- Activity 1: Updated list of suppliers.
- Activity 2: Supplier auditing.
- Activity 3: Inspection of the raw materials delivered to the company.

*Verification of the implementation of the supplier control actions and their efficiency*
The documents generated by the implementation of the program activities will be revised to verify that the activities are being performed and that they fulfil the aspects of quality required by the legislation and established in the contract with the supplying company.

Table 7.11. Program of activities for supplier control.

| What | How | When | Who |
|---|---|---|---|
| List of suppliers and raw materials. If the suppliers are only intermediaries, include in the list the product manufacturer also. | Collect the following data:<br>■ Identification of the supplier (name, surname, Fiscal code)<br>■ Complete address, telephone, e-mail, etc.<br>■ Health record number or Health authorisation<br>■ Raw material delivered (flour, salt, oil, yeast, containers, etc.)<br>■ Contract date with the supply enterprise | The list will be drawn up once and will be updated when there are any changes (new or cancelled suppliers) | Quality technician |
| Supplier auditing | Visit the enterprise and revise some parameters within which the suppliers should work, such as:<br>■ Legal and health authorization affecting their activity<br>■ Verify that they work properly and that they have a management system | Usually an annual visit is advisable | Quality manager |
| Inspection of the raw materials delivered to the company. | For each raw material, state which aspects should be taken into account to be accepted. Table 7.12 shows an example. | With each lot delivered to the company | Quality technician |

Table 7.12. Example of control for flour delivery in a bakery company.

| Product | Supplier | Entry date | Inspection | Evaluation | Incidences | Person responsible for control |
|---|---|---|---|---|---|---|
| Wheat flour in 25 kg sacks. | Harineras SL | 2-4-02 | Conditions of the container | Non-compliant | One sack broken on the outside with no flour leak | Juan Perez |
| | | | Label, expiration date | Compliant | | |
| | | | Revision of quantity and weight | Compliant | | |
| | | | Visual Aspect of the flour | Compliant | | |

The list of suppliers will be revised once a year to check if all the suppliers have been included. On the other hand, the supplier audit should be revised by verifying that it has been conducted correctly and that the report has been adequately fulfilled, indicating who was responsible for the audit, the date, and if the verifications were satisfactory and all possible incidences written down.

The inspection of the raw materials will also be registered, and it will be necessary to perform a comparative control of the invoices and receipts to verify that this task has been performed.

In addition, it will be checked that the results, from the inspection to the raw materials delivered to the company, are satisfactory.

*Corrective action*
When the delivery controls detect incidences, the corrective measures will aim at sending back the product that does not fulfil the aspects agreed upon and sending a complaint to the supplier. If this situation persists or the audit detects non-compliant aspects, the contract with the supplier may be cancelled. In any of these situations, the corresponding corrective action report should be filled in (Table 7.1).

### 7.3.12 Traceability control plan

Given the administration and consumer demands on agri-food safety, it is essential that the companies involved in the food chain guarantee the composition and origin of their products so that it will be easy to identify the source and cause of problems when these occur. In this sense, traceability is the capacity to follow the path of the raw materials and of the ingredients of a product up to the stages of production and processing, as well as the path of the end product during its distribution and delivery. In this process, it is also very important to know how the raw materials were acquired, that is, how the animals have been fed, which plant protection products have been used for treating the vegetables and fruits in the field, etc.

The strength of the agri-food chain depends on each of its links, as the weakest link affects everthing. Therefore, there should be a bottom-up analysis: from the farmers, the industries, suppliers to the consumers; and a top-down analysis, from the consumer to the farmer. Traceability is an information tool that will provide the company with a twofold benefit: on one hand, it will allow the company to know the source of the raw materials it is going to work with, and on the other hand, it will improve the relationship with the consumer, since the information provided to the consumer will increase confidence in the product purchased.

In the agri-food sector, traceability systems are the prerequisites and self-sufficient procedures that provide knowledge about the background and the path of the product or lot of products along the supply chain at a given moment. The precise and accurate identification of the products, as well as their position within the supply chain will facilitate recall from the market and enable the required measures to be taken in order to avoid risks to the consumers.

The most widely used techniques for collecting the required data go from the conventional method of labels and cards to the most sophisticated techniques, like the Internet.

*Traceability control program activities*
The main activities developed to carry out the program, are the following (Table 7.13):
- Activity 1: Systematic traceability of the raw materials when delivered to the company and entered in the process.
- Activity 2: Product traceability process.

## Design and implementation of an HACCP system

The traceability process developed by the company for each of the raw materials aims at knowing the origin and when each lot-product of that raw material starts being processed. Therefore, it is essential to know all possible information about the origin and amount purchased of the raw material. This can be developed by filling in a form designed for this purpose, such as the record shown in Table 7.14. In this way, if the supplier reports any incidence with a certain lot, it will be possible to know quickly the amount remaining in stock and the proportion already used.

On the other hand, it will be necessary to know which specific lots of each raw material have been used to elaborate the product. For this end, it is advisable to fill in a card like the one illustrated in Table 7.15. It will include the date and time that a specific lot of a given raw material has been processed.

There may be concern that one lot involves some risk for consumer safety, and therefore it should be necessary to identify the exact situation of that lot. If the lot were still in the warehouse, its location would be fast, but if the product had already gone out of the company it would be necessary to rely on the traceability system. The traceability system must be prepared to identify the client to whom the product has been delivered. To this end, a card like the one shown in Table 7.16 needs to be filled in.

Table 7.13. Program of activities for traceability control.

| What | How | When | Who |
|---|---|---|---|
| Traceability of each raw material when received and when processing begins. | In both cases, fill in the sheet or card designed for this end | End of each reception and start of processing | Processing and storing official |
| Traceability of the processed product | Fill in the sheet or card designed for this end | Each order delivered | official |

Table 7.14. Sample identification card of a raw material stored.

| Raw material | Reception date | Lot of raw material | N° boxes/sacks | Person responsible for reception of goods |
|---|---|---|---|---|
| Flour | 2-4-02 | 254 | 500 | Juan Pérez |
| Salt | 5-3-02 | 378 | 20 | Juan Pérez |

Table 7.15. Sample identification card for raw material processed.

| Raw material | Reception date | Lot of raw material | Type of product | Lot of processed product |
|---|---|---|---|---|
| Flour | 20-4-03 (11:30 a.m.) | 254 | French bread | 9085 |

*Isabel Escriche, Eva Doménech and Katleen Baert*

Table 7.16. Sample traceability control record.

| Date | client | Type of bread | Lot of processed product | N° boxes |
|---|---|---|---|---|
| 25-4-03 | Mercater | Baguette | 9085 | 2 |
| 25-4-03 | Siatur | Hamburger bun | 9089 | 5 |

*Verification of the implementation of the traceability control actions and their efficiency*
The control for the correct implementation of the traceability procedure will be by visual observation and completing the cards provided in the program, and verifying that all the fields have been documented and there are no omissions.

*Corrective actions*
In the case of detecting deficiencies or errors in the revision, proper measures will be taken and documented in writing by filling in the corresponding card (Table 7.1). More specifically, if a dangerous lot were identified, this lot would be recalled from the warehouse as soon as possible if it were still in the company, or will be collected from the client if it has already been delivered.

## 7.4 Development of an HACCP plan

The HACCP Plan is a protocol with the required information for the implementation of the HACCP management system. The HACCP Plan consists of two main components, the flow chart and the HACCP Control Table. The flow chart is used to represent in a schematic and sequential way the steps of the process to be analysed, and the HACCP Control Table shows the key information for each step of the process, including the CCP's, with a detailed explanation of how, when to proceed with the controls, and who (quality control manager, operator, HACCP team, etc.) should conduct them. In addition, the HACCP Plan will include any additional information required to reach the programmed quality and safety objectives.

International organisations, such as Codex, FAO-WHO, FDA, NACMS, UNESCO, etc, aware of the advantages and benefits of the use of HACCP as a management system in the agri-food industry, have published guidelines, recommendations and case studies to promote HACCP principles and to help businesses to implement the Plan (NFI, 1991; ICMSF, 1992; NACMCF, 1992; NFPA, 1992; NACMCF, 1998; FDA, 1999; FDA, 2001). This section will introduce students to the theoretical principles of the HACCP Plan (preliminary tasks and principles), so as to help them understand the philosophy of the HACCP system and design the correct plan for each situation. However, those students willing to know the HACCP principles in more detail can refer to the bibliography published by these organisations (Seafood HACCP Alliance, 1997; Department of Agriculture, 1999; Codex 1997).

## Design and implementation of an HACCP system

### 7.4.1 The seven principles of the HACCP concept

The HACCP system consists of the following 7 basic principles:
Principle 1:   Conduct a hazard analysis;
Principle 2:   Determine the Critical Control Points (CCP's);
Principle 3:   Establish critical limit(s);
Principle 4:   Establish a system to monitor control of the CCP;
Principle 5:   Establish the corrective action to be taken when monitoring indicates that a particular CCP is not under control;
Principle 6:   Establish procedures for the verification to confirm that the HACCP system is working effectively;
Principle 7:   Establish documentation concerning all procedures and records appropriate to these principles and their application.

### 7.4.2 The 12 stages of HACCP

To establish an HACCP plan, it is advisable to follow a logically structured sequence of steps. In several books and articles, different types of plans with 12 or 14 stages can be found. However, most of them are based on the internationally accepted system as described in the Codex Alimentarius, which includes 12 steps:
Step 1   Assemble HACCP team;
Step 2   Describe product;
Step 3   Identify intended use;
Step 4   Construct flow diagram;
Step 5   On-site confirmation of flow diagram;
Step 6   List all potential hazards, conduct a hazard analysis, and consider control measures;
Step 7   Determine CCP's;
Step 8   Establish critical limits for each CCP;
Step 9   Establish a monitoring system for each CCP;
Step 10   Establish corrective actions;
Step 11   Establish verification procedures;
Step 12   Establish documentation and record-keeping.

### 7.4.3 Step 1: Assembling the HACCP Team

The first task in developing a HACCP Plan is to gather the appropriate personnel to carry it out. HACCP Plan development should never be done by only one person, if the purpose is to elaborate a useful document to meet expectations. For this reason, a multidisciplinary team should be established, including experts in certain specific areas, of certain experience and training, and covering different sectors of the business: engineering, production, sanitation, and quality assurance, etc. In addition, in some cases, the HACCP team may require support from personnel of other departments or external experts. The number of people in the team should not exceed five or six, and to enhance participation, the team should not be structured according to the company's hierarchy.

The selection of the team members will be done by a coordinator, appointed by the company's management. This position usually corresponds to the Quality Control manager of the company. First, the coordinator will enhance the training of the team regarding the tasks they will develop in the elaboration and implementation of the HACCP Plan. This activity can be developed by himself or by an external training company. Proper training will guarantee a positive attitude in the team as regards understanding the advantages of the HACCP system, working in coordination and using the same terminology (NACMCF 1997). In addition, the leader will be in charge of organising the meetings, a task that includes activities such as finding the proper place to hold the meeting and providing the required means such as blackboard, audiovisual equipment, as well as scheduling meetings and asking for permission of the business management for workers to attend those meetings.

During the meetings, the coordinator should lead the team and chair the meetings, coordinate the team work so that all the members can freely give their opinions, avoid conflicts and favour the communication of the decisions adopted in the meeting.

After the meetings, the team leader will be responsible for the meeting proceedings, for registering all the documents generated in the meeting, for informing the management about the decisions made by the team, and for developing the written documents for the HACCP Plan.

### 7.4.4 Step 2: Description of the product

One of the key issues for the correct development of the HACCP Plan is to collect all the necessary information about the food to be processed and about the raw materials used in the process and their distribution. In order to design a safe product, all intrinsic factors should be taken into account, both of the raw materials and of the ingredients to use. These intrinsic factors may serve to control micro-organism growth. The most important factors are: pH, acidity, organic acids, preservatives, water activity and the ingredients. Elaborate a complete description of the characteristics of the product and of the raw materials used. The description is normally performed by filling in some checklists with the required information:

- raw materials:
  - definition/type of ingredients, packaging materials, etc.; providing information on their origin, shipping method, packaging, etc.;
  - percentage in final product;
  - physical and chemical features;
  - official microbiological or chemical criteria;
  - storage conditions before use, etc.
- finished product:
  - product properties (composition and physical, chemical and microbiological features of finished product);
  - process information (e.g. heat-treatment, freezing, brining, smoking, etc.);
  - packaging;
  - storage and distribution conditions;
  - special distribution conditions;

*Design and implementation of an HACCP system*

- legislative requirements;
- labelling instructions, etc.

In this section, describe also the circumstances that may alter food safety and quality during distribution. In addition, indicate whether controls are necessary in the food distribution, and also consider possible further handling of the product, such as retail, food service, further manufacturing, etc.

### 7.4.5 Step 3: Identify the intended use and the consumers of the food product

The type of intended consumer of the product is also an important factor to take into account. It is necessary to envisage whether the product will be consumed by people belonging to risk groups, such as infants, immune-compromised individuals or the elderly (FDA, 1998). If the product is not suitable for use by sensitive or high-risk groups, this should be indicated with proper labelling.

In addition, it is important to describe handling and preparation procedures for use of food (raw, cooked, refrozen, reconstituted, etc) at home by consumers. Also consider if the food will be handled by processors, including retailers, gourmet shops or catering.

### 7.4.6 Step 4: Construct a flow diagram

A flow chart is a simple schematic picture which represents the exploitation, configuration and operation procedures of a process or part of a process. In the development of the flow diagram, on which the HACCP Plan will be based, it is important to include all the additional information relevant for the process to be real. Among the data that can be included in the flow diagram, the most important are: information about raw materials and packaging (indicating the format of reception, characteristics, storage conditions, treatments prior to processing, etc.); details about the process (specifying the sequence of the steps to follow, time and temperature conditions, dead times, process cycles, etc.); data about finished product storage and distribution conditions.

One of the advantages of developing a flow chart is that it enhances the process of understanding and provides an overall vision of the HACCP system, since all the process steps and their relationships are represented in the flow chart.

The HACCP team can choose the type of chart to use, but it is advisable to follow a standard pattern such as those indicated by ISO 10628 (1997). According to this standard, there are three main types of flow charts depending on the information to be represented: block diagrams, process diagrams and pipe or equipment diagrams:

- Block diagrams represent a process or a plant in a simple way by using rectangles for the processes or steps of the process, basic operations, plants, sections or equipment, which may be complemented with further information like material or energy flow rates.
- Process diagrams represent a plant or process but use graphical symbols for the type of equipment, in/out flow direction, materials, energy flows, and operating conditions.

- Pipe and equipment diagrams represent the technical implementation of the process using graphical symbols for the equipment and pipes, and symbols representing the measurement and control functions of the process.

### 7.4.7 Step 5: On-site confirmation of the flow diagram

The HACCP team should perform an on-site review of the operation to verify the accuracy and completeness of the flow chart during all stages and hours of operation. Modifications should be made to the flow chart as necessary, and documented.

### 7.4.8 Step 6: List all potential hazards associated with each step, conduct a hazard analysis, and consider any measure to control identified hazards (principle 1)

The Codex Alimentarius (1997) describes the first of the 7 HACCP principles as:

> 'List all potential hazards associated with each step, conduct a hazard analysis, and consider any measures to control identified hazards.
>
> The HACCP team should list all of the hazards that may be reasonably expected to occur at each step according to the scope from primary production, processing, manufacture, and distribution until the point of consumption. The HACCP team should next conduct a hazard analysis to identify for the HACCP plan which hazards are of such a nature that their elimination or reduction to acceptable levels is essential to the production of a safe food.
>
> In conducting the hazard analysis, wherever possible the following should be included:
> - the likely occurrence of hazards and severity of their adverse health effects;
> - the qualitative and/or quantitative evaluation of the presence of hazards;
> - survival or multiplication of micro-organisms of concern;
> - production or persistence in foods of toxins, chemicals or physical agents; and,
> - conditions leading to the above.'

This document shows that a hazard analysis consists of 2 parts:
- an inventarisation of the hazards;
- an analysis of the identified hazards.

*Inventarisation of the hazards*
A hazard is a biological, chemical or physical agent in or condition of a food with the potential to cause an adverse health effect.

To be able to perform a hazard analysis, some knowledge is required about the potential hazards. Three groups of hazards can be distinguished: biological, physical and chemical hazards.

**Biological hazards** - In this type of hazard one can differentiate between two originating agents, the microbiological and macrobiological agents. Micro-organism agents are the main cause of health alteration on a short-term basis (ICMSF, 1978). The presence of these in a finished product can be due to the raw material, faulty cleaning and production process or incorrect handling by the workers. Within these, four large groups stand out: pathogenic bacteria (this group includes bacteria such as *Salmonella* spp., *Escherichia coli, Clostridium botulinum and perfringens, Staphylococcus aureus, Listeria monocytogenes*, etc.); viruses (among the most important are the Norwalk virus, Vibrio spp. and the Hepatitis A virus); parasites (some of the most common in fish and livestock are *Toxoplasma gondii, Trichinella spirales, Taenida saginata*); and moulds (the producers of mycotoxins have a special importance).

With respect to macrobiological agents, the most important within this type of biological hazard are insects like flies, beetles, spiders, etc, the presence of which is not only unpleasant but they can also be carriers of pathogenic micro-organisms.

**Physical hazards -** There are a large variety of particles or physical entities, that can appear in food, the presence of which can cause cuts, wounds, broken teeth, choking, etc. According to their origin they can be grouped as follows: faulty packaging, and non-edible parts of the food product, for example: bones, fish-bones, skin, splinters, etc.

**Chemical hazards -** The effects of chemical hazards on consumers usually appear after a long period, such as those of a cumulative or cancerous type, although they may also occur in the short term, e.g. allergic reactions. These hazards can be present in food, either because they have their origin in the raw materials or because they have been incorporated during the production process (Escriche and Serra, 1997).

The great number of chemical substances that can be present in the raw materials may have different origins:
- The composition of the food itself (this is the case for fungus toxins like the Amanita genus or in fish like the Tetrodon).
- Environmental pollution (heavy metals, polychlorinates, biphenyls, etc.); added by man but incorrectly used (like phytosanitary products, drugs and medicines, preservatives, etc.).
- By improper conservation of the raw materials, (biogenic amines, mycotoxins, etc).

Other substances can be incorporated during the production process:
- Generated during the production process, such as the Polycyclic Aromatic Hydrocarbons; voluntary addition of authorised substances, but in larger amounts than permitted, or the addition of non-permitted substances: preservatives.
- Substances from the cleaning and disinfection products: detergents, disinfectants, insecticides; substances used in installations and equipment maintenance: lubricants, equipment oils, etc.

Hazard identification is one of the most important points of the HACCP analysis, since the omission of any type of hazard further prevents the possibility of taking the appropriate measures to counteract it (Serra *et al.*, 1999). In the literature there are a great variety of techniques for developing this task in the agri-food industry (Doménech *et al.*, 1999), with different scopes and complexity, such as:

- Log of process deviations. This technique is oriented to search past data relative to the deviations from the good operation of the process. This information may be obtained from the company or from other companies belonging to the same industrial sector. The analysis will cover information regarding actual past process deviations as well as situations which were likely to have caused deviations but which were tackled in time with no subsequent effects.
- What would happen if...? This technique consists of defining the parameters and conditions in which the stage is developed, to later asking what would happen if the situations changed with respect to the parameters previously established.
- Brainstorming. This technique, widely used in the industrial sector, consists in allowing the members of the HACCP team to freely present their ideas and suggestions. To obtain better results and for complete identification, the team will use the flow chart and product description and develop a systematic analysis of what could occur at each step in the process.

The use of these three techniques is not exclusive and, in fact, it is important to point out that better results are obtained when the team members use a combination of all three techniques.

*Control measures*
Once the possible health hazards are known, control measures have to be developed. Here also a reference to the Codex Alimentarius (1997):

'Consideration should be given to what control measures, if any exist, can be applied to each hazard.

*More than one control measure may be required to control a specific hazard and more than one hazard may be controlled by a specified control measure.*'

Control measures are measures that are necessary to guarantee the safety and the suitability of the final products, with regard to the most probable use.

Depending on the content and the effect of the measures a distinction can be made in 3 groups:
- Once-only measures that are used to eliminate the hazard.
  In a lot of cases these measures concern the infrastructure and demand little or no observation. The hazard is eliminated and the control measure needs little or no observation. These measures need to be evaluated periodically with regard to the actual situation.
- Measures that are used to limit the risk, by controlling the hazards.
  This covers a lot of general measures that are not process or product related. The objective of these measures is to avoid or keep to a minimum:
  - additional contamination (chemical, physical and biological);
  - cross-contamination.
  These measures need to be maintained continuously and verified on a regular basis to guarantee food safety. These are the more general PRP measures.
- Measures to limit the risks by eliminating the hazards or by reducing them to an acceptable level.
  These measures are more specific and more directed towards the process. The hazards that can lead to a high risk are kept under control. The maintenance of measures should be followed closely. In most cases, these are critical control points.

## Design and implementation of an HACCP system

*Hazard analysis*
When all potential hazards have been identified, they should be analysed. It is necessary to assess the risk of each identified hazard. In a lot of HACCP plans, established in the starting period, this part of the hazard analysis is missing or is insufficiently established. In these cases, there is no additional step between the inventory of the hazards and the identification of the critical control points.

For all listed hazards, the food industry should investigate the chance or probability that the mentioned hazard is occurring and if it occurs, what the effect on public health may be. In other words, the risk depends on the probability and effect:

risk = probability × effect.

What is the possibility of a negative effect, when the control measure does not work well?

Probability = the risk that the hazard occurs in the final product when the control measures taken are not present or are failing

Effect = the effect of this hazard on the final product (early spoilage) or on the health of the consumer

The literature describes a whole range of approaches to hazard analysis. A simple method is the use of a matrix, in which a figure is given to the assessed probability and effect and in which the risk category can be seen on a scale from 1 to 7. The effect of the hazard can be assessed by indicating that the effect is very limited, moderate, serious or very serious. The chance that the hazard is occurring is indicated as very small, small, real or high.

Risk Assessment: scale from 1 to 7

|  |  | | | | |
|---|---|---|---|---|---|
| | high | 4 | 5 | 6 | 7 |
| | real | 3 | 4 | 5 | 6 |
| **Probability** | small | 2 | 3 | 4 | 5 |
| | very small | 1 | 2 | 3 | 4 |
| | | very limited | moderate | serious | very serious |
| | | | **Effect** | | |

A possible interpretation of the different gradations of probability and effect can be:

## Probability
1 = very small
- theoretical chance - the hazard has never occurred before
- there is a step further in the process to eliminate the hazard
- when the hazard occurs or when the control measures for the hazard fail, production is not possible or no useful final products are produced
- there is a very limited and/or local contamination

2 = small
- the presence of the hazard in the final product through failure/shortcomings of the control measure is almost unlikely
- the control measures for the hazard are general (GMP/PRP); these control measures are already implemented and are followed

3 = real
- failure or shortcomings of the control measure does not result in a systematic presence of the hazard in the final product but the hazard will be present in a certain amount of the final products of the party in question

4 = high
- failure of the control measure results in a systematic error; there is a high risk that the hazard occurs in almost all final products of the party in question

**Effect**

1 = very limited
- there is no problem for the consumer related to public health (paper, soft plastic, large foreign objects such as a knife)
- the hazard can never reach a dangerous concentration (colorants, *S. aureus*,..)
- the hazard is not present anymore at the moment of consumption (rice, pasta, ...)

2 = moderate
- there is an additional development/contamination resulting in early spoilage
- limited, no serious injuries and/or symptoms or only when exposed to an extreme high doses over a long period of time
- a temporary but clear corporal inconvenience or a continuous small inconvenience

3 = serious
- a clear corporal inconvenience with short-term or long-term symptoms which results rarely in mortality
- the hazard has a long-term effect; the maximal doses are not known (dioxins, residues of pesticides, mycotoxins)

4 = very serious
- the consumer group belongs to a risk category and the hazard may result in mortality
- the hazard results in serious symptoms which may result in mortality
- irreparable injuries

When assessing the probability and effect, it is important to know what will happen with the product, raw material or ingredient or for which purpose it is used.

### 7.4.9 Step 7: Determine the critical control points (CCP's) (principle 2)

*Introduction*
A critical control point is generally defined as any point, process step or activity where a potential hazard for food safety can be eliminated, prevented or reduced to an acceptable level. This definition combined with the application of the decision tree can result in a high number of CCP's, making the HACCP plan unclear. The number of critical control points depends on

the presence of PRP, the nature of the product, the complexity of the process and the accepted risk. A company can choose to indicate each point, process or process step that influences a certain hazard, as a CCP. This makes sense only when the company is able to control each point and this is most often not the case. Furthermore, in this system, the most important critical control points are not getting the needed attention.

For this reason, there is a trend in the identification of CCP's to consider as a CCP only these steps, points or processes where loss of control results in an unacceptable risk for public health and where by means of concrete measures an efficient and quick control is possible. The other points, where loss of control doesn't result in an unacceptable risk for public health and where no immediate adjustment of the product happens, is considered as a control point (CP). However, at these control points, inspection is still needed and at regular times, control should be performed. Furthermore, the probability of the occurrence of a serious hazard can only be kept under control, when at these points good preventive measures are present, such as a detailed cleaning and disinfection plan, rules for hygiene, clear work instructions, etc. ...

In summary:
CCP's are points where continuous control is necessary to eliminate or reduce the hazard to an acceptable level. When control of these points is lost, (1) there is a high probability that the products are a risk for public health or are not of good quality, or (2) the effect of a certain hazard is serious. The performed controls should be demonstrable by means of registrations. Points, which are only controlled once a month, are not real CCP's.

CCP's are points which need continuous attention, but the risks can be controlled by general preventive measures, belonging to the basic rules for hygienic and safe operation in a food company (PRP). When the observation of these preventive measures is controlled frequently, the risks are considered as being sufficiently under control.

The identification of CCP's is a complex and critical process. Some production lines are rather extensive and are processing a high number of ingredients. However, the number of CCP's should be limited to 5-10. At a higher number, control becomes too complex. Different companies, producing the same product, can differ with respect to their hazards, risks and also their CCP's (as a consequence of different layout, equipment, ingredients, work conditions). A general HACCP plan can be used as a guide. However, it is still necessary to consider the specific conditions belonging to a specific production line, and each company must identify it's own CCP's.

*Determination of CCP's*
**Using a decision tree -** To help determine where the CCP's are located, there is a tool known as a "decision tree". A decision tree consists of a number of logical questions that have to be answered for each health hazard identified (first HACCP principle) and for each stage of the process at which the health hazard may occur. The answer to each question leads the HACCP team through a given path in the tree to a point of considering whether that phase or stage is a CCP or not.

The use of the decision tree helps the team to think in a structured way (Bryan, 1996). In addition, it forces and facilitates discussion within the HACCP team and improves teamwork. The interpretation of the tree is rather simple, although the correct answer will depend on the experience of the HACCP team (Mortimore and Wallace, 1998). Figure 7.2 shows an example of a decision tree based on the tree proposed by the Codex Alimentarius in 1993.

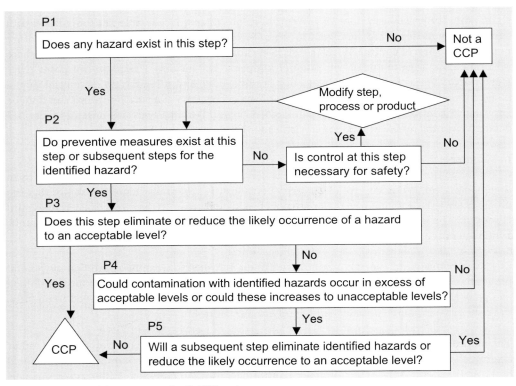

Figure 7.2. Example of decision tree to identify CCP's at the process stages.

- First question: "Does any hazard exist in this step?" may seem obvious, since the tree starts from the hazards previously identified. However, the purpose of this question is to consider only the analysis of the health hazards included in the scope of the study, leaving out, for example, those hazards already included in the prerequisites. The health hazards already identified should be excluded and followed with the HACCP analysis.
- Second question: "Do preventive measures exist at this step or subsequent steps for the identified hazard?" aims at knowing the existence of preventive measures to control the hazard: whether they already exist in the company or if they do not exist but can be implemented.

### Design and implementation of an HACCP system

- Third question: "Does this step eliminate or reduce the likely occurrence of a hazard to an acceptable level?" The answer to this question is the key issue for hazard analysis. For this reason, the HACCP team will have to dispose of the information required, and if necessary ask for the opinion of external technicians and experts in simulation models of microbial growth, which may be very helpful for deciding whether there is the possibility of eliminating or reducing risks to acceptable levels.
- Fourth question: "Could contamination with identified hazards occur in excess of acceptable levels or could these increase to unacceptable levels?" At this step it is important to take into consideration not only the conditions (times, temperatures, etc.) under which the stage under study develops, but also the overall process. In this way, the possible cumulative effect of each factor along the process will be considered.
- Fifth question: "Will a subsequent step eliminate identified hazards or reduce the likely occurrence to an acceptable level?" This question is intended to focus attention only on the strictly necessary critical points. If a subsequent step eliminates the risk of health hazard, this step will not be considered as critical. This is important when the production process is highly complex, when there are a high number of critical points or when control actions make it difficult to go on with normal operation. Nevertheless, as far as practical and in our opinion, it should not focus on the safety of a product but only on one point of the process. Because any undetected cumulative failure could jeopardise the safety management system.

The decision tree will be applied to each identified hazard at each stage of the process until defining all possible CCP's. The HACCP team will signal all the CCP's on the Flow chart of the Process and will elaborate the HACCP Control Table.

**Using hazard analysis -** The identification of CCP's can also be done by using the hazard analysis and the obtained risk categories.
For example:
Risk category 1 & 2:      No specific actions, controlled through PRP
Risk category 3 & 4:      Additional question: Is the general control measure, as it is described in the PRP principles, sufficient to control the identified hazard?
       When Yes: PRP
       When No: CP
Risk category 5, 6 and 7:      critical control point (CCP)

### 7.4.10 Step 8: Establish critical limits for each CCP (principle 3)

This principle imposes the critical limit specification for each preventive measure. Once the CCP's have been identified, the next step will be to decide how they will be controlled to keep the process within the safety limits (Bryan, 1990). These critical limits are the levels or tolerances specified which should not be surpassed to ensure that the CCP is effectively controlled. If such limits are surpassed, the CCP will be out of control and risks of health hazard may occur in the end product.

For some critical limits, it is convenient to provide the upper and lower critical values: upper limit, where a set amount or level cannot be exceeded, and lower limit where a minimum amount is required to produce the safe effect. It may be that for controlling one process stage, different critical values should be established, e.g. pasteurisation time and temperature.

Since step 8 (Critical Limits) establishes the limits between safe and unsafe products, it is fundamental to set accurate values. The HACCP Team, with the purpose of accurately defining the Critical Limits, will have to know the criteria relative to the safety of each CCP. For this end, the Team will require all the data available, expert advice, experimental data, and any other useful information.

A critical limit is associated with a measurable factor and should comply with two characteristics: to allow for routine monitoring and to provide immediate results, in order to predict and anticipate possible loss of control in the process. The Critical Limits can be chemical (e.g. salt content 2%, nitrite content:150ppm, $a_w$ < 4.5, absence of allergens), physical (e.g. milk pasteurisation (72°C for 15 sec, ambient temperature and relative humidity in the cutting plant: at a maximum of 12 °C and 70-80 % RH, intact filters, etc.), or microbiological (absence of *Salmonella* spp./ 25 g raw materials, < $10^2$ ufc/$cm^2$ total amount in utensil surfaces and installations, etc.) and will be related to the type of health hazard to control in the CCP. The microbiological Critical Limits are useful when fast detection techniques are used. Do not use the conventional methods for microbiological growth analysis, as they require several days in the laboratory just to check the efficiency of the other techniques. Once the HACCP Team has established the Critical Limits for all the CCP's, they will be included in the HACCP Control Table.

### 7.4.11 Step 9: Establish a monitoring system for each CCP (principle 4)

Monitoring consists of a number of measurements and observations for each CCP that show that a CCP is under control, within the Critical Limits, and thus that the process is working properly. In this way, it will be ensured that the product is safely manufactured. It is one of the most important parts of the HACCP System.

Monitoring presents three important benefits. It tracks the system's operation so that a trend toward a loss of control can be recognised and corrective action can be taken to bring the process back into control before a deviation occurs; also, it indicates when loss of control and a deviation have actually occurred, and corrective action must be taken; and it provides written documentation for use in verification of the HACCP plan.

The monitoring system for each CCP will depend on the Critical Limits. It is essential that the monitoring system chosen is able to quickly detect a loss of control at a CCP (i.e. a CCP out of the Critical Limits), so that appropriate measures can be taken as soon as possible. The best monitoring system is an on-line continuous system that can be calibrated to detect process deviations and thus to make the appropriate modifications to avoid a loss of control at the CCP. For example, the temperature and time for an institutional cook-chill operation can be recorded continuously on temperature recording charts. If the temperature falls below the scheduled

*Design and implementation of an HACCP system*

temperature or the time is insufficient, as recorded on the table, the batch must be recorded as a process deviation and reprocessed or discarded. Instrumentation used by the food establishment for measuring critical limits must be carefully calibrated for accuracy. Records of calibrations must be maintained as a part of the prerequisites documentation. When it is not possible to monitor a critical limit on a continuous basis, it is necessary to establish that the monitoring interval will be reliable enough to indicate that the hazard is under control. Statistically designed data collection or sampling systems lend themselves to this purpose. When statistical process control is used, it is important to recognise that violations of critical limits must not occur.

Most monitoring procedures for CCP's will need to be done rapidly because the time frame between food preparation and consumption does not allow for lengthy analytical testing. Microbiological testing is seldom effective for monitoring CCP's because of its time-consuming nature. Therefore, physical and chemical measurements are preferred (i.e. temperature recording, relative humidity, pH, salt contents, level of free Cl in water) because they may be done rapidly and can indicate whether microbiological control is occurring. Microbiological testing has limitations in an HACCP system, but is valuable as a means of establishing and verifying the effectiveness of control at CCP's (such as through challenge tests, random testing, or testing that focuses on isolating the source of a problem).

The HACCP Plan must establish the type of monitoring procedures to carry out, where to perform them, frequency and who is responsible for the monitoring tasks. Assignment of responsibility for monitoring is an important consideration. The most appropriate employees for such assignments are often directly associated with the operation. In addition, it should also be indicated how to perform the control to ensure that the monitoring process has been efficient and properly performed.

All records and documents associated with CCP monitoring must be signed or initialled by the person doing the monitoring. This control provides the necessary information to adopt corrective measures and ensure the control of the CCP.

### 7.4.12 Step 10: Establish corrective actions (principle 5)

At this point, the HACCP Team has identified the CCP's of the process and has established the limits that indicate whether a process stage is under control or not. In addition, the HACCP Team has established a monitoring system to check that the process operation is within the correct limits and that there are no process deviations; but what happens when the person responsible for monitoring detects a non-compliance, or when the monitoring results show a trend towards loss of control at one CCP? In such cases, the HACCP Team should have devised a number of corrective measures in the HACCP Plan, which have to be specified in the HACCP Control Table. These corrective actions will be established at two levels: 1. Take all the necessary actions to ensure that the CCP is back under control, and to ensure the production of safe food. Establish what to do with faulty products manufactured during the period of process deviation.

*Safety in the agri-food chain*

After the application of the appropriate corrective action and when the CCP is back under control, it will be necessary to start the revision of the system in order to prevent the failure from recurring. As corrective actions are a tool for system improvement, a record of all possible causes and measures taken will be necessary.

### 7.4.13 Step 11: Establish verification procedures (principle 6)

Verification is defined as those activities that determine the validity of the HACCP plan and that the system is operating according to the plan. An effective HACCP system requires little end product testing, since sufficient validated safeguards are built in early in the process.

The first verification action of the HACCP Plan is done after system implementation; the results obtained can be a ratification or modification of the Plan after checking that some of the criteria adopted are not adequate and/or some important criteria have been omitted.

In addition, a periodic comprehensive verification of the HACCP system should be conducted in order to ensure that it is kept operational. This is logical, since an HACCP Plan elaborated a while ago cannot properly reflect the present situation. This can be due to new scientific data, changes in the legislation, application of new technologies, introduction of new ingredients, lower amounts of preservatives, etc.

### 7.4.14 Step 12: Establishing documentation and record-keeping (principle 7)

In order to apply the system in an effective and efficient way it is necessary to have an accurate and efficient record-keeping system. All the information, which is generated should be gathered and correctly registered with the aim of having written proof for the Administration, the consumer and the company itself of the development of all company activities. In addition, this documented evidence will allow for back and forth monitoring of the process and food, and plan verification, and is the basis for any improvement policy developed. For this reason, record-keeping is an essential feature of an HACCP system and must be planned and carried out as carefully as any other element. The records should be kept legible, and easy to identify and retrieve.

The HACCP Team should decide which data to record and which is the best procedure to follow in order to ensure efficient planning, operation and control of the process.

The length of the documents of the management system may differ depending on the company, type of activity and competence of the personnel.

The documents related to the HACCP Plan will consist basically of:
- Proceedings of the meeting held prior to the elaboration of the HACCP Plan, with all the information about the team: members of the team, responsibilities, training, field of work in or out the company.

## Design and implementation of an HACCP system

- Procedures followed for hazard identification, preventive and/or corrective actions, etc., and all the data generated prior to the elaboration of the plan.
- Flow chart of the process.
- Summary table of the HACCP plan.
- Log of the modifications to the HACCP Plan: changes in the process, in the ingredients or in the plan itself.
- Records of the CCP monitoring actions as well as of the deviations and corrective actions taken.
- Analysis results and checking.
- Reports of validation, verification and audits including information about how it was done.
- The records of the audits, etc.

This documentation will be kept together with the documents of the prerequisites, mentioned in section 7.3. The data records will be kept for as long as required by the regulations of the country.

## 7.5 Implementation

Implementation is the application of the HACCP plan developed for a specific process. Effective implementation is absolutely imperative to ensure food safety (Corlett, 1989, Bryan et al., 1991; MFSC, 1993; Stevenson, 1990). The best plan will not be worth the paper it is written on unless it is transformed into effective action at the operation plant. The successful implementation of an HACCP plan should be facilitated by commitment from top management.

### 7.5.1 Establishment of the implementation team

For small companies, the HACCP team that developed the plan is usually the one in charge of its implementation. Nevertheless, it is advisable to establish another team for the implementation, formed by people that contribute their knowledge in this phase, some of whom may have been part of the team that designed the Plan. The implementation team will carry out a number of tasks that will be the key to achieving the efficient application of the HACCP plan:

- Execute a well thought-out plan, by developing a timeline for the necessary activities, in order to use resources effectively.
- Develop operator procedures for monitoring, to teach officials to execute and register the control sheets correctly, thus avoiding errors, omissions or questions.
- Carry out an appropriate distribution of responsibilities and assignment of responsibilities among all the personnel of the company.
- Inform the company's management on a regular basis of the progress and obtain the means so that these tasks are carried out and therefore the objectives envisaged are successfully fulfilled.

- Check compliance with the facilities and appropriate equipment for the implementation of the preventive actions foreseen in the plan. One of the first and more important activities to develop in the implementation process is the tuning up of the control measures foreseen as necessary to control the hazards. Remember that the team that designed the HACCP plan already made a list (in the 1st principle) of these control measures. In this sense, it is necessary to act at two levels: 1. Carrying out an assessment of the correct operation of all the control measures already in operation prior to the design of the plan, and 2. Ensuring that those new actions that the team considered indispensable to introduce in the process are applied. Naturally, after introducing these new actions, it will be essential to check their effectiveness.

Once the implementation team is convinced that all previous aspects are controlled, the following step towards implementation will involve the monitoring of the CCP's and the registration of the data.

### 7.5.2 Application of the monitoring and recording system

Monitoring is a key element in the rationale of the HACCP (fourth principle). Its main objective is to check that work is carried out under the appropriate conditions and therefore, that the critical points are under control. People in charge of the monitoring of the CCP's play a key role and therefore it is essential that they understand their responsibility and are motivated. Sometimes the people in charge of monitoring CCP's need specific training. The implementation team will be aware of this need and will provide the appropriate training. Later on, it will be essential to check that the people monitoring the CCP's understand the specific tasks to carry out and that they carry them out correctly.

As a consequence of the monitoring of critical points, a large amount of data will be produced, which should be recorded, as stated in principle 7 of the HACCP plan. These records should include detailed information on the critical limits and the corrective actions. Documents should be plain and simple, and must include the date, and the name of the person responsible for monitoring. When the result of the monitoring is an analytical value, the records will not only include the final data, but also enough additional information to ensure the traceability of the analytical result. Some of the required data are: date of the analyses, place, time and person responsible for the sampling, value obtained for each parameter, reference values and non-acceptable values, identification of the laboratory or person in the company that has carried out the analysis, etc.

The subsequent processing of registered data will enable conclusions to be drawn about the suitability and efficiency of the Plan, and will help to guide the management of the company on the possible modifications that need to be carried out in the process to reduce the out-of-control situations. Currently, the statistical analysis of data, especially by means of control charts, is one of the most frequently used tools in agri-food companies.

## 7.6 Verification and maintenance

Even when there is assurance that the CCP's are being monitored, the process is not yet finished. For the designed plan to continue being effective it is necessary to keep it updated and verify it, as with any other management system.

The verification (described in principle 6 of the Plan) will be useful for finding the weaknesses of the system, which will allow for further appropriate actions. Verification is carried out by means of an internal audit. An auditing team will be put together to prepare and carry out the revision, verification and evaluation activities of the available data according to the approaches established. Table 7.17 shows a possible structure for the verification list of an HACCP system.

Table 7.17. Verification lists of a HACCP system.

| Aspects to cover | Compliance |
|---|---|
| HACCP plan | |
| ■ Written HACCP plan prepared for each kind or group of products. | |
| ■ Written HACCP plan implemented. | |
| ■ Written HACCP plan identifies all food safety hazards that are reasonably likely to occur. | |
| ■ Personnel adequately trained to administer the firm's HACCP Plan. | |
| ■ HACCP plan signed and dated as required. | |
| Raw materials | |
| ■ Critical ingredients have been identified | |
| ■ Storage conditions are as established | |
| ■ All the raw materials have been included in the flow chart | |
| ■ There are specifications agreed upon | |
| Hazard analysis | |
| ■ Hazard analyses conducted and written for each kind or group of product processed. | |
| ■ Written hazard analysis identifies all potential food safety hazards and further determines those that are reasonably likely to occur | |
| ■ Hazard analysis reassessed after changes in raw materials, formulations, processing methods/systems, distribution, intended use or consumers. | |
| ■ Personnel adequately trained to administer the firm's hazard analysis. | |
| ■ Written hazard analysis signed and dated as required. | |
| Critical Control Points (CCP) | |
| ■ HACCP plan lists Critical Control Point(s) for each food safety hazard identified as reasonably likely to occur. | |
| ■ Critical Control Point(s) identified in the HACCP plan are adequate control measures for the food safety hazard(s) identified. | |
| ■ Control measures associated with critical control point(s) listed in the HACCP plan are appropriate at the processing step identified. | |

Table 7.17. Continued.

| Aspects to cover | Compliance |
|---|---|
| **Critical Limits (CL)** | |
| ■ HACCP plan lists critical limits for each critical control point(s). | |
| ■ Critical limits defined in the HACCP plan are adequate to control the hazard identified. | |
| ■ Critical limits defined in the HACCP plan are achievable with existing monitoring instruments or procedures. | |
| ■ Critical limits are met. | |
| **Monitoring** | |
| ■ HACCP plan defines monitoring procedures for each critical control point. | |
| ■ HACCP plan defines what will be monitored at each critical control point. | |
| ■ HACCP plan defines how monitoring procedures will be performed at each critical control point. | |
| ■ HACCP plan defines the frequency at which monitoring will be performed at each critical control point. | |
| ■ HACCP plan defines by whom the monitoring will be performed at each critical control point. | |
| ■ Monitoring procedures as defined in the HACCP plan followed. | |
| ■ Monitoring procedures as defined in the HACCP plan adequately measure critical limits at each critical control point. | |
| ■ Monitoring record data consistent with the actual value (s) observed during the audit. | |
| ■ Employees trained in monitoring operations. | |
| **Corrective action** | |
| ■ Predetermined corrective actions defined in the HACCP plan ensure product which may be injurious to health or otherwise adulterated as a result of a deviation do not enter commerce. | |
| ■ Predetermined corrective actions defined in the HACCP plan ensure that the cause of the deviation is corrected. | |
| ■ Appropriate corrective action taken for products produced during a deviation from critical limits defined in the HACCP plan. | |
| ■ Corrective actions defined in the HACCP plan were observed and followed when deviations occurred. | |
| ■ Affected product produced during the deviation segregated and held. | |
| ■ A review to determine product acceptability performed. | |
| ■ Corrective action taken to ensure that no adulterated and/or product that is injurious to health enters commerce. | |
| ■ Cause of deviation was corrected. | |
| ■ Reassessment of HACCP Plan performed and modified accordingly. | |
| ■ Corrective actions documented. | |

Table 7.17. Continued.

| Aspects to cover | Compliance |
|---|---|
| Verification | |
| ■ HACCP plan defines verification procedures. | |
| ■ HACCP plan defines the frequency of verification. | |
| ■ Reassessment of HACCP plan conducted annually, OR after changes that could affect the hazard analysis, OR after significant changes in the operation including raw materials and/or source, product formulation, processing methods/systems, distribution for intended use or intended consumer. | |
| ■ HACCP plan reassessment performed by trained individual as required. | |
| ■ Program in place to review consumer complaints. | |
| ■ Verification records reviewed as required - including date and signature. | |
| ■ Verification records review performed by trained individual. | |
| ■ Critical control point monitoring records reviewed to verify and document that values are within critical limits. | |
| ■ Monitoring record review performed as required. | |
| ■ Corrective action record reviewed as required. | |
| ■ Records of calibration of process control instruments and end product or in-process testing results listed as verification activities in the HACCP plan reviewed. | |
| ■ Verification records or documents are present that validate the effectiveness of the control measure and established critical limit in controlling the identified hazard. | |
| Records | |
| ■ Required information included in the record, e.g. name/location of processor and/or date/time of activity and/or signature/initials of person performing operation and/or identity of product/product code. | |
| ■ Processing/other information entered in the record at time observed. | |
| ■ Records retained as required, e.g. one year for refrigerated products/ two years for preserved, shelf-stable or frozen products. | |
| ■ Records relating to adequacy of equipment or processes retained for 2 years. | |
| ■ HACCP records available for official review and/or copying. | |
| ■ Personnel adequately trained to administer the firm's HACCP System. | |
| Name of auditor(s) (Please Print) | |
| Signature(s): | |
| Date: | |

Once the revision activities have been carried out, the team will meet to agree upon the conclusions and write the audit report. Thus, all the steps and generated information will be recorded in writing, from the objectives intended to the conclusions. In summary, the audit of the HACCP system can provide evidences for further modifications.

Although everything works apparently correctly, a periodic verification of this management system will be carried out, with the purpose of ensuring that it is always operative. Effective maintenance depends largely on regularly scheduled verification activities. The frequency of verification should be enough to ensure the detection of possible deficiencies or possible aspects for improvement. Overall, the periodic maintenance is a continuous and cyclical process that leads to the continuous improvement of the system (Figure 7.3).

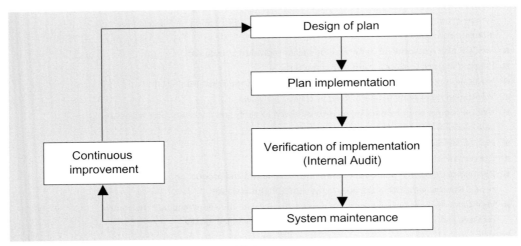

Figure 7.3. Process of continuous improvement of a HACCP System.

## 7.7 Future proposal

The HACCP system used by the agri-food industry, as it has been summarised in this chapter, should be considered as a big step forward for ensuring food safety. As the HACCP concept focuses on food safety, the hazards related to public health should have priority. However, it is advisable to include hazards related to food spoilage as well. Some companies also prefer to include hazards that only affect food quality, in order to provide an adequate response to the quality objectives imposed by the market. However, this is not advisable as the HACCP plan is then unnecessarily extended, especially as far as the first HACCP plan is concerned. If a company chooses to include quality aspects, it is necessary to be able to distinguish these aspects from the other safety hazards.

## Acknowledgements

We would like to thank Asunción Jaime, Miguel Angel Candel and the Foreign Language Co-ordination Office (ACLE) at the Polytechnic University of Valencia for their help in translating this text.

# References

Bryan, F.L., 1990. Hazard analysis critical control point (HACCP) systems for retail food and restaurant operations. J. Food Prot. 53(11):978-983.

Bryan, F.L., C.A. Bartelson, C.O. Cook, P. Fisher, J.J. Guzewich, B.J. Humm, R.C. Swanson and E.C.D. Todd, 1991. Procedures to Implement the Hazard Analysis Critical Control Point System. Int. Assoc. of Milk, Food, Environ. Sanitarians, Ames, IA, 72 pp.

Bryan, F.L., 1997. Another decision tree approach for identification of critical control points. Journal of food protection. Vol 59, nº 11, 1242-1247.

Campanella, J., 2000. Los costes de la calidad. Principios, implantación y uso. Ed. AENOR, Madrid.

CCE (Comisión de las Comunidades Europeas), 2000a. Libro blanco sobre la seguridad alimentaria. Bruselas 12-1-2000. COM (1999) 719 final.

CCE, Comisión de las comunidades europeas, 2000b. "Propuesta de Reglamento del Parlamento Europeo y del Consejo, relativo a la higiene de los productos alimenticios". Bruselas 14/07/2000. COM 2000. 438 final.

Codex Committee on food hygiene Guidelines for the application of the hazard analysis and critical control point (HACCP) system, 1993. In training considerations for the application of the HACCP System to food processing and manufacturing, WHO/FNU/FOS/93. 3II. 1993, WHO, Geneve.

Codex Committee on food hygiene HACCP system and guidelines for its application, 1997. Annex to (CAC/RCP 1- 1969, Rev.3 (1997)) in Codex Alimentarius food hygiene basic texts. FAO/WHO.

Codex alimentarius, 2001a. "Informe de la 34ª edicion del comité del codex sobre higiene de los alimentos". ALINORM 03/13. Programa conjunto FAO/OMS sobre normas alimentarias. 8-13 October 2001.

Codex alimentarius commission, 2001b. Report of the Thirty-Third Session of the Committee on Pesticide Residues The Hague, the Netherlands, 2 - 7 April 2001 JOINT FAO/WHO FOOD STANDARDS PROGRAMME CODEX ALIMENTARIUS COMMISSION Twenty-Fourth Session Geneva, 2-7 July 2001.

Corlett, D.A., 1989. Refrigerated foods and use of hazard analysis and critical control point principles. Food Technol. 43(2):91-94.

Department of Agriculture, 1999. Guidebook for the Preparation of HACCP Plans United States. Food Safety and Inspection Service, September.

Doménech, E., I. Escriche, S. Martorell and J.A. Serra, 1999. Técnicas de identificación de peligros. Adaptación en al industria de alimentos. Alimentaria, septiembre, Vol 305, págs: 19-22.

Escriche, I. and J.A. Serra, 1997. Toxicologia industrial de alimentos. Servicio de publicaciones de la Universidad Politécnica de Valencia. SPUPV 96856.

FDA (Food and Drug Administration Center for Food Safety and Applied Nutrition Managing), 1998. Food Safety: A HACCP Principles Guide for Operators of Food Establishments at the Retail Level Draft: April 15, 1998.

FDA (Food and Drug Administration), 1999. U. S. Department of Health and Human Services Public Health Service 1999 Food Code.

FDA (Food & Drug Administration Center for Food Safety & Applied Nutrition Bacteriological Analytical), 2001. U. S. Department of Health and Human Services Manual Online January 2001.

ICMSF, 1978. Microorganisms in Foods 1, Their significance and methods of enumeration, Método 4, 2nd. Ed..

ICMSF, 1988. El sistema de análisis de riesgos y puntos críticos. Su aplicación a las industrias de alimentos. Ed. Acribia, S.A. Zaragoza.

ICMSF, 1992. HACCP in microbilogical safety and quality. Ed: Backwell Scientific Publications.

ISO 10628, 1997. Flow diagrams for process plants General rules. Asociación Española de normalización y certificación.

ISO 9000, 2000. Gestión de calidad, AENOR. Asociación Española de normalización y certificación.

ISO 15161, 2001. Guidelines on the application of ISO 9001:2000 for the food and drink industry.

Juran, J.M., A. Blanton and A. Blanford, 1999. Juran's quality handbook. 5$^{th}$ edition, 1872 pp, Portland OR, Book News, Inc.

LEY, 2002. 20/2002, de 5 de julio, de Seguridad Alimentaria BOE núm. 181 Martes 30 julio 2002.

MFSC (Microbiology and food safety committee) of the national food Processors association, 1993. Implementation of HACCP in a food processing plant. Journal of Food Protection, vol 56, n° 6, 548-554.

Mortimore, S. and C. Wallace, 1998. HACCP: A Practical Approach, third edition, Gaithersburg, MD: Aspen.

NACMCF, (National Advisory Committee on Microbiological Criteria for Foods), 1992. Hazard Analysis and Critical Control Point (HACCP) system. International Journal of Food Microbiology. Vol 16, 1-23.

NACMCF, (National advisory committee on microbiological criteria for foods), 1998. Hazard Analysis and Critical Control Point principles and application guidelines. adopted august 14 J. of food protection. Vol. 61 n° 6, 762-775.

NFI (National Fisheries Institute), 1991. Seafood industry, hazard analysis critical control point, HACCP, training manual. Arlington, VA,.

NFPA (National Food Processors Association), 1992. HACCP and total quality management winning concepts for the 90's: A review. J. Food Prot. 55:459-462.

Regulation (EC) No 178/2002 of the European Parliament and of the Council laying down the general principles and requirements of food law, establishing the European Food Safety Authority, and laying down procedures in matters of food. Bulletin EU 1/2-2002 Health and consumer protection (1/16), 2002.

Real Decreto 140/2003, de 7 de febrero, por el que se establecen los criterios sanitarios de la calidad del agua de consumo humano. 21-02-2002 BOE 45, 2003.

Seafood HACCP Alliance for training and education, 1997. HACCP: Hazard Analysis and Critical Control Point training curriculum. Third edition. http://nsgd.gso.uri.edu/ncu/ncue98001.pdf.

Serra, J.A., E. Doménech, I. Escriche and S. Martorell, 1999. Risk Assessment an Critical Control Points from the Production Perspective. International Journal of Food Microbiology. Vol 46, 9-27.

Stevenson, K.E., 1990. Implementing HACCP in the food industry. Food Technol. 42(5):179-180.

# Definition of terms used

CCP Decision Tree: A sequence of questions to assist in determining whether a control point is a CCP.

Control: (a) To manage the conditions of an operation in order to maintain compliance with established criteria. (b) The state where correct procedures are being followed and criteria are being met.

Control Measure: Any action or activity that can be used to prevent, eliminate or reduce a significant hazard.

### Design and implementation of an HACCP system

Control Point: Any step at which biological, chemical, or physical factors can be controlled.

Corrective Action: Procedures followed when a deviation occurs.

Criterion: A requirement on which a judgement or decision can be based.

Critical Control Point: A step at which control can be applied and is essential to prevent or eliminate a food safety hazard or reduce it to an acceptable level.

Critical Limit: A maximum and/or minimum value to which a biological, chemical or physical parameter must be controlled at a CCP to prevent, eliminate or reduce to an acceptable level the occurrence of a food safety hazard.

Deviation: Failure to meet a critical limit.

HACCP: A systematic approach to the identification, evaluation, and control of food safety hazards.

HACCP Plan: The written document which is based upon the principles of HACCP and which delineates the procedures to be followed.

HACCP System: The result of the implementation of the HACCP Plan.

HACCP Team: The group of people who are responsible for developing, implementing and maintaining the HACCP system.

Hazard: A biological, chemical, or physical agent that is reasonably likely to cause illness or injury in the absence of its control.

Hazard Analysis: The process of collecting and evaluating information on hazards associated with the food under consideration to decide which are significant and must be addressed in the HACCP plan.

Monitor: To conduct a planned sequence of observations or measurements to assess whether a CCP is under control and to produce an accurate record for future use in verification.

Prerequisite Programs: Procedures, including Good Manufacturing Practices, that address operational conditions providing the foundation for the HACCP system.

Severity: The seriousness of the effect(s) of a hazard.

Step: A point, procedure, operation or stage in the food system from primary production to final consumption.

Validation: That element of verification focused on collecting and evaluating scientific and technical information to determine if the HACCP plan, when properly implemented, will effectively control the hazards.

Verification: Those activities, other than monitoring, that determine the validity of the HACCP plan and that the system is operating according to the plan.

# 8. Steps in the risk management process

*Adriane Mack, Thomas Schmitz, Gereon Schulze Althoff, Frank Devlieghere and Brigitte Petersen*

## 8.1 Introduction

This chapter presents contents from various texts on risk management with consistent use of nomenclature. It selects only the most important definitions of measures. Furthermore, it is shown that databanks and communication systems that are built up in monitoring processes can provide a reliable basis for more research into links between food composition and quality and different types of human health problems. Such research and development can go beyond the limits of human and animal health into occupational health, environmental issues and disaster management. However the chapter explains that this development can only be achieved if there is closer standardisation of terminology and harmonisation of methods in the field of risk management.

The aim of Chapter 8 is to give definitions of risk management, risk communication and risk assessment. In addition, examples of risk management process will be explained. It is to be clarified that the implementation of this process is not only the task of the state but also of the agrarian and food production companies. The learning target is to recognise the principles of the risk management processes and their meaning for consumer protection and food safety.

Safe production of food and drinking water is an important issue and the subject of legislation in most countries. Legislation governs the establishment of safe levels of chemical substances to be allowed in food and drinking water, whether occurring naturally, as deliberate additions or as contaminants. It also defines hygiene measures to minimise the spread of pathogenic agents and the diseases that they may cause. Despite these measures, foodborne illnesses are among the most widespread health problems in the developed world. They may be caused by toxic (chemical) or infectious (microbiological) agents entering the body during ingestion of contaminated food or water. To ensure that resources are used most effectively, it is essential to have a sound basis for assessing the risks of the many different chemical and microbiological agents that may be ingested via food and water (Benford, 2001). Therefore, the European regulation No 178/2002 laying down the general principles and requirements of food law, stated that food legislation needs to be based on risk analysis, except where this is not appropriate to the circumstances or the nature of the measure ((EC) No 178/2002).

Risk assessment is a task of traditional epidemiology. Epidemiology is not only required for research purposes but also for the rational use of resources in disease promotion and control. Epidemiology is one of several disciplines including economics, management, science and medical sociology that support population medicine. Because epidemiology is both population and holistic science it can serve as a bridging mechanism to help focus these disciplines so that they can achieve the more pragmatic objectives of population medicine.

*Adriane Mack, Thomas Schmitz, Gereon Schulze Althoff, Frank Devlieghere and Brigitte Petersen*

While many applications of risk assessment in epidemiology already exist many more links need to be established for the benefit of the health and the welfare of humans and animals. These include economic planning, human health, social sciences, ecology, environmental issues and the provision and management of related services. Such new orientations should take the form outlined below and should lead to actions of risk management on health policy as well as enterprise level which are expressed as recommendations in the following subsections.

Intensive animal and plant breeding, livestock husbandry and food processing, worldwide trade in animals, plants, animal feed stuffs and food, and the development of international tourism have changed our way of life and many other conditions require new development of worldwide epidemiological methods and measures. Interdisciplinary thinking and action are required. Risk assessment is not only a matter of statistics or mathematics. On the contrary it is essential that the data and further findings can be put into practice at interdisciplinary level.

### 8.1.1 Different views of risk and improvement management

Risk and improvement management have two points of view:
- the EU food safety and health policy;
- the enterprises in a food production chain.

Both aspects will be introduced in this chapter.

The formal-normative concept of risk in natural and technical sciences is oriented towards effects and damage. A complete understanding of cause and effect relationships is necessary, however it is normally not available and uncertainties are inadequately taken into account. Risk is distinguished from safety in the formal-normative context. Safety itself can only be understood, however, in relation to uncertainty and therefore to damage. In the legal context, risk is defined by the triad of danger, risk and residual risk. Risk in the legal context is limited to the domain between danger and residual risk but uncertainty has to be considered as distinct from residual risk. In social sciences risk is distinguished from danger. A difference is made between people who decide, i.e. regulators and those who are affected by the decision. People who decide attribute consequences of the decision and the involvement of risks to their own decision. However, concern is grounded in the decisions of others, in non-participation, and possible damage is therefore realised as danger and not as risk (Jung, 2003).

Environmental and health risks are subject to controversy in today's society. Every day the media reports on new scientific findings about risks, on the possibilities for individual risk prevention or on controversies between the various societal actors (e.g. public authorities vs. citizen interest groups). Obviously risk perception, i.e. the individual and subjective risk assessment of those affected or of the societal actors, is of major importance for the individual and society's handling of risks (Schütz and Wiedemann, 2003).

Facing intensive food scandals such as BSE, nitrofen, or acrylamide risk management cannot be successful without a dialogue involving the people in charge and relevant stakeholders. Certainly, communication cannot reduce risks for everyone, but has an important impact on whether different risk assessments lead to a societal or economic crisis. In this regard the performance of responsible authorities is often lacking (Carius and Renn, 2003). From this aspect food safety can be defined as product safety and agri-food system safety as it is described in Table 8.1.

*Table 8.1. Definition of food safety.*

| Food safety concepts | |
|---|---|
| **Product safety *** | **Agri-food-system safety** |
| Safety, non-toxicity of the food | Safety of supply |
| Safety, nutritious food | Safety of distribution |
| Safety of the declaration (all components of the food are shown on a declaration) | Safety of transparency and proximity |
| | Safety of consumer influence on food production |
| Safety of the label (the organic food is truly organic) | Safety of information on the whole food production process (e.g. by using labels) |
| | Safety, no negative impacts of production practices on humans and other living organisms, the environment, climate etc. |

* The traditional definition (e.g. given by the Danish authorities)

### 8.1.2 Historical development

In the 1970's and 80's risk management started to gain momentum having derived its origins from the insurance industry. Its early focus was on protecting against catastrophe and it evolved to protecting unaffordable potential losses. Insurers found that their results were enhanced by encouraging customers to exercise reasonable care and by rewarding good performance. Therefore, risk management evolved from natural intuition and analytical thinking into a more formal process of communication of the controls in place to influence outcomes.

In the 1980s total quality management became accepted as a means for improving the quality of business processes. Today risk management is accepted as a means of protecting the bottom line and assuring long-term performance. Risk management has become a universal management process involving quality of thought, quality of process and quality of action.

There were and still are many different approaches and methods of analysing and managing risk. In the early 1990's a risk management standard was developed in Canada. In 1995 a pre-eminent group of leading business thinkers developed the Australian and New Zealand Standard for risk management - AS/NZS 4360:1995. This standard has received a wide degree of international interest and is widely used as a guideline for implementing risk management. The increased level of focus and formalisation of risk management as a business process has created the opportunity for experienced practitioners and innovative thinkers to capitalise on the latest technology and break new barriers in developing business solutions (webpage 1).

### 8.1.3 Risk classification

From a scientific point of view risks are traditionally characterised by the *extent of damage* and the *probability of occurrence*. The product of both characteristics can be a measure for the classification of risks (Klinke and Renn, 1999). It is also useful to include other criteria of evaluation:
- Incertitude (related to statistical uncertainty).
- Ubiquity defines the geographic dispersion of potential damages.
- Persistency defines the temporal extension of potential damages.
- Reversibility describes the possibility of restoring the situation to the state before the damage occurred (possible restoration practices are reforestation and cleaning of water).
- Delay effect characterises a long time of latency between the initial event and the actual impact of damage. The time of latency could be of physical, chemical or biological nature.
- Potential of mobilization is understood as violation of individual, social or cultural interests and values generating social conflicts and psychological reactions by individuals or groups who feel inflicted by the risk consequences. In particular, it refers to perceived iniquities in the distribution of risks and benefits.

Traditionally, three categories of risks - normal area, the intermediate area and intolerable area - are distinguished (Figure 8.1).

The normal area is characterised by little statistical uncertainty, low catastrophic potential, low risk numbers damage when the product of probability and damage is taken, low scores on the

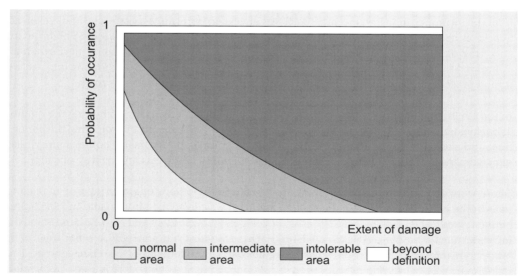

Figure 8.1. Risk areas: normal, intermediate and intolerable area (WBGU, German Scientific Advisory Council on Global Change, 1999) with permission.

criteria: persistency and ubiquity of risk consequences and reversibility of risk consequences, i.e. normal risks are characterised by low complexity and are well understood by science and regulation. The intermediate area and the intolerable area cause more problems because the risks touch areas that go beyond ordinary dimensions. Within these areas the reliability of assessment is low, the statistical uncertainty is high and the catastrophic potential can reach alarming dimensions. The risks may also generate global, irreversible damages, which may accumulate over a long time, or mobilise or frighten the population. In this case, the attitude of risk aversion is absolutely appropriate because the limits of human knowledge are reached.

Theoretically, a huge number of risk classes can be deduced from the eight criteria. Such a huge number of cases would not be useful for the purpose of developing a comprehensive risk classification. Therefore, a classification was developed where similar risk candidates are classified into risk classes in which they reach or exceed one or more of the possible extreme qualities with respect to the eight criteria (Figure 8.2). This classification leads to six risk classes that were given names from Greek mythology.

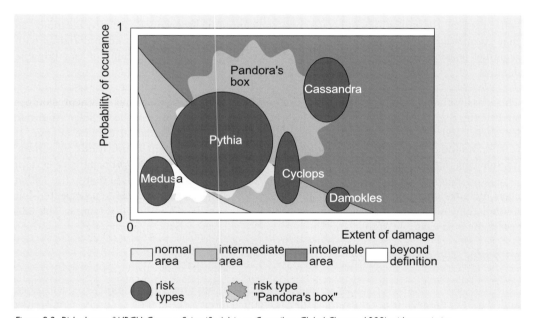

Figure 8.2. Risk classes (WBGU, German Scientific Advisory Council on Global Change, 1999) with permission.

## 8.1.4 Risk class Sword of Damocles

Many sources of technological risks have a very high disaster potential, although the probability that this potential manifests itself in damage is extremely low. Nuclear power plants, large-scale chemical facilities, dams and meteorite impacts are typical examples. A prime characteristic of this risk class is its combination of low probability with high extent of damage. Theoretically, the damage can occur at any time, but due to the safety measures implemented this is scarcely to be expected.

### 8.1.5 Risk class Cyclops

In the risk class *Cyclops* the probability of occurrence is largely uncertain whereas the maximum damage can be estimated. It is often the case that these risks are underestimated. A number of natural events such as volcanic eruptions, earthquakes and floods belong in this category. There is often little knowledge about causal parameters or little observation time in which to identify cyclic regularities. In other cases human behaviour influences the probability of occurrence so that this criterion becomes uncertain. Therefore, the appearance of AIDS and other infection diseases as well as nuclear early warning systems also belong to this risk class.

### 8.1.6 Risk class Pythia

This risk class refers to risk potentials for which the extent of damage is unknown and, consequently, the probability of occurrence also cannot be ascertained with any accuracy. To that extent we must assume for risk potentials of this class that there is great uncertainty with regard to possible adverse effects, and thus also with regard to the probability of ascertainable damage.

This class includes risks associated with the possibility of sudden non-linear climatic changes such as the risk of self-reinforcing global warming or of the instability of the Western Antarctic ice sheet, with far more disastrous consequences than those of gradual climate change. It further includes far-reaching technological innovations in certain applications of genetic engineering, for which neither the maximum amount of damage nor the probability of certain damaging events occurring can be estimated at the present point in time. Finally, the *Pythia* class includes chemical or biological substances for which certain effects are suspected but neither their magnitude nor their probability can be ascertained with any accuracy. The BSE risk is the best example of this.

### 8.1.7 Risk class Pandora's Box

This risk class is characterised by both uncertainties in the criteria probability of occurrence and extent of damage (only presumptions) as a consequence of the high ubiquity, persistency and irreversibility. Besides persistent organic pollutants and biosystem changes, endocrine disruptors can be quoted as examples.

### 8.1.8 Risk class Cassandra

This risk class refers to risk potentials characterised by a relatively lengthy delay between the triggering event and the occurrence of damage. This case is naturally only of interest if both the probability and magnitude of damage are relatively high. If the time interval were shorter the regulatory authorities would certainly intervene because the risks are clearly located in the intolerable area. However, the distance in time between trigger and consequence creates the fallacious impression of safety. Above all, the belief that a remedy will be found before the actual damage occurs can be taken as an excuse for inactivity. Typical examples of this effect are anthropogenic climate change and the loss of biological diversity.

### 8.1.9 Risk class Medusa

The risks belonging to this class refer to the potential for public mobilisation. This risk class is only of interest if there is a particularly large gap between risk perceptions and expert risk analysis findings. The probability of occurrence is low and also the damage is limited. A typical example is the irradiation of foods.

The essential aim of risk classification is to locate risks in one of the three risk areas in order to be able to derive effective and feasible strategies, regulations and measures for the risk policy on the different political levels. The characterisation provides a knowledge base so that political decision makers have better guidance on how to select measures for each risk class. The strategies pursue the goal of transforming unacceptable into acceptable risks, i.e. the risks should not be reduced to zero but moved into the normal area, in which routine risk management and cost-benefit-analysis becomes sufficient to ensure safety and integrity. Table 8.2 shows different management strategies for the different risk classes.

*Table 8.2. Overview of the management strategies (Klinke and Renn, 1999).*

| Management | Risk class | Extent of damage | Probability of occurrence | Strategies for action |
|---|---|---|---|---|
| Science-based | Damocles | ■ High | ■ Low | ■ Reducing disaster potential<br>■ Ascertaining probability |
| | Cyclops | ■ High | ■ Uncertain | ■ Increasing resilience<br>■ Preventing surprises<br>■ Emergency management |
| Precautionary | Pythia | ■ Uncertain | ■ uncertain | ■ Implementing precautionary principle<br>■ Developing substitutes<br>■ Improving knowledge |
| | Pandora | ■ Uncertain | ■ Uncertain | ■ Reduction and containment<br>■ Emergency management |
| Discursive | Cassandra | ■ High | ■ High | ■ Consciousness-building<br>■ Confidence-building<br>■ Public participation<br>■ Risk communication |
| | Medusa | ■ Low | ■ Low | ■ Contingency management |

### 8.1.10 Principle of risk analyses

Regardless of which approaches we have in view - the politically or the company-oriented - the phases of risk analysis are the same.

Quantitative risk assessment started as a formal process to aid in decision making for managing the uncertainties involved in determining the risk of carcinogenic hazards. The U.S. Food, Drug and Cosmetic Act requires that foods and drugs are safe. Therefore, procedures were evolved, during the 1960s and 1970s, for performing animal bioassays to identify carcinogens so these hazards could be eliminated. However, as analytical methods increased in sensitivity from parts per million to parts per trillion and beyond, it became necessary to determine at what levels exposures to potential hazards no longer resulted in measurable risks (Potter, 1996).

In 1983, the National Research Council (NRC) released *Risk Assessment in the Federal Government: Managing the Process*, which forms the basis of a common understanding of risk assessment (NRC, 1983). This document uses definitions that are sufficiently broad to be generally applicable and sufficiently narrow to provide precision in communication. It also establishes a generally accepted risk assessment nomenclature for carcinogenicity and other non-infectious processes (Potter, 1996). Nowadays, this approach is also used for other hazards like microbial hazards and non-carcinogenic chemicals.

It is generally agreed that risk assessment should be an independent scientific process, distinct from measures taken to control and manage the risk. The overall risk analysis process includes risk assessment, risk management and risk communication (Figure 8.3) and involves political,

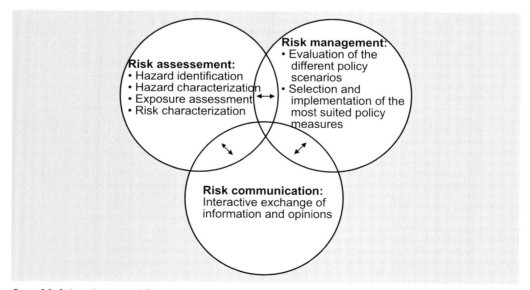

Figure 8.3. Risk analysis: general framework.

## Steps in the risk management process

social, economic and technical considerations. It also needs to take into account the public's perception of risk. Risk management strategies may be regulatory, advisory or technological and take into account factors such as the size of the exposed population, resources required, costs of implementation and the scientific quality and certainty of the risk assessment. Risk communication includes interactive exchange of information and opinions among risk assessors, risk managers, consumers and all other interested parties, often called stakeholders (Benford, 2001).

The following, Table 8.3, Table 8.4 and Table 8.5, give an overview of the definitions of risk assessment, risk management and risk communication of different authors.

*Risk assessment*
Since 1993 WTO and other organisations have tried to find a common definition of risk assessment. Table 8.3 gives an overview of the current definitions of risk assessment.

*Table 8.3. Definition of risk assessment of different authors.*

| Definition risk assessment | Author |
|---|---|
| The evaluation of the likelihood of entry, establishment or spread of a pest or disease within the territory of an importing member according to the sanitary or phytosanitary measures which might be applied and of the associated potential biological and economic consequences; or the evaluation of the potential for adverse effects on human or animal health arising from the presence of additives, contaminants, toxins or disease-causing organisms in food, beverages or feedstuffs. | (WTO GATT, 1994) |
| Identification and quantification of the risk resulting from a specific use or occurrence of a chemical or physical agent, taking into account possible harmful effects on individual people or society of using the chemical or physical agent in the amount and manner proposed and all the possible routes of exposure. Quantification ideally requires the establishment of dose-effect and dose-response relationships in likely target individuals and populations. | (IUPAC Recommendations, 1993) |
| The objective of risk assessment is the provision of scientific advice. Extensive information gathering and analysis is a prerequisite for sound and up-to-date scientific advice. | (White Paper on Food Safety, 2000) |
| A process of evaluation including the identification of the attendant uncertainties, of the likelihood and severity of an adverse effect (s) /event(s) occurring to man or the environment following exposure under defined conditions to a risk source(s). A risk assessment comprises hazard identification, hazard characterisation, exposure assessment and risk characterisation. A process intended to calculate or estimate the risk for a given target system following exposure to a particular substance, taking into account the inherent characteristics of a substance of concern as well as the characteristics of the specific target system. The process includes four steps:<br>■ hazard identification<br>■ dose-response assessment<br>■ exposure assessment<br>■ risk characterisation | (CAC ProcM, 1998); (EU risk, 2000); (EFA, 2000); ((EC) No 1488/94) |

Table 8.4. Definition of risk management of different authors.

| Definition risk management | Author |
|---|---|
| Risk management is understood as the process of weighing policy alternatives and other factors, and where required, selecting and implementing appropriate prevention and control options including regulation measures. | (CAC ProcM, 1998); (EMEA/CVMP/187/00-CONSULTATION) |
| Decision-making process involving considerations of political, social, economic, and engineering factors with relevant risk assessments relating to a potential hazard so as to develop, analyse, and compare regulatory options and to select the optimal regulatory response for safety from that hazard. Essentially risk management is the combination of three steps: risk evaluation; emission and exposure control; risk monitoring. | (IUPAC Recommendations, 1993) |

Table 8.5. Definition of risk communication of different authors.

| Definition risk communication | Author |
|---|---|
| The interactive exchange of information and opinions throughout the risk analysis process concerning hazards and risks, risk-related factors and risk perceptions, among risk assessors, risk managers, consumers, industry, the academic community and other interested parties, including the explanation of risk assessment findings and the basis of risk management decisions | (CAC ProcM, 1998) |
| Risk communication is a key element in ensuring that consumers are kept informed, and in reducing the risk of undue food safety concerns arising. It requires scientific opinions to be made widely and rapidly available, subject only to the usual requirements of commercial confidentiality, where applicable. | (White Paper on Food Safety, 2000); (EFA, 2000) |
| Interpretation and communication of risk assessments in terms that are comprehensible to the general public or to others without specialist knowledge. | (IUPAC Recommendations, 1993) |

*Risk management*
Legislation and control are the two components of risk management (White Paper on Food Safety, 2000). In Table 8.4 different definitions of risk management are shown.

The results of the conducted risk assessment are weighed with related social, economical and political factors to take policy decisions. Normally the executers of a risk assessment are not involved in the decision making so that they are as objective as possible. However, it is important to have a good interaction between risk managers and risk assessors to determine whether a certain risk assessment is feasible or not.

*Risk communication*
Risk communication is the interactive exchange of information and opinions during the entire risk analysis process, concerning hazards and risks, risk-related factors and risk perceptions, between risk assessors, risk managers, consumers, the population, companies, universities and

other interested persons, including the explanation of the results of the risk assessment and the foundation of the risk management decisions.

In Table 8.5 different definitions of risk communication are shown.

## 8.2 Risk assessment

In this under chapter the most important elements, which belong to the risk assessment are described: hazard identification, exposure assessment and risk characterisation.

Risk assessment is a scientific process, conducted by scientific experts, which may begin with a statement of purpose intended to define the reasons why the risk assessment is required and support the aims of the subsequent stages of risk management (Figure 8.4) (Benford, 2001).

A risk assessment consists of 4 parts: hazard identification, hazard characterisation, exposure assessment and risk characterisation.

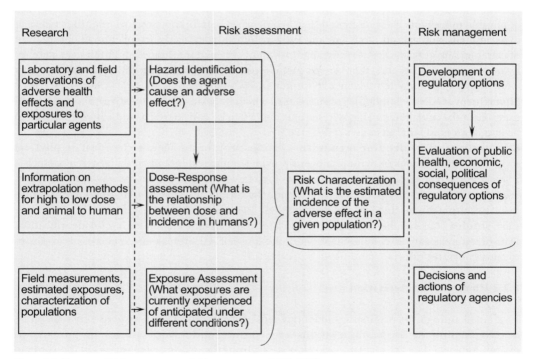

Figure 8.4. Elements of risk assessment and risk management (NRC, 1983).

*Adriane Mack, Thomas Schmitz, Gereon Schulze Althoff, Frank Devlieghere and Brigitte Petersen*

## 8.2.1 Hazard identification

Hazard identification is defined as the process of determining whether exposure to an agent can cause an increase in the incidence of a health condition (cancer, birth defect, etc.). It involves characterising the nature and strength of the evidence of causation. Although the question of whether a substance causes cancer or other adverse health effects is theoretically a yes-no question, there are few chemicals on which the human data are definitive. Therefore, the question is often restated in terms of effects in laboratory animals or other test systems, e.g., 'Does the agent induce cancer in test animals?' (NRC, 1983).

To carry out a hazard identification, different sources of information can be used:
**Epidemiological data -** Well-conducted epidemiological studies that show a positive association between an agent and a disease are accepted as the most convincing evidence about human risk. However, this evidence is often difficult to accumulate. The evidence is low, the number of persons exposed is small, the latent period between exposure and disease is long, and exposures are mixed and multiple. Thus, epidemiological data requires careful interpretation (NRC, 1983).
**Animal-Bioassay data -** The most commonly available data in hazard identification are those obtained from animal bioassays. The inference that results from animal experiments are applicable to humans is fundamental to toxicological research; this premise underlies much of experimental biology and medicine and is logically extended to the experimental observation of carcinogenic effects. Despite the apparent validity of such inferences and their acceptability by most cancer researchers, there are no doubt occasions in which observations in animals may be of highly questionable relevance to humans (NRC, 1983).
**Short-Term studies -** Sometimes tests are used which indicate certain negative effects. A typical example is the mutagenicity assay. These tests need to be considered as indicative and need to be supplemented with bioassays (NRC, 1983).
**Comparison of molecular structure -** Comparison of an agent's chemical or physical properties with those of known carcinogens provides some evidence of potential carcinogenicity. Experimental data support such associations for a few structural classes; however, such studies are best used to identify potential carcinogens for further investigation and may be useful in priority setting for carcinogenicity testing (NRC, 1983).
**Case studies -** Case studies are also an important source of information for hazard identification. Data of outbreaks and accidents can identify the cause of an outbreak or the necessary exposure to cause negative effects.

## 8.2.2 Hazard characterisation

In a small number of instances, epidemiological data permit a dose-response relation to be developed directly from observations of exposure and health effects in humans. If epidemiological data are available, extrapolations from the exposures observed in the study to lower exposures experienced by the general population are often necessary. Such extrapolations introduce uncertainty into the estimates of risk for the general population. Uncertainties also arise because the general population includes some people, such as children, who may be more

## Steps in the risk management process

susceptible than people in the sample from which the epidemiological data were developed (NRC, 1983).

The extrapolation is carried out by fitting a mathematical model to dose-response data that were collected during animal testing or outbreaks of foodborne illnesses. At present, the true shape of the dose-response curve at doses several orders of magnitude below the observation range cannot be determined experimentally. However, regulatory agencies are often concerned about much lower risks (1 in 100,000 to 1 in 1,000). This problem is illustrated in Figure 8.5 for a microbial risk when foods are consumed (NRC, 1983).

In extrapolating from animals to humans, the doses used in bioassays must be adjusted to allow for differences in size and metabolic rates. Several methods are currently used for this adjustment, and they assume that animal and human risks are equivalent when doses are measured as milligrams per kilogram per day, as milligrams per square meter of body surface

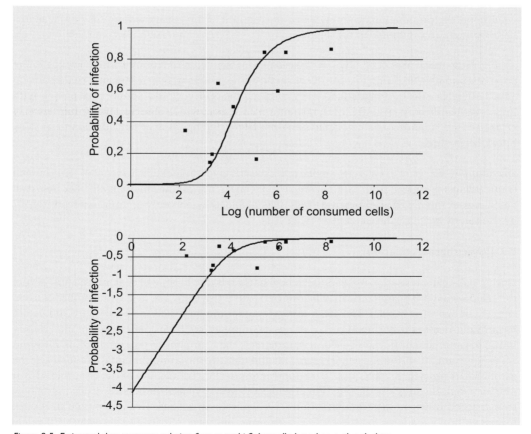

Figure 8.5. Estimated dose-response relation for non-typhi Salmonella based on outbreak data.

area, as parts per million in air, diet, or water, or as milligrams per kilogram per lifetime. Although some methods for conversion are used more frequently than others, a scientific basis for choosing one over the other is not established (NRC, 1983).

### 8.2.3 Exposure assessment

The first task of an exposure assessment is the determination of the concentration of the chemical to which humans are exposed. This may be known from direct measurement but more typically exposure data are incomplete and must be estimated. Models for estimating exposure can be very complex.

When the exposure is determined it is necessary to take into account that the (sub)population is not always equally exposed and that the exposure can vary in time. When, for example, the risk of fumonisins via the consumption of maize containing foods is estimated the eating habits of the population at research have to be taken into account. These eating habits can differ from person to person. Also, when the level of an agent in a food can be determined, the storage conditions, the preparation, frequency and the amount of the consumed product will lead to a high variation in the intake of the agent.

Another important aspect of exposure assessment is the determination of which groups in the population may be exposed to a chemical agent; some groups may be especially susceptible to adverse health effects. Pregnant women, very young and very old people and people with impaired health may be particularly important in exposure assessment. The importance of exposures to a mixture of carcinogens is another factor that needs to be considered in assessing human exposures.

For example, exposure to cigarette smoke and asbestos gives an incidence of cancer that is much greater than anticipated from carcinogenicity data on each substance individually. Because data detecting such synergistic effects are often unavailable, they are often ignored or accounted for by the use of various safety factors (NRC, 1983).

### 8.2.4 Risk characterisation

During risk characterisation it is estimated how likely it is that harm will be done and how severe the effects will be. The outcome may be referred to as a risk estimate, or the probability of harm at given or expected exposure levels. It brings together the information on exposure and health hazards defined in the earlier stages of risk assessment and outlines the sources of uncertainty in the data on which they are based. It summarises the estimates of potential risk in different ways, for example, different scenarios, and should identify the strengths and weaknesses of the estimates. If the data are adequate to support numerical quantification of the risk, then this is also included. The risk characterisation provides the primary basis for making decisions on how to manage the risk in different situations, i.e. whether it is necessary or feasible to reduce exposure, provide advice for susceptible subgroups, etc. The information provided will relate directly to the stated purpose of the risk assessment (Benford, 2001).

## 8.3 Risk management and food safety issues

In this subsection, principles are described, which aim at the accomplishment of suitable preventive measures if a risk is recognized, or that aim at minimising risk if food safety is acutely endangered. The improvement of crisis management and early warning systems will be described on the basis of examples.

### 8.3.1 Precautionary principle

The three components of risk analysis should be applied within an overriding framework for management of food-related risks to human health. There should be a functional separation of risk assessment and risk management, in order to ensure the scientific integrity of risk assessment, to avoid confusion over the functions to be performed by risk assessors and risk managers, and to reduce any conflict of interest However, it is recognised that risk analysis is an iterative process, and interaction between risk managers and risk assessors is essential for practical application (CAC, 1998).

*Control of food hazards*
Food business operators should control food hazards through the use of systems such as HACCP. They should:
- identify any steps in their operations which are critical to the safety of food;
- implement effective control procedures at those steps;
- monitor control procedures to ensure their continuing effectiveness; and
- review control procedures periodically, and whenever the operations change.

These systems should be applied throughout the food chain to control food hygiene throughout the shelf-life of the product through proper product and process design.

Control procedures may be simple, such as checking stock rotation calibrating equipment, or correctly loading refrigerated display units. In some cases a system based on expert advice, and involving documentation, may be appropriate. A model of such a food safety system is described in Hazard Analysis and Critical Control (HACCP) System and Guidelines for its Application (Annex) (Recommended international code of practice for general principles of food hygiene).

The precautionary principle has been invoked to ensure health protection in the European Community, thereby giving rise to barriers to the free movement of food or feed. Therefore, it is necessary to adopt a uniform basis throughout the Community for the use of this principle. In those specific circumstances where a risk to life or health exists but scientific uncertainty persists, the precautionary principle provides a mechanism for determining risk management measures or other actions in order to ensure the high level of health protection chosen in the Community. In specific circumstances where, following an assessment of available information, the possibility of harmful effects on health is identified but scientific uncertainty persists, provisional risk management measures necessary to ensure the high level of health protection chosen in the Community may be adopted, pending further scientific information for a more comprehensive risk assessment. Adopted measures shall be proportionate and no more restrictive of trade than is required to achieve the high level of health protection chosen in the Community, regard being had to technical and economic feasibility and other factors regarded as legitimate in the matter under consideration. The measures shall be reviewed within a reasonable period of time, depending on the nature of the risk to life or health identified and the type of scientific information needed to clarify the scientific uncertainty and to conduct a more comprehensive risk assessment. Food safety and the protection of consumer's interests are of increasing concern to the general public, non-governmental organisations, professional associations, international trading partners and trade organisations.

It is necessary to ensure that consumer confidence and the confidence of trading partners is secured through the open and transparent development of food law and through public authorities taking the appropriate steps to inform the public where there are reasonable grounds to suspect that a food may present a risk to health ((EC) No 178/2002).

### 8.3.2 Improvement of crisis management

Managers should ensure effective procedures are in place to deal with any food safety hazard and to enable the complete, rapid recall of any implicated lot of the finished food from the market. Where a product has been withdrawn because of an immediate health hazard, other products which are produced under similar conditions, and which may present a similar hazard to public health, should be evaluated for safety and may need to be withdrawn. The need for public warnings should be considered.

Recalled products should be held under supervision until they are destroyed, used for purposes other than human consumption, determined to be safe for human consumption, or reprocessed in a manner to ensure their safety (Recommended international code of practice general principles of food hygiene).

*Facing a product recall*
The past decade has witnessed the development and implementation of elaborate systems in order to guarantee that food products and raw materials for food products meet all the demands made by governmental agencies, manufacturers, food distributors or consumers. Despite all these efforts, from time to time defects in food products and raw materials occur. Examples are sensory deviations, microbiological spoilage, contamination with product-foreign compounds or contamination with product-foreign particles. Sometimes the defects can be so severe that a product recall is inevitable.

Frequently, the decision-making process in a product recall is hampered by a lack of information as to the nature of the defect and the possible consequences of the defect. With respect to a product recall important decisions are to be made in a very short time (Kersten, 2000).

## Steps in the risk management process

*Finding the cause and judging the consequences*
Long-standing experience in examining quality defects and providing trouble-shooting strategies is a prerequisite to answering questions like 'what is the nature of the defect?' and 'what is the health risk, if any?' For providing relevant information widespread experience is necessary as well as the highest quality equipment that can identify e.g. a minute quantity of a contaminant that causes deviation. The compound is to be isolated from the food and is to be identified by one of a range of the most advanced analytical techniques available. Sometimes defects are of microbiological origin and the expertise of microbiologists who are familiar with product and process parameters that determine the microbiological quality of foods is necessary.

Once a suspect compound has been identified, top toxicologists need to be consulted to assess the health risk, if any, resulting from consumption of the substance causing the defect.

In this section, some cases of products recalls will be presented. One conclusion that can be drawn is, that in cases of alleged or damaged products, the judgement of well-trained experts is essential (Kersten, 2000).

*General plan for crisis management*
The Commission shall draw up, in close cooperation with the Authority and the Member States, a general plan for crisis management in the field of the safety of food and feed (hereinafter referred to as 'the general plan'). The general plan shall specify the types of situation involving direct or indirect risks to human health deriving from food and feed which are not likely to be prevented, eliminated or reduced to an acceptable level by provisions in place or cannot adequately be managed solely by way of the application of Articles 53 and 54.

The general plan shall also specify the practical procedures necessary to manage a crisis, including the principles of transparency to be applied and a communication strategy ((EC) No 178/2002).

*Crisis unit*
Without prejudice to its role of ensuring the application of Community law, where the Commission identifies a situation involving a serious direct or indirect risk to human health deriving from food and feed, and the risk cannot be prevented, eliminated or reduced by existing provisions or cannot adequately be managed solely by way of the application of Articles 53 and 54, it shall immediately notify the Member States and the Authority. The Commission shall set up a crisis unit immediately, in which the Authority shall participate, and provide scientific and technical assistance if necessary ((EC) No 178/2002).

*Tasks of the crisis unit*
The crisis unit shall be responsible for collecting and evaluating all relevant information and identifying the options available to prevent, eliminate or reduce to an acceptable level the risk to human health as effectively and rapidly as possible. The crisis unit may request the assistance of any public or private person whose expertise it deems necessary to manage the crisis effectively. The crisis unit shall keep the public informed of the risks involved and the measures taken ((EC) No 178/2002).

### 8.3.3 Rapid alert system

A rapid alert system for the notification of a direct or indirect risk to human health deriving from food or feed is hereby established as a network. It shall involve the Member States, the Commission and the Authority. The Member States, the Commission and the Authority shall each designate a contact point, which shall be a member of the network. The Commission shall be responsible for managing the network. Where a member of the network has any information relating to the existence of a serious direct or indirect risk to human health deriving from food or feed, this information shall be immediately notified to the Commission under the rapid alert system. The Commission shall transmit this information immediately to the members of the network. The Authority may supplement the notification with any scientific or technical information, which will facilitate rapid, appropriate risk management action by the Member States. Without prejudice to other Community legislation, the Member States shall immediately notify the Commission under the rapid alert system of:

- any measure they adopt which is aimed at restricting the placing on the market or forcing the withdrawal from the market or the recall of food or feed in order to protect human health and requiring rapid action;
- any recommendation or agreement with professional operators which is aimed, on a voluntary or obligatory basis, at preventing, limiting or imposing specific conditions on the placing on the market or the eventual use of food or feed on account of a serious risk to human health requiring rapid action;
- any rejection, related to a direct or indirect risk to human health, of a batch, container or cargo of food or feed by a competent authority at a border post within the European Union.

The notification shall be accompanied by a detailed explanation of the reasons for the action taken by the competent authorities of the Member State in which the notification was issued. It shall be followed, in good time, by supplementary information, in particular where the measures on which the notification is based are modified or withdrawn. The Commission shall immediately transmit to members of the network the notification and supplementary information received under the first and second subparagraphs.

Where a batch, container or cargo is rejected by a competent authority at a border post within the European Union, the Commission shall immediately notify all the border posts within the European Union, as well as the third country of origin. Where a food or feed which has been the subject of a notification under the rapid alert system has been dispatched to a third country, the Commission shall provide the latter with the appropriate information. The Member States shall immediately inform the Commission of the action implemented or measures taken following receipt of the notifications and supplementary information transmitted under the rapid alert system. The Commission shall immediately transmit this information to the members of the network. Participation in the rapid alert system may be opened up to applicant countries, third countries or international organisations, on the basis of agreements between the Community and those countries or international organisations, in accordance with the procedures defined in those agreements. The latter shall be based on reciprocity and shall include confidentiality measures equivalent to those applicable in the Community ((EC) No 178/2002).

## 8.4 Risk communication and food safety issues

This subsection makes clear that communication represents one of the basic columns of the efforts around food security. A differentiation must be made on one hand between communication between national and EU institutions and on the other hand between public mechanisms and the consumer. Finally, safe food can only be produced, processed and sold if communication is possible between all actors of the value-added chain. This aspect is the basis of the "stable to table approach".

Government, consumer and industry/trade have an important influence on food safety (Figure 8.6).

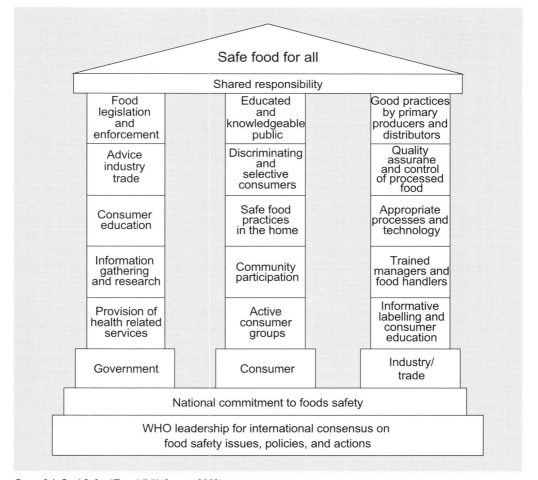

Figure 8.6. Food Safety "Temple" (Käferstein, 2003).

Adriane Mack, Thomas Schmitz, Gereon Schulze Althoff, Frank Devlieghere and Brigitte Petersen

Facing intensive food scandals such as BSE, nitrofen, or acrylamide, risk management cannot be successful without a dialogue involving those in charge and relevant stakeholders. Certainly, communication cannot reduce risks for everyone, but has an important impact on whether different risk assessments lead to a societal or economic crisis. In this regard, the performance of responsible authorities is often lacking (Carius and Renn, 2003).

## 8.4.1 Communication with public authorities

The scientific and technical issues in relation to food and feed safety are becoming increasingly important and complex. The establishment of a European Food Safety Authority, hereinafter referred to as 'the Authority', should reinforce the present system of scientific and technical support which is no longer able to respond to the increasing demands on it. Pursuant to the general principles of food law, the Authority should take on the role of an independent scientific point of reference in risk assessment and in doing so should assist in ensuring the smooth functioning of the internal market. Furthermore, the Authority should provide a comprehensive independent scientific view of the safety and other aspects of the whole food and feed supply chains, which implies wider-ranging responsibilities for the Authority. These should include issues having a direct or indirect impact on the safety of the food and feed supply chains, animal health and welfare, and plant health.

The Authority should contribute through the provision of support on scientific matters, to the Community and Member States' role in the development and establishment of international food safety standards and trade agreements. The confidence of the Community institutions, the general public and interested parties in the Authority is essential. For this reason, it is vital to ensure its independence, high scientific quality, transparency and efficiency. Cooperation with Member States is also indispensable. The Authority should cooperate closely with competent bodies in the Member States if it is to operate effectively. An Advisory Forum should be created in order to advise the Executive Director, to constitute a mechanism of exchange of information, and to ensure close cooperation in particular with regard to the networking system. Cooperation and appropriate exchange of information should also minimise the potential for diverging scientific opinions.

The Authority should take over the role of the Scientific Committees attached to the Commission in issuing scientific opinions in its field of competence. It is necessary to reorganise these Committees to ensure greater scientific consistency in relation to the food supply chain and to enable them to work more effectively. A Scientific Committee and Permanent Scientific Panels should therefore be set up within the Authority to provide these opinions. In order to guarantee independence, members of the Scientific Committee and Panels should be independent scientists recruited on the basis of an open application procedure. The lack of an effective system of collection and analysis at Community level of data on the food supply chain is recognised as a major shortcoming. A system for the collection and analysis of relevant data in the fields covered by the Authority should therefore be set up, in the form of a network coordinated by the Authority. A review of Community data collection networks already existing in the fields covered by the Authority is called for.

Improved identification of emerging risks may in the long term be a major preventive instrument at the disposal of the Member States and the Community in the exercise of its policies. It is therefore necessary to assign to the Authority an anticipatory task of collecting information and exercising vigilance and providing evaluation of and information on emerging risks with a view to their prevention. The Commission remains fully responsible for communicating risk management measures. The appropriate information should therefore be exchanged between the Authority and the Commission. Close cooperation between the Authority, the Commission and the Member States is also necessary to ensure the coherence of the global communication process. The independence of the Authority and its role in informing the public mean that it should be able to communicate autonomously in the fields falling within its competence, its purpose being to provide objective, reliable and easily understandable information ((EC) No 178/2002).

*Communications from the Authority*
The Authority shall communicate on its own initiative in the fields within its mission without prejudice to the Commission's competence to communicate its risk management decisions. The Authority shall ensure that the public and any interested parties are rapidly given objective, reliable and easily accessible information, in particular with regard to the results of its work. In order to achieve these objectives, the Authority shall develop and disseminate information material for the general public. The Authority shall act in close collaboration with the Commission and the Member States to promote the necessary coherence in the risk communication process. The Authority shall publish all opinions issued by it in accordance with Article 38. The Authority shall ensure appropriate cooperation with the competent bodies in the Member States and other interested parties with regard to public information campaigns ((EC) No 178/2002).

*Responsibilities*
Food and feed business operators at all stages of production, processing and distribution within the businesses under their control shall ensure that foods or feeds satisfy the requirements of food law which are relevant to their activities and shall verify that such requirements are met. Member States shall enforce food law, and monitor and verify that the relevant requirements of food law are fulfilled by food and feed business operators at all stages of production, processing and distribution. For that purpose, they shall maintain a system of official controls and other activities as appropriate to the circumstances, including public communication on food and feed safety and risk, food and feed safety surveillance and other monitoring activities covering all stages of production, processing and distribution. Member States shall also lay down the rules on measures and penalties applicable to infringements of food and feed law. The measures and penalties provided for shall be effective, proportionate and dissuasive ((EC) No 178/2002).

*TRACES (TRade Control and Expert System)*
The European Commission has just introduced TRACES, a new IT system designed to improve the management of animal movements both from outside the EU and within the EU. TRACES will consolidate and simplify existing systems and create better tools for managing animal disease outbreaks.

*Adriane Mack, Thomas Schmitz, Gereon Schulze Althoff, Frank Devlieghere and Brigitte Petersen*

All those concerned by animal trade authorities and economic operators will benefit from this modern web-based system, the first EU-wide e-government application in the field of food safety (Table 8.6). In Table 8.6 you will find more detailed information about TRACES.

*Table 8.6. Traces (webpage 3).*

| | |
|---|---|
| Objectives | The Integrated Computerised Veterinary System - TRACES - aims at providing an integrated structure for the existing 'computerised network linking veterinary authorities' - ANIMO - and the upcoming 'database covering import requirements' SHIFT. |
| Policy areas | ■ Animal Health<br>■ Animal Welfare<br>■ Public Veterinary Health |
| Legal obligations: animo | ■ Council Directive 90/425/EEC of 26 June 1990 concerning veterinary and zootechnical checks applicable in intra- Community trade in certain live animals and products with a view to the completion of the internal market (Article 20)<br>■ 92/438/EEC: Council Decision of 13 July 1992 on computerisation of veterinary import procedures (Shift project)<br>■ 92/563/EEC: Commission Decision of 19 November 1992 on the database covering the Community's import requirements, envisaged by the Shift project |
| Institutional requirements | ■ Report 1/2000 of the Court of Auditors on Classical Swine Fever (CSF)<br>■ Resolution A5-396/2000 of the EP on CSF<br>■ Report A5-0405/2002 of the EP on Foot-and-Mouth Disease (FMD)<br>■ Commitment towards the European Parliament (EP) for the development of SHIFT |
| Cascais conclusions | ■ Integration of ANIMO & SHIFT:<br>    1. User interface and application levels (avoid double data entry)<br>    2. European server (statistics, traceability, risk assessment)<br>■ Automated receipt notification of messages<br>■ Acknowledgment of the merchandise (transit, animal waste, animal welfare)<br>■ Risk assessment tool<br>■ Extension to animal products<br>■ Introduction of animal identification<br>■ Messages for complementary information (to be standardised)<br>■ Warning messages on EU imports (wrong destination, non conformity of consignment)<br>■ Rejected consignments database extended (rerouted, destroyed and/or transformed consignments). |
| Goals | ■ Control and traceability of animals and animal products<br>■ Assistance in decision-making for imports of animals and animal products<br>■ Central risk assessment warning<br>■ Reduction in administrative workload |

*Table 8.6. Continued.*

| | |
|---|---|
| System's main characteristics | ■ Electronic certification<br>■ Updated information and alert awareness<br>■ Interoperability<br>■ Risk assessment<br>■ Statistics automatisation<br>■ Databases (legislation, safeguard measures, transport, operators, animal products, animals, staging points)<br>■ Multi-criteria search engine<br>■ Multilingualism<br>■ Custom nomenclature (ND) as common codification |
| Statistics | ■ 94/360/EC: On the reduced frequency of physical checks of consignments of certain products to be implemented from third countries<br>■ 97/794/EC: Laying down certain detailed rules for the application of Council Directive 91/496/EEC as regards veterinary checks on live animals to be imported from third countries<br>■ 97/152/EC: Concerning the information to be entered in the computerized file of consignments of animals or animal products from third countries which are re-dispatched<br>■ 97/394/EC: Establishing the minimum data required for the databases on animals and animal products brought into the Community |
| Risk assessment | ■ Reinforced Controls<br>■ Increased frequency of physical checks<br>■ (94/360/EC & 97/794/EC) |
| Benefits | ■ Avoidance of resources redundancy in the EU's 285 Border Inspection Posts (BIP)<br>■ Standard application of laws and procedures<br>■ Simplification of tasks for official services through a reduction of data entry and automatisation<br>■ Increase in security through data entry in national languages<br>■ Speed of data transmission<br>■ Cooperation between services |

### 8.4.2 Communication with public authorities and consumer/enterprise

*Communication*

The ability to communicate directly and openly with consumers on food issues will give the Authority a high public profile. The Authority will need to make special provision for informing all interested parties of its findings, not only in respect of the scientific opinions, but also in relation to the results of its monitoring and surveillance programmes. The Authority must become the automatic first port of call when scientific information on food safety and nutritional issues is sought or problems have been identified. It will also need to ensure that appropriate information on these issues is published, as part of its commitment to re-establishing consumer confidence. Clearly the Commission will continue to be responsible for communicating risk management decisions (White Paper on Food Safety, 2000).

*Reacting to crises*
Where a food safety emergency occurs, the Authority will collect, analyse and distribute relevant information to the Commission and Member States, and will mobilise the necessary scientific resources to provide the best possible scientific advice. The Authority will have to respond rapidly and effectively to crises, and will take a key role in supporting the EU response. This will promote improved planning and handling of crisis situations at the European level, and will demonstrate to consumers that a pro-active approach is being taken to deal with problems. The Authority will operate the Rapid Alert System, which allows the identification and rapid notification of urgent food safety problems. The Commission will be part of the network and will therefore be informed on a real-time basis. Depending on the nature of a crisis, the Authority may be requested to carry out follow-up tasks, including monitoring and epidemiological surveillance (White Paper on Food Safety, 2000).

### 8.4.3 Communication in food chains

Specialisation in the agri-food sector has led to an immense growth in productivity. However, it has also led to an increase in complex customer-supplier relationships in agri-food chains.

In recent years producers and consumers have become more and more aware that quality of agricultural products depends on the performance of individual links within the complete production process. In contrast, quality management efforts of the various organisations of a supply chain are separated rather arbitrarily. Various boundaries exist between:
- Organisations - quality management and efficiency improvement activities are limited to internal processes at specific links of a netchain.
- Production stages - diverse organisational cultures exist at the different levels of production.
- Nations - different languages, different production methods and quality and information standards, differences in executing EU legislation.

A Dutch-German initiative (GIQS) turns chain management concepts and existing IT tools into practical solutions for quality management in pork netchains. The aim of the research for the findings presented here was to develop a reference information model for a quality information system in pork netchains. This reference model forms the baseline for the functional design of the system currently implemented in three pork netchains (one Dutch and two German), in the following referred to as pilot chains. The aim was to cover the organisational heterogeneity of netchain structures in the region.

*A. Netchain*
A netchain has been defined as a set of networks comprised of horizontal ties between companies within a particular industry or group, such as these networks (or layers) are sequentially arranged based on the vertical ties between firms and different layers (Lazzarini *et al.*, 2001). In the agri-food sector, a variety of network and supply chain structures exist, forming different

## Steps in the risk management process

netchains with a broad diversity of inter-enterprise relationships between its mostly legally independent links. This heterogeneous and complex netchain structures are schematically described in Figure 8.7. They render chain-wide information exchange and coordination activities rather challenging.

*B. Process model*

Quality Management is a structured system for satisfying internal and external customers and suppliers by integrating business environment, continuous improvement and maintenance cycles. EN ISO 9000:2000 ff. has been widely accepted as a framework for implementation of quality management systems and demands, amongst other things, the principles of
- process approach;
- continual improvement;
- mutually beneficial supplier relationships,

formulating the cornerstones for chain-wide cooperation and coordination of quality management activities.

The process model in Figure 8.8 describes product and optimal information flows between clients and suppliers for their individual decision making. Information from internal, client-oriented and supplier-oriented inspections as well as feedback from clients are considered for the analysis and assessment of a company's process management. In addition, structured and regular feedback can be addressed to its suppliers.

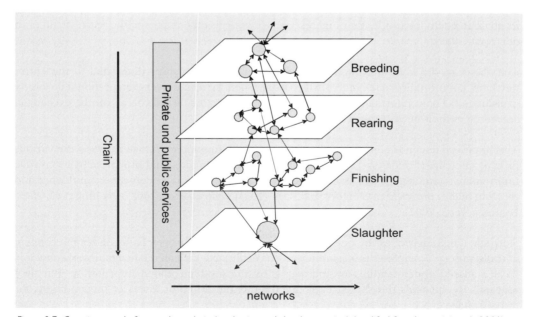

Figure 8.7. Generic example for a pork netchain (production and slaughter section) (modified from Lazzarini et al., 2001).

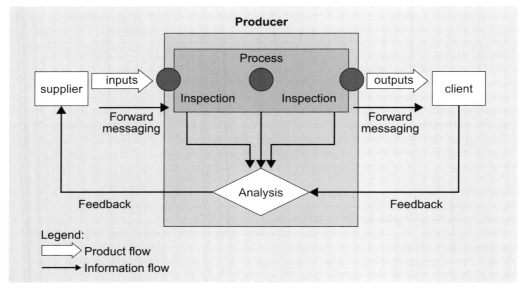

Figure 8.8. Process model with product and information flows.

A quality management system encompassing the horizontal and vertical dimension of a netchain can provide a means to better steer its complexities. Driving force is the recognition that each organisation in a netchain can enhance its performance and the product quality by integrating its goals and activities with other organisations to optimise the results of the entire chain (Van der Vorst, 2000).

The above model is limited to single client-supplier relationships. Extended to the entire netchain, a multitude of such relationships exist. Their information exchange should be jointly coordinated. Then, information even between distant actors of the chains can be exchanged according to their needs (Figure 8.9).

This approach depends greatly on knowledge- and information exchange between the various links of a netchain (Petersen et al., 2002). Effectively implemented and supported by innovative information technology, it can assist in improving productivity, raise consumer confidence and result in higher profits (e.g. Clemons and Row, 1992; Lazzarini et al., 2001; Amelung et al., 2002; Petersen, et al., 2002).

Hofstede (2003) describes the need for information exchange in netchains under the concept of transparency. He defines transparency as "the extent to which all the netchain's stakeholders have a shared understanding of, and access to, the product related information that they request". He specifies the shared understanding as the different levels of preconditions for transparency:

## Steps in the risk management process

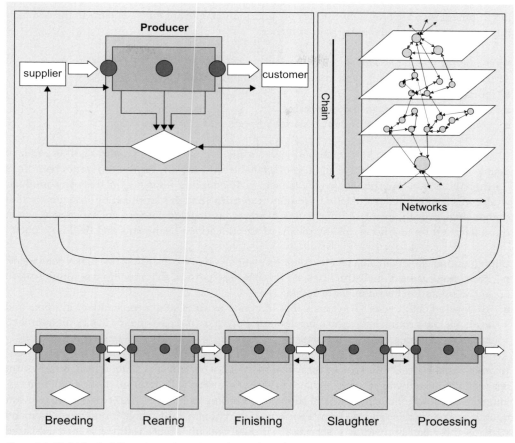

Figure 8.9. Model for the information exchange in netchain wide quality management.

- a shared language;
- shared interpretation of key concepts;
- shared standards for product quality;
- shared reference information models;
- shared technological infrastructure.

The concept of transparency can not only be seen as a defensive marketing tool, but a concept of sharing operational and even strategic information for better coordination of interdependent processes along the whole production cycle (Hofstede, 2003).

Additional workload and investments for chain actors can only be justified by a return on their investment. Dependent on their collaborative and contractual arrangements, in the current legal

*Adriane Mack, Thomas Schmitz, Gereon Schulze Althoff, Frank Devlieghere and Brigitte Petersen*

environment they are only willing to supply information relevant for other parts of the netchain if the system can provide added value for them.

*Inter-enterprise data warehouse*
Three possibilities of chain wide information exchange exist:
- portable data files;
- distributed data storage and processing;
- centralised data storage and processing (Vernède *et al.*, 2003).

It is essential to exchange information by electronic documents via a central hub in order to enable traceability and improve the use of available information for the different chain links (Luttighuis, 2000). Another major advantage of de-coupling information from the product, while preserving a link to detailed product properties though identification systems, is that exchanging parties prevent an information overload, as detailed data are not exchanged, while these data still remain accessible by means of identification (Trienekens and Beulens, 2001).

Implementing and organising chain wide information systems constitutes two major challenges:
- To set up organisational structures as "Trusted third parties" that moderate communication and cooperation between chain actors.
- To develop chain wide IT systems that integrate existing solutions, standard systems and available data sources to reduce implementation time and cost (Beers, 2002); an approach to the latter is described in this paper.

In order to enable cost effective and practical solutions, a central food chain information system has to build on standard systems and link existing data sources. Therefore the concept of inter-enterprise data warehousing has been adopted. According to Devlin's (1997) definition, relevant quality information is selected from a variety of sources along the whole production chain and structured for decision support purposes. Through specific access rights and internet-based interfaces information is made available to various end users, who can use it in their individual business context (Devlin, 1997). Similar to a backbone in the body, the "GIQS backbone" is developed to organise a centralised data management for chain actors including data on the converging and diverging product flows (Figure 8.10). Technical components used in this project are a data model of Chainfood® developed together with Wageningen University, implemented in an Oracle® database environment and combined with Cognos® Business intelligence technology.

*The Reference Information Model*
The Chainfood® data model generically describes the organisation of a database for information management in food chains, applicable to diverse sectors of the food industry. It can become operational for pork chains by applying a specific information model (Verdenius, 2003). An information model is here defined as an implementation guideline suggesting specific information sets, data sources and access rights which are needed to fulfil chain actors' information needs about products or processes.

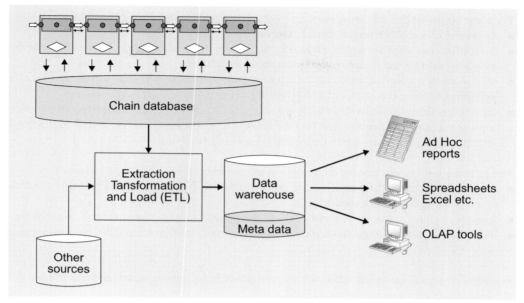

Figure 8.10. Structure of the GIQS Backbone.

Following the process model (Figure 8.8), each netchain link can retrieve data from their control points at the onset, in the middle and end of their part of the production process. Three main categories of information, related to the smallest identifiable unit, can be distinguished:
- Product information - inherent product properties (i.e. birth, provenance, size, components, etc.)
- Process information - production means and history of what has happened to a unit (i.e. farm information, treatments, processing, shipping, sales)
- Extended information - condensed information that allow for more detailed information on quality standards (i.e. QS, IKB), input specifications (i.e. pharmaceuticals, feed products) and means of production (i.e. machines, labour). This section is an extension to the process information.

It is apparent that there is a complex design aspect in developing models for standardised information exchange in netchains. Once in place, reference information models can reduce the workload and design needs for food chains and accelerate implementation activities in the industry (Hofstede, 2003).

Such models provide a basis for communicating information requirements, performing database design and enhancing user interaction with the information management system.

They are useful for the following reasons:
- Provision of a schema for database designers.
- Simplification of a complex situation for users. Individual end users must not be overwhelmed by the complexity of an integrated database.
- The reference information model enables specific chain actors to choose information sets according to their objective, related information needs and existing data sources.
- Effective communications tool visible for everyone. A reference information model provides a picture of the architecture and vision behind the "GIQS Backbone". It is thus anticipated that the model will increase in detail and complexity with time.

The information system developed here for pork netchains shall serve three purposes:
- Improved exchange and analysis of product-related information for preventive health management in primary production.
- Improved chain management and analytical support for chain managers and consultants.
- Provision of new EC Hygiene regulation required information for public meat inspection services.

Research is carried out in the following steps to develop the reference information model:
1. Analysis of inter-enterprise relationships.
2. Analysis of product and process quality information exchange and additional chain information needs by different stakeholders.
3. Mapping of the location, frequency of data collection and administration of quality information sources (database, paper based) in order to suggest potential application integration or interfaces with a central system.
4. Development of an information model for pork netchains.
5. Implementation of a food chain information system according to the model.
6. Validation, review and finalisation of the reference information model.

*Results*
Inter-enterprise relationships and organisational set-up vary in each of the three pilot chains. They could be categorised as:
- Closed system - stakeholders define quality specifications and information standards in contractual arrangements.
- Open system - external organisations (i.e. IKB, QS) determine production requirements and provide an organisational framework for customers and suppliers in the netchain.
- Mixed system - quality specifications and information standards of open systems combined with extra arrangements of the specific netchain.

Though the three pilot chains are differently structured, they have set up similar information exchange systems between their stakeholders. Up to now, it is organised point to point and paper based, though computerised databases are capturing information (especially at abattoir and extension services). The two pilot chains with closed quality systems carry out a more advanced information exchange between the links, where it is used as a resource for planning and controlling. Though defined as very useful by the producers, a structured feedback to their suppliers (breeder, etc.) is not established in any of the three chains.

Additionally, individual links of the chains process further details on the products and processes for their own use. Reports of veterinarians, extension services and auditors are available in paper form in formats provided by the relevant quality systems. Since all this information is stored separately it serves rather for documentation purposes and ad hoc corrective measures, instead of being a means to steer the continuous improvement process. Therefore, it is so far only of minor assistance for chain wide information exchange, reporting, efficient analysis and early warning activities.

Based on currently existing information sources a concept for a maximal useful data input into a central pork chain information system has been developed. A number of specific types of information have been defined as valuable for chain-wide and individual planning and controlling purposes and is listed in the reference information model. The information can be retrieved from different stages of production and is grouped as:
- Product Information: Identification, Provenance, Smallest identifiable unit, Supplier assessments, Public Meat inspection results, Organ results, Classification.
- Process Information: Information on production sites, Breeding programme, Transports, Slaughter, Vaccinations, Parasitical regimes, Feeding, Health and hygiene status, Laboratory results, Quality programme audit results.
- Extended information: Quality programme specifications, Salmonella antibody level calculations and trends, Supplier categorisation calculated on supplier assessments.

This reference model can be seen as a blueprint for a comprehensive food chain quality information system. However, the content is adjustable to meet the specific organisational arrangements, technical environment and information needs of each netchain.

*Conclusions*
Information exchange in netchains is more and more established, but often paper based and point-to-point. A variety of databases contain valuable information. To set up a cost effective quality information system, which meets the demands described above, existing information sources should be linked or integrated into a comprehensive database. A reference information model for pork netchains can assist in reducing costs and handling the complexity of the implementation of such systems. However, this can only act as a recommendation. Chain actors themselves ultimately have to decide on their level of information exchange.

Additional information does not necessarily lead to knowledge gain. For an efficient use of a central information system, available data has to be specifically processed and made available for the various stakeholders. Data warehousing technology is a means of integrating this information to support effective information exchange, continuous improvement process and provide an added value for all stakeholders of a netchain in a chain. Specific tools for pork chain actors will be developed in the next phase of this project (Schulze Althoff and Petersen, 2004).

## 8.5 Principles of enterprise risk management

While the preceding subsection deal with the role of the state with regard to the improvement of food safety, this subsection will describe the role of companies. Again the characteristic elements of risk management can be differentiated. The chapter deals with the alternatives of the companies to protect themselves against different risks and which methods are already used in food production to analyse and evaluate risks.

Under the aspect of quality management risk can be defined as the combination of the probability of an event and its consequences (ISO, 2002). In all types of enterprise there is the potential for events and consequences that constitute opportunities for benefit or threats to success. Risk Management is concerned with both positive and negative aspects of risk. This chapter considers the food safety risks. From this point of view consequences are only negative and therefore the management of food safety risk focuses on prevention and mitigation of harm (webpage 2).

### 8.5.1 Challenges for enterprise risk management

Risk management is a central part of any organisation's strategic management. The focus of good risk management is the identification and treatment of these risks. It increases the probability of success and reduces both the probability of failure and the uncertainty of achieving the organisation's overall objectives.

Risk management should be a continuous and developing process which runs throughout the organisation's strategy and the implementation of that strategy. It should address methodically all the risks surrounding the organisation's activities past, present and in particular, future (webpage 2).

The focus of identifying and analysing the risks may be due to a variety of reasons, such as customer and statutory requirements. The challenges for enterprise risk management are shown in Figure 8.11, which elucidates eight main factors on perceived risks.

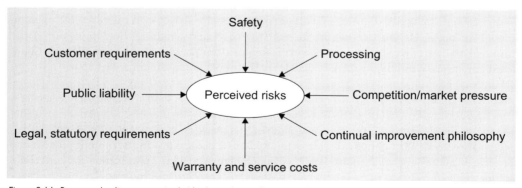

Figure 8.11. Pressures leading to perceived risks (according to Stamatis, 1995).

## 8.5.2 The enterprise risk management process

The steps of the risk management process are demonstrated in Figure 8.12.

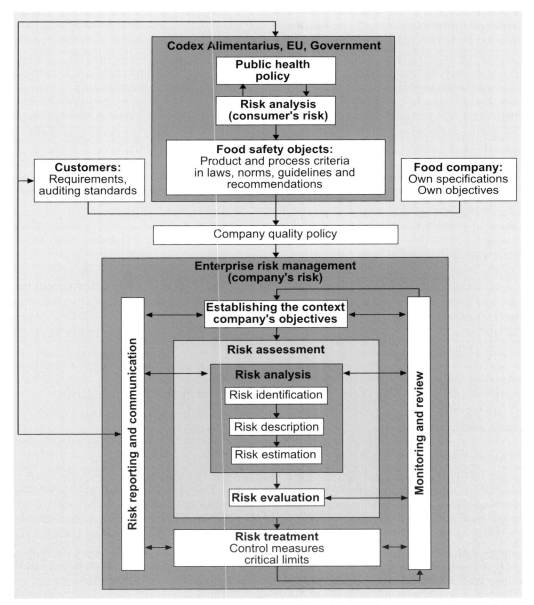

Figure 8.12. The steps of the risk management process (according to DGQ, 1998; webpage 2; Hoornstra and Notermans, 2001).

*Adriane Mack, Thomas Schmitz, Gereon Schulze Althoff, Frank Devlieghere and Brigitte Petersen*

In this chapter the connection between scientific risk analysis and enterprise risk management is explained (Figure 8.12). The following overview gives a detailed description of the steps of the enterprise risk management process. Every step is represented in Figure 8.12.

*Public health policy and scientific risk analysis*
One part of risk assessment in the food production chain is the scientific process by which the hazards and risk factors are identified and the risk estimation is determined (See subsection 0). Risk assessment is an especially important tool for governments when food safety objectives have to be developed in the case of 'new' contaminants in known products or known contaminants causing trouble in 'new' products (Hoornstra and Notermans, 2001).

The public health policy defines food safety objects. Food companies have to translate these food safety objects, customer requirements and auditing standards into enterprise-internal objects, the quality policy and specific own product and process criteria.

As a consequence of this risk assessment, support for food companies is also an important factor:
- during product development;
- during (hygienic) process improvement and;
- as an extension of the more qualitative HACCP-System.

*Establishing the context*
Establishing the context provides an understanding of the environment in which the risk assessment is taking place. This step includes:
- specifying the enterprise-internal risk policy and strategic objectives;
- identifying and analysing product and process criteria and;
- defining key elements for structuring the risk assessment process (e. g. structure of processes and products).

*Risk assessment*
Risk Assessment is defined as the overall process of risk analysis and risk evaluation.
The following steps in the risk analysis are risk identification, risk description and risk estimation (ISO, 2002).

*Risk identification*
Each of the key elements established in the previous step are systematically examined to identify what the risks are and how they may happen. Risk identification sets out to identify an organisation's exposure to uncertainty. This requires an explicit knowledge of the organisation, the market in which it operates, the legal, social, political and cultural environment in which it exists, as well as the development of a sound understanding of its strategic and operational objectives and processes. Risk identification should be approached in a methodical way to ensure that all significant activities within the organisation have been identified and all the risks flowing from these activities defined (webpage 1).

*Risk description*
The objective of risk description is to display the identified risks in a structured format, for example, by using a table. The risk description table can be used to facilitate the description and assessment of risks. The use of a well-designed structure is necessary to ensure a comprehensive risk assessment process. By considering the consequence and probability of each of the risks set out in the table it should be possible to prioritise the key risks that need to be analysed in more detail (webpage 2).

*Risk estimation*
Risk estimation can be quantitative, semi-quantitative or qualitative in terms of the probability of occurrence and the possible consequence. For example, probability and consequences may be high, medium or low. These estimation levels require enterprise-internal definitions of estimation criteria in a matrix. Different organisations will find that different measures of consequence and probability will suit their needs best.

A range of methods can be used to analyse risks. Risk analysis can be carried out by outside consultants especially in small and medium-sized companies. Nevertheless, in-house approach with well-communicated, consistent and co-ordinated processes and methods is fundamental. In-house 'ownership' of the risk management process is essential.

*Risk evaluation*
When the risk analysis process has been completed, it is necessary to compare the estimated risks against risk criteria, which the organisation has established. The risk criteria may include associated costs, legal requirements, socio-economic and environmental factors, concerns of stakeholders, etc. Risk evaluation, therefore, is used to make decisions about the significance of risks to the organisation and whether each specific risk should be accepted or treated (webpage 2).

*Risk reporting and communication*
Different levels within an organisation need different information from the risk management process.

The Board of Directors should:
- know about the most significant risks facing the organisation;
- ensure appropriate levels of awareness throughout the organisation;
- know how the organisation will manage a crisis;
- be assured that the risk management process is working effectively;
- publish a clear risk management policy covering risk management philosophy and responsibilities (webpage 2).

Business units should:
- be aware of risks which fall into their area of responsibility, the possible impacts these may have on other areas and the consequences other areas may have on them have performance indicators which allow them to monitor the key business activities, progress towards objectives and identify developments which require intervention and;
- report systematically and promptly to senior management any perceived new risks or failures of existing control measures (webpage 2).

Individuals should:
- understand their accountability for individual risks;
- understand how they can enable continuous improvement of risk management response;
- understand that risk management and risk awareness are a key part of the organisation's culture and;
- report systematically and promptly to senior management any perceived new risks or failures of existing control measures(webpage 2).

A company needs to report to its stakeholders on a regular basis setting out its risk management policies and the effectiveness in achieving its objectives. Increasingly stakeholders look to organisations to provide evidence of effective management of the organisation's non-financial performance in such areas as community affairs, human rights, employment practices, health and safety and the environment.

The arrangements for the formal reporting of risk management should be clearly stated and be available to the stakeholders. The formal reporting should address:
- the control methods - particularly management responsibilities for risk management;
- the processes used to identify risks and how they are addressed by the risk management systems;
- the primary control systems in place to manage significant risks;
- the monitoring and review system in place.

Any significant deficiencies uncovered by the system, or in the system itself, should be reported together with the steps taken to deal with them (webpage 2).

*Risk treatment*
Risk treatment is the process of selecting and implementing measures to control the risk. Risk treatment includes risk avoidance, risk mitigation, risk transfer or bearing all risks.
Any system of risk treatment should provide as a minimum:
- effective and efficient operation of the organisation;
- effective internal controls and;
- compliance with laws and regulations.

The risk analysis process assists the effective and efficient operation of the organisation by identifying those risks, which require attention by management. Effectiveness of internal control is the degree to which the risk will either be eliminated or reduced by the proposed control measures (Pfeifer, 2001). Table 8.7 points out relevant action alternatives.

*Table 8.7. Action alternatives for risk treatment (according to Pfeifer, 2001).*

| Control measures | Action alternatives |
|---|---|
| Risk avoidance | ■ Abandonment of a product<br>■ Abandonment of a sales market<br>■ Choice of another process technology<br>■ Choice of another recipe |
| Risk mitigation | Technological measures<br>Organisational measures<br>to reduce the probability of occurrence, the probability of detection and the possible consequence of risk |
| Risk transfer | To take out insurance:<br>■ Commercial third party liability insurance<br>■ Product third party liability insurance<br>■ Products recall cover<br>Form of contract:<br>■ General terms and conditions<br>■ Quality agreements |
| Bearing all risks | Do not insure risk<br>Non-insurable risk<br>Deductible |

### 8.5.3 Risk assessment methods and supporting software tools

This subsection provides an overview of some of the risk assessment tools that are used in food companies (Krieger *et al.*, 2003). There are many hazard and risk assessment tools. The choices that follow provide a brief overview of eight commonly used tools that are well suited for combination with the HACCP system (See Chapter 6).

*Failure Mode and Effect Analysis (FMEA)*
The Failure Mode and Effect Analysis (FMEA) is a logical, structured process which evaluates products, processes, devices, or services for possible ways in which hazards (risks, failures, problems) can occur. The FMEA focuses on preventing or minimising risk in a company's operations. It is one of the most commonly used safety analysis methods. For each of the known or potential failures identified, an estimate is made of its occurrence, severity and detection. The estimation is used to produce a risk profile. The Risk Priority Number (RPN) is the product of the Severity (S), Occurrence (O), and Detection (D) ranking:

RPN = (S) x (O) x (D)

The RPN is a measure of risk. The RPN is used to rank in order the concerns in processes (e.g. in Pareto fashion). Subsequently, an evaluation is made of the necessary action to be taken and planned, or ignored. The emphasis is on minimizing the probability of failure or the effect of failure (Stamatis, 1995; Lehnert et al., 2000).

*Fault Tree Analysis*
Fault Tree Analysis (FTA) is a graphical display similar to the shape of a tree. It is an analysis method that provides a systematic description of the combinations of possible occurrences in a system, which can result in an undesirable outcome. This method can combine hardware failures and human failures. FTA begins with the definition of an undesirable event and traces that event through the system to identify basic causes - a top-down appraisal. It is generally applicable for almost every type of risk assessment application (DGQ, 1998; Pfeifer, 2001).

*Event Tree Analysis*
Event Tree Analysis (ETA) is a method that logically develops visual models of the possible outcomes of an initiating event. It uses decision trees to create the models. The models explore how safeguards and external influences (called lines of assurance) affect the path of accident chains. ETA is valuable in analysing the consequences arising from a failure or undesired event. An event tree begins with an initiating event, such as a component failure, increase in temperature/pressure or a release of a hazardous substance. The consequences of the event are followed through a series of possible paths. Each path is assigned a probability of occurrence and the probability of the various possible outcomes can be calculated (DGQ, 1998; Pfeifer, 2001).

*Ishikawa diagram (fishbone diagram, cause-and-effect diagram)*
The Ishikawa diagram helps to find the causes of an effect (problem) and to reveal potential root causes. It graphically shows, in the form of a fishbone, the relationship of causes and sub-causes to an identified effect (DGQ, 1998; Pfeifer, 2001).

*Brainstorming*
Brainstorming is an intentionally uninhibited method for generating the greatest number of possible solutions to a problem, for later evaluation and development using group dynamics. The method relies on team participation and interaction (Stamatis, 1995).

*Surveys and questionnaire*
Surveys are an investigative questioning method. Through a programmed questioning of supplier and customer a picture is formed of problems encountered, customer desires, shape of the process and so on (Stamatis, 1995).

*Checklist analysis*
Checklist analysis is a systematic evaluation against pre-established criteria in the form of one or more checklists. It is used for high-level or detailed analysis, including root cause analysis. It is generally performed by an individual trained to understand the checklist questions and sometimes performed by a small group, not necessarily risk analysis experts. Checklist analysis generates qualitative lists of conformance and non-conformance determinations, with recommendations for correcting non-conformances (DGQ, 1998; Pfeifer, 2001).

*Pareto analysis*
Pareto analysis is a prioritisation method that identifies the most significant items among many. The Pareto Diagram is an X-Y-bar chart with the bars prioritised in descending order (from left to right) and distinguished by a cumulative percentage line. This method employs the *80-20* rule, which states that about 80 percent of the problems or effects are produced by about 20 percent of the causes. The prioritisation of the inputs (causes) indicates those that should be considered first (in other words, those on the left of the chart) (Stamatis, 1995; Pfeifer, 2001).

*Supporting software tools*
In order to increase effectiveness and efficiency, it is important to support the sequential as well as the cooperative handling of the individual combined methods. Computer Supported Cooperative Work (CSCW) enables efficient support especially for modern forms of cooperation in and between groups as well as companies along a food production chain. Risk management tools combine risk assessment methods with a common database. Software-supported methods lead to a less complex and more flexible application of combined risk assessment methods. An integrated electronic guideline as an action manual for problem-orientated and goal-orientated application of methods assist the users. For example, the international software company PLATO AG (www.plato-ag.com) developed a team-oriented solution integrating risk assessment and quality management methods into an easily managed software tool. This integration of the methods in one database provides a real-time interaction through hazard networks, product and process definitions and actions (Schmitz and Petersen, 2001).

# References

Amelung, C., S. Kiefer, T. Scherb and J.G. Schwerdtle, 2002. Qualitätssicherung bei Schweine- und Geflügelfleisch - Konzepte und praktische Umsetzung. Lebensmittelsicherheit und Produkthaftung, Rentenbank Schriftenreihe, 16, 43-91 (in German).

Beers, G, 2002. State of the art of tracking and tracing in Dutch agribusiness, In: K. Wild, R.A.E. Müller and U. Birkner (eds.), Referate der 23. GIL-Tagung in Dresden 2002 pp. 15-19.

Benford, D., 2001. Principles of risk assessment of food and drinking water related to human health, ILSI Europe concise monograph series.

CAC (Codex Alimentarius Commission), 1999. Principles and guidelines for the conduct of microbiological risk assessment, CAC/GL-30.

CAC ProcM, 1998. Codex Alimentarius Commission: Definitions for the Purposes of Codex Alimentarius. In: Procedural Manual. 11 ed, FAO/WHO, Rome 2000, pp 45-49CCFH 31 Codex Alimentarius Commission: Draft principles and guidelines for the conduct of microbiological risk assessment (At Step 8 of the Procedure), Appendix II of ALINORM 99/13A: Report of the thirty-first session of the Codex Committee on Food Hygiene, Orlando, United States, 26 - 30 October 1998, FAO/WHO, 1998 Datei im BgVV K:\General_Subject_Committees\ccfh\CCFH_31_session\ALINORM99_13A.pdf, 1998

Carius, R. and O. Renn, 2003. Participatory risk communication. Ways towards risk maturity. Bundesgesundheitsblatt-Gesundheitsforschung-Gesundheitsschutz, Jahrgang 46:578-585.

Clemons, E.K. and M.C. Row, 1992. Information technology and industrial cooperation: The changing economics of coordination and ownership, Journal of Management Information systems, 9 (2), 9-28.

Devlin, B., 1997. Data Warehouse: From Architecture to Implementation, Addison-Wesley.

DGQ - Deutsche Gesellschaft für Qualität e.V., 1998. Qualitätslenkung in der Lebensmittel-wirtschaft DGQ-Band 21-12, 1. Auflage, Beuth Verlag, Berlin, Köln.

EC No 1488/94, 1994. Commission Regulation of 28 June 1994 laying down the principles for the assessment of risks to man and the environment of existing substances in accordance with Council Regulation (EEC) No 793/93 (Text with EEA relevance). Official journal NO. L 161, 29/06/1994 P. 0003 - 0011, Document 394R1488.

EC No 178/2002, 2002. Regulation of the European Parliament and of the council of 28 January 2002, laying down the general principles and requirements of food law, establishing the European Food Safety Authority and laying down procedures in matters of food safety.

EFA, 2000. Commission of the European Communities: Proposal for a regulation of the European Parliament and of the Council laying down the general principles and requirements of food law, establishing the European Food Authority, and laying down procedures in matters of food safety (presented by the Commission), COM

EMEA/CVMP/187/00-CONSULTATION, 2000. EMEA, The European Agency for the Evaluation of Medicinal Products, Veterinary Medicines and Information Technology, Committee for Veterinary Products: Note for Guidance on the risk analysis approach for residues of veterinary medicinal products in foods of animal origin, (2000)716 - final, Brussels, 8.11.2000.

EU risk, 2000. European Commission - Health & Consumer Protection Directorate-General, Directorate C - Scientific Opinions: First Report on the Harmonisation of Risk Assessment Procedures, Part 2: Appendices. 26-27 October 2000.

European Commission, 2002. Proposal: Regulation of the European Parliament and of the Council of laying down specific rules for the organisation of official controls on products of animal origin intended for human consumption, 2002/0141 (COD).

Hofstede, G.J., 2003. Transparency in Netchains, In: Proceedings of EFITA 2003 Conference, 5.-9. July 2003, Debrecen, Hungary, pp 17-29.

Hoornstra, E. and S. Notermans, 2001. Quantitative microbiological risk assessment. International Journal of Food Microbiology 66, 21-29.

ISO, 2002. ISO/IEC Guide 73, Risk management - Vocabulary - Guidelines for use in standard.

IUPAC Recommendations, 1993. Glossary for chemists of terms uses in toxicology, in: Pure & Appl. Chem, Vol 65, No. 9, p 2003-2122.

Jung, T., 2003. Scientific and public definition of risk. Bundesgesundheitsblatt-Gesundheitsforschung-Gesundheitsschutz Juli 2003, Jahrgang 46:542-548.

Käferstein, F.K., 2003. Actions to reverse the upward curve of foodborne illness. Food control 14, 101-109.

Kersten, J., 2000. An integral approach to practical food hygiene, Conference Food Hygiene Europe, Amsterdam RAI, 6-8 June 2000, Product contamination, analysis and evaluation of the risks, (TNO Nutrition and Food Research, the Netherlands).

Klinke, A. and O. Renn, 1999. Prometheus unbound, challenges of risk evaluation, risk classification and risk management, ISBN 3-932013-95-6.

Krieger, S., B. Petersen and T. Schmitz, 2003. EDV-Bedeutung steigt - Einsatz von Qualitätstechniken sowie deren Unterstützung durch Software. Lebensmitteltechnik 35 (3) 65-67.

Lazzarini, S.G., F.R. Chaddad and M.L. Cook, 2001. Integrating supply chains and network analyses, the study of netchains. Journal on Chain and Network Science, 1 (1), 7-22.

Lehnert S., T. Schmitz and B. Petersen, 2000. Risk and Weak Point Analysis in the Range of Chain-oriented Data Acquisition, In: J.H. Trienekens and P.J.P. Zuurbier (eds.), Chain Management in Agribusiness and Food Industry, Wageningen Pers, Wageningen, the Netherlands.

Luttighuis, P.H.W.M.O, 2000. ICT Service Infrastructure for Chain Management. In: J.H. Trienekens and P.J.P. Zuurbier (eds.), Proceedings of 4th International Conference on Chain Management in Agribusiness and the Food Industry. Wageningen Pers, pp 275-282.

NRC (National Research Council), 1983. Risk assessment in the federal government: managing the process, National Academy Press, Washington D.C.

Petersen, B., S. Knura-Deszczka, E. Pönsgen-Schmidt and S. Gymnich, 2002. Computerised Food Safety Monitoring in Animal Production. Livestock Prod. Science. 76: 207-213.

Pfeifer, T., 2001. Qualitätsmanagement - Strategien - Methoden - Techniken, 3. Auflage, Carl Hanser Verlag, München, Wien.

Potter, M.E., 1996. Risk assessment terms and definitions. J. Food Protect., Suppl, 6-9.

Schmitz, T. and B. Petersen, 2001. Denken in Prozessketten - Transparenz durch IT-gestützte präventive Qualitätstechniken. Die Ernährungsindustrie 5, 56-57.

Schulze Althoff, G. and B. Petersen, 2004. Chain quality information system - development of a reference information model to improve transparency and quality management in pork netchains along the Dutch German border. In: H.J. Bremmers, S.W.F. Omta, J.H. Trienekens and E.F.M. Wubben (eds.), Dynamics in Chains and Networks; Proceedings of the sixth International Conference on Chain and Network Management in Agribusiness and the Food Industry, Wageningen Academic Publishers, Wageningen, the Netherlands.

Schütz, H. and P.M. Wiedemann, 2003. Risk perception in society. Bundesgesundheitsblatt-Gesundheitsforschung-Gesundheitsschutz, Jahrgang 46:549-554.

Stamatis, D.H., 1995. Failure Mode and Effect Analysis - FMEA from Theory to Execution, ASQC Quality Press, Milwaukee, Wisconsin.

Trienekens, J.H. and A.J.M. Beulens, 2001. The implications of EU food safety legislation and consumer demands on supply chain information systems. In: Proceedings of 2001 Agribusiness Forum and Symposium International Food and Agribusiness Management Association, Sydney, Australia.

Van der Vorst, J.G.A.J., 2000. Effective food supply chains. Generating, modelling and evaluating supply chain scenarios. Dissertation, Wageningen University, Netherlands.

Verdenius, F., 2003. Foodprint: Structured development of Traceability systems, in: Proceedings of International Foodtrace Conference, Sitges, Spain, October 30 - 31, 2003 pp. 73-75.

Vernède, R., F. Verdenius and J. Broeze, 2003. Traceability in food processing chains, state of the art and future developments, KLICT position paper www.klict.org, 46 pp.

WBGU, German Scientific Advisory Council on Global Change, 1999. Welt im Wandel. Handlungsstrategien zur Bewältigung globaler Umweltrisiken. Jahresgutachten 1998, Springer, Berlin.

White Paper on food safety, 1999. Brussels, 12 January 2000, COM, 719 final.

WTO GATT, 1994. Uruguay Round of Multilateral Trade Negotiations (1986- 1994) - Annex 1 - Annex 1A - Agreement on the Application of Sanitary and Phytosanitary Measures WTO-"GATT 1994", Official Journal L 336, 23/12/1994 p. 0040 0048. Document 294A1223(05).

webpage 1, Sesel, J., http://www.siliconrose.com.au (04.06.2004)

webpage 2, www.theirm.org, A Risk Management Standard, published by IRM the Institute of Risk Management, ALARM The National Forum for Risk Management in the Public Sector, AIRMIC the Association of Insurance and Risk Managers, 2002

webpage 3, Didier CARTON, Animal health and welfare, zootechnics Uni, http://europa.eu.int/comm/food/animal/diseases/animo/tracesdc.ppt

# 9. Modelling food safety

*Frank Devlieghere, Kjell Francois, Bruno De Meulenaer and Katleen Baert*

## 9.1 Introduction

Some phenomena in food safety need a quantitative insight into the behaviour of chemical or microbial contaminants in food products. Mathematical models can be useful tools in creating this insight. In this chapter, the use of mathematical models to describe the behaviour of contaminants in two domains is discussed: predictive microbiology and migration of chemical compounds out of packaging materials. The mathematical approach to risk assessment is also described. However, it is not the aim of this chapter to overload the reader with mathematical equations. Equations are only given to illustrate the different approaches applied in the fields described above.

## 9.2 Modelling microbial behaviour: predictive microbiology

### 9.2.1 Introduction

Recent crises in the food industry have increased public awareness about the food they eat. Over the last few decades, dioxins and PCB's, but also microbial hazards like *Listeria monocytogenes* or *Bacillus cereus*, have reached the news' headlines. The consumer has become more critical of food demanding more fresh, healthy, safe and nutritionally rich food products.

One way to prove the microbial safety of a food product is by using laborious and time-consuming challenge tests. In these tests the shelf life of a specific food product can be assessed as to spoilage and pathogenic microorganisms in a specific set of storage conditions. These methods have been criticised for their expensive, time-consuming and non-cumulative character (McDonald and Sun, 1999). As most food companies have an increasing number of different products, and storage conditions are different at each stadium in the food chain, it is almost impossible to cover all these product/condition combinations using the classic challenge tests.

Predictive microbiology can be an answer to these problems. Once a model is developed it will be the fastest way to estimate the shelf life according to microbial spoilage and to estimate the microbial safety of a food product. And within no time, this information can be delivered for some different storage conditions or for some slight or more profound recipe changes. Moreover, in the last decade, the power of the computer for both hardware and software has increased exponentially, while the price of a powerful machine has come down, making it possible for everyone to access the current modelling methodologies.

Predictive food microbiology can be defined as the use of mathematical expressions to describe microbial behaviour in a food product (Whiting and Buchanan, 1994). Predictive microbiology is based on the premise that the responses of populations of microorganisms to environmental factors are reproducible and that, by characterising environments in terms of those factors that most affect microbial growth and survival, it is possible from past observations to predict the responses of those microorganisms in other, similar environments (Ross et al., 2000). This knowledge can be described and summarised in mathematical models which can be used to predict quantitatively the behaviour of microbial populations in foods, e.g. growth, death, toxin production from a knowledge of the environmental properties of the food over time.

Classically, the models can be divided into three groups: the primary models depicting the evolution during time in the amount of bacteria under a given set of conditions. When these conditions are favourable for the bacteria, the primary model will be a growth model, while under stressful conditions, the primary model will be an inactivation model. The secondary models deal with the effect of environmental conditions, like temperature, pH, $a_w$ or gas conditions, on the primary model. In a tertiary model the previous two were brought together and implemented in a user-friendly software package.

In this section, there will be a summary of the current modelling tools: the different types of primary and secondary models will be discussed, as well as some tertiary models, the model-developing procedures will be highlighted, the current shortcomings in predictive modelling will be disclosed and some illustrating examples will be presented.

### 9.2.2 Modelling the growth of microorganisms

Several different classification schemes are possible to group the models in the area of predictive modelling. The system of Whiting and Buchanan (1993) is most often used, dividing the models into three groups: primary, secondary and tertiary models, although some other segmentation criteria are also common. There are differences between kinetic and probability models, between empirical and mechanistic models, and between static and dynamic models. In this section, these different groups will be discussed thoroughly and some examples will be given.

*Kinetic vs. probability models*
One way to split up model categories within the area of predictive modelling is to look at the difference between kinetically-based models and probability-based models first. The choice of approach is largely determined by the type of microorganisms expected to be encountered and the number of variables. Kinetic models can predict the extent and rate of growth of a microorganism, while probability-based models consider the probability of some event occurring within a nominated period of time. They are often used to model spore-forming bacteria, and the probability of survival and germinating of spores in canned products. Probability models can also be used to estimate the likelihood that a pathogen can be present at dangerous levels in a food product given a certain set of environmental conditions, or the likelihood that a pathogen is able to grow in a certain environment or not.

Kinetic modelling focuses on the bacterial cell concentration as a function of time. Normally this can be depicted by a sigmoidal growth curve (Figure 9.1), which is characterised by four

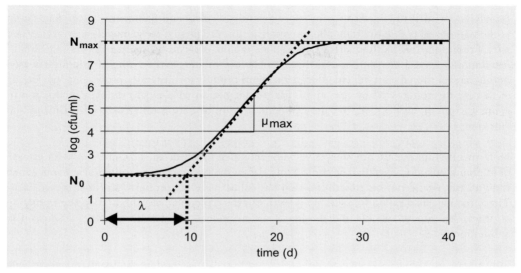

Figure 9.1. Sigmoidal growth curve.

main parameters: (1) the lag phase - $\lambda$ (h) - is the time that a microorganism needs to adapt to its new environment; (2) the growth speed - $\mu$ (h$^{-1}$) - is correlated to the slope of the log-linear part of the growth curve; (3) $N_0$ (cfu/ml), the initial cell concentration, and (4) $N_{max}$ (cfu/ml) the maximal cell concentration.

The effect of different environmental factors like temperature, pH and/or $a_w$ on the parameter values of those sigmoidal curves are then modelled. Evaluation of this fitted sigmoid curve may allow researchers to make predictions about the studied microorganisms in a particular food system. In both modelling approaches, models are constructed by carefully evaluating a lot of data collected on increases in microbial biomass and numbers, under a studied criterion of intrinsic and extrinsic parameters. This allows researchers to make predictions about the studied microorganism's lag phase, generation time or exponential growth rate, and maximum cell density. Traditionally, the studied environmental variables are temperature, pH and $a_w$, but other important conditions like the composition of the gas atmosphere, the type and concentration of acid, relative humidity, concentration of preservatives and/or antimicrobials or the redox potential can also affect microbial growth, and thus those factors can be included in a model.

Kinetic models are useful as they can predict changes in microbial cell density as a function of time, even if a controlling variable, which can affect growth, is changing. However, the main drawback is that kinetic models are difficult to develop as they require a lot of data to be collected to model the interaction effects between the different environmental factors (McDonald and Sun, 1999). The main types of kinetic models will be discussed further in the chapter.

Probability models are used more for spore-forming bacteria and pathogens, as the focus here is on the probability that a certain phenomenon occurs. Originally, these models were concerned with predicting the likelihood that organisms would grow and produce toxins (e.g. for *C. botulinum*) within a given period of time. Probability models can be a helpful tool for a food manufacturer to make an informed decision about product formulation, processing, packaging or storage of products. They give appropriate information with regard to toxin production or spore germination in a food, but they provide little information on growth rate (Gibson and Hocking, 1997).

More recently, probability models have been extended to define the absolute limits for growth of microorganisms in specified environments, e.g. in the presence of a number of stresses which individually would not be growth limiting but collectively prevent growth (Ross *et al.*, 2000). These models, also called "growth/no-growth interface models", are situated on the transition between the growth models and the inactivation models. They will be further discussed in subsection 9.2.3.

Ross and McMeekin (1994) suggested that the traditional division of predictive microbiology into kinetic models and probability models is artificial. They argued that the two types of modelling approach represent opposite ends of the spectrum of modelling requirements, with research at both ends eventually coming together. This can be proved by the growth/no-growth models, which are a logical continuation of the kinetic growth models. When environmental conditions become more stressful, growth speed slows down and the lag phase is extended until growth speed drops to zero, or the lag phase rises to eternity. Within that twilight zone, the modelling needs to shift from a kinetic model, when the environmental conditions still support growth, to a probability model when the growth/no growth interface is reached, demonstrating the near link between the two modelling types.

*Empirical vs. mechanistic models*
The difference between empirical models and mechanistic models might be even more difficult to define than the difference between kinetic models and probability models. An empirical model, or black-box model, can be described as a model where the fitting capacities are the only criteria used. The aim is to describe the observed data in the best possible way, using a convenient mathematical relationship, without any knowledge about biochemical backgrounds or underlying cellular processes. Polynomial models or response-surface models are the best-known representatives of this group. Other examples of empirical models are the Gompertz equation, and artificial neural networks but also square-root models and their modified versions (see later).

Mechanistic models are situated at the other end of the spectrum. Here, the model is based on an understanding of the underlying biochemical processes, and the intercellular and intracellular processes that are controlling the cell behaviour. Hence, the actual microbial knowledge is insufficient to generate a complete mechanistic model. Most of the time a model will be a combination of some mechanistic and some empirical components.

# Modelling food safety

Most researchers agree that (semi-) mechanistic models are inherently superior to empirical models as they give a better understanding of cellular behaviour, so these models are preferable as long as the fitting capacities are not endangered.

*Primary, secondary and tertiary models*
The most used, and maybe the most evident classification within the predictive modelling area is the distinction between primary, secondary and tertiary models, first presented by Whiting and Buchanan (1993).

*Primary models*
Primary models describe microbial evolution as a function of time for a defined set of environmental and cultural conditions. This section will focus on primary growth models, but the same structure can be followed for inactivation modelling.

Classically, a sigmoidal growth curve is used to model cell growth as a function of time, although a three-phase linear model is sometimes used as a good simplification (Buchanan *et al.*, 1997a). In Figure 9.2, a standard growth curve is depicted, plotted over the original dataset. The symbols show the collected data, while the full line represents the sigmoidal growth model. Growth curves are mainly figured on a $\log_{10}$-based scale for the cell density as a function of time, as microbial growth is an exponential phenomenon, but sometimes an ln base is preferred.

Traditionally, the sigmoidal growth curve is described using four parameters: the exponential growth rate $\mu_{max}$, the lag phase $\lambda$, the inoculum level $N_0$ and the maximal cell density $N_{max}$. The exponential growth rate is defined as the steepest tangent to the exponential phase, so the tangent to the inflexion point, while the lag phase is defined as the time at which that extrapolated tangent line crosses the inoculum level (McMeekin *et al.*, 1993). The generation time (GT), the time necessary for a cell to multiply in hours or days, can be calculated from the $\mu_{max}$:

$$GT = \frac{\log_{10}2}{\mu_{max}} \tag{9.1}$$

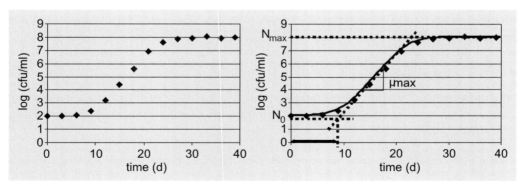

Figure 9.2. Sigmoidal growth curve fitted to experimental data set.

During the short history of predictive modelling, several primary models were developed. The first growth model was described by Monod, based on a purely empirical observation that microbial cell growth is an exponential system.

$$N = N_0 e^{kt} \qquad (9.2)$$

This rather rudimentary model, where $N$ = cell density (cfu/ml), $N_0$ = initial cell level (cfu/ml), $k = \ln 2/GT$ (h$^{-1}$) and $t$ = time (h), is limited as only one growth parameter, the growth rate $\mu_{max}$, can be revealed, while no lag phase can be modelled. Nor is the maximal cell density taken into account, even though cells can't grow infinitely.

The non-linear models were proposed in the eighties. Gibson et al. (1987) introduced the Gompertz function into food microbiology, making it possible to express the log(cfu/ml) as a function of time using the sigmoidal shape.

$$\log N(t) = A + D \exp\{\exp[-B(t-M)]\} \qquad (9.3)$$

where,
$N(t)$ = the cell density at time t;
$A$ = value of the lower asymptote = $N_0$;
$D$ = difference in value between the upper and the lower asymptote = $N_{max} - N_0$;
$M$ = time at which the growth rate is maximal;
$B$ = related to the growth rate.

The parameters from the Gompertz model can be reparameterised to the classic primary growth curve characteristics:

$N_0 = A$;
$N_{max} = A + D$;
$\mu_{max} = B \cdot D / \exp(1)$;
$\lambda = M - 1/B$.

In line with the Gompertz model, a logistic sigmoidal curve was proposed by Gibson et al. (1987) to predict microbial growth.

$$\log N(t) = \frac{D}{A + \exp[-B(t-M)]} \qquad (9.4)$$

The fitting results are similar to the Gompertz function, but the logistic model is a symmetrical model, while most growth curves are not. Therefore, the Gompertz model was preferred over the logistic model.

In the early nineties, the focal point moved from the static primary models to the dynamic primary models. Van Impe et al. (1992) suggested a dynamic first-order differential equation to predict both microbial growth and inactivation, with respect to both time and temperature. This was one of the first models developed that was able to deal with time-varying temperatures over the whole temperature range of growth and inactivation. The previous history of the product can also be taken into account. In the special case of constant temperature, the model behaved exactly like the corresponding Gompertz model.

A second dynamic growth model was developed by Baranyi and Roberts (1994). It was based on a first-order ordinary differential equation:

$$\frac{dx}{dt} = \mu(x)x \qquad (9.5)$$

where $x(t)$ is the cell concentration at time $t$ and $\mu(x)$ is the specific growth rate. If $\mu(x) = \mu_{max}$ = constant, then the equation describes the pure exponential growth.

This first-order differential equation was extended with two adjustment functions: an adaptation function $\alpha(t)$ describing the smooth transition from the inoculum level to the exponential growth phase, and an inhibition function $u(x)$ describing the transition from the exponential growth phase to the stationary phase. Therefore, the structure for the growth model of Baranyi and Roberts is:

$$\frac{dx}{dt} = \mu_{max}\, \alpha(t)\, u(x)x \qquad (9.6)$$

**The model of Baranyi and Roberts**

The basic model of Baranyi and Roberts is given by the equation 9.6, but to apply the model, the adjustments functions $a(t)$ and $u(x)$ have to be defined. The adaptation function is based on an additional parameter $q(t)$, representing the physiological state of the cells introduced. It creates an adaptation function that describes a capacity-type quantity expressing the proportion of the potential specific growth rate (which is totally determined by the actual environment) that is utilised by the cells. The process of adjustment (which is the lag period) is characterised by a gradual increase in $\alpha(t)$ from a low value towards 1.

$$\alpha(t) = \frac{q(t)}{1 + q(t)} \qquad (9.7)$$

The inhibition function $u(x)$ can be based on nutrient depletion, which results in a Monod model (eq. 9.8). But nutrient depletion is only limiting at high cell concentrations and sometimes growth inhibition starts much earlier. Therefore, a simple inhibition function was created, based on a maximum cell density parameter $x_{max}$ and a curvature parameter m, characterising the transition of the growth curve to the stationary phase (eq. 9.9).

$$u(x) = \frac{S}{K_s + S} \qquad (9.8)$$

$$u(x) \approx 1 - \left(\frac{x(t)}{x_{max}}\right)^m \qquad (9.9)$$

For a specific set of conditions, the differential form of the Baranyi model can be solved, resulting in an explicit, deterministic and static model, reparameterised to the classic growth curve parameters (Poschet and Van Impe, 1999):

$$N(t) = N_0 + \mu_{max}A(t) - \frac{1}{m} \ln\left(1 + \frac{e^{m \mu_{max}A(t)-1}}{e^{m N_{max} - N_0}}\right)$$

$$A(t) = t + \frac{1}{\mu_{max}} \ln\left(\frac{e^{-\mu_{max}t} + (e^{\lambda \mu_{max}-1})^{-1}}{1 + (e^{\lambda \mu_{max}-1})^{-1}}\right) \quad (9.10)$$

Nowadays, the growth model of Baranyi and Roberts is still one of the most widely used primary models. Many researchers have used the Baranyi model in specific microbial modelling applications and found in comparison to the Gompertz function, and other models, that it gives satisfactory results (McDonald and Sun, 1999).

This success can be partly attributed to the program Microfit, with which researchers and producers can fit the Baranyi model easily to their datasets. The program is freeware that can be downloaded from the Internet. More info, and the program can be found at www.ifrn.bbsrc.ac.uk/microfit.

*Secondary models*

Secondary models model the effect of intrinsic (pH, $a_w$, etc.) and extrinsic (temperature, atmosphere composition, etc.) factors on the growth of a microorganism within a food matrix. Several modelling approaches have been designed ranging from complete empirical models to more mechanistic-oriented attempts with a lot of variations in model complexity. Four groups will be highlighted here: (1) the Arrhenius model and its modifications, (2) the Belehradek models or square root models, (3) the cardinal value model, and (4) the response surface models or polynomial models.

The *Arrhenius models* are based on the empirical expression that relates the growth rate to the environmental temperature. The basis for this expression was taken from the work of van 't Hoff and Arrhenius from the end of the 19th century, examining the thermodynamics of chemical reactions.

$$k = A\, e^{\frac{-E_a}{RT}} \quad (9.11)$$

where,
$k$ = growth rate ($h^{-1}$);
$A$ = "collision factor" ($h^{-1}$);
$E_a$ = "activation energy" of the reaction system ($kJ.mol^{-1}$);
$R$ = universal gas constant (= 8,31 $kJ.mol^{-1}.K^{-1}$);
$T$ = absolute temperature (K).

The Arrhenius equation was extended by Davey (1989). They modelled the growth rate, here represented by k, as a function of temperature and water activity. The factors $C_0$ to $C_4$ are the parameters to be modelled. When water activity is not a limiting value, the last terms can be removed.

$$\ln k = C_0 + \frac{C_1}{T} + \frac{C_2}{T^2} + C_3 a_w + C_4 a_w \tag{9.12}$$

A second group of secondary models consists of the family of *Belehradek models* or *square root models*. The first attempt was made by Ratkowsky et al. (1982) who suggested an easy liner relationship between the square root of the growth rate constant and temperature. The equation was only to be applied to the low temperature region.

$$\sqrt{k} = b \cdot (T - T_{min}) \tag{9.13}$$

$T_{min}$ is an estimated extrapolation of the regression line derived from a plot of $\sqrt{k}$ as a function of temperature. By definition, the growth rate at this point is zero. It should be stressed that the $T_{min}$ is usually 2-3°C lower than the temperature at which growth is usually observed, so $T_{min}$ has to be interpreted as a conceptual temperature (McMeekin et al., 1993). This basic square root model has been adjusted several times by several authors. Some of them will be summarily highlighted here. A first adaptation was made by Ratkowsky et al. (1983), enlarging the scope to the whole temperature range.

$$\sqrt{k} = b(T - T_{min})\{1 - \exp[c(T - T_{max})]\} \tag{9.14}$$

Other adaptations aimed to include growth factors other than temperature. McMeekin et al. (1987) added an $a_w$ parameter to the basic model, while Adams et al. (1991) included a pH term. McMeekin et al. (1992) suggested that the two additional factors could be included at the same time. Later on, Devlieghere et al. (1999) replaced the pH terms with terms of dissolved $CO_2$ to describe the behaviour of spoilage flora in modified atmosphere packaged meat products. The main form as described by McMeekin et al. (1992) is mentioned here.

$$\sqrt{k} = b\sqrt{(a_w - a_{w_{min}})} \cdot \sqrt{(pH - pH_{min})} \cdot (T - T_{min}) \tag{9.15}$$

A third group of secondary models is formed by the *cardinal models*. The first model was developed by Rosso et al. (1995). It describes the effect of temperature and pH on the growth rate of a microorganism, based on the cardinal values: the optimal, the minimal and the maximal temperature and pH at which growth is possible.

$$\mu_{max}(T, pH) = \mu_{opt} \cdot \tau(T) \cdot \rho(pH) \tag{9.16}$$

where,

$\tau(T) = \begin{cases} 0 & T < T_{min}, \\ \dfrac{(T - T_{max})(T - T_{min})^2}{(T_{opt} - T_{min})[(T_{opt} - T_{min})(T - T_{opt}) - (T_{opt} - T_{max})(T_{opt} + T_{min} - 2T)]} & T_{min} < T < T_{max}, \\ 0 & T > T_{max}, \end{cases}$

$$\rho(pH) = \begin{cases} 0 & pH < pH_{min} \\ \dfrac{(pH - pH_{min})(pH - pH_{max})}{(pH - pH_{min})(pH - pH_{max}) - (pH - pH_{opt})^2} & pH_{min} < pH < pH_{max} \\ 0 & pH > pH_{max} \end{cases}$$

All used parameters within this model do have a biological interpretation, although the model structure itself is a pure empirical relationship.

A forth group of secondary models consists of the *response surface models* or *polynomial models*. Those are pure empirical black-box models, where fitting capacities are the most beneficial properties, and the model structure is simple and easy to fit. Classically, a second order polynomial function is used, so the model will consist of thee groups of terms: the first order terms, the second order terms, and the interaction terms. Here an example is given of a second order model describing the growth rate as a function of temperature, pH, $a_w$ and lactic acid concentration (Devlieghere et al., 2000)

$$\mu = \begin{aligned} & I_\mu + m_1.T + m_2.a_w + m_3.NaL + m_4.CO_2 \\ & + m_5.T^2 + m_6.aw^2 + m_7.NaL^2 + m_8.CO_2^2 \\ & + m_9.T.a_w + m_{10}.T.NaL + m_{11}.T.CO_2 + m_{12}.a_w.NaL \\ & + m_{13}.a_w.CO_2 + m_{14}.CO_2.NaL \end{aligned} \qquad (9.17)$$

When working with polynomial models, a lot of parameters have to be estimated from the data, and the amount rises exponentially when more environmental factors are added to the model. Not all of these parameters are statistically relevant, so often a backward regression is executed to eliminate the non-relevant parameters.

Sometimes, these models are sneeringly called *"bulldozer models"* because of the high computational capacities that are needed to estimate the large number of model parameters and the total absence of a mechanistic approach. Although less elegant, the polynomial models can result in very good fitting capacities and are nevertheless very often used in practice.

*Tertiary models*
Tertiary models are software packages in which the previously described models are integrated into a ready-to-use application tool. As such, the discipline of predictive modelling can find its way to the food industry.

The most well-known tertiary model is the PMP; the pathogen modelling program (Figure 9.3), which was developed by the USDA in the nineties. It describes the behaviour of several pathogenic bacteria as a function of the environmental conditions. The model contains growth curves, inactivation curves, survival, cooling and irradiance models. It also makes predictions for the time to turbidity or the time to toxin production of some pathogens. The model is regularly

## Modelling food safety

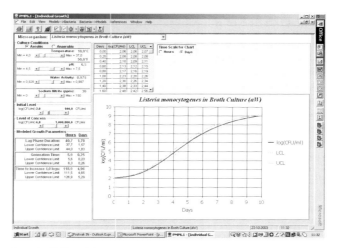

Figure 9.3. Print screen from the PMP model.

updated to include current modelling knowledge. The PMP can be downloaded free from the USDA website (www.arserrc.gov/mfs/pathogen.htm).

Another well-known program is the Seafood Spoilage Predictor (SSP), which focuses on the spoilage of marine, fresh fish, as a function of temperature and gas atmosphere (Figure 9.4). A realistic time-temperature combination can be defined and the subsequent microbial growth will be predicted, together with the remaining shelf life. The program can be downloaded free from the Internet (www.dfu.dk/micro/ssp).

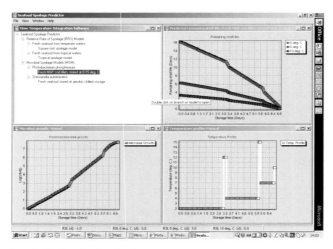

Figure 9.4. Printscreen from the Seafood Spoilage Predictor.

### 9.2.3 Modelling the inactivation of microorganisms

The modelling of inactivation, more specifically modelling of thermal inactivation, was one of the first applications of the discipline that was later called predictive microbiology. Apart from thermal inactivation, non-thermal inactivation can also be modelled. Chemical inhibition is one of the recent research topics within the area of predictive modelling. Other non-thermal inhibition factors can be high pressure or irradiation.

*Modelling thermal inactivation*
Classically, the D & Z values are used to model the thermal resistance of a microorganism. When the cell count is pictured as a semi-logarithmic graph with the logarithm of the cell count on the Y-axis and time on the X-axis (Figure 9.5), the thermal reduction at a certain temperature has a linear relationship. The D-value is then defined as the time needed to obtain one decimal reduction of the cell count.

The linear relationship between the cell count and the heating time is an ideal situation. In practice, there can be some shoulder or tailing effects, especially when dealing with low heating temperatures. The shoulder effects can be explained by a limited heat stability of the bacteria: they can survive the heat treatment for a short period, but after a certain time, a linear decrease in cell count can be observed. The tailing effects can be explained by a heat stable sub-population within the total population. Most bacteria will not survive the heat treatment, although a small group might have a higher resistance against the temperature stress, causing a tailing effect in the curve. More information about the modelling of these inactivation curves can be found in the next section where the different curve-shapes are discussed (Figure 9.7)

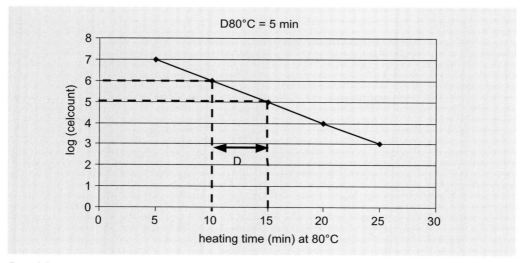

*Figure 9.5. D-value.*

## Modelling food safety

When the D-value is determined at several temperatures, the Z-value can be calculated. When the D-value is depicted as a function of temperature on a semi-logarithmic scale, the Z-value can be defined as the temperature raise that is needed to decrease the D-value by a factor 10 (Figure 9.6).

*Modelling non-thermal inactivation*
The concept of D-values known from thermal heat treatments, assuming a semi-logarithmic survival curve to be linear, has been extended to chemical inactivation. Obviously, the concept of a D-value becomes problematic when experimentally determined semi-logarithmic survival curves are clearly non linear (Peleg and Penchina, 2000) which is often the case for chemical inactivation. Six commonly observed types of survival curves have been distinguished by Xiong *et al.* (1999): log-linear curves (A), log-linear curves with a shoulder (D), loglinear curves with a tailing (C), sigmoidal curves (E), biphasic curves (C) and biphasic curves with a shoulder (F). Two other shapes of chemical inactivation curves have been reported by Peleg and Penchina (2000), those being concave downward (G) and concave upward curves (H). The eight different inactivation curves are shown in Figure 9.7.

Most model development in the field of chemical inactivation has focused on the development of primary models, and to a much lesser extent on secondary model development. In most cases, the influence of environmental factors of the microbial cell on the primary model parameters are described by black box polynomial equations. In many cases the time for a 4D reduction ($t_{4D}$) is described by a polynomial equation as a function of factors, including temperature, pH and concentration of the applied chemical(s). This was, for example, the case for an expanded model for non-thermal inactivation of *Listeria monocytogenes* (Buchanan *et al.*, 1997b). They examined the effect of temperature, salt concentration, nitrite concentration, lactic acid concentration and pH on the inactivation of *Listeria monocytogenes*.

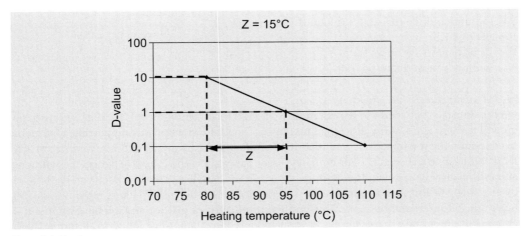

*Figure 9.6. Z-value.*

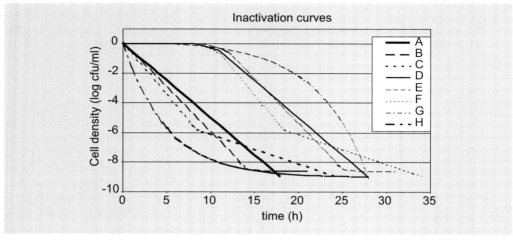

Figure 9.7. Different types of inactivation curves.

$$\begin{aligned}
\text{Ln} (t_{4d}) = \ & 0.0371T + 0.0575S + 3.902P - 1.749L - 0.0547N \\
& - 0.0012TS - 0.00812TP - 0.0131TL + 0.000323TN \\
& + 0.0103SP + 0.00895SL + 0.0001582SN \\
& + 0.1895PL + 0.00356PN + 0.00209LN \\
& - 0.00168T^2 - 0.00749S^2 - 0.3007P^2 \\
& + 0.1705L^2 + 0.0000871N^2 \\
& - 3.666
\end{aligned}$$ (9.19)

where,
$T$ = temperature (4-42 °C)
$S$ = sodium chloride (0.5-19.0% aqueous phase)
$P$ = pH (3.2-7.3)
$L$ = lactic acid (0-2% w/w)
$N$ = sodium nitrite (0-200 µg/ml)

*Modelling the growth/no growth interface*
On the interface between the growth models and the inactivation models there is a growth/no growth twilight zone. Many large food producers choose for non-growth supporting conditions to guarantee the microbial safety of their food product, i.e. they do not allow any growth of any food pathogen in the product. For them it is essential to know the interface between conditions supporting growth and conditions where growth is not possible. This interface can be defined by one single factor, such as temperature (cardinal value), when other factors remain constant, but more often a combination of factors like temperature, pH, $a_w$ and concentrations of a chemical compound is considered. Several models have been developed to describe these multiple factor interfaces.

When modelling the growth/no growth interface, there is a lot of data to be collected near the growth boundaries, as the abruptness of the transition between growth and no growth is often quite steep. Examples can be found in Tienungoon et al. (2000) where a pH decrease of 0.1 to 0.2 units can cause growth cease for Listeria monocytogenes, while Salter et al. (2000) showed that an $a_w$ decrease of 0.001-0.004 forms the growth/no growth boundary of E. coli.

A first attempt to model the growth/no growth interface was made by Ratkowsky and Ross (1995). They transformed the square-root model (*a kinetic model*) from McMeekin et al. (1992), adjusted with a nitrite factor (eq. 9.19), to a logit-based model (*a probability model*) (eq. 9.20).

$$\sqrt{\mu_{max}} = a_1 \cdot (T - T_{min}) \cdot \sqrt{(pH - pH_{min}) \cdot (aw - aw_{min}) \cdot (NO_{2max} - NO_2)} \qquad (9.19)$$

From both sides of the equation, the natural logarithm was taken, and the left side was replaced by the logit (p) = ln (p/(1-p)), where p is the probability that growth occurs. The data set is considered to be binary: p = 1 if growth can be observed for that defined combination of parameters, while p = 0 if no growth is observed. The new equation is:

$$\text{logit (p)} = b_0 + b_1 \ln(T - T_{min}) + b_2 \ln(pH - pH_{min}) \\ + b_3 \ln(a_w - a_{w\,min}) + b_4 \ln(NO_{2\,max} - NO_2). \qquad (9.20)$$

The coefficients $b_0, b_1, b_2, b_3$ and $b_4$ have to be estimated by fitting the model to experimental data. The other parameters ($T_{min}, pH_{min}, aw_{min}$ and $NO_{2\,max}$) should be estimated independently or fixed to constant values, so the coefficients can be estimated using linear logistic regression. Once the b-values are predicted, the position of the interface can be estimated by choosing a fixed p. Often p = 0.5 is preferred, corresponding to a 50:50 chance that the organism will grow under those fixed environmental conditions. Then logit (0.5) = ln (0.5/0.5) = ln (1) = 0.

A second type of model was developed by Masana and Baranyi (2000). They investigated the growth/no growth interface of Brochotrix thermosphacta as a function of pH and water activity. The $a_w$ values were recalculated as $b_w = \sqrt{(1 - a_w)}$. The model could be divided into two parts: a parabolic part (eq. 9.21) and a linear part at a constant NaCl level.

$$pH_{boundary} = a_0 + a_1 \cdot b_w + a_2 \cdot (b_w)^2 \qquad (9.21)$$

A third type of modelling was proposed by Le Marc et al., (2002). The model, describing the L. monocytogenes growth/no growth interface, is based on a four-factor kinetic model estimating the value of $\mu_{max}$. Functions are based on the cardinal models of Rosso et al. (1995).

$$\mu_{max} = \mu_{opt} \cdot \rho(T) \cdot \gamma(pH) \cdot \tau([RCOOH]) \cdot \xi(T, pH, [RCOOH]) \qquad (9.22)$$

The growth/no growth interface was obtained by reducing the $\mu_{Max}$ to zero. This can be done by making the interaction term $\xi(T, pH, (RCOOH))$ equal to zero.

### 9.2.4 Model development procedures

The development of a predictive model in the field of food microbiology is basically performed in four steps (Figure 9.8): (1) planning, (2) data collection and analysis, (3) function fitting, and (4) validation and maintenance (McMeekin *et al.*, 1993). Each step is essential in the model development.

In the planning step, it is essential to know the requirements of the final user of the model. The planning will differ when a model is developed to estimate the direct effect and interactions of independent variables on the growth of a microorganism compared to a model that is developed to predict quantitatively the growth of the microorganism in a specific food product. The decisions to be made in the planning step are summarised in Table 9.1.

Preliminary experiments are often of major importance for achieving the postulated objective with a minimum of time and resources. Reduction of the number of independent variables incorporated in the developed model, for example, can lead to very significant reductions of labour and material costs. Adequate predictive models are indeed difficult to develop when more than four independent variables have to be included. It is therefore necessary to determine the relevant intrinsic and extrinsic parameters for a specific food product and to develop specific food related predictive models based on this knowledge (Devlieghere *et al.*, 2000). The determination of the relevant independent variables can be based on present knowledge combined with preliminary experiments.

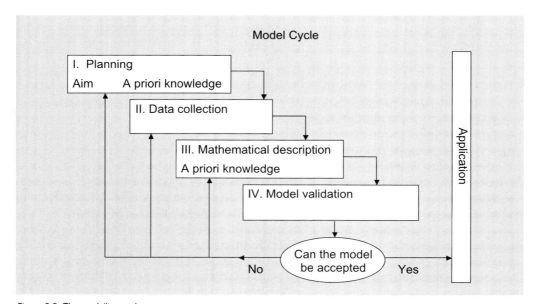

Figure 9.8. The modelling cycle.

## Modelling food safety

Table 9.1. Important questions for the planning step in model development.

- What is the objective?
- What are the most important independent variables for this objective?
- What is the (practical) range of these independent variables?
- What type of inoculum is going to be applied (physiological status, inoculum size, mixture or single strain, way of inoculation)?
- Is a background flora important for the defined objective?
- What dependent variables are going to be determined?
- What substrate is going to be used?
- How much and which combinations of the independent variables will be incorporated in the experimental design?
- How are the data collected?

The development of a model requires the collection of data from experimentation. Growth of microorganisms can be followed by different enumeration methods. *Viable count methods* are generally applied. The advantages of plate counting methods are that they can be made very sensitive (they measure only viable cells), and can be used to determine microbial loads in both real foods and laboratory liquid media (McMeekin *et al.*, 1993). Their chief disadvantage, however, is the amount of labour and materials required to obtain a single determination. A number of workers have therefore chosen to construct their initial models on data obtained from *optical density determinations* in laboratory media. The advantages of this method are speed, simplicity and non-invasiveness, but many disadvantages are reported: for example, limitation to clear media, limited range of linearity between concentration and absorbance/turbidity, low detection limit ($10^6$ cells/ml), low upper limit ($10^{7.5}$ cells/ml) of the linear part, lag determinations on dense populations only(McMeekin *et al.*, 1993) and a higher risk of misinterpretation (McDonald and Sun, 1999).

Optimisation between the quantity of the collected data and the costs involved with data collection is necessary. To fit primary models and subsequent secondary models, data must be collected over the entire growth period and often more than a hundred primary growth curves have to be produced (Mc Donald and Sun, 1999). Bratchell *et al.* (1989) suggested that 15 to 20 observations per growth curve are sufficient, but taking only 10 can lead to big differences between surfaces fitted with data obtained with high and low numbers of observations per growth curve. Gibson *et al.* (1987) stated that up to 10-15 data points are required to obtain growth curves, which can be modelled reliably by sigmoid growth models. Poschet and Van Impe (1999) concluded from their study, in which the influence of data points on the uncertainty of a model was analysed by means of Monte Carlo Analysis, that the uncertainty of the model parameters decreases with an increasing number of data points. However, after a certain number of data points, the uncertainty did not decrease anymore with an increasing number of data points.

The obtained data are fitted to one of the existing primary models expressing the evolution of the cell density as a function of time (i.e. growth curves) to obtain growth parameters (such as maximum specific growth rate, lag phase) for each combination of environmental conditions. In a second step, the obtained combinations of environmental factors with their respective growth parameters are fitted to a secondary model function, as described elsewhere in this chapter. Function fitting is performed by regression analysis that can be linear or non linear, depending on the character of the applied model. The actual process of fitting a function to a data set, i.e. to determine the parameter values that best fit the model to those data, is based on the principle of least squares. This criterion aims to derive parameter values that minimise the sum of the squares of the differences between observed values and those predicted by the fitted model, i.e. the residuals (McMeekin *et al.*, 1993).

Predictive microbiology aims at the quantitative estimation of microbial growth in foods using mathematical modelling. To determine whether predictions provide a good description of growth in foods, models should be validated to evaluate their predictive ability (Te Giffel and Zwietering, 1999). Validation should be performed in media but especially in real food products.

The accuracy of models can be assessed graphically by plotting the observed values against corresponding predictions of a model. Furthermore, mean square error (MSE) and $R^2$ values can also be used as an indication of the reliability of models when applied in foods. Ross (1996) proposed indices for the performance of mathematical models used in predictive microbiology. The objective of those performance indices was to enable the assessment of the reliability of such models when compared to observations *not* used to generate the model, particularly in foods, and hence to evaluate their utility to assist in food safety and quality decisions (Baranyi *et al.*, 1999). A further objective was to provide a simple and quantitative measure of model reliability.

Those indices were termed bias factor ($B_f$) and accuracy factor ($A_f$) and are calculated as follows:

$$B_f = 10^{(\Sigma \log(GT_{predicted}/GT_{observed})/n)} \qquad (9.23)$$

$$A_f = 10^{(\Sigma |\log(GT_{predicted}/GT_{observed})|/n)} \qquad (9.24)$$

The $B_f$ and $A_f$ factor defined by Ross (1996) were refined by basing the calculation of those measures on the mean square differences between predictions and observations (Baranyi *et al.*, 1999). Here, the most general form is given comparing the accuracy factor and the bias factor for a model $f$ and $g$. When comparing a model with the observed values, the prediction values of $g$ are replaced by the observed values.

$$B_{f,g} = \exp\left(\frac{\int_R (\text{Ln } f(x_1,...,x_n) - \text{Ln } g(x_1,...,x_n)) \, dx_1...dx_n}{V(R)}\right) \qquad (9.25)$$

$$A_{f,g} = \exp\left(\sqrt{\frac{\int_R (\text{Ln } f(x_1,...,x_n) - \text{Ln } g(x_1,...,x_n))^2 \, dx_1...dx_n}{V(R)}}\right) \qquad (9.26)$$

## Modelling food safety

where,

$f(x_1,...,x_n)$ = predicted value of the growth rate $\mu$ as a function $f$ determined by $n$ environmental parameters $x$.

$g(x_1,...,x_n)$ = predicted value of the growth rate $\mu$ as a function $g$ determined by $n$ environmental parameters $x$.

$V(R)$ = the volume of the region, $\int_R 1 \, dx_1,...,dx_n$, e.g. if R is a temperature interval $(T_1 - T_2)$, then $V(R) = T_2 - T_1$

### 9.2.5 Current shortcomings in predictive microbiology

*Interactions between microorganisms*
Competition between microorganisms in foods is not considered in most predictive models and very few have been published regarding modelling of microbial interactions in food products. Recently, more attention has been given to incorporate interactions in predictive models, especially describing the effect of lactic acid bacteria on the behaviour of foodborne pathogens.

Vereecken and Van Impe (2002) developed a model for the combined growth and metabolite production of lactic acid bacteria, a microbial group that is often used as a starter culture for fermented foods, like dairy products or fermented meat products. Within the food product, the lactic acid bacteria are used as a protective culture - or biopreservative - against pathogens and spoilage organisms. The model of Vereecken is based on the production of lactic acid by lactic acid bacteria. The transition from the exponential phase to the stationary phase, which is the growth inhibition, could be modelled by the concentration of undissociated lactic acid and the acidification of the environment. Later on, a co-culture model was proposed by Vereecken *et al.* (2003), based on the same theory, describing the mono- and co-culture growth of *Yersinia enterocolitica*, a pathogen, together with *Lactobacillus sakei*, a lactic acid bacterium. The inhibition of the pathogen could be modelled by the lactic acid production of the *Lactobacillus*.

*Structure of foods*
Over the last decade, a large variety of predictive models for microbial growth have been developed. Most of these models do not take into account the variability of microbial growth with respect to space. In homogeneous environments, like broths and fluid foods, this variability does not exist or may be neglected (Vereecken *et al.*, 1998). However, in structured (heterogeneous) foods, growth determinative factors, such as temperature and pH, and thus microbial growth itself may be highly related to the position in the food product. Moreover, due to the solid structure of the food, microorganisms can be forced to grow in colonies, in which competition and interaction effects play a more important role in comparison with fluid products.

Dens and Van Impe (2000) applied a coupled map lattice approach to simulate the spatio-temporal behaviour of a mixed contamination with *Lactobacillus plantarum* and *Escherichia coli* on a food with a very low diffusion coefficient, which is the case in solid foods. They demonstrated pattern formation, making global coexistence of both species possible, which was not the case in a homogeneous liquid medium. Recently, the model was extended (Dens and

Van Impe, 2001) to describe two phenomena: (1) the local evolution of the biomass, and (2) the transfer of biomass trough the medium. In this situation, predictive models based on experiments in broth overestimate the safety of a solid food product.

*Diversity of natural strains*
Natural occurring strains of food pathogens as well as spoilage microorganisms differ in their response to stress conditions. Begot *et al.* (1997) compared the growth parameters of 58 *L. monocytogenes* and 8 *L. innocua* strains at different combinations of pH (5.6 and 7.0), temperature (10 and 37°C) and water activity (0.96 and 1.00). Wide variations between the different strains were observed at the same conditions, especially for the lag phase. In one combination ($a_w = 0.96$; T = 10°C; pH = 7.0), the minimum lag phase was 3.9h while the maximum lag phase amounted to 97.9h. In the same conditions, the maximum generation time was 2.3 times the minimum generation time. Moreover, strains exhibiting the longest generation time of the entire population did not show the maximal lag time. By cluster analysis, it was demonstrated that the group of the fastest growing strains in all conditions was in the majority composed of strains isolated from industrial sites. Such results demonstrate the difficulties in choosing a strain on which to build a predictive model. If the fastest strain is considered, the model will always predict shorter lag time and lower growth rate than those found for the majority of strains, and would thus always provide safe growth estimates. On the other hand, questions could arise about the reality of the predictions made by models developed with the fastest growing strain. The ideal model should also quantify the uncertainty of the predictions, caused by the diversity of natural strains.

*Modelling of the lag phase*
The variability of a population lag phase, i.e. the time needed for a bacterium to adapt to a new environment before it starts multiplying, is an important source of variability in the exposure assessment step when performing a risk assessment concerning *L. monocytogenes*. Research on modelling the behaviour of bacteria in foods has repeatedly shown that the lag phase is more difficult to predict than the specific growth rate.

Pre-adaptation to inimical growth conditions can shorten lag times dramatically and the magnitude of this effect is difficult to predict. Also, the inoculum level has in important impact on the lag phase. It has been observed that the variability of detection times of *L. monocytogenes* increases when a lower inoculum level is applied and, when dealing with constant inoculum levels, an increase in variation can be observed when more salt stress is applied (Robinson *et al.*, 2001). Therefore, when dealing with realistic low inoculum levels of pathogens like, e.g. *L. monocytogenes*, a high variability can be expected in the population lag phase, especially when severe stress conditions are applied. Higher initial densities of bacteria are theoretically associated with a higher likelihood of including at least one cell in the proper physiological state for immediate growth, with only a limited lag time for adjustment (Baranyi, 1998). An individual cell approach is therefore needed to quantify these distributions adequately.

Even when inoculum effects have been minimised, it has still proved difficult to obtain a clear picture of the way lag time varies as a function of the external environment. Several studies have demonstrated a relationship between lag time and growth rate but the general validity of

*Modelling food safety*

this relationship has not been fully explored. A clear understanding about the mechanisms of action of stress conditions on the microbial cell, as well as the cellular adaptation systems, can probably lead to better predictions of the lag phase. Until then, however, the lag phase should not be excluded for the prediction of spoilage or microbial safety of food products, but careful interpretation of the model predictions is necessary.

*Modelling behaviour of spoilage microorganisms*
As the criteria for the shelf life of most food products are set by a combination of criteria for food pathogens and for spoilage microorganisms, mathematical models to predict the behaviour of the major groups of spoilage microorganisms would be useful. The development of comprehensive models for spoilage organisms has not received much attention. The development of mathematical models to predict microbial spoilage of foods is normally very product and/or industry specific which is very limiting. However, in recent years research into predicting spoilage has gained more interest.

*Dealing with unrealistic modelling predictions*
When working with predictions, unrealistic predictions are sometimes made. This is mostly the case for black box models, that contain no mechanistic elements, have unrealistic high $R^2$-values but impossible results as unrealistic curvatures or even negative lag phases or negative growth rates can occur. Geeraerd *et al.* (2004) developed a novel approach for secondary models incorporating microbial knowledge in black box polynomial models. By inserting some a priori microbial knowledge into the black box models, the goodness of fit might be a little less perfect, but the model will be more rigid, and realistic model outputs can be expected.

### 9.2.6 Applications of predictive microbiology

The use of predictive models can quickly provide information and, therefore, it is important to appreciate the real value and usefulness of predictive models. It is, however, important to point out that their applications cannot completely replace microbial analysis of samples or the sound technical experience and judgement of a trained microbiologist (McDonald and Sun, 1999). This can be described by a quote from te Giffel and Zwietering (1999): *a model is a good "discussion partner" giving you good ideas, pointing you in the right direction, but like other discussion partners is not always right*. It has to be considered as one of the tools that decision makers, active in the field of food, have at their disposal to consolidate their decisions. Many applications, which are summarised in Table 9.2, have been proposed for predictive food microbiology.

## 9.3 Modelling of migration from packaging materials

### 9.3.1 Introduction

From the previous chapter on chemical risks in foodstuffs, it can be concluded that an enormous amount of chemicals may induce health problems in humans. The origin of these compounds can be very diverse. Mathematical modelling can sometimes be used to predict the concentration

*Table 9.2. Possible applications of predictive food microbiology.*

| |
|---|
| **Hazard Analysis Critical Control Point (HACCP)** |
| ■ preliminary hazard analysis |
| ■ identification and establishment of critical control point |
| ■ corrective actions |
| ■ assessment of importance of interaction between variables |
| **Risk assessment** |
| ■ estimation of changes in microbial numbers in a production chain |
| ■ assessment of exposure to a particular pathogen |
| **Microbial shelf life studies** |
| ■ prediction of the growth of specific food spoilers |
| ■ prediction of growth of specific food pathogens |
| **Product research and development** |
| ■ effect of altering product composition on food safety and spoilage |
| ■ effect of processing on food safety and spoilage |
| ■ evaluation of effect of out-of specification circumstances |
| **Temperature function integration and hygiene regulatory activity** |
| ■ consequences of temperature in the cold chain for safety and spoilage |
| **Education** |
| ■ education of technical and especially non-technical people |
| **Design of experiments** |
| ■ number of samples to be prepared |
| ■ defining the interval between sampling |

of these chemicals in food. For compounds which are produced in the food during e.g. processing and for compounds which migrate from a food contact material, this technique can be particularly fruitful. For the former type of compounds, classical chemical kinetics can be applied. In this contribution, the focus is on the migration from polymeric food contact materials. This migration phenomenon can be modelled mathematically since the physical processes which govern this process are very well studied and understood. Therefore, initially some of these fundamentals will be discussed in more detail.

### 9.3.2 Physical processes which govern the migration process.

As already mentioned in the chapter on chemical risks, migration from food contact materials should be considered as a sub-microscopic mass transfer from the polymeric food contact material to the food. Three sub-processes should be distinguished in the whole migration process. First of all, a low molecular weight compound will diffuse in the polymer in the direction of the food due to the presence of a concentration gradient (diffusion process). Subsequently, reaching the food-plastic interface, the migrant will be desorbed by the polymer and absorbed by the food (sorption process). Finally the migrant, currently dissolved in the food, will diffuse into the total food matrix, again due to the presence of a concentration gradient.

Alternatively, the latter mass transfer can be accelerated due to a convection process inside the food matrix as will be discussed in more detail later.

So in conclusion, the migration process is governed by diffusion and sorption, respectively. The sorption and diffusion process can be described quantitatively by using the partition coefficient $K_{P/F}$ and the diffusion coefficients $D_P$ and $D_F$, where the indexes P and F refer to the polymer or plastic and the food respectively.

*Partition coefficient and the sorption process*
The partition coefficient of the component 'a' between a polymer and a food can be defined as follows

$$K_{P/F(a)} = \frac{C_{P,a,\infty}}{C_{F,a,\infty}} \tag{9.27}$$

where $C_{P,a,\infty}$ and $C_{F,a,\infty}$ are respectively the equilibrium concentration of the component 'a' in the polymer and the food. Basically this definition is derived from the assumption that in equilibrium conditions (time $t=\infty$) the chemical potential of the migrating substance in the polymer, $\mu_{P,a}$, is equal to the chemical potential of the same compound in the food, $\mu_{F,a}$.

The partition coefficient depends mainly on the polarity of the substance and on the polarity of the two phases involved. The following simple example illustrates the importance of the partition coefficient.

At equilibrium conditions, the amount of migrated substance 'a' into the food, $m_{F,a,\infty}$ can be calculated as follows. Supposing that initially no migrating substance was present in the food ($m_{F,a,0}=0$) and that $m_{P,a,0}$ represents the initial amount of migrant present in the polymer, then it can be concluded from the mass balance that

$$m_{P,a,0} = m_{F,a,0} + m_{P,a,\infty} \tag{9.28}$$

From equation (9.27) however

$$m_{P,a,0} = \frac{m_{F,a,0}}{V_F} \times K_{P/F(a)} \times V_P \tag{9.29}$$

where $V_P$ and $V_F$ are respectively the volume of the polymer and the food.
Consequently,

$$m_{F,a,0} = \frac{m_{P,a,0}}{1 + K_{P/F(a)} \times \frac{V_P}{V_F}} \tag{9.30}$$

In general, it can be assumed that $V_P/V_F \ll 1$. From equation (9.30) it can be concluded that in the case $K_{P/F(a)} \ll 1$, the amount of migrated substance in the food in equilibrium conditions ($m_{F,a,\infty}$) equals the initial amount of substance present in the polymer ($m_{P,a,0}$), which implies that total migration of the substance out of the polymer occurred. If on the contrary, a hydrophobic substance is applied in a hydrophobic polymer contacted with water, it is clear that $K_{P/F(a)} > 1$ and from equation (9.30), it can be concluded that

$$m_{F,a,\infty} << m_{P,a,\infty} \qquad (9.31)$$

indicating that migration remains restricted. Since most polymers used are (fairly) hydrophobic and most of the low molecular weight compounds present in a plastic are hydrophobic as well, it follows from the above example that migration will be especially important in apolar food matrices, like fatty foods. Of course, polar migrants, such as the antistatic polyethylene glycol, will preferentially migrate to more polar, so-called aqueous foods.

Partition coefficients can be determined experimentally, but this approach is often very tedious and prone to experimental errors. Therefore, empirical methods were developed to estimate the partition coefficients for given polymer-migrant food systems. A detailed discussion of these methods falls out of the scope of this chapter.

*Diffusion coefficient and the diffusion process*
The diffusion coefficient $D_a$ of a compound 'a' in a particular matrix follows from Fick's first law, stating that the mass flux of a compound a in the direction x, $J_{x,a}$, during a time 't' through a unit area is proportional to the gradient of the concentration of the compound, $C_a$, considered in the x direction. Mathematically, this gives the following

$$J_{x,a} = -D_a \times \frac{\partial C_a}{\partial x} \qquad (9.32a)$$

Of course, fluxes in the other directions can be defined similarly

$$J_{y,a} = -D_a \times \frac{\partial C_a}{\partial y} \qquad (9.32b)$$

$$J_{z,a} = -D_a \times \frac{\partial C_a}{\partial z} \qquad (9.32c)$$

Due to this flux however, the concentration in a unit cell of the matrix in which the diffusion is taking place, with dimensions $\Delta x \Delta y \Delta z$, will vary accordingly as a function of the time (Figure 9.9).

The net change of the concentration of the compound 'a' in this unit cell as a function of time can be found from

$$-\frac{\partial C_a}{\partial t} = \frac{[J_{x,a}(x+\Delta x) - J_{x,a}(x)]\Delta y \Delta z}{\Delta x \Delta y \Delta z} + \frac{[J_{y,a}(y+\Delta y) - J_{y,a}(y)]\Delta x \Delta z}{\Delta x \Delta y \Delta z} + \frac{[J_{z,a}(x+\Delta z) - J_{z,a}(z)]\Delta x \Delta y}{\Delta x \Delta y \Delta z} \qquad (9.33)$$

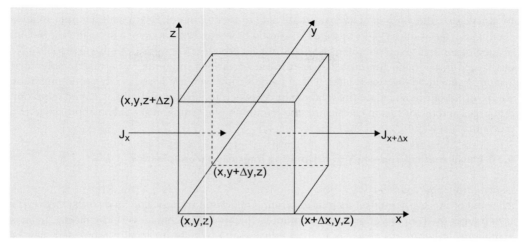

Figure 9.9. Diffusion through an elementary volume.

If the unit cell becomes infinitely small, then equation (9.33) becomes

$$-\frac{\partial C_a}{\partial t} = \frac{\partial J_{x,a}}{\partial x} + \frac{\partial J_{y,a}}{\partial y} + \frac{\partial J_{z,a}}{\partial z} \tag{9.34a}$$

or if the diffusion coefficient is constant

$$\frac{\partial C_a}{\partial t} = D_a \left( \frac{\partial^2 C_a}{\partial x^2} + \frac{\partial^2 C_a}{\partial y^2} + \frac{\partial^2 C_a}{\partial z^2} \right) \tag{9.34b}$$

Equation (9.34a) is known as the general Fick's second law of diffusion.

Various models which are able to predict the diffusion coefficient for a given thermoplastic polymer-substance system have been developed over the years. It is not our intention to review those models in detail in this chapter. It should be emphasised, however, that the diffusion mechanisms in thermoplasts below and above the glass transition temperature of the polymer, are totally different. Basically three different cases can be considered depending on the diffusion rate of the migrant and the relaxation rate of the polymer. Case I diffusion, or Fickian diffusion, occurs when the diffusion rate is much less then the relaxation rate of the polymer. Case II diffusion is characterised by a rapid diffusion in comparison with the relaxation of the polymer. The relaxation rate of the polymer is related to the time the polymer needs to adjust itself to a new equilibrium. Finally, case III or anomalous diffusion occurs when both the diffusion and relaxation rates are comparable. Rubbery polymers correspond quickly to changes in their physical condition, consequently the diffusion of low molecular weight compounds is considered to be Fickian. Due to the limited mobility of the polymer below its glass transition temperature, diffusion obeys mostly case II or III in these circumstances. Therefore, estimation of diffusion coefficients in glassy polymers is even more difficult compared to rubbery polymers.

In most cases, the various models used are able to predict the diffusion coefficient of relative small molecules such as gases or water. Currently however, estimation of a diffusion coefficient of a migration in a given polymer system, is not yet possible. Therefore, empirical models have been used to estimate the diffusion coefficient of various migrants in particular polymeric systems, as will be discussed later. These models, however, are based on dynamic migration studies. In order to elucidate the relationship between the experimentally observed migration dynamics and the diffusion coefficient, the migration phenomenon should be approached in a mathematical way.

### 9.3.3 Mathematical approach to estimating migration from plastics

*General transport equation*
The goal of the mathematical models describing migration from plastic food contact materials is to predict the concentration of the migrant in the food after contact with the plastic. In such a manner, lengthy and costly migration experiments which are legally requested, can be avoided. In addition, from this mathematical approach, relevant parameters controlling the migration process can be identified. Knowledge of these parameters is of prime importance in order to allow realistic simulations of the migration phenomenon in the laboratory and to interpret the obtained results in a correct manner.

A reliable model should consider all mass transfer phenomena and other processes affecting the concentration of the migrant in the food. Basically, the following processes are taken into account:
- diffusion of the migrant;
- convection of the medium in which the migrant is dissolved;
- chemical reactions in which the migrant is involved.

It is important to realise that from a theoretical point of view all these processes can take place in both the food and the polymer. Practically, however, the following processes mainly control the migration behaviour:
- diffusion of the migrant in both the polymer and the food;
- chemical reaction in both the polymer and the food.

Convection of the polymer is very much restricted in normal conditions of use, so will be of no influence with regard to the migration. In liquid foods, convection will cause a quick distribution of the migrant in the food, favouring a uniform concentration of the migrant. In solid foods or highly viscous foods, diffusion of the migrant will be of greater importance compared to convection fluxes of the food itself.

The chemical reactions that a migrant is subjected to, are also an interesting feature. An antioxidant present in the polymer undergoes partial degradation during the production of the plastic contact material, lowering the potent migration of the compound considered. In the food itself, migrants may undergo reactions as well. Bisphenol A diglycidyl ether (BADGE), a cross-linking agent used in epoxy coatings for food cans, is an important example.

Mathematically, the predominant processes affecting the concentration C of the migrant at a particular place with coordinates (x,y,z) in the food-polymer system, can be written as follows:
For diffusion

$$-\frac{\partial C}{\partial t} = \frac{\partial J_x}{\partial x} + \frac{\partial J_y}{\partial y} + \frac{\partial J}{\partial z} \qquad (9.34c)$$

For the chemical reaction

$$\frac{\partial C}{\partial t} = -k \times C^n \qquad (9.35)$$

Equation (9.34c) represents the previously introduced second diffusion law of Fick, and in equation (9.35), n represents the order of the chemical reaction and k is the reaction rate constant.

Combining the two gives the general transport equation in which the convection in both the polymer and the food is supposed to a have minor influence

$$\frac{\partial C}{\partial t} = \left( \frac{\partial^2 DC}{\partial x^2} + \frac{\partial^2 DC}{\partial y^2} + \frac{\partial^2 DC}{\partial z^2} \right) - k \times C^n \qquad (9.36)$$

Of course, the main problem in solving this equation is situated in the second order partial differential equation. Such an equation only has an analytical solution in some special cases. In addition, the diffusion coefficient should be constant. In all other cases, numerical methods should be used in order to solve the equation. Once this equation is solved, however, the problem of a reliable estimate of the diffusion coefficient of the migrant remains. As discussed previously, mechanistic and atomistic diffusion models are currently unable to give reliable predictions for diffusion coefficient of migrants in plastics.

The second part of this equation is rather specific for particular migrants and will not be discussed here in more detail. Therefore, only solutions to the second order differential equation will be presented.

As indicated, a number of assumptions should be made in order to analytically solve the second order partial differential equation given in equation (9.36). Primarily, the diffusion coefficient is supposed to be constant in both the food and the polymer. In addition, it is assumed that diffusion takes place in only one direction, perpendicular to the surface of the polymer. Consequently, the partial differential equation becomes

$$\frac{\partial C}{\partial t} = D \frac{\partial^2 C}{\partial x^2} \qquad (9.37)$$

Solutions to this equation for finite and infinite polymers will be presented in the following section.

*Diffusion from finite polymer*
Further basis assumptions, in addition to those mentioned above, include the following:

- there is one single migrant, which is uniformly distributed in the polymer at t = 0 at a concentration $C_{P,0}$;
- the concentration of the migrant in the food at a particular time, $C_{F,t}$ is everywhere the same;
- a constant distribution of the migrant between the polymer and the food takes place according to

$$K_{P/F} = \frac{C_{P,t}}{C_{F,t}} = \frac{C_{P,\infty}}{C_{F,\infty}} \tag{9.38}$$

- the contact material is a flat sheet;
- the mass transfer is mainly controlled by diffusion taking place in the polymer.

The following solution to equation (9.37) was developed for a polymer in contact with a finite food

$$\frac{m_{F,t}}{m_{F,\infty}} = 1 - \sum_{n=0}^{\infty} \frac{2\alpha(1+\alpha)}{1+\alpha+\alpha^2 q_n^2} \times e^{\left(-\frac{D_P q_n^2 t}{L_P^2}\right)} \tag{9.39}$$

in which $m_{F,t}$ is the amount of migrant in the food at a particular time, t, $q_n$ is the positive root of the trigonometric identity $tg(q_n) = -\alpha q_n$, $L_P$ is the thickness of the polymer, $D_P$ the diffusion coefficient of the migrant in the polymer, and $\alpha$ is given by the following formula

$$\alpha = \frac{1}{K_{P/F}} \times \frac{V_F}{V_P} = \frac{m_{F,\infty}}{m_{P,\infty}} \tag{9.40}$$

in which $V_F$ and $V_P$ represent the volume of the food and the polymer, respectively, and $m_{P,\infty}$ is the mass of the migrant present in the polymer at equilibrium conditions ($t = \infty$).

This rather complex equation (9.38) can be simplified by assuming the finite polymer is in contact with an infinite food. This implies that the concentration of the migrant in the food equals zero, since mathematically speaking, $V_F \to \infty$. Consequently, from equation (9.40) it follows that $\alpha >> 1$. Equation (13) can be simplified into:

$$\frac{m_{F,t}}{m_{F,\infty}} = 1 - \sum_{n=0}^{\infty} \frac{8}{(2n+1)^2 \pi^2} \times e^{\left(-\frac{(2n+1)^2 \pi^2}{4L_P^2} D_P t\right)} \tag{9.41}$$

Equation (9.41) is reported to give the same results as equation (9.39) if the volume of the food ($V_F$) exceeds 20 times the volume of the polymer ($V_P$), which is usually achieved in practice, as well as in migration tests.

Equation (15) can further be simplified for the following two cases:
- long contact time ($m_{F,t}/m_{F,\infty} > 0.6$)

$$\frac{m_{F,t}}{m_{F,\infty}} = 1 - \frac{8}{\pi^2} e^{\left(-\frac{\pi^2}{L_P^2} D_P t\right)} \tag{9.42}$$

## Modelling food safety

- short contact time ($m_{F,t}/m_{F,\infty} < 0.6$)

$$\frac{m_{F,t}}{m_{F,\infty}} = \frac{2}{L_P}\sqrt{\frac{D_P t}{\pi}} \qquad (9.43)$$

For all these models, it was assumed that diffusion in the polymer is the main factor controlling the migration phenomenon. If other processes, such as the dissolution and the diffusion of the migrant in the food, are also important factors to consider, analytical solutions of the diffusion equation (9.37) are not available. Numerical methods for some cases have been described.

A further simplification of the problem, assuming the polymer is infinite, allows for the dissolution and the diffusion of the migrant in the food as explained in the following.

### Diffusion from infinite polymer

The assumption of infinite polymer implies that the concentration of the migrant in the polymer is constant as a function of time ($C_{P,0} = C_{P,t}$). Of course, this does not correspond to reality since it is known that the concentration of the migrant in the polymer is affected by migration. Again, several solutions of equation (9.35) have been proposed for a number of cases, taking into account the following supplementary boundary conditions:

- there is one single migrant, which is uniformly distributed in the polymer at t = 0 at a concentration $C_{P,0}$;
- a constant distribution between the polymer and the food takes place according to

$$K_{P/F} = \frac{C_{P,t}}{C_{F,t}} = \frac{C_{P,\infty}}{C_{F,\infty}}$$

- the contact material is a flat sheet.

Two major cases can be distinguished, depending on the concentration gradient of the migrant in the food.

### No concentration gradient of the migrant in the food

If no concentration gradient is present in the food, this implies that the food is well mixed or that the diffusion of the migrant in the food proceeds much faster compared to the diffusion in the polymer. The general solution of equation (11) is given by:

$$m_{F,t} = \frac{C_{P,0} \times A}{K_{P/F}}\left(1 - e^{z^2}\,\text{erfc}(z)\right) \qquad (9.44)$$

in which A is the contact surface between the polymer and the food, and z is given by

$$z = \frac{K_{P/F} \times \sqrt{D_P t}}{A} \qquad (9.45)$$

This equation is valid for infinite polymers contacted with finite foods, indicating that the migrant slowly dissolves in the food a. Consequently, diffusion is mainly governed by solvatation. If the migrant is very soluble in the food, however, ($K_{P/F} << 1$), equation (9.38) can be simplified into:

$$\frac{m_{F,t}}{A} = 2C_{P,0}\sqrt{\frac{D_P t}{\pi}} \qquad (9.46)$$

Equation (20) represents the migration from an infinite polymer in contact with an infinite food ($C_F = 0$). In this case, diffusion of the migrant in the polymer will dominate the migration process.

*Concentration gradient of the migrant in the food*
If a concentration gradient of the migrant is present in the food, the following equation has been proposed as a solution to equation (9.37)

$$m_{F,t} = 2C_{P,0}\sqrt{\frac{D_P t}{\pi}}\left(\frac{\beta}{\beta+1}\right) \qquad (9.47)$$

in which

$$\beta = \frac{1}{K_{P/F}}\sqrt{\frac{D_F}{D_P}} \qquad (9.48)$$

As can be seen, diffusion of the migrant in both the food and the polymer are taken into account. If in this case, diffusion in the food is high ($\beta > 1$), equation (9.47) is turned into equation (9.46), indicating that due to the high diffusion in the food, the concentration gradient of the migrant in the food is negligible.

If on the other hand, $\beta << 1$, because of the poor solubility of the migrant in the food, migration will be especially dominated by the migration in the food as indicated in the following equation derived from equation (9.46)

$$m_{F,t} = \frac{2C_{P,0}}{K_{P/F}}\sqrt{\frac{D_F t}{\pi}} \qquad (9.49)$$

### 9.3.4 Estimation of material constants

As can be concluded from all the analytical solutions to the general diffusion equation (9.37), diffusion coefficients and the partition coefficient of the migrant should be known in order to apply these equations practically. As will be explained later, from a regulatory point of view, the "worst case" scenario predicting migration is of primary interest. Therefore, it is most frequently assumed that the solubility of the migrant in the polymer is very high, which implies that $K_{P/F} = 1$, thus avoiding difficulties for the estimation of the partition coefficient for a given migrant-polymer food system.

The main problem of a realistic estimate of the diffusion coefficient remains. Diffusion coefficients of migrants range from about $10^{-7}$ cm$^2$/s down to about $10^{-18}$ cm$^2$/s. From the above equations it can be concluded that this large difference in magnitude will play a major role in the final migration result for most cases. Realistic estimates are therefore considered indispensable because underestimated diffusion coefficients will underestimate migration and overestimates will make the practical use of these migration models impossible.

However, the mechanistic models on diffusion currently available cannot be applied for the estimation of diffusion coefficients of migrants in polymers. Alternatively, empirical formulas such as equation (9.50) and (9.51) can be used.

$$D_P = D_0 e^{\left(\xi\sqrt{M_r} - \psi\frac{\sqrt[3]{M_n}}{T}\right)} \qquad (9.50)$$

in which $D_0$, $\xi$ and $\psi$ are constants related to the activation energy of diffusion, the molecular weight of the migrant, T is the absolute temperature and $M_r$ is the molecular weight of the migrant considered (45).

$$D_P = 10^4 e^{\left(A_P - \frac{\tau}{T} - 0.1351 M_r^{2/3} + 0.003\, M_r - \frac{10454}{T}\right)} \qquad (9.51)$$

in which $A_P$ is a so-called polymeric specific diffusion conductance parameter and $\tau$ is a polymer-specific activation energy parameter. For polyolefins, these polymer- specific parameters are defined in the Practical Guide issued by the European Commission (EC. Food contact materials, a practical guide. http://cpf.jrc.it/webpack/downloads /PRACTICAL%20 GUIDE-010302.pdf).

These empirical equations are based on the link between a migration-dynamics database and the general migration model previously presented (equation 9.39)). Therefore, it can be stated that no real independent validation is made, since the diffusion coefficients obtained, using these empirical equations, are used again in the migration models, from which the diffusion coefficients were derived. Although from a scientific point of view this is a serious drawback, equations (9.50) and (9.51) are believed to be very important from a practical point of view, since currently these are the only means by which to obtain diffusion coefficients in a simple manner.

Reliable diffusion coefficients for migrants having a molecular weight of up to 4000 could be calculated in such a way for selected polyolefins between the melting and glass transition temperature of the polymer.

For non-polyolefins, however, which are characterised by a higher $T_g$ (frequently between 50-100°C), such models are not currently available due to a lack of experimental data. Therefore, no useful diffusion coefficient estimates are available yet for these polymers.

### 9.3.5 Practical use of mathematical models

There is a general consensus about the usefulness of mathematically modelling the migration phenomenon, in order to limit laboratory tests which are tedious and costly. This is reflected by the possibility of using mathematical modelling in order to prove compliance with legislation recently accepted within the EU and long accepted in the US. In addition, extensive information on the use of these models is available in the Practical Guide, issued by the European Commission. In this document, reference is made to two software packages which are available on the market (MIGRATEST Lite 2001, FABES GmbH, Munich, Germany) or which can be downloaded free from the Internet (SMEWISE, INRA, Reims, France). Both programs basically use the same migration model and the same empirical equation to determine the diffusion coefficient of a particular polymer-migrant system. The numerical methods to solve equation (9.39) may be different. Nevertheless, both software packages are reliable if they are properly used by trained personnel, with a more than basic knowledge of the migration phenomenon.

Mathematical models, however, are prone to a number of limitations which are important to consider. As indicated before, an important aspect in the evaluation of the models is the correspondence between the calculated and experimental data. Because of the necessity of a reliable estimate of the diffusion coefficient of the migrant, it can be concluded from the above discussions that only models applicable to polyolefins are currently available. It should be noted, however, that polyolefins are currently the most frequently applied polymers for food contact. On the other hand, in the applied migration models it is assumed that the diffusion coefficient of the migrant is constant. If fatty foods are contacted with polyolefins, however, negative migration of the tri-acylglycerols will occur, resulting in a time-dependent change of the migrants' diffusion coefficient in the polymeric system. Since it was concluded previously that migration to fatty foods is of particular concern, this fact can be considered as a major disadvantage.

The models described are only able to predict the migration of known and well-characterised migrants. Consequently, the models will not be able to predict the total amount of substance migrated from a contact material, since the contact material may contain, apart from the additives, a number of other compounds whose identity is not entirely known (e.g. ethylene oligomers or their breakdown products present in polyethylene).

Care should be taken in using too simplified models such as the one discussed above. The use of a more general equation such as equation (9.39) is considered to be better. Specific migration of various additives from polyolefins to olive oil at various time temperature conditions with the predicted values of this migration model were compared in Figure 9.10. For polypropylene, almost all the estimated values were higher then those experimentally observed. From a safety point of view this is interesting. Results for polyethylene were a bit less promising. In order to obtain a more accurate estimation, it was believed that the use of a more realistic $K_{P/F}$, especially at lower temperatures, would be appropriate.

## 9.4 Risk assessment: a quantitative approach

### 9.4.1 Introduction to risk assessment

A risk can be defined as a function of the probability of an adverse health effect and the severity of that effect, consequential to a hazard in food (Codex Alimentarius, 1999). During a risk assessment an estimate of the risk is obtained. The goal is to estimate the likelihood and

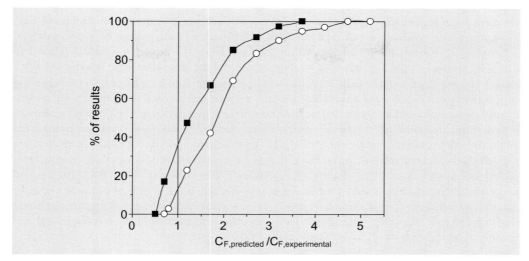

Figure 9.10. Comparison between the predicted and experimental migration levels in olive oil for several additives out of high density PE (○) and PP (■) (Based on the experimental data reported in Adams et al., 1991; Baranyi and Robberts, 1994).

the extent of adverse effects occurring to humans due to possible exposure(s) to hazards. Risk assessment is a scientifically-based process consisting of the following steps: (i) hazard identification, (ii) hazard characterisation, (iii) exposure assessment, (iv) and risk characterisation (Codex Alimentarius, 1999).

During hazard identification, biological, chemical, and physical agents that are capable of causing adverse health effects and which may be present in a particular food or group of foods, are identified (Codex Alimentarius, 1999). In the second step, hazard characterisation, the nature of the adverse health effects associated with the hazards are evaluated in a qualitative and/or quantitative way (Codex Alimentarius, 1999). Therefore, a dose-response assessment should be performed. The dose-response assessment is the determination of the relationship between the magnitude of exposure (dose) to a chemical, biological or physical agent and the severity and/or frequency of associated adverse health effects (response). The overall aim is to estimate the nature, severity and duration of the adverse effects resulting from ingestion of the agent in question (Benford, 2001). Exposure assessment is defined as the qualitative and/or quantitative evaluation of the likely intake of the hazard via food as well as exposure from other sources, if relevant (Codex Alimentarius, 1999). For food, the level ingested will be determined by the levels of the agent in the food and the consumed amount. The last step, risk characterisation, integrates the information collected in the preceding three steps. It interprets the qualitative and quantitative information on the toxicological properties of a chemical according to the extent to which individuals (parts of the population, or the population at large) are exposed to it (Kroes et al., 2002). In other words, estimating how likely it is that harm will be done and how severe the effects will be. The outcome may be referred to as a risk estimate, or the probability of harm at given or expected exposure levels (Benford, 2001).

*Frank Devlieghere, Kjell Francois, Bruno De Meulenaer and Katleen Baert*

Quantitative risk assessment (QRA) is characterised by assigning a numerical value to the risk, in contrast with qualitative risk analysis, that is typified by risk ranking or separation into descriptive categories of risk (Codex Alimentarius, 1999). During QRA a model is used to calculate (estimate) the risk based on the exposure and the dose-response. Besides the QRA model, the exposure and the dose-response can also be described by a model. To calculate the exposure to a microbiological hazard for example, a model can be used that predicts the growth during storage. Several methods can be used to estimate the risk, namely: (i) point estimates or deterministic modelling; (ii) simple distributions, and (iii) probabilistic analysis (Kroes *et al.*, 2002). In a deterministic framework, inputs to the exposure and effect prediction models are single values. In a probabilistic framework, inputs are treated as random variables coming from probability distributions. The outcome is a risk distribution (Verdonck *et al.*, 2001). The method chosen will usually depend on a number of factors, including the degree of required accuracy and the availability of data (Parmar *et al.*, 1997). No single method can meet all the choice criteria that refer to cost, accuracy, time frame, etc. Therefore, the methods have to be selected and combined on a case-by-case basis (Kroes *et al.*, 2002).

## 9.4.2 Deterministic risk assessment

Deterministic modelling (point estimates) uses a single 'best guess' estimate of each variable within the model to determine the model's outcome(s) (Vose, 2000). This method is illustrated with the calculation of the exposure to a contaminant, e.g. the mycotoxin patulin. A fixed value for food consumption (such as the average or maximum consumption) is multiplied by a fixed value for the concentration of the contaminant in that particular food (often the average level or permitted level according to the legislation) to calculate the exposure from one source (food). The intakes from all sources are then totalled to estimate the total exposure.

Point estimates are commonly used as a first step in exposure assessments because they are relatively quick, simple and inexpensive to carry out. Inherent in the point estimates models is the assumption that all individuals consume the specified food(s) at the same level, that the hazard (biological, chemical or physical) is always present in the food (s), and that it is always present at an average or high concentration. As a consequence, this approach does not provide an insight into the range of possible exposures that may occur within a population, or the main factors influencing the results of the assessment. It provides limited information for risk managers and public. The use of this method also tends to significantly over- or under-estimate the actual exposure (Finley and Paustenbach, 1994). Using high-level values to represent either the food consumption or hazard level may lead to high and often implausible overestimates of the intake. Point estimates are generally considered to be most appropriate for screening purposes (Parmar *et al.*, 1997). If they demonstrate that the intake is very low in relation to the accepted safe level for the hazard or below a general threshold value, even when assuming high concentrations in the food and high consumption of this food, it may be sufficient to decide that no further exposure assessments are required (Kroes *et al.*, 2000). In order to refine estimates of exposure, more sophisticated methods for integrating the food consumption and hazard level and more detailed data from industry, monitoring programmes, etc., are needed (Kroes *et al.*, 2002).

## 9.4.3 Probabilistic risk assessment

In probabilistic analysis, the variables are described in terms of distributions instead of point estimates. In this way all possible values for each variable are taken into account and each possible outcome is weighted by the probability of its occurrence.

Probabilistic methods may be used under a classical or a Bayesian statistical paradigm (IEH, 2000). This text will only discuss the classical view (Monte Carlo simulation).

> **Classical and Bayesian probabilistic risk assessment**
>
> In 'classical' (also called 'frequentist') probabilistic risk assessment, probability is regarded as a frequency, i.e. how likely it is, based on repeated sampling, that an event will occur. Monte Carlo simulation is the method most commonly used for classical probabilistic risk assessment (IEH, 2000). It is a computational method developed over 50 years ago for military purposes, and which is used to achieve multiple samples of the input distributions, analogous to doing repeated experiments (IEH, 2000; Vose, 1996). Recently, attention has been paid to the Bayesian probabilistic approach. Bayesian statistical methods are based on Bayes' theorem, which was originally stated in 1763 by Thomas Bayes and was one of the first expressions, in a precise and quantitative form, of one of the modes of inductive inference. According to the Bayesian view, probability is regarded as a 'degree of belief', or how likely it is thought to be, based on judgement as well as such observational data as are available, that an event will occur. Bayesian theory requires a 'prior distribution' which is a probability density function relating to prior knowledge about the event at issue. This is combined with the distribution of a new set of experimental data in a likelihood framework (analogous to a probability distribution) to obtain the so-called 'posterior distribution' of probability, the updated knowledge. Obtaining the posterior distribution can involve solving a complex integral, which is increasingly dealt with by means of Markov chain Monte Carlo methods. Thus, Monte Carlo methods are used both in Bayesian and classical probabilistic methods, although Bayesian analysis extends the use beyond simply simulating predictive distributions (IEH, 2000).

Monte Carlo simulation has been extensively described in literature (Cullen and Frey, 1999; Vose, 1996). The principle is illustrated in Figure 9.11. One random sample from each input distribution is selected, and the set of samples is entered into the deterministic model. The model is then solved, as it would be for any deterministic analysis and the result is stored. This process or iteration is repeated several times until the specified number of iterations is completed. Instead of obtaining a discrete number for model outputs (as in a deterministic simulation) a set of output samples is obtained (Rousseau et al., 2001; Cullen and Frey, 1999). This method is described as a first order Monte Carlo simulation or a one - dimensional Monte Carlo simulation.

The process of setting up and running the models requires appropriate modelling software and a high level of computer processing power. There are a variety of risk analysis software products on the market, which include software for modelling and distribution fitting. Examples of software products are @RISK, Crystal ball and Fare Microbial (free software).

An important advantage of probabilistic risk assessment (PRA) is that it allows consideration of the whole distribution of exposure, from minimum to maximum, with all modes and percentiles. In this way, more meaningful information is provided to risk managers and the

*Frank Devlieghere, Kjell Francois, Bruno De Meulenaer and Katleen Baert*

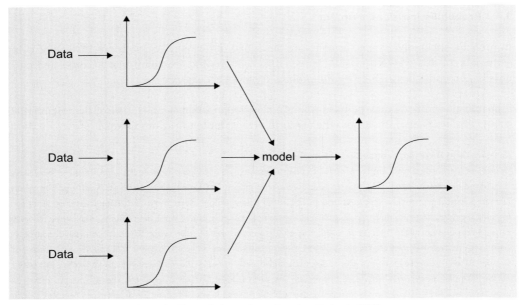

*Figure 9.11. The principle of a first order Monte Carlo simulation (based on Verdonck, 2003).*

public. A second important advantage is the possibility to carry out a sensitivity analysis (9.4.5.) (Finley and Paustenbach, 1994). An important disadvantage of the current PRA procedures is the need for accurate prediction of the tails in a distribution (e.g. the 5- and 95-percentile). These tails are very important in performing a PRA because the largest exposure concentrations (e.g. 95-percentile) will have first effects on the most sensitive population (e.g. 5-percentile) (Verdonck *et al.*, 2001). Also, the high degree of complication and time-consumption are a disadvantage (Finley and Paustenbach, 1994). While probabilistic modelling confers many advantages, Burmaster and Anderson (1994) point out that the old computer maxim of 'garbage in, garbage out' (GIGO) also applies to this technique, and that GIGO must not become 'garbage in, gospel out'. The reliability of the results of a probabilistic analysis is dependent on the validity of the model, the software used and the quality of the model inputs. The quality of the model inputs reflects both the quality of the data on which the input will be based (see frame) and the selection of the distribution to represent the data in the model.

The distributions that are used as a model input can be a distribution function (parametric approach) or the data as such (non-parametric approach). For the parametric approach the data are fitted to a distribution function, like the normal distribution, the gamma distribution, the binomial distribution, etc. The distribution function that gives the best fit is used as a model input. For the non-parametric approach the original data are used as an input of the model.

In the context of exposure assessments, 'simple distributions' is a term used to describe a method that is a combination of the deterministic and probabilistic approach. It employs a

*Modelling food safety*

> **Quality of data**
>
> Different sources of dietary information exist. In this context two methods, namely food supply data and individual dietary surveys, will be discussed.
>
> Food supply data are calculated in food balance sheets, which are accounts, on a national level, of annual production of food, changes in stocks, imports and exports, and agricultural and industrial use. The result is an estimate of the average value per head of the population, irrespective of, for instance, age or gender. Food supply data refer to food availability, which gives only a crude (over-estimated) impression of potential average consumption. When these data are used in a risk assessment, there is high chance that the risk is overestimated (Kroes *et al.*, 2002).
>
> In contrast to food balance sheets, data from individual surveys provide information on average food and nutrient intake and their distribution over various well-defined groups of individuals. These data reflect more closely actual consumption (Kroes *et al.*, 2002). However, it is important to note that individual dietary surveys will be associated with some degree of under-reporting, since being engaged in a dietary survey affects customary consumption. This may lead to some degree of under-estimation of exposure and risk.

distribution for the food consumption, but it uses a fixed value for the level of the hazard. The results are more informative than those of the point estimates, because they take account of the variability that exists in food consumption patterns. Nonetheless, they usually retain several conservative assumptions (e.g. all soft drinks that an individual consumes contain a particular sweetener at the maximum permitted level; 100% of a crop has been treated with a particular pesticide, etc.), and therefore can only usually be considered to give an upper limit estimate of exposure (Kroes *et al.*, 2002).

### 9.4.4 Variability versus uncertainty

In most current probabilistic risk assessments, variability and uncertainty are not treated separately, although they are two different concepts. Variability represents inherent heterogeneity or diversity in a well-characterised population. Fundamentally a property of nature, variability is usually not reducible through further measurement or study (e.g. biological variability of species sensitivity or the variability of the consumed amount of food). As explained in subsection 9.4.2. and 9.4.3., deterministic risk assessment does not take variability into account, while probabilistic risk assessment includes variability in the calculations. In Figure 9.11, the cumulative variability distributions are visualised as a full line (Verdonck *et al.*, 2001).

Uncertainty represents partial ignorance or a lack of perfect information about poorly characterised phenomena or models (e.g. measurement error), and can partly be reduced through further research (Cullen and Frey, 1999). In Figure 9.12. the uncertainty is visualised as a band around the cumulative variability distribution function (dashed lines). For each percentile of the variability distribution, an uncertainty or confidence interval can be calculated (i.e. the uncertainty distribution) (Verdonck *et al.*, 2001). For example, techniques such as bootstrapping can be used to get a measure of the uncertainty associated with parameters of a

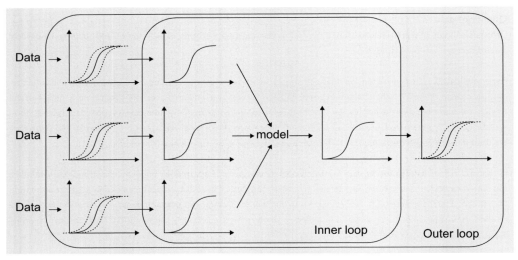

*Figure 9.12. The principle of a second order Monte Carlo simulation (based on Verdonck, 2003).*

distribution due to sample size (Cullen and Frey, 1999). The more limited the data set, the wider the confidence intervals will be.

To separate variability and uncertainty in a PRA, a second order or 2-dimensional Monte Carlo simulation is developed (Burmaster and Wilson, 1996; Cullen and Frey, 1999). It simply consists of two Monte Carlo loops, one nested inside the other (Figure 9.12). The inner deals with the variability of the input variables, while the outer one deals with uncertainty (Verdonck, 2003). Firstly, a distribution is selected randomly from every uncertainty band. Then. this distribution is used in the inner loop and a first order Monte Carlo simulation is carried out. The result is a distribution for the output of the model. This process is repeated several times. In this way a high number of output distributions is obtained and an uncertainty band can be constructed (outer loop).

### 9.4.5 Sensitivity analysis

A sensitivity analysis is performed to determine which variables in the model have the biggest influence on the results. This is achieved by changing the value for each input variable in turn by a fixed amount (e.g. ten percent) while keeping the others constant (IEH, 2000).

The results of sensitivity analysis permit risk managers to consider the relative merits of different strategies for reducing exposure in cases where levels of exposure are deemed unacceptably high. Because the probabilistic analysis provides information on the full distribution of exposures, the exposure assessor can determine how different scenarios will affect different sections of the distribution. For example, different scenarios of modelling nutrient supplementation or fortification can be applied to evaluate the impact at the lower tail of the distribution (possible inadequate nutrient intake) and the upper tail of the distribution (possible nutrient toxicity) (Kroes *et al.*, 2002).

## References

Adams, M.R., C.L. Little and M.C. Easter, 1991. Modelling the effect of pH, acidulant and temperature on the growth rate of *Yersinia enterocolitica*, J. Appl. Bacteriol., 71, 65-71.
Baranyi, J. and T.A. Roberts, 1994. A dynamic approach to predicting bacterial growth in food, Int. J. Food Microbiol., 23, 277-294.
Baranyi, J., 1998. Comparison of statistic and deterministic concepts of bacterial lag, J. Theor. Biol., 192, 403-408.
Baranyi, J., C. Pin and T. Ross, 1999. Validating and comparing predictive models, Int. J. Food Microbiol., 48, 159-166.
Begot, C., I. Lebert and A. Lebert, 1997. Variability of the response of 66 *Listeria monocytogenes* and *Listeria innocua* strains to different growth conditions, Food Microbiol., 14, 403-412.
Benford, D., 2001. Principles of risk assessment of food and drinking water related to human health. ILSI Europe concise monograph series.
Bratchell, N., A.M. Gibson, M. Truman, T.M. Kelly and T.A. Roberts, 1989. Predicting microbial growth: the consequences of quantity of data, Int. J. Food Microbiol., 8, 47-58.
Buchanan, R.L., M.H. Golden and J.G. Phillips, 1997b. Expanded models for the non-thermal inactivation of *Listeria monocytogenes*, J. Appl. Microbiol., 82, 567-577.
Buchanan, R.L., R.C. Whiting and W.C. Damert, 1997[a]. When simple is good enough: a comparison of the Gompertz, Baranyi, and three-phase linear models for fitting bacterial growth curves, Food Microbiol., 14, 313-326.
Burmaster, D.E. and P.D. Anderson, 1994. Principles of good practice for the use of Monte Carlo techniques in human health and ecological risk assessment, Risk analysis, 14, 477-481.
Burmaster, D.E. and A.M. Wilson, 1996. An introduction to second-order random variables in human health risk assessments, Human and Ecological risk assessment, 2, 892-919.
Codex Alimentarius, 1999. Principles and guidelines for the conduct of microbiological risk assessment, CAC/GL-30.
Cullen A.C. and H.C. Frey, 1999. Probabilistic techniques in exposure assessment. A handbook for dealing with variability and uncertainty in models and inputs, Plenum, New York.
Davey, K.R., 1989. A predictive model for combined temperature and water activity on the microbial growth during the growth phase, J. Appl. Bacteriol., 67, 483-488.
Dens, E.J. and J.F. van Impe, 2000. On the importance of taking space into account when modelling microbial competition in structured foods. Math. Comput. Simulat. 53 (4-6), 443-448.
Dens, E.J. and J.F. van Impe, 2001. On the need for another type of predictive model in structured foods, Int. J. Food Microbiol., 64, 247-260.
Devlieghere F., B. van Belle and J. Debevere, 1999. Shelf life of modified atmosphere packed cooked meat products: a predictive model, Int. J. Food Microbiol. 46, 57-70.
Devlieghere, F., A.H. Geeraerd, K.J. Versyck, H. Bernaert, J. van Impe and J. Debevere, 2000. Shelf life of modified atmosphere packaged cooked meat products: addition of Na-lactate as a fourth shelf life determinative factor in a model and product validation. Int. J. Food Microbiol., 58, 93-106.

Finley B. and D. Paustenbach, 1994. The benefits of probabilistic exposure assessment: three case studies involving contaminated air, water and soil, Risk analysis, 14, 53-73.

Geeraerd, A.H., V.P. Valdramidis, F. Devlieghere, H. Bernaert, J. Debevere and J.F. van Impe, 2004. Development of a novel approach for secondary modelling in predictive microbiology: incorporation of microbiological knowledge in black box polynomial modelling, Int. J. Food Microbiol., 91, 229-244.

Gibson, A.M. and A.D. Hocking, 1997. Advances in the predictive modelling of fungal growth, Trends Food Sci. Tech., 8 (11), 353-358.

Gibson, A.M., N. Bratchell and T.A. Roberts, 1987. The effect of sodium chloride and temperature on the rate and extent of growth of *Clostridium botulinum* type A in pasteurized pork slurry, J. Appl. Bacteriol., 62, 479-490.

IEH (The Institute of Environment and Health), 2000. Probabilistic approaches to food risk assessment.

Kroes R., C. Galli, I. Munro, B. Schilter, L. Tran, R. Walker and G. Wurtzen, 2000. Threshold of toxicological concern for chemical substances present in the diet: a practical tool for assessing the need for toxicity testing, Food Chem Toxicol, 38, 255-312.

Kroes, R., D. Müller, J. Lambe, M.R.H. Löwik, J. van Klaveren, J. Kleiner, R. Massey, S. Mayer, I. Urieta, P. Verger and A. Visconti, 2002. Assessment of intake from the diet, Food Chem Toxicol, 40, 327-385.

Le Marc, Y., V. Huchet, C.M. Bourgeois, J.P. Guyonnet, P. Mafart and D. Thuault, 2002. Modelling the growth kinetics of *Listeria monocytogenes* as a function of temperature, pH and organic acid concentration, Int. J. Food Microbiol., 73, 219-237.

Masana, M.O. and J. Baranyi, 2000. Growth/no growth interface of *Brochotrix thermosphacta* as a function of pH and water activity, Food Microbiol., 17, 485-493.

McDonald, K. and D.-W. Sun, 1999. Predictive food microbiology for the meat industry: a review, Int. J. Food Microbiol., 53, 1-27.

McMeekin, T.A., R.E. Chandler, P.E. Doe, C.D. Garland, J. Olley, S. Putro and D.A. Ratkowsky, 1987. Model for the combined effect of temperature and water activity on the growth rate of *Staphylococcus xylosus*. J. Appl. Bacteriol., 62, 543-550.

McMeekin, T.A., J.N. Olley, T. Ross and D.A. Ratkowsky, 1993. Predictive microbiology - theory and application, John Wiley & sons inc., New York, 340p.

McMeekin, T.A., T. Ross and J. Olley, 1992. Application of predictive microbiology to assure the quality and safety of fish and fish products, Int. J. Food Microbiol., 15, 13-32.

Parmar B., P.F. Miller and R. Burt, 1997. Stepwise approaches for estimating the intakes of chemicals in food, Regul Toxicol Pharmacol, 26, 44-51.

Peleg, M. and C.M. Penchina, 2000. Modeling microbial survival during exposure to a lethal agent with varying intensity, Crit. Rev. Food Sci. Nutr., 40 (2), 159-172, 2000.

Poschet, F. and J.F. van Impe, 1999. Quantifying the uncertainty of model outputs in predictive microbiology: a monte carlo analysis, Med. Fac. Landbouwwet.Universiteit Gent, 64 (5), 499-506.

Ratkowsky, D.A., J. Olley, T.A. McMeekin and A. Ball, 1982. Relationship between temperature and growth rate of bacterial cultures, J. Bacteriol., 149 (1), 1-5.

Ratkowsky, D.A. and T. Ross, 1995. Modelling the bacterial growth/no growth interface, Lett. Appl. Microbiol., 20, 29-33.

Ratkowsky, D.A., R.K. Lowry, T.A. McMeekin, A.N. Stokes and R.E. Chandler, 1983. Model for bacterial culture growth rate throughout the entire biokinetic temperature range, J. Bacteriol., 154 (3), 1222-1226.

Robinson, T.P., O.O. Aboaba, M.J. Ocio, J. Baranyi and B.M. Mackey, 2001. The effect of inoculum size on the lag phase of *Listeria monocytogenes*, Int. J. Food Microbiol., 70, 163-173.

Ross, T. and T.A. McMeekin, 1994. Predictive microbiology, Int. J. Food Microbiol., 23, 241-264.
Ross, T., P. Dalgaard and S. Tienungoon, 2000. Predictive modelling of the growth and survival of *Listeria* in fishery products, Int. J. Food Microbiol., 62, 231-245.
Ross, T., 1996. Indices for performance evaluation of predictive models in food microbiology, J. Appl. Bacteriol., 81, 501-508.
Rosso, L., J.R. Lobry, S. Bajard and J.P. Flandrois, 1995. Convenient model to describe the combined effects of temperature and pH on microbial growth, Appl. Env. Microbiol., 61 (2), 610-616.
Rousseau, D., F.A.M. Verdonck, O. Moerman, R. Carrette, C. Thoeye, J. Meirlaen and P.A. Vanrolleghem, 2001.Development of a risk assessment based technique for design/retrofitting of WWTPs, Water Science & Technology, 43, 287-294.
Salter, M.A., D.A. Ratkowsky, T. Ross and T.A. McMeekin, 2000. Modelling the combined temperature and salt (NaCl) limits for growth of a pathogenic *Escherichia coli* strain using nonlinear logistic regression, Int. J. Food Microbiol., 61, 159-167.
Te Giffel, M.C. and M.H. Zwietering, 1999. Validation of predictive models describing the growth of *Listeria monocytogenes*, Int. J. Food Microbiol., 46, 135-149.
Tienungoon, S., D.A. Ratkowsky, T.A. McMeekin and T. Ross, 2000. Growth limits of *Listeria monocytogenes* as a function of temperature, pH, NaCl and Lactic acid, Appl. Env. Microbiol., 66 (11), 4979-4987.
Van Impe, J.F., B.M. Nicolaï, T. Martens, J. De Baerdemaeker and J. Vandewalle, 1992. Dynamic mathematical model to predict microbial growth and inactivation during food processing, Appl. Env. Microbiol., 58 (9), 2901-2909.
Verdonck, F, 2003. Geo-referenced probabilistic ecological risk assessment, PhD thesis, Ghent University.
Verdonck, F., C. Janssen, O. Thas, J. Jaworska and P.A. Vanrolleghem, 2001. Probabilistic environmental risk assessment, Med. Fac. Landbouw. Univ. Gent, 66, 13-19.
Vereecken, K., K. Bernaerts, T. Boelen, E. Dens, A. Geeraerd, K. Versyck and J. van Impe, 1998. State of the art in predictive food microbiology, Med. Fac. Landbouwwet.Universiteit Gent, 63 (4), 1429-1437.
Vereecken, K.M. and J.F. van Impe, 2002. Analysis and practical implementation of a model for combined growth and metabolite production of lactic acid bacteria. Int. J. Food Microbiol., 73, 239-250.
Vereecken, K.M., F. Devlieghere, A. Bockstaele, J. Debevere and J.F. van Impe, 2003. A model for lactic acid-induced inhibition of *Yersinia enterocolitica* in mono- and coculture with *Lactobacillus sakei*. Food Microbiol. 20, 701-713.
Vose D., 1996. Quantitative risk analysis. A guide to Monte Carlo simulation modelling, Wiley and Sons, New York.
Vose D., 2000. Risk analysis: a quantitative guide, Wiley and Sons, New York.
Whiting, R.C. and R.L. Buchanan, 1993. A classification of models in predictive microbiology - Reply, Food Microbiol, 10 (2), 175-177.
Whiting, R.C. and R.L. Buchanan, 1994. Microbial modeling, Food Tech., 48 (6), 113-120.
Xiong, R., G. Xie, A.E. Edmondson and M.A. Sheard, 1999. A mathematical model for bacterial inactivation, Int. J. Food Microbiol., 46, 45-55.

# 10. Traceability in food supply chains

Jacques Trienekens and Jack van der Vorst

## 10.1 Introduction

The European consumer has become increasingly concerned about the safety of food and the negative effects of bio-industrial production. This concern has been heightened by several sector-wide crises in the last decade (such as the BSE crisis, the dioxin crisis, classical swine fever, and foot and mouth disease in Europe). Governments, both national and international, respond by imposing new legislation and regulations for the safety and quality of food products. Retailers react by imposing new demands on their suppliers. To comply with the new demands, companies are forced to introduce sophisticated information systems that focus on identification and registration, and tracking and tracing capabilities. In this chapter tracking and tracing systems, the results of an international benchmark study, and current ICT applications in the global food supply chain will be discussed.

The modern consumer demands products of high and consistent quality, in broad assortments, throughout the year, and at competitive prices. Society also imposes constraints on companies in order to economise on the use of resources, ensure animal-friendly and safe production and restrict pollution. Unfortunately, today's consumer has become increasingly concerned about the quality and safety of food, and the negative effects of bio-industrial production. It is estimated that millions of Europeans get sick every year from food contamination. Important causes are salmonella, campylobacter and *E. coli* O 157. Moreover, consumers will find recall announcements almost weekly in the newspapers. Even though food products are now safer than ever before, from a technical point of view and due to many quality control programs, the safety perception of consumers has decreased significantly.

Since the discovery of BSE in cattle as the probable cause of the deadly human form, known as new variant Creutzfeldt-Jakob disease, there has been a large-scale crisis in the European cattle sector. Between 1990 and 1999 there was a reduction of 6% in sales of cattle meat in the EU (with peaks and falls). The British meat sector suffered the most from the crisis in this period. In 2000 several new discoveries of BSE were made in other European countries, like France and Germany. By mid-February 2001, the consumption of cattle meat had dropped by as much as 80% in several parts of Germany.

Market demand is no longer confined to local or regional supply. The food industry is becoming an interconnected system with a large variety of complex relationships. This is changing the way food is brought to the market. Currently, even fresh produce shipped from halfway around the world can be offered at competitive prices. This has spurred an enormous growth of product assortment in the supermarkets (in many European supermarkets in the 1990s the number of articles more than tripled from 10,000 to more than 30,000).

Together with safety and quality demands of consumers, these developments have changed the production, trade, and distribution of food products beyond recognition. Governments, both national and international, are responding to this by imposing new legislation and regulations to ensure safe and animal-friendly production, restrict pollution, and to economise on the use of resources. Examples are the Codex Alimentarius standards (FAO/WHO), the General Food Law (EU 2002/178) and the EU-BSE regulations.

For food businesses this implies placing more emphasis on quality and safety control, on traceability of food products and on environmental issues and, at the same time, shifting from bulk production towards production of specialities with high added value. Furthermore, because of their the way they are embedded in the network economy, collaboration with other parties becomes important for all businesses to achieve safe and high quality food products for the consumer. This means that business strategies must now move their focus from traditional economical and technological interests to topical issues such as the safety and healthiness of food products, animal friendly procedures, the environment, etc. These processes are affecting the entire food chain, from producer through to retailer. To effectively address (paradoxical) demands facing businesses, many problems and opportunities must be approached from a multi-disciplinary and farm-to-fork perspective, and trade-offs must be made between different aspects of production, trade and the distribution of food.

Our approach in this chapter is multi-disciplinary and farm-to-fork, and takes a supply chain management perspective on traceability and food safety in food chains and networks.

### 10.1.1 Multidisciplinary and farm-to-fork approach

We recognise four major areas of influence, covering a large number of research disciplines (Figure 10.1). Although they represent different perspectives on the production, trade and distribution of food, they are complementary in food chain explanation.

- The economics area of influence is related to efficiency (in cost-benefit terms) and consumer orientation. To be effective and profitable companies must form alliances with other parties in the production column resulting in supply chains and networks. Examples of important issues are:
  - compliance of production and distribution systems with consumer values;
  - cost/benefit relationships of quality and safety control systems in supply chains and networks;
  - the impact of (EU) standards for health and animal welfare on the external competitiveness of businesses.
- The environment area of influence is related to the way production, trade and distribution of food is embedded in its (ecological) environment. Examples of important issues are:
  - use of energy and energy emissions in production and distribution of food products;
  - recycling of waste and packaging materials throughout supply chains from consumer to farm;

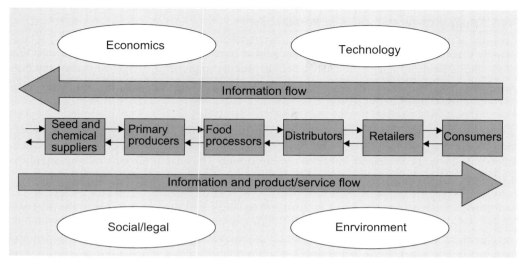

Figure 10.1. Food system dimensions.

- sustainable food production systems (also referring to issues like bio-diversity and landscape architecture).
- The technological area of influence is related to the way technology (product and process technology, transport technology, information and communication technology) can be applied to improve production and distribution of high quality and safe food products. Important issues are:
  - standards in relation to national and international legislation and regulations;
  - systems to guide and control processes and flow of goods throughout the (international) supply chain (such as tracking and tracing, HACCP);
  - provision of product and process information to the public (information attached to the product and general information to the public).
- The social and legal area of influence (norms and values) is related to societal constraints on production, distribution and trading of food, and to issues like human well-being, animal welfare and sustainable socio-economic development. Important issues are:
  - national/EU legislation and secondary legislation (international business norms, conventions) regarding food products;
  - consumer expectations and behaviour with regard to food products (e.g. GMO's);
  - the international legal/regulatory framework and the possibility of differentiating food products complying with higher health and animal welfare standards.

All four areas of influence cover food chains from farm to fork.

*Jacques Trienekens and Jack van der Vorst*

## 10.1.2 Supply Chain Management Perspective

Supply Chain Management approaches can help us in the analysis and redesign of product and information flow throughout the chain from the primary producer up to the consumer, and thus has a direct influence on traceability and food safety issues. Figure 10.2 depicts a typical supply chain (Lambert and Cooper, 2000).

Figure 10.2 shows that SCM views a company as being in the centre of a network of suppliers and customers in different tiers/links of the chain-network with whom the company has to cooperate in order to deliver value to the consumer. The match of demand and supply is central in this approach.

An important aspect of SCM is the exchange of (quality) information throughout the different stages of the chain. In the complex and time-critical network of supplier and customer relations, guaranteeing high quality and safe products is essential. A system that collects, stores and analyses (quality and safety) data from all the relevant processes is of major importance for good functioning of the food chain.

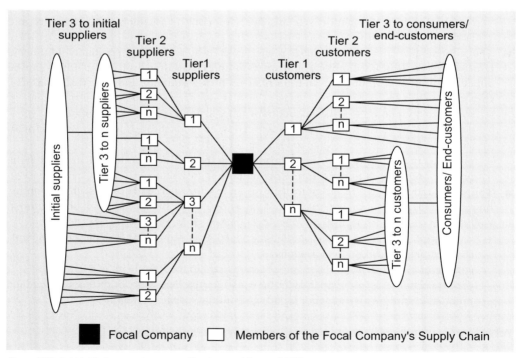

*Figure 10.2. Supply Chain network structure (Lambert and Cooper, 2000).*

Successful SCM quality programs will evolve from coercive to collaborative relationships. The integration of process information is the target. This includes both feed-forward and feedback information flows. The enabling data handling and reporting systems will incorporate greater quality management and traceability capabilities.

The next section will go into the relationship between quality systems and traceability. Section 10.3 deals with traceability systems and section 10.4 will deal with technological aspects.

## 10.2 Relationship between quality systems and traceability in food chains

### 10.2.1 Quality systems in food chains

Companies around the world are increasingly using quality assurance systems to improve their product and production processes and to protect themselves against liability claims. This section will describe the major quality systems used in companies throughout Europe and the way they pay attention to supply chain aspects and traceability.

Quality assurance systems enable the application and verification of control measures intended to assure the quality and safety of food. They are required at each step in the food production chain to ensure safe food and to show compliance with regulatory and customer requirements. Governments have an important role in providing policy guidance on the most appropriate quality assurance systems and verifying/auditing their implementation as a means of regulatory compliance (FAO, 2002). There is a definite move away from the old end-of-line product inspection approach to a new environment of a quality assurance approach where the supplier assumes responsibility for safety. This means that food safety needs to be managed along the entire supply chain.

In Europe producers, food industries and NGO's (Non-Governmental Organisations) in different countries have taken various initiatives for quality assurance systems. In Sweden research has been carried out into European quality assurance in the food sector (Tuncer, 2001). Through desk research and interviews with experts 103 quality assurance systems were identified throughout Europe. The UK showed the largest number of initiatives, followed by Germany and The Netherlands. Below five groups of initiatives are discussed.

Quality assurance initiatives in different European countries (Based on Tuncer, 2001):
- **Certification systems for sustainable agriculture (30 initiatives)** - These systems focus on environment-friendly production and the use of specific quality standards. Examples of such systems are "EKO" in The Netherlands and "CRAE" in Spain. Traceability takes place through documentation. Farmers must keep receipts and product documentation for monitoring and control purposes. A disadvantage of these systems is the heavy administrative load for the farmers.

- **National or sectoral quality assurance systems (26 initiatives)** - These systems aim at control of the primary production phase by defining standards and monitoring performances. Examples are the "Farm Assured British Beef and Lamb" (FABBL) and "National Dairy Farm Assured Scheme" (NDFAS) in the UK, and "Integraal Keten Beheer" (IKB, 'Chain management' in English) in The Netherlands. Contrary to the previous category these standards do not refer to sustainable production methods but aim at healthy and safe food products. They comply with government standards and have an important focus on animal welfare and traceability systems. This explains why traceability tools are fairly advanced in these systems, i.e. the use of ear-labels, 'animal passports' and ICT systems.
- **Quality assurance systems initiated by food industries (8 initiatives)** - These are certificates of branches that are managed by national or international businesses that aim for specific and distinct processes (e.g. SAI: Sustainable Agriculture Initiative). A traceability example is "Hipp's traceability system", which allows the producers of baby food to trace the origin of all raw materials in every jar based on a production code.
- **Retailer systems (17 initiatives)** - These systems are controlled by European retailers. The most important systems aim at sustainable production and high quality and are based in countries where the large European retailers have an important market share, e.g. Germany, UK and The Netherlands. Examples in the UK are the collaboration between TESCO and The Royal Society for the Prevention of Cruelty to Animals (RSPCA) in their "Freedom Food Scheme" and "EUREPGAP" (see also above) introduced by Euro-Retailer Group. In The Netherlands there is Albert Heijn's "Earth and Value" programme. Sometimes products are marketed through a branch name. Examples are KF's "Änglamark" in Sweden and Tengelmann's "Naturkind" in Germany.
- **Regional or traditional quality assurance systems (22 initiatives)** - This category includes all initiatives that refer to regional or local production and have implemented their own standards. An example in the Netherlands is "Nautilus".

In the USA quality systems have existed for many years too. Traceability related issues, however, have attracted hardly any attention (Bredahl *et al.*, 2002). The systems aim in particular at the physical health of animals on the farm and not on issues such as animal welfare. Examples are the Beef Quality Assurance (BQA) program of the National Cattlemen's Beef Association (NCBA) aiming for the reduction of residues in veal and the National Pork Producers Council Pork Quality Assurance (PQA) program aiming at 'good management practices'. Although in the last year more attention has been given to issues like animal welfare and the environment, traceability has only recently reached the top management agenda.

In general, systems initiated by retailers cover the largest part of the chain, in contrary to other initiatives that only include a few links. Good Manufacturing Practice (GMP), Hazard Analysis and Critical Control points (HACCP) and ISO certificates are currently important instruments in assuring food safety and quality on a company level. Until now, however, most quality assurance systems have not included traceability covering the whole food chain. Produce and half-fabricates can be traced with HACCP separately, but without giving a fork-to-farm overview. Risks, so far, are tackled through supplier and chain audits and through monitoring programs.

## Traceability in food supply chains

An exception is found in meat chains in countries where, as a result of recent events, much attention is given to traceability issues.

Many manufacturing systems, including food manufacturing, have sought registration to the ISO 9001 Quality Standards. These require that the product should be able to be traced from the current stage back through all its stages of manufacture through accurate and timely record keeping. The requirement for paper documentation has recently been changed; computer records alone can now be used as evidence of compliance (Food standards agency, 2002).

Within food manufacturing it is also common to see traceability systems used alongside HACCP to provide verifiable documentation, which monitors the critical control points and allows remedial action to be taken if a product falls below quality. Some manufacturers consider their traceability systems (dominantly linked to process control) to be separate from HACCP (linked to quality management). But others consider traceability and HACPP to be inextricably linked as part of a product quality management system. These may not necessarily be opposing views, but represent different viewpoints related to how the systems have been implemented in practice (Food standards agency, 2002).

Traceability forms an important element in the quality system of the supply chain. The European Commission's White Paper on Food Safety states: "A successful food policy demands the traceability of feed and food and their ingredients. Adequate procedures to facilitate such traceability must be introduced. These include the obligation for feed and food businesses to ensure that adequate procedures are in place to withdraw food and feed from the market where a risk to the health of the consumer is posed. Operators should also keep records of suppliers of raw materials and ingredients so that the source of a problem can be identified".

### 10.2.2 Traceability requirements

For organisations in the food and agribusiness the interest in traceability is growing. First, a good tracking and tracing system offers possibilities to follow the product and the processes it undergoes. This leads to more transparency, which makes it possible to offer specific information to buyers and consumers. This again can play a major part in (re)gaining the trust of the consumer. Moreover, by sharing information between partners, information and goods flow can be better managed, resulting in lower costs and more flexibility throughout the chain. Table 10.1 depicts traceability requirements from industry, government and consumers.

Not only the players in the market but also governments are demanding more transparency in the chain. The new law on product accountability implies that traceability in the chain should be guaranteed. Currently, tracking and tracing is high on the agenda of policy makers.

*Identification and registration of GMOs*
In the last few years a discussion on the labelling of GMOs (genetically modified organism) has arisen between the EU and the USA. Labelling of GMOs is obligatory in the USA only if the product differs essentially from the "original", e.g. if the nutritional value differs, or if the product contains

## Jacques Trienekens and Jack van der Vorst

Table 10.1. Traceability requirements for different stakeholders (Food Standards Agency, 2002).

| Stakeholders | Traceability requirements |
|---|---|
| Industry | ▪ To comply with relevant legislation<br>▪ To be able to take prompt action to remove products from sale and protect brand reputation<br>▪ To minimise the size of any withdrawal and hence the costs incurred in recovering, disposing or reconditioning products already placed on the market<br>▪ To diagnose problems in production and pass on liability where relevant<br>▪ To minimise the spread of any contagious disease amongst livestock<br>▪ To protect the food chain against the effects of animal disease<br>▪ To assure meat and meat products and maintain markets and consumer confidence<br>▪ To create differentiated products in the market place because of the way they have been produced<br>▪ Enable avoidance of specific foods and food ingredients, whether because of allergy, food intolerance or lifestyle choice |
| Government | ▪ Protect public health through the withdrawal of food products<br>▪ Help to prevent fraud where analysis cannot be used for authenticity<br>▪ Control zoonotic disease e.g. salmonella<br>▪ Enable control with regard to human and animal health in emergencies<br>▪ Control livestock diseases through the rapid identification of disease sources and dangerous contacts<br>▪ Monitor and control livestock numbers for subsidy claims |
| Consumers | ▪ Protect food safety by effective product recall<br>▪ Enable avoidance of specific foods and food ingredients, whether because of allergy, food intolerance or lifestyle choice |

an allergen that is not present in the original. The EU demands that all GMO products, with a GMO contamination of > 0,9%, must be labelled as such. Another difference between the EU and the USA is reflected by the following example: the US dairy industry may label milk of cows not treated with the bST hormone as such only if the label also states that there is no indication of significant differences in milk from cows treated with bST (Folbert and Dagevos, 2000).

Labelling requirements and retailers' action for products produced from, or containing, GMO's have, so far, been mostly limited to food products intended for human consumption. However, if the EU were to pass the EC's proposed legislative measures, which extend labelling requirements to animal feed ingredients, the impact on countries exporting to the EU, notably the USA, could be very substantial. Retailers could require poultry and livestock producers to raise animals on non-GM diets. Since the EU greatly relies on animal feed ingredients imported from the USA (particularly soybean and MGF, a by-product of ethanol production), and if the USA is to continue supplying the EU, methods that allow the delivery of non-GM products will have to be developed. A considerable increase in the demand for non-GM products (at the 0,9% contamination threshold) might lead to a substantial disruption of the market. So far, the USA

## Traceability in food supply chains

market for animal feed seems to be suited to the delivery of bulk, undifferentiated products, but it could hardly respond to a more significant demand for non-GM products.

The EU General Food Law Regulation contains clear requirements for traceability (see the description of Article 18 in Chapter 13 on modern European food safety law)

This general traceability requirement is non-prescriptive but encompasses all food and feed business operators including primary producers. Retailers of goods to the final consumer are exempt from the requirements of forward traceability.

Although the statements are clear, the GFL is unclear about the required performance of chain traceability. For example, what is the maximum time available for the tracing and what is the required level of detail? The specific requirements for the extent of traceability, in other words how much information is carried, will vary and depend among other things on the nature of the product, on farm practices, customer specifications or legal requirements.

Although a large number of legislative, economic and technological bottlenecks still exist that prevent smooth implementation of traceability systems, many stakeholders recognise the benefits of traceability in food chains (Table 10.2).

Table 10.2. Benefits of traceability for different stakeholders (derived from: Food standards agency, 2002).

| Consumer | Business | Government |
|---|---|---|
| Protect food safety by effective product recall. | Protect public health through the withdrawal of food products. | Comply with relevant legislation. |
| Enable avoidance of specific foods and food ingredients, whether because of allergy, food intolerance or lifestyle choice. | Help to prevent fraud where analysis cannot be used for authenticity. | Be able to take prompt action to remove products from sale. |
| | Enable control with regard to human and animal health in emergencies. | Be able to diagnose problems in production and pass on liability where relevant. |
| | Monitor and control for subsidy claims. | Assure food products and maintain market and consumer confidence. |

## 10.3 Traceability systems

### 10.3.1 Traceability perspectives

If we zoom in on the relationship between food chains and the network of processes as described in section 1, we can identify relationships as depicted by Twillert (1999). He has developed a framework to describe different traceability views for looking at a (food) supply chain process. He identified four perspectives (see Figure 10.3):

- Enterprise
  This perspective focuses on functions and different levels of activities of a single organisation within the supply chain. Precedence relationships between processes within a single organisation are needed to guarantee traceability.
- Multi-site
  It is possible that an organisation in the chain has more than one location. As a result, the flow of materials and information comprises a complex network of related sister organisations. Not only the precedence relationships between processes at one location, but also the relationships between the processes of different sites should be recorded to guarantee traceability.
- Supply chain
  This perspective includes the entire supply chain from primary producer to consumer. Therefore, the precedence relations between the processes of all actors involved in the entire FSCN should be recorded to guarantee traceability.
- External environment
  This also includes the stakeholders of the supply chain that are not directly involved in processing the products. Examples are consumer organisations, authorities and society as a whole.

For the purpose of traceability, information needs to be exchanged between the different locations of a single enterprise, but also between the different chain participants in the food chains and stakeholders in the external environment. Therefore, information exchange should facilitate both horizontal and vertical communication within and outside the chain.

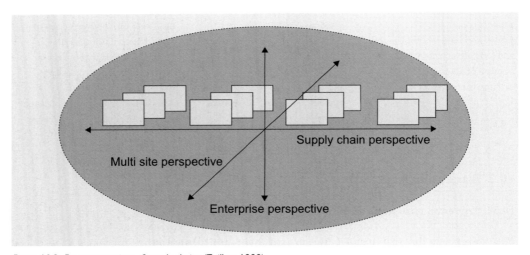

Figure 10.3. Four perspectives of supply chains (Twillert, 1999).

## 10.3.2 Definition of traceability

Traceability is of importance at chain level, as well as at company level. At company level a system should provide information on the location of the product and on the history of the product (product and process information). At chain level, besides information on the location of products, information on the origin of the product is also of importance. In this regard it is important to identify the current unique characteristics of lots (components) and the historical relationship between lots. Traceability systems are used for recall-management, but also for consumer information, for logistic management in distribution centres, for quality management, for risk management in food chains, for an efficient sales process at the supermarkets, etc.

Given the demands of stakeholders (industry, government, consumers), it is extremely important for companies in the food chain to be able to guarantee the composition and origin of their products by the establishment of a collaborative chain information system. The following demands can be put on such chain traceability information systems (see also Figure 10.4):
- Identification of produce and products throughout the food chain. Identification aims at recognising an item as a unique set of data. The identification function in a company provides items with unique codes (barcode, label, tag, etc.).
- Tracking of items: the determination of the ongoing location of items during their way through the supply chain.
- Traceability of items throughout the food chain. Tracing aims at defining the composition of an item and the treatments that item has received during the various stages in the production life cycle. Chain upstream (backward) tracing aims at determining the history

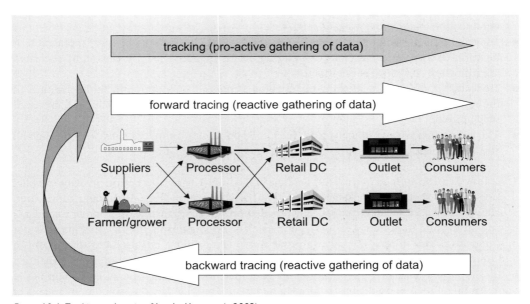

Figure 10.4. Tracking and tracing (Van der Vorst et al., 2003).

of items and is used to determine the source of a problem of a defective item. Chain downstream (upward) tracing aims at the determination of the location of items that were produced using, for example, a contaminated batch of raw materials.

Product identification and product tracking and tracing each refer to a different set of requirements imposed on products and materials. They also have different drivers.
The importance of identification is to be able to clearly distinguish one material or product from another during the manufacturing process, by means of tags, labels, routing sheets, colours, etc. Identification is typically done according to the supplier's established procedures.

The basic idea of tracking and tracing is the possibility to determine where a certain item is located and to trace the history of that item. On the basis of that information, it should also be possible to determine the source of any (quality) problem of an item, and it should be possible to find out where the other items with the same problem are located in the supply chain.

Some definitions of traceability:
- Lot traceability deals with the identification of a lot or batch of material, the tracking (location and quantity), and the tracing ('where from' and 'where used') information of material. A lot or a batch is a quantity produced together and sharing the same production costs and resultant specifications (Van Rijn *et al.*, 1993).
- Food traceability can be defined as that information necessary to describe the production history of a food, and any subsequent transformations or processes that the food might be subject to on its journey from the grower to the consumer's plate (Wilson and Clark, 1997).
- Tracking and tracing is a modern tool that gives insight into the origin of products to all links of the chain. That insight is used to optimise the processes in the separate links and to enhance the total chain (Weigand, 1997).
- Traceability is the quality management system's ability to trace the history, application or location of an item or activity, or similar items or activities, by means of recorded identification (ISO 9000 series)(ISO,2001).
- Traceability: a) attribute allowing the ongoing location of a shipment to be determined, b) the registering and tracking of parts, processes, and materials used in production, by lot or serial number (Apics).
- The ability to trace and follow a food, feed, food producing animal or substance through all stages of production and distribution (EU regulation - Food Law).

Traceability can be defined in a narrow and in a broad way. The narrow option involves tracking & tracing in its original and most basic form. The basic flow of tracking and tracing information is usually created for recall purposes. The broad approach positions the function of tracking & tracing far beyond the original scope. The registering of data during the tracking phase is not only done to be able to carry out backward and forward traceability, but also to be able to control and optimise the process within a company or a supply chain. An example of enhanced optimisation possibilities by means of using data that flows through the supply chain is the optimisation of recipes on the basis of lot characteristics provided by the supplier. By means of a tracking and tracing functionality in this extended form, companies will be able to decrease failure costs, increase productivity and guarantee quality.

Tracking & tracing in a narrow sense allows people to see where (food) products are at all times. The on-line tracking function creates a historical record by means of recorded identification, which allows traceability of components and usage of each end product.

Tracking and tracing in a broad sense implies that process and product information could also be used for the optimisation and control of processes in and between the separate links of the supply chain to be able to decrease failure costs, to increase productivity and/or to guarantee quality. Based on this insight, Van der Vorst (2004) defined three chain traceability strategies:

- Compliance-oriented strategy: the organisation only complies with legal regulations with the help of end-of-pipe techniques. It focuses on the registration of incoming and outgoing materials and leaves the process as a black box. No optimisation activities take place. The chain is usually fragmented since each company individually complies with (legal) demands.
- Process improvement-oriented strategy: the organisation strives for control of product traceability within its link by means of production-integrated measures in order to achieve compliance with legal regulations as well as improved process efficiency and thus a better return. An example of a process-oriented measure is the introduction of local ICT that provides traceability and also enables a more efficient handling of logistics flows.
- Market-oriented (branding) strategy: with this, organisations aim for the establishment of full traceability within the supply chain to achieve competitive advantage (by creating added value in the market place). This may require the redesign of processes to separate small production lots, standardisation of information carriers, adjusted planning and control of production processes, and so on. The traceability performance is the result of the joint effort to design and produce a product. This requires a chain structure in which the individual links work intensively together to open new markets. Integration and common goals are key aspects.

## 10.3.3 Types of traceability data

Product-related information, in the context of food chains, can refer to various types of information from various sources. There will be particularities for each individual food chain. Yet, a general-purpose classification can serve as a starting-point for finding information that is to become transparent in a supply chain. All these types of information pertain to the smallest homogeneous product unit (Beulens *et al.*, 2004).

1. Inherent properties of a product. These can be seen or otherwise measured in the product. Most inherent properties can change over time, e.g. taste, content of chemical components, bacteriological status, and visual attractiveness. Some are not likely to change, e.g. size.
2. Process properties. These constitute the history of what has happened to a unit. If units are combined, for instance, only on the basis of equal inherent properties, then you get units with non-homogeneous process properties. That may become a problem when tracing product provenance through the supply chain on process properties becomes necessary, e.g. in the case of a recall.

3. Properties of means of production used on the product. This is in fact an extension of the previous category. It includes machines and labour. In the case of machines, contamination issues can arise. For instance, if a machine is used for non-GM seed after having been used for GM seed, very rigorous cleaning is necessary. In the case of labour, ethical issues can arise. Customers may, for instance, refuse to buy products if they suspect that child labour is involved.
4. Origin or provenance data of a product. This deals with information about the processes, associated actors, resources, raw materials and intermediate products that have been used to produce the product.
5. Actors involved. They execute processes, 'produce' products, services and information. Properties of actors may be 'inherited' by the processes and products produced.
6. Relationships between the entities previously mentioned with associated properties of interest. For example if a process consumes inputs, uses machines and produces according to some specification, properties may be associated with that combination of entities.

### 10.3.4 Information decoupling point

As described in the previous sections, traceability information regarding product and process characteristics is linked to products in every part of the chain. However, when goods are exchanged between enterprises that are part of a supply chain, in most cases the majority of information that can be linked to products is left at the supplying company. In other words, the information is de-coupled from the products and only aggregated information accompanies the product further through the chain. (In this sense labels on products only represent a potentially small part of the information available).

Traceability, then, is guaranteed via the coupling of aggregated product properties to detailed product properties by codes or certificates. These codes or certificates linked to products must give access to the information left behind at the links upstream of the chain. For companies this means that they must implement information systems that are able to identify, register and track the product throughout the chain, while preserving the link between aggregated and detailed product information.

The major advantage of de-coupling information from the product, while preserving a link to detailed product properties, is that exchanging parties prevent an information overload, as detailed data are not exchanged, while these data still remain accessible by means of identification. The point at which this de-coupling occurs is called the information de-coupling point (Beulens *et al.*, 1999). Figure 10.5 depicts the information de-coupling point (IDP).

To be able to guarantee product safety and quality, it is essential that IDP's are placed and defined carefully. Especially in case of mistakes or suspension, well-defined information de-coupling points will make it possible to recall products and to identify causes in a relatively accurate and easy way.

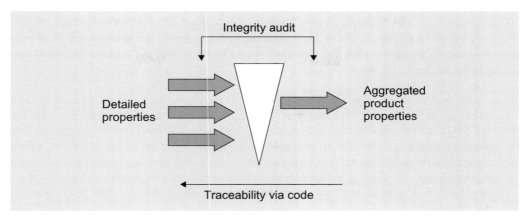

Figure 10.5. Information de-coupling point (IDP).

In the context of defining IDP's, batch-generating characteristics play an important role. A generating characteristic could be seen as 'something' that is applicable to a certain (group of) products. An example at the supplier's site could be found in the production of raw materials according to the purchase specifications of the manufacturers. The manufacturer assumes that all batches, of a certain product, that are delivered are produced according to the given specifications. In case of, for example, receiving raw materials from certified suppliers, the certificate that accompanies raw material lots 'assures' that all delivered lots are manufactured according to the certification standards.

The above-mentioned example of generating characteristics could be called "general generating characteristics". They are called general, as the characteristics are assumed to be applicable to all delivered batches of a certain product. In the IDP, a name is assigned to (a set of) generating characteristics, for example, 'chlorine-free' or 'organic'. Detailed information with respect to the general generating characteristics is usually kept at the supplier's site. By means of identification codes the data remain accessible.

In addition to general generating characteristics, however, there are also 'specific characteristics'. The difference between specific and general characteristics is that specific characteristics discriminate between batches. If a supplier delivers six boxes of a specific type of meat according to the purchase identifications, the manufacturer assumes that all boxes contain the specific type of meat that is manufactured according to his specifications. If purchase specifications do not change, this will also be applicable to next deliveries. So, the purchase specifications represent one or a set of general generating characteristics.

From a quality point of view, however, the six boxes might be different. If, for example, three boxes were processed on a certain day in the week before delivering and the three other boxes were produced two days later, two different lots are delivered. So, to be able to distinguish between both lots, the supplier should add extra information to the batches. Examples of extra

## Jacques Trienekens and Jack van der Vorst

information/data are the unique lot numbers, data on quality characteristics per batch, information about the origin of raw materials that were used to manufacture a specific batch, etc. These characteristics, which discriminate between batches, form the 'specific characteristics' of batches (Beulens *et al.*, 1999).

From a tracking & tracing point of view, the specific characteristics are more interesting than general ones, as the specific characteristics form the discriminating factors between batches. On the basis of the specific characteristics, manufacturers are able to create batches in warehousing and in production. In the above-described case of receiving six boxes with a specific type of meat, the manufacturer knows that two batches should be created on the basis of exchanged data on the specific generating characteristics. The relationships between a concrete batch of materials and data on its characteristics should be established by means of lot identification.

We may conclude that in designing traceability systems a key item is the separation of lots aiming at identity preservation (IP) of lots. In some sectors this is so important that product flows go through dedicated facilities, for example, organic or eco-biological produce.

*Tactical and strategic Identity Preservation, in the case of GMOs*
So far, the attitude of consumers in the EU and other countries towards the application of biotechnology in the food sector, and the legislative measures that have been introduced to regulate GMOs and derived products, have not had substantial implications for the traditional bulk commodities marketing system. However, recent and forthcoming biotechnology and food safety legislation, requiring either monitoring or labelling, or both, may foster the development of a system that will have to respond to the increasing requirements of product differentiation and market segregation (Mochini, 2001). Future developments in both market demand and EU regulation in the food and animal feed sector will require the establishment of a supply chain capable of separating GMOs and non-GMOs.

According to Buckwell *et al.* (1998), Identity Preservation (IP) is a system of crop or raw material management that preserves the identity of the source or nature of the material. This system should enable the supply chain: a) to keep non-GMO and GMO products separate, and b) to demonstrate the presence or absence of modified material in compliance with either market specifications or regulatory requirements. Rial (1999) defines IP as "a system of production and delivery in which the grain is segregated based on intrinsic characteristics (such as variety or production process) during all stages of production, storage and transportation".

However, we propose a distinction between "tactical/operational" IP and "strategic" IP. Tactical IP utilises the same resources to move and process conventional undifferentiated grains and aims at preventing grains and the processed products commingling throughout the various processes and steps of the supply chain, by requiring careful monitoring of operations, cleaning of machinery and rigorous testing. Strategic IP refers to a system where the crops and processed products go exclusively through dedicated facilities, thus reducing the operational costs involved

*Traceability in food supply chains*

in ensuring that no commingling occurs, but requiring infrastructural investments. It also usually involves containerised shipping and testing only prior to containerisation.

The choice between and feasibility of tactical and strategic IP, as well as the overall economics, depends upon several factors, but primarily (Coppola, 2002):
- the magnitude of the demand for non-GM organisms;
- the level of GM organism contamination allowed;
- the likely premiums consumers would be willing to pay.

### 10.3.5 Typical traceability demands on the food industry

As we have seen in the previous sections, registration of process and product characteristics (e.g. composition, history of products) is essential for the food industry, for the purpose of traceability, for production management and to comply with new rules and regulations. The special characteristics of the food industry, in combination with the implementation of EU legislation on identification and registration, have specific implications regarding the use of data (on products and processes) in various management processes.

Many authors have provided an overview of typical food industry characteristics, sometimes categorised for different chain actors, and/or the activities they perform (van Rijn *et al.*, 2003; Trienekens, 1999; Hvolby and Trienekens, 1999; Van der Vorst, 2000; Beulens *et al.*, 2004). Examples are the unpredictable supply of produce, the quality variation between different suppliers and between different lots of produce, the perishable character of fresh produce etc. These characteristics influence the implementation of traceability systems. A summary of the factors that have the most effect on the complexity of traceability are presented below:
- Diverging and converging product streams that make it difficult to follow the different raw materials that go into the product, and the resulting end products.
- The inhomogeneity of raw materials and intermediate products, due to, for example, weather conditions, biological variation and seasonality, but also as a possible result of variations in production. A typical characteristic of food products is their variability in form, shape, taste, etc.
- Contamination of (many) different batches of raw materials. Because batches in many food industries are mixed, cross-contamination of batches is a general problem in the food industry.
- Batch or continuous production. If production takes place in batches, identification can be retained by batch. However, in the case of continuous production (e.g. milk) identification can only be retained by time of production.
- Many sources of batches of raw materials (home and abroad). Because of the internationalisation of food chains and networks, sourcing becomes more and more international. This makes traceability hard to achieve.
- Many actors with formal and informal relationships in the chain. In the food chain transactions often take place at arm's length. Sound transaction administration is often lacking.

- Lack of connections between physical and administrative product flows. In general, one could say that food chains and (chain) processes within them are complex systems. Consequently, implementing traceability and transparency systems is also complex.
- Variable and multi-level recipes. Products can be based on more than one recipe, for example, different raw materials and use of different production means can lead to similar products.
- Presence of active material. Active material, for example, protein, is contained in other material components of the product and determines the value of the end products. The concentration, amount or percentage can vary. Registration of the total product quantity is therefore not sufficient, but rather the active component should be identified separately.
- Perishable character of products. For certain materials storage life constraints apply. As a consequence, using up materials according to 'first in first out' (FIFO) may not apply and different batches of the same product, but of different age, cannot be grouped and must be handled separately.

Research projects in the food industry in the Netherlands identified an extended list of product and process data of importance in the food industry (Trienekens and Beulens, 2001). In this section two examples of traceability requirements at companies in different food chains are given. Both case studies have been developed in projects in practice with the aim of arriving at functionality demands for (traceability) systems.

The first example illustrates traceability requirements linked to processes at an average distribution company in the fruit and vegetables chain (Table 10.3).

The second example depicts more complex traceability requirements for the food industry (Table 10.4).

Table 10.3. Important information system functions at a trader of fruit and vegetables.

| Supply | Storing of purchase specifications |
| --- | --- |
| | Coupling of lot number to supplier data and production data |
| | Coupling of lot number to entry control data |
| Storage | Coupling of lot number to storage location and conditions (e.g. temperature, time) |
| Assembly | Registration of new lot numbers linked to preceding lot numbers |
| Packaging/labelling | Coupling of lot number with label |
| Sales | Coupling of lot numbers to invoice data |
| | Coupling of lot number to distribution data |

Table 10.4. Demands regarding identification and registration in various business processes in the food industry.

| Production planning | Registration of lot characteristics |
|---|---|
| Order-management (purchasing) | Registration of data during order entry (supplier, delivery data, delivery time, etc.) |
| | Having insight into the location of the ordered goods in the supply chain |
| | Registration in case of purchasing raw materials for the production of samples in R&D |
| Warehouse management | Links between batch numbers of suppliers and lot numbers of food industry |
| | Registration of data on lot characteristics during receipt |
| | Lot traceability in case of splitting or mixing lots |
| | Location control per lot |
| | Reverse logistics: identification of raw materials lots that are returned from production |
| Manufacturing | Registration of actual lot numbers that are used in production |
| | Registration of process variables per "batch" |
| | Lot traceability in case of using more batches for one packaging order |
| Order management (sales) | Registration of actual lot numbers that are sent to the customers |
| | Registration of complaints |
| | Using shelf-life restrictions in sales |
| Freight management | Registration of data during order picking and truck loading (employee who picked the order, actual lot numbers per sales order line, departure time, departure temperature, etc.) |
| | Being able to have insight into the distribution of trays over retail outlets |
| | Tracking of pallets (or other returnable packaging materials) for finished goods |

## 10.4 International benchmark study on traceability systems

### 10.4.1 Introduction

In the autumn of 2002 an international benchmark study was carried out initiated by the Dutch Ministry of Agriculture, Nature Management and Fisheries and the Ministry of Economic Affairs (Van der Vorst *et al.*, 2003). The research aimed at gaining insight into current international practices in food supply chains regarding the use of Information and Communication Technologies (ICT) to support traceability of food products to guarantee food safety. In this paper we will present the main results and some best practices.

The benchmark was carried out in The Netherlands, Germany, Spain, Sweden, the United Kingdom, Australia and the United States of America. In each country the following supply chains were investigated (from feed supplier - via the primary producer, processing industry and retail, to consumer): meat products, dairy products, fruit & vegetables, and grain/bread products.

The research approach was twofold. First of all, an electronic questionnaire was developed on the use of ICT for traceability systems to guarantee food safety, and it was sent to best practice organisations in each supply chain in each country. In each country local experts selected companies and supply chains that were seen as innovative with respect to the use of ICT and traceability.

This questionnaire was based on the research model depicted in Figure 10.6. The model proposes that the realised chain performance concerning traceability in food supply chains is based on the following elements:
- the restrictions put on the supply chain by government (laws) or societal organisations;
- the strategy of the company and supply chain concerning food safety and traceability;
- the organisation of the supply chain (degree of coordination between the stages in the supply chain); and
- the use of ICT in the supply chain (use of standard bar-codes, presence of scanning devices, planning modules, etc.).

The chain performance on traceability is characterised by: the number of links in the supply chain that can be traced back and forward; the tracing unit that defines the level at which the traced object is uniquely identified (e.g. a farmer, a delivery, a cow); the time needed for tracing the products; and the reliability of the tracing.

Second, the benchmark was supported and enriched by interviews and desk-research. We searched in the public domain (the Internet) for traceability initiatives and approached software suppliers, food companies, sector organisations, governmental institutes, and so on.

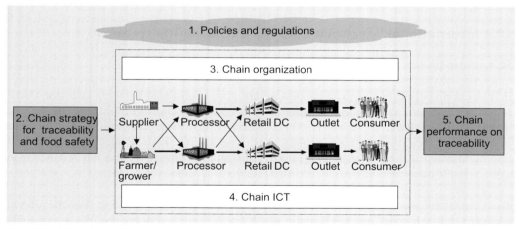

Figure 10.6. Research model.

## 10.4.2 Benchmark results

The following general results were obtained from the project:
- There is a huge diversity of performance concerning traceability in the supply chain. In many cases companies still focus on their own business instead of the complete supply chain. Most companies have arranged some kind of traceability between company walls. However, they are unaware of the potential traceability of their suppliers, mainly because they have multiple alternating suppliers with whom they do not exchange strategic information.
- There are a number of best practices that have nearly full traceability in the supply chain. These best practices are usually completely integrated firms or highly coordinated supply chains that comprise feed suppliers, farmers, processing firms and retail concerns, that have agreed on the use of specific standards and systems. Most of these best practice chains achieve more than the present legal food safety demands and try to increase their market share and/or margin by distinguishing themselves from competition, hence following the compliance-oriented strategy.
- Legislation is an important incentive for companies to comply with traceability demands. The European standards are most strict in this regard. As of 1 January 2005 in the EU, the origin and destination of all animal feed, raw materials, and products in all stages of production and distribution must be registered (on a transaction basis) as discussed in a previous section (General Food Law). Countries that export to Europe must conform to these standards. Because of the international market demands, companies often go further than obliged by their national legislation.
- Most companies focus on prevention instead of traceability. Legislation so far gives no clear rules for the required performance of traceability systems (speed of traceability, detail level, etc.). Companies, therefore, focus on GMP, HACCP, ISO, etc. However, these systems provide in-company traceability but no chain traceability. To cope with this shortcoming, companies perform supplier audits and install monitoring programmes. The demand for traceability is usually higher when the risk of incidents is higher and consumer trust is at stake (or competitive advantage can be created).
- There are only a few ICT applications specifically designed for traceability. In most cases traceability is established via the linking of existing registration systems destined for other purposes, such as purchasing, production, sales, laboratory, financing, etc. (such as ERP, WMS and LIMS systems). The research also shows that there are only minor differences between ICT applications in different countries. This confirms the idea that the ICT market is a global market (e.g. ERP systems, bar-code, RFID, internet, XML are globally comparable).
- When traceability systems are in place, companies cannot always profit fully. When incidents occur, retailers often remove all articles from the shelves and not just the articles from the specific lot concerned. One of the main reasons is that retailers have discovered that in time the number of products related to the incident often grows and consumer confidence in the product decreases. Furthermore, incidents often lead to general import restrictions without taking account of existing traceability systems.

### 10.4.3 Comparing chains and countries on traceability

One of the main conclusions of the research is that the differences between chains are larger than the differences between countries, concerning the use of ICT for traceability. The research shows that food supply chains have become global chains. Regulations are different in different parts of the world, but because retailers and processors have become global players the same rules apply for everyone. For example, British retailers have amongst the strictest requirements for food safety and they are supplied by the Spanish, Australian, Dutch, and German amongst others. So, the British requirements concern all their suppliers, and accordingly all have to take actions to fulfil them.

However, Table 10.5 shows a comparison of the best practices of the seven selected countries regarding chain management and traceability,. The United Kingdom and the Netherlands are in general trendsetters, while Australia and Germany are the main followers (both very active in the field of meat production; Germany because of high consumer demands, Australia because of export interests to Europe). The UK leads with regard to legislation and the setting up of institutes and has the most examples of traceability systems. In Sweden and the USA there is a lot of focus on quality assurance, but much less on traceability (one explanation could be that there has either been no food incident yet (Sweden) or traceability is not an issue for the consumer (USA)).

As stated, the differences between chains are larger than between countries concerning the use of ICT for traceability. Most traceability initiatives are found in the meat supply chain, mainly because of the recent incidents in these chains. Table 10.6 shows some major characteristics of ICT applications found in the different chains.

*Table 10.5. Qualitative comparison between countries.*

|  | Government commitment | Chain organisation | Use of ICT for traceability | Traceability performance |
|---|---|---|---|---|
| Trendsetter | United Kingdom<br>The Netherlands | United Kingdom<br>The Netherlands | Australia<br>The Netherlands | United Kingdom<br>Australia<br>The Netherlands |
| Follower | Sweden<br>Australia<br>Germany<br>Spain | Australia | United Kingdom<br>Sweden<br>USA | Sweden<br>Germany |
| Stay behind | USA | Sweden<br>Germany<br>Spain<br>USA | Spain<br>Germany (except for meat) | Spain<br>USA |

Table 10.6. Characteristics of typical food supply chains.

| | Meat | Dairy | Fruit & vegetables | Grain-bread |
|---|---|---|---|---|
| Characteristics | Long and complex chain. Development towards integration. Much attention to traceability. | Long, in general integrated and controlled chains. Mingling of milk flows makes traceability complicated. | Very diverse chain structures. Development towards retail integration and branding. | Complex product flows (commodity). GMO issue gets more attention than traceability for food safety. |
| Incentives for working on traceability | Law, customer demands, branding. | Prevention; focus on quality assurance. | Law, customer demands, more and more branding. | Law (GMO issues), few branding initiatives. |
| Best practice | Integrated chain from feed supplier to retail. | Cooperative chain from farmers to processor. | Coordinated chain with certified suppliers and contractual relationship with retail. | Chain with preferred certified suppliers with contractual relationships. |
| Chain performance traceability | Full chain within 36 hours and individual animal. | Within 24 hours but to large number of farms. | Within few hours to specific harvest grower. | Within 24 hours to number of farmers. |
| Use of ICT | Ear labels (some electronic), barcodes (EAN), transponders, DNA identification, central databases, EDI, Internet. | No chain systems; functional silos. Use of barcodes (EAN), RFID, EDI, Internet. | Little use of ICT; functional silos, custom-made software, use of barcodes (EAN), RFID. | Little use of ICT, many custom-made single systems; use of barcodes (batches). |
| Initiatives for traceability | Many in many countries. | Few, focus on prevention. | Many in many countries; focus on product coding. | Most initiatives aim at lot separation (e.g. GMOs). |
| Conclusion | Many developments but more international cooperation required. More and more branding. | Traceability is not an issue. | Development from spot market to chains; more differentiated pre-packed products. | Development from commodity to specialty (branding). |

### 10.4.4 Bottlenecks and success factors

During the research a number of bottlenecks and success factors were identified for full, fast and reliable traceability in food supply chains. The most important *bottlenecks* that were identified are the following (Table 10.7 presents an overview of the bottlenecks that were identified in the four supply chains worldwide):

- Indefinite and differentiated performance levels concerning traceability resulting in a follow-and-wait policy of actors in the supply chain.
- Little economical incentive: it is unclear what the exact benefits of traceability will be (especially as long as consumers are unwilling to pay more for a traceable product), and it is also unclear what the costs of traceability are.
- High investments in infrastructures for 100% traceability: agribusiness is characterised by a large commingling of products due to expensive facilities and infrastructures that are based on commodity flows. Changing infrastructures to a system of separation of flows costs enormous amounts of money.
- Lack of chain organisation and chain transparency.
- Traceability of products in quality assurance schemes is restricted: these schemes are usually focused on parts of the supply chain and not the complete supply chain.
- Lack of standardisation: information for establishing traceability is registered in all links of the chain. The problem is that each actor does it according to his own standards, which reduces the ease and ability with which data is exchanged. This holds true within one country as well as between countries.
- Businesses in food supply chains have such specific characteristics that each supply chain has its own specific elements: standardisation is therefore difficult.

*Table 10.7. Overview of the bottlenecks concerning traceability in the four chains.*

| | |
|---|---|
| Meat | Registration of lots and handling is insufficient, especially in first part of the chain |
| | Insufficient standardisation in registration of data |
| | Insufficient integration of systems |
| | High costs and inadequate incentives for traceability |
| | Unclear division of authority between institutions |
| | Quality assurance systems aim at certification, not traceability |
| Dairy | Registration of lots and handling insufficient, especially in first part of the chain |
| | Combination of lots at processing stage prevents traceability to suppliers |
| | Rest streams are input for production processes without traceability options |
| | No forward traceability of products from processor to supermarket |
| | Traceability systems are not integrated and are partly paper systems |
| Vegetable & fruit | Traceability is lost at retailers and traders |
| | Unit of traceability is strongly dependent on the packaging (insufficient traceability through auction and stores) |
| | Legislation is organised per link, leading to traceability being unsystematic |
| | Lack of standards for coding |
| | Batches are too small to make traceability cost effective |
| Grain | Purchase of grain on the world market |
| | Insufficient quality and safety assurance systems for food additives |
| | Traceability is organised separately per chain link; start and end of the chain are weak points |
| | In-company information systems are not integrated |
| | Traceability demands for adjustment of production and storage facilities |

## Traceability in food supply chains

The most important success factors were identified as follows:
- Unmistakable definition of the legally required functionalities of traceability systems in food supply chains and the minimal performance requirements.
- Identification, registering and exchange of data in all links of a supply chain according to a uniform standard and at the same level of detail (tracing unit).
- Implementation of a risk assessment of the supply chain and a focus on the main risks.
- Making the added value of traceability visible to everyone depending on the functionality (product recall, logistical optimisation, etc.).
- Use of a joint approach by all chain participants in the development of a functional and modular basic design for a traceability system that is suitable for a large number of specific situations in food supply chains.
- Reduction in the number of suppliers and the commingling of products in the supply chain; make the chain transparent!

## 10.5 Technology for traceability systems

### 10.5.1 Introduction

Systems that register tracking and tracing data must, next to the identification and collection of data via AIDC (Automatic Identification and Data Collection/Capture) technology, be integrated on a company level with administrative systems, to simplify the analysis and exchange of data. These data can also then be used for the planning and evaluation of production. We can refer here to examples such as Warehouse Management Systems (WMS), Laboratory Information Management Systems (LIMS) and Enterprise Resource Planning (ERP) systems. Furthermore, systems for the communication of traceability data are needed, for example, Electronic Data Interchange (EDI) or eXtended Markup Language (XML). For an efficient and effective data exchange in the supply chain, one also needs a common and standardised infrastructure to facilitate the exchange of data in the supply chain (Figure 10.7). The European

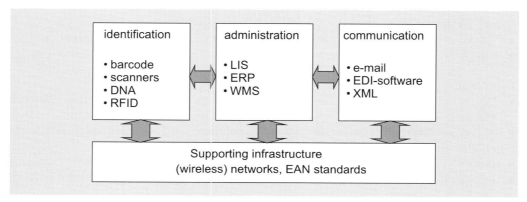

*Figure 10.7. Overview of ICT-elements related to tracking and tracing (Van der Vorst et al., 2003).*

*Jacques Trienekens and Jack van der Vorst*

Article Number (EAN) Association develops standards for use in logistical systems, in close cooperation with the American equivalent, the Uniform Code Council (UCC). For example, they define global unique numbers for industrial producers. In the remainder of this section we will focus on the available AIDC technologies.

### 10.5.2 AIDC applications

AIDC (automatic identification and data communication) technologies have been developing rapidly and are widely applied in all industries. In its simplest form the identification may be a numeric or alphanumeric string in read-only format, which gives access to data stored elsewhere. However, the amount of information that can be carried within the identification system has been expanding rapidly. AIDC technologies make automatic and fast data collection and registration of large quantities of products possible.

Applications and benefits of AIDC technology in the chain (DTI, 2001):

*In the Factory*
- Detailed information stored on the item to maintain a comprehensive product history.
- Box contents verified and reconciled with purchase orders prior to shipping.
- Problems fixed at the source, increasing overall quality assurance.
- Picking and packing productivity, throughput, and use of capital equipment increased.
- Decreased labour-hour requirements.

*In the Distribution Centre*
- Inventory control, quality assurance, and decreased time for count cycles and inventory audits.
- Items inside boxes identified and inventoried without opening.
- Contents confirmed, information updated, and manual re-ticketing for special handling eliminated.
- Streamlines order picking and packing, while also eliminating human error.

*During Transportation*
- Tracking the location of the products allowing on-time delivery and theft prevention.
- Controlling conditions during transportation, decreased product spoilage and damage.

*In the Store*
- Nearly 100 percent of merchandise available for purchase resulting in increased sales and customer service.
- Real-time visibility of product gaps on the sales floor; improved accuracy, and manual reads eliminated.
- Prevents customers "walking the sale" or leaving the store without making a purchase.
- RFID as secure payment device speeds transactions, increases loyalty and enables one-to-one CRM.

## Traceability in food supply chains

Figure 10.8 gives an overview of AIDC applications. Some of these are now widely used for product identification and registration in the food chain.

Included under the automatic identification/registration umbrella are the following technologies (Hill and Cameron, 2000; www.aimglobal.org):
- **Barcode** - A barcode is a machine-readable code consisting of a series of bars and spaces printed in defined ratios. In the same way the human eyes see an object, a bar code scanner sees a bar code and converts the visual image into an electrical signal. The information encoded in this electrical signal is then processed by a decoder analogous to the processing of information from the human eye by the brain (bar code product handbook). In short, barcode technology encompasses the symbologies that encode data to be optically read, the printing technologies that produce machine-readable symbols, the scanners and decoders

**Bar Code Standards**

There are many different barcode symbologies, or languages. Each symbology has its own rules for character (e.g. letter, number, punctuation) encodation, printing and decoding requirements, error checking, and other features. The various barcode symbologies differ both in the way they represent data and in the type of data they can encode: some only encode numbers; others encode numbers, letters, and a few punctuation characters; still others offer encodation of the 128-character, and even 256-character, ASCII sets. The newest symbologies include options to encode multiple languages within the same symbol; allow user-defined encodation of special or additional data; and can even allow (through deliberate redundancies) reconstruction of data if the symbol is damaged. At the last count, there were about 225 known barcode symbologies but only a handful of these are in current use and fewer still are widely used (www.aimglobal.org). In the food sector the most commonly used codes are EAN 13 (article number) and EAN 128. EAN 128 allows inclusion of extra information, such as trade units, best before dates, product origin, lot numbers, etc. EAN Europe has recently developed an application of the 128 code for the beef chain, which includes information such as country of birth, country of slaughter, reference number of cutting firm, etc.

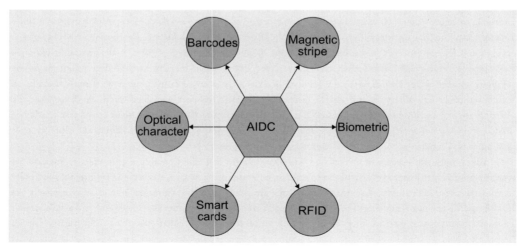

*Figure 10.8. Overview of AIDC applications.*

that capture visual images of the symbologies and convert them to computer-compatible digital data, and the verifiers that validate symbol quality (www.aimglobal.org). Typically, laser scanners can read barcodes from near contact to 12 inches distance, whereas very powerful scanners reach as far as 4 feet (Joshi, 2000).

- **Biometric Procedure -** In the context of identification systems biometrics is the general term for all procedures that identify people by comparing unmistakable and individual physical characteristics. In practice, these are fingerprinting procedures, voice identification, retina (or iris) identification, and DNA analysis. Biometrics are best used for applications that require unique, absolute and secure user identification. An application in the food sector is DNA identification of animals, which is (still) too expensive for large-scale applications. Biometric identification uses a physiological trait digitally encoded and stored to accomplish this identification. Biometric systems may simply identify the individual or allow a system to tap into a whole range of "rules" regarding that individual.

- **RFID Systems -** Radio frequency identification (RFID) covers a range of data-carrying technologies, for which transfer of data from the identifier to the reader is achieved by a radio-frequency link. International standards for the use of RFID systems are being put into place, but are not as developed as those for barcodes.

As the graphic shows, when a transponder enters a read zone, its data is captured by the reader and can then be transferred through a standard interface to the host computer, printer or

**An example of developments in RFID technologies.**

Motorola: BiStatix (http://www.motorola.com)
Motorola, Inc. signed a memorandum of understanding with Dai Nippon Printing Co., Ltd., one of Japan's largest business forms manufacturers, to jointly develop and market "smart" label solutions incorporating Motorola's breakthrough BiStatix(tm) radio frequency identification (RFID) technology. The new paper-based solutions will contain information stored in small computer chips that can be read and modified through a wireless interface, making the smart labels and forms an ideal tool for the tracking and efficient routing of materials. BiStatix tags can be read without clear line of sight and are not rendered unreadable by the effects of moisture, dirt, dust or paint. In addition, while bar code technology is "read-only," BiStatix RFID technology is read/write capable, meaning that the data it contains can be changed, updated and/or locked. Because BiStatix RFID technology uses silicon and printed ink, BiStatix smart labels can be created more effectively than earlier generations of RFID technology-enabled labels, which required the incorporation of a costly metal coil and resonant capacitor. Furthermore, BiStatix smart labels can be read after being folded, crumpled and even ripped, and are also fully disposable. Additionally, BiStatix tags can be printed easily on paper and other non-conductive surfaces and integrated into existing printing processes.
The BiStatix tags and readers in conjunction with smart forms and labels provide a cost-effective solution for companies seeking to improve their business operations in the area of tracking objects, monitoring inventory and other supply-chain related processes. Applications of BiStatix are intended for books, magazines and other products like packaging. The printing products provided by Dai Nippon will become exceptionally "intelligent" and enable various business operations to increase their efficiency. Our vision for BiStatix is that everything from baggage tags to inventory labels will soon have wireless communications and digital intelligence, ushering in a new era of productivity and efficiency growth.

programmable logic controller for storage or action. The electromagnetic field pattern where the transponder is read is affected by such things as the size and shape of the readout antenna, the orientation of the transponder as it passes through the field, and other electronics in the environment. RFID is increasingly used in warehouse environments for logistics purposes and in food chains, for example, to track animals from birth through processing, and register/transmit data on feed, antibiotic use, health data, weight, etc. For this purpose animals can have an eartag or, as we see for example in some pork chains, a tag can be injected in the animal.

The AIDC technologies described below are not (yet) found in many food companies for product identification.

- **Optical Character Recognition (OCR)** - OCR was first used in the 1960s. Nowadays, OCR is applied in production, service and administrative fields as well as in the banking world for the registration of cheques. It is commonly used for high-volume document management in the insurance and healthcare industries. OCR is also used in heavy-duty manufacturing environments for reading direct-marked, human-readable part numbers. However, OCR systems have failed to become universally applicable because of their high price and the complex reader that is required.
- **Magnetic Stripe** - Magnetic stripe technology is one of the more widespread and established AIDC technologies. The magnetic stripe can be found on the back of bank and credit cards, and customer loyalty cards. Credit cards were first issued in 1951, but it was not until the establishment of standards in 1970 that the magnetic stripe became a factor in the use of the cards. Today financial cards all follow the ISO standards to ensure read-reliability worldwide. Credit cards constitute the largest users of magnetic stripe cards together with transit cards. With the advent of new technologies the demise of the magnetic stripe is predicted. However, with the investment in the current infrastructure this is not likely to be any time soon.
- **Smart Cards -** A smart card is an electronic data storage system with or without additional computing capacity (microprocessor card) that is incorporated into a plastic card the size of a bankcard. Smart cards differ from magnetic stripe cards in two ways: the amount of information that can be stored is much larger and some smart cards can be reprogrammed to add, delete or rearrange data.

AIDC technology is developing fast. Further extension of the possibilities of the different technologies for product identification and registration will be a strong incentive for the further development of tracking and tracing systems in food supply chains. Developments in barcoding, RFID technology and biometric procedures are particularly promising in this regard. But new applications of other AIDC technologies may also be expected.

## 10.6 Major research themes and questions for discussion

For the successful implementation of supply chain information and traceability systems, much research still needs to be done and many developments have to take place. This chapter will end by formulating some important research issues in the field of traceability in food supply chains.

*Jacques Trienekens and Jack van der Vorst*

Important criteria for a sound identification and registration system are that it must be fraud-insensitive, and that data on the origin and history of a certain product must be able to be traced quickly and effectively. Fraud-insensitivity refers to linking an identification code to the product (e.g. ear-tags on pigs can get lost; in the EU often up to 5%) and to exchanging correct data between chain parties. Changes in information systems alone cannot bring the necessary integrity and auditability of information gathered and provided. They must be accompanied by other organisational changes in business processes and management.

*What are the combined technological and organisational requirements for guaranteeing the integrity of information on company, national and international chain levels?*

Internationalisation of the food chain makes it imperative that identification and registration systems also become standardised and internationalised. Possibilities for fast, real-time accessibility to data are, however, restricted because of the large number of different, and sometimes still manual, databases in the different countries. Most countries still do not have a central database, and there are no automatic coupling systems between national databases yet, let alone international coupling systems. Tracing of the origin of a product is, therefore, still a long and difficult process in many cases. A complicating factor is that many companies/chains are reluctant to provide product and process information to third parties, for strategic reasons.

*What are the technological requirements and the demands of the international legislative/regulative system for achieving transparency of food production on company, national and international chain level?*

Collaboration between national governments to achieve an international regulatory system is crucial in obtaining transparency in cross-border supply chains. However, current debates and disagreements in the European Parliament and Council of Ministers about labelling shows that a consensus between different EU countries about the registration is difficult to reach. The many differences between the US, EU and other nations have also been the subject of negotiations within the Codex Alimentarius Commission.

Supply chain companies in developing countries face particular difficulties in implementing safety and quality regulations, as these companies are often faced with poor infrastructural and institutional facilities. Governments in these countries rarely actively encourage cross-border supply chain development and do not compel the implementation of food safety and quality regulations. It could be that the sanitary and phytosanitary conditions under the WTO accelerate the development of standards and compliance across trading countries.

*How to arrive at global standards and regulations, and at the same time offer developing countries balanced and sustainable socio-economic growth?*

Although in most current European supply chains data are related to origin of produce and products, in the future new consumer demands may require the registration of new data. For example, in an animal chain, data on feed, medication, husbandry and welfare conditions may all have to be registered. An increasing number of firms are offering various guarantees to their customers, concerning the use of pesticides or medications, animal welfare, and other "ethical" variables, especially in Europe. For example, the Label Rouge program in France provides consumers with a guarantee that its poultry is raised under free range conditions. Others are the various organic certification protocols guaranteeing consumers that no artificial chemicals have been used in production.

*What are the economic and technological trade-offs in the design of chain systems that allow for future, yet unknown, demands for chain transparency from food chain stakeholders?*

Concern about the potential long-term health effects of (e.g. GM) foods underlie a lot of the food safety regulations that are being promulgated for that specific segment of the food industry. As with pharmaceutical products, some have suggested that subjecting foods to post-market surveillance could improve the level of safety by ensuring that any late emergence problem will be identified and addressed by the health system. The preceding discussions on food safety technologies in the food industry provide a base infrastructure for the implementation of a post-market surveillance system by providing information on the components of food items and chain of custody information. If such systems are extended to encompass consumers (Figure 1) and the public health system, they will provide the direct link between food and health, facilitating rapid access to information on food products that produce unexpected adverse health effects in consumers.

*How to link long-term public-health effects with (short-term) industry data-capture systems?*

From the previous discussion in this chapter it appears that the business community at large faces a great number of challenges. Challenges derived from the need to satisfy consumer demands, and to comply with rapidly changing legal requirements and business requirements. All this while ensuring low costs. Parts of these requirements are such that meeting them is a "licence to produce". Not being able to meet them means loosing this licence. Finally, it must be noted that changes in information systems alone cannot bring the necessary integrity and auditability of information gathered and provided. These changes must be accompanied by other, organisational, changes in business processes and management.

# References

Beulens, A.J.M., M.H. Jansen and J.C. Wortmann, 1999. The Information Decoupling Point, In: K. Mertins, O.Krause and B.Schallock (eds.), Global Production Management. Boston: Kluwer Academic Publishers.

Beulens, A.J.M., L.W.C.A. Coppens and J.H. Trienekens, 2004. Traceability requirements in food supply chain networks, Submitted to Computers in Industry.

Buckwell, A., G. Brookes and D. Bradley, 1998. Economics of Identity Preservation for Genetically Modified Crops, CEAS, Wye, England.

Bredahl, M.E., James R. Northen, Andreas Boecker and Mary Anne Normile, 2002. Consumer Demand Sparks the Growth of Quality Assurance Schemes in the European Food Sector, In: Anita Regmi (ed.), Changing Structure of Global Food Consumption and Trade. Market and Trade Economics Division, Economic Research Service, U.S. Department of Agriculture, Agriculture and Trade Report. WRS-01-1.

Coppola, L., 2002. Non-GM maize gluten feed and soybean meal; strategies for the USA supply chain, Dutch Ministry of Agriculture, Nature Management and Fishery, The Hague The Netherlands.

DTI, Department of Trade and Industry, 2001. It's more than just a barcode! A guide to AIDC technologies, London, March 2001.

FAO, 2002. www.fao.org/es/ESN/food/foodquality_en.stm

Folbert, J.P. and J.C. Dagevos, 2000. Veilig en Vertrouwd - Voedselveiligheid en het verwerven van consumentenvertrouwen in comparative context. Den Haag, the Netherlands: LEI.

Food Standards agency, 2002. Traceability in the food chain, Food Chain Strategy Division, Food Standards Agency.

Hill, J.M and B. Cameron, 2000. Automatic Identification and Data Collection: Scanning into the future, Montgomery Research Sites, Ascet volume 2, 4/15/00.

Hvolby H.H. and J.H. Trienekens, 1999. Manufacturing Control Opportunities in Food Processing and Discrete Manufacturing Industries. International Journal for Industrial Engineering, May.

ISO, 2001. Guidelines on the application of ISO 9001: 2000 for the food and drink industry, first edition, reference number ISO 15161: 2001 (E).

Joshi, Y.V., 2000. Information Visibility and Its Effect on Supply Chain Dynamics, thesis Master of Science at the Massachusetts Institute of Technology.

Lambert D.M. and M.C. Cooper, 2000. Issues in Supply Chain Management, Industrial Marketing Management, 29, 65-83.

Moschini, G., 2001. Biotech-Who Wins? Economic benefits and costs of biotechnology innovations in agriculture, The Estey Centre Journal of International Law and Trade Policy, 2(1), 93-117.

Rial, T., 1999. Containerised Oilseed, Grain, and Grain Co-Products Exports, Des Moines, Iowa.

Rijn, Th.M.J. van and B.V.P. Schyns, 1993. MRP in process; the applicability of MRP-II in the semi-process industry, Assen, Van Gorcum.

Trienekens J.H., 1999. Management of Processes in Chains, a research framework. Doctoral Thesis. Wageningen University.

Trienekens J.H. and A.J.M. Beulens, 2001. The implications of EU food safety legislation and consumer demands on supply chain quality assurance and information systems, IAMA Symposium, Sydney Australia June 2001, www.ifama.org

Tunçer, B., 2001. From Farm to Fork? Means of Assuring Food Quality; An analysis of the European food quality initiatives, MSc thesis, Lund, Sweden, IIIEE Reports 2001:14

Twillert, J. van, 1999. Tracking & Tracing in Semi-Process Industries, Msc Thesis. Wageningen University.

Van der Vorst, J.G.A.J., 2000. Effective food supply chains; generating, modelling and evaluating supply chain scenarios, PhD thesis Wageningen University.

Van der Vorst, J.G.A.J., J. van Beurden and H. Folkerts, 2003. Tracking and tracing of food products - an international benchmark study in food supply chains, Rijnconsult, The Netherlands.

Van der Vorst, J.G.A.J., 2004. Performance levels in food traceability and the impact on chain design: results of an international benchmark study, In: H.J. Bremmers, S.W.F. Omta, J.H. Trienekens and E.F.M. Wubben (eds.), Dynamics in Chains and Networks, Wageningen Academic Publishers, Wageningen, the Netherlands.

Weigand, A., 1997. Tracking and tracing in agri & food, Deventer, Kluwer Bedrijfsinformatie.

Wilson, T.P. and W.R.Clarke, 1997. Food safety and traceability in the agricultural supply chain; using the internet to deliver traceability, s.l., Food Track plc.

# 11. Microbial analysis of food

*Anna Maraz, Fulgencio Marin and Rita Cava (cases by Andreja Rajkovic)*

## 11.1 Introduction

Modern civilization with its large population (ca. 6.5 billion in the year 2005) could not be supported without methods for providing food safety. The experiences and outbreaks that have taken place in the history of mankind made us think of hygiene importance in food production, handling and processing. All foods can be contaminated anywhere along the production and distribution conduit before reaching the consumer. Beef can be contaminated by livestock colonisation at the front end of the production process; lettuce can be contaminated as a result of improper handling by field workers or consumers. Proper preparation and sanitation methods along the entire food chain are essential for preventing foodborne illnesses.

Though only a minority of microorganisms are pathogenic (disease-producing), practical knowledge of these microbes is necessary for food safety. In the field of microbial food safety, the development and evaluation of microbial analysis methods for the detection of food pathogens or their toxins has a central role. Many different techniques were developed and applied with varying degrees of success in research and/or routine microbial food analyses. Indicator organisms were tested first and the information gained was used to assume the probable occurrence of pathogens. Recent decades have brought new techniques and methods to light, in particular immunological, molecular and modern microscopic techniques for pathogens that are difficult to detect with the classical culturing techniques, e.g. for viruses, viable but non-culturable cells and species taxonomically hard to identify.

## 11.2 Objectives

The objective of this chapter is to discuss the basis of traditional (culture-based) and newly developed rapid and molecular techniques that are routinely used to determine the microbiological state of food and raw materials. Special emphasis will be placed on elaborating the advantages and disadvantages of the different microbiological techniques used in food microbiology.

The first part of this chapter describes briefly the characteristics of culture-based techniques and the great progress that has been made in this field by developing and commercialising new selective and indicator culture media that are also suitable for screening and detecting foodborne pathogens. The second part of this chapter introduces the rapid methods, including both immunological and molecular techniques. The significant progress made in the latter technique in the recent past has made it necessary to go more deeply into the scientific and technical bases of nucleic acid-based microbiological methods.

## 11.3 Detection, identification and typing of foodborne pathogens

Microbiological analysis is carried out on raw materials, processing lines, and final products. The main aims of microbiological controls are:
- determination of the total number of microorganisms present;
- determination of the number of indicator microorganisms;
- determination of the levels of specific spoilage microorganisms;
- determination of the levels of specific pathogens and toxins;
- checking for the presence or absence of specific pathogens.

Microorganisms can be detected and investigated by several microbiological techniques, which belong to one of the following broad groups (Notermans *et al.*, 1997; De Boer and Beumer, 1999):
- microscopy;
- culture-based techniques;
- immunological techniques;
- molecular techniques.

Not only are the microbiological state of the raw material and the food itself important, but also that of the food processing environment, e.g. the microbiological quality of the air and water used, the level of microbial contamination of the surfaces, etc. The tests that are actually carried out will vary according to the type of raw materials, the food-processing technology applied, and the criteria used to assess quality (Garbutt, 1977).

As pathogenic microbes are generally present at very low concentration, very sensitive assays are needed to be able to detect pathogens in raw materials, during food processing, and in the final product. At the same time the techniques used for detection, identification and enumeration of foodborne pathogens should be as rapid as possible. As a consequence of the generally low levels of pathogens in food, a pre-enrichment or enrichment step is generally unavoidable.

Rapid detection of microorganisms for monitoring involves (Fung, 1995, De Boer and Beumer, 1999):
- automated instrumental analysis;
  - impedimetry;
  - turbidimetry;
- bioluminescence assay;
- direct epifluorescent (DEFT) microscopy;
- flow cytometry;
- immunological methods;
- nucleic acid-based methods.

### 11.3.1 Validation of microbiological techniques

Microbiological tests are important in the food production chain and also in laboratories and agencies responsible for controlling food. Microbiological tests are applied by:

- Governmental food inspection authorities to enforce legal regulations.
- International trade authorities to determine compliance with a microbiological standard.
- The food industry to maintain quality control and process requirements.
- Academic laboratories to conduct research.
- Reference laboratories to provide surveillance data and epidemiological information.

Microbiological methods should be technically reliable and importantly all parties involved should agree with and accept the methods employed (Debevere and Uyttendaele, 2003). It is indubitably helpful if standardised methods are used. The objective comparison and acceptance of data obtained by different laboratories is made possible if the standardisation of microbiological methods is carried out by international organisations (e.g. ISO = International Standard Organization, AOAC International = Association of Official Analytical Chemists International, CEN = Comité Européen de Normalisation), by national agencies (e.g. AFNOR = Association Française de Normalisation, DIN = Deutsches Institute für Normung) or by trade organisations (e.g. IDF = International Dairy Federation). Standardised methods, which may be considered as reference methods, serve as guidelines for the reliable analysis of food. They are generally culture-based techniques and could be accepted as official methods by national agencies or production companies.

If a new method is intended to be used for food analysis it is necessary to carry out a validation procedure. Validation is the demonstration that the technical performance of a method is comparable to the existing standard method. Validation provides confidence to the end user that the method has proven ability to detect and enumerate the organism or group of organism specified (Debevere and Uyttendaele, 2003).

The validity of microbiological data depends on the microbiological test applied and on the performance of the laboratory carrying out the test. Validation involves determination of the performance characteristics of the method by use of reference material inside the laboratory and in inter-laboratory tests.

During validation the following performance characteristics of the methods are to be determined (Notermans et al., 1997; De Boer and Beumer, 1999; Debevere and Uyttendaele, 2003):
- **Accuracy** - Accuracy is the closeness of agreement between the results obtained and the true value. The accuracy of test results also depends on the precision of the method.
- **Precision** - This is defined as the closeness of agreement between the independent test results obtained by repeating the experimental procedure several times under similar conditions. Precision is generally expressed by determining the standard deviation or standard error. Precision can measure both the repeatability and reproducibility.
- **Repeatability** - This is defined as variability of the test results obtained within the same laboratory under similar conditions
- **Reproducibility** - Indicates the variability observed in an inter-laboratory study when the same method is applied under similar conditions.
- **Detection limit** - The smallest number of microorganisms or the smallest quantity of toxin that can be detected in the sample.

- **Sensitivity** - Sensitivity of the method is the proportion of microorganisms that can be detected. It can be calculated using the following equation:

$$\text{Sensitivity (\%)} = \frac{\text{number of true positives } (p)}{p + \text{number of false negatives}} \times 100$$

Failure to detect the target when it is present is a false-negative result and will lower the sensitivity of the test.

- **Specificity** - The specificity of a method is its ability to discriminate between the target organism or substance (e.g. toxin) and other organisms or substances.
It can be calculated using the following equation:

$$\text{Specificity (\%)} = \frac{\text{number of true negatives } (n)}{n + \text{number of false positives}} \times 100$$

A positive result in the absence of the target is a false positive result and will lower the specificity of the method.

- **Robustness/ruggedness** - This may be defined as the sensitivity of the method to (small) changes in the environmental conditions or parameters during execution (e.g. temperature of incubation, sources of supplies and reagents, etc.).

An appropriate procedure should be developed for the validation of each type of methods. The validation protocol is different for qualitative and quantitative methods. In 2002 the European standard "Protocol for the validation of alternative methods" was accepted by CEN. Despite that fact that different validation schemes have been worked out and accepted, there are still a number of issues to be addressed in the preparation of a solid validation scheme such as:

- Choice of reference method: the reference method should be an internationally standardised method.
- Number of food types to be tested and number of samples to be analysed: the number of food types to be included depends on the applicability of the method. A sufficient number of identical food samples should be analysed to be able to use appropriate statistics for the interpretation of data.
- Naturally contaminated or artificially contaminated food samples are used for testing: whenever possible naturally contaminated food samples should be applied because they represent best samples encountered in real life. If it is not possible to acquire a sufficient number of naturally contaminated foods, artificially contaminated (spiked) samples may be allowed.
- The source and number of inoculum strains: reference materials, containing appropriate but well-defined levels of target organism may be used for spiking the samples. Strains that have been isolated from the same type of food products are generally preferred over clinical isolates for spiking.

Some performance characteristics of different conventional and alternative methods are compared in Table 11.1.

Table 11.1. Comparison of different conventional and alternative methods (de Boer and Beumer, 1999).

| Method | Sensitivity (detection limit) cfu/ml or gram | Specificity | Time before results |
|---|---|---|---|
| Plating technique | 1 | Good | 1-3 days |
| Bioluminescence | $10^4$ | No | 0.5 hours |
| Flow cytometry | $10^2 - 10^3$ | Good | 0.5 hours |
| DEFT | $10^3 - 10^4$ | No | 0.5 hours |
| Impedimetry | 1 | Moderate/Good | 6-24 hours |
| Immunological methods | $10^5$ | Moderate/good | 1-2 hours |
| Nucleic acid-based methods | $10^3$ | Excellent | 6-12 hours |

## 11.3.2 Conventional culture-based methods (Holbrook, 2000)

*Plate count*

The traditional plate count technique is based on the fact that living microbial cells or clumps of cells grow and form colonies inside or on the surfaces of solid culture media. The concentration of cells in a food sample can be determined by counting the colonies that appear after incubation.

Application of this technique requires the production of homogenous food samples, in which the microbial cells present in the food are evenly dispersed in a liquid that can be easily pipetted and spread. After disintegration of samples a series of dilution are made, followed by the inoculation of samples from each dilution on agar plates. After incubation the growing colonies, which appear inside or on the surface of the plates, are counted for the estimation of cell concentration in the samples. The concentration of microorganisms is defined as the number of colony-forming units (cfu) per ml or gram of the original sample. The most commonly used plating techniques are 'pour and spread' plating. The information obtained from the plate count depends on the following:

- The choice of diluent. Liquid is generally used for the preparation of homogenous food samples and dilution series. Diluents that give the highest possible recovery of microorganisms present in the food sample should be used. Water generally does not fit this requirement because it may cause cell damage. Common diluents are 0.85% sodium chloride with or without 0.1% peptone, 0.15% peptone water, or phosphate buffered peptone water. Special diluents may be required, e.g. low redox diluents for anaerobes or artificial seawater for marine bacteria. Addition of 1% TWEEN 80 surfactant can improve microbial yield, especially in the case of food with moderate to high fat concentration.
- Methods used for quantitative removal of microbes. Different swabs are used to remove microbes from surfaces but they rarely give quantitative and reproducible results. Sprays (e.g. a spray gun operated with high-pressure air) followed by washing can result in a quantitative and non-destructive removal of microbes from surfaces. Ultrasound sonication is also an effective technique for removal of microorganisms from hard surfaces, resulting in suspensions of microorganisms with very low debris. Cavitation, however, can be lethal

to some cells. The small sample sizes handled and needed to sterilise containers can also be problematic. For a long time blenders were considered as the most effective homogenisers for food samples. However, noise, the need to clean and sterilise after use, overheating, and the high levels of debris generated are notable drawbacks. The sheer force and heat generated by the blades may kill microbes and reduce the estimated total count. Stomachers are currently the method of choice as they use paddle action and therefore do not have the drawbacks of blenders; they also have the added advantage that sterile disposable plastic bags can be used during homogenisation. The homogenisation time needs to be standardised, as microcolonies continue to break down into smaller units that increase the count as the homogenising time increases. Pulsifiers are a recently introduced type of homogeniser. In this case the food sample is placed in a sterile disposable bag and is beaten at a high frequency (35000 rpm), producing a combination of shock waves and intensive stirring, which results in separation of the microbial cells from the food matrix.

- The medium used. Special selective and differential or elective media are used to count specific organisms, especially pathogens. The physiological state of the target organisms must be taken into account. Treatment used during food processing and conditions present in the food (e.g. low water activity, pH, or the presence of antimicrobial compounds of food) can sublethally injure microbial cells. Therefore, it is often necessary to allow the organisms to repair the injury before transferring them to selective media. In the case of presence/absence tests, food samples are incubated first in non-selective enrichment/resuscitation media. Organisms present in a dry sample may be sensitive to osmotic shock and can be damaged as a result. To avoid this, samples must be hydrated gradually. Pathogens are generally present in much lower concentrations than other microorganisms. Therefore detection of pathogens in food requires the use of selective media to separate the target organisms from the others. Selectivity can be achieved by addition of certain agents to the medium and/or by using specific culture conditions (e.g. aerobic or anaerobic, or a particular incubation temperature). Most selective agents show a degree of toxicity toward the target organisms, and therefore substrates may be added to stimulate specifically the growth of the target organisms. Differential media contain substances that allow differentiation of the target organisms from others present in the food.
- The plating method. The spread plate technique gives surface colonies that develop normally and are easy to count. The pour plate method, however, involves the use of agar poured at 45 °C which may kill some psychrotrophs or psychrophiles, or damage cells and reduce the count. The technique results in growth of both surface and sub-surface colonies. The latter can show restricted growth that can be difficult to count and can be confused with food particles on low dilution plates. When pathogens are to be tested quantitatively, plating can be carried out on a non-selective agar medium, which is then overlaid with the selective agar.
- Incubation conditions. The temperature, time, and gaseous atmosphere for incubation need to be carefully selected in relation to the organism or group of organisms to be enumerated, e.g. plate counts for mesophiles are carried out at 37 °C, for thermophiles at 50-55 °C, and those for psychrotrophs at 18-25 °C.

*Most probable number (MPN) technique*
Most probable number (MPN) technique is used to estimate the number of microbial cells or cfu in food samples, if the criterion for a particular microorganism requires detection of low numbers. This technique is frequently used for estimating the number of indicator bacteria or *E. coli*. It is also occasionally used if the criterion for a pathogen does not require absence and the numbers are too low. The typical MPN technique involves general or selective or differential liquid media, which are inoculated with samples from a dilution series. Positive tubes show growth (thus becoming turbid) often in conjunction with another characteristic of the microorganisms, e.g. the ability to produce gas and/or acid from a sugar. Assessment of the numbers of cfu per g in a food is based on a statistical method, where the number of positive tubes in the dilution series is determined and the MPN is calculated using a statistical table. The main disadvantages of the MPN method are that it is labour intensive and has poor precision unless there are a large number of replica tubes per dilution.

*Presence/absence tests (qualitative methods)*
In several cases the acceptable limit of pathogens in food is so low that it is not possible to analyse this number in food quantitatively.

*Modified and automated conventional methods*
Traditional agar-based methods are expensive and labour-intensive. Many attempts have been made to improve laboratory efficiency and to reduce material and labour costs (Table 11.2):
- Gravimetric diluters automatically add the correct amount of diluents to the test sample before homogenisation.
- The spiral plater application reduces the quantity of agar plates used by eliminating the need for serial dilutions before plating. It deposits a small volume of the sample in a spiral line such that there is a dilution ratio of $10^4$ from the centre to the edge of the plate. Colonies appearing along the spiral line can be counted either manually or automatically. As the volume dispensed at any point is known, this technique provides a time-consuming version of colony counting.
- Dipslides are simplified agar plates, which are suitable for the direct detection and estimation of microbes without preparation and dilution of samples. Selective or non-selective agar culture media are layered onto flexible plastic slides and enclosed in sterile plastic tubes. The slides are pressed onto the surface or dipped into the liquid being tested. After incubation, colonies appearing on the slide are counted and can be isolated. Dipslides are useful for testing contamination of surfaces or estimation of the total number of colonies or group of special microbes (e.g. coliform) in milk, water and other fluids.
- The Petrifilm is an alternative to agar-poured plates. In this system the plastic Petri-dish contains rehydratable nutrients that are embedded in a film along with a gelling agent, soluble in cold water. A given volume of liquid sample is placed on the film, and the growth of any cells present is supported by rehydrated media. After incubation, colonies can be counted directly on the film as with conventional plates. Petrifilms are available for total aerobic plate counts, yeast and mould counts, and for the selection of different microorganisms such as *E. coli* and *E. coli* O157:H7.

- The Hydrophobic Grid Membrane Filter (HGMF) technique works by confining colony growth to a grid of 1600 cells. This technique has the advantage of removing inhibitors and concentrating organisms, as well as having a three-log counting range. In this procedure the homogenised and pre-filtered food sample is filtered through a membrane filter, which traps the target microorganisms on the cells of the grid. Then the HGMF is placed on an appropriate agar medium and microcolonies are counted after a short incubation time.
- Chromogenic or fluorogenic substrates in selective culture media combine the enumeration, detection and identification steps thereby eliminating the use of subculture media and further biochemical tests. These substrates yield brightly coloured or fluorescent products as a result of the action of different bacterial enzymes or metabolites. Fluorogenic media containing 4-methylumbelliferyl-β-D-glucuronide (MUG) as a substrate are widely applied for the detection and enumeration of *E. coli*. This medium uses the enzyme β-D-glucuronidase (GUD) as an indicator for *E. coli*, as this enzyme is present in 94-96% of *E. coli* strains. However, it is not present in *E. coli* O157:H7. GUD causes the release of 4-methylumbelliferone from MUG, giving rise to a fluorescent blue colour under UV light. Liquid fluorogenic media can be applied in MPN enumeration of coliforms and *E. coli*.
- Numerous colony-counting devices have been developed for automation of this time-consuming step. Various image analysis systems have been shown to be useful and cost effective in large laboratories, especially those in the dairy industry.

Table 11.2. Modification and automation of conventional methods in food microbiology (de Boer and Beumer, 1999, modified).

| Method | Application |
|---|---|
| Dilution of sample | |
|     Gravimetric dilutor | Diluent addition |
| Plating techniques | |
|     Spiral plate | Enumeration |
|     Dipslides | Enumeration |
|     Petrifilm | Detection + Enumeration |
|     HGMF | Detection + Enumeration |
|     Chromogenic/fluorogenic culture media | Detection + Enumeration/Colony counting |
| Counting | |
|     Automation | Colony counting |
| DEFT | Enumeration |
| Fluorescent antibody (FA)-DEFT | Detection + Enumeration |
| Bioluminescence | Detection + Enumeration |
| Flow cytometry | Detection + Enumeration |
| Impedimetry | Detection + Enumeration |
| Turbidimetry | Enumeration |
| Miniaturised biochemical tests | Identification |
|     API tests (bioMerieux) | Identification of Enterobacteriaceae, non-fermenters, anaerobes, etc |
|     BBL Crystal (Becton Dickinson) | |
|     Vitek (bioMerieux) | Identification of Enterobacteriaceae, Gram negatives, Gram positives |

## 11.3.3 Light microscopy

Although this technique is rapid and requires little in the form of consumables, it can rarely be used for food analysis. An example is if known samples of liquid food or food homogenate are spread evenly in a 1 cm² area, dried and stained with a suitable dye, e.g. methylene blue, the microbial cells present can be counted directly with the use of an optical light microscope. However, it has several drawbacks such as:
- It can not distinguish between dead and living cells.
- The presence of cell debris in most foods makes counting very difficult.
- The sensitivity range is very limited. The minimum number of cells that can be detected is in excess of $10^5$/ml.

Modern microscopic techniques can overcome some of these difficulties and automated instruments are also appearing. Moreover, image analysers are continually improving in their performance and affordability. Microscopy can shorten the incubation period if the growing micro-colonies are counted directly by a microscope, optical scanner or coulter counter.

## 11.3.4 Fluorescent microscopy (Sharpe, 2000)

Direct epifluorescent filter technique (DEFT) is a direct method used for enumeration of microorganisms, which bind the fluorochrome, acridine orange. In this technique food samples are first filtered thorough a membrane filter, thus permitting concentration of the microorganisms, then the cells are stained directly with acridine orange on the filter. The filter is then mounted under a cover glass and viewed using epifluorescent microscopy. The main advantage of DEFT is that cells are easy to detect and it can improve the limits of detection. It is also able with certain limitations to discriminate viable and dead cells. This is based on the fact that acridine orange binds both to cellular RNA and DNA, causing the DNA and RNA to have a green and an orange fluorescent colour respectively. Orange fluorescence masks green fluorescence so that cells that are considered viable, and have a high RNA content have an orange fluorescent colour, whereas those with a low RNA content fluoresce green. Although cells with a high and low RNA content appear orange and green respectively, this does not necessarily equate with viability. Dormant cells with low RNA content will appear green although viable. Newly dead cells with high RNA content will fluoresce orange. This may not be a problem for the analysis of fresh food in which the organisms present are all likely to be active, but for processed foods the situation is more problematic. This technique has been developed to monitor the microbiological quality of raw milk but it can also be very useful for hygiene monitoring. When higher numbers of samples are to be analysed an automated version with image analyser has been developed. Combination of DEFT with fluorescent antibodies (FA) enables the detection of specific foodborne microbes as in the case of Salmonellae. E. coli O157:H7 can be enumerated directly by the FA-DEFT method within 1 hour at concentrations above $10^3$ per ml in milk or juice, or in beef at levels down to 16 cfu per g when using trypsin/Triton X-100 treatment before filtration.

## 11.3.5 Bioluminescence (Neufeld et al., 1985; Kyriakides and Patel, 1994; Sharpe, 2000)

This technique was first developed based on the reaction of adenosine triphosphate (ATP) and luciferin in the presence of the enzyme luciferase yielding luminescence light according to the following reaction scheme:

$$\text{luciferin} + \text{ATP} + O_2 \xrightarrow[Mg^{2+}]{\text{luciferase}} \text{oxyluciferin} + \text{AMP} + h\upsilon \text{ (light)}$$

If luciferin and luciferase are in excess, the maximum intensity of light emitted is proportional to the ATP concentration (1 photon for each ATP molecule) and can be quantified by luminometers. Data for ATP content strictly relate to the biomass, but can also be related to the number of target cells if the system has been previously calibrated for that organism.

Microbial cells contain different but characteristic amounts of ATP; bacterial cells contain about $10^{-3}$ pg (=1fg) of ATP and yeast cells about 0.1 pg (=100fg) of ATP. Luminometers can detect about 0.1-1 pg ATP per ml, therefore theoretically the technique is able to detect $10^2$-$10^3$ bacterial cells per ml or 10-100 yeast cells per ml in suspensions. Under practical circumstances, $10^4$-$10^5$ bacterial cells per ml and $10^2$-$10^3$ yeast cells per ml may be more achievable levels of sensitivity.

However, raw materials of food like plant and animal tissues are rich in ATP, which degrades slowly and incompletely when cells decompose. Therefore, if bacterial biomass is to be estimated, the bacterial cells have to be separated efficiently and the non-bacterial ATP must be removed (e.g. by ATPase (apyrase) treatment) before the assay.

ATP-based bioluminescence is mainly used for measuring biofilms and food residues on surfaces and utensils and therefore gives an indication of the cleaning and sanitation efficiency. Food residues left on the surfaces of equipment and other places may bind to microbial cells that can contaminate products during processing. After a long or short time biofilms containing food material and microorganisms can be formed, which are generally very resistant to cleaning and disinfection. Measuring the ATP luminescence of surfaces provides information about both impurity and microbial contamination. For these reasons it has become an excellent technique for on-line hygiene monitoring in HACCP programs. Although the ATP measurement may not discriminate between food matrix and microbial cells, the existence of positive readings proves that cleanup and sanitation were not done properly. The test can be applied directly to surfaces or by taking swabs, and gives results in a matter of minutes.

A number of instruments and kits are available for hygiene monitoring food processing lines. The instruments are portable and contain disposable sampling units and reagent kits for continuous and safe use. Some instruments are automated and programmed to direct the user to critical control or other testing points and can store and analyse a huge number of test data. These test systems have some limitations since several factors including pH and the presence of enzyme inhibitors may influence the reaction. They also cannot be used in the case of dry-cleaning. Further development of these tests may solve some of the problems (e.g. application of buffered systems).

## Microbial analysis of food

Correlation is generally poor between the ATP luminescence and plate count. This could arise from the following situations:

- Sub-lethally damaged or injured cells contain less ATP than normal cells which can lead to the underestimation of the number of microbial cells present in a sample.
- A filtration step may be required to remove the organisms from the food before measurement if the food contains materials that absorb light (quenching factors). However, filtration may be very difficult if the food contains particulate matter. A combination of HGMF and ATP luminescence can overcome this difficulty. After filtration the hydrophobic membrane is sprayed with bioluminescent reagents and the developed microcolonies are counted using a microscope equipped with an image analyser.
- Difficulties may arise if both bacterial and yeast cells are present in the sample because of the much higher amount of ATP in yeast cells.

The main steps in the bioluminescence analysis of food samples using ATP photometry are:
1. Breaking down the non-microbial cells in the food to release their ATP.
2. Elimination of the non-microbial ATP by ATPase (apyrase) treatment.
3. Releasing the microbial ATP from the microbial cells.
4. Assaying the amount of microbial ATP by adding the luciferin/luciferase system and recording the intensity of the light emitted.
5. Calculation of the cfu per ml by using the correlation between the intensity of the light and ATP content or using a correlation curve between light intensity and cfu per ml.

ATP-based luminometry has been successfully used to assess the quality of fresh meat and milk, measure the activity of starter cultures, and to test UHT liquid foods for sterility.

In addition to bioluminescence tests based on the measurement of ATP released from the cells, non-luminescent bacteria have been genetically engineered into luminescent cells. The *lux* gene of a naturally luminescent bacterium *Vibrio fischerii* has been cloned in several pathogenic bacteria or their phages giving rise to luminescent bacteria. In this case the energy source for luminescence is $FMNH_2$ instead of ATP. Specific phages transfer the *lux* gene to bacteria during infection, which become luminescent within a short period of time. The specificity and sensitivity of phage infection enable this procedure to be used instead of the selective enrichment of bacteria. Luminescent bacteria provide an excellent experimental system for studying the propagation and survival of cells under different circumstances. Moreover, the efficiency of disinfection and activity of biocide preparations or the presence of antimicrobial residues can be tested for the lack of luminescence as stressed, injured or killed bacteria become dark.

### 11.3.6 Flow cytometry (Hadley et al., 1985; Sharpe, 2000)

The technique of flow cytometry is based on the principle of light microscopy together with the possibility that cells can be detected automatically when passed through a focused laser beam. The ratio of absorbed and scattered laser light depends on several factors such as the size, shape, and fluorescence of the cells and their constituents. Scattered light is analysed by a system consisting of lenses and photosensors via the induced optical changes.

The power of this technique is derived from its ability to detect and count different cell types in a suspension as it can measure different parameters for each cell simultaneously. By the application of fluorescent dyes and antibodies, it is possible to measure the total nucleic acid or DNA or RNA contents of cells, assess cell viability, or detect the presence of specific surface antigens. This technique is both rapid and sensitive; results can be obtained within a few minutes and the detection limit is between $10^2$-$10^3$ cells per ml. Sample preparation is quite simple, but only liquid material containing particles smaller than 100 μm can be analysed. If the nucleic acid staining fluorescent dye ethidium bromide is used, the total number of microorganisms present in a sample can be determined quickly.

Flow cytometry has been successfully applied in the detection of yeast cells in soft drinks or bacterial cells in milk. Fluorescent labelled antibodies produced against a series of pathogens (e.g. *Salmonellae, E. coli, Campylobacter*) have provided the possibility of detecting a few cells in a high background of other microbial or somatic cells. Although the simple fluorescent staining of cells is not suitable for discriminating between viable and dead cells, a number of fluorochromes have been developed for the flow cytometric assessment of cell viability. This is based on the assumption that the non-fluorescent (fluorogenic) enzyme substrate is taken up and hydrolysed by the living cells only and this results in the accumulation of the fluorochrome inside the cells. This technique is suitable for recognising the "viable but non-culturable" cells of some pathogens such as *Campylobacter jejuni* and *Vibrio vulnificus*. In spite of several theoretical advantages in food microbiological control, flow cytometry has not achieved widespread application. The biggest drawbacks are the high cost and complexity of the equipment used. However, this is now changing as cheaper bench-top equipment is becoming commercially available.

### 11.3.7 Impedimetry or electrical impedance measurement (Bolton and Gibson, 1994)

Aqueous solutions conduct electricity to different extents. Impedance is resistance to the flow of an alternating current (AC) through a conducting material, e.g. culture medium. Conductance not only depends on the concentration of the charged ions present, the size of the molecules (because size influences mobility) are also an important factor. Conductance is the reciprocal of resistance to the flow of current. Culture media with high conductance have low resistance and vice versa. Capacitance is associated with the accumulation of charged ions at the interface between the electrodes and the culture medium. Impedance can be determined by the combination of the conductive and capacitative resistance.

Changes in the impedance of a culture medium are associated with the conversion of uncharged or weakly charged substrates into smaller and/or highly charged molecules by the metabolic activities of organisms. For example, amino acids are deaminated with the release of ammonium ions; this causes a decrease in resistance to the flow of current (impedance) through the culture medium, thereby progressively increasing the medium's conductance. Increase in conductivity is greatly influenced by the chemical composition of the medium and special components are added to maximise the effect.

## Microbial analysis of food

Impedimetry is becoming a more and more popular method for estimating the level of microbial contamination of food and for detecting pathogens. It can be used for both qualitative and quantitative applications: to detect the bioburden load, the quantity of selected groups of microorganisms, or the presence/absence of selected pathogens. Several instruments are available, differing in the sample size, number of samples to be measured simultaneously, complexity and automation of measurement. All are completely computer controlled and can be connected to laboratory automation systems. They automatically record and interpret data, and determine conformance with specifications. The scheme of conductometric microbiological analysis with a Bactometer is shown in Figure 11.1.

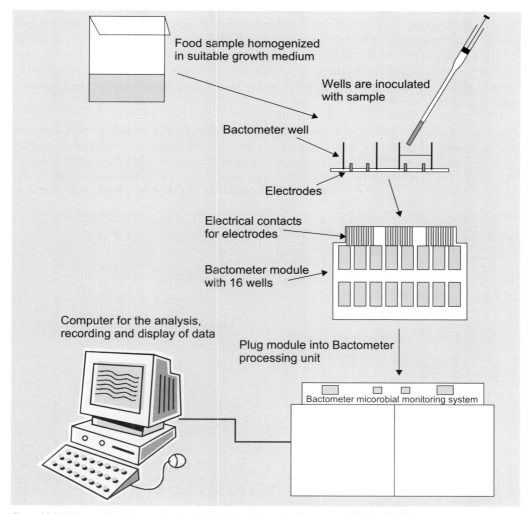

Figure 11.1. Scheme of conductometric microbiological analysis with a Bactometer. (Garbutt, 1977).

Equipment for impedance analysis contains a processing unit, which monitors change in impedance and also serves as an incubator. It is linked to a computer with software designed to analyse, record and display information produced by the processing unit. Samples, e.g. food homogenised in a suitable culture medium, are pipetted into the wells of a module, each well having two electrodes. When the module is plugged into the processing unit a low AC is passed between the electrodes, and any change in the impedance is monitored. Impedance changes due to microbial growth only become detectable when cell concentrations reach $10^6$ to $10^8$ per ml. If further changes in the impedance are plotted against time, a typical impedance curve is obtained as illustrated in Figure 11.2. Although it looks like a growth curve, the first part of the curve is not a lag phase but a combination of the lag phase and the time to reach the threshold of the detectable cell concentration. The time taken to reach this threshold is called the impedance detection time (DT or IDT). The impedance detection time is dependent on both the initial cell numbers and their metabolic activity during growth, i.e. its duration depends both on the number of cells in the original sample and their metabolic and multiplicative energies. Theoretically, it is possible to obtain an IDT for one cell in a well. The IDT is influenced by several factors, such as the incubation temperature, medium used, types of the organisms present and their lag phase.

The number of colony-forming units can be estimated by calibration. If impedance measurements and standard plate counts are determined simultaneously with a large number of samples under the same conditions, and IDT versus plate count data are plotted to produce a scattergram, then the line of the best fit can be calculated by regression analysis (see Figure 11.3). Because of

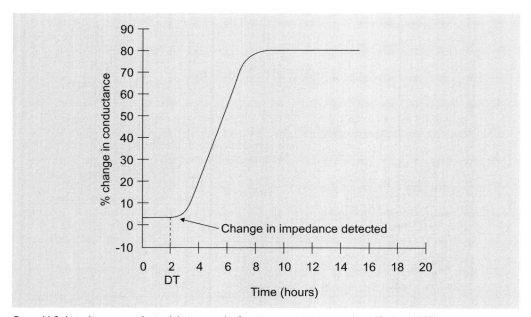

Figure 11.2. Impedance curve obtained during growth of a microorganism in pure culture (Garbutt, 1977).

## Microbial analysis of food

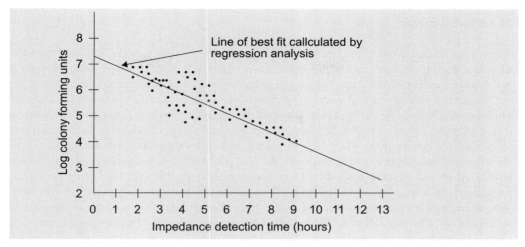

Figure 11.3. Calibration curve for determination of colony-forming units in food sample by a conductometer (Garbutt, 1977).

variations in microbiota and the possibility that the cells present are stressed, each food type requires a different calibration curve, e.g. frozen vegetables require a different curve from meat or pasteurised milk. Since impedance data reflect both the initial cell number and metabolic activity, they can be more suitable for predicting shelf life than plate counts. Using appropriate media, impedimetry can be applied as an alternative to plate counts, coliform counts, and yeast and mould counts. The time necessary to get results, on average 6 to 12 hours, is considerably shorter by using impedimetry. Sterility testing of UHT dairy products and fruit juices generally requires less than 48 hours. Specific detection of the most important pathogens can be performed with currently developed protocols and media in a lot of food types by impedimetric analysis. The first step in this assay is generally a non-selective pre-enrichment followed by selective enrichment in media containing a compound, which is converted to a highly charged molecule.

Indirect conductometry has been developed for those cases where selective media of high ionic strength has to be used (e.g. detection of salt-tolerant microbes) or if the metabolites of the organism will not increase conductance (e.g. detection of yeasts and moulds). In the case of indirect conductometry, electrodes are immersed into a small vial of potassium hydroxide solution placed above the culture medium. Carbon dioxide evolved during growth reacts with the potassium hydroxide, thereby reducing its conductance.

Electric impedance is one of the most versatile of all alternative microbial methods. Examples of commercially available equipments are the Bactometer (BioMerieux), Malthus System (Malthus Instrument Ltd.), BacTrac and µ-Track (Sy-Lab) and RABIT (Don Whitley Scientific Ltd). All are highly-automated systems able to measure multiple samples and present analysis reports automatically. The capacity of the instruments is variable, with instruments of a high capacity (around 700 samples/day) for big companies as well as those of a small capacity (around 20 samples/day) being available (e.g. BacTrac and µ-Track, respectively).

## 11.3.8 Turbidimetry

Suspending microbial cells in a clear solution will cause the solution to become turbid. Turbidity increases during propagation of cells or decreases when cells lyse. Turbidity or absorbance of visible light can be measured by a (spectro)photometer when cell concentration exceeds $10^6$ per ml. Generally, the microbial content of homogenised food samples cannot be assessed photometrically because the background turbidity - caused by the food particles - is too high or the level of background turbidity changes during incubation. Changes in the background turbidity are a consequence of settling, utilisation of food by the cells, enzymic digestion, etc. Direct turbidimetry can, however, be used if the food is added in a small quantity as an inoculum or if 10 or 100 fold dilutions are made before the measurement. Turbidity measurement during propagation of cells in liquid culture media is useful for the determination of the growth curve, thus providing a means for the analysis of the effect of growth conditions on the multiplicative activity of cells (length of the lag phase, generation time). Predictive microbiology is based on setting up predictive microbial growth under different conditions. To derive predictive models of microbial growth, a lot of combinations of growth conditions have to be set up. Turbidity measurement is a simple, reproducible and cheap technique for deriving data used in predictive microbiology.

Automated instrumental analysis of microbial growth based on turbidimetry can be performed using the Bioscreen Analysis System, which measures microbial growth in a vertical optical pathway. In this system cells are suspended in microplate vessels and incubated under controlled conditions. The instrument then measures the turbidity in a regular way, and stores and analyses the data. Similarly, as in the case of impedimetry, the time required to achieve the detection limit is a function of the inoculum concentration, therefore it allows estimation of the initial cell concentration present in the sample. The system has been used to predict the total cell count in a variety of food products.

## 11.3.9 Immunological techniques (Barbour and Tice, 1997; Radcliffe and Holbrook, 2000)

Immunological techniques are based on the specific binding of the antibodies to antigens. Antibodies are produced by the B-lymphocytes of warm-blooded animals and human beings as an immunological response to infectious agents such as bacteria, fungi, parasites and viruses. Most antigens are large complex molecules with different chemical structures. The ability of a molecule to function as an antigen depends on its size and the complexity of its structure. Proteins, glycoproteins and nucleoproteins are considered as the best antigens but some glycolipids and polysaccharides also solicit good immune responses. Cellular, extracellular and viral antigens induce the production of heterogeneous (polyclonal) antibodies, which bind specifically to the antibody binding sites of the antigens called epitopes. Antigen-antibody immune complexes are therefore formed resulting in inactivation of the infectious agents. Antibodies belong to the globular protein fraction of blood serum and are classified as IgA, IgD, IgE, IgG and IgM according to the type of heavy chains they have; $\alpha$, $\beta$, $\varepsilon$, $\gamma$ and $\mu$, respectively. Isolated immunoglobulins are able to react with the antigens not only *in vivo* but also *in vitro* in what are known as serological reactions. Most of the immunoglobulins participating in serological reactions belong to the IgG class. The IgG molecule is made up of two long (or heavy, H) γ chains and two short (or light, L) κ or λ chains that are connected by disulphide linkages, each of which contains a variable antigen binding site (or paratopes) at the N-terminal region. Because each IgG molecule has two antigen binding sites they form a complex with antigens having more than one epitope. The binding reactions of the antigens and antibodies are basically determined by the nature of the antigens. Microbial cells and their extracellular polymeric products, such as exotoxins, capsular polymers and enzymes, as well as viruses have good antigenic properties. Pathogens are first recognized by the phagocytes, which then decompose into smaller unique antigen molecules. Phagocytosis is non-specific, and the target can be any foreign substance, including pathogens and their components. The phagocytes "present" the pathogen-derived antigens to antigen-specific immune lymphocytes (this is called antigen presentation) and B-lymphocytes respond by producing the antigen-specific antibodies. This means that every infectious cell contains billions of antigenic molecules, the most important among them being the surface and the extracellular antigens.

During infection or artificial immunisation of animals (generally rabbits, goats or sheep) a mixed population of IgG molecules are formed called polyclonal antibodies; that is, many B cells are stimulated to produce antibodies to a complex antigen. Antisera are prepared from the blood of the immunised animals and used for performing serological reactions (Table 11.3).

Polyclonal antisera, however, have several disadvantages such as the variability in the antibody spectra during the immunological response, which causes results that are not entirely reproducible. Problems may also occur as a result of cross-reactions among the antibodies. For these reasons polyclonal antibodies are difficult to standardise for immunodiagnostic procedures. However, the advent of the hybridoma technique at the end of seventies overcame these disadvantages. In the hybridoma technique the isolated B-lymphocytes of an immunised mouse (each B-lymphocyte clone secreting only one specific antibody) were fused with culturable tissue (myeloma) cells and the selected hybrid cells (called hybridomas) were used for the production of monoclonal antibodies. Involvement of the monoclonal antibodies in the immunoassays greatly enhanced development of the immunodiagnostic procedures by providing a consistent and reliable source of characterised antibodies of high affinity. Therefore, immunodiagnostic procedures applying polyclonal antibodies have been succeeded by monoclonal antibody techniques.

Further advancement in the production of antibodies with enhanced specificity and affinity has been achieved since the elaboration of the methodology of recombinant vaccine production.

Table 11.3. Types and components of basic serological reactions.

| Serological reaction | Antigen | Antibody | Type of Ig |
|---|---|---|---|
| Agglutination | Cells | Agglutinin | IgG, IgM, IgA |
| Precipitation | Soluble (colloidal) macromolecules | Precipitin | IgG, IgM, IgA |
| Cell lysis | Cell membrane | Lysin | IgG, IgA |
| Toxin-antitoxin reaction | Extracellular toxin | Antitoxin | IgG, IgM |

In this technique DNA sequences of selected IgG molecules coding for the antigen binding sites are cloned in *E. coli*, which as a consequence of gene expression produce recombinant vaccines. This cloning system made it possible to produce "mutant antibodies", which have even greater specificity and affinity.

Bacterial antigens may be divided into three main types according to their location:
1. Cell surface antigens, which are readily accessible to antibody reactions such as the somatic (O), flagellar (H), capsule (K), fimbriae antigens and outer membrane proteins (OMP);
2. Non-exposed cellular antigens, present inside the cell wall or in the cytoplasm and inaccessible in an intact organism to antibody reactions;
3. Extracellular antigens such as exotoxins, bacteriocins and enzymes (e.g. streptokinase, haemolysins, collagenase, DNase, etc.)

**O antigens** are lipopolysaccharides (LPS) present on the outer surface of bacterial cell walls in Gram-negative bacteria. They are heat stable and are also resistant to several chemicals like ethanol. LPS molecules of *Enterobacteriaceae* are complex molecules composed of three distinct moieties, a lipid component, a core oligosaccharide and a side chain that confers the O serotype specificity.
**H antigens** are proteins of flagellae and ciliae that are usually highly immunogenic. The major protein component of flagella is flagellin. H antigens are heat-labile and are inactivated by acids and alcohol.
**K antigens** are capsular polysaccharides that can be removed by heat treatment although they are not typically heat-labile. Although they are rarely involved in serotyping they can inhibit the O-antigen reaction and therefore it may be necessary to use heat treatment before O serotyping. K antigen of *Salmonella typhi* is called Vi antigen.
**Exotoxins** of bacteria are generally highly immunogenic (there are exceptions); the antibody is the antitoxin.
**Extracellular polysaccharides (EPS)** are produced by moulds during growth in food and culture can be easily detected immunologically, thus providing a suitable assay for detecting *Aspergillus* and *Penicillium* spp. An immunogen must be of a certain minimum molecular weight (approximately 5000 Da) to induce a response. Although the actual area of the antigen that is recognised by the antibody is relatively small, antibodies can distinguish between proteins that differ by only a single amino acid. Many small organic molecules are not antigenic by themselves but can become antigenic and a specific antibody is produced if they are covalently bound to a larger carrier molecule such as a protein. Such small molecules capable of eliciting antibody production are termed hapten. An antibody specific to the hapten moiety of the antigen will bind not only to the conjugated antigen but also to the free hapten. Different immunoassays have bee developed for very sensitive and specific detection and quantification of haptens like antibiotics, growth hormones, mycotoxins, preservatives, additives, etc in food.
**Immunodiagnostic tests** are one of the main techniques in the identification and characterisation of pathogenic microorganisms and their metabolic products. As for all the diagnostic methods, sensitivity and specificity are also the most important characteristics of these techniques. In this case specificity is the ability of an antibody preparation to recognise a single antigen. A desirable level of specificity implies that the antibody is specific for a single antigen,

## Microbial analysis of food

and will not cross-react with any other antigen, and therefore will not provide false positive results. Sensitivity defines the lowest amount of antigen that can be detected. The most desirable level of sensitivity implies that the antibody in the test is capable of identifying a single antigen molecule. High sensitivity prevents false negative reactions. The sensitivity order of some common immunological tests is as follows:

Precipitation reactions < immunofluorescence < direct agglutination < indirect agglutination < radioimmunoassay (RIA) < enzyme linked immunoassay (ELISA)

Sensitivity of different immunoassays suitable for detection of foodborne pathogens is shown in Table 11.4.

Immunoassays can be classified as either homogeneous or heterogeneous. In homogeneous (or "marker-free") immunoassays only the native antibody is added to the test material, thus it is not necessary to eliminate the free (unbound) antibodies during the assay. The result is then obtained directly. These immunoassay techniques involve the oldest and most simple tests such as agglutination and precipitation. They are generally simple tests with short reaction times, and are therefore easier to automate than heterogeneous assays. Whole cells or soluble molecules of the cells are applied in the test and agglutination or precipitation will be the result of specific binding of the antigens to the antibodies, respectively. If concentrations of the participating components are equally high, the reaction can be examined visually. Antigens and antibodies are then quantified by titration. The titre is the highest dilution of the antigen or antibody

*Table 11.4. Sensitivity of immunoassays for foodborne pathogens (Barbour and Tice, 1997).*

| Format | Sensitivity |
|---|---|
| Direct microscopic detection (fluorescent antibody) | |
| ■ Salmonellae | not available |
| ■ E. coli O157:H7 | 16 CFU/g |
| Immunodiffusion | |
| ■ Staphylococcus enterotoxins | 0.3 µg/ml |
| ■ Salmonellae | not available |
| Latex agglutination | |
| ■ E. coli heat-labile enterotoxin | 32 ng/l |
| ■ Salmonellae | not available |
| Enzyme-linked immunosorbent assay (ELISA) | |
| Toxins | |
| ■ Aflatoxin | 10-250 pg/ml |
| ■ Botulinum toxin | 50-100 mouse $LD_{50}$ |
| ■ Staphylococcus enterotoxins | 0.1 ng/ml |
| Cells | |
| ■ Salmonellae | $10^5$-$10^6$/ml |
| ■ Listeriae | 1 CFU/g |

solution that is able to produce a serological reaction. The general procedure of titration is to set up a two-fold dilution series of the antisera or the antigen (1:2, 1:4, 1:8, 1:16, 1:32, and so on) and determine the highest dilution producing a serological reaction. The titres are labelled 2, 4, 8, 16, and 32, respectively.

In heterogeneous assays a conjugated (labelled or enzyme-linked) specific antibody is detected, which is why it is necessary to eliminate (wash out) the unbound antibodies. It requires longer assay times than the homogenous tests but the separation of free labelled antibodies from the bound ones gives better potential for high sensitivity, and the removal of food matrix during the process can improve specificity. ELISA is the most common example of this type of assay.

*Agglutination*
Agglutination involves the binding of a particulate antigen (generally whole cells) to the antibody. Many agglutination reactions are used for the detection of antigens or antibodies in microbial diagnosis, as a simple, inexpensive, highly specific and rapid immunoassay. Despite this they are not as sensitive as some other immunoassays. In slide agglutination tests immunosera containing species-specific or strain-specific antibodies are added to a dense cell suspension of the microorganisms on a microscopic slide, and clump formation is evaluated as a positive reaction. For some pathogens different antisera are available for testing the virulence factors. These antisera contain antibodies that are specific to the polymorphic antigens present in the cells. This is called serotyping and is still very valuable for the rapid detection and discrimination (typing) of the most frequent pathogens (e.g. *Salmonella enteritidis* serotype 4b). The limitations of the technique are that a large series of antisera should be available for typing, cross-reaction between the antisera may happen and experience is necessary for the evaluation. It is mostly used for confirmation of selective or biochemical tests, to characterise a pathogen and for epidemiological studies. Several enteropathogenic Gram-negative bacteria like *Salmonella* are tested by the presence of O (cell wall) and H (flagellar) antigens using poly"O" and poly"H" antisera, respectively. For confirmation of presumptive positive *Salmonella* colonies, positive reactions are required for both antisera, which cover most of the *Salmonella* serovars likely to be encountered in foods. Further serotyping includes application of O- and H-monospecific antisera, which enables identification of the serovars and subtyping of a given isolate. In *E. coli* a particularly pathogenic strain was classified, which is identified by the presence of two specific antigens (number 157) in its cell wall (designated O157) and antigen number 7 on its flagella (designated H7). This strain, known as *E. coli* O157:H7 has emerged in the recent past as one of the most dangerous enteropathogenic bacteria.

Instead of using soluble antibodies present in the immunosera, antibody-coated particles can be applied in the agglutination reaction. Latex beads and magnetic beads are most frequently used for such purposes. In latex agglutination small (0.8 µm) synthetic latex beads coated with the antibody(s) specific to the cellular antigen(s) are mixed with small amount of cells (e.g. a colony) on a microscope slide and incubated for a short period of time. If the organism is present, the milky-white latex suspension will become visibly clumped, indicating a positive agglutination reaction. A series of latex bead agglutination assays has been developed for the rapid detection and identification of pathogens like *Staphylococcus aureus, Streptococcus pyogenes, Campylobacter* spp. *E. coli* O:157:H7, *Vibrio cholerae* O1, etc. Another type of latex agglutination is based on the competition of the hapten in the sample with the same hapten molecule immobilised on the latex beads for the binding site on the antibody. If there is no hapten in the sample, the antibody can react only with the latex particles. This results in an agglutination reaction that can be seen visually or can be measured turbidimetrically. If there are hapten molecules in the sample, they will compete for the antibody binding sites and the agglutination reaction slows or fails. Therefore, the rate of change in turbidity is inversely related to the concentration of hapten molecules in the sample.

Magnetic particles coated with antibodies are used as immunocaptures in the immunomagnetic separation (IMS) technique: a homogenised food sample, pre-enriched or enriched culture is mixed with the antibody-coated magnetic particles. After agitation the magnetic beads are collected with the aid of an external magnet, then the supernatant is removed. A new technique has been developed recently where the food sample is circulated by a pump which throws the magnetic beads, thus enhancing the efficacy of separation. When cells are collected ("captured") the beads are plated on a medium and incubated for growth. As an alternative to cultural enrichment procedures, immunomagnetic separation provides a rapid and targeted concentration of the pathogens.

*Enzyme-Linked Immunosorbent Assay (ELISA)*
The enzyme-linked immunosorbent assay (ELISA) has become one of the most widely used serological tests for antigen detection. This method employs enzymes to label antibody molecules used for antigen detection. Adding chromogenic substrate to the antibody-enzyme conjugate - antigen complex, induces an enzymatic reaction which produces a coloured product that is detected or measured by a colorimeter. Typically linked enzymes include horseradish peroxidase, alkaline phosphatase and β-galactosidase. Two basic ELISA methods are used: the direct (sandwich) assay and the indirect assay. The first technique is used for detecting antigens and the second for detecting antibodies. In the sandwich assay (the first specific "capture") antibody is immobilised on a solid support, usually plastic (polystyrene) microtiter plates, tubes, nitrocellulose membranes or magnetic beads. The test antigen is then added, which reacts with the capture antibody. After washing away the unbound antigens, the second enzyme-conjugated "reporter" antibody is added. Unbound antibodies are again removed by washing. A three-layered complex is formed which looks like a sandwich. The reporter antibody is detected by adding a substrate specific to the enzyme. If the enzyme-linked antibody is present, an enzymatic reaction will occur resulting in a coloured product. The intensity of the colour developed is proportional to the amount of antigen present. Figure 11.4 illustrates the scheme of *Salmonella* antigen detection by a sandwich ELISA assay.

In the indirect ELISA assay the "sensitising" antigen is adsorbed onto a solid support (generally microtiter wells or latex beads), and free antigens are then washed away. Test antiserum is then added, and antibody-antigen complexes (binding) are formed if antibodies specific to the antigen are present. Unbound antibodies are then washed away followed by the addition of enzyme-bound antibodies. The conjugate binds to the test antibodies and after unbound conjugate is washed away, the attached ligands are visualised by the addition of a chromogenic substrate.

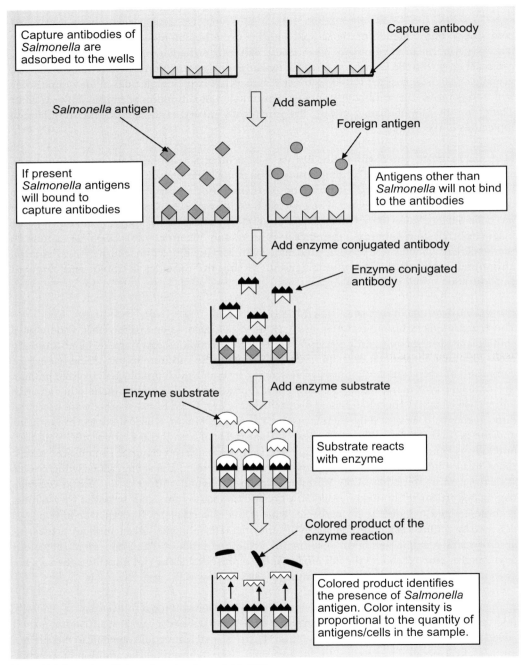

Figure 11.4. Detection of Salmonella antigen by sandwich ELISA.

The amount of test antibodies present is quantified in the same way as antigens in the double sandwich method.

Although ELISA techniques give results within a short period of time (generally between 1 and 3 hours), pre-enrichment and enrichment stages are required to increase cell numbers to a level that can be reliably detected (about $10^5$ cells per ml). Overall there is a gain of 2 days over the traditional method. A number of approaches have been used to speed things up even further, e.g. a system that captures *Salmonella* cells on a dipstick. The dipstick is dipped into the pre-enrichment broth and captures *Salmonella* cells. The stick is then transferred to a growth medium enabling the cell numbers to increase before ELISA is carried out. This allows for the selective enrichment stage to be dispensed with, saving another day.

*Radioimmunoassay (RIA)*
Radioimmunoassay (RIA) is similar in principle to ELISA, only differing in that radioisotopes are used for labelling the antibodies instead of enzymes. In the original RIA technique a radioisotope-labelled antigen is used that competes for antibody with unlabelled standard or antigen in experimental samples. The radioactivity associated with the antibody is then detected by means of radioisotope analysers. If there are a lot of antigens in the experimental sample, they will compete with the radioisotope-labelled antigens for antigen binding sites on the antibodies, and little reactivity will be bound. The isotope used most commonly is $^{125}$I. The use of radioisotopes has some disadvantages such as the instability of isotopes, associated health risks and the problem of waste disposal. For these reasons isotope labelling is less frequently used, and other alternative "cold" labelling systems such as fluorescence have been developed and commercialised.

*Automated immunoassays*
Automated immunoassays are becoming very common in food microbiology laboratories. They have been developed basically for testing large series of samples but now equipment for testing a low number of samples are also available. Beside ELISA, fluorescent immunoassay (FIA) and chemiluminescent immunoassays (CLIA) have also been automated. All assay steps (adding of reagents, washing, detection and evaluation) are performed automatically by the instrument. In addition, some commercial immunoassay systems are fully automated and include barcode specimen ID. For foodborne microbial antigens, the VIDAS system (bioMerieux), Opus (Tecra) and EIAFoss (Foss Electric) are examples of automated immunoassays.

## 11.3.10 Nucleic-acid based assays (Barbour and Tice, 1997)

*Nucleic acid hybridisation assays*
Hybridisation of nucleic acid molecules is based on the rule that single stranded (ss) nucleic acids reconstitute the double stranded (ds) form via hydrogen bonds under appropriate conditions, if extended complementary sequences are present in the combined nucleic acid molecules. Hybridisation of complete genomes is rarely used in food microbiology, but in taxonomy determination of the percentage of homology it is a widely used approach for delineating or merging species. Theoretically, closely related species or strains generally have

a high percentage of homology but distantly related species have a low percentage. However, taxonomic distance cannot be estimated from the level of homology. Not only DNA-DNA but also DNA-RNA hybridisation can take place between the transcribed RNA molecules (mRNA, rRNA and tRNA) and the template DNA strand. In food microbiology nucleic acid hybridisation is basically used to detect pathogens by targeting their DNA or RNA sequences using labelled complementary DNA molecules known as probes (Falkinham, 1994). There are different hybridisation assay formats, many of which have been commercialised and used as rapid molecular techniques for detecting pathogens or toxins. The sensitivity of the assay is generally in the range of $10^4$ to $10^6$ copies of the target gene, which can also be expressed as cell concentration if the copy number of the target gene is constant.

Preparation of the target DNA samples includes cell permeabilisation or lysis, and a certain level of DNA purification is generally necessary. As the food matrix can interfere with hybridisation or detection reactions, proper dilution of the food sample is a general basic requirement before the assay.

As mentioned earlier, gene probes have been developed for the specific detection of particular genes by hybridization. Probes are shorter or longer ssDNA molecules and the most suitable oligonucleotide probes are in the range of 15 to 30 base pairs (bp). Probes have to be labelled for recognition of the target gene by hybrid formation, thus providing mainly qualitative (presence or absence) or sometimes quantitative results.

The specificity of hybridisation is influenced by several conditions, which have to be optimised for each detection assay. The stringency of hybridisation (a measure of the degree of base pair mismatch) increases by increasing the length of probes, but in some cases a certain level of mismatch pairing is necessary between the target gene and probe, because some sequence variation may occur among the target genes. The other major factors influencing the stringency of hybridisation are temperature and salt concentration. Once a probe sequence is determined, optimal conditions for the favourable stringency are generally empirically determined.

Labelling of probes for detection is a basic requirement. Radioactive labelling ($^{32}P$ or $^{35}S$) was the first technique used because of its high sensitivity and stability, however, similarly stable and sensitive non-radioactive ("cold") labelling techniques have now been developed. There are a number of advantages to non-isotopic probes; 1/ The non-isotopically labelled probe can be detected more rapidly; 2/ The sensitivity can be equal to the radioactive labelling or even higher; 3/ Non-isotopic probes can be more stable; 4/ There is no radioactive waste produced and personnel are not exposed to the hazards associated with the use of radioisotopes. As a consequence of the widespread use of nucleic acid probes, non-isotopically labelled probes are becoming more and more popular. Probes for routine purposes are always labelled non-radioactively.

Non-radioactive labelling offers either direct or indirect detection. In the direct system the capture probe and the reporter probe are covalently linked. The advantage of this technique is that only a single reaction is required for detection, however, the labelled reporter has to maintain its activity during the whole procedure. In the indirect system detection of the hybridised probe (e.g. by enzymatic reaction) is delayed after hybridisation is completed. Labelling groups can be incorporated into the probes basically in the same manner as DNA molecules are labelled for laboratory purposes (e.g. by nick translation, random priming, reverse transcription), but the labelling of probes by PCR amplification seems a more reliable and convenient technique.

Hybridisation assays can be classified in different ways. One possible means of classification is by determining whether hybridisation of the probes and NAs takes place in solution or if one of them is anchored to a solid support. Solid supports are typically different membranes or polymer matrices. The first DNA detection system developed was Southern blotting. In this technique isolated DNA is first cleaved by a restriction endonuclease to small fragments (< 50 kbp), it is then separated by gel electrophoresis and transferred to a nitrocellulose membrane. Finally, a radioactively labelled probe is hybridised to the denatured membrane-bound DNA fragments. Presence of the target gene is checked by autoradiography. Colony hybridization is a quite simple technique where colonies developed on an agar medium are first transferred to a membrane. In the next step, cells are lysed to release DNA, denatured by alkali treatment and fixed to the membrane. Labelled probe is added and the colonies containing the target gene are detected. Colony hybridisation is generally not sensitive or reliable enough for direct detection of pathogenic microorganisms in food. Such cases occur when, 1/ the level of the pathogen is generally too low to be detected by direct plating, 2/ the level of the background microorganisms is too high and they overgrow the pathogens, and 3/ if the virulence gene is present on a plasmid, which has been lost during the enrichment procedure. Another method using immobilised target DNA is the dot blot. The only difference between the colony and dot blot is that in the latter case the origin of the target cells is a liquid. Besides this all the other steps are similar in both assays. Sandwich hybridisation offers more sensitivity than the formerly introduced techniques. The concept of sandwich formation is similar to ELISA, because the capture probe is immobilised to a solid phase, after binding of the target NA to the capture probe, the reporter probe will then hybridise to an adjacent sequence of the same target molecule. Unbound probes are washed away and the reporter is detected according to the labelling technique.

Besides radioactive labelling the other labelling techniques are similar to those described in antibody labelling assays. Biotin-avidine or biotin-streptavidine, and digoxigenin-antidigoxygenin reactions can be used for labelling if a very high sensitivity is not required. Antibodies covalently bound to enzymes such as horseradish peroxidase or alkaline phosphatase are added after hybridisation of the probe and target gene take place. Enzyme reactions can be followed by adding colorimetric, fluorescent, chemiluminescent or bioluminescent substrates. Assay sensitivity is immensely improved in the last two assays mentioned.

Some examples of nucleic acid hybridisation-based assays suitable for the detection of foodborne pathogens are shown in Table 11.5.

Table 11.5. Examples of nucleic acid hybridisation-based assays for foodborne pathogens (Hill et al., 2001).

| Bacteria | Target gene |
|---|---|
| Escherichia coli | Heat-stable toxin (ST) |
| ETEC | Heat-labile enterotoxin (LT) |
| EHEC | Shiga-like toxins (SLT) |
| EIEC | Invasive genes |
| EHEC O157:H7 | O157:H7 |
| Listeria species | r-DNA |
| L. monocytogenes | Listeriolysin O |
|  | Major secreted polypeptide (msp) |
| Salmonella species | r-DNA |
| Shigella species | Invasive genes |
| Campylobacter jejuni | r-DNA |
| Vibrio cholerae | Cholera enterotoxin |
| V. parahaemolyticus | Thermostable direct hemolysin (tdh) |
| V. vulnificus | Cytotoxin-hemolysin |
| Yersinia enterocolitica | Cytotoxicity/Sereny |
|  | Invasive gene (ail) |
| Staphylococcus aureus | Enterotoxin B |

## 11.3.11 PCR-based assays (Atlas and Bej, 1994)

In the late eighties the Polymerase Chain Reaction (PCR) technique was developed and has since revolutionised several areas of molecular biology, among them molecular diagnosis. Many of the molecular biology techniques used before PCR were labour intensive, time consuming and required a high level of technical expertise. Additionally, working with only trace amounts of DNA made it difficult to apply basic techniques such as NA hybridisation.

*Basics of PCR*
The objective of PCR is to produce a relatively large amount of a specific piece of DNA from a very small amount *in vitro*. Technically, it is performed by the controlled enzymatic amplification of a template DNA molecule containing the specific DNA sequence of interest. The template can be any form of dsDNA such as genomic DNA, plasmids, or DNA from bacteriophages and other viruses. In theory, only one copy of the template DNA is needed to generate millions of copies of new DNA molecules. The ability to amplify precise sequences of DNA for further analysis or manipulation is the true power of PCR. Some of the main reasons that PCR is such a powerful tool are its simplicity and specificity. All that is required for the reaction are relatively inexpensive reagents. Specificity of PCR comes from the ability to target and amplify one specific segment of DNA (or gene) out of a complete genome or isolated DNA.

The PCR reaction consists of repetitive cycles of the following 3 steps: denaturation of template DNA, annealing of oligonucleotide primers and synthesis of complementary DNA strands (see Figure 11.5).

The basic steps of PCR are illustrated in Figure 11.6 and are also detailed below:
1. **DNA Denaturation -** Separation of the two strands of the DNA template by heat denaturation - usually around 94 °C.
2. **Primer annealing -** Annealing the primers to the template strands - cooling the system to around 40 to 60 °C. Primers are short oligonucleotides that flank the gene or DNA segment of interest. The oligonucleotide primers are designed to hybridise to regions of DNA that flank the desired target gene sequence, annealing to complementary strands of the target sequence. The appropriate temperature for primer annealing depends on the base composition, length and concentration of the primers. Generally, the ideal annealing temperature is 5 °C below the melting temperature ($T_m$) of the primers. Stringent annealing temperatures help to increase specificity. If the temperature is lower than the optimum, the PCR products frequently include additional non-target DNA fragments. Such non-specific background products can be further target sequences during the next PCR cycles and result in false DNA fragments.
3. **Extension -** DNA polymerase extends the primers - usually done at around 72 °C. Primers are extended across the target sequence by using a heat-stable DNA polymerase. As a result of which two new complementary DNA strands are synthesized.

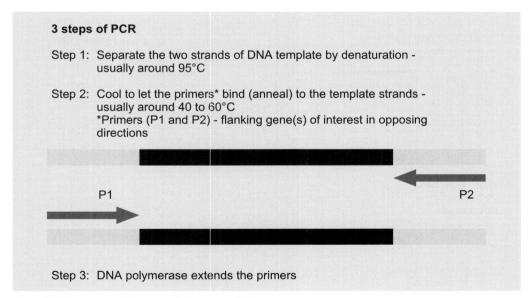

*Figure 11.5. Basic steps of Polymerase Chain Reaction (PCR).*

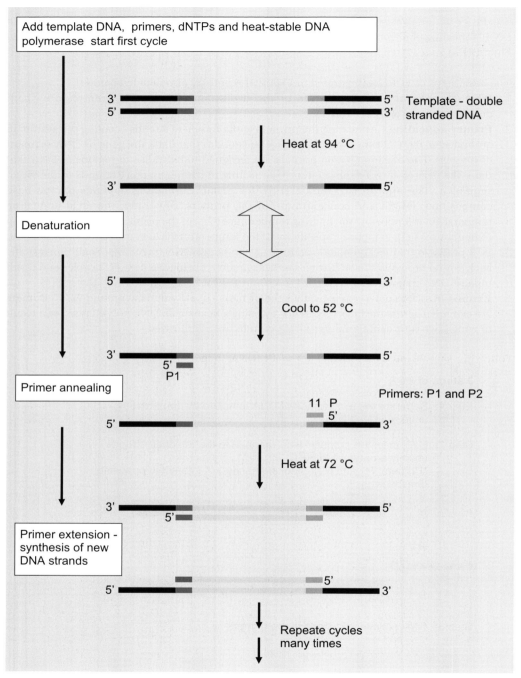

Figure 11.6. The schematic simplified principle of Polymerase Chain Reaction (PCR).

## Microbial analysis of food

By repeating the three-stage process many times over, a nearly exponential increase in the amount of target DNA results. The product of each PCR cycle is complementary to and capable of binding the primers, and therefore the amount of DNA is exponentially doubled in each successive cycle. Twenty cycles theoretically could produce 1 million copies of target sequence ($2^{20} = 10^6$), but for practical purposes 30 to 40 cycles are usually used, producing at least a $10^6$-fold increase in the concentration of the target DNA sequence. The power of PCR is shown by the fact that if each cycle takes about 5 minutes, $2^{12}$ copies can be made in one hour.

The essential reagents for PCR are:
- reaction buffer (containing $Mg^{2+}$);
- template DNA;
- two primers that flank the fragment of DNA to be amplified;
- deoxyribonucleotides (dNTPs as dATP, dTTP, dGTP, dCTP);
- heat-stable DNA polymerase.

The DNA polymerase used in PCR must be thermostable and thermoactive, that is, it should not be inactivated by the repeated heat treatments of PCR step 1 (at 92-94 °C) and it must also be active at the temperature of primer extension (at around 72 °C). The DNA polymerase from the thermophilic archea *Thermus aquaticus* (*Taq* polymerase) meets these requirements. Cloning and subsequent modification of the *Taq* polymerase gene has resulted in cheap and multipurpose variants of the enzyme.

Following sample preparation, the template DNA, oligonucleotide primers, thermostable polymerase, the four deoxyribonucleotides and reaction buffer are mixed in a single micro test-tube. The tube is placed into a thermal cycler. The thermal cycler contains a block that holds the samples and can be rapidly heated and cooled across extreme temperature differences. The rapid heating and cooling of this thermal block is defined as "temperature cycling" or "thermal cycling" (see Figure 11.7).

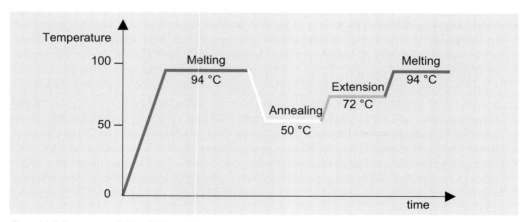

*Figure 11.7. Temperature profile of PCR reaction.*

The most simple and general detection of PCR products is based on the observation of the DNA bands of the expected size in an ethidium bromide stained agarose gel. For diagnostic purposes, other more complicated end point detection techniques have been developed, which can detect PCR products not only after gel separation but also in solution. The most important detection systems are:
- agarose or polyacrylamide (PAGE) gels;
- fluorescent labelling of PCR products and detection with laser-induced fluorescence;
- plate capture and sandwich probe hybridisation.

*Variants of the basic PCR method*
Beside the amplification of a target DNA sequence by typical PCR, several specialised types of PCR have been developed for specific applications.
- **Nested PCR** - PCR with nested primers applies two sets of primers that are used in two subsequent PCR runs. The first PCR is performed with a set of primers that produce a product larger than the targeted sequence. The stringency of the first amplification condition is generally low. This is achieved by using a lower annealing temperature than the optimum. A second PCR reaction is then performed with another set of primers that anneal to an internal region of the first PCR product and amplify the targeted sequence. The second PCR is run under stringent conditions, which result in an increase in the fidelity of the reaction. Nested PCR can also reduce the effect of PCR inhibitors.
- **Multiplex PCR** - Multiplex PCR is that type of PCR amplification when more than one DNA segment is amplified simultaneously by using multiple sets of primers. In this case primers with similar annealing temperatures should be chosen. The length of the target DNA sequences should also be similar, because the shorter sequence could be favoured over the longer one during the amplification, resulting in different yields of the amplified products.
- **MAMA-PCR** - Mismatch Amplification Mutation Assay (MAMA-PCR) can distinguish between targets that differ by only one nucleotide. Primers are constructed so that the 3' end of one of the primers contains a single mismatch with that allele to be detected and a double mismatch with that allele to be excluded. Thus, genes that differ by a single nucleotide in the amplification region can be detected and differentiated.
- **Real Time Quantitative PCR** - One of the drawbacks of conventional PCR is that estimation (quantification) of the initial concentration of the template in the reaction is difficult. There were several attempts to solve this problem, but the early quantitative PCR techniques were only suitable for the quantification of PCR products and not for the estimation of the initial target sequence quantity. The recently developed Real Time Quantitative PCR provides a very accurate and reproducible quantification of gene copies and has quickly revolutionised this field (Heid *et al.*, 1996). The method measures PCR product accumulation through a dual-labelled fluorogenic probe (TaqMan Probe). It uses the 5' nuclease activity of Taq polymerase to cleave a non-extendible hybridisation probe during the extension phase of PCR. The approach uses dual-labelled fluorogenic hybridisation probes. A fluorescent dye (e.g. FAM) serves as a reporter and its emission spectra are quenched by a second fluorescent dye (e.g. TAMRA). The nuclease degradation of the hybridisation probe releases the quenching of the FAM fluorescent emission, resulting in an increase in peak fluorescent emission at 518 nm. The use of a sequence detector (ABI prism) allows for measurement of the fluorescence

spectra of all 96 wells of the thermal cycler continuously during PCR amplification. Therefore the reactions are monitored in real time (Nogva and Lillehaug, 1999). Certain instruments allow detection of more than one pathogen simultaneously by using multiple TaqMan primer and probe combinations in the same tube, each being detected in a unique optical channel at the respective wavelength. The TaqMan assay is a versatile technique for the detection and identification of different pathogens in diverse food matrices at levels as low as $10^1$ cfu/ml, although pre-enrichment and DNA template preparation are necessary before the assay (McKillip and Drake, 2004). In recent years detection of *E. coli* O157:H7 in raw milk and other foods, *Salmonella* in meat (Chen *et al.*, 1997), *Clostridium botulinum* in modified atmosphere packed fish, enterotoxigenic *Bacillus cereus* in milk (Kim *et al.*, 2000) were reported as applications of the TaqMan approach.

- **Reverse transcriptase (RT)-PCR** - Not only DNA but also RNA can be the target sequence for amplification. However, it has to be converted to DNA first. The enzyme reverse transcriptase (RT) technique is used for synthesising a copy of the complementary DNA strand (cDNA), which can then serve as a template for standard PCR. RT-PCR is useful for detection of RNA viruses and also the expression of genes can be studied by this method via detection of mRNA transcripts.

As DNA can be quite persistent in dead cells, simple PCR that is based on a DNA target is not suitable for distinguishing between viable and dead cells. Moreover, traditional culture-based techniques used for the detection and enumeration of viable but non-culturable (VNBC) cells of pathogenic bacteria are not accurate. This is due to the fact that the selective media employed generally prevent growth and colony formation of these types of cells. Pathogenic bacteria can become sub-lethally injured or stressed during specific stages of food processing or storage, enter into a VBNC state, but still pose a threat to the consumer. Therefore, in order to detect and identify VBNC cells of pathogenic bacteria, as well as to assess virulent gene expression, RNA-based methods must generally be used. mRNA molecules are transcribed only in viable cells and are quickly decomposed after being used in translation. This means that mRNA transcripts are good indicators of either the metabolic status of bacteria or the presence of VBNC pathogens. The first assay for the detection and measurement of gene expression in foodborne pathogenic bacteria implemented reverse transcriptase PCR (RT-PCR). This technique involves total RNA extraction from enrichment cultures followed by DNase-I treatment to eliminate the interfering genomic DNA, thereby preventing false positive results. Reverse transcriptase converts mRNA to cDNA with sequence specific primer and dsDNA is synthesised in a conventional PCR cycle by a second primer flanking the sequence of interest. Although RT-PCR monitoring of cell viability proved applicable for bacterial foodborne pathogens, this methodology has not been widely applied in food control laboratories. This is attributed to the fact that the method is labour intensive and has poor reproducibility results.

A developed version of RT-PCR, NASBA (nucleic acid sequence-based amplification) can overcome the problems mentioned before (Compton, 1991). The first steps of NASBA are identical to RT-PCR, but during second-strand DNA synthesis the specific primer used contains the T7 RNA polymerase promoter sequence (Figure 11.8). The result is that the T7 RNA polymerase applied in the next step transcribes dsDNA, thereby generating many mRNA copies

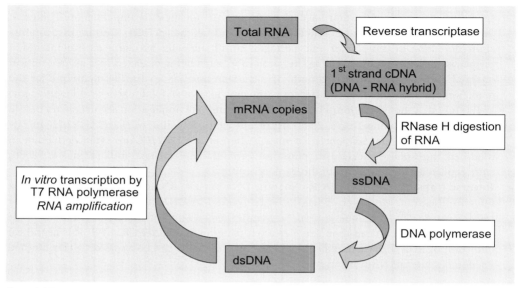

*Figure 11.8. Nucleic Acid Sequence-Based Amplification (NASBA) - amplification of mRNA.*

of the original transcript template. The amplified RNA can be visualised via subsequent RT-PCR or by incorporating fluorophores during its synthesis. The process continues at a single temperature (about 42 °C), so neither thermostable polymerase nor a thermocycler is required. Another advantage is that a high DNA background does not generate false positive results.

*Variants of DNA separation by gel electrophoresis*
Conventional gel electrophoresis of nucleic acid (NA) molecules is carried out by placing the samples that contain dissolved NA molecules in a solid matrix. Most commonly agarose or polyacrylamide gels are used as solid holders. In a static electric field NA molecules migrate through the pores of the gel, at rates dependent on their sizes and shapes. Small or compact molecules migrate more rapidly than large or loose molecules. Migration of linear molecules is in accordance with their molecular masses. After a defined period of time of electrophoresis (usually a few hours), the locations of the NA molecules in the gel are assessed by staining with a fluorescent dye, generally ethidium bromide is used. NA bands are then visualised under a UV light. A molecular size standard is run under the same condition for the estimation of the sizes of NA molecules.

Several parameters affect the separation and mobility of NA molecules in gel electrophoresis. Under ordinary conditions molecules up to 20 kilobase pairs (kb) can be separated according to their sizes. Above this size all molecules will show essentially the same mobility in a static electric field and, thus, will not be separated from each other. Using special conditions, conventional electrophoresis can separate NA molecules as large as 50 kb. Generation of DNA fragments of these sizes is generally achieved by using restriction endonucleases (often called restriction enzymes) for digestion of the DNA. Special types of restriction enzymes recognise and cut within specific short sequences of DNA. Therefore, the mixture of fragments obtained after the digestion of a particular sample of DNA is non-random and reproducible. Some information on the most frequently used restriction endonucleases are shown in Table 11.6.

Table 11.6. Examples of restriction endonucleases, their specific recognition and cutting sites.

| Enzyme designation | Producing bacteria | Recognition and cutting sequence* |
|---|---|---|
| Bam HI | Bacillus amyloliquefaciens H | ↓<br>5'-G-G-A-T-C-C-3'<br>3'-C-C-T-A-G-G-5'<br>↑ |
| Eco RI | Escherichia coli R | ↓<br>5'-G-A-A-T-T-C-3'<br>3'-C-T-T-A-A-G-5'<br>↑ |
| Eco RII | Escherichia coli R | ↓<br>5'-C-C-T-G-G-3'<br>3'-G-G-A-C-C-5'<br>↑ |
| Hae III | Haemophilus aegyptius | ↓<br>5'-G-G-C-C-3'<br>3'-C-C-G-G-5'<br>↑ |
| Hin dIII | Haemophilus influenzae d | ↓<br>5'-A-A-G-C-T-T-3'<br>3'-T-T-C-G-A-A-5'<br>↑ |
| Not I | Nocardia otitidis-caviarum | ↓<br>5'-G-C-G-G-C-C-G-C-3'<br>3'-C-G-C-C-G-G-C-G-5'<br>↑ |
| Pst I | Providencia stuartii | ↓<br>5'-C-T-G-C-A-G-3'<br>3'-G-A-C-G-T-C-5'<br>↑ |
| Sau 3A | Staphylococcus aureus | ↓<br>5'-G-A-T-C-3'<br>3'-C-T-A-G-5'<br>↑ |

*Arrows indicate the sites of enzymic attack. Note that in the double-stranded conditions, the base sequence will base pair with the same sequence (a palidome) running from the opposite direction.

If the enzymes recognise short (4-6) nucleotide sequences they will cut large DNA molecules into many specific fragments. Separation of the DNA fragments by electrophoresis results in a typical and reproducible pattern of DNA. This method is called REA (Restriction Enzyme Analysis) or RFLP (Restriction Fragment Length Polymorphism) and can be suitable for discriminating bacterial strains by molecular typing. The major limitation of this method is the difficulty of comparing complex DNA patterns, which consist of many bands that may be unresolved or overlapping. Hybridisation of specific labelled probes to the separated DNA fragments after transferring to a membrane (blotting) could overcome this problem by improving the resolution of bands in the patterns. The best results were obtained by using labelled probes of rDNA sequences for hybridisation.

*Pulsed Field Gel Electrophoresis (PFGE) techniques*
PFGE allows separation of large DNA fragments by a special gel electrophoresis technique in which the direction of the electric current switches according to a predetermined pattern. Large DNA fragments are produced by using infrequently-cutting restriction endonucleases. PFGE RFLP patterns contain less bands than the conventional RFLP patterns and are highly discriminatory and reproducible. There are some disadvantageous of this technique such as the complicated and laborious sample preparation procedure and the need for relatively expensive, specialised electrophoresis equipment.

*Denaturing Gel Electrophoresis (DGE) techniques*
Separation of nucleic acid molecules by simple agarose gel electrophoresis results in migration of the molecules according to their sizes. A reproducible pattern will be generated if the DNA solution contains molecules of distinct sizes. Differences in the nucleotide sequences of DNA molecules of the same size will not influence mobility. However, the denaturation properties of dsDNA molecules are influenced by the DNA base composition and sequence. The temperature when complete denaturation (melting) of the dsDNA molecules takes place is dependent on their G+C contents, and their melting points ($T_m$) are in direct proportional to their G+C contents. Partial denaturation of the DNA, however, is influenced not only by the G+C content but also by the nucleotide sequence of the DNA molecule under study. If a DNA molecule is melted in distinct domains, its mobility in the gel will be influenced by the extension of partial denaturation. Many electrophoretic techniques have been developed for the purpose of analysing the presence of sequence alterations in DNA molecules containing the same genes. The most common methods include different DGE techniques, such as *Denaturing Gradient Gel Electrophoresis (DGGE), Temporal Temperature Gradient Gel Electrophoresis (TTGE) and Constant Denaturing Gradient Gel Electrophoresis (CGGE)*. DGE techniques are able to identify small case, even one base-pair alterations (i.e. point mutations), in short (some hundreds bp long) DNA molecules. Such kinds of DNA fragments are produced by PCR amplification using specific primers and separated by polyacrylamide gel electrophoresis (PAGE).

A separation technique on which DGGE is based was first described by Fischer and Lerman (1983). In DGGE the denaturing environment is created by a linear denaturant gradient formed with urea and formamide at a uniform run temperature in the range of 50-65 °C. In this polyacrylamide gel, the dsDNA molecules are subjected to an increasing denaturant environment and will melt in discrete segments called "melting domains". The Tm of these domains is sequence specific. When the melting temperature of the lowest domain is reached, the DNA will become partially melted, creating branched molecules and reducing its mobility in the gel. DNA containing sequence alterations will encounter mobility shifts at different positions in the gel. If the fragments completely denature, then migration again becomes a function of size (Figure 11.9). The denaturing gradient may be formed perpendicular or parallel to the direction of electrophoresis (Figure 11.10). Parallel DGGE is more convenient for comparison of the mobility of PCR products because it allows better separation of fragments.

In TTGE DNA molecules migrate in a gel containing a constant concentration of urea. During electrophoresis, the temperature is increased gradually and uniformly. The result is a linear temperature gradient over the length of the electrophoresis run.

In CDGE the denaturant concentration that gives optimal resolution from a parallel or perpendicular DGGE gel is held constant. The optimal concentration of denaturant is determined from the concentration that causes the maximum split between the PCR products under analysis.

*Microarrays*
In living organisms thousands of genes and their products function in a complicated and orchestrated way that creates the mystery of life. However, traditional methods in molecular biology generally test one gene in one experiment, thereby giving only limited information and

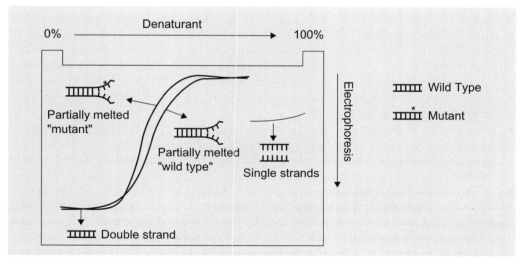

*Figure 11.9. DNA melting properties of DNA fragments (wild type and mutant alleles are shown here) containing sequence alterations and run in a denaturing gradient gel. At low concentration of denaturant the DNA fragments remain double-stranded, but as the concentration of denaturant increases, the DNA fragments begin to melt. Then, at very high concentrations of denaturant, the DNA fragments will completely melt, creating two single strands.*

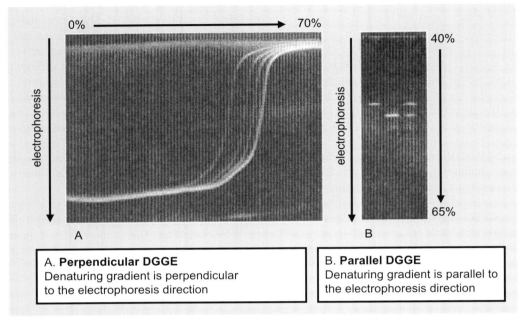

*Figure 11.10. Types of Denaturing Gradient Gel Electrophoresis (DGGE): perpendicular and parallel.*

the whole picture of gene function is therefore hard to obtain. DNA microarrays or chips make it possible to detect and type different pathogens simultaneously or analyse virulence gene expression in food-associated pathogens (Hong et al., 2004). DNA chips contain large arrays of oligonucleotides in solid support and provide a medium for matching known and unknown DNA samples on the basis of DNA hybridisation. The process and identification of the unknown is fully automated. The detection of the probe - sample hybrid is achieved by direct fluorescence scanning or by enzyme mediated detection (Gerhold et al., 1999). DNA microarray technology is implemented in two major formats:

- Identification of the target gene by DNA-DNA hybridisation. In this case a cDNA probe (converted from RNA by reverse transcriptase) is immobilised on a solid surface such as glass by a robot in a special array and exposed to the target DNA samples either separately or in mixture.
- Determination of the expression level/abundance of genes. In this variation, an array of cDNA oligonucleotides is immobilised on a solid support. The array is exposed to labelled sample cDNA and hybridised, and the identity/abundance of complementary sequences is determined with quantitative instrumentation. Such an approach is suitable for global analysis of gene expression and is a promising tool for future application in food microbiology.

Some examples of DNA amplification-based assays suitable for detection of foodborne pathogens are shown in Table 11.6. Immunological and molecular assays are frequently combined aiming to increase sensitivity and/or specificity of detection and typing. Table 11.7 shows the

Table 11.7. Relevant characteristics of immunological and DNA-based assays targeting pathogenic microorganisms in food (Barbour and Tice, 1997, modified).

| Sample | Relevant characteristics | | | | |
| --- | --- | --- | --- | --- | --- |
| | Target concentration | Food interference | Direct assay | Applicable method | Accuracy |
| Food | very low | high | not amenable | immunocapture after dilution | dependent on sensitivity |
| Food, pre-enriched culture | low | high | amenable | PCR testing immunocapture | dependent on sensitivity and specificity |
| Food, selective enrichment culture | high | low | amenable | Immunoassay DNA hybridisation PCR testing | dependent on specificity |
| Colony (pure culture) | highest | no | amenable | Immunoassay DNA hybridisation PCR testing | dependent on specificity (final confirmation) |

applicability of immunological and molecular techniques in the direct detection of microorganisms in food, (pre)-enriched food or pure culture.

Table 11.8 and Table 11.9 show examples of commercially available immunological and nucleic acid-based assays for the detection of foodborne pathogens and toxins, respectively. These methods are validated and generally have one or more advantageous performance characteristics in comparison with the conventional microbiological tests.

## 11.3.12 Methodology for identification and typing of microorganisms

Taxonomy is defined as the science of biological classification. It consists of three tools: classification, nomenclature and identification. Classification is the orderly arrangement of taxonomic units into larger groups or taxa, e.g. species into genera. Nomenclature concerns the naming of the classified units according to the internationally accepted rules. The rules for the naming of bacteria are described in the *International Code of Nomenclature of Bacteria* (Lapage *et al.*, 1992). Identification of an unknown unit means that its characteristics are compared with that of the known units in order to name it appropriately; e.g. characteristics of an isolated strain should be determined, after which the name of the species where it belongs is specified. The basic taxonomic group in microbiology is the species, which could be characterised as a collection of strains which share several common properties and differ significantly from other groups of strains. A strain is a population of organisms that arise from a single organism and share practically identical characteristics (often also called a pure culture). Strains of a given species can differ in some characteristics from each other. Biovars are variant microbial strains that differ in biochemical or physiological characteristics whereas serovars have different

Table 11.8. Examples of commercially available immunological assays for the detection of foodborne pathogens and toxins (Feng, 2001).

| Assay | Organism | Test name | Manufacturer |
|---|---|---|---|
| Immunoprecipitation | E. coli O157:H7 (EHEC) | PetrifilmHEC | 3M |
| | Listeria | Clearview | Unipath |
| | Vibrio cholerae | CholeraSMART | New Horizon |
| Latex agglutination | E. coli O157:H7 (EHEC) | Prolex | PRO-LAB |
| | Listeria | Listeria latex | Unipath |
| | Campylobacter | Campyslide | Becton Dickinson |
| | Salmonella | Bactigen | Wampole Labs |
| | Staphylococcus aureus | Staph Latex | Difco |
| RPLA (reverse passive latex agglutination) | E. coli ETEC- labile toxin | VET-RPLA | Oxoid |
| | E. coli Shiga toxin | Verotox-F | Denka Seiken |
| | Vibrio cholerae enterotoxin | VET-RPLA | Unipath |
| Immunomagnetic separation | E. coli O157:H7 (EHEC) | Dynabeads | Dynal |
| | E. coli ETEC- stabile toxin | E. coli ST | Oxoid |
| | Listeria | Listertest | VICAM |
| ELISA (enzyme-linked immunosorbent assay) | E. coli O157:H7 (EHEC) | TECRA | TECRA |
| | Listeria | Listeria-TEK | Organon Teknika |
| | Bacillus cereus | TECRA | TECRA |
| | Clostridium botulinum | ELCA | Elcatech |
| | Salmonella | Salmonella TEK | Organon Teknika |
| | Staphylococcus aureus | S. aureus VIA | TECRA |
| | S. aureus enterotoxin | SET-EIA | Toxin Technology |
| ELFA (Enzyme-linked fluorescent assay) | E. coli O157:H7 (EHEC) | VIDAS | bioMERIEUX |
| | Listeria | VIDAS | bioMERIEUX |
| | Campylobacter | VIDAS | bioMERIEUX |

antigenic properties. Every species has its type strain, which is generally among the first strains characterised and exhibits a majority of the characteristics typical to the species. *Bergey's Manual of Determinative Bacteriology* (Holt et al., 1994) contains the most comprehensive and up-to-date system of bacterial taxonomy.

In food microbiology species-level identification of foodborne pathogens and frequently that of the food-poisoning microorganisms is very important. Sub-species identification relies on the ability of the techniques to recognise and discriminate strains or a series of isolates that belong to the same species. This is commonly referred to as typing. Descendants of a cell produced by vegetative multiplication (simple or mitotic division) harbour genetically (clonally) identical cells, as it is in the case of the single-cell colonies of bacteria, yeasts and moulds.

Table 11.9. Examples of commercially-available, nucleic acid-based assays used in the detection of foodborne bacterial pathogens (Feng, 2001).

| Organism | Trade Name | Format | Manufacturer |
|---|---|---|---|
| Clostridium botulinum | Probelia | PCR | BioControl |
| Campylobacter | AccuProbe | probe | GEN-PROBE |
| | GENE-TRAK | probe | Neogen |
| Escherichia coli | GENE-TRAK | probe | Neogen |
| E. coli O157:H7 | BAX | PCR | Qualicon |
| | Probelia | PCR | BioControl |
| Listeria | GENE-TRAK | probe | Neogen |
| | AccuProbe | probe | GEN-PROBE |
| | BAX | PCR | Qualicon |
| | Probelia | PCR | BioControl |
| Salmonella | GENE-TRAK | probe | Neogen |
| | BAX | PCR | Qualicon |
| | Probelia | PCR | BioControl |
| Staphylococcus aureus | AccuProbe | probe | GEN-PROBE |
| | GENE-TRAK | probe | Neogen |
| Yersinia enterocolitica | GENE-TRAK | probe | Neogen |

In food microbiology and epidemiology it is necessary to identify the pathogenic agents and discriminate or prove identity of the isolates. The most important requirements for ideal identification and typing systems are summarised below (De Boer and Beumer, 1999):
- universally applicable;
- independent of growth conditions;
- specific;
- sensitive;
- cheap;
- rapid;
- simple.

Characterisation of organisms for identification purposes is basically based on either the phenotypic or genotypic approaches. Phenotypic approaches (e.g. morphological and biochemical characters) rely on the expression of the examined characteristics that are coded for by the genes of the organism. However, the expression of genes is influenced by several environmental factors. Therefore, to obtain reliable results, the growth conditions during characterisation must be standardised. In contrast to the phenotypes of the organisms, genotypes are essentially an invariant nature of the nucleic acid sequences that comprise their genomes. Genomes are built up from nucleic acids, which consist of genetically determined, unique sequences. Nucleic acid sequences harbour either conservative (identical or highly similar) or variable genes, as well as non-coding sequences. Nucleic acid analysis provides a powerful tool for the classification and identification of microbes. Highly related organisms share highly similar nucleotide sequences, while organisms of distant taxonomic groups show extended variations in their genome sequences. Thus, gene or genome sequence determination and comparison offers a reliable tool for the evolution-based identification and classification of organisms. In many cases, nucleic acid analysis has confirmed classical taxonomic hierarchies. In other cases, new organisms and relationships that were not obvious from classical methodologies were discovered.

Classification based on the evolutionary relationships of the organisms is considered as the phylogenetic system. If organisms are grouped together based on the mutual similarity of their phenotypic characteristics, the system is considered as a phenetic one.

Phenotypic identification assays are based on the determination of several morphological and biochemical characteristics, that are - as we described earlier - highly influenced by the growth conditions. For identification purposes generally the most constant characteristics typical to the taxonomic unit, are selected. The cultivation conditions (e.g. culture media, pH, temperature, atmospheric gas composition, etc) should also match those described in the identification protocol. Miniaturised phenotypic systems have been developed for identification purposes that have highly improved the conventional, culture tube- and Petri dish-based techniques.

Genome-based identification assays meet almost all the requirements, mostly because of the rapid developing instrumental and automated molecular techniques. The invention of gene probes and polymerase chain reaction (PCR)-based techniques allowed for the design and development of tests that can rapidly detect and identify specific bacterial strains and/or virulence genes without the need to establish pure cultures. Also, the price of the molecular diagnostic kits tends to be comparable with those of culture-based facilities.

In the following we describe the main characteristics, advantages and disadvantages of the identification techniques most frequently used by routine diagnostic laboratories and also show the prospects of the currently developed identification schemes.

*Phenotypic identification systems*
**Conventional identification systems -** Conventional (often called classical) identification systems are based on the characterisation of microorganisms by colonial and cellular morphology and also a series of physiological and biochemical traits, which are obtained by using conventional microscopic and culture-based techniques. Using a dichotomous key method and testing for several individual characteristics, the list of possible species is narrowed until a positive identification is made and the name of the species is found. Computerisation can help the evaluation procedure. The applicability of this type of identification for routine microbiological analysis of food is limited, because standardisation of the assays is problematic and personal experience is generally required for correct and reliable identification.
**Miniaturised phenotypic assays** - Several companies have developed miniaturized identification systems, which have greatly improved conventional assays. The main advantages of miniaturised tests are that they are standardised, simple and rapid. Several assays are automated with automatic reading and data processing, containing a large comprehensive

library for correct identification. Standardisation of the inoculation step (age and size of the inoculum), incubation conditions and interpretation of the results, is a necessity. In the tests preformed culture media are incorporated to the wells of a plastic strip, offering a method for automatic reading. Results are based on enzymatic reactions, detection of growth or metabolic activity. Increase in cell density (turbidity) or change in colour are the general positive signals. Miniaturised tests for routine analysis are mostly designed for a particular group of microorganisms (e.g. *Enterobacteriaceae, Lactobacilli*), thus pre-identification could be necessary. Several small-scale identification kits are available for routine analysis such as the API Systems, Minitek, Enterotube, BBL-Crystal, etc. Some companies have developed large-scale automated miniaturised identification systems with a big library for species identification (e.g. Microbial Identification System by MIS Newark, DE, USA), these are, however, generally too expensive for a small microbiological laboratory. If such a service is available for a limited time, it can also be useful for routine identification purposes.

*Genotypic identification systems (Hill and Jinneman, 2000)*
A series of genes coding for ribosomal RNAs (rDNA genes) has proved to be ideal for genotypic identification purposes for all kinds of living organisms. Although ribosomal DNA genes are organised in a very similar manner in the living world, the consequent differences found among the bigger or smaller taxonomic units has provided a suitable tool for phylogenetic studies. In bacteria 16S (SSU = small subunit) and 23S (LSU = large subunit) rDNA cistrons and the internal transcribed spacer (ITS) regions have been particularly presented as ideal targets for molecular identification as they contain extremely conservative (stable) domains (having extremely low mutational and recombination rates) and less conservative domains. Moreover, a large number of copies are present, around one hundred per cell. Several molecular hybridisation and PCR-based techniques, as well as direct DNA sequencing have been used for developing molecular identification systems. The most direct method, based on the polymorphic character of the ITS region, is to carry out PCR amplification using primers of highly conservative flanking sequences. The PCR product can be digested with a restriction enzyme and the resultant products can be resolved electrophoretically. Not only the ITS region but other rDNA sequences can be amplified by PCR and digested with restriction enzymes for molecular identification and typing. These techniques have the common name ARDRA (Amplified rDNA Restriction Analysis) or RFLP of rDNA. A second approach is to use DNA probes which contain a sequence that is unique to the target of interest. The hybridisation of the PCR product with such a probe can be used to identify the target organism. The third approach is to carry out DNA amplification using general prokaryotic rDNA primers (especially for amplification of SSU sequences), then the PCR products are sequenced and the isolate of interest is identified using molecular databanks of rDNA sequences. rDNA databanks comprise of complete or partial rDNA sequences. The Ribosomal Database Project (RDP, http://rdp.cme.msu.edu) provides ribosome related data and services, including online data analysis and aligned and annotated 16S rRNA sequences. The bacterial SSU database now contains around 20,000 entries. Databases are generally available for free. All these rDNA-based identification techniques are often called ribotyping.

The use of fluorescently-labelled oligonucleotide hybridisation probes to detect and identify whole cells by Fluorescence In Situ Hybridisation (FISH) has been particularly popular during the last decade. In this technique fluorescently labelled probes target stable RNAs, generally the cytoplasmic rRNA molecules. The high degree of evolutionary conservation makes bacterial 16S and 23S rRNA molecules very suitable for phylogenetic studies above the species level. Nevertheless, the sequence variations found between the relatively conserved rRNA sequences form the basis for designing probes specific for most bacteria and other microbes. Furthermore, the high cellular abundance of rRNA allows individual cells to be identified by the labelled probes. FISH with rRNA-specific probes followed by observation with epifluorescence microscopy allows for the identification of a specific microorganism in a mixture with other bacteria.

This technique is relatively simple because no special treatment of cells or nucleic acid isolation is necessary before hybridisation, which takes place on a microscopic slide. Smears of pure or enriched cultures and food or clinical samples containing enough target cells for a microscopic examination are suitable for evaluation by this technique (Amann et al., 1995; DeLong et al., 1989).

Despite the success of FISH in many applications, it had several drawbacks that made the improvement of this technique necessary. Problems related to the decreased sensitivity and specificity of FISH are non-uniform cell permeability, variable target site accessibility, poor sensitivity if the rRNA content of target cells is low, and background fluorescence derived from autofluorescence of samples or unspecific binding of probes to compounds present in the samples (Theron and Cloete, 2000). FISH with PNA (Peptide Nucleic Acid) probes that target rRNA overcomes the majority of these problems. PNA is a DNA analogue with a polyamide backbone instead of a sugar phosphate backbone, which obeys Watson-Crick's rule of base-pairing rules for hybridisation to complementary nucleic acid targets, DNA or RNA (Egholm et al., 1993). Due to their uncharged neutral backbones, PNA probes have more favourable hybridisation characteristics than DNA probes, such as high specificity, strong affinity, and faster hybridisation kinetics that result in the improved hybridisation characteristics of highly structured targets such as rRNA. In addition, the relatively hydrophobic character of PNA compared to that of DNA, enables PNA probes to penetrate the hydrophobic cell walls of bacteria and fungi following preparation of standard smears.

The unique properties of PNA enable the development of assay formats, which go above and beyond the possibilities of DNA probes. PNA probes targeting specific rRNA sequences of bacteria and other microorganisms of food microbiological and clinical value, have recently been developed and applied to a variety of rapid assay formats. Some simply incorporate the sensitivity and specificity of PNA probes into traditional methods, such as membrane filtration and microscopic analysis; others involve recent techniques such as real-time and end-point analysis of amplification reactions (Stender et al., 2002). Ribosomal RNA sequences of closely related strains, sub-species, or even different species are often identical and therefore cannot be used as differentiating markers. One possibility of overcoming this limitation was by using a small stable cytoplasmic RNA molecule of high copy number that is tmRNA. (Schönhuber et al., 2001).

The general applicability of the PNA FISH method made the simultaneous identification of multiple species in a mixed population an attractive possibility, something never accomplished using DNA probes. Four colour images using differently labelled PNA probes provided simultaneous identification of *E. coli, P. aeruginosa, S. aureus and Salmonella*, thereby demonstrating the potential of multiplex FISH for various diagnostic applications (Perry-O'Keefe et al., 2001).

A commercial hybridisation-based molecular approach is the RiboPrinter Microbial Characterisation System (DuPont Qualicon, USA, www.qualicon.com/riboprinter.html). It is an automated and standardized ribotyping instrument, which generates characteristic "fingerprint" patterns of specific rDNA sequences. Characterisation refers to species-level identification and sub-species or strain characterisation (typing). It provides this information quickly (within 8 hours) and reliably, with a compact, fully automated instrument that requires no special skills to operate. It provides the speed, accuracy and resolution needed to identify microorganisms and characterise them efficiently and consistently.

Main steps of the analysis involve:
1. Picking the samples (typically pre-enriched or enriched).
2. Heat treatment of samples followed by cell lysis.
3. Released DNA is digested by a restriction enzyme for fragmentation.
4. DNA fragments are separated according to size by agarose gel electrophoresis.
5. Transfer of fragments to membrane.
6. Hybridisation with DNA probes and chemiluminescent labelling.
7. Pattern detection with a digitizing camera.
8. Data processing and generation of a RiboPrint pattern.
9. Identification and characterisation by using electronic RiboPrint Database.

The main advantages of the system are:
- Up to eight isolates can be tested at one time and up to 32 samples can be loaded in a workday.
- Characterisation and identification is performed less than 8 hours after pre-enrichment or enrichment of the sample.
- This analysis provides rapid and accurate results with great confidence.
- RiboPrint patterns characterise environmental isolates, pathogens, spoilage organisms, and beneficial organisms that are important with regard to food safety and other purposes.
- The RiboPrinter Database contain over 6000 RiboPrint patterns, covering almost 200 bacterial genera and over 1400 species and serotypes.

Large numbers of specific primers have been developed for the detection of strains belonging to a given species or certain serotypes. Figure 11.11 illustrates examples of when European serotypes of *Yersinia enterocolitica* strains were detected by a specific primer-pair (A) and when general primers were used for detection of the species *E. coli* and specific primers for detection of *E. coli* O:157 (B).

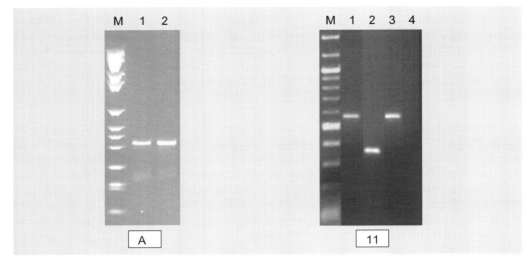

Figure 11.11. PCR detection of Yersinia enterocolitica (A) and E. coli O:157 (B) by specific primers. A. Y. enterocolitica strains amplified by a primer-pair specific to the European serotypes. B. E. coli; Lanes 1 and 2: O157:H7, lanes 3 and 4: non-pathogenic strain. Lanes 1 and 3: PCR products by species-specific primers. Lanes 2 and 4: PCR products by O157:H7-specific primers.

The BAX Microbial Detection System (DuPont Qualicon USA, www.qualicon.com/bax.html) is the first commercial PCR-based instrument developed for the screening and detection of pathogenic bacteria in food and environmental samples. The characterisation of food contaminating pathogenic protozoa like *Cryptosporidium parvum* is also possible. The BAX System provides a reliable pathogen detecting method for quality assurance and HACCP implementation. It targets a unique DNA sequence specific to the microorganism for detection by PCR. After sample preparation PCR quickly amplifies that DNA fragment to a level where it can be detected, therefore the results are available within a few hours. Including the pre-enrichment step (ca. 20 hours), results are obtained after one day.

Main steps of the analysis:
1. Collection of enriched sample.
2. Preparation of target DNA by heating the cells in a lysis reagent solution to rupture their cell walls and release the DNA (no DNA isolation is necessary).
3. PCR tablets, which contain all the reagents necessary for PCR (plus fluorescent dye for fluorescent detection), are hydrated with the lysed sample and processed in a thermocycler/detector.
4. Within a few hours the polymerase chain reaction (PCR) amplifies a DNA fragment that is specific to the target.
5. The PCR product is detected by gel electrophoresis or by a fluorescent signal.
6. Interpretation of the results: targeted microorganisms are indicated by the presence or absence test - presence or absence of a single band in the gel or measuring the activity of the labelled PCR product.

The main advantages of the system are:
- The equipment is highly automated and because it has a single program for all bacteria, this allows for tests of multiple targets in the same run and up to 96 samples per batch.
- The simplified sample preparation procedure and automated detection process minimise human errors and false positives.
- All the PCR components are combined into a single tablet that is pre-packaged in the sample tubes.
- Results are available the next day and are clearly displayed as a simple positive or negative report.
- User friendly screen prompts guide the user through the entire procedure, reducing the need for highly skilled technicians and expensive training.
- Electronic files enable users to print, share and store the results for easy archiving and retrieval.
- False positive results, due to frequently encountered DNA homology between closely related organisms, are minimized because the BAX system can detect segments of genes typical to the target species or strains.

Extensive validation shows that the BAX system is equal to or better than culture based methods in detecting pathogens. The speed and accuracy of the system allows for the determination of product safety before shipment, therefore avoiding shipping delays and recalls of suspected products.

*Molecular subtyping techniques*
Molecular subtyping aims to identify different strains within a species via techniques generating highly specific fingerprints of strains or isolates. Molecular fingerprints using statistical software are suitable for the comparison of a large number of isolates. The data obtained is suitable for taxonomic or epidemiological purposes.

Restriction Fragment Length Polymorphism (RFLP) was among the first molecular techniques developed for assessing strain-relatedness by molecular fingerprinting of organisms. In this technique DNA extracted from the cells or amplified by PCR is digested with sequence specific DNA splitting enzymes known as restriction endonucleases. DNA fragments generated by the enzymic digestion are then separated by gel electrophoresis. If rare cutting endonucleases are used for digestion of total cellular DNA, large fragments are generated that can be separated by pulsed field gel electrophoresis (PFGE). RFLP patterns of the same isolate obtained with a unique enzyme are completely identical because no sequence polymorphism exists that would generate new restriction sites or eliminate them. Different strains belonging to the same species generally have altering RFLP patterns. RFLP patterns of PCR amplified *flaA* gene sequences of *Campylobacter* strains are shown in Figure 11.12.

Separation of PCR products by DGGE is also suitable for discriminating strains belonging to the same or closely related species. Figure 11.13 illustrates such an example for typing *Campylobacter jejuni* and *C. coli* strains.

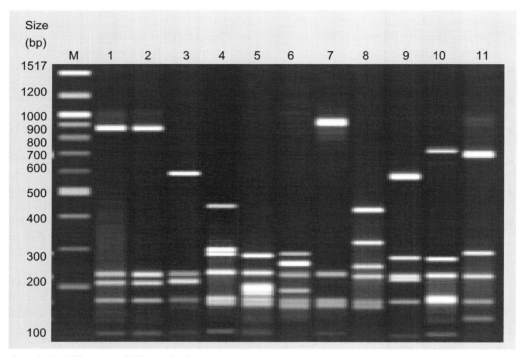

Figure 11.12. RFLP patterns of PCR amplified flaA gene sequences. Fragments were generated by DdeI restriction enzyme, separated by agarose gel electrophoresis and stained with ethidium bromide. Campylobacter jejuni type strain is in lane 8, C. coli type strain is in lane 11 and Campylobacter food isolates are running in the other lanes (M is molecular size standard).

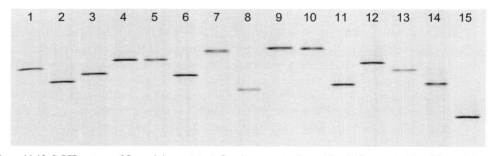

Figure 11.13. DGGE patterns of Campylobacter jejuni, C. coli type strains (lanes 14 and 15, respectively) and Campylobacter food isolates (lanes 1-13) based on flaA gene typing.

Random Amplified Polymorphic DNA (RAPD) fingerprinting is a PCR-based typing technique where patterns are generated by the amplification of an arbitrary nucleotide sequence and the amplified DNA is separated by simple agarose gel electrophoresis. These patterns arise from using a single, short primer (about 10-12 nucleotides) and the PCR reaction is conducted under non-stringent conditions that allow for mispairing of primer and template DNA, as the consequent primer will anneal at many different sequences throughout the genome. Some primers will be randomly positioned so that amplification can occur in conjunction with other nearby primers. Several regions will be amplified that are bound by template DNA sequences showing similarity to the primers. If typable strains contain polymorphic DNA (i.e. they are different in some genomic sequences), they exhibit distinguishable patterns. The major drawback of this technique is the poor reproducibility, evident in the extended variations that are observed at both intra- and interlaboratory level. This problem can be partly resolved by standardisation of the PCR conditions. Figure 11.14 shows RAPD patterns of strains belonging to different *Campylobacter* species. Some strains generate distinguishing patterns (e.g. lanes 3 and 9), but others do not give characteristic banding patterns.

The Amplified Fragment Length Polymorphism (AFLP) technique combines the simplicity of RFLP and the specificity of the PCR reaction by selective amplification of restriction fragments. An AFLP reaction typically results in fingerprint patterns consisting of 50-100 bands. Application of an automated laser fluorescent sequencer may help to simplify and enhance standardisation of this technique (Aarts *et al.*, 1999).

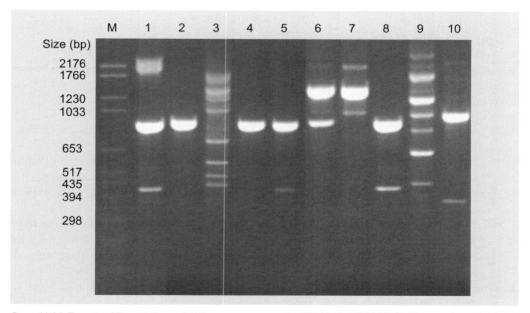

*Figure 11.14. Example of Campylobacter RAPD patterns, generated by OPE 15 (5'-ACGCACAACC-3') primer. Lanes 1-5: RAPD patterns of Campylobacter jejuni strains. Lanes 6-7: RAPD patterns of Campylobacter coli strains. Lanes 8-9: RAPD patterns of Campylobacter lari strains Lane 10: RAPD pattern of a Campylobacter faecalis strain.*

## 11.4 Case studies in detection and quantification/enumeration of pathogens and their toxins

### 11.4.1 Detection of Bacillus cereus emetic toxin (cereulide) in solid food using boar semen motility bioassay, cell cultures and HPLC-MS

*Bioassay*
Bioassay as such represents a suborganismic biological test where one does not use an organism as a whole, but rather certain functional elements of an organism. In this test Boar semen (commercially produced boar semen used for *in vitro* insemination with ca $15 \times 10^6$ cells ml$^{-1}$) is used. During the measurements, the Total motility of the semen, as well as Progressive motility, is followed. The total motility is visually observable movement of boar spermatozoa in space, while progressive motility designates movements that are characterized by certain values of motility: average path velocity (AVP), straight line velocity (SLV) and curvilinear velocity (CLV). Both parameters are measured using computer-aided semen analyser.

The test utilises the biological effect of cereulide on mitochondria and, via a computer, measures the change in boar semen motility as a toxic response parameter. It can give qualitative and semi-quantitative information of the toxin presence.

*Specific consumables and equipment*
- Leja two-chamber slides (one slide provides space for 2 analyses) (Figure 11.15).
- Hamilton Thorne CEROS using adequate software package.

*Practical procedure*
- Extraction:
  - Weigh 50 g of the food sample (wet wt) in a flat bottom glass flask and allow to evaporate to dryness.
  - After adding ethanol to the dry mass, let the submerged food incubate at room temperature for 1h with horizontal shaking and than boil for 15 minutes at 100°C.
  - Collect the ethanol phase in a glass phial, evaporate to dryness. Dissolve the residue in a suitable volume (1 to 5 ml) of methanol or ethanol.
  - Positive and negative control should come from the same type of food (or even the same food) inoculated with cereulide producing and non-producing strains. Extra negative control should be foreseen in order to make sure that the food intrinsic factors do not interfere with the semen motility.
- The test operation procedure:
  - Boar semen is motile at 37°C and therefore this temperature must be provided on the microscope stage during motility observation.
  - Boar semen and the sample should be mixed in ratio 40:1 and the mixture injected quickly to the Leja slide.
  - Follow the behaviour of the total and progressive motility.

## Microbial analysis of food

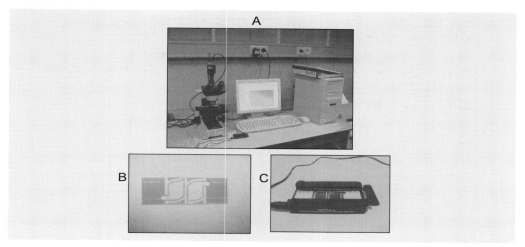

Figure 11.15. Hamilton-Thorne Ceros 12.1 semen analyser (A), Leja slide with 2 chambers (B) and portable stage slide heater (C).

*Data processing and report*

One should calculate integral change of motility with the time. This value should be compared with that of the positive control and with the standard curve. This will give a semi-quantitative estimate of the amount of the toxin present.

For only qualitative observation no data calculations have to be made, only observation that the progressive motility dropped for more than 20-30% of the negative control after 5 minutes of monitoring. If the progressive motility remains within 20% variation of that of the negative one may conclude that the sample was negative with the test detection limit (ca. 20 ng cereulide per g of sample). An example of the processed data, where progressive motility change is plotted against the time needed for a given change is shown in the Figure 11.16.

*Cell culture*
- Extraction procedure can be done in the same way as for the bioassay.
- The simplified test operation procedure (Finlay et al., 1999)

3-(4,5-Dimethylthiazol-2-yl)-2,5-diphenyltetrazolium bromide (MTT) is a yellow, water-soluble tetrazolium salt which is converted to insoluble purple formazan by metabolizing cells. The emetic toxin has been shown to affect mitochondrial function, and as MTT is regarded as an indicator of cell viability, and therefore cytotoxicity, it was considered that the toxin might affect MTT conversion.
- Serially diluted potato puree extracts is to be mixed with trypsinised Hep-2 (laryngeal carcinoma) at $10^6$/ml and the mix in microtiter plates incubated for 40 h at 37 °C in 5% $CO_2$.
- After removal of the medium MTT is applied and plates further incubated at 37 °C for 3 h.
- Solubilisation of intracellular formazan with dimethyl sulfoxide

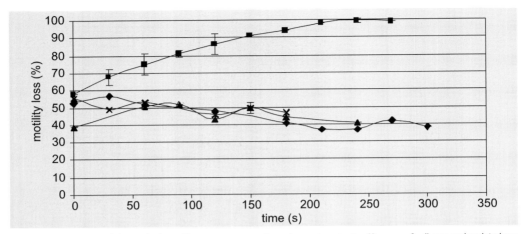

Figure 11.16. Evolution of the motility loss of boar spermatozoa after exposure to extracts of B. cereus foodborne outbreak isolates 5958c (♦), 5964a (■), 5969a (▲) and 5972a(×).

- Reading of absorbance at 570 nm with a microtiter plate reader.
- The mean endpoint titer should be recorded as the reciprocal of the highest dilution giving a colorimetric reading lower than that of the negative control.

*Chemical assay (HPLC-MS)*
- Extraction procedure can be done in the same way as for the bio-assay.
- The simplified test operation procedure (Häggblom et al., 2002).
  - Standards used in the test are valinomycin (commercially obtained) and cereulide purified from *B. cereus*, both dissolved in methanol.
  - Column used is $C_8$ column (100 by 2.1 mm, 5-μm particle size).
  - The solvent: mixture of 95% acetonitrile, 4.9% $H_2O$, and 0.1% trifluoroacetic acid at a flow rate of 0.15 ml min$^{-1}$, with a sample injection volume of 1.0 μl.
  - $A_{210}$ was monitored with a Diode array detector.
  - After chromatographic separation, the sample is introduced to the electrospray ion trap mass analyser. A mass range from 500 to 1,300 $m/z$ is to be collected.
  - For detection and quantification of low concentrations (< 100 pg per injection) close to the detection limit, single ions 1129 and 1171 ($NH_4^+$ adducts) were monitored for valinomycin and cereulide, respectively.

## 11.4.2 Detection of Salmonella spp. in foods and animal feed stuffs

More performance data is currently available for *Salmonella* isolation and detection methods than for any other pathogen. This and the fact that *Salmonella* are well established foodborne pathogens of an omnipresent nature led to the meticulous standardised methods for their detection. However, not only standard methods are in use, but also new rapid techniques providing significant shortening of analyses have found their application in modern food

## Microbial analysis of food

microbiology. Figure 11.17 gives a simplified overview of some of the possible detection methods, indicating standard method as the most time consuming and PCR based method as the shortest

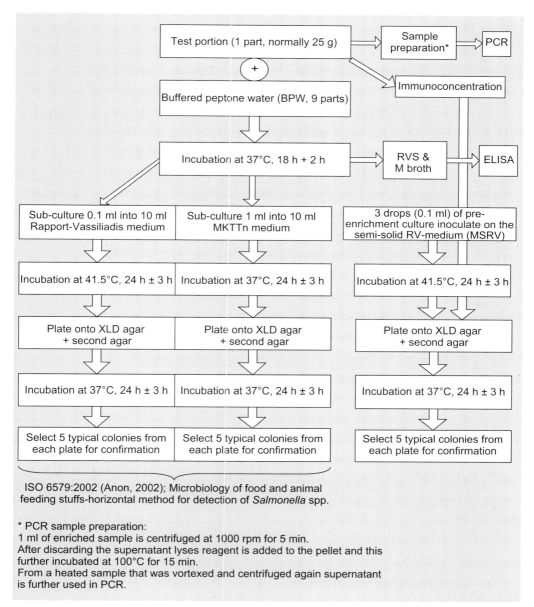

Figure 11.17. Simplified overview of different procedures in standard versus rapid methods for detection of Salmonella spp. in foods and feed stuffs.

ISO 6579:2002 (Anon, 2002); Microbiology of food and animal feeding stuffs-horizontal method for detection of *Salmonella* spp.

* PCR sample preparation:
1 ml of enriched sample is centrifuged at 1000 rpm for 5 min.
After discarding the supernatant lyses reagent is added to the pellet and this further incubated at 100°C for 15 min.
From a heated sample that was vortexed and centrifuged again supernatant is further used in PCR.

one. In all scenarios presented samples are first incubated in an entirely non-selective primary enrichment (pre-enrichment) medium. This step allows for resuscitation of stressed *Salmonella* cells. From there on differences between methods occur, as outlined in Figure 11.17.

# References

Aarts, H.J., L.E. Hakemulder and A.M. van Hoef, 1999. Genomic typing of *Listeria monocytogenes* strains by automated laser fluorescence analysis of amplified fragment length polymorphism fingerprint patterns. Int. J. Food Microbiol. 49:95-102.

Amann, R.I., W. Ludwig and K.H. Schleifer, 1995. Phylogenetic identification and in situ detection of individual cells without cultivation. Microbiol. Rev. 59:1919-1925.

Anon, 2002. ISO 6573:2002. Microbiology of food and animal feeding stuffs-Horizontal method for detection of *Salmonella* spp. Geneva: International Organization for Standardization.

Atlas, R.M. and A.K. Bej, 1994. Ch 19. Polymerase chain reaction. In: P. Gerhard, R.G.E. Murray, W.A. Wood and N.R. Krieg (eds.), Methods for general and molecular microbiology. pp. 418-435; Am. Soc. Microbiol., Washington.

Barbour, W.M. and G. Tice, 1997. Ch. 39. Genetic and immunological techniques for detecting foodborne pathogens and toxins. In: M.P. Doyle, L.R. Beuchat and T.J. Montwille (eds.), Food microbiology. Fundamentals and frontiers. pp. 710-727; ASM Pres, Washington.

Bolton, F.J. and D.M. Gibson, 1994. Automated electrical techniques in microbiological analysis. In: P.D. Patel (ed.), Rapid analysis techniques in food microbiology. pp. 131-169. Blackie Academic & Professional, London.

Chen, S., A. Yee, M. Griffits, C. Larkin, C.T. Yamashiro, C. Behari, C. Paszki-Kolva, K. Rahn and W.A. De Grandis, 1997. The evaluation of fluorogenic polymerase chain reaction assay for the detection of *Salmonella* species in food. Int. J. Food Microbiol. 35:239-250.

Compton, J., 1991. Nucleic acid sequence-based amplification. Nature 350:91-92.

Debevere, J. and M. Uyttendaele, 2003. Validating detection techniques. In: A. Thomas and A. McMeekin (eds.), Detecting pathogens in food. pp 69-92. Voodhead Publ., Cambridge, England.

De Boer, E. and R.R. Beumer, 1999. Methodology for detection and typing of foodborne microorganisms. Int. J. Food Microbiol. 50, 119-130.

DeLong, E.F., G.S. Wickham and N.R. Pace, 1989. Phylogenetic stains: ribosomal RNA-based probes for the identification of single cells. Science 243:1360-1363.

Egholm, M., O. Buchardt, L. Christensen, C. Behrens, S.M. Freier, D.A. Driver, R.H. Berg, S.K. Kim, B. Norden and P.E. Nielsen, 1993. PNA hybridizes to complementary oligonucleotides obeying the Watson-Crick hydrogen-bonding rules. Nature 365:566-568.

Falkinham, J.O. III, 1994. Ch. 28. Nucleic acid probes. In: P. Gerhard, R.G.E. Murray, W.A. Wood and N.R. Krieg (eds.), Methods for general and molecular microbiology. pp. 701-710; Am. Soc. Microbiol., Washington.

Feng, P., 2001. Rapid methods for detecting foodborne pathogens. Bacteriological Analytical Manual Online. htpp://cfsan.fda.gov/~ebam/bam-a1.html

Finlay, W.J.J., N.A. Logan and A.D. Sutherland, 1999. Semiautomated metabolic staining assay for Bacillus cereus emetic toxin. 65, 1811-1812.

Fischer, S.G. and L. Lerman, 1983. DNA fragments differing by single base-pair substitutions are separated in denaturing gradient gels: correspondence with melting theory. Proc. Natl. Acad. Sci., 80: 1579-1583.

Fung, D.Y.C., 1995. What's needed in rapid detection of foodborne pathogens. Food Technol. June 64-67.

Garbutt, J., 1977. Essentials in food microbiology. Arnold Publ., London Doyle.

Gerhold, D., T. Rushmore and C.T. Caskey, 1999. DNA chips: promising toys have become powerful tools. TIBS 24:168-173.

Hadley, W.K., F. Waldman and M. Fulwyler, 1985. Rapid microbiological analysis by flow cytometry. In: W.H. Nelson (ed.), Instrumental methods for rapid microbiological analysis. pp 67-89. VCH Publishers.

Haggblom, M.M., C. Apetroaie, M.A. Andersson and M.S. Salkinoja-Salonen, 2002. Quantitative analysis of cereulide, the emetic toxin of *Bacillus cereus*, produced under various conditions. 68, 2479-2483.

Heid, C.A., J. Stevens, K.J. Livak and P.M. Williams, 1996. Real Time Quantitative PCR. Genome Res. 6, 986-994.

Hill, W.E., A.R. Datta, P. Feng, K.A. Lampel and W.L. Payne, 2001. Identification of foodborne bacterial pathogens by gene probes. Bacteriological Analytical Manual Online. htpp://cfsan.fda.gov/~ebam/bam-a1.htm.

Hill, W.E. and K.C. Jinneman, 2000. Principle and applications of genetic techniques for detection, identification and subtyping of food-associated pathogenic microorganisms. In: B. Lund, T.C. Baird-Parker and G.W. Gould (eds.), The microbiological safety and quality of food, pp. 1813-1851; Aspen Publ. Gaithersburg, Maryland.

Hong, B.X., L.F. Jiang, Y.S. Hu, D.Y. Fang and H.Y. Guo, 2004. Application of oligonucleotide array technology for the rapid detection of pathogenic bacteria of foodborne infections. J. Microbiol. Methods 58:403-411.

Holbrook, R., 2000. Detection of microorganisms in foods - Principles of culture methods. In: B. Lund, T.C. Baird-Parker and G.W. Gould (eds.), The microbiological safety and quality of food, pp. 1761-1790; Aspen Publ. Gaithersburg, Maryland.

Holt, J.G., N.R. Krieg, P.H.A. Sneath, J.T. Staley and S.T. Williams, 1994. Bergey's Manual of Determinative Bacteriology, 9th ed. Williams & Wilkins, Baltimore.

Kim, Y.-R., J. Czaijka and C.A. Batt, 2000. Development of fluorogenic probe-based PCR assay for detection of *Bacillus cereus* in nonfat dry milk. Appl. Environ. Microbiol. 66:1453-1459.

Kyriakides, A.L. and P.D. Patel, 1994. Luminescent techniques for microbiological analysis of food. In: P.D. Patel (ed.), Rapid analysis techniques in food microbiology, pp. 196-231. Blackie Academic & Professional, London.

Lapage, S.P., P.H.A. Sneath, E.F. Lessel, V.B.D. Skerman, H.P.R. Seelinger and W.A. Clark, 1992. International Code of Nomenclature of Bacteria, 1990 revision. American Society for Microbiology, Washington, DC.

McKillip, J.L. and M. Drake, 2004. Real-time nucleic acid-based detection methods for pathogenic bacteria in food. J. Food Protect. 67: 823-832.

Neufeld, H.A., J.G. Pace and R.W. Hutchinson, 1985. Detection of microorganisms by bio- and chemiluminescence techniques. In: W.H. Nelson (ed.), Instrumental methods for rapid microbiological analysis. pp. 51-65. VCH Publishers.

Nogva, H.K. and D. Lillehaug, 1999. Detection and quantification of *Salmonella* in pure cultures using 5'-nuclease polymerase chain reaction. Int. J. Food Microbiol. 51, 191-196.

Notermans, S., R. Beumer and F. Rombouts, 1997. Detecting foodborne pathogens and their toxins. Conventional versus rapid and automated methods. In: M.P. Doyle, L.R. Beuchat and T.J. Montwille (eds.), Food microbiology. Fundamentals and frontiers, pp. 697-709; ASM Pres, Washington.

Perry-O'Keefe, H., S. Rigby, K. Oliveira, D. Sørensen, H. Stender, J. Coull and J.J. Hyldig-Nielsen, 2001. Identification of indicator microorganisms using a standardized PNA FISH method. J. Microbiol. Methods 47:281-292.

Radcliffe, D.M. and R. Holbrook, 2000. Detection of microorganisms in food - Principles and application of immunological techniques. In: B. Lund, T.C. Baird-Parker and G.W. Gould (eds.), The microbiological safety and quality of food, pp. 1791-1812; Aspen Publ. Gaithersburg, Maryland.

Schönhuber, W. G. Le Bourhis, J. Tremblay, R. Amann and S. Kulakauskas, 2001. Utilization of tmRNA sequences for bacterial identification. BMC Microbiology, 1:20.

Sharpe, A.N., 2000. Detection of microorganisms in foods: Principles of physical methods for separation and associated chemical and enzymological methods of detection. In: B. Lund, T.C. Baird-Parker and G.W. Gould (eds.), The microbiological safety and quality of food, pp. 1734-1760; Aspen Publ. Gaithersburg, Maryland.

Stender, H., M. Fiandaca, J.J. Hyldig-Nielsen and J. Coull, 2002. PNA for rapid microbiology. J. Microbiol. Methods. 48:1-17.

Theron, J. and T.E. Cloete, 2000. Molecular techniques for determining microbial diversity and community structure in natural environment. Crit. Rev. Microbiol. 26:37-57.

# 12. Analysis of chemical food safety

*Carmen Socaciu*

## 12.1 Introduction

### 12.1.1 General aims and objectives of analysis chemical food safety

Food Quality Control (FQC) is a process designed to evaluate the performance of a product or production chain and to take corrective action in order to maintain and assure food quality. A food product or the food production process is controlled across the whole technological chain, from the raw material up to the final product.

The aim of chemical food safety is to provide a general picture of the main steps and components which are integrated in this concept (as showed is Figure 12.1).

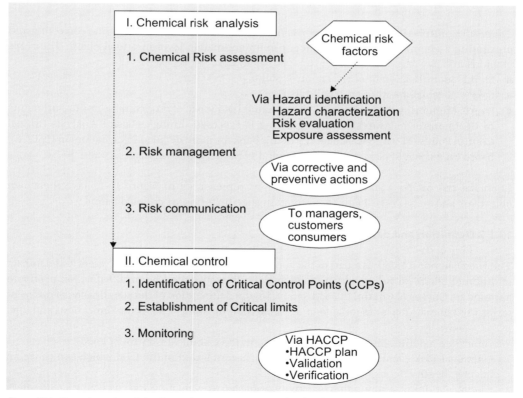

Figure 12.1. General overview of the aims and components integrated in the chemical control of food.

*Carmen Socaciu*

The general objective of the Chemical Control in Food concept is to supervise the negative impact of chemical hazards on food safety. To accomplish this, it is necessary to make an appropriate chemical risk analysis (I) and to provide a system of Chemical Control (II).

According to Figure 12.1, chemical risk analysis (I) is a process which consists of three components: risk assessment (1), risk management (2) and risk communication (3). These components represent 3 successive steps to be followed if one aims to know how to evaluate and how to correct and prevent the occurrence of different chemical hazards in food.

- First step is the chemical risk assessment (I.1), based on the evaluation of known chemical agents in the food and establishment of the risks induced by consumers exposure to the food contaminated with specific chemicals having toxic potential and harmful effects on health. This step includes four points to follow: identification of chemical hazard, characterisation of chemical hazard, exposure assessment and risk characterisation.
- Second step is represented by risk management (I.2), which includes the alternatives to accept, minimise or reduce the assessed risks, and select and implement appropriate options by corrective and preventive actions.
- Third step is represented by risk communication (I.3), which represents an interactive way to exchange information and opinions on chemical risks among managers or to communicate results to customers and consumers.

Chemical Control (II) includes all actions and activities dedicated to evaluating, correcting and preventing a chemical hazard or reduce it to an acceptable level in a food chain. The steps included are:
- Step 1: Identification of critical control points (CCPs) (II.1).
- Step 2: Establishment of critical limits (II.2).
- Step 3: Monitoring the control (II.3) by a planned sequence of observations or measurements of CCP control measures. This monitoring is executed through Hazard Analysis Critical Control Points (HACCP, see Chapter 7 of this book for more detailed information on HACCP), based on a specific plan, and validation and verification measures.

Chemical risk analysis and chemical control, as represented in Figure 12.1, have their own objectives and steps (1-3). They are defined in specific terms, as detailed below.

### 12.1.2 Definition and delimitation of terms

The Food Chemical Safety is a part of the general concept of Food Safety and refers to the impact of different chemical risk factors on food quality. To assure the Chemical Safety we need and appropriate Chemical Control of Food Safety. So, the delimitation of the significance of different terms used in this area is essential to understand their meaning, their interconnections and role.

We offer here a definition of 14 key words used in the Chemical Control of Food
1. Chemical Risk Factor = a chemical agent or hazard found in the food which can cause an adverse heath effect.
2. Chemical Control = any action or activity dedicated to evaluating and correcting a food safety risk(by accidental exposure to a chemical, chemical degradation during processing, etc.).

3. Chemical Risk Analysis = a process consisting of three components: risk assessment (1), risk management (2) and risk communication (3).
4. Chemical Risk Assessment = a scientific and practical evaluation of known or potential adverse health effects, resulting from human exposure to chemical hazards.
5. Chemical Hazard Identification = the identification of known or potential health effects associated to a specific chemical agent.
6. Chemical Hazard Characterisation = qualitative and / or quantitative evaluation (by hazard analysis) of the adverse effects associated with a specific chemical agent.
7. Hazard Analysis = the process of collecting and evaluating information on chemical hazards and conditions leading to their presence, in order to decide their significance for food chemical safety and security.
8. Food Chemical Safety = an assurance process which prevents harm to a consumer due to chemical risk factors found in food when it is prepared and/or eaten
9. Food Chemical Security = assurance that a food does not contain chemicals which affect people's lives or does not represent a terrorist threat.
10. Exposure Assessment = qualitative and / or quantitative evaluation of the degree of intake by humans.
11. Risk Assessment = the integration of hazard identification, characterization, risk evaluation and exposure assessment, in order to estimate the adverse effects of different specific chemicals to a given population.
12. Risk Management = the process of choosing policy alternatives to accept, minimise or reduce assessed risks, and to select and implement appropriate options.
13. Corrective Action = any action to be taken in order to re-establish good food quality in the light of a chemical risk factor.
14. Risk Communication = interactive process of exchange of information and opinion on risk among food industry managers or to communicate to customers and consumers.

## 12.1.3 Specific aims and objectives of this chapter

According to the general aims and objectives of the Chemical Control of Food (as presented in Figure 12.1), the following aspects are addressed in this and other chapters of the book:
- Aspects regarding chemical risk analysis (I), mainly about chemical risk assessment (I.1.) Via hazard identification, characterisation and exposure assessment- Chapter 3.
- Aspects regarding risk management and risk communication (I.2. and I.3), which are common for the control of many risk factors (including chemical ones)- Chapter 8.
- Aspects regarding Chemical Control (II) via identification of Critical Control Points, establishment of critical limits and monitoring via the HACCP system- Chapters 6 and 7.

Table 12.1 integrates the specific objectives of Chemical Risk Assessment and the activities which are carried out to achieve these objectives.

In this chapter the specific aim is to present an overview of the evaluation of chemical risk via the chemical analysis of food. We focus on the most specific activity linked to chemical control

Table 12.1. Objectives and activities which are included in the Chemical Risk Assessment.

| Objective of action | Activity (what is determined) |
|---|---|
| Hazard Identification | Type(s) of chemical(s) involved |
|  | Which/how foodstuff(s) is/are affected |
|  | The nature of harmful effects of chemicals to human health |
| Hazard characterisation | Association of a specific chemical with specific toxicological end-points |
|  | Relationship between the individual toxicity to dose and target organ/organism. |
|  | Identification of „zero effect" |
| Risk evaluation | Qualitative food analysis: isolation and identification of risky chemical(s) |
| a. Risk identification by exposure to a specific chemical | Quantitative food analysis: isolation, identification and quantitative determination of risky chemical(s) |
| b. Risk intensity by exposure to a certain quantity of a specific chemical |  |
| Exposure analysis | The amount of chemicals which have contaminated the food |
|  | The quantity of food which has been consumed in relation to the risk of chemical intoxication |

(known in its restricted meaning as "verification by analysis"), namely the determination of chemical hazards in food.

Our objectives are focused on the application of different methodologies and techniques in both qualitative and quantitative determination of chemical hazards in food matrices.

## 12.1.4 Structure of the chapter

The first part of the chapter describes chemical risk evaluation via food analysis and it gives an insight into:
- Methods and techniques used according to their aim (qualitative or quantitative).
- Methods and techniques used according to their performance.
- Specificity of some methods according to the specific chemicals to be determined.
- Legislative requirements and method standardisations.

The second part includes practical examples of new and advanced protocols of food analysis for the identification and quantification of specific chemical risk factors.

The references are important sources for further investigations and are divided into:
- reviews and relevant books and articles on Food Analysis;
- specific references for different classes of chemical hazards found in food;
- specific legislation;
- world wide web sources;
- scientific journals of interest for this field.

## 12.2 Chemical risk evaluation via food analysis

Chemical risk evaluation includes and prescribes all analytical methodologies and techniques dedicated to the identification and quantification of chemicals found in food products, taking into account the prescriptions of general quality systems within which every laboratory of analysis must operate.

### 12.2.1 General methodologies used in chemical analysis of food

Methodology represents a general concept in analysis, describing the application of one or a succession of several methods, which includes a chain of techniques, used in order to satisfy the requirements of a precise and accurate determination. Food samples can be controlled either by direct inspection and measurements (i), or by analysis procedures (ii). This is a first classification of methodologies.

Direct inspection and physical measurements determine the sensorial and some physical attributes of food. For example, the sensorial evaluation refers to: colour, odour/smell, shape/appearance, taste/flavour, mouth feel, texture. The physical direct measurements refer to pH, temperature, humidity, and viscosity.

The analysis procedures include three successive, necessary steps: sampling (I), sample preparation (II) and evaluation by qualitative or quantitative measurements (III). Finally, the performance of the analysis or measurement has to be considered in order to assure confidence in the results.

Sampling (I). There are precise criteria established for carrying out correct sampling: the food sample has to be identical to the bulk and to have the same intrinsic properties (homogeneity of distribution in the food matrix). According to Pomeranz and Meloan (1994) cited by Luning et al. (2002), the causes which induce a variation in sampling of agri-food products include irregular shapes, water loss, evaporation of specific volatiles, enhancement of enzymatic reactions after mechanical damage. A statistical sampling plan has to be respected in order to obtain a representative sample from the food under investigation. This plan should contain data about the purpose of inspection, and nature of the sample (history, homogeneity, type of laboratory procedures).

Sample preparation for analysis (II). This is a critical step since it is needed to minimise the modifications of sample (enzymatic and/or oxidative degradations), the prevention of microbial contamination of the sample, and the preparation of a homogeneous sample from which the relevant target compounds have to be extracted and analysed.

The preparation procedures involve mechanical grinding of dry materials or homogenisation of moist materials (1), followed by a physical, chemical or enzymatic treatment (2) (e.g. sonication, immersion in liquid nitrogen, enzymatic hydrolysis or inactivation by heat or by inorganic salts).

An important precaution has to be considered: avoidance of oxidative and microbial spoilage by storage of the sample in the deep freezer or refrigerator, under nitrogen or by addition of preservatives.

Evaluation by analysis (identification or quantitative measurements of risk factors) (III). The analyses include at least three essential steps:
- extraction, purification and separation procedures;
- identification;
- quantitative determination of marker molecules.

The methods and techniques which are used will be described in section 12.2.3. The performance of measurement represents the most important step and includes many requirements and precautions, such as specificity, sensitivity, accuracy and precision (Horwitz, 1988; Pomeranz and Meloan, 1994).
- Specificity represents the ability of the procedure to extract and analyse the target compound. The specificity of the method used is crucial since the method should not be affected by interfering compounds which could react with the same reagents as the target compound.
- Sensitivity represents the ability of the method or the instrument to react to a low concentration of compound. The more sensitive the method and/or instrumentation are, the lower quantities of the target compound that can be identified.
- Accuracy reflects the degree of superposition between a mean estimation of a concentration of target compound and its real determination. The deviation between these two concentrations may be due to the method, the effect of interfering substances or alterations of the target compound during analysis.
- Precision of the determination reflects the coincidence between the real concentration and the practical measurement of target compound, made by the same laboratory (by repetition at different times) or in parallel by different laboratories (in this case the repetition is called reproducibility).

## 12.2.2 Methods and techniques used for qualitative and quantitative evaluation of food quality and safety

The analytical methodologies are classified and divided generally into qualitative and quantitative methods. A method is designed to offer information about the ways to identify or to quantify a chemical or a class of chemicals from the product to be analysed. The method, e.g. physical, chemical, immunological, radioimmunological, enzymatic, etc.) is complemented with more techniques, such as instrumental, chromatographic, and electrophoretic procedures. Figure 12.2 includes a general view and classification of different techniques and methods used in Food Control.

The qualitative methods are designed to identify a molecule, a class of compounds or a food characteristic. They may include:
- Sensorial evaluation: colour, taste/flavour, odour/smell, appearance, texture.

## Analysis of chemical food safety

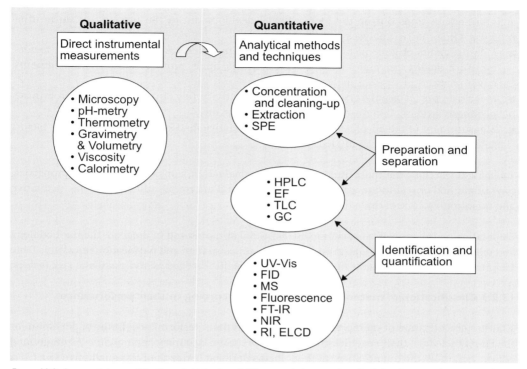

Figure 12.2. A general view and the flow of application of different techniques and methods (qualitative and quantitative) used in food control.

- Physical evaluation: instrumental identification of some parameters which may be related to chemical risk factors: pH, pressure, temperature, water activity, rheological (viscosity) properties.
- Biochemical evaluation of microorganism development; identification of enzymatic and metabolic specific biomarkers.
- Direct identification of the presence of risk molecules or modification of different food components affected by chemical risk factors: carbohydrates, lipids, amino acids and proteins, enzymes, vitamins, pigments, flavours, inorganic elements, etc.
- Separation followed by identification of chemical risk factors: residues (pesticide, veterinary drugs, antibiotics, hormones), environmental contaminants (nitrates/nitrites, heavy metals, polyaromatic hydrocarbons, dioxins), mycotoxins, plant toxins, endogenic toxins derived by food degradation or by migration from packages.

The quantitative methods give a better performance than the qualitative ones, and are more costly. There are parallel tests involved in method validation via standard curves, recovery testing and limits of detection, reproducibility, specificity and sensitivity levels. These types of analyses can be applied at different points of the food chain, from the beginning, in raw materials, and

at intermediate steps during the technological development of the product, or in the final products. Quantitative analyses include:
- Measurements of general food quality parameters which can be affected by risk factors: moisture, acidity, ash and minerals, carbohydrates, lipids, proteins, amino acids, pigments, flavours, etc.
- Measurement of regulatory factors: enzyme activity (e.g. lipase, amylase, proteinase, pectinase, oxidase, invertase), hormones, vitamins, etc.
- Measurement of the absolute quantity and the relative distribution of different risk factors (mentioned above) found in food.

Table 12.2 describes the most important types of evaluation, considering the food properties, type of method or technique which are appropriate and are frequently used for the qualitative and quantitative control of food.

Depending on the type of risk factor to be identified or monitored in the food, the methodology and techniques of sample preparation, extraction, concentration and purification may vary. Table 12.3 reveals the most frequent methods used for specific categories and classes of risk factors.

## 12.2.3 Classification of methods and techniques according to their performance

Another classification of methods takes into account their performance: limits of detection for the target molecule, their specificity, sensibility, precision, accuracy and rapidity. As mentioned before, generally the quantitative methodologies perform better than the qualitative ones, but of course they involve more laborious steps and are more expensive. Choosing to apply a certain method is depends on the level of confidence in it and the need of the investigator. According to performance level, the methods of analysis may be also classified as classic & routine methods, reference & advanced methods, rapid & recent methods. Each of the food components can be determined by different, specific and average performance-level methods, depending on the rapidity, accuracy and sensibility required for the target molecule determination.

*Classic and routine methods*
Many common methods used for years in laboratories to determine food quality and safety are still classic and routine. Their characteristics are: long-term usage, their validation by many laboratories and their use in many countries as standardied methods. These methods are used by small laboratories belonging to factories, farms or standardisation bodies.

Common methods include different techniques used for the determination of nitrogen, ash, moisture, sugar content in water, meat, confectionery, etc, described in detail in many books and some of them designated as national/international standards (see Kjeldahl method for proteins and total nitrogen, butyrometric determination of fat in milk, sucrose by double polarisation). The most relevant common, routine methods used in Food analysis (Pearson, 1973; Meloan and Pomeranz, 1973; Kirk and Sawyer, 1991), as well some indications about the standardised ones (by AOAC) are the basis of many standardised methodologies which refer to:

## Analysis of chemical food safety

Table 12.2. General types of methods and techniques used frequently for qualitative and quantitative evaluation of risk factors in food.

| Type of evaluation | Food property/ risk factor | Type of methods | Techniques |
|---|---|---|---|
| Sensory evaluation | colour, appearance taste/flavour, odour/smell, texture | Consumer acceptance tests | Panel difference techniques: visual inspection, ranking, paired comparison, numerical scoring |
| Physical properties | pH | Instrumental | Potentiometry |
|  | Temperature | Instrumental | Thermometry |
|  | Moisture and water activity | Gravimetry | Water evaporation in oven Conductivity |
|  | Texture and rheological (viscosity) properties | Instrumental | Tenderometry, Penetration Texturemetry |
| Chemical evaluation of food composition | Identification of food components affected by risk factors: carbohydrates lipids proteins and amino acids enzymes vitamins pigments flavors inorganic elements | Separations coupled with identification Direct Instrumental determination Chemical and enzymatic methods (substrate determination or inhibition studies) | Chromatography (OC, TLC, HPLC, GC) based on adsorption, size exclusion, ion exchange Electrophoresis Atomic absorption and Flame photometry Conductivity Polarimetry Electrometry NMR and MS spectroscopy Colorimetry UV-Vis, IR, NIR, fluorescence spectroscopy Radio(immuno) assays |
| Identification of chemical risk factors | Toxic residues Natural/endogen toxins Environmental contaminants Food additives Toxins derived by food degradation and which migrate from packages | Separations coupled with identification Instrumental measurements Chemical and enzymatic methods (substrate determination or inhibition studies, metabolic biomarkers (lactic acid, H2S, ATP, etc) | Chromatography (OC, TLC, HPLC, GC) based on adsorption, size exclusion, ion exchange Electrophoresis Atomic absorption (AAS) and Flame photometry Conductivity Polarimetry Electrometry NMR and MS spectroscopy Colorimetry UV-Vis, IR, NIR, fluorescence spectroscopies Immunoenzymatic assays (ELISA) |

*Table 12.3. Specific methods used for different risk factors.*

| Category of chemical risk factor | Class of risk factor | Methods of separation and identification | Quantitative methods |
|---|---|---|---|
| Residues | Pesticides | TLC | GC-FID, GC-MS, HPLC, LC-MS |
| | Antibiotics | TLC | HPLC |
| | Hormones (steroids) | TLC | GC-FID, HPLC |
| | Veterinary drugs | TLC | HPLC |
| Naturally/ endogen toxins | Alkaloids and phenols | TLC | HPLC |
| | Biogenic amines | TLC | HPLC |
| | Mycotoxins | TLC | |
| Environmental contaminants | Fertilizers(nitrates/nitrites) | TLC | Colorimetric, HPLC |
| | Mycotoxins | TLC | HPLC, Immunoassay |
| | Heavy metals | TLC | AAS |
| | PAHs, PCBs and dioxins | precipitation | GC, HPLC |
| Food additives | Colorants | TLC | HPLC |
| | Antioxidants | TLC | HPLC |
| | Sweeteners | - | HPLC |
| | Emulsifiers | - | HPLC |
| Toxins formed during processing and migration | PAHs and acrylamide | TLC | GC-FID, HPLC |
| | Lipid peroxides | TLC, | HPLC, UV-Vis spectrometry |

- determination of moisture (AOAC 950.46, Pearson methods - p.31-39);
- determination of volatile oils (Pearson methods - p.45-49);
- determination of nitrogen and fibre (Pearson methods - p.52-54);
- determination of ash (AOAC 920.153, Pearson methods - p.57-58);
- determination of sugars (AOAC 989.05, Pearson methods - p.62,69);
- determination of acidity (Pearson methods - p.70-71);
- determination of food additives (AOAC 948.15, Pearson methods - p.75-94);
- analysis of inorganic groups and metal ions (AOAC 985.15-52, Pearson methods - p.98-114);
- analysis of oils and fats (Pearson methods - p.119-127);
- analysis of milk and milk products (Pearson methods - p.132-162);
- analysis of meat and flesh products (Pearson methods - p.169-208);
- analysis of eggs (Pearson methods - p.278-280);
- analysis of flour and bread (Pearson methods - p.216-235);
- analysis of sugars, pectins, juices and confectionery (Pearson methods - p.237-266);
- analysis of vegetables and fruits (Pearson methods - p.274-276);
- analysis of water (Pearson methods - p.270-273).

In spite of the progress in developing new methods and techniques, many of these methods are applied and requested by national or international legislative bodies.

Determination of enzyme activities is also included in routine methods as a reflection of food quality. As long as enzymes are labile proteins they are good quality and act as degradation markers for raw material quality, for technological chain performance and for the quality and safety of the final product. In this context, the analysis includes the measurement of rheological properties of the enzyme substrate monitored either by physical measurements (viscosity), and analysis of the specific end products or by-products resulting from enzyme activity. Common techniques used for all routine analysis include volumetric titration, gravimetry, colorimetry, polarimetry, flame-photometry, atomic absorption, UV-Vis spectrophotometry. More and more, the recent methods, that are rapid, precise and cheap (bioassays), are replacing the classic ones. One example is identification via immunoassays, using Enzyme Linked ImmunoSorbent Assay (ELISA) (Bergmeyer, 1983; Fox, 1991).

Many more common screening methods are now considered to be "classic" since they offer a lot of information about the existence of some degradation molecular markers in food. For example, Thin Layer Chromatography (TLC) is an old and well known, classical method which has been used for many years to separate and identify all kinds of food molecules by specific colouring or UV-absorption or fluorescence evaluation. Nowadays, there is a continuing interest in it as a simple screening technique for different classes of food contaminants and additives. TLC can use different stationary phases (silica, alumina, cellulose or reverse phase C18 RP F254 plates), coupled or not with TLC scanning densitometers, for a more precise qualitative or semi-quantitative evaluation (Sherma, 2000).

*Reference & advanced methods*
Due to the continuing progress and evolution in equipment design and production, advanced methods of analysis were created and are now performed. These methods generally use more sophisticated procedures applied by central quality control laboratories or by governmental agencies with qualified personnel. They are also used to verify the reliability of "rapid" methods, and of classic methods, or to calibrate the instruments. They can be used as well for food quality assessment, and for analysis of food contaminants and additives.

The main characteristics of these methods are: high selectivity and specificity, good precision and accuracy, low limits of detection and the possibility to make a multi-analyte evaluation. Many of these methods have been validated through collaborative studies, in order to obtain laboratory ISO accreditation, which involves a *sine qua non* utilisation of validated methods.

New instrumental techniques are applied in this category of methods. They include chromatography, electrophoresis, enzymatic determinations and bioassays via spectrometric techniques, e.g. High Performance Liquid Chromatography (HPLC) or LC-MS, Gas Chromatography (GC-FID or GC-MS) capillary electrophoresis (CE), ELISA, Fluorimetry, UV-Vis spectrometry, NIR or FT-IR spectrometry.

In time, such techniques will become routine procedures and recently some of them have been adopted by the official international bodies like AOAC, CEN, and ISO. A review of the most important reference and research methods is presented below.

*Carmen Socaciu*

The sampling and preparation of samples before analysis is essential for such methods. Sample extraction and cleaning-up are time-dependent, crucial procedures for isolation and concentration of different analytes from such complex matrices as agrifood products.

As shown in Figure 12.1, there are 3 possible steps for preparing the sample: by extraction, concentration or cleaning-up. Liquid-liquid extractions with an appropriate solvent, cleaning-up on open columns, and concentration of the analyte under vacuum, are commonly used methods. Recently these steps were replaced by a Solid-Phase Extraction (SPE) with C18-SPE cartridges as an improved extraction and cleaning-up technique. Another recent innovation is the dialysis unit (automated liquid-liquid extraction procedure) which can be coupled directly to the HPLC injector.

The separation techniques include gas- or liquid-chromatographic or electrophoretic procedures. These procedures can concomitantly separate, identify and quantify the target molecules from a mixture/extract by coupling the separation device to different detectors for UV-Vis light absorption or fluorescence, or by mass spectrometry (MS), light scattering (ELSD), refractometry (RI detectors), or flame ionization (FI detectors). ELSDs are more sensitive than RIs and allow gradient elution or biosensor detection.

These general steps in the analytical process are applicable for food components as well as quality markers (pigments, proteins, amino acids, sugars, oils, volatiles, etc), environmental residues or contaminants (pesticides, antibiotics, mycotoxins, etc).

These procedures assure high sensitivity (low limits of detection, under 1 ppm), selectivity, specificity, high accuracy and precision. These methods are obtained after long-term research studies and validation procedures, are reproducible and accurate, but labour-intensive and costly.

Some characteristic applications for these techniques are presented below:
- High Performance Liquid Chromatography (HPLC) coupled to different detection types, allows the separation of polar (using NP columns) or non-polar molecules (RP-C8 or C18 columns). In the main, unpolar molecules are separated by NP-HPLC (when no water is contained in the sample or eluent) or by RP-HPLC. Usually plant lipids, food glycerides, and phospholipids, pigments, unpolar vitamins, steroids, mycotoxins, can be separated and identified/quantified by these methods, while polar molecules may be separated only by RP-HPLC, where the column support (C18) is resistant to water. The detection is usually done by DAD or LS or RI. Some examples of HPLC separations are presented in section 12.3. Natural or synthetic polymers (proteins, pectins, carbohydrate hydrolysates) can be separated and purified by LPLC (Low Pressure Liquid Chromatography), based on gel filtration and size-exclusion mechanisms (Nollet, 2000; Hewlett Packard application, 1996).
- Ion chromatography (IC) is a type of liquid chromatography, more appropriate for anions and cations, as well other polar molecules (sugars, complex carbohydrates, water-soluble food colorants, amino acids, sweeteners, and antioxidants). It is based on ion-exchange columns, aqueous effluents or buffers and the detection is usually made by RI detectors. Recently, the development of high performance anion-exchange chromatography (HPAEC) with

## Analysis of chemical food safety

CarboPac columns and pulsed amperometric detection (PAD) has provided an excellent procedure for separating mono-, di- and oligosaccharides in a single run (Weiss, 1995; Henshall, 1997).

- Gas chromatography (GC) is still a well-known and validated method in food analysis, and well-documented (Nollet, 1996; Pare and Belanger, 1997). The main limitation of this technique is the stability of the evaporated components of a liquid mixture or that of volatile compounds up to 300°C. Due to the extensive application of HPLC, GC techniques are used less nowadays. The coupling of GC to FID/MS, MS, MS/MS detectors or the use of capillary GC columns has significantly improved the accuracy of determinations (Cserhati and Forgacs, 1999).

- Capillary electrophoresis (CE) is a new analytical technique that allows rapid and efficient separation of mixture components due to their different electrodynamic properties, as they migrate through narrow bore capillary tubes under an electrical potential (Frazier *et al.*, 2000). CE is widely used in the pharmaceutical industry but less in food analysis, due to the absence of well-validated analytical procedures for extraction, cleaning-up and separation protocols. CE includes different technical variants such as capillary zone electrophoresis (CZE), micellar electrokinetic chromatography (MEKC), capillary isoelectric focusing (CIEF), capillary gel electrophoresis (CGE). The main problem that needs solving is the poor detection limit obtained with UV and DAD. Recently, the coupling of biosensors, as specific detectors, to CE has been reported (Bossi *et al.*, 2000).

- Flow-injection analysis (FIA) was applied initially in water, soil and feeding stuff analysis but has been used more recently in food analysis, using biosensors for detection (Galensa, 1998). Recently, there have been reports of electro-chemiluminescence-based optic biosensors or bulk acoustic biosensors coupled to FIA for the analysis of beverages and soft drinks.

- Inductively-coupled plasma atomic emission spectrometry (ICP-AES) is gradually replacing the old routine method of flame atomic absorption (flame photometry) for analysis of metal cations (Na, K, Ca, Mg) in food and beverages. The sample preparation is a limiting factor in transforming the food product into a suitable solution for ICP-AES analysis. The classical dry-combustion method of disintegrating the samples is still preferred to microwave acid digestion or wet digestion with nitric/perchloric acid. For heavy metals, atomic absorption spectrometry (AAS) is still in use, but with improvements, using plasma or acetylene-flame instead.

- Fourier-Transformed Infrared Spectroscopy (FT-IR) is widely known and applied to identify different functional groups of target molecules or as detection equipment coupled to HPLC. Generally, the "Infrared" techniques (FT-IR and also Near Infrared Spectroscopy - NIR) are being applied more and more in food analysis, as has been recently reviewed (Davies and Giangiacomo, 2000; Wetzel, 1998). They are used not only as precise methods of identification but also as screening methods (for fingerprinting food matrices) and quantitative rapid methods to evaluate some chemical markers of food (protein in grains and seeds, water, etc).

*Rapid & recent methods*
These groups of methods are extending, and focus the attention of scientists and food companies or laboratories. They can be applied to routine quality control analyses or process control of food additives. Many publications describe the new developments in this area, but few validated

procedures are available as yet in the literature. The most important methods included in this category are the immunoassays, the methods based on biosensors, the enzymatic kits, pH-differential methods, X-ray fluorescence methods and NIR or FT-IR techniques.

Generally, these methods need as initial steps rapid and accurate extraction and cleaning-up procedures, such as the solid-phase extraction (SPE), based on solid/liquid partition on polypropylene or polystyrene columns or discs, containing one or more specific adsorbents (Pimbley and Patel, 2000). This procedure assures a higher speed of separation, better reproducibility, versatility, and a reduction in time and price.

**Immunoassays** - Immunoassays provide a powerful analytical tool in food analysis. They can be divided into those that are based on primary reactions, i.e. those that measure directly the antibodyantigen reaction (e.g. radioimmunoassays and enzyme-labelled assays) and those based on secondary reactions, i.e. those that measure the antibodyantigen reactions indirectly (e.g. immunoprecipitation). These direct and indirect terms relate to whether it is the actual binding of the antibody to antigen that is being measured or whether it is some property of the antibodyantigen complex, once binding has occurred, that is being measured.

The most frequently used immunoassay is Enzyme-Linked ImmunoSorbent Assay (ELISA). Although ELISA itself is a relatively simple and straightforward technique, preparation of a food sample prior to the test requires optimisation. The sample preparation frequently requires analyte extraction and concentration from complex food matrices, where there is a danger of low analyte recovery and/or rendering the analyte undetectable. There are several formats of ELISA that can be used, but essentially they all follow the same principal and rely on the specificity of antibodies with the sensitivity of simple enzyme assays. There are two main variations on this method: firstly, detection of the presence of antigens that are recognised by an antibody, and secondly it can be used to test for antibodies that recognise an antigen. An ELISA is a five-step procedure: 1) coat the microtiter plate wells with antigen; 2) block all unbound sites to prevent false positive results; 3) add antibody to the wells; 4) add anti-mouse IgG conjugated to an enzyme; 5) reaction of a substrate with the enzyme to produce a coloured product, thus indicating a positive reaction. There are many different types of ELISAs, but with all their variety and versatility they will depend upon having one component of a ligand pair fixed to a solid surface. The choice of which component is bound and whether the detection is by a direct or indirect means, and whether there are two or more components in the immunoassay, will be dependent on the research needs.

The first step in the development of an immunoassay is the preparation of a suitable antibody. Animals are immunised with the antigen, and subsequently serum containing *polyclonal* antibodies is collected. Assay performance can often be improved by the preparation of *monoclonal* antibodies that exhibit greater specificity for the target antigen. Spleen cells derived from immunised animals are used to produce hybridoma cells that secrete monoclonal antibodies whilst in culture.

## Analysis of chemical food safety

The production of antibodies for a given antigen also requires several steps, such as extraction, concentration and purification of the target molecule, and if the target molecule has no intrinsic antigenic properties it must be to linked to the molecule of an antigenic protein to form a conjugate. This conjugate is than used to induce the production of specific antibodies "anti-conjugate" which will belong to one of the classes of immunoglobulins (e.g. immunoglobulin class G (IgG); immunoglobulin class M (IgM)).

ELISA has found an application in many domains of food safety related analyses, e.g. detection of pathogenic microbes or their metabolites (bacterial toxins, mycotoxins, etc.), hormones, antibiotics, and growth promoters.

**Biosensors** - The term sensor is defined as a device or system (electronics, software) that responds to a specific physical or chemical property, which can be identified or measured quantitatively. A biosensor is an analytical device which contains a biological sensing element connected or integrated within a transducer, which converts a biological event into a response to be processed. The biological sensing element can be either catalytic (enzyme, microorganism) or non-catalytic, like an affinity sensor (antibody receptor). Some reviews focus on biosensor technology (Cunningham, 1998; Kress-Rogers, 1998, Ramsay, 1998; Scott, 1998), but so far there is still a lack of understanding of the requirements and the specific conditions for the use of biosensors in food analysis. A biosensor comprises two elements: a biological recognition element (antibody, enzyme, receptor, cell) and a signal transduction element (optical, amperometric, acoustic, electrochemical) connected to a data acquisition and a processing system (Patel, 2000). The application of different chemiluminescence biosensors for the determination of different food analytes was recently reported (Aboul-Enein et al., 2000). Immunosensors, initially developed for medical purposes, may have great potential for development in food analysis, as it was mentioned recently (Bostrom and Lindeberg, 2000) for vitamins B in fortified food, using an automated continuous flow system based on surface plasmon resonance (SPR) technique (Table 12.4).

Table 12.4. Different food components or contaminants, measured via biosensors (Watson, 2002). The mechanism of transduction and the examples of sample types are also mentioned.

| Food components | Type of biosensor | Mechanism of determination | Sample type |
|---|---|---|---|
| Glucose, sucrose | Enzymatic/glucose oxidase | Membrane electric potential | Vegetables, cereals, dairy products |
| Lactate | Enzymatic/lactate oxidase | Membrane electric potential | Meat and cooked food |
| Glutamate | Enzymatic/glutamate oxidase | Potentiometric | Food and additives |
| Choline | Enzymatic/choline oxidase | Microwave hydrolysis and enzyme | Infant formulas |
| Sulphite | Enzymatic | Amperometric measurement of cytochrome c reduction | Beverages |
| Vitamin C | Enzymatic /ascorbate oxidase | Potentiometric | Vegetables and fruit juices |

**Molecularly imprinted polymer-based methods (MIPs)** - Another alternative to the use of immunosensors via immunoaffinity, is the Molecular Imprinted Polymers (MIP) method, based on synthetic polymers built with specific "sites" as (bio)sensors for specific target (print) molecules (PM) (Figure 12.3). The procedure involves the preparation of a "mould" made of functional monomers (Mo) around PM (antibiotic, steroid, etc) (Sellergren, 2000). The print molecule PM and the complex formed is then co-polymerised with cross linkers (e.g. ethyleneglycol dimethacrylate) into a rigid polymer.

Then the PM is extracted, leaving its specific recognition sites free. Using this technique, "plastic antibodies" are created, as particulate MIPs "mould" into affinity sites for catching specific PMs, which are then able to act as biological antibodies (Yano and Karube, 1999, Ye *et al.*, 1999).

In practice, the MIP technique involves 4 steps:
- the print molecule is dissolved in a porogen together with the monomer(s) and forms a non-covalent soluble complex;
- an initiator of polymerisation is added, forming a rigid polymer;
- the mass of polymer is ground and wet-sieved with a solvent, forming particles of 10-25 micrometers (called MIP) which may act as supports for SPE, electrophoresis or are packed in HPLC columns;
- PM is extracted from the MIP "moulds", which represent affinity specific sites (biosensors) for "catching" other different print molecules (drugs, herbicides, sugars, nucleotides, amino acids, proteins) from mixtures.

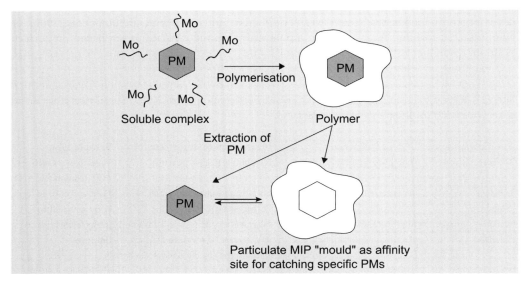

*Figure 12.3. The mechanism of the formation of Molecular Imprinted Polymers (MIPs) and its application as a method of identification and measurement of different food organic contaminants.*

## Analysis of chemical food safety

Some examples of MIPs used for contaminant analysis are: identification and quantitative evaluation of triazine and chlorotriazine herbicides, penicillin, chloramphenicol using MIP-based SPE, fluorometry and HPLC, at detection levels of 0.02-1000 ppm (Watson, 2002). MIPs are also used in combination with transducer elements (optical fiber, electrochemical, piezoelectric or acoustic sensors) for a rapid "biomimetic sensorial" detection and quantitative evaluation, and also for contaminants (e.g. o-xylene, PAHs) at 0.01-1 ppm.

The applications of MIPs have been directed mainly at organically soluble contaminants but their applications in agrifood and the environment will increase. With new approaches like surface imprinting, MIPs can also be created in aqueous media for aqueous-soluble analytes, extending the application to proteins, DNA, carbohydrates and microorganisms. Recently, new synthetic materials were found to be capable of recognising proteins (Shi and Ratner, 2000). Protein-binding nano-cavities were prepared on a carbohydrate-like surface using a novel radiofrequency plasma deposition of thin films. Protein-specific nanometer-sized "pits" were used as templates to catch specific proteins like albumin, immunoglobulin, fibrinogen, and lysozyme.

Immunomagnetic MIPs were also recently created for the separation of (S)-propanol, which when bound to the template exhibits super-paramagnetic properties. Since MIPs are heat stable, they can be sterilised and used in different biotransformations. Such applications mimic the catalytic reaction of an enzyme, the MIP acting as a "plastizyme". It was demonstrated that MIP can act as a site for a specific substrate, such as S-L-aspartic acid, to react with L-phenylalanine methyl ester, resulting in aspartame. Using the MIPs, the yield of the reaction increased 1.4 times more than in a classical reaction (Watson, 2002).

Until now, MIPs have not been widely used for routine applications due to some limitations: preparation of MIP is costly, if used as an HPLC stationary phase; the printing yield is still low; the MIP-based sensors still have long response times; and they are still produced on a laboratory scale, and not yet available as commercial products. But the combination of MIPs with transducers to produce efficient biosensors is expected to become a powerful real-time analytical technology. To be of commercial value, the application of MIP-based sensors and ligand-binding assays still needs to be demonstrated on agrifood matrices, and the performance and cost of these applications need to be at least equivalent to modern, high-performance analytical techniques (HPLC, GC-MS, ELISA).

**X-Ray fluorescence (XRF) -** This is a non-destructive technique, very suitable for rapid in-line analysis of inorganic components in food (Pomeranz and Meloan, 1994; Price and Major, 1990), and easy to use for dry materials, pellets or fluid samples.

Specific instruments were created for this measurement, ED-XRF or WD (Energy Dispersive or Wavelength Dispersive, respectively). Being a comparitive measurement, a calibration curve for each type of food is necessary. Typical applications are the determination of salt content, iron and calcium in infant formulas, etc. Measurement time is only 5-10 minutes per sample.

**Near-Infrared (NIR) -** NIR and Fourier-Transformed Infrared Spectroscopy (FT-IR). The "Infrared" techniques, as mentioned before, are widely known and applied for scientific purposes to identify specific functional organic groups, but their applications in food analysis were only recently reviewed (Davies and Giangiacomo, 2000; Wetzel, 1998).

The NIR techniques allow the direct determination of components (water, starch) in the seeds of cereals, dry food, and FT-IR allows the identification of oil authenticity, rancidity via oxidation, etc. NIR and FT-IR also became good methods for rapid checking of the fortification of food with vitamins, using a total reflectance FT-IR accessory (ATR). After a correct calibration of the instrument, the precision of measurements was in the range of 4-8% compared to those obtained by the reference HPLC method.

**Rapid enzymatic and test kit methods -** Specialised companies like Merck or Beckmann deliver different kits based on enzymatic assay techniques. For example, BIOQUNAT is a kit for the determination of aspartame sweetener and nitrates in food. RQ flex kits use test strips for the semi-quantitative determination of ascorbic acid in fortified food. The ferrospectral kit is used to identify iron salts in food. Colorimetric rapid assays for lecithin and choline, using enzymes and co-immobilised dye conjugates in polyacrilamide gels, have recently been applied. The differential pH techniques may also offer rapid ways to analyse fruit acids (like ascorbate, lactate, malate or citrate) in food.

Other rapid methods for identifying and quantifying food additives or contaminants have been described recently, for example, photo-chemiluminescence (Hermann, 2000). This rapid method involves the generation of free radicals and their detection, making possible the determination of both fat-soluble or water-soluble antioxidants.

**Bioassays -** Bioassays are an old way of determining the safety of food and the environment. The overall tendency is to perform fewer animal tests, whereever possible, and replace them with precise and repeatable analytical measurements. This, however, offers no information on the biological activity of the detected analyte. Animal bioassays are, therefore, being replaced more and more by *in vitro* models employing mammalian and prokaryotic cells. Despite the rapid improvements in analytical chemistry, these methods are not sufficient to deal with the complex mixtures of chemical hazards found in food. They also do not meet the requirements for rapid screening assays for monitoring programs.

Bioassays with pro- and eukaryotic cells capable of detecting compounds *in vitro*, based on their effects, offer an alternative solution. For example, for the detection of antibiotics in milk and meat, there are some screening methods in use that don't require the identification of a single responsible substance.

Recent advances in cell biology and in biotechnology have induced the development of new bioassays, based on the induction of gene expression and specific effects in cellular systems. This "new generation" of bioassays may be used in the future in combination with sensitive analytical methods. For example, the DR-CALUX assay for dioxins (see below, in subsection 12.3.3.).

## Analysis of chemical food safety

Another interesting application of bioassays is the analysis of acetylcholinesterase inhibitors like organophosphate and carbamate pesticides. These rapid screening tests allow the *on-site* testing of fruits and vegetables. The increasing knowledge about the effects of compounds on signal transduction pathways in cells, including the transcription of specific genes, should lead to the development of new bioassays. Concomitantly, specific and rapid cleaning-up procedures are needed and are sure to be available soon.

In conclusion, the current trend in analytical chemistry applied to the control of food quality and safety is directed towards "user-friendly" miniaturised instruments, "laboratory-on-a chip" applications (Myers, 2000; Cefai, 2000).

Due to the tremendous evolution in PC technology, many instrument accessories (detectors, CE or HPLC capillaries, pumps, etc) will be housed in a "chip" to improve the rapid and commercial application of such methods. Biosensors, being easy to miniaturise will use more specific detectors and will be adapted to a large variety of target molecules. Immunoassay tests and immunosensors, like MIPs, will allow the detection of contaminants and allergens in food, authenticity determinations, as well as food additives monitoring.

Regarding the reference methods, HPLC is developing rapidly towards better detection limits and accuracy, replacing UV-Vis or DAD detection with combined fluorescence/DAD but especially MS or MS/DAD, as well as by detection with biosensor chips. This technique will always require specialised chemists, and the cost will be high. Improvements in CE techniques, especially with better detection using MS and biosensor chips, will convert this technique into a good analytical system for the food industry.

### 12.2.4 Legislative requirements for food chemical surveillance

In 1990 the EU commission recommended the establishment of Quality Standards for all laboratories involved in inspection and sampling under the OCF Directive. This recommendation was adopted in the Directive on Additional Measures concerning the Food Control of Foodstuffs (AMFC).

Upon adopting these directives, national laboratories have to respect all the requirements regarding their organisation, how they carry out the analyses, which analyses are used, etc. All these aspects have to be accredited to an internationally recognised standard, included in proficiency schemes and using validated methods.

By virtue of the EN 45001 Standards and ISO/IEC Guide 17025 and AMFC Directives, food control laboratories should be accredited to the EN 45000 series of standards. Codex Alimentarius and its standards are integrated into the food laws of all developed or developing countries.

The "Codex Committee on Methods of Analysis and Sampling" (CCMAS) developed criteria for assessing the competence of testing laboratories involved in food control. They reflect the EU recommendations for laboratory quality standards and methods of analysis. The EU requirements are based on accreditation, proficiency testing, use of validated methods of analysis and quality control procedures.

The Food Standards Agency (FSA) undertakes food chemical surveillance. The requirements for a laboratory to be recognised and authorised are classified on three levels:
A. Quality assurance system (GLP), Internal Quality Control and Validation Methods
B. Information regarding the sample storage conditions, methods of analysis, limits of detection (LOD), limits of quantification (LOQ)
C. Information flow from the laboratory to the customers
D. Details about these surveillance levels are described elsewhere (Watson, 2002).

*Standardisation of analysis methodologies*
There are many organisations which publish standardised methods of analysis for food components and contaminants. These methods are validated through collaborative trials and by internationally accepted protocols. Among these organisations are the European Committee for Standardization (ECS), the American Organization for Analytical Chemistry (AOAC), and the Nordic Committee for Food Analysis (NMKL). ECS has a technical committee dealing with horizontal methods for contaminants and food additives (TC 275). This committee includes working groups (numbered from 1 to 11), responsible for specific chemical determinations from food products, as shown in Table 12.5.

*Future directions for applying different methods of analysis*
Future applications of the standardised methods of analysis have to take into account the development of new methods, the use of more automation, the measurement uncertainty, the in-house method validation, and the recovery percentage. There is some concern within the food analytical community as to the most appropriate way of applying a method and estimating its variability and its validation. As a result, IUPAC guidelines have been developed to give information on the acceptable procedures.

Food analysts should recognise that all new issues have to be addressed on an international basis, established until now by the European Union (Council Directives 89/397/EEC, 93/99/EEC) and European Committee for Standardization CEN/CENELEC 1989). As for the introduction of new countries and scientific/laboratory entities to the EU, this problem has to be solved uniformly and harmonised on a European scale.

At the international level, it is difficult to defend the selection of a particular method of analysis, for a class of food contaminants or for a certain food analyte, for inclusion in the legislation.

Specifying a single method, the analyst loses the choice of freedom or uses an inappropriate alternative method. This procedure may inhibit automation or make it difficult to replace an unsatisfactory method with another, more available method. As a consequence, the use of an alternative method, or criteria to which methods have to apply, is now considered and adopted in some sectors of food analysis. Some important issues have to be considered by the food analytical community, such as:

Table 12.5. Working groups and specific standardised methods for different food contaminants (Watson, 2002).

| N° | Working group/aim | Work item | Type of determination | Methods used |
|---|---|---|---|---|
| 1 | Sulfites | 1988-1:1998 | Direct sulfite concentration | Optimized Monier-Williams procedure, Enzymatic |
| 2 | Sweeteners | 1996/1999 | Saccharin | Spectrometry |
| | | | Acesulfam K | Spectrometry |
| | | | Aspartam, Ciclamate, Acesulfam and Saccharin | HPLC |
| 3 | Pesticides in Fatty Food | A | Pesticides and Polychlorinated biphenyls (PCBs) | Extraction of fats, cleaning up and confirmation methods (TLC) |
| 4 | Pesticides in non-fatty food | A | Multiresidue methods | Extraction, cleaning-up, GC |
| | | B | Dithiocarbamates and thiuram disulphides | Spectrometry |
| | | C | | GC |
| | | D | Bromine residues | Xanthogenate method |
| | | E | N-methyl carbamate residues | Inorganic and total bromide methods |
| | | | Veterinary drugs: Thiabendazole, benomyl, carbenadazim | HPLC HPLC |
| 5 | 5 Mycotoxins | A | Aflatoxins B1, B2, G1, G2/cereals | HPLC with derivatisation and immunoaffinity |
| | | B | Ochratoxin A/cereals | HPLC/silica cleaning up |
| | | C | Ochratoxin A/cereals | HPLC/immunoaffinity |
| | | D | Patulin | HPLC/ |
| | | E | Fumonisin | HPLC |
| | | F | Criteria of analysis for mycotoxins)(CEN report) | HPLC |
| | | | Okadaic acid, dinophysis toxin/mussels | HPLC/milk |
| | | | Aflatoxin M1/milk | |
| 6 | Nitrates | | Nitrates/nitrites | HPLC/IC Enzymatic |
| 7 | Food irradiation | | Irradiated food containing bones, cellulose, fat | ESR-spectroscopy, thermoluminescence, GC-MS |
| 8 | Vitamins | | Vitamin A | HPLC |
| | | | B-carotene | HPLC |
| | | | Vitamin D | HPLC |
| | | | Vitamin E (all tocopherols) | HPLC |
| 9 | Trace elements | A | General considerations | CVAAS - pressure digestion |
| | | B | Hg | AAS - dry ashing |
| | | C | Pb and Cd | AAS - pressure digestion |
| | | F | Pb, Cd, Cr, Mo | AAS - dry ashing |
| | | G | Pb, Cd, Cr, Mo, Ni | AAS - microwave digestion |
| | | H | Pb, Cd | |
| 10 | Genetically Modified Organisms | A and B | Detection | Sampling and nucleic acid extraction |
| | | C and D | Detection | Qualitative nucleic acid and protein-based methods |

- *The uncertainty of a measurement.* Any result can be an "estimate" and the true value is spread around this result. How to estimate this variability? From validation studies for the method? Or from the standard deviations of individual measurements?
- *The validation way.* Although a method should be validated within a collaborative trial, practical and economic reasons render this procedure difficult to fulfil in practice. As a consequence, new IUPAC guidelines have been developed for a single laboratory validation method.
- *Recovery.* International guidelines show how recovery information should be handled by different countries and organisations, in order to correct the discrepancies.

## 12.3 Some examples of chemical analysis protocols for the identification and quantification of chemical risk factors

To demonstrate the application of the concepts and methodologies/methods/techniques presented above, we have made a selection of some relevant "case studies" - some new, and some older but accepted examples - taking into account the main food chemical risk agents.
Be aware of the fact that in literature and in the European standard office new methods, advanced methods or rapid/cheap methods dedicated to the identification and quantitative evaluation of chemicals as risk factors for food safety, are being reported and registered all the time.

### 12.3.1 Residues of pesticides, antibiotics and veterinary/human drugs, and steroid hormones

*Pesticides*
There are many established methods based on GC-FID or HPLC for different pesticides. In general, the organochlorine and organophosphorous pesticides are separated by GC, and the carbamates by HPLC. We give an example of separation and identification (Figure 12.4)

*Antibiotics and veterinary/human drugs*
Due to their weak stability at high temperatures, antibiotics and many other drugs cannot be separated by GC, so the best method remains HPLC in different variants of elution system and using different types of RP- or NP-HPLC columns (Figure 12.5).

*Steroid hormones*
In general, the synthetic hormones (oestrogens or oestrogen-like molecules) are administered to cows or sows to stimulate their growth rate. The residues of hormones can be found in meat or milk and can affect human metabolism. To measure the presence and the quantity of such hormones or metabolites in food or biological samples, one can make use of either GC or HPLC techniques (Figure 12.6)

Bioassays are also increasingly used for other compounds with hormone (oestrogenic) potential (assays using reporter genes in breast tumour or yeast cells, named ER-CALUX, in analogy with CALUX assay). These tests should allow the detection of estradiol in meat and blood at ppb levels.

### Analysis of chemical food safety

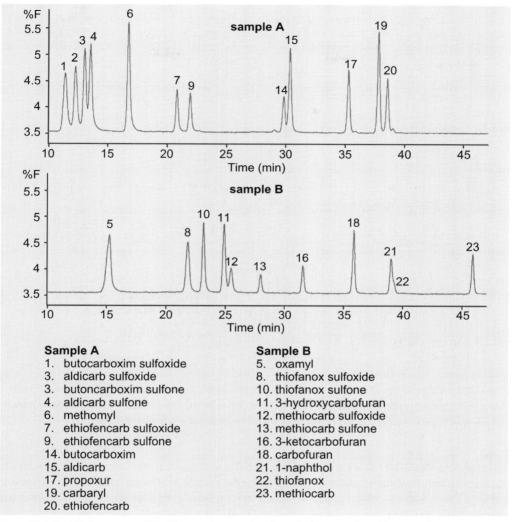

Figure 12.4. The HPLC separation of 23 different carbamates from 2 standard mixtures (A and B), using a $C_{18}$ column, a mobile phase of water/methanol (88:12, v/v) - methanol (gradient for 49 minutes) and a fluorescence detection ($\lambda_{em}$ = 425 nm, $\lambda_{exc}$ = 230 or 330 nm).(Hewlett-Packard appl., 1996). The limit of detection:1 ppb.

### 12.3.2 Environmental contaminants: inorganic contaminants (fertilisers and heavy metals), organic aromatic dioxins and metabolites, and mycotoxins

*Inorganic contaminants in food: nitrate/nitrite fertilisers and heavy metals*
The presence of inorganic ions like heavy metals and nitrates/nitrites with adverse effects on human health is undesirable in food. The major sources of contamination are industrial pollution

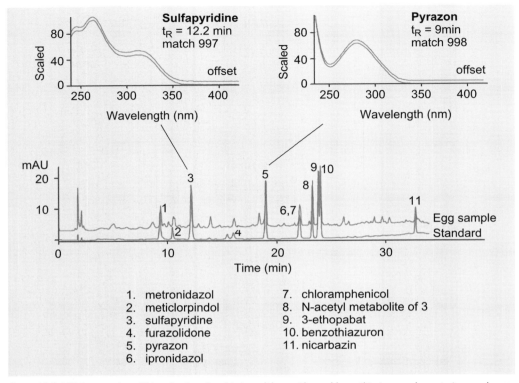

Figure 12.5. HPLC separation of 11 molecules of antibiotics, sulphonamides and benzothiazines used as veterinary or human drugs, from a standard mixture and from eggs. The UV spectra (above) were obtained by DAD detection and demonstrate the identity of HPLC peaks nr. 3 and 5, for the food extract and the standard, as a good means of identification.

and excess of fertilisers in the soil and plants. Lead, cadmium, arsenic, aluminium and mercury have been the main subjects of monitoring in the food chain, with data about the concentration of such elements in food being combined with the intake of foodstuffs. Other metals which are essential for all plants and animals, like Cu, Zn, Sb, Cr, Co, and Ni, can become toxic at high levels of exposure.

The analytical procedures for their evaluation include 2 steps: solubilisation via treatments with acids (mineralisation) and quantification by atomic absorption spectrometry (AAS), which is the best method, validated and calibrated with standard curves specific for each element.

Nitrates occur naturally in the environment as a source of protein synthesis and can be added as fertilisers. From ammonium, a gradual oxidation to nitrites and nitrates is rapid in soil and transferred in plants or water. If an excess of nitrate is retained in the soil, green leafy vegetables concentrate it. Both nitrate and nitrite are used in meat products as antiseptic and red-colouring agents. With the formation of nitroso-derivatives *in vivo* these ions can be considered toxic, above

## Analysis of chemical food safety

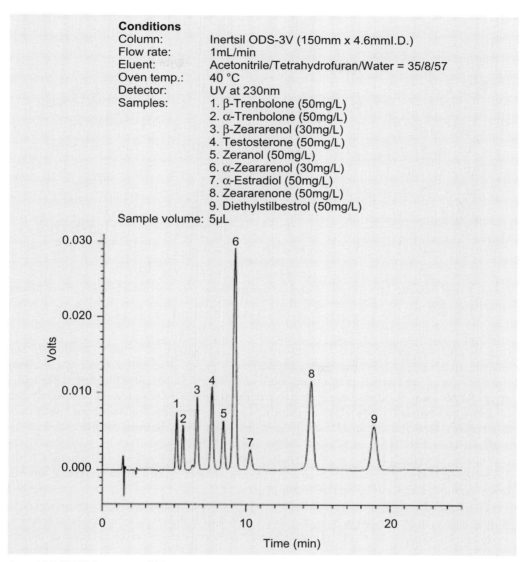

Figure 12.6. RP-HPLC separation of different steroid hormones (natural or synthetic homologues) which can be found as residues in pork or beef.

the average dietary exposure of 1.3 mg/day. The analysis of these ions is classically by means of identification reactions based on red-coloured diazonium salts with sulfanilic acid. Modern procedures include spectrophotometric determinations or HPLC-DAD evaluation (Figure 12.7).

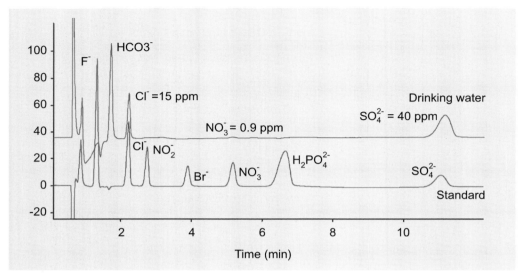

Figure 12.7. HPLC separation of different anions which can be found as contaminants in water and different food products (Hewlett-Packard appl., 1996).

Organic aromatic contaminants in food are products of industrial activity. To identify and quantify such hazards, sensitive direct analytical methods as well as immunochemical analyses were developed. The development of methods to identify and quantify chlorinated pesticides 40 years ago, led to the identification of polychlorinated biphenyls (PCBs), chlorinated dioxins (PCDDs) and furans (PCDFs). Concomitantly, polycyclic aromatic hydrocarbons (PAHs) and other organic chemicals (benzene, naphthalene, phthalates, and chlorinated aliphatic derivatives) were also identified. These compounds are generally very persistent in the environment, in the food chain and in the human body. They have proved to be highly cytotoxic and genotoxic. They are analysed either by classical techniques (TLCs, GC-FID) or by new methods (HPLC, GC-MS, bio- or immunoassays) as illustrated below (Figure 12.8).

After discovering dioxins in the food chain, very laborious and expensive analytical procedures were developed, in order to detect more than 17 different derivatives of chlorinated dioxins (PCDDs and PCDFs). Meanwhile, it was established that other substances like planar PCBs and brominated PAHs should be included in the evaluation. So, new analytical methods were needed.

In response to this limited and expensive analytical capacity, bioassays with mammalian cells were developed, initially based on the effects on cell proliferation, and then on the binding of dioxins to a specific receptor (Ah- receptor). As a consequence, the EROD assay measures the cytochrome P450 activity following the binding of dioxins to this receptor. The specificity of this test increased with the use of a rat hepatoma cell line containing a reporter gene luciferase, which, in response to dioxins, synthesises luciferase in a dose-dependent way and can be

# Analysis of chemical food safety

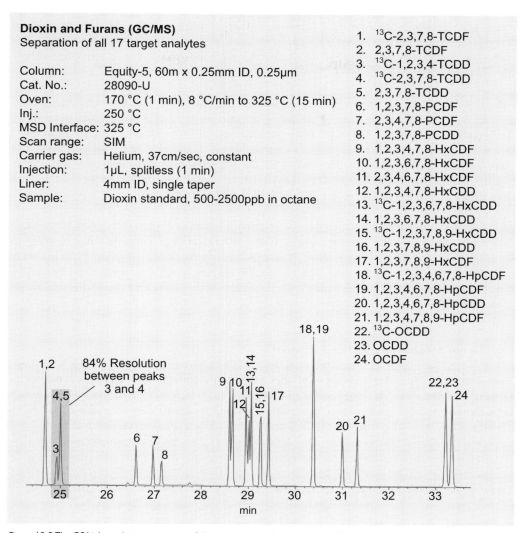

Figure 12.8 The GC high-resolution separation of dioxin and its metabolites as well as furans, using an Equity-5 column and Helium as carrier gas in a temperature range of 170-350°C. (Supelco catalogue, 2004).

quantified by an enzymatic light-producing reaction (CALUX bioassay). This test can identify less than 0.05 ng of TCDD. Meanwhile, a rapid clean-up procedure for fat samples was developed, based on silica column separation, so dioxins can be easily separated and concentrated from different matrices. The false results were compared using three methods, GC-MS, CALUX and EROD assays, with validation tests for different agonists also under evaluation.

Mycotoxins are produced by certain filamentous fungi growing naturally especially in cereals and oilseeds, after harvest and during storage; they are low-molecular weight, non-antigenic secondary metabolites of such fungi, formed via the polyketide or amino-acid route (aflatoxins), the terpene route (trichothecenes), or the tricarboxylic route (rubratoxin).

At least 300 potential mycotoxins have been isolated from raw materials but only about 20 were identified in food and feeds. Human intake of mycotoxins occurs via plant-based and animal-derived foods. Major food/feed sources of mycotoxin residues are milk and dairy products, meat and organs, e.g.liver, kidney, fermented and mould-ripened foods, and food additives. The most important mycotoxins are: aflatoxins B1,B2, G1, G2, ochratoxin A, penicillic acid, rubratoxin, patulin, citrinin, cyclopiazonic acid, fumonisin B1, trichothecene, T-2 toxin. For details about the toxicity and risk assessment, see the chapter in Smith, 2002.

Analysis of mycotoxins. Efficient sampling, sample preparation and methods of analysis for mycotoxins from food and feed form the basis of quality control procedures. Since mycotoxins have diverse chemical structures, there are no uniform methods of analysis and succession of flow patterns,i.e. sampling, extraction in specific solvent, clean-up, separation, detection, quantification and finally, confirmation. The main procedures used for mycotoxin analysis are presented in Table 12.6.

Table 12.6. Main procedures to follow for mycotoxin analysis.

| Step of analysis | Description | Aim |
| --- | --- | --- |
| Preparation of the sample and extraction | Grinding + mixing + blending + extraction in the appropriate solvent | High yield of mycotoxin extraction and separation from non-soluble food |
| Cleaning-up | Liquid-liquid partition (L/L) Open column chromatography Elimination of divalent metals ($Pb^{2+}$, $Fe^{2+}$, $Cu^{2+}$) | Separation of the toxin from other compounds extracted in the solvent |
| Separation and derivatisation | Thin Layer chromatography (TLC) Gas-liquid chromatography (GLC) Liquid-liquid chromatography (LC) | Separation of different types of mycotoxins and from other interfering toxins with similar structures |
| Detection and quantitation | Fluorescence on TLC plate or in solution UV spectrometry GLC-FID HPLC MS | Separation and identification of individual molecules |
| Confirmation | TLC detection MS NMR | Identification of the molecule structure by comparison with pure standards |

Most mycotoxins are non-volatile; GLC is limited in use and is recommended only for trichothecene mycotoxins. TLC is the most widely used quantification technique although HPLC is increasingly used since it offers increased sensitivity, improved accuracy and the ability to be automatic. HPLC does not always produce superior results to TLC but has greater versatility and is more specific by using detection by FT-RID, DAD, MS. The cost of HPLC is significantly higher than TLC (high capital output, highly trained personnel, time consuming, difficult cleaning up to achieve the detection limits, etc). There is now a great need for screening methods that are able to analyse large volumes of food and feed samples, and can be carried out by inexperienced operators. Figure 12.9 shows an example of an HPLC protocol and chromatogram for aflatoxins.

The immunoassay methods are now a major means of analysing mycotoxins, especially aflatoxins, and are different to chromatographic methods. These methods involve reversible binding between antigens (the analyte or mycotoxin) and selective antibodies, resulting in a specific complex antigen-antibody. In commercial applications, the radioimmunoassay (RIA) and Enzyme-linked immunosorbent assay (ELISA) and affinity chromatography are the principal immunochemical methods. Immunoaffinity chromatographic columns and cartridges for specific mycotoxins are now being increasingly used as preliminary cleaning-up procedures rather than HPLC and GLC methods.

Novel methodologies are based on the effect of mycotoxin extracts on primary and established/immortalised animal/human cell lines, determining the protein or DNA synthesis as end-points.. The development of such techniques will serve as standardised bioassays for detecting mycotoxins in food (Lewis *et al.*, 1998).

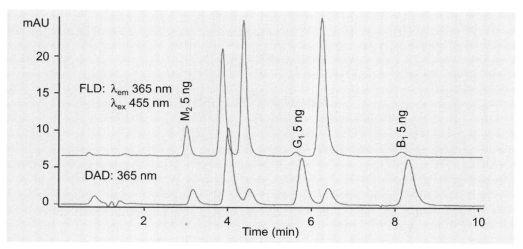

Figure 12.9. HPLC analysis of aflatoxins using C18 columns and UV detection by DAD (365 nm) or fluorescence ($\lambda_{em}$ = 365 nm, $\lambda_{exc}$ = 455 nm) (from Hewlett-Packard appl. 1996). The linearity of the calibration curve is in the range of 1-500 ng by UV detection.

## 12.3.3 Food additives: Synthetic colorants

In confectionery and patisserie, as well as in dairy products and fruit juices, many synthetic colorants are still used. Some of them are potentially toxic, so their identification and quantitative determination is useful. Figure 12.10 shows an HPLC protocol to separate some of these colorants and the way they are identified. UV-Vis spectra are specific for each colorant, so with HPLC-DAD it is possible to have an immediate identification of each peak separated from the HPLC column.

## 12.3.4 Toxic compounds formed during food processing: polyaromatic hydrocarbons (PAHs) and lipid peroxides

Polyaromatic Hydrocarbons (PAHs) are multicyclic compounds which can be formed during heat processing at high temperatures (e.g. frying in oil) and are potentially toxic (carcinogenic effect). So, their separation and identification is vital in meat products, bakery products, etc.

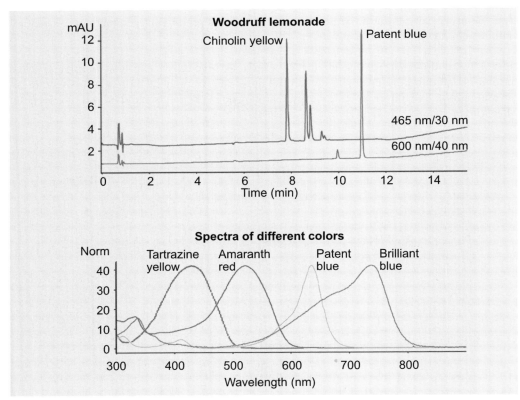

Figure 12.10. HPLC separation of food colorants using DAD detection. The spectra presented below identify the colorants according to their specific $\lambda_{max}$.

## Analysis of chemical food safety

Lipid peroxides. The oxidative degradation of oils and fats is known to take place in light, in the presence of moisture, and during exposure to heat. The oxygenated derivatives are formed via free radical reactions; they are peroxides, hydroperoxides or epoxides, short-life compounds converting finally to aldehydes or ketones with a rancid smell and taste. The unsaturated oleic (O), linoleic acid (L) and palmitoleic (P) acids in particular, but also saturated stearic acid (S), are labile substrates for oxidation and form different oxygenated derivatives, as shown in Figure 12.11, an HPLC chromatogram in which such compounds are separated, and identified in UV, at 215, 240 or 289 nm.

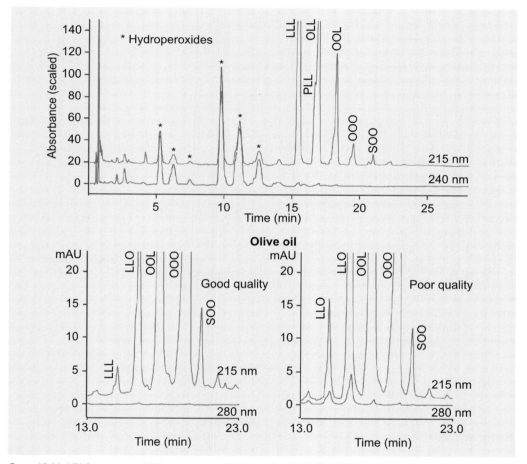

Figure 12.11. HPLC separation of different epoxides (LLO), peroxides (OOL, SOO, OOO) and hydroperoxides (LOOH) resulting from the oxidation of olive oil during thermal processing. Identification was made by UV detection of specific maxima at 215, 240, 280 nm.

*Carmen Socaciu*

## Abbreviations

| | |
|---|---|
| AAS | Atomic Absorption Spectrometry |
| AMFC | Additional Measures for Food Control |
| AOAC | American Organization for Analytical Chemistry |
| ATR | Accessory Total reflectance |
| CCMAS | Codex Committee for Methods of Analysis and Sampling |
| CEN | Community European Norms |
| DAD | Diode Array Detection |
| EF | Electrophoresis |
| ELISA | Enzyme Linked ImmunoSorbent Assay |
| ELSD | Electric Light Scattering Detector |
| FID | Flame Ionisation Detector |
| FT-IR | Fourier-Transformed Infrared (Spectroscopy) |
| GC | Gas Chromatography |
| HPLC | High Performance Liquid Chromatography |
| IR | Infra Red |
| LC | Liquid-liquid Chromatography |
| LPLC | Low Pressure Liquid Chromatography |
| MAA | MethAcrylic Acid |
| MIP | Molecular Imprinting Polymer |
| MS | Mass Spectrometry |
| NIR | Near Infra Red (Spectroscopy) |
| NMR | Nuclear Magnetic Resonance |
| OC | Open Column |
| PAHs | PolyAromatic Hydrocarbons |
| PCBs | PolyChlorinated Biphenyls |
| PCDDs | PolyChlorinated Dioxin Derivatives |
| PCDFs | Poly Chlorinated Dioxin Furans |
| RID | Refraction Index Detector |
| SFE | Solid Phase Extraction |
| SPR | Surface Plasmon Resonance |
| TLC | Thin Layer Chromatography |
| UV-Vis | Ultraviolet-Visible (Spectroscopy) |
| XRF | X-Ray Fluorescence |

## References

Aboul-Enein, H.Y., R.I. Stefan, J.F. van Staden, X.R. Zhang, A.M. Garcia-campana and W.R.G. Baeyens, 2000. Recent developments and applications of chemiluminescence sensors, Crit.Rev.Anal.Chem., 30, 271-289.

Bergmeyer, H.U., 1983. Methods of Enzymatic analysis, Academic Press, New York.

Bossi, A., S.A. Piletsky, P.G. Righetti and P.F. Turner, 2000. Review: Capillary Electrophoresis Coupled To Biosensor Detection, J Of Chromatogr A, 892, 143-53.

Bostrom, C.M. and J. Lindeberg, 2000. Biosensor-based determination of folic acid in fortified food, food Chem., 70, 523-532.

Cefai J., Complexity and miniaturization, LC GC Europe, 13, 752-764.

Cserhati, T. and E. Forgacs, 1999. Chromatography in Food Science and Technology. Lancaster Technomic Publishing Co.

Cunningham, A.J., 1998. Introduction to bioanalytical Sensors. New York, Wiley Press.

Davies, A.M.C. and R. Giangiacomo, 2000. Near Infrared Spectroscopy, Proceedings of the 9th International Conference, NIR Publications.

European Committee for Standardization, 1989. General Criteria for the Operation of Testing Laboratories, European Standard EN 45001, 45002 and 45003, Brussels, CEN/CENELEC.

Fox, P.F., 1991. Food Enzymology, vols. I and II, Elsevier, New York.

Frazier, R.A., J.M. Ames and H.E. Nursten, 2000. Capillary Electrophoresis for Food Analysis - Method Development, Cambridge, Royal Society of Chemistry.

Galensa, R., 1998. Biosensor-coupling with FIA and HPLC systems, Lebensmittelchemie, 52, 15-28.

Henshall, A., 1997. Use of Ion Chromatography in Food and Beverage Analysis, Cereal Foods World, 42, 414-19.

Hermann, H., 2000. Rapid automated analysis of antioxidants, Labor Praxis, 24, 24-27.

Horwitz, W., 1988. Sampling and preparation of samples for chemical examination. J. Assoc. Off. Anal. Chem., 71, 241-245.

Kirk, R.S. and R. Sawyer (eds.), 1991. Pearson's Composition and Analysis of Foods. 9th ed. Longman Group UK Ltd.

Kress-Rogers, E., 1998. Instrumentation and Sensors for the Food Industry, Cambridge, Woodhead.

Lewis, C.W., J.E. Smith, J.G. Anderson and R.I. Freshney, 1998. In: Miraglia *et al.* (eds), Mycotoxins and Phycotoxins: development in Chemistry, Toxicology and Food Safety, Alaken Inc., Fort Collins, Colorado, 159-164.

Luning, P.A., W.J. Marcelis and W.M.F. Jongen, 2002. Food Quality management, a techno-managerial approach, Wageningen Pers.

Meloan, C.E. and Y. Pomeranz, 1973. Food Analysis Laboratory Experiments, The Avi Publishing Company Inc., Westport, CT.

Myers, P., 2000. Hype and chips, LC GC Europe, 13, 744-750.

Nollet, L.M.L., 1996. Handbook of Food Analysis, Vol 1 and 2 New York, Marcel Dekker.

Nollet, L.M.L., 2000. Food Analysis by HPLC, 2nd Ed., New York, Marcel Dekker.

Pare, J.R.J. and J.M.R. Belanger, 1997. Instrumental Methods in Food Analysis, Amsterdam, Elsevier.

Patel, P.D., 2000. Biosensors for measurement of analytes implicated in food safety: a review, Leatherhead Food RA Scientific and Tech. Notes, 125.

Pearson, D., 1973. Laboratory techniques in Food Analysis, London, Butterworths.

Pimbley, D. and P.D. Patel, 2000. Agrifood applications of solid-phase extraction: a review, Leatherhead Food RA Scientific and Tech. Notes, 197.

Pomeranz, Y. and C.E. Meloan, 1994. Food Analysis: Theory and Practice, 3rd Edition, Chapman and Hall, New York.

Price, B.J. and H.W. Major, 1990. X-ray fluorescence proves useful for quality control, Food Technol., 44, 66-70.

Ramsay, G., 1998. Commercial Biosensors, New York, J.Wiley and Son.
Scott, A.O., 1998. Biosensors for Food Analysis. Cambridge, Royal Society of Chemistry.
Sellergren, B., 2000. Molecularly imprinted polymers-Man-made mimics of antibodies and their application in analytical chemistry, London, Elsevier Sci.
Sherma, J., 2000. Review: Thin Layer Chromatography in Food And Agricultural Analysis, J Of Chromatogr A, 880, 129-47.
Shi, H. and B.D. Ratner, 2000. Template recognition of protein-imprinted polymer surfaces, J.Biomed.Mater.Res., 49, 1-11.
Smith, J.E., 2002. Mycotoxins, In: D.H. Watson (ed.), Food Chemical safety, vol. 1 and 2, CRC Press, Woodhead Publ.Ltd.
Watson, D.H., 2002. Food Chemical safety, vol. 1 and 2, CRC Press, Woodhead Publ.Ltd.
Weiss, J., 1995. Ion Chromatography, VCH, Weinheim, Germany.
Wetzel, D.L.B., 1998. Analytical Near Infrared Spectroscopy', In: D.L.B. Wetzel and G. Charalambous (eds.), Instrumental Methods In Food And Beverage Analysis. Amsterdam, Elsevier, 141-94.
Yano, K. and I. Karube, 1999. Molecularly Imprinted Polymers for biosensor applications, Trends Anal.Chem., 18, 199-204.
Ye, L., O. Ramstrom, R.J. Ansell, M.O. Mansson and K. Mosbach, 1999. Use of molecularly imprinted polymers in a biotransformation process, Biotechnol.Bioeng., 64, 650-655.

# 13. Modern european food safety law

*Bernd van der Meulen and Menno van der Velde[1]*

## 13.1 Introduction

### 13.1.1 The ground plan of this chapter

This chapter gives an overview of European[2] food safety law from the perspective of European Union (EU) law.[3] Regulation 178/2002, known as the General Food Law, is the new foundation and centre for a large-scale reformulation of food law in the European Community (EC). This process is driven by several concerns, among which two stand out: the intimate connection of food law with the internal market, and the lessons learned from the major food scares of the 1990's. The General Food Law is also at the centre of this chapter. The development of the European Community's food law is traced over the three decades of its history since the end of the transition period used to create the common market. Knowledge of certain aspects of this development is necessary to understand the present situation of the internal market, harmonisation of national legislation, and food safety law. Readers will see how law is used as an instrument to achieve food safety and to deal with food safety problems. This chapter opens with some aspects of EC law that are crucial for food law. Against this background the turbulent development of food law in Europe from the beginnings of the European Community to the release of the White Paper on Food Safety in 2000 is described in section 13.2. The next two sections are dedicated to the development of European food safety law on the basis of the design

---

[1] This chapter is based on the authors' Food Safety Law in the European Union, Wageningen Academic Publishers, 2004. The authors wish to thank Ronald Verbruggen, Anna Szajkowska and Aaron Chase Underwood.

[2] Strictly speaking the European Union does not represent all of Europe, but we have taken the liberty of writing Europe when we mean a major part of it with great influence on food law, even outside the territory of the European Union, and even outside the European Economic Area. We feel this to be at least as justified as the habit of using the word America when we mean only a part of that continent.

Legally speaking we would have to differentiate clearly when to use the term European Union, and when European Community, as these are quite different entities. This awkwardness will stay with us until the day that Article IV-2 of the Draft Treaty Establishing a Constitution for Europe (!) becomes operative. That will repeal and replace the earlier treaties like the European Community Treaty and the European Union Treaty. From then on even the basic rules will have found one home. Until then we will freely interchange the words European Community and European Union, using European Community especially when the special characteristics of that Community are decisive for the subject matter at hand. Or as the Draft Constitution formulates it in Article 1 when "The Union shall exercise in the Community way the competences they [the Member States] confer on it."

[3] There is an unfortunate ambiguity in the English word "law". Sometimes it signifies the law generally, all legislation, other rules, custom and case law taken together, like in European Union Law. At other times it means a specific law, an Act made by the legislator on a specific issue, for instance the General Food Law. So it is that talking about general food law is something quite different from talking about the General Food Law. Other European languages do not have this ambiguity: French has "droit" and "lois", German "Recht" and "Gesetz", Dutch "recht" and "wet".

laid down in the White Paper. Section 13.3 discusses the general provisions of the General Food Law; section 13.4 some institutional aspects. The sections following that deal with substantive matters of food law: the composition of food (13.5), and the handling of food (13.6). Section 13.7 addresses communication on food from the angle of informed choice.

To curb the size of this chapter we concentrate on substantive food law. We do not take up procedural matters like enforcement. We also pay little attention to institutional aspects such as the internal organisation of the European Food Safety Authority. We are aware of the crucial role this new organisation is expected to perform in the new EC food law, but we have chosen to concentrate on its tasks, leaving a detailed description and analysis of its bodies and internal organisation for another occasion. In substantive law we stick to EC law[4]. We have selected the horizontal legal instruments (the instruments that regulate certain aspects that many foodstuffs have in common, rather than vertical recipe legislation that regulates all aspects of one particular food, from farm to fork), as these horizontal legal instruments are heavily influenced by the new approach of the European Community to food law. We do not describe the national food law of the Member States. Given the necessity to be selective about substantive food law as well, we feel justified in omitting this aspect of food law as it is ruled almost completely by the EC food law that we do describe and analyse. No such justification is available for omitting the tantalising developments of international food law on the level of the entire world. For this we have claimed some additional space in section 13.8, especially because this international food law is influencing EC food law more and more.

### 13.1.2 The European Union and Food Law

The founders of the European Communities[5] wrote into the treaties a system whereby the Communities have power only if a treaty article expressly gives this particular power. If the treaty is silent on a particular power, this power remains in the hands of the Member States. This basic principle is technically called the 'attribution of power'[6]. Article 2 of the European Community Treaty lists the objectives and the means of the EC; Article 3 prescribes at least twenty-one activities that are necessary to reach the goals set in the preceding article. The third part of the Treaty unfolds the Article 3 activities into thirty-one Chapters, each a set of specifications attuned to the characteristics of its policy field. It is with this elaboration of each policy that the necessary powers are attributed. The EC Treaty provides a separate transfer of power for each separate policy. Each transfer takes from the abstract toolkit that which is deemed to fit

---

[4] Literature on EC law generally: Gormley, Laurence W., Introduction to the Law of the European Communities, 3th ed., London, 1998; Mathijsen, P.S.R.F., A Guide to European Union Law, 7th ed., London 1999.

[5] Originally there were three Communities: the European Coal and Steel Community (ECSC), founded in 1952, the European Economic Community and the European Atomic Energy Community, better known as Euratom, both founded in 1958. The ECSC ceased to exist when the fifty years duration of its founding Treaty lapsed. The two remaining Communities were founded for an indefinite period. The EC took over the functions of the ECSC in 2002.

[6] See EC Treaty, Article 5, first sentence and Article 7, paragraph 1, second sentence, and (parts of) all Articles in the Treaty's "Part three: Community Policies", attributing power for a specific policy.

the particular policy, and assigns the institutions that have to carry out these tasks, in accordance with the procedures that are also given.

The attribution of powers for food policy and food law is complicated. It has several bases in the EC Treaty. The articles of the EC Treaty that attribute powers for food policy and food law are the articles for the common agricultural policy, approximation of national law needed for the internal market, the common commercial policy, public health policy and finally consumer protection.[7] Table 13.1 indicates the different parts of the EC Treaty that attribute powers to make food policy and food law.

The basic instrument for food law is Regulation (EC) No 178/2002, usually called the General Food Law (GFL). It is a pivotal legislative instrument that lays the basis for further development of EU food law. It mentions as the basis of its legality the EC Treaty articles presented above as the attribution articles, except the article for consumer protection. In Table 13.1 these articles are indicated as power clauses. The institutions of the Community have to prove that the necessary power resides with the Community each time they want to act. Therefore, each authoritative act of the Communities starts with mentioning the specific treaty article(s) that give(s) the powers that are used: "Having regard to Article ...".

### 13.1.3 From common to internal market

In the 1985 White Paper Completing the Internal Market[8] the EU Commission took the consequences of the harmonisation effort for the common market. Although many directives had been made, there were still large obstacles on the road to one single market. The Commission's White Paper gave an inventory of more than 300 legislative measures that were necessary to have a real common market with no internal borders. The program of the internal market was to achieve that objective by the last day of 1992. The Single European Act Treaty introduced the internal market in 1987 in the EEC Treaty and created a special harmonisation article for the effort to make the 300 measures before the target date. The new Article 100a introduced the possibility to use other instruments for the task of making common rules, and it prescribed majority decision making for these measures.

The White Paper also introduced a new approach to harmonisation. The case law of the Court of Justice prescribed mutual recognition of the laws of the Member States if a good has been lawfully made and put on the market in a Member State.[9] However, Member States could still claim as an overriding necessity "the protection of health and life of humans, animals or plants" (Article 30 EC Treaty) to maintain specific national legislation and thereby create or maintain a barrier to free movement. So the Commission proposed to concentrate the harmonisation effort on the national measures that create barriers to trade allowed by EC law. The legislative effort of the internal market programme had to supplant the legitimate national trade barriers by EC

---

[7] The EC Treaty Articles in the same sequence: 37, 95, 133, 152 and 153.
[8] COM (85) 310 Final, presented to the public by EC Commissioner Lord Cockfield on 15 June 1985.
[9] This case law is discussed for the Cassis de Dijon case in paragraph 13.2.2.

Table 13.1. Attribution of powers: food policy and food law. Food policy and food law have several bases in the EC Treaty.

| European Community Treaty ||
|---|---|
| **Part one** | **Part three** |
| | Community policies |
| Article 1 establish EC | TITLE I Free movement of goods |
| | Chapter 1 Customs Union |
| Article 2 | Article 25 |
| EC tasks, | Customs duties ... and charges having equivalent effect prohibited |
| Establish common market | Power clause: Article 26 |
| | Chapter 2 Prohibition of quantitative restrictions |
| Article 3 | Article 28 and 29 |
| 1. For purposes Article 2, activities of EC timetable: | Quantitative restrictions and all measures having equivalent effect prohibited |
| a. Customs Union | |
| b. Common commercial policy | TITLE II Agriculture |
| c. Internal market | Article 32 (1) common market > agriculture |
| ... | Article 34 (1) |
| e. Common agricultural policy | Power clause: Article 37 ... (2) ... |
| h. Approximation of laws | TITLE VI |
| p. Public health | Chapter 3 Approximation of laws |
| t. Consumer protection | Power clause: Article 94 ... Council acting unanimously on a proposal from the Commission ... issue directives for approximation ... law Member States ... directly affect the establishment or functioning of the common market. |
| ... | Power clause: Article 95 ... derogation from Article 94 ... Council ... procedure referred to in Article 251 ... measures for the approximation ... law Member States ... object establishment and functioning internal market |
| | TITLE IX Common commercial policy |
| | Power clause: Article 133 ... (2). Commission proposals to Council |
| | Power clause: (3) agreements with States or international organisations ... Council authorise Commission to negotiations. Council and Commission responsible |
| | TITLE XIII Public health |
| | Article 152 (1) A high level of human health protection shall be ensured ... |
| | Power clause: (4) The Council procedure Article 251 (b) **derogation from Article 37** measures in the veterinary and phytosanitary fields which have as their direct objective the protection of public health; |
| | TITLE XIV Consumer protection |
| | Article 153 (1) ... promote the interests of consumers and to ensure a high level of consumer protection, the Community shall contribute to protecting the health, safety and economic interests of consumers ... |
| | Power clause: (4) Council, procedure Article 251 ... adopt the measures referred to in paragraph 3(b): (b) measures which support, supplement and monitor the policy pursued by the Member States. |

law to safeguard the interests of safety and health. The approach to regulating these concerns was changed. Instead of making detailed rules for all aspects of the matter, the technical provisions were no longer a part of the EC legislation. The directives or regulations simply refer to standards made by European standardisation organisations. Industry had established these private law legal persons to develop unified European product standards and to rationalise the market. To give an example, CEN[10] and CENELEC[11] are private standardisation organisations for technical consumer products. These organisations keep track of technical progress to keep the European standards up to date. The EC legislation follows the same developments by referring to these standards, without the need to go through the process of lawmaking each time the standards change. The reference in the directive or regulation to the technical standards follows the changes in the private law standard automatically. However, this easy means of legislation carries the consequence that these standards are not part of the binding law. They remain voluntary, but industry is stimulated to apply the standards because then their products get free movement in the internal market. National authorities have to assume that goods made according to the standards meet the requirements of EC legislation. For goods that are not made according to the standards it is up to the manufacturer to prove that the product meets the requirements of the regulation or the directive.

### 13.1.4 Strengthening of substantive EC Treaty law on food

The BSE food crisis echoed in the corridors at the Intergovernmental Conference on the 1997 Treaty of Amsterdam. This treaty changed the 1992 European Union Treaty and the 1992 European Community Treaty. A number of articles were changed to ensure better policies on, among other issues, food. Article 95 EC Treaty, the special harmonisation article introduced by the Single European Act to help the creation of the internal market, already had in its third paragraph the commitment that the Commission would base its harmonisation proposals on a high level of protection of public health, safety, environmental protection and consumer protection. The Amsterdam Treaty added to this: "taking account in particular of any new development based on scientific facts." This reference to new developments based on scientific facts is of obvious importance to food law and the work of the European Food Safety Authority.

The Amsterdam Treaty also changed Article 152 EC Treaty paragraph (1) into: "A high level of human health protection shall be ensured in the definition and implementation of all Community policies and activities. Community action, which shall complement national policies ...". In the original article, EC policies and activities had to contribute to a high level of human health protection. Furthermore, the same article gives the Council and the Parliament acting together the power to overrule agricultural interests for the sake of public health: "Article 152 (4)(b) (adopting:) by way of derogation from Article 37 (giving powers for measures on agriculture), measures in the veterinary and phytosanitary fields which have as their direct objective the protection of public health;".

---

[10] CEN: Comité Européen de Normalisation (the European Committee for Standardisation).
[11] CENELEC: Comité Européen de Normalisation Electrotechnique (European Committee for Electrotechnical Standardisation).

*Bernd van der Meulen and Menno van der Velde*

Finally, the Amsterdam Treaty added to the already existing Article 153 on consumer protection the clause "... the Community shall contribute to protecting the health, safety and economic interests of consumers, ...". Within the context of these developments in primary EC Treaty law, specific developments took place in the area of secondary law that is particularly concerned with food.

## 13.2 Food law: development, crisis and transition

European Food Law has developed in several stages. As with any subject area it is useful to take historical developments into account in order to gain a better understanding of what remains of past structures, of mistakes made, lessons learned and probable future developments.

From the beginnings of the European Community in 1958 until the eruption of the BSE crisis in the mid-1990s, European food law was principally directed towards the creation of an internal market for food products in the EU. This market-oriented phase can be divided into two stages. During the first, emphasis was on harmonisation through vertical directives. This stage ended with the 'Cassis de Dijon' case law. During the second stage emphasis shifted to harmonisation through horizontal directives[12].

The BSE crisis, and other food scares that followed soon after, brought to light many serious shortcomings in the existing body of European food law. It became evident that fundamental reforms would be needed. In January 2000, the Commission announced its vision for the future development of European food law in a "White Paper on Food Safety".[13]

The "White Paper on Food Safety" emphasised the Commission's intent to change its focus in the area of food law from the development of a common market to assuring high levels of food safety. In the years since its publication, a great deal of important legislation has been passed, and more proposals are in development or under consideration. As this chapter looks toward the future, we will focus principally on food law as it has developed since the White Paper, and on proposals for continuing reform in the coming years. See Table 13.2 on the development of European food law.

### 13.2.1 Creating an internal market for food in Europe and vertical legislation

When the six original members of what is today the European Union signed the Treaty of Rome in 1957, they created a community with an economic character. This was reflected not only in

---

[12] The distinction between vertical and horizontal directives will be discussed in paragraphs 13.2.1 and 13.2.2.

[13] COM (1999) 719 def. Commission White Papers traditionally contain numerous proposals for Community action in specific areas, and are developed in order to launch consultation processes at the European level. If White Papers are favourably received by the Council, they often form the basis of later "Action Programs" to implement their recommendations. The White Paper on Food Safety already included an Action Program on Food Safety.

*Table 13.2. Development of European food law.*

| Phase | Turning point | Main orientation | Main legal instruments |
|---|---|---|---|
| First | Cassis de Dijon (1979) | Market | Vertical directives |
| Second | BSE crisis (1997) | Market | Horizontal directives |
| Third | | Safety | Regulations |

its original name - the European Economic Community - but also in the original goal to create a common market. At the heart of the instruments to achieve this goal are the famous four freedoms of the European Community: free movement of labour, free movement of services, free movement of capital and free movement of goods. Free movement of goods (Article 3 (1)(c) EC Treaty) has been vital to the development of food law. Initially, legislation was aimed primarily at facilitating an internal market - food products included. Agreement about the quality of food products was considered relevant. To reach such agreement directives were issued on the composition of certain specific food products. This is called vertical (recipe or compositional standards) legislation.

*Example of vertical legislation*
A well-known example is the ongoing discussion about how much cocoa a food product should contain to be called chocolate, and whether vegetable fats other than cocoa butter may be used. The Directive relating to Cocoa and Chocolate products Intended for Human Consumption[14] set the cocoa percentage at 35 for the entire European Union. Since 3 August 2003[15] up to a maximum of 5% of the total weight may be added in other vegetable fats to certain chocolate products.

Early attempts to establish a common market for foodstuffs in Europe by prescribing harmonised product compositions along such lines faced two substantial obstacles. Firstly, at that time all legislation required unanimity in the Council, which gave each Member State a virtual right of veto over new legislation. Secondly, there was the sheer scale of the task. Browse through a supermarket in any EU Member State and consider the variety of products on the shelves. There are, as the Community institutions soon realised, simply too many food products to deal with. Creating compositional standards for each product would have been a 'mission impossible', and the Commission wisely chose to seek alternatives. Nevertheless, a great many products remain subject to European rules on compositional standards.[16] These, however, will not be considered in this chapter.

---

[14] Council Directive 73/241/EEC of 24 July 1973 on the approximation of the laws of the Member States relating to cocoa and chocolate products intended for human consumption O.J. L 228, 16/08/1973, p. 23 - 35, meanwhile replaced by Directive 2000/36/EC of the European Parliament and of the Council of 23 June 2000 relating to cocoa and chocolate products intended for human consumption, O.J. L 197, 03/08/2000, p. 19 - 25.

[15] The date on which the term for implementation of Directive 2000/36 expired. Since then this rule from the Directive has direct effect if it is not yet transformed into national law.

[16] E.g. sugar, honey, fruit juices, milk, spreadable fats, jams, jellies, marmalade, chestnut puree, coffee, chocolate, natural mineral waters, minced meat, eggs, fish. Wine-legislation is a body of law in itself.

Bernd van der Meulen and Menno van der Velde

### 13.2.2 Advancement through case law

It was the Court of Justice of the European Communities that 'restarted the engine' of European co-operation in the area of food law, in this case through new, broad interpretations of Article 28 of the EC Treaty[17]. Article 28 is a key provision for the free movement of goods in the common market. It prohibits quantitative restrictions on imports and all measures having equivalent effect.[18] See law text box 13.1.

*Law text box 13.1. Article 28 EC Treaty.*

Quantitative restrictions on imports and all measures having equivalent effect shall be prohibited between Member States.

*Dassonville*
In the 1974 Dassonville case[19], the European Court of Justice gave a broad definition of the concept of 'measures having equivalent effect'. The case concerned parallel imports of Scotch whisky into Belgium. Belgian law required that a certificate of origin from the British authorities accompany such products. Mr Dassonville bought Scotch whisky in France for re-importation into Belgium. It was cheaper there, as Scottish exporters had been trying to gain a share in the French market with reduced prices. As Mr Dassonville could not obtain British certificates of origin in France, he created his own and was subsequently charged with fraud by the Belgian authorities. The Belgian criminal court was asked to determine whether, under European law, Mr Dassonville could be convicted for fraud. The Belgian court submitted this question to the EC Court of Justice in Luxemburg. The Court found that this particular Belgian law could not be applied as it constituted a restriction on trade and was thus prohibited under Article 28. The Court held "that all trading rules enacted by Member States which are capable of hindering, directly or indirectly, actually or potentially, intra-Community trade are to be considered as measures having an effect equivalent to quantitative restrictions", and are thus prohibited in the absence of a specific allowable justification. Such justifications could be the protection of human, animal or plant health. From this ruling it follows that mere disparities between national laws cannot be held against products from an exporting Member State as measures having equivalent effect to quantitative restrictions. In later rulings the Court detailed how such situations should be handled in practice.

*Cassis de Dijon*
The Cassis de Dijon[20] ruling was seminal in this respect. A German chain of supermarkets - price fighter Rewe - sought to import Cassis de Dijon, a fruit liqueur, from France. The German

---

[17] At that time numbered Article 30.
[18] See paragraph 13.1.1 Table 1.
[19] EC Court of Justice 11 July 1974, Case 8/74 (Dassonville), European Court Reports (ECR) 1974, page 837.
[20] EC Court of Justice 20 February 1979, Case 120/78 (Cassis de Dijon), ECR 1979, page 649.

authorities, however, refused to authorise the import because the alcohol content was lower than allowed by German national law, which stipulated that such liqueurs should contain at least 25% alcohol. Cassis de Dijon contained just 20% alcohol.

The German authorities acknowledged that this was a restriction on trade, but sought to justify it on the basis that beverages with too little alcohol pose several risks. The German authorities argued that alcoholic beverages with low alcohol content could induce young people to develop tolerances for alcohol more quickly than beverages with higher alcohol content. Also, consumers already familiar with German law might feel cheated if they purchased such products with the expectation of higher alcohol content. Finally, Germany submitted that in the absence of such a law, beverages with a low alcohol content would benefit from an unfair competitive advantage because taxes on alcohol are high, and beverages with lower alcoholic content would be saleable at significantly lower prices than products produced in Germany according to German law.

The Court held that the type of arguments presented by the German authorities would be relevant, even where they did not come under the specific exceptions contained in the EC Treaty, provided that those arguments met an urgent need. This is known as the rule of reason. The Court's broad interpretation of the prohibition against trade restrictions should be applied within reason.

The Court found that Germany's public health argument did not meet this standard of reasonableness. The Court specifically cited the availability of a wide range of alcoholic beverages on the German market with an alcohol content of less than 25%. As to the risk of consumers feeling cheated by lower than expected alcohol content, the Court suggested that such a risk could be eliminated with less effect on the common market by displaying the alcohol content on the beverages label.

*Mutual recognition and horizontal legislation*
For cases such as this one, in which there are no specific justifications for restrictions on the trade between Member States, the Court introduced a general rule: products that have been lawfully produced and marketed in one of the Member States, may not be kept out of other Member States if they do not comply with their national rules. This is called the principle of mutual recognition. In essence, the Court's ruling was that, within the context of the common market, what is good enough for consumers in one Member State is good enough for consumers across the Community. This principle of mutual recognition signalled a giant leap forward in European law, and 'Cassis de Dijon' has become a symbol of this stage in the development of the European understanding of free movement of goods.

With its ruling, the Court in Luxemburg laid the legal foundation for a well-functioning common market. Food products that comply with the statutory (safety) requirements of the Member State where they are brought on the market must, in principle, be admitted to the markets of all other Member States. Several commentators expressed concern that the Cassis de Dijon decision would lead to product standards based on the lowest common denominator. It is clear that manufacturers established in Member States with the most lenient technical requirements or legal procedures do gain a competitive advantage.

*Example*
Such a tendency is visible in imports as well. Partly owing to the know-how of the RIKILT Institute of Food Safety in Wageningen, Dutch authorities are in a better position than many other countries to detect contaminated foodstuffs. Although this benefits safety, it leads several companies to seek refuge in other European ports than Rotterdam.

The limitations and drawbacks of the principle of mutual recognition highlighted the need for further harmonisation of product standards at the European level. For Member States with more stringent national standards, European-level regulations became the best hope for raising their neighbours' standards. The Cassis de Dijon ruling marked a significant change in the perception of the benefits of harmonisation. Before Cassis, harmonisation was seen merely as a condition for the functioning of the common internal market. Afterwards, emphasis shifted to the need to alleviate the consequences of the common internal market. In legal terms, too, the wave of harmonisation that followed Cassis differed from earlier efforts. Emphasis shifted from product-specific, vertical (recipe) legislation, to horizontal legislation, meaning general rules on common aspects for all foodstuffs, or at least for as many foodstuffs as possible.

Mutual recognition remains the rule up to this day. The consequence bears repetition; food products that have legally come to the market in any Member State, may in principle be sold without restrictions across the whole territory of the European Union.

*Example of horizontal legislation*
A prime example of this development is the Labelling Directive,[21] which includes[22] a prohibition on misleading statements on foodstuff packaging[23] and a requirement to display specific indications, including a list of ingredients. Another example is the Hygiene Directive,[24] which requires manufacturers of foodstuffs to identify the critical points in the production process and to draw up procedures to ensure safety.[25] These mandatory self-regulations can be enforced through national legal procedures.

---

[21] Directive 2000/13 O.J. 2000 L 109 p. 29.

[22] Meaning: requires Member States to implement such prohibition.

[23] Article 2 (a).

[24] Council Directive 93/43/EEC of 14 June 1993 on the hygiene of foodstuffs O.J. L 175, 19/07/1993, p. 1-11. In late 2002 the European Commission announced that the Hygiene directive 93/43 and 17 other Directives would be joined and harmonised with other directives containing rules on hygiene for different products (such as fresh meat, milk and fish) and be simplified. A change of instrument was also foreseen. The Community exchanged the directives for one regulation. On 30 April 2004 Regulation 852/2004 O.J. 2004 L 139, 30/04/2004, p.1 was published, which will enter into force on 1 January 2006 at the soonest. This regulation is discussed in 13.6.1.

[25] The so-called HACCP system: "Hazard analysis and critical control point".

## 13.2.3 The consumer in European food law

One consideration in the Cassis de Dijon case deserves further attention. The Court stated that consumers might be adequately protected if they are adequately informed.[26]

Law text box 13.2. *The Court of Justice on the position of the consumer, EC Court of Justice 20 February 1979, Case 120/78 (Cassis de Dijon).*

> As the Commission rightly observed, the fixing of limits in relation to the alcohol content of beverages may lead to the standardisation of products placed on the market and of their designations, in the interests of a greater transparency of commercial transactions and offers for sale to the public.
> However, this line of argument cannot be taken so far as to regard the mandatory fixing of minimum alcohol content as being an essential guarantee of the fairness of commercial transactions, since it is a simple matter to ensure that suitable information is conveyed to the purchaser by requiring the display of an indication of origin and of the alcohol content on the packaging of products.

This passage illustrates the position of the consumer in European food law. Unlike under German law at that time (and under US law to this day), the Court will not treat European consumers as ignorant, highly vulnerable, or in need of protection by legislators who deem that they know what is best for the consumer. Rather, European food law presumes that consumers are reasonably intelligent, responsible and capable of making informed choices. Later on we will see that under European food law, consumer protection requires providing the consumer with the opportunity to make informed choices.

## 13.2.4 Breakdown

The heyday of market-oriented food law based on mutual recognition continued until the BSE crisis became widely known in the latter half of the 1990s. Public awareness of the epidemic, and the time it had taken British and European authorities to address it, presented a major challenge to European co-operation in the area of food safety. When the extent of the crisis became public, the European Union issued a blanket ban on British beef exports. In response, Britain adopted a policy of non-cooperation with the European institutions, and sought to deny the extent and seriousness of the BSE problem.[27]

The European Parliament played a crucial role in defusing this crisis. Although often accused of being a debating society of little consequence, during the BSE crisis the Parliament proved itself capable of political decisiveness and effective democratic oversight. A temporary Enquiry Committee was instituted to investigate the actions of the national and European agencies

---

[26] See also paragraph 13.8.6 Law text box 21 (Emmenthal Cheese case).

[27] To convince the population that there was nothing wrong with British beef the responsible Minister had his young daughter eat a hamburger on TV.

involved in the crisis.[28] The Enquiry Committee presented its report in early 1997.[29] The report strongly criticised the British government as well as the European Commission. The Commission was accused of wrongly putting industry interests ahead of public health and consumer safety. The Enquiry Committee did not confine itself to an analysis and critical comments. The report went on to make concrete recommendations for improvements in the structure of European food law. Paradoxically, this reproachful report provided the Commission with the impetus it had hitherto lacked to take the initiative for restructuring European food law in a way that considerably strengthened its own powers. The Commission's then President, Jacques Santer, undertook far-reaching commitments to implement the Committee's recommendations. As early as May 1997, the Commission published a Green Paper on the general principles of food law in the European Union. Consumer protection was made the first and foremost priority. An inspection agency - the Food and Veterinary Office (FVO) in Dublin - was created and the establishment of an independent food safety authority[30] was announced.

At the European summit in Luxemburg at the end of the same year, the European Council adopted a statement on food safety. The Commission kept the pressure on beyond 1997, eventually gaining the support of the European Court for the measures taken against the United Kingdom at the height of the crisis[31]. Meanwhile, public attention had turned to a new food safety scare, the Belgian dioxin crisis. The Commission proved it had learned a valuable lesson from its experience with BSE, and moved quickly and efficiently to protect consumers from the dioxin crisis. Nonetheless, the dioxin crisis brought to light further shortcomings in European food law.

Despite the resignation of Santer's Commission (which was succeeded by the Commission led by Romano Prodi), food safety remained a priority issue. On 12 January 2000 the Commission published its White Paper on food safety. The Commissioner responsible for food safety was David Byrne.

### 13.2.5 The White Paper: a new vision on food law

The White Paper aimed to restore and maintain consumer confidence. To achieve this it proposed an ambitious legislative programme. Eighty-four laws and policy initiatives[32] were

---

[28] O.J. 1996 C 261/132.

[29] Report of the Temporary Committee of Inquiry into BSE, set up by the Parliament in July 1996, on the alleged contraventions or maladministration in the implementation of Community law in relation to BSE, without prejudice to the jurisdiction of the Community and the national courts of 7 February 1997, A4-0020/97/A, PE 220.544/fin/A. See http://europe.eu.int > EN Gateway to the European Union > tab Institutions > European Parliament > Activities > Plenary sessions > Texts adopted by Parliament > by word in title > date range: 1994-1999 > 1997 > Words in title: Temporary Committee Inquiry BSE > BSE Resolution EP 19/02/1997 fourth indent hyperlink to Report.

[30] It seems that right now more faith is placed in autonomous technocratic agencies than in politically responsible administrative bodies.

[31] See judgments of 5 May 1998, cases C-157/96, C-180/96 and 1 October 1998 C-209/96.

[32] Listed in the Annex to the White Paper. This Annex bears the title Action Plan on Food Safety.

***Modern european food safety law***

scheduled for the near future, and implementation commenced with unprecedented drive. The White Paper focused on a review of food legislation in order to make it more coherent, comprehensive and up-to-date, and to strengthen enforcement. Furthermore, the Commission backed the establishment of a new European Food Safety[33] Authority, to serve as the scientific point of reference for the whole Union, and thereby contribute to a high level of consumer health protection.

*Planning a European Food Safety Authority*
The establishment of an independent European Food (Safety) Authority was considered by the Commission to be the most appropriate response to the need to guarantee a high level of food safety. According to the White Paper, the Authority should be entrusted with a number of key tasks embracing independent scientific advice on all aspects relating to food safety, operation of rapid alert systems, communication and dialogue with consumers on food safety and health issues as well as networking with national agencies and scientific bodies. The European Food (Safety) Authority ought to provide the Commission with the necessary analysis, however, it will remain the Commission's responsibility to decide on the appropriate response to that analysis.[34]

*Planning new food safety legislation*
Besides proposing the creation of an independent Authority, the White Paper called for a wide range of other measures to improve and bring coherence to the corpus of legislation covering all aspects of food products from "farm to fork". The Commission identified a wide range of measures necessary to improve food safety standards. Considering the enormous developments described above, it is clear that, in a number of areas, existing European legislation had to be brought up to date. A new legal framework was proposed to cover the whole of the food chain, including animal feed production, the establishment of a high level of consumer health protection, the attribution of clear primary responsibility for safe food to industry, producers and suppliers, and, finally, setting up appropriate official controls at both national and European level. The ability to trace products through the whole food chain was considered a key issue. The use of scientific advice will underpin food safety policy, whilst the precautionary principle[35] will be used where appropriate. The ability to take rapid, effective safeguard measures in response to health emergencies throughout the food chain was recognised as an important element. Proposals for the animal feed sector will ensure that only suitable materials are used in its manufacture, and that the use of additives is more effectively controlled. Certain food quality issues, including food additives, flavourings and health claims, will be addressed, and controls over novel foods improved.[36]

---

[33] In the White Paper the Commission speaks of a European Food Authority. The word 'safety' has been inserted later in the General Food Law. See paragraph 13.4.1.

[34] As we will see in paragraph 13.4.1, the European Food Safety Authority has been called in to being in Regulation 178/2002.

[35] This principle prescribes caution in the face of scientific uncertainty. See paragraph 13.3.8.

[36] Two years after the publication of the White Paper new pieces of legislation started to appear in the Official Journal. The flow is continuing until this day.

*Planning improvement of food safety controls*
The experience of the Food and Veterinary Office, the Commission's own inspection service, which visits Member States on a regular basis, has shown that there are wide variations in the manner in which Community legislation is implemented and enforced. According to the White Paper, this means that consumers cannot be sure of receiving the same level of protection across the Community, which makes it difficult to evaluate the effectiveness of national measures. The White Paper proposed developing a Community framework for the development and operation of national control systems, in co-operation with the Member States. This would take account of existing best practices, the experience of the FVO, and be based on agreed criteria for the performance of these systems, leading to clear guidelines on their operation. In support of Community level controls, more rapid, easier-to-use, enforcement procedures in addition to existing infringement actions would be developed. Controls on imports at the external borders of the Community should be extended to cover all feed and foodstuffs, and action taken to improve co-ordination between inspection posts.

*Planning improvement of consumer information*
If consumers are to be satisfied that the actions proposed in the White Paper will lead to a genuine improvement in food safety standards, they must be kept well informed. The Commission, together with the new European Food Safety Authority, will promote a dialogue with consumers to encourage their involvement in the new food safety policy. At the same time, consumers need to be kept better informed of emerging food safety concerns, and of risks to certain groups from particular foods. Consumers have the right to expect information on food quality and constituents that is helpful and clearly presented, so that informed choices can be made. Proposals on the labelling of foods, building on existing rules, have been brought forward in line with the White Paper proposals.[37] The importance of a balanced diet, and its impact on health, should also be presented to consumers.

*International dimension*
The European Community is the world's largest importer and exporter of food products. The actions proposed in the White Paper will need to be effectively presented and explained to the EU's trading partners. An active role for the Community in international bodies will be an important element in explaining European developments in food safety law.[38]

*Follow up*
The turn of the millennium saw the beginning of the planned reversal of European food law. It does not seem likely that we shall see the end anytime soon.[39] Only two years after the White Paper was published, the corner stone of new European food law was laid: Regulation 178/2002. This regulation is often referred to in English as the 'General Food Law'. The Germans speak of it as a 'Basisverordnung' - perhaps a more precise phrase given that the regulation is in fact

---

[37] The principle of informed choice will be discussed in paragraph 13.7.

[38] See paragraph 13.8 Globalisation.

[39] Not withstanding the claim that Commissioner Byrne made at the end of his term that some 90% of the work has been done. See the press release 'Achievements of Commissioner David Byrne, Commissioner for Health and Consumer Protection 1999-2004', Brussels 10 October 2004, MEMO/04/240.

*Modern european food safety law*

the basis upon which European and national food laws are now being re-constructed.[40] The main objective of the General Food Law is to secure a high level of protection of public health and consumer interests with regard to food products.

### 13.2.6 An analytical approach to food law

The history of European food law has resulted in a vast body of heterogeneous legislative measures. Its development was driven by incident rather than planning and - as pointed out in the White Paper - reflects little in the way of a coherent design. Nevertheless, it is possible to approach food law analytically.

Firstly, a distinction can be made between substantive and procedural food law. Substantive law focuses on the content of legal relations. Procedural law focuses on how to realise this content. Procedures for pre-market approval of food products, legal protection against measures of the authorities and, most of all, inspection, monitoring and enforcement - including provisions that attribute the necessary powers to public authorities - belong to procedural law. It appears that most of the substantive rules and regulations can be categorised in a limited number of ways. These categories can be compared to 'atoms', which when combined in different ways form a huge variety of 'molecules'. Students and scholars who grasp the 'atoms' will find it easier to cope with the 'molecules'. Three[41] atoms are easily distinguished: (1) rules concerning the substance of food, (2) rules concerning the handling of food, and (3) rules concerning communication about food. See Figure 13.1 on the structure of food law. The distinction between these three types is not watertight. Often one can argue whether a certain rule is of one type or another,[42] but that does not diminish the usefulness of the distinction. The account in the following sections will be based on this distinction, but first we will turn to the General Food Law. The General Food Law is intended to introduce the overall design that had previously been lacking in European food law. The General Food Law gives a context to the 'molecules' by assigning responsibilities and formulating goals and principles.

## 13.3 The General Food Law: general provisions of food law

### 13.3.1 The General Food Law

The first step in the realisation of the reform of food law, as planned in the White Paper on Food Safety, was the passage of 'Regulation (EC) No 178/2002 of the European Parliament and of the

---

[40] New European food law displays several characteristics in which it is different from its predecessor: more emphasis on horizontal legislation (than on vertical legislation), more emphasis on regulations that formulate the goals that have to be achieved, so-called objective regulations, than on means regulations, increased use of regulations (rather than directives) and thus increasing centralisation.

[41] In its definition of food law, German literature distinguishes some more different types: extraction, production, composition, quality, labelling, packaging and designation.

[42] For instance, the rule that a product must comprise of at least 35% cacao to be called chocolate can be considered a rule on content (35%) or a rule on communication, 'if you want to call it chocolate...'.

*Safety in the agri-food chain*

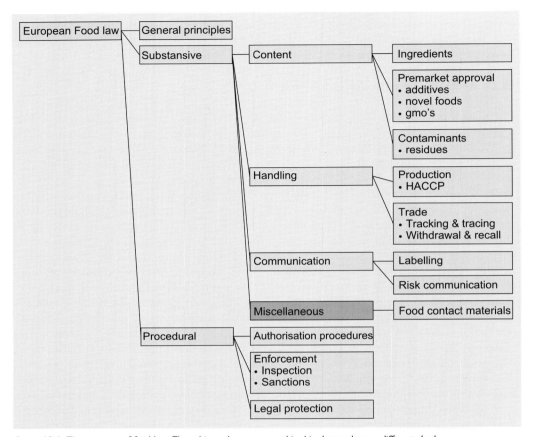

Figure 13.1. The structure of food law. The subjects that are treated in this chapter have a different shade.

Council of 28 January 2002 laying down the general principles and requirements of food law, establishing the European Food Safety Authority and laying down procedures in matters of food safety'.[43] The popular name of this regulation is: the General Food Law (GFL). As is apparent from the regulation's official title, the General Food Law seeks to accomplish three objectives:[44]
1. to lay down the principles on which modern food legislation should be based in the European Union as well as in the Member States;
2. to establish the European Food Safety Authority;
3. to establish procedures for reactions to food safety crises including the so-called Rapid Alert System.

---

[43] O.J. L 31, 1.2.2002, p. 1.

[44] These are the first three of the eighty-four legislative initiatives mentioned in the Action Plan on Food Safety, the Annex to the White Paper on Food Safety.

*Modern european food safety law*

To avoid confusion it should be noted that the General Food Law is not a code encompassing all food legislation. It is the fundament to a general part of food law. Next to it many (hundreds) of other European and national rules and regulations continue to play their role. This section treats the general provisions laid down in the first two chapters of the GFL. The rest of the GFL, the institutional aspects, are covered in section 13.4.

### 13.3.2 Aim and scope

The General Food Law provides a general framework for both national and Community food law in the European Union. In Article 5 - discussed here below - the GFL states the objectives of these national and European provisions of food law. In Article 1 it gives the aim and scope of the GFL itself.

*Law text box 13.3. Article 1 GFL on aim and scope of the GFL.*

1. This Regulation provides the basis for the assurance of a high level of protection of human health and consumer interest in relation to food, taking into account in particular the diversity in the supply of food including traditional products, whilst ensuring the effective functioning of the internal market. It establishes common principles and responsibilities, the means to provide a strong scientific base, efficient organisational arrangements and procedures to underpin decision-making in matters of food and feed safety.
2. For the purposes of paragraph 1, this Regulation lays down the general principles governing food and feed in general, and food and feed safety in particular, at Community and national level. It establishes the European Food Safety Authority. It lays down procedures for matters with a direct or indirect impact on food and feed safety.
3. This Regulation shall apply to all stages of production, processing and distribution of food and feed. It shall not apply to primary production for private domestic use or to the domestic preparation, handling or storage of food for private domestic consumption.

There are some basic notions in this article that have already been mentioned before:
- The principles and responsibilities are common to all food law in the EU.
- The principles and responsibilities are common to all stages of food and feed production and distribution ('from farm to fork'). This is sometimes called the 'holistic' approach to food law.

### 13.3.3 Definitions

The first step that the GFL takes in integrating the whole body of national and community food law in the EU, is the development of a common 'language', by providing definitions for the most important notions. It is important to note that the definitions are given 'for the purpose of this Regulation'. By consequence the GFL definitions do not apply automatically in the legislation of the Member States - although the GFL is a regulation - nor in other pieces of community legislation. The legislators will have to ensure that the GFL definitions are introduced in all relevant legislative acts. An example can be found in the new Hygiene Regulation for food of animal origin (853/2004):[45]

*Law text box 13.4. Inclusion of GFL definitions in Article 2 Regulation 853/2004.*

The following definitions shall apply for the purposes of this Regulation:
1. the definitions laid down in Regulation (EC) No 178/2002;
2. the definitions laid down in Regulation (EC) No 852/2004;
3. the definitions laid down in Annex I;
4. any technical definitions contained in Annexes II and III.

The definition of 'food' provided in the second article of the General Food Law is essential. Its fulfilment is a precondition for the applicability of the GFL. If a product meets this definition, it is a food in the sense of the GFL and the GFL applies to it. The same holds true for all the other laws and regulations that use this definition. In due course that should be the whole body of food law in the European Union and its Member States.

*Law text box 13.5. Definition of 'food' in Article 2 General Food Law, Regulation (EC) no 178/2002 O.J. L 31, 28 01 2002.*

For the purposes of this Regulation, 'food' (or 'foodstuff') means any substance or product, whether processed, partially processed or unprocessed, intended to be, or reasonably expected to be ingested by humans.
'Food' includes drink, chewing gum and any substance, including water, intentionally incorporated into the food during its manufacture, preparation or treatment. It includes water after the point of compliance as defined in Article 6 of Directive 98/83/EC and without prejudice to the requirements of Directives 80/778/EEC and 98/83/EC.
'Food' shall not include:
a. feed;
b. live animals unless they are prepared for placing on the market for human consumption;
c. plants prior to harvesting;
d. medicinal products within the meaning of Council Directives 65/65/EEC and 92/73/EEC;
e. cosmetics within the meaning of Council Directive 76/768/EEC;
f. tobacco and tobacco products within the meaning of Council Directive 89/622/EEC;
g. narcotic or psychotropic substances within the meaning of the United Nations Single Convention on Narcotic Drugs, 1961, and the United Nations Convention on Psychotropic Substances, 1971;
h. residues and contaminants.

---

[45] Regulation (EC) No 853/2004 of the European Parliament and of the Council of 29 April 2004 laying down specific hygiene rules for food of animal origin O.J. L 139, 30/04/2004, p. 55 - 205.

## Modern european food safety law

Other important definitions are those of 'food business' and 'food business operator'. These definitions are important because many provisions address these businesses (operators). If someone meets the definition, s/he has to comply with food law, which is also defined. These definitions are given in Law text box 13.6.

*Law text box 13.6. Definitions of 'food business', 'food business operator', and 'food law', Article 3 General Food Law, Regulation (EC) no 178/2002 O.J. L 31, 28 01 2002.*

> 'food business' means any undertaking, whether for profit or not and whether public or private, carrying out any of the activities related to any stage of production, processing and distribution of food;
>
> 'food business operator' means the natural or legal person responsible for ensuring that the requirements of food law are met within the food business under his control;
>
> 'food law' means the laws, regulations and administrative provisions governing food in general, and food safety in particular, whether at Community or national level; it covers any stage of production, processing and distribution of food, and also of feed produced for, or fed to, food-producing animals.

### 13.3.4 General principles

The General Food Law is not entirely clear on the meaning it attaches to the word 'principle'. In its second chapter, the General Food Law identifies six 'principles' of food law; four of these address food law as such and two address transparency. These principles are followed by three general obligations of the food trade. The second chapter ends with eight Articles containing general requirements of food law.

In our view the 'general obligations' and the 'general requirements' are of the same basic nature as the 'principles'. For this reason we will call them all 'principles'. These principles are presented in Table 13.3. We will discuss most principles briefly and elaborate a little bit more on the precautionary principle because it is controversial, and on the food safety principle (or duty of care) because it is the most important principle for the subject of this book. The principle on presentation (labelling) will be discussed in section 13.7.1. The responsibilities of food business operators for traceability, withdrawal and recall will be discussed in subsection 13.7.2.

### 13.3.5 Focused objectives

The GFL provides a limited set of objectives for food legislation.[46] In other words, a principle of focus applies. Food law shall pursue one or more of the general objectives of a high level of protection of human life and health and the protection of consumers' interests, including fair practices in the food trade, taking account of, where appropriate, the protection of animal health

---

[46] Article 5 GFL.

Table 13.3. Principles of food law (Chapter II GFL).

| Principle | Content |
|---|---|
| **General principles (Section 1)** | |
| General objectives (or: Focus) (art. 5) | Protection of: <br>■ human life & health; <br>■ consumers interests; <br>■ fair trade; <br>■ animal health & welfare, plant health, environment; <br>■ free movement of feed & food products; <br>■ application of international standards. |
| Consumer protection (art. 8) | ■ provide a basis for informed choice; <br>■ prevent fraudulent or deceptive practices; <br>■ prevent adulteration of food; <br>■ prevent other misleading practices. |
| Science based (art. 6) | ■ based of risk analysis; <br>■ risk assessment based on all scientific evidence, independent, objective, transparent; <br>■ risk management takes into account results of risk assessment, other legitimate factors, precautionary principle. |
| Precautionary principle (art. 7) | If the possibility of harmful effects on health are identified, but scientific uncertainty persists: <br>■ provisional risk management measures may be taken; <br>■ pending further research; <br>■ proportionate; <br>■ no more restrictive to trade than required; <br>■ with regard to technical and economic feasibility and other legitimate factors; <br>■ measures shall be reviewed within a reasonable time. |
| **Principles of transparency (section 2)** | |
| Public consultation (art. 9) | ■ the public shall be consulted in food legislation; |
| Public information (art. 10) | ■ risk communication; <br>■ the general public shall be informed of suspected risks and of measures taken. |
| **General obligations of food trade (section 3)** | |
| Import (art. 11) | ■ imports shall comply with EU food law |
| Export (art. 12) | ■ exports shall comply with EU food law |
| International standards (art. 13) | ■ contribute to development; <br>■ promote international organisations; <br>■ contribute to agreements; <br>■ special attention to needs of developing countries; <br>■ promote consistency. |

Table 13.3. Continued.

| Principle | Content |
|---|---|
| **General requirements** | |
| Food safety (art. 14) | No placing unsafe food on the market |
| | Food is unsafe: |
| | ■ injurious to health; |
| | ■ unfit for consumption (contamination, putrefaction, deterioration, decay). |
| | Applies to entire batch unless proven safe |
| | Regarding: |
| | ■ normal use; |
| | ■ information. |
| | Regarding: |
| | ■ long term effects; including future generations; |
| | ■ cumulative toxic effects; |
| | ■ sensitive category of consumers; |
| | Food complying with food law is deemed safe for the aspects covered: |
| | ■ compliance does not bar authorities from taking measures. |
| Feed safety (art. 15) | ■ No placing unsafe feed on the market; |
| | ■ No feeding unsafe feed to food producing animal. |
| Presentation (art. 16) | ■ No misleading consumers |
| Responsibilities (art. 17) | ■ Business operators from farm to fork for following food law; |
| | ■ Member States for enforcement from farm to fork (controls, communication & penalties). |
| Traceability (art. 18) | ■ from farm to fork; |
| | ■ one step up one step down; |
| | ■ operators have systems in place; |
| | ■ information available for authorities; |
| | ■ adequate identification & labelling. |
| Responsibilities for food (art. 19) | ■ operator who has reason to doubt safety, must withdraw food; |
| | ■ if food has reached consumer: inform consumer and if necessary recall; |
| | ■ inform and co-operate with competent authorities. |
| Responsibilities for feed (art. 20) | ■ operator who has reason to doubt safety, must withdraw feed; |
| | ■ inform and co-operate with competent authorities. |
| Liability (art. 21) | ■ general provisions on product liability apply. |

and welfare, plant health and the environment. Next, the free movement of food products is mentioned, as is adherence to international standards. Although this range of objectives is rather wide, it is not unlimited. For instance, there does not seem to be much room for legislators to make food law provisions aiming specifically at objectives like the interests of the food industry, or of food industry employees.

### 13.3.6 Consumer protection

Article 8 GFL states that food law shall aim at the protection of the interests of consumers and shall provide a basis for consumers to make informed choices in relation to the foods they consume. It shall aim to prevent:
a. fraudulent or deceptive practices;
b. the adulteration of food; and
c. any other practices which may mislead the consumer.

Although this provision has not been placed right next to article 5, it seems clear that here objectives are also at stake.

### 13.3.7 Risk analysis

Food law is science based or, in the words of the GFL, "food law shall be based on risk analysis except where this is not appropriate to the circumstances or the nature of the measure".[47] This principle is at the core of the new European approach to food law as 'food safety law'. The GFL regards risk analysis as a process consisting of three interconnected components: risk assessment, risk management and risk communication.[48] As these issues are addressed in this book's chapter on modelling and risk management, we will not discuss risk analysis any further.

### 13.3.8 Precautionary principle

The precautionary principle adds subtlety to the risk-analysis principle.[49] In specific circumstances, following an assessment of available information, where the possibility of harmful effects on health is identified but scientific uncertainty persists, provisional risk management measures necessary to ensure the high level of health protection chosen in the Community may be adopted, pending further scientific information for a more comprehensive risk assessment.[50]

*Background*
The precautionary principle is defined in Article 7 of the General Food Law. The GFL's preamble states that, in some cases, scientific risk assessment alone cannot provide all the necessary information on which risk management should be based. The GFL invokes the precautionary

---
[47] Article 6 GFL.
[48] Article 3 (10) GFL.
[49] Many thanks to Anna Szajkowska who wrote this paragraph.
[50] Article 7 GFL.

principle to ensure health protection in the Community and to set an uniform basis for the use of precautionary measures within the Community, in order to avoid arbitrary decisions giving rise to unjustified barriers to the free movement of food and feed. The preamble states that in circumstances where "a risk to life or health exists but scientific uncertainty persists, the precautionary principle provides a mechanism for determining risk management measures or other actions in order to ensure the high level of health protection chosen in the Community".

*Precaution and risk assessment*
There is a close link between the articles on risk analysis, including science-based risk assessment, which are placed in Article 6, and the precautionary principle. Art. 6 (3) mentions the elements which risk management should take into account in order to achieve the general objectives of food law: "the results of risk assessment, and in particular, the opinions of the Authority referred to in Article 22,[51] other factors legitimate to the matter under consideration and the precautionary principle". This means that the precautionary principle is used in the framework of risk analysis.

The precautionary principle article is divided in two parts. The first defines the circumstances under which the principle may be invoked:
- Potentially harmful effects on health must have been identified.
- Scientific evaluation does not determine the risk with sufficient certainty.

The second part deals with limitations of the use of the principle, which must be:
- proportionate and no more restrictive to trade than is required to achieve the high level of health protection chosen in the EU;
- reviewed within a reasonable period of time, depending on the nature of the risk to life or health identified and the type of scientific information needed to clarify the scientific uncertainty and to conduct a more comprehensive risk assessment.

*Provisional measures*
The precautionary principle provides a basis for measures of a provisional character. Article 7 GFL states that "provisional risk management measures" may be adopted, "pending further scientific information for a more comprehensive risk assessment". The measures based on the precautionary principle should be maintained as long as the scientific data are inadequate, imprecise or inconclusive and as long as the risk is considered too high to be imposed on society. However, this is not always linked to a time factor, but to the development of scientific knowledge.

Measures resulting from the precautionary principle may take various forms. When a precautionary measure seems to be appropriate to a particular situation, a decision has to be taken as to the nature of the action. Besides the adoption of legal measures, there are other possible actions that can be inspired by the precautionary principle, such as funding a research program, or informing the public about the adverse effects, etc. Precautionary measures must be proportional to the desired level of protection. A total ban may not be a proportional response

---
[51] I.e. the European Food Safety Authority, discussed in paragraph 13.4.1.

to a potential risk in all cases. Risk can rarely be reduced to zero. They must also be non-discriminatory: comparable situations should not be treated differently, and incomparable situations should not be treated in the same way.

The precautionary principle includes a notion of "cost-effectiveness". The precautionary measures are not simply there for banning products that are deemed not to be entirely safe. The central element of the principle is the process of weighing risks and benefits, comparing the costs of action and non-action, before accepting a product supposed to be dangerous. It means that where there is no obvious gain from taking the risk, it is better not to take it. This is not simply a cost-benefit analysis from an economical viewpoint, its scope is wider, and includes non-economic factors, like the efficacy of possible options and their acceptance by society. Consumers may require more efforts to protect health.

### 13.3.9 Transparency: Public consultation and Public information

Two principles fall under the heading 'transparency'. The first of them is that the public must be consulted in the development of food law.[52] The second is that authorities have a duty to inform the public of foodborne health risks, and the measures that are taken to counteract them.[53] This is the concept of 'risk communication'.

### 13.3.10 International trade: Import, Export and International Standards

European food safety requirements apply to imported food.[54] As we will see hereafter, these requirements on food safety not only address the condition of the product when it arrives at the European border, but also the way it has been handled in processing and trade and even in the choice of raw materials. This principle therefore implies considerable extra-territorial ambitions of EU food law.

In principle,[55] exported food must also comply with the European food safety standards.[56] This principle on the one hand makes the European origin of a food product a quality guarantee. On the other hand it facilitates controls and enforcement, as all production in the EU in principle has to meet the same safety standards, regardless of the market for which they are produced. The EU and the Member States contribute to the development and application of international standards.[57]

---

[52] Article 9 GFL.
[53] Article 10 GFL.
[54] Article 11 GFL.
[55] Third countries can explicitly agree to exceptions.
[56] Article 12 GFL.
[57] Article 13 GFL.

## 13.3.11 Food safety: a duty of care

The most important notion in the General Food Law is the central place it accords to safety. The GFL imposes on food business operators the responsibility for the safety of the food they bring to the market. Article 17 which bears the title 'Responsibilities', states in its first paragraph that food business operators must ensure adherence to food law.

*Law text box 13.7. Article 17 (1) GFL on adherence to food law.*

Food and feed business operators at all stages of production, processing and distribution within the businesses under their control shall ensure that foods or feeds satisfy the requirements of food law which are relevant to their activities and shall verify that such requirements are met.

This sounds like stating the obvious: 'one should adhere to the law', but there is more to it than that. Article 14 GFL forbids bringing food to the market if it is unsafe. Food is deemed to be unsafe if it is injurious to health or unfit for human consumption.[58] This imposes a general responsibility to ensure the safety of any food brought to the market. Furthermore, this implies a general risk for the food business operator in case a food that has been brought to the market turns out to have been injurious to health. Even the effects on the health of subsequent generations are taken into consideration. In other words adverse effects of food may constitute an offence committed by the food business operator.

The conveyance of responsibility to industry also has consequences for the legislator. Where possible, food law should define the aims to be achieved rather than the methods to be applied.[59] The Member States watch over the food business operators on behalf of the EC. See Law text box 8 for the obligations the Member States have in this respect.

*Law text box 13.8. Article 17 (2) GFL on the Member States' responsibility to enforce food law.*

Member States shall enforce food law, and monitor and verify that the relevant requirements of food law are fulfilled by food and feed business operators at all stages of production, processing and distribution.
For that purpose, they shall maintain a system of official controls and other activities as appropriate to the circumstances, including public communication on food and feed safety and risk, food and feed safety surveillance and other monitoring activities covering all stages of production, processing and distribution.
Member States shall also lay down the rules on measures and penalties applicable to infringements of food and feed law. The measures and penalties provided for shall be effective, proportionate and dissuasive.

---

[58] Interestingly, this is not part of the definitions of Article 3, but part of Article 14.

[59] In other words, the GFL prefers so-called objective-regulation to means-regulation. This is expressed by the Commission in the draft regulation, which later became the GFL. See COM (1999) 719 final, O.J. C 96 E/247, § 1.7.

Although article 17 of the General Food Law holds the Member States responsible for the enforcement of food law, European food law increasingly sets standards for national enforcement and provides for supervision. On 30 April 2004 a new Regulation was published in the Official Journal of the European Union: 'Regulation (EC) No 882/2004 of the European Parliament and of the Council of 29 April 2004 on official controls performed to ensure the verification of compliance with feed and food law, animal health and animal welfare rules'. Food that complies with specific Community provisions governing food safety - or in the absence of these with national provisions - shall be deemed to be safe insofar as the aspects covered by the specific Community provisions are concerned.[60]

## 13.4 The General Food Law: institutional aspects

### 13.4.1 The European Food Safety Authority

In article 22 the GFL establishes the European Food Safety Authority (EFSA). The EFSA is an independent agency responsible for risk assessment. The GFL gives the EFSA (called 'the Authority') the following mission:

Law text box 13.9. Article 22 (2) GFL on the mission of EFSA.

> The Authority shall provide scientific advice and scientific and technical support for the Community's legislation and policies in all fields which have a direct or indirect impact on food and feed safety. It shall provide independent information on all matters within these fields and communicate on risks.

The quality and accessibility[61] of scientific advice is considered to be of paramount importance to ensure effective, timely and appropriate decision-making. The EFSA must cover all parts of the food chain. Scientific matters are considered as a continuum from primary production through to the consumer. The operation of the EFSA as an independent entity is intended to ensure that there is a functional separation of the scientific assessment of risk from risk management decisions. The reason is that scientific risk assessment should not be swayed by policy or other external considerations. This is designed to guarantee impartiality and objectivity. Risk management on the other hand, in its widest possible sense, remains within the domain of the Commission, the Parliament and the Council. Risk management decisions take into account all relevant aspects and are considered to be the function of accountable, political structures. Risk managers, therefore, have to take into consideration not only science but also many other matters - for example, economic, societal, traditional, ethical or environmental factors, as well as the feasibility of controls. For this reason there should be close collaboration, interaction and exchanges between the EFSA and those charged with the responsibility of

---

[60] Article 14 (7) GFL.
[61] See www.efsa.eu.int.

managing risk. Through an Advisory Forum, the EFSA provides central co-ordination to the efforts and resources of the national food authorities and agencies in Europe. It also provides a central focus for such networks.

*The organisation of the European Food Safety Authority*
EFSA is composed of four bodies: the Management Board, the Executive Director (heading the staff), the Advisory Forum and the Scientific Committee with Scientific Panels.[62] The Management Board is composed of one representative of the Commission and fourteen members who are appointed by the Council in consultation with the European Parliament from a list drawn up by the Commission. The fourteen members are selected on the basis of their competence and expertise. Four of the members must have a background in organisations representing consumers and other interests in the food chain.[63]

The Management Board has to keep EFSA on the course set by its mission statement, to adopt the annual work programme and the revisable multi-annual programme. Both work programmes must be consistent with the Community's legislative and policy priorities in the area of food safety.

The Executive Director is the legal representative of EFSA.[64] She or he is responsible for the day-to-day administration of EFSA, implementing the decisions of the Management Board. The Executive Director draws up a proposal for EFSA's work programmes in consultation with the Commission, and acts as the liaisons officer between EFSA and the European Parliament, ensuring a regular dialogue with its relevant committees.

The Advisory Forum is the link between EFSA and the Member States.[65] It is composed of representatives from competent bodies in the Member States, who undertake tasks similar to those of EFSA. There will be one representative designated by each Member State. The Forum has to meet at least four times a year. The Advisory Forum gives the Executive Director advice on all aspects of his or her tasks, and in particular on drawing up a proposal for EFSA's work programme. The Advisory Forum is a mechanism for an exchange of information on potential risks and the pooling of knowledge. It ensures close cooperation between EFSA and the competent bodies in the Member States.

The Scientific Committees were attached to the EU Commission in the period before the GFL. To enhance their independency they are now a part of the EFSA. The Scientific Committee and permanent Scientific Panels are responsible for providing the scientific opinions of the EFSA, each within their own spheres of competence. The Scientific Committee is the institutionalised meeting of the Chairs of the Scientific Panels and six independent scientific experts who do not belong to any of the Scientific Panels. The six independent experts are appointed by the

---

[62] Article 24 GFL.
[63] Article 25 GFL.
[64] Article 26 GFL.
[65] Article 27 GFL.

Management Board. The Scientific Panels are composed of independent scientific experts appointed by the Management Board. Eight Panels have been established with a specified mandate so far.[66] The Scientific Committee is responsible for the general co-ordination necessary to ensure the consistency of the scientific opinion procedure, in particular with regard to the adoption of working procedures and harmonisation of working methods. It provides opinions on multi-sectoral issues falling within the competence of more than one Scientific Panel, and on issues which do not fall within the competence of any of the Scientific Panels. The GFL lays great stress on the openness and transparency of the activities of EFSA, as far as confidential aspects of its work permit. Members of the Scientific Committee and the Scientific Panels have to sign a commitment to independence, and have to report each year on their direct or indirect interests that are relevant to ESFA.

*Tasks of EFSA*
The following seven tasks will be briefly described: to present scientific opinions; to find a solution for diverging scientific opinions; to provide scientific and technical assistance to the Commission; to build a data collection; to establish monitoring procedures; and to promote networking and commission scientific studies. EFSA has to process the information delivered by the rapid alert system as a part of the monitoring procedures.

**Task 1 -** A main task of EFSA is to prepare and present scientific opinions. This work is done by the Scientific Committee or the Scientific Panels. Requests for a scientific opinion are assigned by the Executive Director to one of the Scientific Panels according to its mandate, or to the Scientific Committee. Article 29 GFL specifies that opinions can be requested by the Commission acting within the sphere of its authority and in all cases where Community legislation grants this right. Opinions can also be requested by the European Parliament and the Member States. EFSA can give opinions on its own initiative. Article 18 (4) MB Decision specifies the items that must be present in an opinion.

*Law text box 13.10. Article 18 (4)(5), EFSA's Management Board Decision on the content of a scientific opinion.*

... 4. A scientific opinion shall comprise the question posed by the Commission, the Parliament, a Member State or the EFSA itself, the background to the request, the information considered, the scientific reasoning and the opinion of the Scientific Committee or Panel.
5. ... The full scientific opinion shall be published without delay on the EFSA's Web site.

**Task 2 -** In the case of diverging scientific opinions, the Advisory Forum has to contribute to finding a solution when disagreement occurs between EFSA and an organisation working on the same topics in a Member State. When both parties identify a substantive divergence of opinion over scientific issues, Article 30 (4) GFL orders them to tackle this problem. Their co-

---
[66] See for this list Article 28 GFL, combined with Article 18 MB Decision.

operation has to result in resolving the differences or in a joint document clarifying the contentious scientific issues and identifying the relevant uncertainties in the data.

**Task 3 -** Providing scientific and technical assistance to the Commission if requested. This assistance will consist of scientific or technical work involving the application of well-established scientific or technical principles. The scientific evaluation by EFSA is not needed. The assistance can be given without compromising the independent position of EFSA. The Commission will enlist this assistance especially for the establishment or evaluation of technical criteria and the development of technical guidelines.

**Task 4 -** ESFA shall build up a collection of data in particular relating to:
a. food consumption and the exposure of individuals to risks related to the consumption of food;
b. incidence and prevalence of biological risk;
c. contaminants in food and feed;
d. residues.

EFSA will collect data in close co-operation with all organisations operating in the field of data collection, including those from applicant countries, third countries or international bodies.

**Task 5 -** ESFA will establish monitoring procedures for systematic collecting of information to identify emerging risks in the fields within its mission. Where EFSA has information leading it to suspect a serious risk, the Member States, other Community agencies and the Commission have to co-operate to find and deliver to EFSA any relevant information. To enable it to perform its task of monitoring the health and nutritional risks of foods as effectively as possible, EFSA shall receive all messages forwarded via the rapid alert system. EFSA analyses this information and gives it to the Commission and the Member States for risk analysis.

**Task 6 -** EFSA has to promote networking between European organisations that work in the same fields as EFSA. The aim of such networking is, in particular, to facilitate a scientific co-operation framework by the co-ordination of activities, the exchange of information, the development and implementation of joint projects, and the exchange of expertise and best practices in the fields within EFSA's mission. [67]

**Task 7 -** EFSA has to commission scientific studies necessary for the performance of its mission, using the best independent scientific resources available. EFSA has to avoid duplication with Member State or Community research programmes and shall foster co-operation through appropriate co-ordination. The Advisory Forum has to assist the EFSA in this respect, giving substance to its role of ensuring the pooling of knowledge and close co-operation between EFSA and the competent bodies in the Member States.

---

[67] Article 36 GFL

## 13.4.2 Standing Committee on the Food Chain and Animal Health

Most EU policies are executed by the administrations of the Member States, working in these cases under EU law. Only a part of the EU policies is administrated at the Union level by the institutions of the Community. Then two institutions are involved, the Commission and the Council. The EC Treaty is built on the assumption that the Council will delegate most of its powers for the day-to-day administration of EC policies to the Commission.[68] Indeed the Council has made Council Decision 1999/468/EC of 28 June 1999 laying down the procedures for the exercise of implementing powers conferred on the Commission.[69] In addition, the Council makes a basic legal instrument for each important policy field, laying out the ground rules. In this basic document the Council specifies the powers it wants to keep for itself and the powers it delegates to the Commission. In the case of a delegation of power, the Council chooses from Decision 1999/468 EC one of the five mechanisms to exert its influence during the execution by the Commission of the powers delegated to it. Most policies require the formation of a Committee composed of civil servants as representatives of the Member States, and chaired by a civil servant representing the Commission. For food law this is the "Standing Committee on the Food Chain and Animal Health", established in article 58 GFL.[70]

The Committee "assist[s] the Commission in the exercise of implementing powers", as the beautiful euphemism of Council Decision 1999/468/EC formulates it. The rules supporting this system have become known as Comitology. They enable the Council to delegate some of its powers to the Commission, and at the same time keep the final say for itself. The Standing Committee is a Regulatory Committee, the most powerful of the committees and decision mechanisms that can be created on the basis of the Council Decision. The regulatory procedure is to be used for measures of general scope, designed to apply essential provisions of basic instruments. The decisions the Regulatory Committee has to make include measures concerning the protection of the health or safety of humans, animals or plants. The essence of the Regulatory Committee procedure is that the Commission can decide to implement its proposal only when it is in accordance with the opinion of the Committee. When the Committee disagrees, or is not able to take any decision, the Commission has to submit a proposal to the Council relating to the measures to be taken. The Council must adopt the Commission's proposal or indicate its opposition within three months from the date of referral. If on the expiry of that period the Council has not taken either one of these decisions (unable to find a majority saying "Aye", and unable to find a majority saying "Nay"), the Commission can transform its proposal into law. However, if within that same period the Council has indicated by qualified majority that it opposes the proposal, the Commission has to re-examine it. The Commission may submit an amended proposal to the Council, re-submit its original proposal or present a legislative proposal on the basis of the Treaty to change the basic regulation.

---

[68] Article 202 and 211 EC Treaty.

[69] Council Decision 1999/468/EC of 28 June 1999 laying down the procedures for the exercise of implementing powers conferred on the Commission, also known as the Comitology Decision, O.J. L 184, 17/07/1999 p. 23 - 26

[70] Articles 58 and 59 GFL.

With the Regulatory Committee procedure several additions can be made to the rules provided by the GFL. To give a few examples: provisions for the application of the traceability requirements,[71] the adaptation of the number and names of the Scientific Panels,[72] the implementing rules for scientific opinions,[73] the rules to decide which organisations can be included in the scientific network,[74] and finally, the measures to implement the rapid alert system.[75]

### 13.4.3 Rapid Alert System

*Information*
The General Food Law introduces a new rapid alert system for feed and food.[76] The system foresees obligatory notification of any direct or indirect risk to human health, animal health or the environment within a network consisting of national competent authorities, the EFSA and the European Commission. It builds upon the former rapid alert system for food, extending it to include the feed sector and feed and food imports from outside the EU. The European Commission is entrusted with managing the system and ensuring immediate transmission of information to all contact points. Participation in the rapid alert system is in principle open to candidate countries, third countries and international organisations - subject to negotiated agreements. The EFSA's role is to supply scientific and technical information that will be helpful to Member States in deciding follow-up steps.

*Emergency measures*
If an alert is given through the network, the authorities of the Member States must take appropriate steps to inform the public when there are reasonable grounds to suspect a risk. The GFL confers special powers on the European Commission for taking emergency measures. Such measures can be taken where it is evident that feed or food originating in the EU, or imported from a third country, is likely to constitute a serious risk to human health, animal health or the environment, and that such a risk cannot be satisfactorily contained by measures taken by the Member States. Such action can be initiated by the Commission itself or be requested by a Member State. Depending on the gravity of the situation, emergency measures can include the suspension of a feed or food from the market, or special conditions or other appropriate interim measures restricting the product's marketing or use.

---

[71] Article 28 (4) GFL.
[72] Article 18 (5) GFL.
[73] Article 29 (6) GFL.
[74] Article 36 (3) GFL.
[75] Article 51 GFL.
[76] Articles 50-52 GFL.

## 13.5 The composition of food

European food law now applies 'from farm to fork'. It seems self evident that the ultimate test of the effectiveness of food law lies in the quality and safety of the food product when it reaches the consumer's fork. It may be surprising then that only a relatively small part of food legislation is concerned with the composition and quality of food products at the time they reach the consumer.

For the purpose of analysis, a distinction can be made between components that the producer adds purposefully (ingredients/raw materials), and contaminants that are undesirable materials of either natural origins (such as mycotoxins of microbial origin) or due to human activity (e.g. cadmium of industrial origin, or acrylamide due to cooking and processing).

Concerning the former, the paramount question is to what extent raw materials are subject to pre-market approval. Concerning the latter, it is important to what extent their presence in food is restricted. Table 13.4 gives a graphic representation of these distinctions.

As far as food safety is concerned, European food law is lenient towards foodstuffs used as raw materials that have an evident tradition of (supposedly) safe use within the European Union, and no prior market authorisation is required for these.[77] Foodstuffs not having a history of use in the European Union are called novel foods, and must be authorised before being brought to

Table 13.4. Categories of food components (between brackets the applicable sections).

| Intentional presence Ingredients | | Unintentional presence Contaminants (13.5.5) | |
|---|---|---|---|
| No pre-market approval | Pre-market approval required (13.5.1) | Natural | Human |
| | Additives (13.5.2) | Micro-organisms | Environmental pollution <br> ■ e.g. cadmium |
| | Novel foods (13.5.3) | Mycotoxins | Food-processing <br> ■ residues of processing aids <br> ■ acrylamide |
| | GMOs (13.5.4) | | Residues (13.5.5) <br> ■ pesticides <br> ■ feed additives <br> ■ veterinary medicine |
| Other ingredients (13.5.1) | | | |

---

[77] However, even for such foodstuffs there is the limitation that they cannot be used in specific final foods where vertical standards prohibit that. Examples are vegetable oils in butter, or an additive in a food for which there is no authorisation for that additive, e.g. a sweetener in a fruit juice. It should be stressed that the prohibition is a trade prohibition, not a prohibition related to food safety.

the market. Raw materials that are not normally consumed as food are called additives. The use of additives is forbidden unless explicitly authorised.

### 13.5.1 Pre-market approval: Ingredients

As stated above, food producers are given considerable freedom to choose their ingredients provided that they have been used traditionally. No prior market authorisation is required unless explicitly stated otherwise. Such exceptions apply to additives and novel foods, as well as genetically modified foods, which only recently have been distinguished from other novel foods.[78]

### 13.5.2 Additives[79]

*The Additives Framework Directive - additives definition and general principles*
At present, the basic provision setting the rules and procedures on additives is the Additives Framework Directive (89/107) on the approximation of the laws of the Member States concerning food additives authorised for use in foodstuffs intended for human consumption.[80] There, the definition of additives is given.

*Law Text Box 13.11. Definition of 'additive', Article 2 Framework Directive 89/107 on the approximation of the laws of the Member States concerning food additives, O.J. 1989 L 40 p. 27.*

any substance not normally consumed as a food in itself and not normally used as a characteristic ingredient of food whether or not it has nutritive value, the intentional addition of which to food for a technological purpose in the manufacture, processing, preparation, treatment, packaging, transport or storage of such food results, or may be reasonably expected to result, in it or its by-products becoming directly or indirectly a component of such foods.

---

[78] In addition, in the near future, flavouring substances will also require formal authorisation by positive list prior approval after scientific scrutiny. A draft framework regulation has been under discussion for quite a while: Commission working document WGA/004/03 of 28 AUG 2003: "Draft Proposal for a Regulation of the European Parliament and of the Council on food additives authorised for use in foodstuffs intended for human consumption".

[79] Many thanks to Mr. Ronald Verbruggen (director Orbyte bvba; consultant Food regulatory and scientific affairs), who wrote this paragraph.

[80] See also European Parliament and Council Directive 94/34/EC of 30 June 1994 amending Directive 89/107/EEC on the approximation of the laws of Member States concerning food additives authorised for use in foodstuffs intended for human consumption O.J. L 237, 10/09/1994, p. 1 - 2. A consolidated version of the Additives Framework Directive can be found at http://europa.eu.int/eur-lex/en/consleg/pdf/1989/en_1989L0107_do_001.pdf. The Decision with the list of 'traditional' foodstuffs however (292/97) is not part of that file. See note 85 for the publication of the list.

In its Annex I, the Additives Framework Directive sets out a comprehensive list of 24 categories of food additives.[81]

Some examples of coined names of additives categories are: Anti- oxidants; Preservatives; Colours; Sweeteners; Emulsifiers; Gelling agents, and Anti-caking agents.

The food additives category descriptors refer to the technological function that the members of that category generally perform in food. The legislation on additives is built on so-called positive lists. These are exhaustive lists that indicate which additives may be used in which food product and - many times - at which maximum level.[82] Any material not listed as authorised in the Food Additive Directives is prohibited as a food additive. Before a substance is added to the list of additives it is subject to a safety assessment by EFSA. It must be demonstrated that there is a technological need, that there is no safety hazard for the consumer, and that the consumer is not misled when the additive is being used. If a substance is approved it is assigned an E number. A food additive acquires its E number by being mentioned with an E number in an additives directive.[83] The E number can be used to draw up the ingredient list for the label of food products that contain the additive. However, the full name may be used as well instead of the E number.

An amendment to the Additives Framework Directive (94/34) authorises Member States to maintain a prohibition of certain additives in the production of certain foodstuffs considered traditional. Such an exception was requested by Germany with a view to preventing the use of intense sweeteners in beers brewed according to the Reinheitsgebot.[84] Vigorous opposition from many sides led finally to the permission for all Member States to make exemptions for 'traditional' foodstuffs from the horizontal legislation on additives in general, not just on sweeteners. In order to cope with the problem that EU law did not define 'traditional', the Commission published a Decision (292/97) listing once and for all those foodstuffs that can claim the 'traditional' exemption from the additive rules.[85]

*The Additive Authorising Directives*
Historically, additives authorisation directives, after the framework directive, were drawn up in three packs. First came sweeteners, then colours, and finally all others were issued under the common name of "miscellaneous additives" ("food additives other than colours and sweeteners"). The "Miscellaneous Additives" Directive comprises many functions. The two

---

[81] It shall be noted that the concept of "additive" is being used for other products, such as for feed. Yet, for feed, the meaning of "additive" is quite different, and - at variance with food additives - includes a great many enzymes, micro-organisms, and coccidiostats.

[82] See, however, hereafter the concept of QS.

[83] The E numbers therefore are neither granted by a separate legislative act, nor by a scientific advisory committee (such as EFSA, the former SCF, etc.).

[84] German law on "purity" of beer, in fact a historic law on taxes levied on beer.

[85] Decision No 292/97/EC of the European Parliament and of the Council of 19 December 1996 on the maintenance of national laws prohibiting the use of certain additives in the production of certain specific foodstuffs O.J. L 048, 19/02/1997, p. 13 - 15.

former groups each hold only one function. Some materials can occur in more than one function. For example, the polyols (such as sorbitol and xylitol) are listed both as sweeteners and as humectants in the "Miscellaneous Additives" Directive.

*Sweeteners Directive*
The Sweeteners Directive (94/35) has been amended twice.[86] The first amendment (96/83 O.J. 1996 L 48 p. 16) essentially filled gaps left in the first directive, and clarified some of the concepts. Recently, a second amendment was issued (2003/115), essentially in order to introduce two new sweetener compounds: sucralose, and the acesulfame-aspartame salt.

*Colours Directive*
The Colours Directive (94/36) has never been updated, and is not likely to be amended in the near future.[87]

*"Miscellaneous Additives" Directive*
The "Miscellaneous Additives" Directive holds many more additives than the two other directives. Where 14 sweeteners are authorised for use, and some 32 colours, "Miscellaneous Additives" authorises the use of 261 additives. The original Directive (95/2) has been amended five times by 96/85 - 98/72 - 2000/15 - 2003/52 - 2003/114[88]. The first three were the usual type of amendments, granting additional authorisations. Directive 2003/52 is a bit unusual in that it prohibits one specific additive - E 425 konjac - in jelly confectionery, in order to prevent the risk of choking, due to the significant swelling of konjac when it absorbs water. One of the novelties in the latest one of these updates (2003/114) is that it introduces the additives necessary for preparation, storage, and use of flavourings into the system. This satisfies a longstanding request from the flavourings industry, to prevent problems existing earlier. The flavourings industry felt handicapped for years by the fact that, in flavourings, a number of additives were used for which no formal EU authorisations were ever given. Consequently, this entailed a grey area for their labelling in the ingredient list.

*Specifications - criteria of purity and identity*
For most additives authorised, the EU has issued specifications (criteria of purity), in a set of several directives with amendments.

*Additional general issues, safety, and principles*
To the practitioner, the tables annexed to the directives, holding the lists of authorisations, are the main part. The Additives Framework Directive is the right place to establish the principles

---

[86] European Parliament and Council Directive 94/35/EC of 30 June 1994 on sweeteners for use in foodstuffs O.J. L 237, 10/09/1994, p. 3 - 12. A consolidated version of the Sweeteners Directives can be found in http://europa.eu.int/eur-lex/en/consleg/pdf/1994/en_1994L0035_do_001.pdf.

[87] European Parliament and Council Directive 94/36/EC of 30 June 1994 on colours for use in foodstuffs O.J. L 237, 10/09/1994, p. 13 - 29.

[88] European Parliament and Council Directive No 95/2/EC of 20 February 1995 on food additives other than colours and sweeteners O.J. L 061, 18/03/1995, p. - 40. A consolidated version of the "Miscellaneous Additives" Directive can be found in http://europa.eu.int/eur-lex/en/consleg/pdf/1995/en_1995L0002_do_001.pdf.

and give explanations of the concepts. Nevertheless, as the authorising directives were written, complements on concepts were added to these directives as well. Two concepts clarified in that way are:
- "Carry over" which means that an additive may lawfully occur in a compound foodstuff to the extent that it is permitted in one of its ingredients. Under "reverse carry over" an additive may not occur in a food for direct consumption, or not occur at the level under consideration, but will be lawfully accepted if it is used in a raw material or in an intermediate, used solely in the preparation of a compound foodstuff, and to the extent that the resulting compound foodstuff conforms to the authorisations given in the directives. Those two carry over notions are expressed in the Colours Directive (Art. 3), as well as in the "Miscellaneous Additives" Directive (Art. 3 (1)). Notice though that the term "carry over" itself is not even mentioned in those Articles. The term appears however in the tables, for specific additives.
- "Quantum satis" (QS): The legislators have accepted this concept from Belgian law for a number of authorisations. It means that "...no maximum level is specified. However additives shall be used in accordance with Good Manufacturing Practice (GMP), i.e. at a level not higher than is necessary to achieve the intended purpose and provided that they do not mislead the consumer." The notion is defined in Article 2 (8) of the "Miscellaneous Additives" Directive and identically in Art. 2 (7) of the Colours Directive. Remarkably, QS is also used, but not defined, in the Sweeteners Directive. There is no doubt, however, that the same interpretation is to be applied there.

*Future developments*
A new framework measure on food additives has been in preparation for quite a while in the EU administration. It will be a regulation. That framework regulation has already been announced in the White Paper on Food Safety. It is to replace (i.e. repeal) the Additives Framework Directive 89/107, which is now in force (2004).

The core issues of the new approach are as follows:
1. At present, additives for feed are authorised by mere decisions of a standing committee of the Commission. Additives for food, however, require full parliamentary approval. This is considered to be a painstakingly long and expensive process, rarely shorter than two years. The ensuing political infighting may lead to irrational compromises, and the result in the law can be disappointing. The Commission hopes to align in the new framework regulation the food additives authorisation procedure with the one already in place for feed additives for many years.
2. At present, open-ended authorisations for additives are still granted - in principle. In the new framework, it is planned that all additives will get a temporary 10-year authorisation, with renewals upon re-evaluation.
3. Additives consisting of, containing, or produced from GMOs are to be authorised under Regulation 1829/2003 on GMO food and feed.

An additional future amendment to the "Miscellaneous Additives" Directive (95/2) is in the pipeline of the European legislation process. The original Additives Framework Directive

(89/107) has already announced that the food additives legislation will be merged into a ( = one) "comprehensive directive". However, it may still take a while before that materialises.

### 13.5.3 Novel foods

Food products and ingredients that have not been used to a significant degree for human consumption within the EU prior to passage of the Novel Foods Regulation 258/97 (27 January 1997 O.J. 1997 L 43, 14 February 1997, p. 1), are so-called novel foods. They have to pass a safety assessment before they may be brought to the market. Article 1 of the Novel Foods Regulation specifies four categories of novel foods, see Law text box 13.12 for the specifications.

*Law text box 13.12. Definition of four categories of novel foods, Article 1 (2) Novel Foods Regulation 258/97. Until 18 April 2004, genetically modified foods were mentioned as (a) and (b). As from this date the lemmas (a) and (b) are deleted and a separate regime applies for GM foods.*

This Regulation shall apply to the placing on the market within the Community of foods and food ingredients which have not hitherto been used for human consumption to a significant degree within the Community and which fall under the following categories:
a. [deleted];
b. [deleted];
c. foods and food ingredients with a new or intentionally modified primary molecular structure;
d. foods and food ingredients consisting of or isolated from micro-organisms, fungi or algae;
e. foods and food ingredients consisting of or isolated from plants and food ingredients isolated from animals, except for foods and food ingredients obtained by traditional propagating or breeding practices and having a history of safe food use;
f. foods and food ingredients to which has been applied a production process not currently used, where that process gives rise to significant changes in the composition or structure of the foods or food ingredients which affect their nutritional value, metabolism or level of undesirable substances.

This definition aims to cover all new food products, whether originating from animals, plants or micro-organisms, but only those new processing methods that have the effects mentioned sub c and f. New breeds of animals are exempt if they are bred in a normal way from 'conventional' parents.

Novel foods must not present a danger to the consumer, mislead the consumer or be nutritionally disadvantageous compared with the products they replace.

Before being placed on the market, novel foods must undergo Community assessment after which an authorisation decision may be taken. Under the assessment procedure, the competent body of the Member State that receives an application for pre-market approval must make an initial assessment and determine whether an additional assessment is required. If it is found that no additional assessment is required, and neither the Commission nor a Member State raises an objection, the applicant will be informed that s/he may place the product on the market. In other cases the Commission must take an authorisation decision with the assistance of the Standing Committee on the Food Chain and Animal Health.

The authorisation decision defines the scope of the authorisation and specifies, as appropriate, the conditions of use, the designation of the food or food ingredient, its specification and the specific labelling requirements. The regulation lays down specific requirements concerning the labelling of these foodstuffs. The following must be mentioned:
- Any characteristics such as composition, nutritional value or the intended use of the new foodstuff which renders it no longer equivalent to an existing food.
- The presence of materials which may have implications for the health of some individuals.
- The presence of materials which give rise to ethical concerns.

This is in addition to the general requirements on labelling, discussed hereafter.

*Example*
A well-known example of an approved novel food is Becel Pro Activ; a spreadable fat with a blood cholesterol lowering* effect. See Commission Decision 2000/500/EC of 24 July 2000.[89]
*Hence bcl which abbreviation turned into a name.

### 13.5.4 Genetically modified foods

A special category of novel foods, which since 2004 has been subject to a separate regulatory framework, are genetically modified organisms (GMOs) (including micro-organisms) used for human consumption (GM food). GM foods need an authorisation on the basis of a double safety assessment before they may be brought to the market. They need an authorisation for the deliberate release of a GMO into the environment, under the criteria laid down in Directive 2001/18 (O.J. 2001 L 106 p.1), and an authorisation for use in food and/or feed under the criteria laid down in Regulation 1829/2003[90].

The European legislator is proud to have introduced a "one door-one key" principle. This means several things. First, an authorisation for GM food is valid throughout the Community. Unlike pharmaceutical products, for instance, there is no need to acquire a separate authorisation for each Member State where the product is brought to market. Secondly, Regulation 1829/2003 makes it possible to file a single application for obtaining both the authorisation under Directive 2001/18 and the authorisation under Regulation 1829/2003. This single application is followed

---

[89] 2000/500/EC: Commission Decision of 24 July 2000 on authorising the placing on the market of 'yellow fat spreads with added phytosterol esters' as a novel food or novel food ingredient under Regulation (EC) No 258/97 of the European Parliament and of the Council (notified under document number C(2000) 2121) O.J. L 200, 08/08/2000, p. 59 - 60.

[90] Regulation (EC) No 1829/2003 of the European Parliament and of the Council of 22 September 2003 on genetically modified food and feed, O.J. L 268, 18/10/2003, p. 1 - 23.
Regulation 1829/2003 applies to GMOs for food use, food containing or consisting of GMOs and food produced from or containing ingredients produced from GMOs (Article 3). See also Article 4 (4).

*Modern european food safety law*

by a single risk assessment process for which EFSA is responsible, and a single risk management process involving the Commission and the Member States through a regulatory committee procedure. Finally, if a product is likely to be used both as a food and as a feed it can only be authorised for both, or not at all.

The role of national authorities is limited. Essentially, they act only as a box office to receive applications, though Member States do retain the right to make comments or raise objections. EFSA may ask national authorities to carry out a risk assessment. Finally, Member States may exercise some influence through the committee or through the Council.

Applications are to be submitted first to the competent authority of the Member State where the GM food product is to be marketed first. The application must define the scope of the application, indicate which parts are confidential and must include a monitoring plan, a labelling proposal and a detection method for the new GM food or feed. The applicant must present copies of the available studies that have been carried out and any other available material to demonstrate that the GM food complies with the following criteria. GM food/feed must not:

- have adverse effects on human health, animal health, or the environment;
- mislead the consumer;
- differ from the food/feed it is intended to replace to such an extent that its normal consumption would be nutritionally disadvantageous for the consumer/animals.

If the market authorisation for GM food is granted, it is valid throughout the Community for a renewable period of 10 years. Conditions and restrictions may be connected to it. The obligation to implement a monitoring plan can be imposed on the applicant.

### 13.5.5 Restricted substances

*Contaminants*
Beside raw materials that the producer intentionally includes in a food product, all kinds of chemicals or (micro) organisms may find their way into the final product unintentionally before it reaches the consumer. This situation is covered to a certain extent by the general rules on food safety, which insist that no dangers to the consumer be present. However, there are also more specific rules. First amongst these is the Framework Regulation 315/93.[91] This regulation opens the possibility of establishing maximum content levels for certain substances in food. If these levels are surpassed, the food may not be brought to the market. The regulation opens in the first paragraph of article 1 with a definition of 'contaminant'.

---

[91] Council Regulation (EEC) No 315/93 of 8 February 1993 laying down Community procedures for contaminants in food O.J. L 037, 13/02/1993, p. 1 - 3.

*Bernd van der Meulen and Menno van der Velde*

*Law text box 13.13. Definition of a contaminant, Article 1 (1) Framework Regulation 315/93.*

> Contaminant means any substance not intentionally added to food which is present in such food as a result of the production (including operations carried out in crop husbandry, animal husbandry and veterinary medicine), manufacture, processing, preparation, treatment, packing, packaging, transport or holding of such food, or as a result of environmental contamination. Extraneous matter, such as, for example, insect fragments, animal hair, etc, is not covered by this definition.

Regulation 466/2001 gives maximum levels for a whole series of contaminants.[92] For instance, for nitrates in lettuce and spinach, aflatoxins in nuts, dried fruit, grain, herbs, and milk.

*Residues*
A particular kind of contaminant is residue from veterinary medicine, additives in feeding stuffs (in animal products like meat, milk or eggs), or crop protection products (in plants). For these products Maximum Residue Levels (MRLs) can be established. If these levels are surpassed, the food may not be brought to the market.

*Pesticide residues*[93]
The legislation on pesticide residues is a good example of the EU approach to the presence of undesired substances in food. The regulatory framework on pesticides consists of two parts: the pre-market approval of pesticides for use in agriculture and the establishment of maximum residues limits in food commodities.

Pre-market approval
In 1991, the Council adopted Directive 91/414/EEC concerning the placing of plant protection products on the market.[94] This Directive provides for the Commission to examine pesticide active substances under *all* their safety aspects (consumer protection, worker protection, fate and behaviour in the environment, ecotoxicology).
The setting of MRLs is very closely coordinated with the parallel activities on this evaluation of pesticides' active ingredients.

MRLs
Legislation for pesticide residues, including the setting of MRLs in food commodities is a shared responsibility of the Commission and the Member States.[95] To date (2004), more than 17,000 Community MRLs have been set for various commodities for 133 pesticide active substances.

---

[92] Commission Regulation (EC) No 466/2001 of 8 March 2001 setting maximum levels for certain contaminants in foodstuffs O.J. L 077, 16/03/2001, p. 1 - 13.

[93] In writing this paragraph, use has been made of: European Commission, Introduction to EC Pesticides Residues Legislation, http://europa.eu.int/comm/food/plant/protection/resources/intro_en.pdf.

[94] Council Directive 91/414/EEC of 15 July 1991 concerning the placing of plant protection products on the market O.J. L 230, 19/08/1991, p. 1 - 32.

[95] For those pesticide/commodity combinations where no Community MRL exists, the situation is not harmonised and the Member States may (subject to satisfying their obligations under the Treaty) set MRLs at national level to protect the health of consumers.

*Modern european food safety law*

Current pesticide MRL legislation is derived from/based on four Council Directives:
a. Council Directive 76/895/EEC establishing MRLs for selected fruits and vegetables[96];
b. Council Directive 86/362/EEC establishing MRLs for cereals and cereal products[97];
c. Council Directive 86/363/EEC establishing MRLs in products of animal origin[98]; and
d. Council Directive 90/642/EEC establishing MRLs in products of plant origin, including fruits and vegetables.[99]

MRLs are not maximum toxicological limits. They are based on good agricultural practice and they represent the maximum amount of residue that might be expected in a commodity if good agricultural practice was adhered to during the use of a pesticide. Nonetheless, when MRLs are set, care is taken to ensure that the maximum levels do not give rise to toxicological concerns.

In principle, MRLs are set on the basis of the following:
a. Supervised agricultural residue trials establish the residue level in or on an agricultural crop treated with a pesticide under specified use conditions (Good Agricultural Practice = GAP).
b. Using appropriate consumer intake models, the daily residue intake under normal and worst case scenarios can be estimated for the European population and for national populations and sub-populations (e.g. infants).
c. Data from toxicological tests on the pesticide allow for the fixing of an "acceptable daily intake" (ADI). Usually this involves finding the highest dose that would produce *no* adverse effects over a lifetime (chronic) exposure period and then applying appropriate safety factors.
d. If the estimated daily consumer intake for all commodities calculated under (b) is lower than the ADI calculated under (c) then the residue level under (a) is set as the MRL. In cases where the calculated intake is higher, the use conditions described need to be modified to reduce the residue level in the commodity. If this is not possible the use of that pesticide on that crop cannot be tolerated and the MRL is set at the limit of determination (effectively zero). MRLs for processed products and composite foodstuffs are normally calculated on the basis of the MRL set for the agricultural commodity by application of an appropriate dilution or concentration factor. For composite foodstuffs MRLs are calculated taking into account the relative concentrations of the ingredients in the composite foodstuff. Only exceptionally may specific MRLs be determined for certain processed products or certain composite foodstuffs.

---

[96] Council Directive 76/895/EEC of 23 November 1976 relating to the fixing of maximum levels for pesticide residues in and on fruit and vegetables O.J. L 340, 09/12/1976, p. 26 - 31.

[97] Council Directive 86/362/EEC of 24 July 1986 on the fixing of maximum levels for pesticide residues in and on cereals O.J. L 221, 07/08/1986, p. 37 - 42.

[98] Council Directive 86/363/EEC of 24 July 1986 on the fixing of maximum levels for pesticide residues in and on foodstuffs of animal origin O.J. L 221, 07/08/1986, p. 43 - 47.

[99] Council Directive 90/642/EEC of 27 November 1990 on the fixing of maximum levels for pesticide residues in and on certain products of plant origin, including fruit and vegetables O.J. L 350, 14/12/1990, p. 71 - 79.

## 13.6 Food handling

If a food business operator starts a new production process, or engages in new trade relations, s/he has to ascertain that both the handling of the food by the company and between companies is in agreement with the applicable provisions of food law. According to Article 17 of the General Food Law, food business operators are responsible for meeting the relevant (food safety) requirements at all stages of production, processing and distribution of food. Under the new Hygiene Regulation they are even held responsible for formulating these requirements.

There are several kinds of rules addressing the handling of food. Some aim to prevent food safety problems from occurring, some aim to be prepared in case food safety problems occur, and others give obligations and instruments to deal with food safety problems when they occur. See Table 13.5 on food handling.

Table 13.5. Food handling.

|  | Production | Trade |
|---|---|---|
| Preventing problems (13.6.1) | HACCP (13.6.1) | (Tracking & tracing) |
| Preparing for problems (13.6.2) | (HACCP) (13.6.2) | Tracking & tracing (13.6.2) |
| Dealing with problems (13.6.3) | (HACCP) | Withdrawal & recall (13.6.3) |

### 13.6.1 Preventing problems: hygiene

The safety of food products on a consumer's plate depends largely on the way they have been produced. For this reason rules have been made to ensure that safe methods of production are used. At this moment a whole range of prescriptions exists on the hygienic production of food. This involves a general Directive (93/43)[100] and several vertical (product specific) directives for food products of animal origin (meat, fish, eggs, etc.). On 14 July 2000 the European Commission introduced a proposal for a horizontal Regulation on the Hygiene of Foodstuffs.[101] This proposal aimed to codify, harmonise and simplify the existing legislative framework. The choice of a regulation to replace directives means that food hygiene legislation is lifted from harmonised national law to uniform European law. The result is Regulation (EC) No 852/2004 of the European Parliament and of the Council of 29 April 2004 on the Hygiene of Foodstuffs,[102] published in the Official Journal of the European Union on 30 April 2004. Its entry into force depends on some legislative measures that have to be taken, but will not be sooner than 1

---

[100] Council Directive 93/43/EEC of 14 June 1993 on the hygiene of foodstuffs O.J. L 175, 19/07/1993, p. 1 - 11.
[101] COM(2000) 438 final / 2000/0178(COD), O.J. 19.12.2000 C 365 E/43.
[102] Regulation (EC) No 852/2004 of the European Parliament and of the Council of 29 April 2004 on the hygiene of foodstuffs O.J. L 139, 30/04/2004, p. 1 - 54; and Regulation (EC) No 853/2004 of the European Parliament and of the Council of 29 April 2004 laying down specific hygiene rules for food of animal origin O.J. L 139, 30/04/2004, p. 55 - 205.

January 2006.[103] The word 'hygiene' is used in a broad sense. It means measures and conditions necessary to control hazards and ensure fitness for human consumption of a foodstuff taking into account its intended use.

The regulation imposes a general obligation on food business operators to ensure that in all stages of production, processing and distribution, food under their control satisfies the relevant hygiene requirements. At the heart of these requirements is the "Hazard analysis and critical control point" (HACCP) system. The HACCP principles are defined in Article 5 of Regulation 852/2004 on the Hygiene of Foodstuffs. Several chapters of this book will deal with these principles, so we will not elaborate on them here.

It is interesting to note that the legal definition leaves it to industry to formulate the specific standards that must be adhered to. We see here a sort of self-regulation, but not a voluntary sort. The application of the HACCP principles can be facilitated and encouraged by national or Community guidelines for good practices. In its annexes the regulation gives general hygiene requirements for primary production and for all food business operators. Most of these requirements are concerned with cleanliness and prevention of cross-contamination.

### 13.6.2 Preparing for problems

*HACCP*
The HACCP system discussed above is not only relevant for the prevention of food safety problems. The principle mentioned in article 5 (1)(e) of the Hygiene Regulation requires that the HACCP plan include corrective action to be taken in case of problems. To a large extent the legislator leaves it to the food business operator to prepare for problems that may occur within the business. However, for problems that may affect other stages in the food chain, some mandatory requirements are in place.

*Tracking and tracing*
The General Food Law requires that food and food ingredients are traceable. Recitals 28 and 29 read:
'Experience has shown that the functioning of the internal market in food or feed can be jeopardised where it is impossible to trace food and feed. It is therefore necessary to establish a comprehensive system of traceability within food and feed businesses so that targeted and accurate withdrawals can be undertaken or information given to consumers or control officials, thereby avoiding the potential for unnecessary wider disruption in the event of food safety problems.'
'It is necessary to ensure that a food or feed business including an importer can identify at least the business from which the food, feed, animal or substance that may be incorporated into a food or feed has been supplied, to ensure that on investigation, traceability can be assured at all stages.'

---

[103] Thus action 8 of the Action Plan on Food Safety (Annex to the White Paper on Food Safety) will be implemented.

*Bernd van der Meulen and Menno van der Velde*

The intention of the traceability system is therefore to enable food safety problems to be identified at the source, and across the food chain. To this end food business operators must keep comprehensive records of exactly where their food material originated and where it went. For food products of animal origin traceability is nothing new. The General Food Law, however, broadens the scope of traceability to all foods. The relevant provision in the General Food Law is article 18, see Law text box 13.14.

*Law text box 13.14. Traceability, Article 18 General Food Law Regulation 178/2002.*

1. The traceability of food, feed, food-producing animals, and any other substance intended to be, or expected to be, incorporated into a food or feed shall be established at all stages of production, processing and distribution.
2. Food and feed business operators shall be able to identify any person from whom they have been supplied with a food, a feed, a food-producing animal, or any substance intended to be, or expected to be, incorporated into a food or feed.
   To this end, such operators shall have in place systems and procedures, which allow for this information to be made available to the competent authorities on demand.
3. Food and feed business operators shall have in place systems and procedures to identify the other businesses to which their products have been supplied. This information shall be made available to the competent authorities on demand.
4. Food or feed which is placed on the market or is likely to be placed on the market in the Community shall be adequately labelled or identified to facilitate its traceability, through relevant documentation or information in accordance with the relevant requirements of more specific provisions.
5. Provisions for the purpose of applying the requirements of this Article in respect of specific sectors may be adopted in accordance with the procedure laid down in Article 58(2).[104]

So far the Commission has not produced the more detailed provisions mentioned in paragraph 5. The burden to interpret and apply this provision therefore remains on industry.

It is clear, however, that Article 18 does not require an intact paper trail to accompany each individual food ingredient from the farm to the fork. Traceability requirements go only one step up and one step down the food chain. Food business operators must be able to identify their own sources and customers (except the final consumer). The burden to reconstruct the whole food chain is on the authorities and to that end traceability information has to be made available to those authorities on demand. A heated discussion is taking place as to what extent 'identity preservation' is required.

*Example*
Imagine a severe poisoning case caused by a pizza. Food safety demands that all pizzas of the same production lot be taken off the market. If it is uncertain which ingredient caused the problem, all ingredients will be suspect. It is said that a pizza consists of about 70 ingredients. If the same type of raw material is purchased from several producers, all those producers may

---
[104] A regulatory committee according to the Council's Comitology Decision, see paragraph 15.4.2, note 69.

be affected if the pizza baker is unable to identify which ingredient was used in which pizza. If on the other hand the source of the problem can be identified as being a specific salami, then only the pizza baker and the supplier of that salami (and maybe some of his other customers) would suffer losses.

The more precise the identity preservation, the smaller the losses in case of a food safety problem, and the smaller the number of food business operators affected. For this reason each food business operator should define which level of identity preservation is achievable. It remains to be seen to what extent the courts will hold food business operators liable for damages suffered by suppliers unjustly implicated in food safety problems, particularly where better tracking systems might have excluded them.

*GMO traceability*

A different traceability regime applies to GMOs. Regulation 1830/2003 defines 'traceability' as 'the ability to trace GMOs and products produced from GMOs at all stages of their placing on the market through the production and distribution chains'.[105] Unlike conventional products, for GM products a paper trail is required. Regulation 1830/2003 is explicit on the information that must accompany GM food. The European Commission is to devise a system of unique identifiers that will be assigned to each GMO. From the time a product consisting of or containing GMOs is placed on the market, suppliers shall ensure that the following information is transmitted in writing to the recipient of the product:
a. that it contains or consists of GMOs;
b. the unique identifier(s) assigned to those GMOs.

At every next stage this information must be passed on for each ingredient or additive that it concerns. Small traces - no more than 0.9% - are exempted from the traceability requirements if they are adventitious and unavoidable.[106]

Article 1 of Regulation 1830/2003 states that traceability has the objectives of facilitating accurate labelling, monitoring the effects on the environment and, where appropriate, on health, and the implementation of the appropriate risk management measures including, if necessary, withdrawal of products.

### 13.6.3 Dealing with problems: Withdrawal and Recall

As stated above, food business operators may not bring food to the market if it is unsafe (Article 14 GFL). The product must be withdrawn from downstream businesses or recalled from the

---

[105] Regulation (EC) No 1830/2003 of the European Parliament and of the Council of 22 September 2003 concerning the traceability and labelling of genetically modified organisms and the traceability of food and feed products produced from genetically modified organisms and amending Directive 2001/18/EC O.J. L 268, 18/10/2003, p. 24 - 28.

[106] Art. 7 Regulation 1830/2003, amending art. 21 of Directive 2001/18/EC of the European Parliament and of the Council of 12 March 2001 on the deliberate release into the environment of genetically modified organisms and repealing Council Directive 90/220/EEC - Commission Declaration O.J. L 106, 17/04/2001, p. 1 - 39.

consumer if unsafe food nonetheless is discovered to have made it to market. The General Food Law deals with this situation in article 19 shown in Law text box 13.15.

*Law text box 13.15. Responsibilities for food: food business operators, Article 19 GFL Regulation 178/2002.*

1. If a food business operator considers or has reason to believe that a food which it has imported, produced, processed, manufactured or distributed is not in compliance with the food safety requirements, it shall immediately initiate procedures to withdraw the food in question from the market where the food has left the immediate control of that initial food business operator and inform the competent authorities thereof. Where the product may have reached the consumer, the operator shall effectively and accurately inform the consumers of the reason for its withdrawal, and if necessary, recall from consumers products already supplied to them when other measures are not sufficient to achieve a high level of health protection.
2. A food business operator responsible for retail or distribution activities which do not affect the packaging, labelling, safety or integrity of the food shall, within the limits of its respective activities, initiate procedures to withdraw from the market products not in compliance with the food-safety requirements and shall participate in contributing to the safety of the food by passing on relevant information necessary to trace a food, cooperating in the action taken by producers, processors, manufacturers and/or the competent authorities.
3. A food business operator shall immediately inform the competent authorities if it considers or has reason to believe that a food, which it has placed on the market may be injurious to human health. Operators shall inform the competent authorities of the action taken to prevent risks to the final consumer and shall not prevent or discourage any person from co-operating, in accordance with national law and legal practice, with the competent authorities, where this may prevent, reduce or eliminate a risk arising from a food.
4. Food business operators shall collaborate with the competent authorities on action taken to avoid or reduce risks posed by a food, which they supply or have supplied.

It is likely that a considerable amount of discussion will arise as to what extent the information that business operators are obliged to provide, may be used by the authorities when imposing sanctions on the operator. The German application of this provision - art. 40a of the Lebensmittel- und Bedarfsgegenständegesetz (Act on food and equipment) - explicitly states that the information provided by the food business operator may not be used against him in criminal proceedings. This provision will undoubtedly stimulate operators to come forward with problems they discover within their organisation. However, there is also the risk that this provision will be misused to escape punishment for intentional neglect.

If the concerned food business operator has an adequate traceability system in place, withdrawal should not be too much of a problem. All the information on which product to withdraw from which customer should be present in the system. However, it is not required that traceability systems encompass the consumer. Therefore, in most cases the business operator will not have in his possession information on the identity of the concerned consumers. For this reason recall-actions need to resort to publicity in the media.

## 13.7 Informed choice: presentation of food products

It is important for food businesses to communicate with their customers and consumers. Probably the most important means of communication are the words and pictures attached to the food product on the label. From the point of view of the business, the most important message will usually be the brand name under which the product is sold. This brand name can be protected from use by other businesses by registering it as a trade mark.

The freedom of speech that food business operators exercise on the label is limited in food law. Certain information is mandatory, other information is prohibited. The General Food Law lays down the principles on consumer information in Articles 8 and 16 (See Law text box 13.16 and 13.17). Their function is to give consumers the opportunity to make informed choices and to protect them from misleading practices.

*Law text box 13.16. Protection of consumers' interests, Article 8 GFL Regulation 178/2002.*

Food law shall aim at the protection of the interests of consumers and shall provide a basis for consumers to make informed choices in relation to the foods they consume. It shall aim at the prevention of:
a. fraudulent or deceptive practices;
b. the adulteration of food; and
c. any other practices which may mislead the consumer.

*Law text box 13.17. Presentation, Article 16 GFL Regulation 178/2002.*

Without prejudice to more specific provisions of food law, the labelling, advertising and presentation of food or feed, including their shape, appearance or packaging, the packaging materials used, the manner in which they are arranged and the setting in which they are displayed, and the information which is made available about them through whatever medium, shall not mislead consumers.

This principle has been elaborated in the (horizontal) Labelling Directive. Several other pieces of legislation give additional standards on consumer information. A proposal on health claims is currently under debate.

### 13.7.1 Labelling

An important issue in consumer protection is to ensure that the consumer is aware of what he or she consumes. Many rules exist concerning the obligation of food business operators to provide the consumer with adequate information by labelling.[107] The most important codification of these rules is to be found in Directive 2000/13 of the European Parliament and of the Council of 20 March 2000 on the approximation of the laws of the Member States relating to the labelling,

---
[107] See also the rules on additives and novel foods mentioned above.

presentation and advertising of foodstuffs: the so-called 'Labelling Directive'.[108] Labelling means 'any words, particulars, trade marks, brand name, pictorial matter or symbol relating to a foodstuff and placed on any packaging, document, notice, label, ring or collar accompanying or referring to such foodstuff'. Labelling may not be misleading. All pre-packaged foodstuffs must be labelled in a language that is easily understood. Usually this means in the national language of the Member State. Other information is mandatory, restricted or forbidden. (Forbidden information will be discussed below under the heading 'claims'.) There are ten required (mandatory) pieces of information:

1. the name under which the product is sold;
2. the list of ingredients;
3. the quantity of certain ingredients or categories of ingredients;
4. in the case of pre-packaged foodstuffs, the net quantity;
5. the date of minimum durability or, in the case of foodstuffs which from the microbiological point of view are highly perishable, the 'use by' date;
6. any special storage conditions or conditions of use;
7. the name or business name and address of the manufacturer or packager, or of a seller established within the Community;
8. particulars of the place of origin or provenance where failure to give such particulars might mislead the consumer to a material degree as to the true origin or provenance of the foodstuff;
9. instructions for use when it would be impossible to make appropriate use of the foodstuff in the absence of such instructions;
10. with respect to beverages containing more than 1,2 % by volume of alcohol, the actual alcoholic strength by volume.

Most of these requirements are self-evident. We refer the reader to our book for more details.[109]

## 13.7.2 List of ingredients

The list of ingredients shall include any substances, including additives, used in the manufacture or preparation of a foodstuff and still present, even if in altered form, in the finished product. The ingredients must be listed in descending order of weight. The quantity of an ingredient (as a percentage) must be mentioned[110] only:

- Where the ingredient or category of ingredients concerned appears in the name under which the foodstuff is sold.
- Where it is usually associated with that name by the consumer.
- Where the ingredient or category of ingredients concerned is emphasised on the labelling in words, pictures or graphics.

---

[108] O.J. L 109, 6.5.2000, p. 29. A consolidated version can be found at http://europa.eu.int/eur-lex/en/consleg/pdf/2000/en_2000L0013_do_001.pdf.

[109] Bernd van der Meulen and Menno van der Velde, Food Safety Law in the European Union, Wageningen (Academic Publishers) 2004.

[110] This is known as quantitative ingredients declaration, or: QUID.

*Modern european food safety law*

- Where the ingredient or category of ingredients concerned is essential to characterise the foodstuff and distinguish it from products with which it might otherwise be confused due to its name or appearance.

New legislation requires the list of ingredients to mention the presence - however minute - of substances known for their ability to spark allergic reactions.[111]

### 13.7.3 Labelling of food additives[112]

Additives are ingredients. Article 6 of the Labelling Directive makes labelling of their use compulsory in the ingredient list. There are, however, some exemptions; the most notable one being drinks with an alcohol content of more than 1.2%. So far, these need not carry ingredient labelling.[113] Consequently, alcohol-free beers routinely carry a list of ingredients. Alcoholic drinks do not show an ingredients list, except in countries where national law requires that, such as in Germany.

Additives whose presence in a foodstuff is merely due to carry over[114] and that have no technological function in the final foodstuff need not be indicated in the ingredient list. "Solvents" or "media" for additives or for flavouring need not be labelled either.[115] Additives in the ingredient list must be preceded by the name of their additive labelling category (i.e. their function[116]), as laid down in Annex II of 2000/13. If a substance exerts more than one function, it is the principal function that should be indicated.

There are additional special labelling requirements for certain categories of additives, for sweeteners and colours. Agents causing severe adverse reactions have to be indicated by a warning on the label. The only additives to which this requirement applies are sulphites. Irrespective of whether it is for the purpose of intolerance warnings or for ingredient listing, sulphites are considered not present when their total quantity from all sources is less than 10 mg/kg. The additives must be designated by their specific name or by their E number. The E number is a number officially assigned by the EU.[117] In addition to the rules on important

---

[111] Directive 2003/89, amending Directive 2000/13 O.J. L109, 2000, p. 29.
[112] Many thanks to Mr. Ronald Verbruggen who wrote this paragraph.
[113] Article 6 Directive 2000/13.
[114] For the notion of carry over, see: 13.5.2.
[115] Article 6 § 4 (c) ii of Directive 2000/13.
[116] See 13.5.2.
[117] Complementary lower case specifiers sometimes have to be added to distinguish very similar substances with specifications somewhat different from the ones of the main parent compounds. These specifiers can be of two kinds. Specifiers "i, ii, iii, etc." indicate that the ADI is not different and therefore different conditions of use do not have to be established. And, in ingredient labelling, E-i numbers can be put together with the parent substance (i.e. neither a separate item nor a "i, ii, iii. .." specifier is required in the ingredient list). Specifiers "a, b, c ... etc.," on the other hand, indicate that a separate risk evaluation was carried out, which resulted in an independent ADI. Consequently, such substances are assigned independent conditions of use, and should therefore be labelled separately - with their specifier - as different ingredients. Examples of the first kind are the diphosphates: E 450 i through vii. Examples of the second kind are E 407 (carrageenan) (10) and E 407a (alternatively refined carrageenan).

*Bernd van der Meulen and Menno van der Velde*

labelling for additives in pre-packaged food, there are also rules for the labelling of food additives sold as such to manufacturers (additives as ingredient raw materials by themselves), as well as to the ultimate consumer (in baking powder, or in table-top sweeteners).[118] Note that tabletop sweeteners are not foodstuffs but (mixtures of) additives. The latter general rules can be found in the Additives Framework Directive (89/107).[119]

### 13.7.4 GM labelling

In relation to GM food, the principle of informed choice requires that consumers are informed of the use of gene technology in the process that ultimately leads to the food product they are about to buy or consume. Therefore, Regulation 1829/2003 requires in Article 13 that 'the words "genetically modified" or "produced from genetically modified [name of the ingredient]" shall appear in the list of ingredients'.[120] This requirement has been further elaborated in Article 4(6) of Regulation 1830/2003 reproduced in Law text box 13.18.

*Law text box 13.18. Principle of informed choice: GM labelling. Labelling genetically modified organisms.*

Article 13 Regulation 1829/2003
... 'the words "genetically modified" or "produced from genetically modified [name of the ingredient]" shall appear in the list of ingredients'.

Article 4 (6) of Regulation 1830/2003:
For products consisting of or containing GMOs, operators shall ensure that:
a. for pre-packaged products consisting of, or containing GMOs, the words 'This product contains genetically modified organisms' or 'This product contains genetically modified (name of organism(s))' appear on a label;
b. for non-pre-packaged products offered to the final consumer the words 'This product contains genetically modified organisms' or 'This product contains genetically modified (name of organism(s))' shall appear on, or in connection with, the display of the product.

These labelling requirements apply to all products derived from GMOs, even highly refined ones. Exception is no longer made for products in which no protein or DNA is present.

There are, however, a very few exceptions remaining. The labelling requirement does not apply to foods containing material which contains, consists of or is produced from GMOs in a

---

[118] See Article 4 Directive 94/35 O.J. 1994 L 237 p. 3.

[119] Directive 89/107 O.J. 1989 L 40 p. 27.

[120] It should be noted that the word 'genetically' must be in the labelling. It may cause some confusion that there exist additives that must be labelled as 'modified starch'. Here the word 'modified' refers to conventional techniques of modification, not to gene technology. Many examples of such products can be found on the website of Greenpeace: http://weblog.greenpeace.org/ge/map.html.

proportion no higher than 0.9 %[121] of the food ingredients considered individually or food consisting of a single ingredient, provided that this presence is adventitious or technically unavoidable. The burden of proof is on industry: 'In order to establish that the presence of this material is adventitious or technically unavoidable, operators must be in a position to supply evidence to satisfy the competent authorities that they have taken appropriate steps to avoid the presence of such material.'[122] To comply with this requirement, food business operators must have detailed information at their disposal concerning the history of the raw materials they use. Research has shown for instance that it takes a distance of 24.5 metres between a field of GM maize and a field of non-GM maize for the non-GM field to remain below the 0.9% threshold. Other products need even greater distances. Labelling standards will require complete transparency and traceability in the food chain, and a strict segregation of the GM food chain from the conventional food chain for producers on the non-GM side to be able to live up to these standards.

### 13.7.5 Claims

Article 2 (1)(b) of the Labelling Directive states that labelling must not attribute to any foodstuff the property of preventing, treating or curing a human disease, or refer to such properties. It is less clear to what extent labelling or advertising may communicate that a product has certain beneficial qualities. At present, that remains an issue for national legislation. This legislation differs from Member State to Member State. Those differences hinder intra-community trade. On 16 July 2003, the European Commission introduced a draft regulation[123] covering nutrition claims (e.g. "rich in vitamin C" or "low in fat") and health claims (i.e. claims of a positive relationship between a specific food and improved health). It sets rules for making such claims and also allows some health claims that were previously prohibited. Disease-related messages ("reduction of disease risk" claims), which until now were totally prohibited, will be allowed if they can be scientifically substantiated and are authorised at the European level. Nutrition and health claims are only allowed if they are substantiated by generally accepted scientific data. The food business operator must justify any claims used.

Nutrition claims are about what the food products contains, health claims are about what the food product does.

---

[121] Art. 12 (2) Regulation 1829/2003. In case of a GMO that has not yet been authorised, a presence of 0.5% maximum is considered not to constitute an infringement provided that this GMO has benefited from a favourable opinion from the Community Scientific Committee(s) or the EFSA before the date of application of Regulation 1829/2003 (art. 47 Reg. 1829/2003).

[122] Art. 12 (3) Regulation 1829/2003.

[123] COM(2003) 424 final (action 65 in the Action Plan on Food Safety).

*Nutrition claims*
A "Nutrition claim" is any claim which states, suggests or implies that a food has particular nutrition properties due to:
a. the energy (calorific value) it;
   - provides;
   - provides at a reduced or increased rate; or
   - does not provide; and/or
b. the nutrients or other substances it;
   - contains;
   - contains in reduced or increased proportions; or
   - does not contain.

The draft regulation comes with an exhaustive annex listing claims that can be made (e.g. low energy, low fat, fat free, with no added sugars, high fibre, light, etc.) and the conditions applying to them. For instance, a claim stating that a product is "light" or "lite", and any claim likely to have the same meaning for the consumer, must meet the same conditions as those set for the term "reduced". That is, the claim shall be accompanied by an indication of the characteristic(s), which make the food "light" or "lite". A claim that a food is energy-reduced, and any claim likely to have the same meaning for the consumer, may only be made where the energy value is reduced by at least 30%, with an indication of the characteristic(s), which make(s) the food reduced in its total energy value.

*Health claims*
"Health claim" means any claim that states, suggests or implies that a relationship exists between a food category, a food or one of its constituents, and health.[124] Health claims must be authorised by the Commission on the basis of an assessment completed by EFSA. The applicant must provide a dossier containing scientific studies to substantiate the claim and a proposal for the wording of the claim, which must be authorised in every Community language!

*Limitations to nutrition and health claims*
The draft regulation proposes limits on food business operators' rights to communicate the nutritional or health benefits of certain food with undesirable nutritional profiles (e.g. foods that are high in fat or sugar). According to the Commission, such foods can be consumed in moderation as part of a healthy diet, but to associate them with health and nutrition claims may lead consumers that are currently eating them in moderation to consume them in greater quantities. The proposal does not call them "bad food" but prevents them from being marketed as "good food" with positive messages about health and nutritional benefits. Health claims on beverages containing more than 1.2 % alcohol will also not be allowed since alcohol is known to cause other health and social problems. Only claims referring to a reduction in alcohol or energy content will be allowed. It follows from the proposal's new approach to health claims that any information about foods and their nutritional or health benefits used in labelling, marketing and advertising which is not clear, accurate and meaningful, and which cannot be

---

[124] On health claims see: Korver, Onno, Kühn, Martha Celcilia, Richardson, David, The Functional Foods Dossier: Building Solid Health Claims, Bennekom (The Netherlands), 2004.

substantiated, will not be permitted. Furthermore, vague claims referring to general well-being (e.g. "helps your body to resist stress", "preserves youth") or claims making reference to psychological and behavioural functions (e.g. "improves your memory" or "reduces stress and adds optimism") will not be allowed. Slimming or weight control claims are also forbidden (e.g. "halves/reduces your calories intake"). Reference to and endorsement by doctors or health professionals will not be permitted as it might suggest that not eating the specified food could lead to health problems.

## 13.8 Globalisation

### 13.8.1 Introduction

The European Union is not an island. The EU engages in trade on a large scale with third countries. Food is one of the largest product sectors of trade, both in tonnes and in turnover.[125] On the global level the World Trade Organisation (WTO) tries to remove barriers to trade not unlike the way the EU has tried to remove these kinds of barriers from the internal market.

To achieve this, several measures have been taken. First tariff barriers were reduced and to the extent that this was successful, non-tariff barriers became more of a concern.

In the food trade differences in technical standards like packaging requirements may cause problems, but mostly concerns about food safety, human health, animal and plant health induce national authorities to take measures, which frustrate the free flow of trade. To address these concerns two WTO treaties were concluded: the Agreement on Technical Barriers to Trade (the TBT Agreement) and the Agreement on the Application of Sanitary and Phytosanitary Measures (the SPS Agreement).

The SPS Agreement was drawn up to ensure that countries only apply measures to protect human and animal health (sanitary measures) and plant health (phytosanitary measures) based on the assessment of risk, or in other words, based on science. The SPS Agreement incorporates, therefore, safety aspects of foods in trade. The TBT Agreement covers all technical requirements and standards (applied to all commodities), such as labelling, that are not covered

---

[125] The White Paper on Food Safety has the following to say on the agri-food sector in the EU (COM (1999) 719 final, p, 5): "The agri-food sector is of major importance for the European economy as a whole. The food and drink industry is a leading industrial sector in the EU, with an annual production worth almost 600 billion €, or about 15% of total manufacturing output. An international comparison shows the EU as the world's largest producer of food and drink products. The food and drink industry is the third-largest industrial employer of the EU with over 2.6 million employees, of which 30% are in small and medium enterprises. On the other hand, the agricultural sector has an annual production of about 220 billion € and provides the equivalent of 7.5 million full-time jobs. Exports of agricultural and food and drink products are worth about 50 billion € a year. The economic importance and the ubiquity of food in our life suggest that there must be a prime interest in food safety in society as a whole, and in particular by public authorities and producers." Note that these figures do not include the 10 new Member States who joined the EU on 1 May 2004.

by the SPS Agreement. Therefore, the SPS and TBT Agreements can be seen as complementing each other. In this section we will limit our attention to the SPS Agreement.

### 13.8.2 The World Trade Organisation

To a certain extent the WTO is a supranational organisation. The treaties concluded between its members are binding. There is an arbitration procedure to resolve conflicts. The winning party may implement economic sanctions if the decision reached in an arbitration procedure is not implemented by the party found at fault.

### 13.8.3 The Agreement on the Application of Sanitary and Phytosanitary Measures

As stated above, the SPS Agreement is very important from a food-safety point of view.

The SPS Agreement recognises the right of the parties to this Agreement to take sanitary and phytosanitary measures necessary for the protection of human, animal or plant life or health, provided that such measures are not inconsistent with the provisions of the Agreement. The measures must be scientifically justified and they may not be discriminating, nor constitute disguised barriers to international trade.

*Law text box 13.19. Article 2 (2) SPS Agreement on Science.*

> 2. Members shall ensure that any sanitary or phytosanitary measure is applied only to the extent necessary to protect human, animal or plant life or health, is based on scientific principles and is not maintained without sufficient scientific evidence, except as provided for in paragraph 7 of Article 5. ... (paragraphs 3 and 4 omitted)

However, if the measures are in conformity with international standards, no scientific proof of their necessity is required. These measures are by definition considered to be necessary. The most important international standards on food safety are to be found in the so-called Codex Alimentarius.

If no international standards apply or if measures are chosen that are stricter than the applicable international standards, the burden of proof is on the authority taking the measure. Scientific research will have to provide the proof required. The European Union was confronted with this requirement in a dispute about meat from American cattle that had been treated with hormones. The Union refused to admit this meat to its market. The United States filed a complaint under the WTO, and was found to be in the right.[126] The Codex Alimentarius allows under certain restrictions the use of hormones in cattle. The EU could not prove that their concern was science based.

---

[126] The WTO's appellate body ruled on 16 January 1998. The ruling has reference WT/DS48/AB/R and can be found at www.wto.org.

*Law text box 13.20. SPS Agreement on Codex Alimentarius.*

---
Article 3

Harmonization

(paragraph 1 omitted)

2. Sanitary or phytosanitary measures which conform to international standards, guidelines or recommendations shall be deemed to be necessary to protect human, animal or plant life or health, and presumed to be consistent with the relevant provisions of this Agreement and of GATT 1994. ...

(paragraphs 3 - 5 omitted)

SPS Agreement Annex A Definitions

3. International standards, guidelines and recommendations
    a. for food safety, the standards, guidelines and recommendations established by the Codex Alimentarius Commission relating to food additives, veterinary drug and pesticide residues, contaminants, methods of analysis and sampling, and codes and guidelines of hygienic practice.

---

### 13.8.4 Codex Alimentarius

*General*

What is this Codex Alimentarius that provides such important standards for international trade? In 1961 the Food and Agricultural Organisation (FAO) and the World Health Organisation (WHO) established the Codex Alimentarius Commission (CAC). Over the years the CAC established specialised committees. These committees are hosted by Member States all over the world. Some 165 countries, representing 98% of the world's population, participate in the work of Codex Alimentarius.

Food standards are established through an elaborate procedure of international negotiations. All these standards taken together are called 'Codex Alimentarius'. In Latin this means 'food code'. It can be seen as a virtual book filled with food standards. Beside the food standards, Codex Alimentarius includes advisory provisions called codes of practice or guidelines. At present the Codex comprises more than 240 commodity standards, over 40 food hygiene and technological codes of practice, over 1,000 food additives and contaminants evaluations, and over 3,200 maximum residue limits for pesticides and veterinary drugs. Finally, the Codex Alimentarius includes requirements of a horizontal nature on labelling and presentation, and on methods of analysis and sampling.[127]

*Legal force*

The Codex standards do not represent legally binding norms. They do slightly resemble the directives in European law in the sense that they present models for national legislation, however without the obligation to implement them. Member States undertake to transform the

---

[127] Susana Navarro and Richard Wood, Codex Deciphered, Leatherhead Surrey UK, 2001, p. 1. On Codex Alimentarius, see also: FAO/WHO, Understanding The Codex Alimentarius, Rome 1999 (available on the Internet; www.codexalimentarius.net).

Codex standards into national legislation. However, no sanctions apply if they do not honour this undertaking.

One might ask what is the use of such non-binding standards. The answer comprises different elements. Generally, nation states are reluctant to enter into internationally binding agreements because these limit their sovereignty. For this reason it turns out to be easier to agree on non-binding 'soft law' standards than on binding 'hard law'. By agreeing on non-binding standards, the participating states develop a common language. All states and other subjects of international law will mean the same thing when they meet to negotiate about food, they mean food as defined in the Codex. The same holds true for milk and honey and all the standards that have been agreed upon. The notion of HACCP has been developed - and is understood - within the framework of Codex Alimentarius.[128]

The mere fact that national specialists on food law enter into discussion on these standards will influence them in their work at home. A civil servant drafting a piece of legislation will look for examples. In food he will find examples in abundance in the Codex. In these subtle ways the Codex is likely to have a major impact on the development of food law in many countries even without a strict legal obligation to implement.

It turns out more than once that soft law has a tendency to solidify. Once agreements are reached parties tend to put more weight on them than was initially intended. In the following sections it will be shown that this is true for Codex standards as well. Due to several developments they are well on their way to acquiring at least a quasi-binding force.

### 13.8.5 WTO/SPS

As we have seen above, the inclusion of the Codex Alimentarius in the SPS Agreement greatly enhances its significance. WTO members who follow Codex are liberated from the burden to prove the necessity of the sanitary and phytosanitary measures they take. If they cannot base their measures on Codex, they have to prove that their measures are science-based. For the food sector the practical result is that they have access to the majority of the world's markets if their products are up to Codex standard. It is much easier to apply Codex than to apply a whole range of differing national standards. The catch is that industry depends on national governments to take action within WTO if they have to face trade barriers that do not comply with Codex.

### 13.8.6 European Union

In the context of European food law, the Codex is also strengthening its legal impact. Again the European Court of Justice is the engine that propels progress. In its case law it uses Codex standards as an interpretation aid for open standards in European law. In the so-called Emmenthal Cheese case, for instance, the Court was called upon to judge whether the definition of Emmenthal Cheese in French legislation constituted a barrier to trade as interdicted in

---

[128] Recommended International Code of Practice, General Principles of Food Hygiene CAC/Rcp 1-1969, Rev. 3-1997, Amd. (1999).

## Modern european food safety law

article 28 of the EC Treaty.[129] The case concerned a French cheese trader who was prosecuted for trying to sell as Emmenthal cheese a product that did not have a brown rind as required in French legislation. The Court decided that the French law was indeed in breach of article 28 of the EC Treaty. In its reasoning the Court considers that the definition of Emmenthal in Codex does not mention a rind. In other words: in cases like this, the Codex standard helps to define the limits that European law sets for national legislators.

*Law text box 13.21. Codex Alimentarius used by the Court of Justice of the European Communities to define Emmenthal Cheese, judgment 5 December 2000, case C-448/98.*

32. In the case at issue in the main proceedings it should be noted that, according to the Codex alimentarius referred to in paragraph 10 of this judgment which provides indications allowing the characteristics of the product concerned to be defined, a cheese manufactured without rind may be given the name Emmenthal since it is made from ingredients and in accordance with a method of manufacture identical to those used for Emmenthal with rind save for a difference in treatment at the maturing stage. Moreover, it is undisputed that such an Emmenthal cheese variant is lawfully manufactured and marketed in Member States other than the French Republic.
33. Therefore, even if the difference in the maturing method between Emmenthal with rind and Emmenthal without rind was capable of constituting a factor likely to mislead consumers it would be sufficient whilst maintaining the designation Emmenthal for that designation to be accompanied by appropriate information concerning that difference.
34. In those circumstances the absence of rind cannot be regarded as a characteristic justifying refusal of the use of the Emmenthal designation for goods from other Member States where they are lawfully manufactured and marketed under that designation.
35. The answer to the question referred for a preliminary ruling must therefore be that Article 30 of the Treaty precludes a Member State from applying to products imported from another Member State where they are lawfully produced and marketed a national rule prohibiting the marketing of a cheese without rind under the designation Emmenthal in that Member State.

*Codex in the General Food Law*
The significance of the Codex Alimentarius is recognised in the General Food Law. In Article 13 the General Food Law lays emphasis on the importance of the development of international standards. More important however is article 5, paragraph 3 of the General Food Law. Although this paragraph leaves a wide margin of appreciation for the national and EU legislators, it introduces an obligation to take international standards like Codex into account.

This comes close to an obligation under European law for both the Union and the Member States to include standards like the Codex in national and community food legislation. If this Article will indeed be interpreted and applied in this way, this will mean a boost for the legal position of Codex standards in Europe. The General Food Law itself gives the right example. The definition of food, which is the GFL's foundation, is based on the Codex Alimentarius. This

---
[129] At that time numbered 30.

*Law text box 13.22. International standards for food law, Article 5 (3) General Food Law Regulation 178/2002.*

> Where international standards exist or their completion is imminent, they shall be taken into consideration in the development or adaptation of food law, except where such standards or relevant parts would be an ineffective or inappropriate means for the fulfilment of the legitimate objectives of food law or where there is a scientific justification, or where they would result in a different level of protection from the one determined as appropriate in the Community.

example is followed in more recent legislation. The Hygiene Regulation (852/2004), for instance, states in its 15th consideration:

"The HACCP requirements should take account of the principles contained in the Codex Alimentarius."

In addition, the principle laid down in article 6 GFL that food law shall be based on risk analysis can be understood in this context. The whole body of European food safety law can be seen as sanitary measures - e.g. measures to protect human health - in the sense of the SPS Agreement. If requirements of European food safety law are not in conformity with the Codex Alimentarius, sooner or later they will be contested under the SPS Agreement as barriers to international trade. They will only stand in the WTO forum if they are science based, that is, based on risk analysis.

*The EU joins Codex*
For over 40 years, the EU was not a member of the Codex Alimentarius Commission. It exercised its influence through its Member States who were in Codex. On 17 November 2003 [130] the next logical step was taken in the process of increasing recognition of Codex. The Council applied on behalf of the European Communities for membership of the Codex Alimentarius Commission.

## 13.9 Conclusion

Food law cannot be understood in isolation from the general aspects of EU law. Nevertheless, food law is well on its way to becoming an academic specialisation in its own right.

From the 2000 White Paper on Food Safety onwards, a new body of food safety law is rapidly developing in the EU. The (2002) General Food Law provides the general principles. The most important of these is a duty of care for food safety resting on food business operators.

Food business operators have to address the legal limits of their liberty in the choice of raw materials, their obligations in organising their production process and trade relations and the rules concerning their communication with consumers.

---

[130] Council decision 2003/822/EC, O.J. L 309/14, 26.11.2003. In this way point 83 in the Action Plan on Food Safety has been implemented.

To study EU food safety law in further detail, it is helpful to distinguish substantive food law and procedural food law. The main part of substantive food law consists of three types of rules: rules on the composition of food, rules on the handling of food and rules on the communication about food.

Some procedures were mentioned in the context of pre-market approval of food.

Substantive rules have to function in the legal system that produces them. For food law and food safety law the legal environment is becoming increasingly more international, which increases the complexity of the law. Many rules and actions on food safety law originate at the international level outside the EU. These inputs have to be channelled through the complex of the Union and its Member States. The shift in legislative instruments from directives to regulations signals a centralisation of food law at the EU level with less room for national adaptation. How strong this centralisation will turn out to be depends heavily on the substance of the regulations that build on the General Food Law. Less detailed regulation leaves more room for diversity in the Union, more delegation from the Council to the Commission can lead to another style of policy-making going hand in hand with the development of the European Food Safety Authority. EFSA itself is a contribution to centralisation as the organiser and centre of gravity of the European network of independent scientific food safety organisations.

# 14. Consumer perception of safety in the agri-food chain

*Wim Verbeke, Joachim Scholderer and Lynn Frewer*

## 14.1 Introduction

The aim of this section is to describe the scope and objectives of this chapter on consumer perception of safety in the agri-food chain. Furthermore, the rationale for taking consumer behavioural issues into account in agri-food safety debates is provided.

In order to shed some light on consumer behaviour with respect to food safety issues, this chapter provides both some basic principles of consumer behaviour and a selection of topical case studies. First, basic principles of consumer behaviour and consumer decision-making that are applicable in food consumer research are introduced. To this end, consumer motivation for food choice and a classical model of consumer decision-making with related information-processing concepts and influencing factors are set forth. Particular attention is paid to the potential role of risk perception in shaping consumer attitudes and behaviour. Secondly, selected cases about consumer perception of safety in the agri-food chain are presented. Topical cases include discussions on microbiological risk (food poisoning), on chemical risk (Coca-Cola), physical risk (GM food), BSE and the role of traceability and labelling, and the specific case of food allergy. With this broad, rather than in-depth scope, this chapter aims at providing insights and raising readers' interest in the wide range of contemporary consumer behaviour issues relevant to safety debates in the agri-food chain.

Food safety scares have substantially increased consumer concerns about food consumption and potential human health risks. Some of the recent issues of consumer concern and consecutive research focus include BSE, dioxins, genetic modification, or specific outbreaks of food poisoning from microbiological or chemical contamination. Without exception, former real or perceived food safety problems extended into food scares after extensive mass media coverage. A wide diversity of studies consistently report declining consumer confidence, deteriorating perception and decreasing consumption rates after exposure to adverse food-health communication. After discussing the wide range of biological, chemical or physical safety risk factors facing today's agri-food chains, and explaining the various systems established to handle these risks, the present chapter aims at explaining the role of consumers. Ultimately, consumers vote for products with their available budget and, in accordance with perceived product value, consumers pay prices that make up the profit of all previous agri-food chain participants. Hence, understanding consumer behaviour is critical to making the right managerial and marketing decisions, including strategic choices with respect to risk management, risk assessment and risk communication.

Wim Verbeke, Joachim Scholderer and Lynn Frewer

## 14.2 Consumer motivations for food choice

The aim of this section is to discuss the consumer's major motivations for food choice. Motivations are shown to be person-related psychological factors that play a key role in initiating decision-making and directing consumer behaviour, as will be discussed in section 14.3.

Motives or consumer motivations perform a central role in consumer behavioural processes. Motives are defined as enduring predispositions that direct behaviour towards attaining specific goals or objectives (Assael, 1987). Motives function both to arouse behaviour and direct it to certain ends (Engel *et al.*, 1995). The best-known classification of motives was presented by Abraham Maslow (1954), who introduced a hierarchy of motives ranging from physiological, through safety, social, and esteem to self-actualisation. The idea is that consumers will satisfy the basic motivational level first, before trying to satisfy higher levels in the hierarchy of motives.

Von Alvensleben (1997) provided an overview of the major motives for food demand. In the specific case of food choice behaviour, satisfying hunger and thirst emerge as the basic physiological motives. In today's affluent society with plentiful food, the physiological motive mainly pertains to optimum satisfaction of nutritional needs, i.e. avoiding over-nutrition. Physiological motives may help explain consumer choice of low fat diets or functional foods, for instance. Safety constitutes the second level motive, which in the case of food is quite straightforward. Consumers may decide to accept organic foods and reject GM foods for safety motives. The physiological and safety motives in food choice are strongly linked to health. The third level motive, social motive, includes a sense of belonging, love, friendship and affection. Specific food choices for special occasions or food choices in compliance with important reverent persons (social norms, religious motives) fit with this social motive. Environmental or political motives for food choice are also related to the social motive.

Once physiological, safety and social motives are satisfied, consumers will aim at esteem, prestige and status. Food choices in accordance with this type of motivation are, for instance, the purchase of luxury foods (e.g. the so-called Veblen (1899) goods, which are more desirable the higher their price in certain consumer segments), or food choice for specific hedonistic motives (e.g. full fat products). Clearly, conflicts may emerge between esteem motives and lower level motives, like safety and health. The ultimate level is self-actualisation or self-fulfilment. To some extent, consumer choice for convenience foods fits with this motive. Time saving realised in shopping, food preparation and consumption provides more room for activities, like learning, spiritualism or sports that allow consumers to actualise or fulfil the self.

## 14.3 Consumer decision-making and consumer behaviour

The aim of this section is to provide a framework for investigating food consumer behaviour, including the classical stage model of the decision-making process, extended with information-processing concepts and a classification of influencing factors. Furthermore, there is a discussion

## Consumer perception of safety in the agri-food chain

about why consumers need a lot of time and spend quite some effort on some purchasing decisions, whereas others are mainly based on routines or habits.

### 14.3.1 A framework for investigating food consumer behaviour

Consumer behaviour is defined as "those acts of individuals directly involved in obtaining, using and disposing of economic goods and services, including the decision processes that precede and determine these acts" (Engel *et al.*, 1995). From a microeconomics point of view, much emphasis has traditionally been placed on consumer decision-making and choice behaviour. Most of the presented schemes are so-called stage models, which assume that consumers move through a problem solving process, ranging from the recognition of needs, over information search and the evaluation of alternatives, to reach the final stage of choice or purchase. Studying consumer behaviour based on stage models is also referred to as the decision-making perspective in consumer behaviour research. From this decision-making perspective, purchase is considered as one point in a particular course of actions undertaken by a consumer. In order to understand that ultimate point, an examination of preceding events, such as problem or need recognition, the search and processing of information, and the evaluation of product alternatives, is needed. Typically, needs are defined as discrepancies between the actual versus the desired state of being or feeling. Consumers feel thirsty and need a drink. It should be noted that a want is more specific than a need, e.g. a consumer needs a drink, and wants a particular branded soft drink. Demand can be considered as a want that is backed up by spending power and willingness to pay.

After realising a need, consumers can start searching for information about potential solutions to satisfy the need that was recognised. Both internal and external sources can be consulted. Internal sources typically pertain to previous experience and memory, whereas external sources include commercial or non-commercial stimuli in the consumers' environment.

The following step is the evaluation of alternative solutions on criteria that are relevant for the individual consumers in the specific situation. Such criteria are referred to as attributes, about which consumers hold specific beliefs. Beliefs about attributes, combined with attribute importance weights, result in product preference, which is logically translated into purchasing intentions. In most cases, purchasing intentions yield actual purchasing behaviour, unless the consumer is confronted with situations, like out-of-stock products or in-store promotions.

Attributes are product characteristics that are either intrinsic, like taste, texture or colour, or extrinsic, like packaging, brand or label, to the product. Another attribute classification distinguishes between search, experience and credence attributes. Search attributes are available for product evaluation before purchase. Typical examples are price, appearance, brand and packaging. Experience attributes can only be evaluated upon or after purchase and/or product use. Examples are taste and texture. Credence attributes are attributes that consumers cannot evaluate or verify themselves. Instead they have to put trust in people or institutions, like government controls or industry claims. Attributes relating to production (e.g. organic), processing (e.g. free of additives, free of GM ingredients) and product contents (e.g. nutrient or contaminant content) are typically of the credence type. Safety as a product attribute is mainly of the credence type. However, when safety is guaranteed through trustworthy branding or labelling, it may reach the status of a search attribute. Safety can also be of the experience type, e.g. when safety pertains to some type of microbiological risk like Salmonella or *E. coli*, which eventually result in immediate illness. The stage of evaluation of alternatives is also where perception comes into play, since consumers' beliefs about product attributes are strongly determined by their perception.

The classical four-stage model of the decision-making process forms the point of departure in many consumer studies (Engel *et al.*, 1968; Engel *et al.*, 1995). The model can be extended and integrated, first, with a "hierarchy of effects"-model as initiated by Lavidge and Steiner (1961) and revisited by Barry and Howard (1990). Second, concepts related to information processing, as presented by McGuire (1978) and more recently discussed by Scholten (1996), can be supplemented. Finally, a classification of factors or variables that potentially influence the consumer decision-making process is adopted (Pilgrim, 1957; Steenkamp, 1997) (see Figure 14.1).

The "hierarchy of effects" indicates the different mental stages that consumers go through when making buying decisions and responding to commercial or non-commercial information

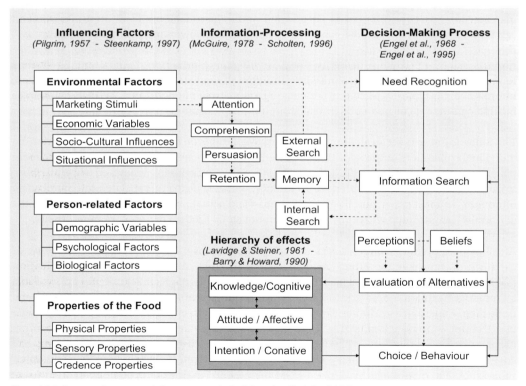

*Figure 14.1. Framework consumer behaviour towards food. Based on Verbeke (2000).*

messages. While its inception dates back to philosophical work by Plato (Holbrook, 1995), the concept was introduced in consumer behaviour literature around the beginning of the 20th century as the "AIDA-model" (Attention, Interest, Desire, Action) (Strong, 1925). The concept is particularly relevant with respect to setting communication objectives. Communications with mainly cognitive objectives aim at cognitive processing at the consumer level, hence increasing consumer knowledge (e.g. providing factual nutrition information to consumers). Affective communications use more emotional arguments, creating feelings and aiming directly at improving consumers' attitudes towards the product. Conative communications try moving consumers to immediate action without the preceding cognitive processing or stimulating attitudinal reactions. Typical examples are in-store sales promotions like premium packs or promotional pricing. The three stages are also referred to as the three components of attitude, with beliefs constituting the cognitive component, brand evaluations constituting the affective component, and the tendency to act or the behavioural intention relating to the conative component. Ever since then, the hierarchy concept has been subject to theoretical and empirical research, as well as to controversy stimulating scientific debates among applied economists, psychologists, sociologists and consumer behaviourists. While it is generally agreed that a structure including a cognitive (learning, knowing), affective (thinking, feeling) and conative (intending, doing) component holds true, no clear-cut evidence about the sequence and inter-distance of these hierarchical steps appears to be available (Barry and Howard, 1990; Ambler, 1998; Vakratsan and Ambler, 1999).

Since, in the current food situation, specific attention is paid to potential influences on consumer decision-making that result from communication and marketing, an information-processing concept is included in the framework. The concept identifies communication effects in terms of ordered stages: exposure and attention to communication, comprehension, persuasion, which refers to attitude change, and finally, retention of a new attitude. This type of model was advanced as a framework for the study of persuasion in the field of social psychology, with a specific focus on the impact from persuasive communication.

At any point in time or throughout the decision-making process, judgements and choices are affected by a variety of stimuli from the environment, as well as by internal processes and characteristics from the consumers themselves. Numerous classifications of stimuli have been set forth in literature. Also, it has generally been recognised that boundaries between groups of stimuli are fuzzy and that factors can be mutually exchangeable between groups. Based on one of the earliest presented models of consumer behaviour towards food (Pilgrim, 1957) and on a review of factors affecting food acceptance and behaviour (Shepherd, 1990), Steenkamp (1997) proposed a classification with three types of influencing factors: environmental factors, person-related factors and properties of the food. Marketing stimuli, the economic and socio-cultural environment, as well as situational influences constitute the consumer environment. Person-related factors or individual difference variables relate to demographic, psychological and biological characteristics of the individual consumer. Classifications of food properties, like intrinsic versus extrinsic, or search, experience and credence attributes, have been discussed before. Combinations of these factors explain why some consumers go through all steps of the decision-making process for particular products or in particular cases, whereas they don't in others. In the next section there will be an explanation of how differences in extensiveness of decision-making yield different types of consumer behaviour.

### 14.3.2 Types of consumer behaviour

Going through the different levels of the afore-mentioned multi-stage model of consumer decision-making is referred to as extended problem-solving behaviour. This type of decision-making requires fully active reasoning, which usually requires quite some time and effort to be put into information search, information processing and the evaluation of alternatives.

The degree of active reasoning is determined by the consumer's level of involvement, the degree of differentiation between alternatives, and eventual time pressures. First, involvement means relevance, importance or pertinence to the individual. In a food context, involvement can result, for instance, from particular hedonic or symbolic values of food or from heightened perceived safety risks. Whereas most food products are regarded as low-involvement products, some foods in specific situations achieve a status near high involvement, e.g. meat during a meat safety crisis, or religious foods. Second, an optimal degree of differentiation between products stimulates active reasoning. Either too much or too little differentiation limits the degree of active reasoning. Fresh food product categories like meat or vegetables usually have a rather low degree of differentiation, whereas the product category of branded soft drinks often shows a too high degree of differentiation to stimulate active reasoning. Third, when time pressure is high, active reasoning may be impossible, or when time pressure is low, active reasoning may be stimulated. The latter may be the case, for instance, for functional foods. The fact that these foods display health benefits after an extended period of use means that consumers have some time to think and reason before making a purchasing decision.

Looking at most of today's consumer decisions about food, it is clear that this extended problem-solving behaviour is not very common in everyday food purchasing decisions; involvement may be low, differentiation not optimum, and time pressure too high. In normal conditions, our food-purchasing decision is not the most important one in our life; an eventual bad choice has no major financial implications and can easily be corrected on our next shopping trip. Hence, food-purchasing decisions mainly resort under restricted strategies, such as limited problem solving, route problem solving (also called habitual purchasing behaviour), and variety seeking behaviour.

When consumers are satisfied about their choice, i.e. product performance has met or exceeded expectations, there is a high likelihood of repeat purchase, which ultimately may result in product or brand loyalty. Later decisions, e.g. when realising the need for mineral water replenishment at home, are easily taken as a routine or out of habit. Consumers prefer routine decision-making for obvious reasons of ease and convenience.

Limited problem solving typically occurs in situations where consumers are confronted with new phenomena, new information or new brands in an otherwise well-known product category. Food safety problems and related media coverage are examples of such new phenomena or information that may distort consumer purchasing habits, and change behavioural strategies from routine to limited problem solving.

Finally, variety-seeking behaviour typically occurs in situations where consumers are minimally involved, sometimes even bored with the familiar, and where product differentiation is significant. Consumers may lack optimal stimulation from the familiar products or brands, and attempt to experiment, often after seeing advertisements or in-store promotional efforts. This type of behaviour explains why even "loyal" consumers sometimes switch brands.

## 14.4 Food safety facts versus consumer perception

This section aims at explaining why consumer perception, e.g. about food safety, often differs from scientific facts or expert views. This section introduces the so-called perception filter that mediates between scientific objectivity and human subjective perception.

In recent years, it seems that consumers are in the main uncertain about the safety and quality of their food, despite the fact that our food has never been safer. Safety is one of the factors that determines food-purchase intentions. Under normal conditions, the majority of consumers are not anxious about product safety, although a certain fear is always present in a latent state. However, the perceived safety can drop dramatically when new information is provided even without medical or scientific evidence as demonstrated by the recent events concerning BSE, GM food, Coca-Cola or acrylamid. Even when the fear appears disproportionate to food scientists, it is not the objective safety that is important to food-quality perception, rather it is the perceived safety that is critical (Cardello, 1995).

Research shows that the public tends to judge the relative risks from food safety issues differently from the way experts expect them to. Furthermore, there is often little relationship between the perceived hazard of a food safety concern and its actual hazard. Consumers often place much importance on factors that are of little or no relevance, whilst ignoring factors that in reality pose a substantial threat to safety (Smith and Riethmuller, 2000; Miles and Frewer, 2001; Miles, 2003). Food- and lifestyle-related heart and coronary diseases, as well as lung cancer from smoking, are relatively large risks, which are, however, largely underestimated by consumers. Simultaneously, newly emerging food-processing technologies or foodborne illnesses caused by microbiological or chemical contamination are examples of overestimations of relatively small actual risks.

A so-called perception filter is held responsible for the bias between reality, scientific evidence or facts on the one hand, and consumer perception of these facts on the other hand (Wierenga, 1983) (Figure 14.2). Facts result from scientific objectivity, and pertain, for instance, to product properties like quality, safety, nutritional value or price. These attributes or characteristics are manageable, measurable and repeatable throughout the agri-food chain, as has been demonstrated in several of the previous chapters in this volume. Consumer perceptions relate to human subjectivity, and as indicated before, they often deviate from the expert view on facts

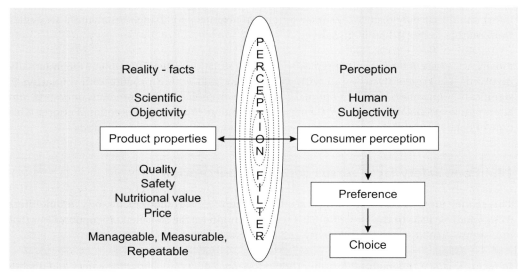

Figure 14.2. Perception filter accounting for the gap between reality and consumer perception. Based on Wierenga (1983).

and reality. The perception filter between reality and perception should be considered as some of kind of mirror that reflects, deflects or distorts factual information. The shape and size of the perception filter is determined by a large number of influencing factors that explain why perception deviates from the objective facts, for some consumers more than for others, or in some situations whereas not in others. Communication, situational and individual factors are the main factors that mediate between scientific objectivity and human subjectivity. Ultimately, the human subjective perception of facts determines the development of attitudes and preferences, based on which buying and consumption choices are made. It should come as no surprise that it is common sense among marketers and communication executives to refer to the existence of "only one reality, namely the perceived one". Clearly, this is another extreme standpoint, but it definitely illustrates that using terms such as consumer *mis*judgement or public *mis*perception is exemplary of highly tentative, speculative and single-sided *mis*reasoning.

## 14.5 Food risk perception

The aim of this section is to explain why some food-safety risks balloon into food scares, whereas other, often more serious risks never attain the "status" of a crisis or scare.

Given the premise that food crises have emerged frequently during the last decade, a logical question relates to the pathways of food-risk perception and food-scare development. For instance, why did smoking never evolve into a crisis whereas BSE, dioxin, Coca-Cola in Belgium or acrylamid in Sweden did? As explained in the previous section, perceptions or beliefs about risks, which often differ from technical risk estimates, drive individual responses to risks. A good starting point for understanding consumer risk perception is provided by the psychometric paradigm developed by Paul Slovic and his co-workers (Fischhoff *et al.*, 1978), which demonstrated that psychological factors determine a person's response to different hazards, including those in the area of food safety. Psychological factors of relevance include, for example, whether the risk is perceived to be involuntary (i.e. in terms of personal exposure), catastrophic (i.e. affecting large numbers of people at the same time), or unnatural (i.e. technological in nature). These psychological factors increase or reduce the threat value of different hazards.

Another key to understanding food scares pertains to the theory of social amplification of risk (Kasperson *et al.*, 1988), providing insight into the problem of why some relatively minor risks elicit often-strong public reactions. Another related key to a better understanding of the pathway of crisis development pertains to a classification of three types of factors that are crucial for any problem to evolve into a crisis (Bennett and Calman, 1999). Besides several psychological fright factors and panic elements, as indicated in the previous section, the classification also includes media triggers. Table 14.1 provides a qualitative assessment of the presence or absence of these facilitating factors or catalysts for the example of risks that were imposed by BSE, dioxin, Coca-Cola and high-fat diets.

Fright factors pertain mainly to the individual's perception of the seriousness of a risk. Fright relates to a risk that is run involuntarily, for instance, for human beings eating in the strict sense is not a voluntary decision. Contrary to generic or unbranded products like most meat, fruit or vegetables, brand choice is a voluntary decision. Hence, risky branded products can more

*Table 14.1. From food safety problem or risk to crisis or scare. Based on Bennett (1999) and Verbeke (2003).*

| Catalysts | BSE | Dioxin | Coca-Cola | High-fat diets |
|---|---|---|---|---|
| **Fright factors** | | | | |
| ■ Involuntary risk | + | | | |
| ■ Inevitable risk | ++ | + | | |
| ■ Contradictory messages | + | ++ | + | |
| ■ Difficult to understand | +++ | + | +++ | |
| **Panic elements** | | | | |
| ■ Universal risk, catastrophic | + | | | |
| ■ New risk | +++ | + | + | |
| ■ Believable risk | + | + | ++ | + |
| ■ Uncertainty | ++ | ++ | ++ | |
| ■ Unnatural, technological | + | +++ | +++ | |
| **Media triggers** | | | | |
| ■ Accused / Suspects | + | ++ | | |
| ■ Personalities | ++ | ++ | | |
| ■ Crime | | + | | |
| ■ Visual impact | +++ | | | + |

easily be avoided. Furthermore, fright increases when the problem is perceived as inevitable, e.g. it cannot be avoided or eliminated through personal precautions like careful cooking. This is the case with BSE or dioxin, though not with most microbiological contaminants such as *Salmonella* and *Campylobacter*. Finally, fright increases when the problem is subject to contradictory messages from different stakeholders, e.g. opposing views held by scientists and politicians, and also resulting from a difficult understanding of the real problem or risk, even by scientists.

Panic elements pertain to the nature of the risk itself, including whether the risk is universal, new, believable and uncertain. Universal does not necessarily refer to a global or worldwide exposure, but to a large potential exposure (probability), as is again more the case with generic or unbranded food products. In the consumer's perception, any beef could be contaminated with BSE, whereas only chicken from Belgium or the one Coca-Cola brand was suspect during the crises in 1999 (see further). Furthermore, newness, believability and uncertainty heightened the panic value of the crises at hand.

Finally, media triggers are crucial in the development of a crisis. Some elements like the presence of accused - or even better - suspects, a link with personalities (e.g. ministers) or crime, and a strong visual impact attract mass media. BSE was exemplary in terms of visual impact, with the mad cow and UK government on stage, whereas the Belgian government and the accused animal feed component suppliers acted as ideal media triggers for the dioxin crisis in 1999. Clearly, few or none of the afore-mentioned fright factors, or panic elements or media triggers are fulfilled in well-known risk behaviours such as smoking, alcohol or the intake of high-fat diets.

Also of interest is the finding that the perceived benefits associated with a particular hazard may offset perceived risk (Alhakami and Slovic, 1994). Flynn *et al.* (1994) have used the psychometric approach to explain the apparent differences between lay and expert perceptions of risk. Lay perceptions are often more complex than perceptions of the same hazards held by experts, and involve a greater number of psychological constructs.

It is useful to distinguish between two categories of potential hazard: those related to technology and its applications, and those related to lifestyle choices (Miles *et al.*, 2004). Perceptions of technology risks are shaped by beliefs that the risks are out of control, are unnatural, and are somehow adding to the existing risk environment. For lifestyle hazards, however, people are frequently over-optimistic about their own risks from hazard exposure (Weinstein, 1980). People tend to rate their own personal risks from a particular lifestyle hazard as being less when they compare them to an 'average' member of society, or even when the comparison is directly with someone with similar demographic characteristics (for a review in the food area see Miles and Scaife, 2003).

In the area of food risks, optimistic biases are much greater for lifestyle hazards (such as food poisoning contracted in the home, or illness experienced as a consequence of inappropriate dietary choices) compared to technologies applied to food production (such as food irradiation or genetic modification of food). At the same time, people perceive that they *know more* about the risks associated with different lifestyle hazards when compared to other people, and are in *greater control* of their personal exposure to specific hazards. In contrast, this does not occur for perceptions of personal knowledge about, and control over, technology-related food risks. As a consequence, the development of effective communication about microbial contamination of food becomes more complicated. People perceive that information about risk reduction is directed towards other consumers who they believe are at more risk from the hazard, who also have less control over their personal exposure to the associated risks, and who possess less knowledge regarding self-protective behaviours. This results in people failing to take precautions to reduce their personal risks from a particular hazard (Weinstein, 1989).

Optimistic bias is reduced for hazards perceived to occur more frequently, (Weinstein, 1987), or which have been explicitly experienced by individuals (Lek and Bishop, 1995). Increased perceptions of personal control influence consumer perceptions such that optimistic bias is increased (Hoorens and Buunk, 1993). If an individual can identify a stereotypical individual, unlike themselves, to whom high levels of risk are attributed, their optimistic bias is increased (Weinstein, 1980), except under circumstances where an individual perceives the stereotype to be rather similar to themselves (Lek and Bishop, 1995). Optimistic bias is likely to act as a barrier to intervention addition; people may have low acceptance of food technologies introduced to alleviate problems associated with microbiological risks, such as food irradiation (Bruhn, 1995) or high-pressure processing.

## 14.6 Selected cases in food safety perception

The aim of presenting case studies is to confront readers with the wide diversity of contemporary consumer research issues related to food safety risks. References to research papers included in the case descriptions allow the search for detailed information about methodologies, results and conclusions.

### 14.6.1 Microbiological risk - "It's not my problem, it's theirs"

In many ways, microbiological risks are the "classical" food safety problem in all aspects related to consumer risk perception, consumer behaviour, and risk communication. Consumers tend to be aware of the ubiquity of the problem: in surveys asking consumers how different food safety problems should be prioritised by regulators, microbiological risks tend to be ranked highest by US consumers (e.g., Food Marketing Institute, 1997), and second to third highest, topped only by pesticide and hormone residues, by EU consumers (e.g., INRA Europe, 1998). Also, many consumers have personally experienced incidents of food poisoning caused by *Salmonella*, *Campylobacter*, or *E. coli* (Miles *et al.*, 1999). In comparative risk perception studies, these microbiological agents tend to fall in that quadrant of the perceptual map where risks are perceived to be relatively dreadful in their consequences, but also as relatively common and well-known to scientists (Fife-Schaw and Rowe, 2000; Kirk *et al.*, 2002).

*Wim Verbeke, Joachim Scholderer and Lynn Frewer*

The reduction of microbiological risks has been a highly prioritised risk management area in practically all UN nations for decades. Strict legislation and extensive controls exist in most countries, often on the basis of HACCP approaches, that have helped achieve a generally high standard of food safety in primary production, food processing and manufacturing, and retail distribution. According to analyses published by the US Centers for Disease Control and Prevention (1994), only 3% of outbreaks of foodborne illnesses can be attributed to food mishandling at processing plants, whilst 77% can be attributed to mistakes made at foodservice sites and restaurants, and a further 20% to food mishandling by consumers in their homes.

Risk communication to consumers has concentrated on the latter issue, with only partial success. Observation studies by Worsfold and Griffith (1997) showed that consumers engage in a wide variety of food mishandling practices at all stages of the meal preparation process (see Table 14.2), and this even though most of the consumers in their study turned out to be quite aware of the risk associated with these practices when questioned afterwards. The obvious question is, of course, why consumers so often fail to act upon their knowledge. Partly responsible for this is a general misattribution of the locus of contamination, resulting in faulty perceptions of the likelihood that microbiological contamination will occur in the household. The likelihood of microbiological contamination at processing plants is generally overestimated by consumers, whilst the likelihood of contamination in their own household or at foodservice sites is generally underestimated by consumers (Centers for Disease Control and Prevention, 1994; US FDA,

*Table 14.2. Frequency of different food handling malpractices that increase microbiological risk. Source: Consumer observation study by Worsfold and Griffith (1997).*

| Preparation stage | Malpractice | Frequency (%) |
|---|---|---|
| Transport | Transported food subjected to temperature abuse | 45 |
| Storage | Chilled ingredients stored above 5°C | 58 |
| Preparation | Hands not washed prior to preparation | 66 |
| Cooking | All ingredients cut on a single board | 60 |
| | Hands not washed after handling raw animal ingredients | 58 |
| | Some/all vegetables not washed | 41 |
| | Raw poultry washed in sink | 33 |
| | Failure to use recommended method of cleaning cutting board | 25 |
| | Meat/poultry packaging not removed from preparation area | 18 |
| | Product not cooked to internal temperature of at least 74°C | 15 |
| Post cooking | Cooked product held at ambient temperature for more than 90 min | 35 |
| | Cooked product stored in refrigerator at a temperature above 5°C | 17 |
| | Product not reheated to an internal temperature of at least 74°C | 11 |
| | Cooked food handled directly | 9 |
| | Cooked food cut on board not cleaned according to a recommended method | 8 |
| | Product re-heated more than once, with intervening holding periods at room temperature | 6 |

1997). A comparison of US data on perceived and actual likelihood of contamination is shown in Figure 14.3.

A second barrier to more responsible food-handling practices in consumer households is the phenomenon known by health psychologists as optimistic bias. The phenomenon is ubiquitous and can be found for virtually all risks that are under high personal control (such as unsafe driving, smoking, high-fat diets, etc.). As already discussed in the previous section, it describes people's tendency to estimate their personal risk as being substantially lower than the risk of the average person in their reference group (Weinstein, 1989; Sjöberg, 2000). Food poisoning through malpractice in the kitchen is a typical example of a lifestyle hazard where optimistic bias plays a crucial role. It is possible to improve public health through consumer adoption of safer domestic hygiene practices, as many cases of food poisoning result from poor domestic hygiene practices. For microbiological risks, the tendency for optimistic bias was found to be exceptionally strong (Redmond and Griffith, 2004). Unfortunately, optimistic bias is known to be utterly resistant to health education and risk communication (Weinstein and Klein, 1995). Hence, it is somewhat questionable whether consumer food-handling practices at home can realistically be improved much further.

Health educators and other communication professionals need to understand how the public perceive different hazards and the associated risk, in order to facilitate the structuring of risk-related messages in such a way that consumers change their risk perceptions and risk-related behaviours. Research has indicated that consumers are more likely to think in depth (and consequently take note of the information in the message) about information which is directly relevant to them specifically, to which they are motivated to pay attention, and which addresses

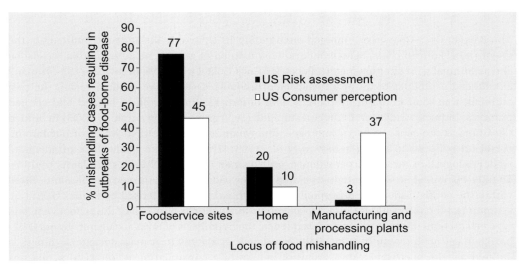

Figure 14.3. Comparison of actual and perceived locus of microbiological contamination of food. Source: US Centers for Disease Control and Prevention (1994).

*Wim Verbeke, Joachim Scholderer and Lynn Frewer*

their interests and concerns. Affective or emotional factors, such as fear about the hazard, may motivate them to think about risk communication focused on a specific issue. This is, however, likely to be rather low in the case of microbial risks (Fischer *et al.*, 2005). Trust in the information source providing the information may also act as a peripheral cue as to the merits of the messages contained (Petty and Cacioppo, 1986). There is, however, some evidence that, in the case of communication about microbial food safety, information source characteristics are less influential than message relevance in influencing risk perceptions associated with food poisoning (Frewer *et al.*, 1997a).

One approach to effective risk communication, which is likely to be an effective way to change consumer behaviours, is to focus on segmenting the population according to their information needs, and developing information with high levels of personal relevance to specific groups of respondents, who may be at greater risk than the rest of the population. Information is more likely to result in attitude change (and subsequent behaviour change) if perceived personal relevance is high (Petty and Cacioppo, 1986). The problem with such an approach is that it is resource intensive, as research firstly needs to be conducted in order to identify individual differences with respect to people's perceptions and behaviours, and then tailored information needs to be delivered.

Simply applying legislative reforms to small sectors of the food chain is unlikely to have a major impact on public health unless consumer behaviour is also addressed (Fischer *et al.*, in press). It is difficult, if not impossible, to legislate for consumer behaviour in the home. The development of an effective and targeted communication strategy is likely to be the only way to produce improvement in public health in the food-safety area.

### 14.6.2 Chemical risk - "Coca-Cola: the luxury of brand power"

The Belgian Coca-Cola crisis emerged on June 8th 1999, two weeks after the outbreak of the major (meat) dioxin crisis. School children complained of general malaise, nausea, headache and abdominal pain after having drunk bottled Coca-Cola. Over subsequent days, the company announced a product recall of all suspected products. Despite the product recall, the first outbreak was soon followed by more school outbreaks, linked both to bottled and canned products. Inquiries indicated hydrogen sulphide ($H_2S$) and carbonyl sulphide (COS) in bottled Coca-Cola as the cause of odour and taste abnormalities, and identified traces of p-chloro-m-cresol on pallets and cans (Demarest *et al.*, 1999). However, due to the lack of adequate epidemiological or toxicological evidence, only a small number of health complaints could be directly attributed to those product-related causes. Additionally, numerous elements fitted within the configuration of epidemic hysteria outbreaks as summarised by Sirois (1999) (i.e. symptoms, school setting, age group, occurrence during the last month of the school year, and type and amount of rumour). As a consequence, the hypothesis of mass sociogenic illness (MSI) has been set forth (Nemery *et al.*, 1999; Van Loock *et al.*, 1999). A mass sociogenic illness is defined as "the occurrence in a group of people of a constellation of physical symptoms suggesting an organic illness but resulting from a psychological cause, with each member of the group experiencing one or more of the symptoms that can not be explained biologically"

(Philen *et al.*, 1989; Wessely and Wardle, 1990). It concerns a social phenomenon involving otherwise healthy people (Boss, 1997). Frequently used synonymous terms are: epidemic hysteria, mass hysteria, and mass psychogenic illness. Only in March 2000, the Belgian Health Council formally confirmed the diagnosis of mass sociogenic illness during the Coca-Cola crisis, with the exception of the first cases.

Following recommendations by Frewer (1999), Bennett (1999), Sirois (1999) and Boss (1997), Verbeke and Van Kenhove (2002) executed behavioural research related to this food safety crisis. The basic aim was to investigate the role of the personality trait, 'emotional stability', and attitude toward the brand on behaviour, i.e. the restoration of consumer trust as exemplified in the decision to cease consumption during and restart after the crisis. The findings of this research were threefold. First, a direct and positive effect of attitude toward the brand on the behavioural response (Coca-Cola consumption) was found. This finding is in line with the most recent research in behavioural and consumer sciences, indicating the existence of a direct link between attitude and behaviour (Engel *et al.*, 1995). Second, no direct effect of the personality trait 'emotional stability' on behaviour was revealed. However, an indirect effect from personality to behaviour was discovered. Emotional stability correlated positively with attitude, which in turn associated with a faster restoration of Coca-Cola drinking. Thus, in this specific case of a premium branded product, the impact of personality was mediated by attitude toward the brand. Third, differences in information perception and importance were found between consumers with low versus high emotional stability scores. Lower emotional stability associated with a higher need for information (or communication), and with higher importance attached to information during this (perceived) food safety crisis. Therefore, these consumers with lower emotional stability require specific attention in future communication. It is in this respect important to realise that the specific target audience of consumers with low emotional stability cannot readily be identified through behavioural or socio-demographic variables as our analyses have shown.

This branded food product crisis revealed that with respect to future risk and health communication, fast provision of adequate, reliable and sufficient information to consumers is crucial. Despite experience with numerous previous crises, this conclusion didn't appear so obvious in practice. In the specific case of Coca-Cola, it took a whole week before the company addressed the public about eventual health risks, possible causes of the problems and the actions undertaken (June 15, 1999). Government and health practitioners (through the Scientific Society of General Practitioners) received detailed information only on June 17, 1999. This meant that reliable and scientifically-based information could be spread to the public only 10 days after the first incident. Uncertainty about the assessment and management of the risk and speculative mass media reports stirred up the crisis in the meantime. Furthermore, fragmentary and incomplete information from the company to the government prevented targeted public health interventions, instead leading to the massive product recall enforcement guided by precautionary principles.

The Belgian Health Council drew three lessons from the Coca-Cola case (Ministry of Public Health, 2000). Firstly, immediate government action should install a cell responsible for crisis management, with a single contact person who receives and spreads information to policy makers, health practitioners and the press. Secondly, epidemiological investigations should start immediately. Thirdly, an ad hoc task force under the auspices of the Health Council or the newly established Federal Agency for Safety in the Food Chain should be installed. It is noteworthy that mainly responsibilities, risk assessment and risk management tasks are dealt with, while the communication task itself is largely ignored. Based on the experience with the crises in Belgium, Anthonissen (2001) indicated the following key elements of crisis communication: background information, details about the incident, actions undertaken, sympathy, and consumer reassurance. Clearly, it should not take days to communicate the first four elements, and doing so raises the chance of avoiding a large-scale scare. Consumer reassurance, however, can hardly be provided immediately since it will only be successful when based on scientific evidence resulting from careful investigation. This is where the previous Health Council recommendations come into play, at least when communication is not ignored in the meantime.

Last but not least, this case of a premium branded food product facing a safety crisis demonstrated that full recovery of market share and consumer acceptance is feasible. Brand and image strength, and appropriate (despite being delayed) marketing communications together with favourable consumer attitudes, accounted for restoration of consumer trust. It is a scenario that many generic fresh food products facing a safety crisis can only dream of.

### 14.6.3 Physical risk - "GM food: the misunderstood food safety problem"

Genetically modified (GM) foods are *the* paradigmatic example of a new type of problem in which consumer perceptions of food safety are driven by highly generalised, socio-political attitudes towards a process (here, genetic modification) rather than a safety evaluation of particular products or the circumstances of their handling and consumption. The GM foods issue is complex and often clouded by political considerations; hence we will treat the case in a more detailed way than the others in this chapter. Historically, the roots of social-scientific research in this area can be traced back to a series of technology assessments that were begun in the late 1980s. In those years, it had become apparent that the emerging field of modern biotechnologies would have a wide-ranging impact on the operation of several industry sectors, including agriculture, food manufacturing, and food processing. In several EU member states, government bodies commissioned social-scientific research to help them formulate a coherent public policy towards these technologies.

In Germany, the Office of Technology Assessment in the German Parliament began monitoring public opinion about modern biotechnology as early as 1985. The Dutch Ministry of Education and Science started in the same year with a number of small-scale qualitative studies, whilst the Ministry of Agriculture commissioned its first consumer survey in 1988. Denmark began in 1987, with the Board of Technology commissioning a panel survey on behalf of the Danish Parliament. In the UK, these policy formation processes started substantially later. In 1991, the UK Department of Trade and Industry commissioned a first research project. None of the remaining EU member states undertook comparable technology assessment exercises before 1998. The European Commission, on a supra-national level, started the Special Eurobarometer series "Europeans and Modern Biotechnology" in 1991.

Since these early days, research has diversified remarkably (for a review, see Frewer *et al.*, 2004), and the paradigmatic orientation of the research has shifted. In the early days, virtually all research understood itself as technology assessment, and communication objectives were framed in terms of a public understanding of science. In current research, both terms have more or less disappeared. Interdisciplinary consumer research has become the new term of reference, and consumer perceptions of food safety are often at the centre of its attention, not least because regulatory efforts, in particular on an EU level, have been shifting away from environmental issues and towards food safety issues (as was particularly evident in EU Regulation 258/97 concerning novel foods and novel food ingredients).

To a certain degree, this shift was motivated by empirical findings. In qualitative as well as survey research, European consumers regularly reported that they were feeling "unsafe" and put "at risk" by GM foods. Regulators as well as stakeholder groups tended to interpret this in classical food safety terms, i.e. as consumer worries about toxicity, allergenicity, and microbiological safety of these foods, and steered research activities in the agricultural and biological sciences as well as in the social sciences towards these issues. Since then, social scientists have been struggling with the question what consumers actually *mean* when they say "unsafe" or "risky". The issue is far from resolved; however, four points of convergence can be identified in the literature.

The first point of convergence is the level of complexity on which GM foods are cognitively represented in consumers. Qualitative research (e.g., Bredahl, 1999; Grunert *et al.*, 2001; Hagemann and Scholderer, 2004; Hamstra and Feenstra; 1989; Scholderer *et al.*, 1999; Schütz *et al.*, 1999) as well as quantitative research (Bredahl, 2001; Brüggemann and Jungermann, 1998; Midden *et al.*, 2002; Scholderer *et al.*, 2000) suggests that the object of consumer attitudes is the technology as such, represented as an abstract concept, rather than as a variety of different products and applications. The matters on which food safety evaluations would normally focus (chemical compounds in the case of toxicity analysis; proteins in the case of allergenicity analysis; micro-organisms in the case of microbiological analysis) are not actually represented in consumer understanding of GM foods.

The second point of convergence, related to the first, is the level of complexity on which consumers make evaluations. Actual beliefs as they are commonly understood in attitude research (Fishbein and Ajzen, 1975) are rarely found in qualitative data. Evaluations appear to be predominantly made through references to even more abstract attitude objects such as health, environment, animal welfare or producer power (Bredahl, 1999; Grunert *et al.*, 2001; Hagemann and Scholderer, 2004; Miles and Frewer, 2001; Scholderer *et al.*, 1999) where the particular implication between one and the other is left unspecified (Schütz *et al.*, 1999). Consistent with this, factor analyses of sets of beliefs usually find that just two dimensions are sufficient to represent the covariation among consumer beliefs (Bredahl, 2001; Hamstra, 1991; Midden *et al.*, 2002; Saba and Vassallo, 2002; Scholderer *et al.*, 2000), suggesting a fairly high level of redundancy in the belief sets. In other words, the highly specific and measurable concepts on which technical food safety evaluations would normally focus are simply not used by consumers in their subjective evaluations of the safety of GM foods.

The third point of convergence is that attitudes towards GM foods are strongly related to other, more general socio-political attitudes, including attitudes towards environment and nature (Bredahl, 2001; Frewer *et al.*, 1997b; Hamstra, 1995; McCarthy and Vilie, 2002), attitudes towards science and modern technology (Borre, 1990; Bredahl, 2001; Hamstra, 1991; McCarthy and Vilie, 2002; Sparks *et al.*, 1994), and social trust, i.e. the willingness to rely on institutions that regulate emerging technologies and manage their risks (Siegrist, 2000). This is highly consistent with the conclusion made above about the level on which evaluations are made: the objects of these general socio-political attitudes are often the same as the abstract concepts to which gene technology is related in investigations of belief content. It appears that consumers perceive the very process of genetic modification to be in conflict with their basic attitudes and values, and that this perceived conflict is simply re-expressed by consumers when they say that GM foods are "unsafe" or "risky".

The fourth point of convergence is that attitudes towards GM foods appear utterly resistant to persuasion. Not a single study reported in the literature has ever been able to change consumer attitudes through communication (Frewer *et al.*, 1996, 1998, 1999, 2003; Peters, 1998; Miles and Frewer, 2002; Scholderer and Frewer, 2003; Søndergaard *et al.*, 2003). The information materials used in the different studies used such different arguments as absence of risks, improved risk assessment methods, introduction of traceability systems, public participation in the risk management process, and a whole range of producer- and consumer-related benefits; all to no avail. The results of Peters (1998) even indicated that the more such communication relies on benefit arguments, the more it prompts processes of active refutation in consumers. Furthermore, the product choice experiments reported by Scholderer and Frewer (2003) suggest that such communication may backfire on the communicator in terms of non-attitudinal aspects of consumer behaviour: reduced purchase probabilities for GM foods were found in all experimental groups where consumers had been exposed to benefit communication *although* no attitude change had occurred. Interpreted in terms of the attitude-structure findings we discussed above, this suggests that consumers perceive pro-GM communications to argue against their basic attitudes and values, and even denying the legitimacy of these basic attitudes and values. Of course, such communication can only fail.

Despite these points of convergence, a number of unresolved questions remain. Given that attitude change by means of persuasive communication appears to be a fruitless task, the managerially most troubling problem is what else can be done to convince consumers of the safety of GM foods. As we have argued in other publications (e.g., Scholderer *et al.*, 2000), the building of an alternative attitude system may be the only way to circumvent the boomerang effects observed by Peters (1998), and Scholderer and Frewer (2003). Such an attitude system should be based on direct experience with high-quality products, instilling a positive *hedonic* experience (Mela, 2001) in consumers, which could then override the effects of the existing attitude system. In a demonstration experiment using high-quality cheese samples (Grunert *et al.*, 2003; Scholderer *et al.*, 2001), we indeed succeeded in creating such a hedonic-based system, and observed positive effects on consumer attitudes towards GM foods in general, as well as a de-coupling of these attitudes from more general socio-political attitudes.

## 14.6.4 BSE - "Power of mass media, potential of traceability and labelling?"

Meat production and consumption have come under heavy criticism over the last decade. Many organisations including consumers, industry, producers and governments, as well as scientists from a plethora of disciplines, have recently been involved in debates that were initiated by numerous occurrences and stirred up by conflicting motivations and influencing factors. Meanwhile, meat has been referred to as the food item in which consumer confidence decreased most during the last decade (Richardson *et al.*, 1994; Becker *et al.*, 1998). Distinct changes at the consumer level increasingly determine the present and future outlook of the meat chain. Meat has traditionally constituted a substantial part of the Western-European diet. Increasing economic and social welfare since the 1950's resulted in increasing amounts of animal protein intake. Top meat consumption levels were noticed during the first half of the nineties in most European countries, but ever since the BSE crisis, fresh meat consumption levels have generally decreased. The Bovine Spongiform Encephalopathy (BSE) epidemic peaked in the UK in 1993 and emerged into one of the major food scares in Europe from 1996 on. Public concern was driven less by the risks of BSE per se, but rather the failure of the UK government to acknowledge the uncertainty about BSE as a potentially causative agent of the human form of the disease, Creutzfeldt-Jakob Disease (nvCJD), prior to 1996. As a consequence, the whole process of how scientific advice was provided by the government was reconsidered. The emphasis on whether or not institutions should communicate about uncertainty shifted to one of how best to do it (Frewer and Salter, 2002; Frewer *et al.*, 2005). Public risk perception was also affected by the failure to provide information relevant to the actual concerns of consumers about food hazards, as was also seen again later with the 1999 dioxin crisis in Belgium.

The impact of mass media communication during the BSE crisis was investigated through two empirical studies in Belgium, one based on cross-sectional survey data, and the other based on time series data. Consumers, who were exposed to mass media coverage of fresh meat issues, reported significantly higher meat consumption decreases with reference to the past as well as stronger intentions for decrease in the future. It was also found that consumers who paid a high level of attention to media reports expressed higher health consciousness, more *mis*perception of health risks and higher levels of concern about potential health hazards that were frequently reported in the mass media. Most importantly, the negative press effect was strongest among younger consumers, who cut their meat consumption after being exposed to mass media coverage in a similar way to 60+ aged consumers (Verbeke *et al.*, 2000).

The negative impact from television publicity was confirmed through econometric time series analysis. Probabilities of cutting fresh meat consumption were boosted as consumers reported to have paid high attention to television coverage of meat issues. Similarly, parameters of television coverage indices were largely significant and negative in an Almost Ideal Demand System for fresh meat, contrary to the estimates of the advertising expenditure variables, which were insignificant. In the case of beef in Belgium during the second half of the nineties, a negative press vs. advertising impact ratio of five to one was found (Verbeke and Ward, 2001). This means that five units of positive news are needed to offset the impact of one similar negative message.

*Wim Verbeke, Joachim Scholderer and Lynn Frewer*

Traceability and beef labelling have been issues in Europe since the BSE crisis starting in 1996 and are currently also prominent in the regulatory debate in the US. The beef safety crises and subsequent decline in beef consumption, particularly in Europe, have forced governments and the meat industry to react and to work toward restoring consumer confidence. Traceability systems and subsequent quality and origin labelling of beef were considered as major instruments for addressing the problem (Gellynck and Verbeke, 2000; Verbeke, 2001). The systems are fully operational and traceability information has been placed on meat labels. Nevertheless, consumer interest in this kind of information cannot be taken for granted. Communication efforts aimed at informing consumers about the existence and meaning of beef traceability failed to evoke an active information search by consumers (Verbeke *et al.*, 2002), and consumers are really not interested in traceability cues on beef labels (Verbeke and Ward, 2006, in press).

Apparently, and despite a considerable number of meat safety crises, consumer interest is low in cues directly related to traceability and product identification while much higher for others like readily interpretable indications of quality such as certified quality marks or seals of guarantee, as well as for mandatory standard information like expiration date. Hence, although traceability has to be in place for legal purposes and in order to help guarantee product quality or origin, consumers are not interested in the traceability information per se. The obvious conclusion is that the role of traceability information is mainly defensive, aimed at guaranteeing a safe product for the next level in the agri-food chain. Its potential as an offensive tool, i.e. its potential usefulness from a marketing perspective, is highly questionable. The obvious conclusion is that, while possibly useful for legal purposes and quality management in the agri-food chain, traceability does not have to be predominant on the food label.

### 14.6.5 Food allergy - "A particular communication challenge"

It is important to communicate information about food allergy to patients in an effective way, which is also relevant to the interests and needs of end users. For example, one might identify different needs linked to different patient groups; children, adolescents, adults and the parents of food- allergic children.

The prevalence of food allergy is highest in young children (Sampson, 2001), yet very little is known about the most effective methods of communication for this group. Given that many parents self-diagnose and self-treat allergy (Eggesbo *et al.*, 2001), inappropriate communication may compromise the health or nutritional status of affected children. However, adolescents and young adults are the groups most at risk from severe food-allergy reactions (Gowland, 2002). Communication difficulties are exacerbated as "risk-taking behaviour and ignoring advice from figures of authority are features of developing independence" (Hourihane, 2001). This implies that specifically segmented communication approaches should be tailored to the needs of food-allergic adolescents. In contrast, adults with self-diagnosed food intolerance need useful, accurate and trustworthy information about getting a clinical diagnosis or seeking further information about their symptoms (Booth and Knibb, 2000).

*Consumer perception of safety in the agri-food chain*

The only treatment currently available for food allergy is avoidance of target foods by sensitive individuals. As a consequence, allergic individuals must learn to cope with constantly exercising extreme vigilance when examining products for potentially problematic ingredients, implying that effective labelling is essential in the prevention of allergic responses. Joshi *et al.* (2002) report that the ability of parents of food-allergic children to accurately read food labels is very poor, primarily because of the complexity and ambiguity of information provided. Indeed, poor labelling practices and so-called precautionary labelling might also lead to unnecessary restrictions in the child's diet (McCabe *et al.*, 2001).

Food-induced anaphylaxis in allergic individuals already aware of their allergy happens mostly at "safe sites", such as homes, schools, workplaces and hospitals (Eigenman and Zamora, 2002). One conclusion that can be drawn is that an important strategy to avoid accidental reactions includes clear labelling of forbidden foods, and increased information provision at all levels (Wood 2002), including the catering sector.

There is potential for novel proteins from conventional or genetically modified sources to produce an allergic response in vulnerable individuals, as crops produced with agricultural biotechnology ultimately result in the introduction of foods containing novel proteins (Taylor and Hefle, 2001). The communication issue here relates to the provision of information about the efficacy of safety testing rather than health communication *per se* (Knibb *et al.*, 1999). Advances in processing techniques may keep labile allergens intact, or, conversely, enable the specific destruction of allergen activity (Soler-Rivas and Wichers, 2001). Other issues of societal concern regarding genetically modified crops may also emerge as a focus of communication activities (van Putten *et al.*, submitted). For example, public perceptions that the integrity of nature is in some way compromised by the introduction of genetically modified crops may increase the complexity of the debate about whether crops with reduced allergenicity are acceptable.

## 14.7 Conclusions

This chapter provided the basic principles of consumer behaviour and consumer decision-making with relevance to safety issues in the agri-food chain. The chapter also introduced a number of selected case studies about consumer perception of safety in the agri-food chain. The major lessons from this chapter are summarised below:

- Food purchasing and consumption decisions can be driven by safety-related motivations. In this case, concerns or uncertainty about food safety may trigger problem or need recognition, information search and information processing. New information, e.g. negative press related to food safety, can change routine or habitual food purchasing into limited problem-solving type behaviour.
- Food safety is one of the product attributes that consumers can use during their evaluation of alternative products. Safety is usually considered a credence attribute, especially when safety is related to the absence of, for instance, chemical residues or GM ingredients that have no immediate health impact. Notable exceptions are when safety is guaranteed and

trusted through control certificates, labels or brands (in that case, safety can become a search attribute, e.g. food allergy labelling), or when safety leads to immediate health problems (in that case, safety becomes an experience attribute, e.g. presence of microbiological contamination).
- A perception filter, shaped by individual, environmental and situational factors, is responsible for the gap between scientific reality and consumer perception. Whereas scientific reality pertains to manageable, measurable and repeatable practices, it is consumer perception that determines beliefs, attitudes, preference and the ultimate choice or behaviour.
- Some food safety or lifestyle risks evolve from a problem into a crisis, while others don't. Catalysts for the evolution from problem to crisis can be classified as psychological fright factors and panic elements, and media triggers. When several of these catalysts are present to a high degree, the problem or risk stands a reasonable chance of turning into a crisis or scare.
- Food poisoning from microbiological contamination largely results from food-handling malpractice by consumers. Optimistic bias is responsible for the underestimation of food contamination in the home. It relates to people's tendency to estimate their personal risk as being substantially lower than the risk of the average person in their reference group. Optimistic bias is utterly resistant to health education and risk communication. A potentially effective - though resource intensive - approach is to focus on segmenting the population according to their information needs, and developing information with high levels of personal relevance to specific target segments.
- The Coca-Cola case with a small amount of initial chemical contamination, and the resulting incidence of mass sociogenic illness, illustrates the strong potential impact of emotions and psychology with respect to food safety perceptions. Relevant lessons were drawn with respect to future risk communication. The case also exemplifies the recovery potential of a strong brand that can build on its past image and mainly favourable consumer attitude.
- Perceived risk from technological innovation can be highly driven by socio-political factors, rather than the safety evaluation of the resulting products, as was illustrated with the case of genetically modified foods. This case demonstrates the complexity of cognitive representations in consumers' minds and the complexity on which consumers make evaluations. Furthermore, the GM case shows how difficult it may be to persuade consumers to change their initial attitudes.
- Although traceability and labelling were expected to solve part of the safety problems confronting the meat chain, it remains debatable whether consumers are after all interested in this type of additional information. Some consumer segments may be, whereas by far the majority prefers more direct indications of food quality and safety, like expiry dates, and even brand names or price information.
- The specific case of risk relating to food allergies underlines again the need for segmented communication approaches, clearer labelling and increased information provision at all levels in the agri-food chain.

As indicated before, this chapter aimed at providing insights into and raising readers' interest in the wide range of contemporary consumer behaviour issues relevant to safety debates in the agri-food chain. As a result, the presented theoretical concepts and frameworks may have remained fragmentary, while the presented case studies lack methodological and empirical details. We respectfully refer readers to the original publications for in-depth discussions and full details.

## References

Alhakami, A.S. and P. Slovic, 1994. A psychological study of the inverse relationship between perceived risk and perceived benefit. Risk Analysis, 14, 1085-1096.
Ambler, T., 1998. Myths about the mind: Time to end some popular beliefs about how advertising works. International Journal of Advertising, 17, 501-509.
Anthonissen, P.F., 2001. Murphy was een optimist: Hoe ondernemingen door crisiscommunicatie in leven bleven [Murphy was an optimist: How companies survived thanks to crisis communication]. Tielt: Uitgeverij Lannoo.
Assael, H., 1987. Consumer behaviour and marketing action. Boston: Kent Publishing Company.
Barry, T. and D. Howard, 1990. A review and critique of the hierarchy of effects in advertising. International Journal of Advertising, 9, 121-135.
Becker, T., E. Benner and K. Glitsch, 1998. Summary report on consumer behaviour towards meat in Germany, Ireland, Italy, Spain, Sweden and the United Kingdom: Results of a consumer survey. Working paper FAIR-CT95-0046. Göttingen: Department of Agricultural Markets and Agricultural Policy, The University of Hohenheim.
Bennett, P., 1999. Understanding responses to risk: some basic findings, In: P. Bennett and K. Calman (eds.), Risk communication and public health, pp. 3-19. Oxford: Oxford University Press.
Bennett, P. and K. Calman, 1999. Risk communication and public health. Oxford: Oxford University Press.
Booth, D.A. and R.C. Knibb, 2000. What's the evidence that your problem comes from that food? Food Allergy and Intolerance 1, 191-196.
Borre, O., 1990. Public opinion on gene technology in Denmark 1987 to 1989. Biotech Forum Europe, 7, 471-477.
Boss, L., 1997. Epidemic hysteria: a review of the published literature. Epidemiologic Reviews, 19, 233-243.
Bredahl, L., 1999. Consumers' cognitions with regard to genetically modified food. Results of a qualitative study in four countries. Appetite, 33, 343-360.
Bredahl, L., 2001. Determinants of consumer attitudes and purchase intentions with regard to genetically modified foods: Results of a cross-national survey. Journal of Consumer Policy, 24, 23-61.
Brüggemann, A. and H. Jungermann, 1998. Abstrakt oder konkret: Die Bedeutung der Beschreibung von Biotechnologie für ihre Beurteilung. Zeitschrift für Experimentelle Psychologie, 45, 303-318.
Bruhn, C.M., 1995. Consumer attitudes and market response to irradiated food. Journal of Food Protection, 58, 175-181.
Cardello, A., 1995. Food quality: Relativity, context and consumer expectations. Food Quality and Preference, 6, 163-170.
Centers for Disease Control and Prevention, 1994. Location of food mishandling. Morbidity and Mortality Weekly Reports (MMWR), November 1, 1994. Atlanta, GA: U.S. Department of Health and Human Services.

Demarest, S., A. Gallay, J. van der Heyden, F. van Loock and H. van Oyen, 1999. Case control study among schoolchildren on the incident related to complaints following the consumption of Coca-Cola products, Belgium, 1999. Brussels: Scientific Institute of Public Health, Unit of Epidemiology.

Eggesbo, M., G. Botten and H. Stigum, 2001. Restricted diets in children with reactions to milk and egg perceived by their parents. Journal of Pediatrics, 139, 583-587.

Eigenman, P.A. and S.A. Zamora, 2002. An internet-based survey on the circumstances of food-induced reactions following the diagnosis of IgE-mediated food allergy. Allergy, 57, 449-453.

Engel, J., D. Kollat and R. Blackwell, 1968. Consumer behavior. New York: Holt, Rinehart and Winston.

Engel, J., R. Blackwell and P. Miniard, 1995. Consumer Behavior. Fort Worth: The Dryden Press.

Fife-Schaw, C. and G. Rowe, 2000. Extending the application of the psychometric approach for assessing public perceptions of food risk: some methodological considerations. Journal of Risk Research, 3, 167-179.

Fischer, A.R.H., A.E.I. de Jong, R. de Jonge, L.J. Frewer and M.J. Nauta, 2005. Improving Food Safety in the Domestic Environment: The Need for a Transdisciplinary Approach. Risk Analysis, 25(3), 503-17.

Fischhoff, B., P. Slovic and S. Lichtenstein, 1978. How safe is safe enough? A psychometric study of attitudes towards technological risks and benefits. Policy Sciences, 9, 127-152.

Fishbein, M.A. and I. Ajzen, 1975. Belief, attitude, intention, and behavior: An introduction to theory and research. Reading, MA: Addison-Wesley.

Flynn, J., P. Slovic and C.K. Mertz, 1994. Gender, race, and perception of environmental health risks. Risk Analysis, 14, 1101-1108.

Food Marketing Institute, 1997. Trends in the United States: Consumer attitudes & the supermarket. Washington, DC: FMI.

Frewer, L., 1999. Public risk perceptions and risk communication, In: P. Bennett and K. Calman (eds.), Risk communication and public health, pp. 20-32. Oxford: Oxford University Press.

Frewer, L. A. Fischer, J. Scholderer and W. Verbeke, 2005. Food safety and consumer behaviour, In: W.M.F. Jongen and M.T.G. Meulenberg (eds.), Innovation in agri-food systems: Product quality and consumer acceptance, pp. 125-145. Wageningen: Wageningen Academic Publishers.

Frewer, L.J. and B. Salter, 2002. Public attitudes, scientific advice and the politics of regulatory policy: the case of BSE. Science and Public Policy, 29, 137-145.

Frewer, L.J., C. Howard and R. Shepherd, 1996. The influence of realistic product exposure on attitudes towards genetic engineering of food. Food Quality and Preference, 7, 61-67.

Frewer, L.J., C. Howard and R. Shepherd, 1997b. Public concerns in the United Kingdom about general and specific applications of genetic engineering: Risk, benefit, and ethics. Science, Technology & Human Values, 22, 98-124.

Frewer, L.J., C. Howard and R. Shepherd, 1998. The importance of initial attitudes on responses to communication about genetic engineering in food production. Agriculture & Human Values, 15, 15-30.

Frewer, L.J., C. Howard, D. Hedderley and R. Shepherd, 1997a. The elaboration likelihood model and communication about food risks. Risk Analysis, 17, 759-770.

Frewer, L.J., C. Howard, D. Hedderley and R. Shepherd, 1999. Reactions to information about genetic engineering: Impact of credibility, perceived risk immediacy and persuasive content. Public Understanding of Science, 8, 1-15.

Frewer, L.J., J. Lassen, B. Kettlitz, J. Scholderer, V. Beekman and K.G. Berdal, 2004. Societal aspects of genetically modified foods. Food and Chemical Toxicology, 42, 1181-1193.

Frewer, L.J., J. Scholderer and L. Bredahl, 2003. Communicating about the risks and benefits of genetically modified foods. Risk Analysis, 23, 1117-1133.

Gellynck, X. and W. Verbeke, 2001. Consumer perception of traceability in the meat chain. Agrarwirtschaft: German Journal of Agricultural Economics, 50, 368-374.

Gowland, M.H., 2002. Food allergen avoidance: risk assessment for life. Proceedings of the Nutrition Society 2002, 61, 39-43.

Grunert, K.G., L. Bredahl and J. Scholderer, 2003. Four questions on European consumers' attitudes to the use of generic modification in food production. Innovative Food Science and Emerging Technologies, 4, 435-445.

Grunert, K.G., L. Lähteenmäki, N.A. Nielsen, J.B. Poulsen, O. Ueland and A. Åström, 2001. Consumer perceptions of food products involving genetic modification: Results from a qualitative study in four Nordic countries. Food Quality and Preference, 12, 527-542.

Hagemann, K. and J. Scholderer, 2004. Mental models of the benefits and risks of novel foods (MAPP project paper no. 06/2004). Aarhus: MAPP.

Hamstra, A., 1995. Consumer acceptance of model for food biotechnology: Final report. The Hague: SWOKA.

Hamstra, A.M., 1991. Biotechnology in foodstuffs. Towards a model of consumer acceptance. The Hague: SWOKA.

Hamstra, A.M. and M.H. Feenstra, 1989. Consument en biotechnologie. Kennis en meningsvorming van consumenten over biotechnologie [Consumers and biotechnology. Consumer knowledge and opinions about biotechnology] (SWOKA onderzoeksrapport nr.85). Den Haag: SWOKA.

Holbrook, M., 1995. Consumer Research: Introspective essays on the study of consumption. Thousand Oaks: Sage Publications.

Hoorens, V. and B.P. Buunk, 1993. Social comparison of health risks: Locus of control, the person-positivity bias, and unrealistic optimism. Journal of Applied Social Psychology, 23, 291-302.

Hourihane, J.O.B., 2001. The threshold concept in food safety and its applicability to food allergy. Allergy, 56, 86-90.

INRA Europe, 1998. Eurobarometer 49: Les Européens et la sécurité des produits alimentaires. Luxembourg: Office for Official Publications of the European Communities.

Joshi, P., S. Mofidi and S.H. Sicherer, 2002. Interpretation of commercial food ingredient labels by parents of food-allergic children. Journal of Allergy and Clinical Immunology, 109, 1019-1021.

Kasperson, R., O. Renn, P. Slovic, H. Brown, J. Emel, R. Goble, J. Kasperson and S. Ratick, 1988. Risk Analysis, 8, 177.

Kirk, S.F.L., D. Greenwood, J.E. Cade and A.D. Pearman, 2002. Public perception of a range of potential food risks in the United Kingdom. Appetite, 38, 189-197.

Knibb R.C., D.A. Booth, R. Platts, A. Armstrong, I.W. Booth and A. Macdonald, 1999. Psychological characteristics of people with perceived food intolerance in a community sample. Journal of Psychosomatic Research, 47, 545-554.

Lavidge, R. and Steiner, G. (1961). A model for predictive measurements of advertising effectiveness. Journal of Marketing, 25, 59-62.

Lek, Y. and G.D. Bishop, 1995. Perceived vulnerability to illness threats: The role of disease type, risk factor perception and attributions. Psychology and Health, 10, 205-219.

Maslow, A.H., 1954. Motivation and personality. New York: Harper & Row.

McCabe, M., R.A. Lyons, P. Hodgeson, G. Griffiths and R. Jones 2001. Management of peanut allergy. Lancet, 357, 1531-1532.
McCarthy, M. and S. Vilie, 2002. Irish consumer acceptance of the use of gene technology in food production, In: J.H. Tienekens and S.W.F. Omta (eds.), Paradoxes in chains and networks, pp. 176-187. Wageningen: Wageningen Academic Publishers.
McGuire, W., 1978. An information processing model of advertising effectiveness, In: H. Davis and A. Silk (eds.), Behavioral and Management Science in Marketing, pp. 156-180. New York: Ronald Press.
Mela, D., 2001. Development and acquisition of food likes and dislikes, In: L.J. Frewer, E. Risvik and H. Schifferstein (eds.), Food, people and society: A European perspective on consumers' food choices, pp. 9-22. Berlin: Springer.
Midden, C., D. Boy, E. Einsiedel, B. Fjæstad, M. Liakopoulos, *et al.*, 2002. The structure of public perceptions, In: M.W. Bauer and G. Gaskell (eds.), Biotechnology: The making of a global controversy, pp. 203-224. Cambridge: Cambridge University Press.
Miles, S. and L. Frewer, 2001. Investigating specific concerns about different food hazards. Food Quality and Preference, 12, 47-61.
Miles, S. and L.J. Frewer, 2002. QPCRGMOFOOD Work Package 6: Socio-economic impact of GM regulation and GM detection (Report to the European Commission). Norwich: Institute of Food Research.
Miles, S. and V. Scaife, 2003. Optimistic bias and food. Nutrition Research Reviews, 16, 3-19.
Miles, S., 2003. Public perception of food safety, In: IAAS (Ed.), Food Quality: A challenge for North and South, pp. 171-180. Heverlee: IAAS Belgium.
Miles, S., D.S. Braxton and L.J. Frewer, 1999. Public perceptions about microbiological hazards in food. British Food Journal, 101, 744-762.
Miles, S., M. Brennan, S. Kuznesof, M. Ness, C. Ritson and L.J. Frewer, 2004. Public worry about specific food safety issues. British Food Journal, 106, 9-22.
Ministry of Public Health, 2000. The Coca-Cola crisis June 1999 in Belgium: evaluation of the events, discussion, conclusions and recommendations. Brussels: Ministry of Public Health, Health Council.
Nemery, B., B. Fischler, M. Boogaerts and D. Lison, 1999. Dioxins, Coca-Cola, and mass sociogenic illness in Belgium. Lancet, 354, 77-78.
Peters, H.P., 1998. Is the negative more relevant than the positive? Cognitive responses to TV programs and newspaper articles on genetic engineering. Paper presented at the 5th International Conference on Public Communication of Science and Technology (PCST). Berlin, September 17-19, 1998.
Petty, R.E. and J.T. Cacippio, 1986. Communication and Persuasion: Central and peripheral routes to attitude change. New York: Springer-Verlag.
Philen, R.M., T.W. McKinley, E.M. Kilbourne and R.G. Parrish, 1989. Mass sociogenic illness by proxy: Parentally reported epidemic in an elementary school. Lancet, 2(8676), 1372-1376.
Pilgrim, F., 1957. The components of food acceptance and their measurement. American Journal of Clinical Nutrition, 5, 171-175.
Redmond, E.C. and C.J. Griffith, 2004. Consumer perceptions of food safety risk, control and responsibility. Appetite, 43, 309-313.
Richardson, N., H. MacFie and R. Shepherd, 1994. Consumer attitudes to meat eating. Meat Science: An International Journal, 36, 57-65.
Saba, A. and M. Vassallo, 2002. Consumer attitudes toward the use of gene technology in tomato production. Food Quality and Preference, 13, 13-21.

Sampson, H.A., 2001. Immunological approaches to the treatment of food allergy. Paediatric Allergy and Immunology, 12, 91-96.

Scholderer, J. and L. Frewer, 2003. The biotechnology communication paradox: Experimental evidence and the need for a new strategy. Journal of Consumer Policy, 26, 125-157.

Scholderer, J., I. Balderjahn, L. Bredahl and K.G. Grunert, 1999. The perceived risks and benefits of genetically modified food products: Experts versus consumers. European Advances in Consumer Research, 4, 123-129.

Scholderer, J., L. Bredahl and L. Frewer, 2000. Ill-founded models of consumer choice in communication about food biotechnology, In: F. van Raaij (ed.), Marketing communications in the new millennium: New media and new approaches, pp. 129-152. Rotterdam: Erasmus University.

Scholderer, J., T. Bech-Larsen and K.G. Grunert, 2001. Changing public perceptions of genetically modified foods: Effects of consumer information and direct product experience. Appetite, 37, 162.

Scholten, M., 1996. Lost and found: The information-processing model of advertising effectiveness. Journal of Business Research, 37, 97-104.

Schütz, H., P.M. Wiedemann and P.C.R. Gray, 1999. Die intuitive Beurteilung gentechnischer Produkte - kognitive und interaktive Aspekte, In: J. Hampel and O. Renn (eds.), Gentechnik in der Öffentlichkeit: Wahrnehmung und Bewertung einer umstrittenen Technologie, pp. 133-169. Frankfurt/Main: Campus.

Shepherd, R., 1990. Overview of factors influencing food choice, in Ashwell, M. (Ed.), Proceedings of the 12th British Nutrition Foundation Annual Conference, pp. 12-30. London: British Nutrition Foundation.

Siegrist, M., 2000. The influence of trust and perceptions of risks and benefits on the acceptance of gene technology. Risk Analysis, 20, 195-204.

Sirois, F., 1999. Epidemic hysteria: School outbreaks 1973-1993. Medical Principles and Practice, 8, 12-25.

Sjöberg, L., 2000. Factors in risk perception. Risk Analysis, 20, 1-11.

Smith, D. and P. Riethmuller, 2000. Consumer concerns about food safety in Australia and Japan. British Food Journal, 102, 838-855.

Soler-Rivas, C. and H.J. Wichers, 2001. Impact of (bio)chemical and physical procedures on food allergen stability. Allergy, 56, 52-55.

Søndergaard, H.A., K.G. Grunert and J. Scholderer, 2003. Consumer attitudes towards different forms of enzyme production and implications for intentions to buy novel food products (Report to the European Commission). Aarhus: MAPP.

Sparks, P., R. Shepherd and L.J. Frewer, 1994. Gene technology, food production, and public opinion: A UK study. Agriculture and Human Values, 11, 19-28.

Steenkamp, J.-B., 1997. Dynamics in consumer behavior with respect to agricultural and food products, In: B. Wierenga, A. van Tilburg, K.Grunert, J.-B. Steenkamp and M. Wedel, (eds.), Agricultural marketing and consumer behavior in a changing world, pp. 143-188. Dordrecht: Kluwer Academic Publishers.

Strong, E., 1925. The psychology of selling. New York: McGraw-Hill Book Company.

Taylor, S.L. and S.L. Hefle, 2001. Will genetically modified foods be allergenic? Journal of Allergy and Clinical Immunology, 107, 765-771.

Vakratsan, D. and T. Ambler, 1999. How advertising works: What do we really know. Journal of Marketing, 63, 26-43.

Van Loock, F., A. Gallay, S. Demarest, J. van der Heyden and H. van Oyen, 1999. Outbreak of Coca-Cola-related illness in Belgium: a true association. Lancet, 354, 680-686.

Van Putten, M.C., L.J. Frewer, L.J.W.J. Gilissen, B.G.J. Gremmen, A. Peinenberg and H.J. Wichers, (submitted). Novel Foods and food allergies. The issues. Trends in Food Science and Technology.

Veblen, T., 1899. The theory of the leisure class. An economic study of the evolution of institutions. New York: Modern Library.
Verbeke, W., 2000. Influences on the consumer decision-making process towards fresh meat: Insights from Belgium and implications. British Food Journal, 102(7), 522-538.
Verbeke, W., 2001. The emerging role of traceability and information in demand-oriented livestock production. Outlook on Agriculture, 30 (4), 249-256.
Verbeke, W., 2003. Consumer reactions to food risks and communication. In: T. Eklund, H. De Brabander, E. Daeseleire, I. Dirinck and W. Ooghe (eds.), Strategies for safe food: Analytical, industrial and legal aspects: Challenges in organisation and communication, Volume 1, pp. 13-18. Heverlee: KVCV.
Verbeke, W. and P. Van Kenhove, 2002. Impact of emotional stability and attitude on consumption decisions under risk: The Coca-Cola crisis in Belgium. Journal of Health Communication, 7, 455-472.
Verbeke, W. and R. Ward, 2001. A fresh meat almost ideal demand system incorporating negative TV press and advertising impact. Agricultural Economics, 25, 359-374.
Verbeke, W. and R. Ward, 2006. Consumer interest in information cues denoting quality and origin: an application of ordered probit models to beef labels. Food Quality and Preference, in press.
Verbeke, W., R. Ward and J. Viaene, 2000. Probit analysis of fresh meat consumption in Belgium: Exploring BSE and television communication impact. Agribusiness: An International Journal, 16, 215-234.
Verbeke, W., R. Ward and T. Avermaete, 2002. Evaluation of publicity measures relating to the EU beef labelling system in Belgium. Food Policy, 27, 339-353.
Von Alvensleben, R., 1997. Consumer behaviour. In: D.I. Padberg, C. Ritson and L.M. Albisu (eds.), Agro-food marketing, pp. 209-244. Wallingford: CAB.
Weinstein, N.D., 1980. Unrealistic optimism about future life events. Journal of Personality and Social Psychology, 39, 806-820.
Weinstein, N.D., 1987. Unrealistic optimism about susceptibility to health problems: Conclusions from a community-wide sample. Journal of Behavioral Medicine, 10, 481-500.
Weinstein, N.D., 1989. Optimistic biases about personal risks. Science, 246, 1232-1233.
Weinstein, N.D. and W.M. Klein, 1995. Resistance of personal risk perception to debiasing interventions. Health Psychology, 14, 132-140.
Wessely, S. and C.J. Wardle, 1990. Editorial: Mass sociogenic illness by proxy: Parentally reported epidemic in an elementary school. British Journal of Psychiatry, 157, 421-424.
Wierenga, B., 1983. Model and measurement methodology for the analysis of consumer choice of food products. Journal of Food Quality, 48, 119-137.
Wood, R.A., 2002. Food manufacturing and the allergic consumer: Accidents waiting to happen. Journal of Allergy and Clinical Immunology, 109, 920-922.
Worsfold, D. and C. Griffith, 1997. Food safety behaviour in the home. British Food Journal, 99, 97-104.

# 15. Ethics in food safety

*Kriton Grigorakis*

Ethics are not just part of philosophy, but also act as regulators in our society defining our limits of acceptance in various issues. Food production and consumption are primary aspects in our lives and therefore subjects to our ethics. Many people would consider food ethics as a philosophical matter with only theoretical relevance to our food chain. But this is not the case, because we often need to take ethical decisions with relevance to food, and especially concerning safety. Moreover, the importance of ethics has increased drastically due to modern biotechnology applications in the food industry.

This chapter aims at analysing food ethics, the implication for food safety and the role of consumers. In the first part main ethical principles, theories and ethical positions will be presented in order to introduce the readers to some philosophical terms and to help them follow the analysis. In the second part of this chapter, food ethics will be viewed from an historical perspective in order to indicate what has historically affected contemporary food ethics. Subsequently, we will try to give answers to how ethical evaluation can take place and how decisions can be driven. Following on from that, the involvement of the consumer in food safety ethics will be studied. In the third part, specific food ethics issues will be presented, like GM food and its ethics, the ethical background of food labelling, and the ethical aspects of food patenting.

This chapter will offer the readers an opportunity to acquire some knowledge in an important field of applied ethics, and to define their own position, and the ethical implications and obligations in the food production / consumption process. Finally and most importantly, it will allow the readers to ask themselves questions on specific topics and to form their own opinions on food ethics, which, in contrast to the exact sciences, does not in many cases provide straight answers.

## 15.1 General introduction to ethics

In this section the reader is introduced to some basic principles in ethics. The first section explains the differences between ethics and morals. Subsequently, basic theories and principles in ethics are explained. The third section discusses how the basic theories in ethics are related to each other and how they are applied in practice. In the last section, the distinction between intrinsic and extrinsic ethical concerns is explained.

### 15.1.1 Ethics and morality, ethical and moral

Ethical and moral are two terms that are very commonly used in ethics philosophy and in everyday life. They can easily be confusing in their meanings, because in everyday language and by various philosophers the terms are often used interchangeably (Peter Singer for example, as mentioned by Reiss and Straughan, 1996). However, the terminology of these two words differs, and therefore certain distinctions can be made between them.

*Moral*: Every human being is considered to have moral views, except maybe in certain neuropsychology pathological situations caused by specific damage in certain brain areas. These moral views can refer to any subject of life, and designate what is right and wrong and what certain actions ought or ought not to be performed. These moral views are diverse and can include any aspect of life. These moral concerns may derive from a lot of deliberation and reflection or from very little; they may be firmly grounded as a consistent set of rational principles, or they may not; their justification may have been analysed and consciously established, or it may not. Many of our moral views are subconsciously transplanted into us through the society in which we were brought up.

*Ethics*: Ethics is normally considered as a narrower concept than morality. The term can be used in several different but related senses. The most general definition that can be made is "...a set of standards by which a particular group or community decides to regulate its behaviour and to distinguish what is legitimate or acceptable in pursuit of their aims from what is not." Thus, we talk about food ethics, and labour ethics, etc.

From a more technical perspective, ethics can also refer to a particular branch of philosophy - 'moral philosophy'. A central task of ethics is a critical investigation of the fundamental principles and concepts that are used in moral debate. Ethics aim to analyse the different arguments in it. Ethics is in a sense a critical 'second order' activity, which puts our 'first order' moral beliefs under scrutiny. Some aspects of life do not have ethical significance, although they generate moral concerns. Many people, for example, may have moral concerns about old and lonely people but it is not clear what ethical validity these have. In another approach (Macer, 1995), ethical is a philosophical concept that prescribes objectivity, while moral is the common sense of morality, of what is right and wrong in society.

What is moral is ultimately (and usually) encoded by the law, and is derived from the general public perception of morality. What is immoral is very often also unethical. However, this is not always the case. Unethical practices or actions are often tolerated by society, and thus, under the definition of moral, we would characterise these actions as 'morally acceptable' actions. As an example, we only have to think of the destruction of the environment, which has been going on for such a long time. Human actions as a result of a lack of purposeful measures often have a negative environmental impact (e.g. pollution). Another example is that in some African societies it is considered as a moral obligation to amputate the genitals of female infants (many of them die because of bleeding or infections). One could say that this is obviously an unethical tradition.

## 15.1.2 Basic ethical theories and principles

*Deontological theory* (Immanuel Kant)
Immanuel Kant, the 18th century German philosopher, introduced the Ethical theory of duty in his book "Grundlegung zur Metaphysik der Sitten " ("Fundamental Principles of the Metaphysic of Morals"). In other words, this is a theory based on our ethical duties. The term 'deontocratic' means, in Greek, 'ruled by duty'.

Kant's ethics are about respecting everyone as an individual without calculating the consequences. The principle of autonomy is based on Kant's notion of rights, which appeals to our major duty 'to treat others as ends in themselves'. In essence, this is the basic rule: 'Do as you would be done by'.

Our duties are very specific in every aspect of our lives, and there is a certain hierarchy among them. An ethical decision is every decision that follows the commands of our most powerful duty every time. Immanuel Kant's theory provides us with a certain set of principles, which when followed provides us with a clear view of what is ethical. On the other hand, the basic objections to his theory are that it is a very strict, rigid and sovereign view of ethics in a world where truths are not just black and white. In practice, a major flaw is that there is no certain rule by which to decide how to prioritise duties. For example, the duties to protect others from harm and to tell the truth: what happens when telling the truth is a cause of harm? How can I decide which of the two contradicting duties is more important?

*Utilitarianism* (Jeremy Bentham, John Stuard Mill)
The Utilitarian theory was developed by two English philosophers of the 18th century, Jeremy Bentham and John Stuard Mill. Their ethical theory can be epitomised as 'The greatest good for the greatest number'. The utilitarian theory suggests that we conclude what is ethical after weighing the costs and benefits and finding the most positive balance among them, i.e. finding the greatest possible good.

This ethical approach provides us with a tool for an ethical assessment, a cost/benefit analysis. It is a practical approach to ethics, which is a very great advantage in a world of rationality. However, several defects in the theory can easily be seen: naïve forms of utilitarianism depend on predictions of outcome (very often wrong) and (in many cases fallible) assessments of who or what counts in the cost/benefit analyses. Subjectivity is the weak point of utilitarianism. Furthermore, utilitarianism can lead to gross inequality, as long as the majority 'are happy', or even crime (stolen money distributed to the needy). Finally, good and harm are often incommensurable. How can we, for instance, weigh the safety of a product, which is designated for humans, against the suffering of the animals that are used to test it? The principle of well-being corresponds to issues prominent in utilitarian theory.

*Contractarianism* (John Rawls)
John Rawls is a contemporary US philosopher who faces the problem of ground and framing ethics based on a concept of overlapping consensus, a contract of society. He introduced the

concept of justice as a key principle of his ethical theory (Rawls, 1972), his notion of 'justice as fairness'.

He was concerned to develop a principle that would allow one to decide fairly in a society how to distribute resources among people. The veil of ignorance, as proposed by J. Rawls, ("people do not know how the various alternatives will affect their own particular case and they are obliged to evaluate principles solely on the basis of general considerations."), allows no one to tailor principles to his own advantage. The weak point in J. Rawls' approach is summarised in the question of how to define what fairness means. For example, does it mean that goods should be distributed according to need, ability or effort?

### 15.1.3 Discussion on ethical theories

It is not clear whether the basic ethical theories oppose each other. Thomson (2001) makes the discrimination between the philosophies that evaluate the moral justifiability of an action in light of its future consequences, and those that evaluate moral justifiability in light of whether it affords adequate respect for the freedom and autonomy of other people. The former are the utilitarian or consequentialist philosophies and the latter are the non-consequentialist, or right-based philosophies. Within this distinction utilitarianism falls into the first category (actually consists of the first category) and the deontological approach (Kantianism) falls into the second category.

Consequentialist philosophies base the outcome of their ethical evaluation on the best balance of benefit and harm, the utilitarian maxim. At a particular time, the action with the best expected value is determined (reflecting both the cost/benefit of the predicted outcome, as well as the anticipated degree of uncertainty or indeterminacy), optimisation is achieved, and this act is considered ethical. Kantianism, on the other hand, implies that ethics is a matter of respecting people's freedoms and rights. Therefore, its rule is that one may not impose risk upon other people without seeking and receiving consent. It demands informed consent.

That these are different philosophical approaches has been analysed previously. The question is whether they are also contradictory in terms of their ethical evaluation. First of all, we saw that each suggests a different decision procedure for determining ethical responsibility. Food safety is one of the practical issues that produces tension between the two approaches. Thomson (2001) has made an inventory of cases where utilitarianism provided opposing final decisions on ethical importance compared to the Kantian approach. Thomson's analysis (which is not one of the present aims) concludes that a Kantian approach would give ethical importance to the matter of food labelling (this will be further discussed), in the matter of pesticide residues, religious matters, food origin/novelty matters, in cases of distrust of the science, and in the light of people's preferences, even if they are not based on a rational decision. Consent criteria (the Kantian approach) demand that consumers have an alternative even when the likelihood of harm is low.

From the other perspective, utilitarianism, according to Thomson, does not demand an ethical stance for any food aspect that does not implicate public safety. The two approaches also differentiate in matters of environmental exposure, risks and hazards. Thus, the Kantian approach presumes a crucial moral difference between risks of human and natural origin, with the latter imposing only prudential and not moral issues. According to Thomson only utilitarianism will consider these in the ethical decision-making process. Indeed, these different approaches seem to produce incompatible results in application, in some cases. The question is, which principle should be applied, and a mixture of pragmatic and philosophical considerations determines the answer. According to Thomson, although scientists may be more comfortable with the outcome-optimising logic of utilitarianism, religious values and scepticism provide grounds for developing an informed consent approach.

The Food Ethics Council Report (Food Ethics Council, 1999) concludes that all these theories in practice are likely to contribute, at varying degrees, to people's attitudes on what should be done in specific circumstances, and that no-one could consistently act as a total utilitarian or a total Kantian. Every position is a blend of these theories. We should, however, agree with Thomson (1998), who supports the theory that utilitarianism is the more dominant of the philosophical streams in the practice of everyday ethics. We will later discuss the evidence that is provided and come out in support of this.

## 15.1.4 Extrinsic versus intrinsic ethical concerns

All moral concerns and all ethical arguments for and against a certain action can be divided into two categories based on their nature. A set of arguments may support that this certain action (or technology) is thought to be intrinsically wrong in itself or extrinsically wrong because of its consequences. Purchase (2002) makes a very apt remark saying that intrinsic versus extrinsic concerns corresponds to opposition of principles versus outcome. This important distinction can be applied to a large number of moral issues and can often help in the identification of the specific grounds of certain concerns. Confusion can arise if this distinction is not drawn because in practice the logical form of one set of arguments (intrinsic arguments) cannot be countered by a set of arguments of a different logical form (extrinsic ones). The intrinsic arguments 'cut deeper than extrinsic ones', as Straughan mentions (Reiss and Straughan, 1996). Purchase confirms it saying, "When there is a major intrinsic concern about a proposed action, extrinsic arguments are not usually convincing counter-arguments." In intrinsic arguments attention is focused upon the precise nature and the distinguishing characteristics of an action (or technology). If I feel that euthanasia is intrinsically wrong, then it is clearly the act of euthanasia *per se* that I consider as a moral concern. It is the act of killing someone that is my concern. On the other hand, if I am afraid that, in some cases, people who might have been saved will die because of its application, then I am objecting to the possible consequences of euthanasia and not to the real action (which would be perfectly okay if we could ensure its proper application).

A major problem arises from the extrinsic ethical concerns. The claim that any activity or process will have undesirable consequences means making predictions about future facts. And at this point uncertainty edges in, as no conclusive proof can ever be provided that a particular set of events will inevitably occur in the future. Extrinsic concerns are in a sense provisional, they carry weight in proportion to the likelihood of the predicted consequences actually occurring. Their validity is strengthened by past experience in similar matters and through experimentation, or case studies. In practice it is possible to make reasonable informed ethical judgments about the consequences of technologies, for instance, genetic modification of plants.

On the other hand, intrinsic objections are much more rigid. No one can convince me to change my mind about something that I consider intrinsically wrong, but someone might convince me that measures will definitely be taken to eliminate certain consequences.

## 15.2 Principles of food ethics and safety

In this section, more principles of especially food ethics in relation to food safety, are explained in more detail. Starting with a concise history of food ethics, we then explain how ethics can be evaluated in a structured way, and, finally, the role of consumers in discussions on food ethics and safety is explained.

### 15.2.1 Food ethics from a historical perspective

Many would regard food ethics as a part of the contemporary philosophy of ethics, mainly developed by needs arising from biotechnology. However, food ethics have a deep historical perspective, having been a part of human practice and philosophy for years. What we consider as modern food ethics originated from these traditional aspects. Zwart (2000) has thoroughly reviewed the traditional food ethics areas that affect today's food ethics, and hereafter we will present them according to his analysis.

In ancient Greece food ethics, or rather *dietetics,* constituted a matter of substantial importance. The basic maxim of Greek morality, *kata physin*, i.e. live and act in accordance with nature, applied to ancient Greek food ethics as well. Man, in contrast to all other animals, is equipped with the faculty of reason when it comes to food intake. This allows him (or forces him) to participate in nature in an actively conscious way. To live *kata physin* basically meant to live a life of temperance. In ancient Greek and Roman ethics the connection between nature and temperance was self-evident. Hippocrates (1923/1957) had already stressed that a truly human life is not a life of passive consumption. Food products yielded by nature have to be improved and refined. Partaking of crude and uncompounded food was the cause of much terrible suffering. Dietetics in ancient Greece was a fully moral task - it was a way of life. In order to maintain health and well-being, the selection and preparation of food became an item of major concern, for which, however, there seemed to be no general recipes. Self-observation, self-inspection and experimentation in every individual person could develop into a moral regime, a moral life-style. The individual patterns of consumption adhered to one and the same basic moral scheme, namely the idea of temperance as a primary condition of human flourishing. Moreover, temperance allowed the moral elite to be distinguished from the masses, the mere consumers. Nowhere do we find an indication that certain food products in ancient Greece were to be regarded as illicit in and by themselves. Everything was allowed as long as one's food practices remained within the limits of temperance.

According to the tradition in the Hebrew Bible, a completely different moral logic was followed. Thus, the Bible's moral logic was guided not by the idea of temperance, but by the idea of a basic distinction between what was allowed and what was not allowed. In the context of food ethics, divine legislation introduced a dichotomy between admissible and inadmissible, legitimate and illicit food. For instance, "These are the creatures you may eat," the Bible says, " of all the larger land animals you may eat any hoofed animal which has cloven hoofs and also chews the cud; those which only have cloven hoofs or only chew the cud you must not eat" (Leviticus 11: 1-4). Many other parts of the bible contain other similar rules regarding food consumption. Countless efforts have been made to explain the why of these stringent directions, notably in terms of health, hygiene, and other utilitarian concerns, but none of them has completely succeeded in effacing their basically arbitrary nature.

Thus, the Hebrew Bible introduced a new, highly significant principle into the history of food ethics, namely the idea that certain food products are to be regarded as contaminated in view of their origin just because they are unlawful in themselves, and not because they are unhealthy, tasteless, and difficult to digest or any other consequential aspect. The Bible was first to introduce the intrinsic value of right or wrong in eating. Instead of the Greek logic of "more" or "less", the Bible came along with a binary logic of prohibited versus permitted.

Although from a general historical perspective Jewish and Christian traditions are often very similar, in the history of food ethics this is not the case. Indeed, what is so striking in the food ethic proclaimed by Jesus, is the basic idea of carelessness that it conveys. Food intake in Christian tradition becomes completely insignificant from a moral point of view. The early Christian food ethic is an ethic of de-problematisation. Very characteristic of this position is the quotation from Matthew: "No one is defined by what goes into his mouth; only by what comes out of it [...] Do you not see that whatever goes in by the mouth passes into the stomach and so is discharged at a certain place? But what comes out of the mouth has its origins in the heart; and that is what defines the person". The teaching of Jesus, in comparison with the stringent food ethic of the Hebrew Bible, shows a tolerant laxity and contains abrupt revisions. Jesus simply urges those who follow him to lose all interest in food production and consumption. This point of view is also found in Paul, who in his First Letter to Corinthians concludes, regarding food consumption, that nothing is unclean in itself. Even meat sacrificed to idols and subsequently sold in the market may be eaten by the Christians (1 Cor 10: 19-11; 8:11-13).

In the medieval years, the elaborated food ethic mainly adhered to the Christian principle of disregard. However, food concern found a place in monastic life, directed at the strengthening of self-discipline and a gradual submission to the *homo naturalis*. Gradually, the monastic concern with food grew into something of an obsession, and abstention from food became an objective in its own right. Food ethics was now aimed at the mortification of the flesh and the extinction of all desire, as well as of all worldly involvement. The ancient Greek moral of temperance was now replaced by "asceticism" in the sense of excessive abstention. In the popular imagination and in real life, however, the official ideology of asceticism was complemented by the counter-image of a gluttonous monk.

Ecclesiastical regulations had a considerable impact on the food practices of the masses, notably by prescribing times of fasting or regulations for special fasting days, such as Friday's meat abstention. Also, from 4AD onwards, Lent, the forty-day period of fasting, (between carnival - *carne vale* means "Adieu meat" - and Easter) was established. This tradition coincided with nature's own cycle of scarcity, and in fact it canonised pagan practices that had been followed since before Christianity.

In the sixteenth century, the monastic food ethic of mortification had already become a target of strong moral criticism. Even within Christian circles, food intake was now regarded with a much more positive point of view. Lutheranism now recommended, through the words of its founder Martin Luther, consuming large quantities of food as a remedy against temptation and melancholy. M. Luther's contemporary Ignatius Loyola, founder of the Jesuit order, also stressed the importance of a healthy and well cared for body. Thus, the Renaissance was characterised by the "rehabilitation of food".

Quite in opposition to monastic life, the medieval worldly elite had always distinguished itself by consuming large quantities of meat. Now (in the Renaissance), however, a civilisation of food intake was to take place. This consisted of the gradual increase of delicacy and sensitivity, notably with regard to meat consumption. Initially, the animal was served whole, clearly recognisable, and the process of skinning and dissecting it was done just before consumption, in public. At a certain point, dissection and consumption came to be discretely separated from one another. In fact, this increase in distance between the preparation and consumption of meat, as well as of other products, continued well into the present. This separation was reinforced by the emergence of moral scruples with regard to the butchering of animals. Subsequently, however, it became a source of suspicion, of moral concern with regard to food production in its own right. This distance between preparation and consumption is both concealing and disquieting. There is continuity between the food ethics, as they were modulated in the Renaissance and the food ethics of the modern era.

The food ethics of modernity are characterised by a new element, which is the scientific point of view. Modern science consists basically of a combination of systematic observation and quantification. The iatrophysics of the $17^{th}$ century, with Santorius as its founding father, can be regarded as the ancestor of modern dietetics. Modern dietetics, rather than being merely a scientific or medical endeavour, has a moral import to it as well. And this becomes noticeable in the famous treatise of Macrobiotic (Makrobiotik), published in Germany in 1796 by Hufeland. The question of how to extend one's life was considered therein, from a scientific as well as from a moral perspective.

Kant, in his Kritik der Urteilskraft (*German*), denied that dietetics could be regarded as a form of ethics. Since it basically consists of the application of scientific knowledge, it is a technical, rather than a "practical" (i.e. "ethical") endeavour. Later, he changed his mind somewhat:

*Ethics in food safety*

dietetics is a moral endeavour insofar as it entails the systematic effort to subject one's sensuality to reason. In the Metaphysic of Morals (Metaphysik der Sitten) Kant had argued that it was morally illicit to benumb one's mind with the intake of food or alcohol, thus depriving oneself of the use of one's intellectual faculties.

The nineteenth century is an important chapter in the history of food ethics, because the importance of the social dimension in food production and consumption emerged and was recognised. Within the economical philosophy of Thomas Malthus, in his famous essay on population growth, food constitutes a major moral problem. In contrast to animal population growth, which is sooner or later restrained by lack of food, mankind can rely on foresight, calculation, and morality to elaborate a more rational solution. By means of temporary sacrifices, a global catastrophe and famine may be prevented. The most powerful of all desires is the desire for food, followed closely by the sexual passion. Both should be subjected to regulation and self-direction. Thomas Malthus supported his argument with demographical and mathematical reasoning. Thus, food ethics provides the model for ethics as such. In absence of a policy of self-constraint, global starvation can be expected. According to Malthus, food intake and lack of self-restraint became problematic for the first time, due to their social rather than individual impact.

Karl Marx, similarly, focused on the social dimension of food. In Marx's work the attention was mostly directed towards the production, rather than the consumption of food. Food products become the basic items of concern. They became the incarnation of social tension and conflict, rendering them materially tangible. The rise of capitalism effected the destruction of the self-providing rural communities of the past and greatly increased the distance between consumption and production. Thus, food products generated by capitalism represent a basic process of estrangement and alienation. Marxist thought also raised for the first time the concern about the machine-like exploitation of animals (Sinclair, 1905 / 1946).

Many aspects that have emerged through the years are still relevant today. What differs is that they are functioning under new conditions. Two of them are the major changes that characterise contemporary food chain systems. Firstly, the fact that the materiality of food has inevitably changed, and food has materialised into industrial food products. And secondly, that we ourselves have become consumers, acting indirectly and from a distance on the systems of food production.

Three basic trends in today's food ethics can be outlined: the dietetics, the idea of moral contamination, and the awareness of the social dimension. Dietetics, as the Greeks initially shaped them in the art of temperance, is still a part of current food ethics. However, a major change has occurred due to the introduction of the scientific point of view. Contemporary dietetics relies not on the subjective experience, but rather on exact measurement, as well as labelling practices, informing consumers about the ingredients and components of the food they consume. It has been quantified and objectified.

The idea of problematic and unproblematic, contaminated and uncontaminated, good and bad food, i.e. the binary distinction that was firstly found in the Bible, still exists and goes side by side with dietetics. Various examples occur, confirming this binary logic, in vegetarianism and the rejection of non-kosher food products. The rejected products are regarded as contaminated, not simply because they are unhealthy, tasteless or hard to digest, but because they are made from animals. This is a form of intrinsic objection against the consumption of certain animals. However, scientific support, today, provides tools for extrinsic objections strengthening this position. For instance, a cow or a pig consumes a much greater amount of energy than they eventually produce (as meat, or other products), so a reduction in meat consumption could help to diminish the global problem of food scarcity.

At this point the importance of labelling is introduced. Due to the increased distance between production and consumption of food and due to the massive production of novel foods, consumer dependence on food providers has increased to a considerable level. Additionally, significant changes in the food production system have been introduced with the use of pesticides, artificial fertilisers, preservatives, genetic modification, and other forms of biotechnology. A whole range of morally dubious and potentially problematic food has been produced. The moral implication of theses developments is that a food ethic based on the above mentioned binary logic would rely more and more on labelling practices.

Finally, the social dimension, as discovered in the nineteenth century, is of considerable importance to contemporary food ethics. A strong tool of criticism in food ethics, for instance, towards biotechnology, is the social impact. This social impact can have more than one face, for example, public safety, public health, the economy, and the environmental effects (biodiversity, extinction of species).

Labelling practices are developed to inform and, if possible, to reassure the public about the moral identity of the food products whose production remains to a large extent "buried out of sight".

### 15.2.2 Ethical evaluation

*Ethical evaluation scheme*
The questions that arise are: how can an ethical evaluation be done in technical terms, taking into account the contributing factors, and how can a conclusion be drawn from it?

The Second Report of the Food Ethics Council made an attempt to develop a scheme to evaluate ethics systematically. The report categorises ethical problems in three basic groups that are ethically affected in combination with three basic ethical values. The three groups, which are ethically relevant, are: 1) the ecosystem (the biota), 2) the producers, and 3) the consumers. The three basic ethical principles which should be taken into account, are: 1) the principle of well-being (corresponding to issues prominent in utilitarian theory), 2) the principle of autonomy (corresponding to the notion of rights advanced by Immanuel Kant), and 3) the principle of justice (based on John Rawls' justice theory).

These counteracting groups and ethical principles give a scheme, a so-called "matrix", which can be called an ethical evaluation scheme, and which is schematically presented as follows:

## Ethics in food safety

|  | Wellbeing | Autonomy | Justice |
|---|---|---|---|
| The biota | Conservation (protection of the biota) | Biodiversity maintenance | Sustainability of biotic populations |
| Producers | Adequate income and working conditions | Freedom to adopt and not to adopt | Fairness in trade and law |
| Consumers | Safety and acceptability of food | Choice (e.g. by labeling) | Universal affordability of food. |

Mepham (2000) adopts the same scheme adding a fourth group that is also ethically affected:

|  | Wellbeing | Autonomy | Justice |
|---|---|---|---|
| The treated organism | Welfare issues (i.e. animal welfare) | Behavioral freedom | telos |

In the above addition, the two first elements (well-being and autonomy) obviously refer only to animals (e.g. there are no "plant welfare issues"). The third one, *telos*, was inspired by the Aristotelian term telos, interpreted as the meaning of life. Telos shares many values with the concept of "intrinsic value", and with what Kant's terms mean "each of us is an end in himself" (if, of course, this Kantian term is expanded not simply to human, but also to animals).

M. Hayry (2000) uses this Food Ethics Council - Mepham matrix to make an analysis of how this can be used in practice. The example he uses is the introduction of BST (Bovine Somatotropin, or Bovine Growth Hormone). Cows are injected with a genetically engineered hormone (rBST) in order to produce an increased milk yield. The ethical analysis of M. Hayry concluded, within the lines of the matrix, that the use of rBST is ethically unacceptable. Hayry also uses an alternative pure consequentialist analysis (though not claimed to be superior to the matrix) to draw attention to problems that the matrix did not address.

Still, questions remain, firstly about the adequacy of the afore-mentioned matrix, and secondly about how straightforward it is to make an evaluation following the scheme. Mepham, with respect to the evaluation and decision-making, wrote:
*"[...], certain ethical principles may appear to be rejected, others are infringed and any national decision procedure will require that the relative weights assigned by different parties to the different impacts be made explicit. [...] It is even conceivable that, in a qualitative sense, opponents and proponents of a prospective technology might agree as to the nature of the impacts (i.e. whether positive or negative) but differ in the significance assigned to them. [...] but even disagreements at this stage can be constructive steps in the overall process of democratic decision-making."*

Conclusively, the scheme is more a method for a spherical approach to the ethical dilemmas that exist, than a decision model. Mepham himself agreed with this conclusion:
*"[...] but the framework should not be viewed as a decision model. Rather it guides ethical deliberation and identifies areas of dispute."*

Hayry admits that, in the end, the analysis and decision will remain reduced to a utilitarian calculus level. The proposed models have not escaped criticism. However, there is a need for a common language, as well as some tools to proceed with. And these are what the Mepham - Food Ethics Council matrix provides.

*Criticism of the Food Ethics Council "matrix"*
Vicky Fraser (2001) made a three-point criticism with respect to the ethical evaluation procedure: ethics application has gone through (1) hyper-commodification, (2) hyper-rationalisation, and (3) hyper-simplification.

According to Korthals (2002): No analysis of principles and norms is sufficient for a synthesis or construction of ethical solutions. It remains a narrow approach.

*General issues in decision-making*
Decision-making is quite a complex and time-consuming process. The first stages of information gathering and processing, and the observation and analysis of possible actions, play a key role in decision-making because they will more or less define the data that need to be assessed. Investigation and information gathering pose problems of (1) access to sources, (2) of information filtering, and (3) of selective diffusion and formalisation. Information is a means of choice.

Beyond the technical difficulties of information collection, the major problem arises from uncertainty, either because information is imperfect, or because the process of acquiring it is costly.

Furthermore, the procedure follows a kind of rationality, which nevertheless depends on procedural subjectivity. Its consequence is the fact that the export of optimisation rules may be inefficient in terms of time and cost, when cognitive capacities are limited by means of calculation and periods of memorising.

The solution, in a situation that seems to be a dead-end, is that the decision should hang upon satisfaction rather than optimisation. This mechanism occurs as a psychological and cognitive tool in every rational individual. The solution that is to be found this way may not be the optimal one, but it always satisfies a minimum standard of acceptability. The only *a priori* demands are those of the satisfaction of certain standards or levels. From the moment these are met, the choice will be the best available (and not the best possible).

Decision-making, although always based on rational procedures and estimations (mostly in a utilitarian context, i.e. a cost / benefit estimation) may conclude in quite opposite outcomes. None of them can be condemned, and it is even impossible to characterise one as better, or possessing more validity than the others. These various and contradictory outcomes can be justified through many mechanisms:
1. Rationality as an idea may be a common thing, but it can always follow different ways of assessing a problem. In other words, there may be one kind of rationality but various procedural rationalities.

## Ethics in food safety

2. Different procedures often gain different data about the same problem. Even scientific truth has a variable degree of subjectivity, and contradictory scientific results are a very common phenomenon. Authoritative reports often present contradictory results on the same subject, e.g. conflicting reports on GM food (John Vidal, The Guardian, Tue 17$^{th}$ Sept. 2002). This may sound strange, but it is rather unusual to conclude similar results when dealing with complicated biological systems, where many unknown or unexpected factors interfere. Thus, elimination of contradictions is unlikely to be just a matter of further empirical research data, but rather a critical discussion of the methodological tools used for the research.
3. Decision-making always indulges the idea of trust or distrust. Whom we trust depends further on our cognitive judgment, and also on emotional mechanisms that are of course completely subjectively defined.
4. A crucial factor in decision-making is the way the values are put into an hierarchy. And the hierarchy of values can never be judged in an objective way, as each one of us prioritises different values in life.
5. Generalisation of rational behaviour, when faced with uncertainty, has one unique but often insufficient tool: risk assessment. And there is in almost every case a situation where uncertainty cannot be translated into probability. Within this situation rationality does not offer any additional tools. Solutions will thereafter be a matter of more than pure cognitive mechanisms, mechanisms that indulge intuition and emotion.

The question that arises is, how do we come to a conclusion with a solution that embraces the collective dimension? How are we going to distinguish and guard the social choice within the terms of democracy?

The procedure includes three general steps:
1. Scientific analysis. The experts' role consists in attributing subjective probabilities to events.
2. Social dialogue. This includes information diffusion, discussion (debates, questionnaires) formation of an informed public choice, public response.
3. Normalisation of the public choice. Formation of regulations, norms and legislation. Development of checking mechanisms.

### 15.2.3 Food safety ethics and consumers

*Introduction*
When it comes to a dialogue, consumers are the crucial entity, for three reasons. The first reason is that consumers are the final recipients. Secondly, because it is the most numerous group involved. Finally, consumers have significantly increased their power over the years.

It is not easy to understand and analyse the role of the consumer in food ethics. And this is due to several reasons. First of all, consumers do not consist of a uniform group, but rather of numerous groups with different education, information access, economic status, different customs and ways of life, and therefore with different concerns about food, different beliefs and different attitudes towards various situations. However, an attempt can be made to make generalisations and cover spherically the role of consumers as a regulating entity in food ethics topology, and to discuss the factors that interfere with their attitudes.

We have already seen that, during the nineteenth century, food ethics were enriched by the social factor. It is a matter of fact that ethics were always mirroring social perception. What has changed is that in the industrial and the high-technology years a new role has emerged, the role of the consumer. Contemporary western societies are characterised and are dependent on consumption. Goods (of all kinds) are produced at incredible rates and are (or at least should be) consumed with corresponding speed. The whole market system is organised in a consumer-centred way; it tries to convince the consumer to consume the products. These result in a consumer that (potentially) has power.

Even within this contemporary period of consumption, consumer attitudes have historically changed. In the 1970's price was all-important when considering which product to buy. With the boom of the 1980's choice was becoming a more important consideration with convenience high on the purchasing decision list. In the late '90s, along with choice and competition, information and loyalty were becoming an important part of the buying decision (Rossiter, 1997). Korthals (2001a) adumbrates the image of the contemporary consumer as the "floating consumer". This image, which he characterises as a normal phenomenon, can be described as "depending on lots of considerations, often difficult for someone from outside to grasp, at different moments the consumer makes ambivalent decisions". This characterisation confirms the difficulties inherent in the effort to understand the consumers' role.

At this point, it is useful to start with consumer concerns and positions about new technologies (like genetically modified food). The reasons for distrust are the technology and its products at various levels. These issues have been extensively discussed and reviewed (Reiss and Straughan, 1996; Frewer and Shepherd, 1998). However, the following generalisation can be made about consumers' distrust. Straughan (1990) analysed consumers' ethical concerns about food production by asking three questions: Is it safe? Is it fair? Is it natural?

Consumers' first concern about the new technologies, according to most analyses, is the matter of safety and the risks that are implicated in every novelty in the technology. This is a natural response, the expression of the self-preservation instinct. Religious taboos are also of equal importance, as long as most people follow a religion, and most of the religions have food-oriented taboos or objections (mostly of an intrinsic nature) about interfering with and modifying the Creation. These two sets of arguments (safety and religious objections) are of capital importance, and this becomes obvious from various research projects into the public perception of GM food (Scott, 1997; Ellahi, 1994), all of which find them to be in first place. Thus, it is not incidental that the whole debate about genetically modified food is focused, on the one hand, on risk and safety, and on the other hand on respect for people's religious beliefs.

The main difference is that the former are extrinsic concerns, and therefore can differentiate and change according to the information that people receive, the influence of the media, incidents that are related to genetic modification (e.g. disease incidents or environmental

## Ethics in food safety

accidents can raise more objections, and group reactions). Moseley (1999) concludes that at the consumer level, there are worries about the future safety of the technology, i.e. not a worry about the technology but rather about its consequences. In contrast, religious taboos are unlikely to change.

*Risk, safety and consumers*
Rossiter (1997) said that since the introduction of genetic engineering in the food chain, all sorts of comments like 'scary science' and 'can we trust the scientist?' emerged. This is proof of two sets of consumer concerns that are inextricably linked; the matter of risk-handling and the matter of trust and distrust.

Regarding food safety, and food-related risks, consumer attitudes towards this matter seem to vary and to depend on various factors. Thus, the way consumers cope with risks is dependent on the social relations of everyday life, of which consumption practices are part (Halkier, 2001). Therefore, Halkier (2001) concludes, risk handling in food consumption is characterised by ambivalences and is socio-culturally broader than the cognitive rationality.

Some research has been conducted into consumers' individual differences concerning the perceived risk (Halkier, 2001; Frewer et al., 1994) and attempts have been made to group them into common attributes. Halkier (2001) distinguishes three types of consumer risk-handling. The first is the worried approach, where consumers follow the public debate and obtain information that they see as problematic. Risk in food is, in this situation, a subject of negotiation and conversation. The worried risk-handling consumers typically identify with being societal actors, capable of acting and making a difference through their role as consumers. The second type of risk-handling, in Halkier's analysis, is the irritated one. Here, risk is primarily associated with risk communication in the media, campaigns and leaflets, but also includes interpersonal risk communication. Consumers are frustrated that risk communication creates dilemmas in their everyday life and they often can't see a solution. In these cases consumers have difficulties in seeing themselves as societal actors through their consumer role. A diffuse lack of trust is expressed in the accumulated amount of knowledge. This attitude leads to a situation where the consumer feels that no matter what he buys, the product will always be related to a certain risk, over which he has hardly any control. The third type is the pragmatic type of risk-handling. Here, risk is understood as a daily-life problem along the lines of all daily-life problems. Food risk, in general, is a potential problem with which one must have a sensible relationship in a busy everyday life. The trust in this case is placed with the authorities or with people that they (the consumers) know personally. A similar type of attitude is to be found in what Gronow (1997) characterises as the 'modern individual with tolerance'.

In the Frewer et al. (1994) analysis, technological and environmental risks are perceived as being greatest by egalitarians (individuals with a positive attitude towards equality of social conditions). This attitude opposes that of the individualists (who express support for continuation of economic growth as a mean of improving the quality of life) or hierarchists (those who trust and have great respect for organised controls and institutions). Fatalists, on the other hand, see themselves as being outside the organised social frameworks, and therefore they feel they have no influence over technological risks.

These generalisations are sometimes over-simplifications, however they can be useful tools in some cases, especially in food policy planning or in educational or marketing planning.

### 15.2.4 Consumers, trust and distrust

The above analyses, about consumers' attitudes towards food risks, raise another relevant question, the question of trust and distrust. Giddens (1994) stresses the value of active trust: "...trust has to be energetically treated and sustained". Active trust ordinarily presumes a process of mutual narrative and emotional disclosure.

Societal structures play a key role in trust-relations formation. In relative terms, Almas Reidar says that we have moved from an "industrial society" to a "risk society". The difference lies in the structure: while the former was structured through social classes, the latter is individualised. Individuals handle risks by composing their own risk identity profile. And it is up to the individual to choose which experts to believe before buying and eating. It is no longer the case that all experts agree and that public food regulatory bodies can tell people what to do. What mainly affects and alters consumers' trust and distrust is: 1) information (information *ex ante*), and 2) food-related incidents (information *ex post*).

Information and trust are intertwined. The magnitude of the information in each case rests upon the trust in the information source. Frewer *et al.* (1996) have argued that source credibility is one of the most important factors in effective risk communication. Conversely, confirmation or rejection of information by everyday reality alters our degree of trust for the information source. Rossiter (1997) reports an increase in information provided to the consumer by opinion-forming organisations (such as non-governmental organisations, consumer organisations and the media), which made consumers feel better informed, but also more insecure, and left them feeling caught up in food scares.

Furthermore, in some areas information and trust are complementary entities. A consumer's right to be informed with respect to his food has certain boundaries. This right should therefore be supplemented with trustworthy procedures. This need becomes even more persistent because the consumer himself / herself cannot judge every scientific claim.

*Consumers' freedom of choice*
Straughan (1995) has mentioned that freedom must always be a relative term, whatever its context may be. Total freedom is an unattainable and unimaginable concept. Therefore, one needs to understand that limitations on freedom will always exist, just like the limitations in all other areas of life. Advertisement and various marketing techniques (such as supermarket lighting, music and location of products) can undeniably be used to affect overall consumer purchase, although it is difficult to claim that these techniques wholly determine the choices of the shopper.

## Ethics in food safety

The questions are: what is the degree of freedom consumers ought to have, and what constraints are justifiable and acceptable? When it comes to consumer choice the conflict is always between paternalism on the one hand and autonomy on the other hand, i.e. between the desirability of making judgments about the best interests of others and of allowing the individuals to make their own judgments. Straughan (1995) correctly claimed that in practice almost nobody would like to rule out all possible restrictions. For instance, the ban on a food containing a lethal ingredient (although restricting one's choice) would encounter no objections. Of course, in everyday life things are not always so obvious and clear-cut when it comes to restrictions on food choice. What does free choice mean?

First of all, the process of choosing in itself has costs, such as gathering and processing relative information which requires specialist knowledge and techniques that require money, time and effort. Apart from the economic costs, there are also psychological costs for the individual consumer. Trying to gain and manipulate relevant data may lead to information overload, or even worse produce concerns arising from the sense of responsibility and associated with worry and anxiety that one may make a wrong/bad choice. On the other hand, being faced with a wide range of alternatives can be disturbing and disorientating. Here, we are confronted with what is called the "fear of freedom", a situation where people are unable to tolerate the uncertainty involved in having to make a choice from so many alternatives. Thus, the conclusion is that freedom of choice cannot be equated with breadth of choice. A reasonable selection of alternatives would cover all aspects of the desired characteristics.

The maximum range of choices would secure the very best buy. However, some economists and philosophers have argued the merits of a policy of "satisfying" rather than optimising (i.e. to find the "good enough" rather than the "best" possible option). This option can be applied to questions about the consumers' choices. The 'satisfier' consumer will pursue the first reasonably satisfactory option, and will not bother finding the optimum alternative. The question that arises, though, is whether consumers are mainly satisfiers or optimisers.

Another problem that is imposed by the freedom of choice is that, if freedom of choice is something to be valued and respected, what is to be done with those who choose to relinquish all or some of this freedom. It is obvious that there cannot be a simple answer, or a single solution with respect to the freedom of choice. However, considering that not all people have the same shopping habits, nor do all have the same criteria and the same attitude when choosing (satisfiers vs. optimisers), freedom of choice as a consumer principle is not concerned solely with the range of alternative products. Sensitivity should rather be shown towards preferences that lie outside the norm.

Various kinds of psychological factors are seen by some people to offer a justification for limiting the consumer's freedom of choice, but this also raises counter-arguments. The foundation of these counter-arguments is that what is needed is not a dilution of the principle, but an educational program to enable all consumers to live with and benefit from the widest possible range of choice.

Practically, consumers' freedom of choice can and should be restricted when it comes to their safety. (Although the question that remains is whether the consumer should be free to choose unsafe or harmful products.). Concerning the safety of a food, two opposite extremes can be distinguished with reference to the consumer's freedom of choice:
1.  all potentially harmful or risky food should be absolutely banned, and
2.  no products whatever should be banned, but instead consumers should be given all available information with respect to safety and risk.

These two are a translation of the paternalism versus freedom dilemma into practical terms. Although the first opinion seems to be attractive, it fails to follow a simple logical point; that there is no such thing as absolutely safe. All foods include a degree of risk. We should always talk about probabilities of danger (or safety). Thus, it is logically impossible to follow the first position in strict terms.

The second viewpoint is also practically impossible to follow. And this is due to the fact that few, if any, consumers would welcome such total freedom (for reasons that have already been mentioned). Nobody could stand an everyday life in which a high probability of potential harm has to be encountered. Therefore, it is not surprising that the overwhelming majority of consumers are in favour of safety regulations.

After demolishing these two extremes, there is a need to establish a framework in which restrictions occur, but freedom of choice is maintained. Therefore, we come to the problem of determining the risk / safety level and evaluating where to put a limitation, a ban from a certain degree of danger and higher.

And automatically, we face the enormous complexities involved in determining the level and extent of risk. First of all, this happens due to the impossibility of certainty about cause and effect. The variability and complexity of the biological systems (in which the food as well as the consumer is included) increases the difficulty.

However, risk should always be estimated, and cause/effect probabilities should always be evaluated, as well as the number of individuals judged to be at risk and the level of intake at which the risk appears to become significant (in terms of probability). The higher the values of the former parameters, the higher the need for a restriction in consumer choice (banning the product). Technically speaking, the source of risk should be evaluated (whether the product is intrinsically "risky", or becomes risky in certain forms, or whether risk implies defects, pollution or poor quality). The strongest cases for restrictions are foods that are included in the third category (defective, polluted, or bad quality food). Legislation and food policy decide the level of risk that is allowed within the food chain. Thereafter, the consumer is the one who will do the calculations, balance the risks and benefits of the food, and make his choice.

*Consumer responsibility*
It is obvious that the role of the consumer, regarded not as a passive being but indulging consumer concerns, and consumer sovereignty, does not simply imply rights but also demands

a certain responsibility. In reference to the consumer grouping from Frewer *et al.* (1994), (see Risk/safety and the consumer), the fatalists make up the group that completely refutes its responsibilities. All other consumer groups show a distinctly strong position against food risk cases.

Responsibility can be defined as a personal commitment in a certain action. It demands presence, attention and vigilance. Our actions can change the course of events; we are responsible for the evolution of the world, for we do (or do not) participate in the choices, actions and deliberations that make the world what it is. As far as health risks are concerned, the notion of legal responsibility is limited. Moral responsibility goes together with rights and duties, imperatives and moral injunctions.

Consumer responsibilities have to do with information and social indulgence. Citizen's information is formalised knowledge, leading to an informed, reasoned judgement, which is socially useful. Concerning risks, focus groups, citizens' conferences and juries have tended through institutional formalisation, to bring about a common language and aimed at achieving a consensus. According to the ethics of discussion (as developed by the German philosophers Appel and Habermas), the truth is the result of a consensus achieved by a discussion of participants. Furthermore, consumers have the power of pressure, and it is their responsibility to use it correctly, to take a reasonable approach to risks, and to make an informed choice. The common informed choice would be the social choice in each case.

Of course, a degree of citizen judgment is relinquished to politicians and experts (or, in other words, to government and scientists). However, no transfer of responsibility can be taken as an excuse for either of the participants in the discussion. The roles are more or less well defined. The scientists and experts propose certain choices; they are not authorised to decide. The public has the duty (and right) to be informed of the imposed risks, to participate in the discussion, to make a commonly accepted informed choice, and to bring forth its common interest. The final decision should be a common consensus. The legislative system (lawyers, country with its food politics) has to ensure and normalise this consensus. It is also up to the legislation to take into account and protect the rights and interests of consumer group minorities, when these are not in conflict with common rights. Then we pass from a situation of risks suffered, to one of acceptable risks, to a notion of risks consented to.

*Consumer sovereignty*
We have already seen the key role that the consumer plays in food ethics. Consumers possess the power to direct things toward certain ethical decisions. The term consumer sovereignty is a new term and is used to describe the power of the consumers' act and how their attitudes are justified. The production system is increasingly confronted with consumers who not only buy (or don't buy) goods, but also demand that those goods are produced conforming to certain ethical standards. These standards may differ incredibly, and may also change depending on the signals the consumers get.

There has always been a traditional distinction between the two roles: that of the citizen, considered as a political and active entity, and that of the consumer that was firstly seen as a passive and apolitical being. This traditional distinction is to be found in many analyses, right from Rousseau up to contemporary analysis (such as the writing of the Norwegian John Elster).

Very characteristic of the way the consumer was formerly regarded is seen in the classical position of Schumpeter in 1911: *"It is the producer who as a rule initiates economic change, and consumers are educated by him [...]. [...the consumers are] taught to want new things, or things which differ in some respect or other from those which they have been in the habit of using."* (Schumpeter, 1999, p. 65). The term consumer sovereignty was introduced in order to overcome this conceptual dichotomy of consumer and citizen, and to view the consumer in a new light.

Korthals (2001b) made an analysis of consumer sovereignty by expounding two concepts of the term. The narrow liberal one conceptualises consumer sovereignty as the right of the individual consumer to get information on food products and to make his or her own choice. Within this approach, there is a very strong emphasis on rules and principles with respect to the autonomy of individuals. According to Korthals (2001b), this traditional concept fails to sufficiently cope with problems of limited possibility to select the relevant information, and with problems of trust in credible food institutions. Furthermore, this approach is centred on views with respect to risk and safety. But consumer concerns also presuppose that consumers regard food as an essential part of theirs *modus vivendi* (way of life). And this means that further to the risk / safety issues, other concerns also come up, like animal welfare issues, or environmental stewardship practices.

In the broad liberal view these initial concepts are enriched with values, preferences, practices of care, and involvement. What changes in the latter view is the level and the way of communication. Public debates and public committees are tools, are energetic acts, expressions of consumer sovereignty. Their aims basically indulge information diffusion, the creation of some value orientation, and the ability to act, to exert social pressure.

Public communication is primarily important, through public debates whose aim is information diffusion. With the concept of public debates a new dimension enters the ethical analysis, and this is because the attention of ethicists moves towards formulation of criteria of successful and rational public debates. Further to the debates, public committees are proposed, in order to create some value orientation to guide the people involved in balancing the different ethical claims.

## 15.3 Special issues in food ethics and safety

In this section the focus is on specific issues that are related to food ethics, like the ethics of genetically modified food, labelling and ethics, and finally some remarks on food ethics and patenting.

*Ethics in food safety*

### 15.3.1 Genetically modified food and ethics

The food industry has been traditionally positive about the application of biotechnology, as far back as ancient history, in the production of bread, cheese, beer and other fermentation processes. What has changed in biotechnology in recent years is the transformation that it underwent by the application of molecular biology. New techniques have evolved for the modification of the genetic material of living organisms. This modification consists of the insertion into the organism of foreign genetic material. The organism that hosts the foreign genetic material (the foreign nucleic acid) is a Genetically Modified Organism (GMO) and carries new characteristics compared to the non-modified counterparts, due to the expression of this foreign genetic material (fore more details on GMOs see Chapter 5).

One can easily accept the fact that genetic modification has caused a worldwide revolution in many traditional systems in the food and agriculture sector. Where then does the problem lie?

As with most new technologies, the potential and future advantages are presented first. This does not happen by chance; this is the tool used to convince us of the positive effects of the innovation (otherwise no introduction of new technologies would ever take place). FAO (2001) states that acknowledging the potential (of genetic modification) is not to ignore the risks it implies, and furthermore that "...like all new technologies, (the GMOs) are instruments that can be used for good and for bad...". The FAO concludes: "The development of GMOs raises perhaps the broadest and most controversial array of ethical issues concerning food and agriculture today".

M. Reiss (2001) made a serious attempt to cover all ethical sensitive considerations with respect to GM food. Firstly, he made a distinction between the separate stages of production and use of GM crops, as follows:
- research by scientists;
- commercial development;
- regulatory approval;
- planting by farmers;
- the stocking of GM food by retailers;
- the purchase and consumption of GM food by consumers.

This separation into stages is a helpful tool with which to make a spherical approach to the ethical dilemmas. Therefore, this scheme of M. Reiss will now be used to analyse the relative ethical dilemmas. The general question that is posed herein is, what are the ethical obligations, so that the GM food is faced from an ethical perspective.

In the research stage, this question is transformed into, whether or not it is ethically correct to proceed with research. Reiss raised four main general ethical reasons for allowing a particular piece of research to go ahead, and respectively four arguments against research continuation.

The arguments for research continuation, i.e. against a research ban are:

- The freedom of scientists is based on the ethical principle of freedom, according to which one must have very strong arguments for banning things. It is noteworthy to mention that we allow people to doing things even when we know that it would be better for them, in some sense, if they didn't (smoking for example).
- The autonomy of scientists is the argument that represents the belief that everyone does the best possible if he /she acts autonomously. This does not imply, of course, that scientists choose their research without any constraints.
- The intrinsic quest for knowledge is a third argument, according to which research, discovery and acquisition of knowledge is the duty of the scientist (and of human beings in general).
- The fourth argument for research continuation is the usefulness of information / ideas gained, which is the ultimate utilitarian argument in this direction.

Respectively four arguments are raised against GM food research:
- It would be wrong in itself to want to do the research. This is an intrinsic objection against research of a certain kind. Maybe this argument seems to carry a paradox, but if the subject is "research on more effective ways of human or animal torture" the argument becomes obvious. If research into genetic modification can be considered as violating the basic principles of life, this argument has a strong validity.
- The process of the research itself would have unacceptable consequences. To conceptualise this argument we should think about research that can be useful, positive from the point of view of its results (for instance, research on human psychology) but harmful, or negative for those involved in the research, negative in its process (research on human psychology which would involve excessive deception of the participants).
- The net consequences of the results would be harmful. Herein we have the utilitarian argument taken from the other side, the utility of the consequences.
- The money could be spent better elsewhere. This is a very widely used argument against many forms of research.

At the stage of commercial development, we encounter a wealth of literature on business ethics (e.g. Jackson, 1995). At one extreme, it can be argued that "the business of business is business"; in other words companies have a duty only to those who own them. At the other extreme, for example from certain utilitarian perspectives, it can be said that companies have a duty to ensure that they operate so as to maximise global happiness/utility. The ethical limitations of one or both of these two positions underlie the existence and function of companies involved in food production (including GM food production).

A rather different approach is to adopt a consumer or stakeholder focus. This pragmatic approach is based on the belief that a company is viable only if it satisfies to a certain degree the customers, and only if it maintains sufficiently good relationships with its stakeholders (customers, employees, regulators, suppliers, etc.). From this perspective, companies should not just consider the "bottom line" of financial performance but should also talk about the "triple bottom line", taking into account their social and ethical responsibilities. This approach offers a less economically focused, narrow point of view. Companies genuinely seeking to fulfil what they feel are their social / ethical responsibilities, e.g. fair treatment of employees, and reduction of pollution, still strive to be financially successful but take on board a wider range of aims and objectives.

At the stage of regulatory approval, the ethical responsibilities of a regulatory/approval system are considered. In other words, what we ask for here are the criteria and the ethical obligations of the law/legislation. The central obligations of a regulatory system are (i) to prevent certain harms, and (ii) to provide especial protection for those unable to take legal actions against those responsible for harms. A further duty of a regulator may be to allow people to act autonomously. A big issue here is the need/obligation around food labelling, which is discussed later in the chapter. At this point we have to give an answer as to when can or when should the Authorities interfere as a regulator against someone's will. An answer in this discussion was attempted with what is called the Harm Principle, an approach that has not been without problems. The EU authorities have, therefore, adopted the so-call Precautionary Principle, for risk assessment of novel foods.

At the stage of the farmers/planters involvement, the particular duty of environment preservation arises. It includes two main aspects: the preservation of soil and the resources in general, and the preservation of wildlife. The preservation of soil, water resources, etc. is pretty much also a duty of self-interest. Furthermore, farmers also have the duty to reduce fluctuation of yields (by proper handling), e.g. two very bad harvests after twenty years of good harvesting would lead to high rates of malnutrition, suffering, and even death.

The producers - retailers of food can be considered as commercial companies and what has been said about their duties / ethical responsibilities also applies here. A central focus here is the obligation of farmers / retailers to provide the consumers the right to choose between GM food and food which is not genetically modified, and therefore to produce and market both of them. Practically speaking, retailers do infringe this right. Of course, the counter-argument is that the market provides the consumers with what the majority asks for. Do producers have the duty to produce and the retailers the duty to stock certain products for a minority of consumers even when it would be commercially better for them not to do so? It is difficult to argue for such a duty, especially as producers are not charities, nor is that their reason for existence. Here, the role of the government/country/legislation system enters again to protect minority food interests.

The duties / obligations of the consumers have already been discussed.

*The Harm principle*
The Harm Principle is based on J. S. Mill's philosophy, according to which: "the only purpose for which power can be rightfully exercised over any member of a civilized community, against his will, is to prevent harm to others." When applied to the issue of GM food, the state's action will be to prohibit manufacturing and commercialisation of these foods in cases where they cause some harm. What has been complicating things is defining the term "harm". Holtug (2001) attempted to characterise harm as *a* quantity of welfare. This quantitative utilitarian interpretation is problematic. Besides the difficulty of quantifying welfare, there is a problem in defining the

meaning of reduction of welfare. In order for a person to be harmed, he must be rendered worse off, but worse off than what? This is called the baseline problem. For example, in the case of GM food, the question is translated as "should we compare GM crops-produced welfare (or harm) with the welfare (or harm) produced by conventional crops or by organic crops?"

A second attempt to define "harm" is as quality of welfare. Here, the reduction of quality of welfare would mean a setback in people's interests in terms of basic desires. This would imply that the state should intervene against GM food only if its implication for human health was of such a degree that it would damage their ability to satisfy basic desires. On the other hand, neither liberty would be protected, e.g. even a book would be banned in case it was considered to pose a threat to some of the basic desires of some people!

A third attempt to characterise "harm" is to make a moralisation of harm. However, in order to incorporate such a moralised concept of harm in the Harm Principle, we need a moral theory. Which means, we need to know what makes acts right or wrong. Thus, even within this interpretation, the Harm Principle does not give straightforward answers. The Harm Principle by itself fails to provide a comprehensive moral basis for regulating GM Food.

*The Precautionary principle*
In February 2000, the European Committee adopted the so-called Precautionary Principle. This principle does not imply that we should not take any risks in the area of new technology, but that in the absence of sound assessments of risks (and contingency plans for dealing with undesired impacts) it is both foolhardy and unscientific to proceed rapidly with the introduction of powerful new technologies. In general terms, significant uncertainties about the effects of novel foods on human safety and the environment suggest that there should be a presumption against their use unless good reasons can be advanced to overrule this. It can be summarised as the "No, unless" principle.

Susan Carr (2002) describes the Precautionary Principle as the balance between scientific and ethical/value-based aspects. She distinguished three stages of precaution. The trigger stage is where initial judgments and initial identification of potentially harmful impacts can take place. The second stage is the decision stage, when it is decided whether or not to adopt the precautionary principle. At this stage a political decision is taken, which may have to respond to varying kinds or degrees of public pressure. Within this stage a balancing act takes place between scientific data and societal values. Decisions should be taken after clarification of how societal values and democratic decision-making can be integrated with scientific risk assessment. The third and final control point is the application of precautionary measures stage. Here, the guidelines to be followed are no different from those that should be followed when there is a complete risk assessment.

The application of Precautionary Principle in the European Union has led to a polarised debate about GM food.

## 15.3.2 Labelling and Ethics

The first question that arises, with respect to food labelling in general, is whether labelling is an ethical issue. And if it is, how is it grounded?

The consumers' demand for food labelling has definite ethical roots. Religious restrictions, for example, are one of the major issues for labelling demand. Hygiene issues are another group of issues. Thus, for instance, allergen-containing food can be lethal for certain people and therefore such foods must be labelled. Also, the legislator and producer, just like the consumer, must be aware that people who lack certain enzymes cannot digest certain food components (for example, phenylcetonuric people that cannot consume any phenyl-alanine source, or people that cannot metabolise milk). This is also an issue for labelling. Food safety also matters, and thus the food label is required to inform the consumer about the time point beyond which the food is unsafe to consume, i.e. the shelf life of the food. All these issues have created specific labelling demands. It is obvious that religious, hygiene and safety issues have an ethical background. These labelling demands increased as the food being consumed became more and more complicated by the use of food technology, that in many cases transformed initial food sources into non-recognisable derivatives, or mixed materials to produce a different final product.

Thomson (2001) gives another dimension to the question, i.e. to what extent is labelling a matter of ethics, claiming that this depends on the ethical approach. He distinguishes between the utilitarian and the deontological - Kantian approach with respect to their position about food labelling. According to the Kantian approach, informed consent is required for food consumption (as for all things in life with an ethical perspective). Therefore, this is translated into obligatory food labelling. The consumer should get adequate information. The criteria for informed consent are complex and can become quite technical. Firstly, they range between diligent efforts to apprise affected parties and the minimum condition for exiting the situation. Secondly, they should respect a broad range of cultural and religious values.

Thomson contrasts this with the utilitarian approach, where labelling is not of ethical importance unless it can differentiate positively the utilitarian maxim (to increase the benefit / harm balance). A utilitarian might think that information should be provided only if it is likely to improve overall public health. Subsequently, it would be unethical to provide information that would lead to less than optimal choices from the standpoint of public health. Or, if not unethical such information would be judged inefficient, since information is costly, and sometimes the marginal costs of more information outweigh its marginal benefit. Thus, arguably, information about a food's conformity with religious dietary law or vegetarian diets might fail to pass the utilitarian test, because such information would have no demonstrable bearing on public health. Conclusively, the ethical background to the food labelling demand is the freedom of choice. And is based on a Kantian - deontological ethical approach.

In this direction, some would obviously conclude that information flow to the consumer should be unhindered and therefore labels should contain all possible information about the products.

Apart from its technical difficulties, this also brings about another counter-argument. Consumers could suffer from information overload. The food industry does not want to drown consumers with information that they do not need or do not want to know in detail (Frewer *et al.*, 1996). The theoretical ideal in information is that which is enough to enable informed choice. This, of course, is not an adequate definition and can therefore give rise to endless disputes about how much information should be considered as adequate.

### 15.3.3 Food and intellectual property

Generally, the need for protection of inventions in food processing is present, because without the chance of reward and profit there would be little or no incentive to make an improvement (which would be pirated sooner or later).

The object behind the granting of patents is the encouragement of innovation and invention. The patent does not permit the patentee to exploit his invention as he wishes, but does stop others using the invention. As Reiss and Straughan (1996) outline, patenting is not a positive right to do something oneself; it is the negative right to exclude others from doing something. The former statement is the negative answer to the question of whether patenting implies ownership. Before the invention can be used, the patentee should also comply with all the rules that control development, manufacturing, and marketing of the product. These rules exist in order to guard public health, safety, ethics, and the environment. Almost any invention can be patented. What cannot be patented are ideas, theories, mathematical algorithms and computer programs, and the laws of nature. Patents for all sorts of inventions are well recognised (chemicals, machinery, processing).

The utilitarian argument is the one that is most often used to justify the patent system. Since society benefits from the disclosure, the use of new inventions as well as from the subsequent economic activity, patents are justified. Any welfare losses due to restrictions are far outweighed by the advantages. The granting of rights is a matter of social convention and not of natural law or distributive justice.

There are many counter-arguments against patenting. It is not within the present scope to analyse them, but a simple reference to them should be made. The patent system can lead to under-investment, over-investment, and the inefficient allocation of resources. Patents encourage "work-arounds" (the development of copycat inventions that only slightly differ from the original patented invention). Also, a serious misallocation of research funds can be a consequence, as applied rather than basic research is profoundly favoured. The legal and administrative costs of the patent system are huge. Disclosure, although it is supposed to be encouraged by patenting, can actually be discouraged, as the possibility of someone stealing your patent greatly increases. This has happened in practice in many cases ("patent first, publish later" mentality has been widespread). Many scientists have learned to keep basic research materials to themselves. The industry development stage inequalities in combination with the patent system enhance inequalities and make it difficult for latecomers to enter.

## Ethics in food safety

Finally, even after strict economical analysis no conclusions can be drawn about the positive or negative effect of the patent system. As Machlup (1958) admitted, "no economist, on the basis of present knowledge, could possibly state with certainty that the patent system, as it now operates, confers a net benefit or a net loss upon the society". Michele Svatos (1996) made an analytical presentation of the above counter-arguments.

*A historical review of food patenting / intellectual property*
- Plants and animals that are used for human food have been developed over the years through natural processes and selective breeding.
- In the mid twentieth century in the UK and even earlier (in the 30s) in the USA, successful plant breeders were given the exclusive right of exploitation of their new varieties by plant variety rights. Animal breeders have not been lucky enough: there has never been an equivalent right for animals.
- Modern technology has allowed more sophisticated and more selective changes to exploited organisms. The establishment of another monopoly, the patent, has protected these changes. Patenting of inventions is an old custom (fifteenth century, Venice).
- Patents in biotechnology were first established for medicinal biotechnology, and food biotechnology followed.
- Biotechnological products were first patented in 1980. The Diamond vs. Chakrabarty case was the first Supreme Court decision where "Biotechnology was recognized as a field of endeavour in which inventions could secure monopoly rewards" (Kenney, 1986; see M. Svatos, 1996).

*GM food patenting and ethical considerations*
- The long-term decline in public research, the privatisation of GM technologies, and the emphasis on crops, needs and priorities of the developed countries, do not bode well for feeding the increasing populations of the developing world.
- The sterility of GM crops (for safety reasons, and for profit reasons), and their high price encouraged by intellectual property rights, may eliminate the ability of farmers of the third world to stick with these crops. This will happen because many poor farmers often obtain next year's crop from previous-year seeds. Thus, the intellectual property rights of the GMO manufacturers and the right of the poor producers to retain and plant second-generation seeds will collide.
- Patenting does not imply ownership. Plant genetic resources have traditionally been considered as common property. But this changed with GM crops, and generated a conflict between the "gene poor" but "technology rich" North and the "gene rich" and "technology poor" south. (Sedjo, 1992). On the other side is the utilitarian approach, where labelling is not of ethical importance unless it can positively differentiate the utilitarian maxim. A utilitarian might think that information should be provided only if it is likely to improve overall public health.

## 15.4 Conclusive remarks

After completing the analysis of food ethics, there is one question that remains: Can we hope for the establishment of international bioethics, with common values? The answer to this question is yes and no! Yes, we can establish what we call cross-cultural bioethics (Macer, 1995). But, no, we cannot apply the same ethical principles in the same way and to any significant degree. Universalism is not achievable in terms of common bioethical truths, and even if it is achievable it won't be correct or respectful of all humans. Indeed, Fraser (2001) condemned this "desire for universal truths".

Instead, it is right to work on the establishment of an international code, or even better, a *carta* (a map) of what is ethical and what is not, of what is acceptable about food and what not, in each civilisation in humanity. Maybe this was the right idea underlying the "Narrative food ethics" that Fraser (2001) proposed, opposing the principled approaches to food ethics.

In the case of Europe, a commonly acceptable food ethics code and policy has already been achieved, and this was relatively easy due to a more or less common history and culture, which shaped common ethical apprehensions. However, even within this general similarity, different food traditions generate oppositions now and then within European cultures. A characteristic case was that of the ban on animal intestine consumption due to the BSE breakout in the mid nineties. This encountered strong opposition in Greece where food ethics traditionally involve various dishes with broiled lamb and caprine intestines.

Furthermore, despite the similarities in ethical views, European countries also wanted to adopt different ethical decisions about GM food. This was the case in 1999, when France and Greece proposed a complete restriction on GM foods in Europe, a proposition that was turned down by the other Members as an unnecessary and illegal action. Instead, the EU decided simply to tighten regulations (Guardian, Fri. 25$^{th}$ June 1999). In conclusion, the EU should proceed with common policies and the establishment of common food ethical codes, but should always take into account the sensitive social individualities and traditions of the Member States.

Purchase (2002) tries to make a future prediction in the area of biosciences ethics. There appears, he says, to be no slow-down in the pace of scientific discovery. Scientific developments will continue to outpace knowledge of their implications for society. The initially calculated risks and benefits will sometimes be proved incorrect. Legislation will lag behind the cutting edge of scientific evolution, because the legislation responds to the public concerns with a delay, due to the fact that it takes time to formulate rules. Thus, Purchase predicts that the scientific evolution-related ethical issues will continue to have an impact. This impact will be negative, when scientists fail to understand the importance of consideration of the ethical aspects of their work before they proceed with it. It is not the public that has to be educated in order to know how important the work of the scientist is. It is more a matter of a partnership of scientists, ethicists and the public, in order to develop an understanding of ethical issues in advance of and simultaneously with scientific development.

*The GM Food Debate - a conflict of food ethics?*
The US and the EU have approached the use of GM foods from different perspectives. The US agricultural community has embraced the use of GM foods while the US public has remained largely silent on the issue. The EU, on the other hand, has adopted a significantly less positive approach. Mostly due to recent public scepticism, the EU has been resistant to the use of GM foods. The differences in approach mirror the different focus in agricultural policy. In the US the agricultural policy is linked to the development of output-enhancement and cost-reduction, whereas in the EU agricultural policy reform provides incentives for lower output and emphasises the quality aspects of the products. Furthermore, food safety concerns have dominated the food-chain system in Europe. In the field of risk management, North Americans adopt a point of view in which the main aim is to educate the consumer, whereas the European viewpoint favours a two-sided communication.

This has led to a polarised debate between the EU and the US focused on two main issues: 1) the approval of GM foods by food safety regulatory bodies, and 2) the labelling of GM foods.
It is not our purpose here to analyse the whole chronicle of this debate. Having gradually tightened precautions, due to public pressure, the EU has been criticised by the US and Canada. The latter viewed the precautions as politically motivated, imposed for the EU's biotechnology sector to allow it to catch up with its North American competitors. Beyond the possible political / economical opposition, there still remain questions. What is the degree of ethical background in such situations, and what is the impact of food ethics on societies and vice versa? And finally, most of all, to what practical extent can food ethics individualities be respected?

## References

Carr, S., 2002. Ethical and value-based aspects of the European commission's precautionary principle, Journal of Agricultural and Environmental Ethics, 15, 31-38.
Ellahi, B., 1994. Genetic Engineering for Food Production - What Is It All About? British Food Journal, 96, 13-23.
FAO, 2001. Building amore equitable and ethical food and agriculture system. Ethics series-1, Rome, pp 26-31.
Food Ethics Council, 1999. Novel Foods: Beyond Nuffield. The second report of Food Ethics Council.
Fraser, V., 2001. What's the moral of the GM Food story. Journal of Agricultural and Environmental Ethics, 14, 147-159.
Frewer, L.J., R. Shepherd and P. Sparks, 1994. Biotechnology and Food Production: Knowledge and Perceived Risk. British Food Journal, 96, 26-32.
Frewer, L.J. and R. Shepherd, 1998. Consumer perceptions of modern food biotechnology, In: Genetic Modification in the Food Industry, Roller, S. and S. Harlander (eds.), pp. 27-46, Blackie, London.
Frewer, L.J., H. Chaya and R. Shepherd, 1996. Effective communication about genetic engineering and food. British Food Journal, 98, 48-52.
Giddens, A., 1994. Risk, trust, reflexivity, In: Reflexive Modernization: Politics, Tradition and Aesthetics in the Modern Social Order, Beck, U., A. Giddens and A. Lash (eds.), 184-197, Polity Press, Cambridge.

Gronow, J., 1997. The Risks and Dangers of Eating: Trust and Legitimacy. Essex: Paper for European Sociological Association Conference.

Halkier, B., 2001. Risk and food: environmental concerns and consumers practices. International Journal of Food Science and Technology, 36, 801-812.

Hayry, M., 2000. How to apply ethical principles to the biotechnological production of food - the case of bovine growth hormone. Journal of Agricultural and Environmental Ethics., 12, 177-184.

Holtug, N., 2001. The Harm Principle and Genetically Modified Food.", Journal of Agricultural and Environmental Ethics, 14, 169-178.

Jackson, J., 1995. Reconciling Business Imperatives and Moral Virtues. In: Introducing Applied Ethics., Almond, B. (ed.), pp. 104-117, Blackwell, Oxford.

Kenney, M., 1986. Biotechnology: The University-Industrial Complex, Yale University Press, New Haven, CT, 257.

Korthals, M., 2002. Ethical Dilemmas of Sustainable Agriculture. Food Ethics Review, 1, 25-35.

Korthals, M., 2001a. Ethical dilemmas in sustainable agriculture. International Journal of Food Science and Technology, 36, 813-820.

Korthals, M., 2001b. Taking consumers seriously: two concepts of consumer sovereignty. Journal of Agricultural and Environmental Ethics., 14, 201-215.

Machlup, F., 1958. "An Economic Review of the Patent System"., Patent Studies no 1, Subcommittee on Patents, Trademarks and Copyrights of the Committee on the Judiciary U.S. Senate, 85th Congress, Second Session, pp. 15.

Macer, D.R.J., 1995. Biotechnology and Bioethics: What is Ethical Biotechnology? In: Biotechnology vol. 12: Legal Economic and Ethical Dimensions. (Vol. editor D. Brauer.), pp. 117-154., Eds. H.J. Rehm and G. Reed., second edition, VCH (Verlagsgesellschaft), Weinheim, D.

Mepham, B., 2000. A framework for the ethical analysis of novel foods: the ethical matrix. Journal of Agricultural and Environmental Ethics, 12, 165-176.

Moseley, B.E.B., 1999. The safety and social acceptance of novel foods. International Journal of Food Microbiology, 50, 25-31.

Purchase, F.H.I., 2002. Review Article: Ethical issues for bioscientists in the new millennium. Toxicology Letters, 127, 307-313.

Rawls, J., 1972. A theory of Justice., Oxford University Press, Oxford.

Reiss, M.J., 2001. Ethical considerations at the various stages in the development, production, and consumption of GM crops. Journal of Agricultural and Environmental Ethics., 14, 179-190.

Reiss, M.J. and R. Straughan, 1996. Improving Nature: The science and ethics of genetic engineering, Cambridge University Press, Cambridge pp. 288.

Rossiter, L., 1997. Consumers attitudes, In: Genetic Modification: Who wants it? Proceedings for an IFST Meeting., Food Science and Technology Today, 11, 247-249.

Scott, J., 1997. Public perception of genetically modified food. Food Science and Technology Today, 11, 237-240.

Sedjo, R.A., 1992. Property Rights, Genetic Resources, and Biotechnological Change. Journal of Law and Economics, 35, 199-213.

Sinclair, U., 1946. The Jungle. Cambridge: Bentley, 1905/1946.

Schumpeter, J., 1999. The theory of Economic Development. Oxford University Press, Oxford.

Straughan, R., 1990. Genetic Manipulation for Food Production: Social and Ethical Issues for Consumers. British Food Journal, 92, 13-20.

Straughan, R., 1995. What's your poison?: The freedom to choose our food and drink. British Food Journal, 97, 13-20.

Svatos, M., 1996. Biotechnology and the utilitarian argument for patents, Social Philosophy and Policy Foundation, USA, pp. 113-144.

Thompson, P.B., 1998. Agricultural Ethics: Research, Training and Public Policy. Iowa State University Press, pp. 239.

Thompson, P.B., 2001. Risk, consent and public debate: some preliminary considerations for the ethics of food safety. International Journal of Food Science and Technology, 36, 833-844.

Zwart, H., 2000. A short history of food ethics., Journal of Agricultural and Environmental Ethics, 12, 113-126.

# Concluding remarks

*Frank Devlieghere, Pieternel Luning and Roland Verhé*

This handbook clearly demonstrates the complexity and multi-disciplinary nature of different aspects of food safety. As a result of several food crises in the last decade, food safety is nowadays an integral aspect of food production besides being an established issue of concern for governments in Europe. Therefore, there is clearly a need for food safety experts at both industrial and governmental level.

Very often, food safety is considered as a cost and indeed a significant financial commitment has to be made by companies to ensure the safety of their products. Moreover, the costs incurred (in the form of recalls, law suits, loss of image etc.) when food products on the market are found to be unsafe, can be very high. The consumers, on the other hand, consider the safety of a food product as an inherent characteristic and are not aware of the great efforts made by the industry and government.

The focus of companies and government has recently shifted towards the relationship between food and health. Although many diseases are linked to the over-consumption of specific food products, food can in contrast be used as an ideal vehicle for the consumption of health-improving agents. The industry has found a new market there, and important research efforts are being made for developing more healthy or health-improving foods. With regard to this important evolution, one should be careful not to overlook safety aspects during the development of new food products. The consequences for safety of changes in food formulations, intended to improve the nutritional or technological properties of a food product, should always be evaluated. Personnel with knowledge of the safety aspects of foods are therefore essential members of any team developing food products with purported beneficial health effects.

When safety aspects are considered, it is essential to have insight into the different potential hazards, and even more importantly to have knowledge of the factors that turn hazards into risks. Decreasing the detection limits of agents causing a hazard, increasing the influence of the media, and product responsibility, are only some of many factors that push companies towards safety control measures that are sometimes extreme. It is therefore essential that decisions can be taken on a quantitative basis and with sound reasoning, as illustrated in this book. Therefore, ideally experts in food safety should not only have knowledge of possible hazards and how to analyse them, but should also be acquainted with related fields such as quality assurance systems, food production systems, modelling, risk analysis, legislation, consumer and ethical aspects.

# Authors

- Anna **Aladjadjiyan**
  Agricultural University, Mendeleev 12, 4000 Plovdiv, Bulgaria, anna@au-plovdiv.bg

- Katleen **Baert**
  Ghent University, Department of Food Safety and Food Quality, Coupure Links 653, 9000 Gent, Belgium, Katleen.Baert@UGent.be

- Rita **Cava**
  Universidad de Murcia, Facultad de Veterinaria, Tecnología de los Alimentos, Nutrición y Bromatología, Campus Universitario de Espinardo-Murcia, 30071 Murcia, Spain

- Anton **Cencic**
  OS Janko Glazer, Lesjakova 4, 2342 Ruse, tone.cencic@guest.arnes.si

- Avrelija **Cencic**
  Department of Microbiology, Biochemistry and Biotechnology, Faculty of Agriculture, Vrbanska C30, 2000 Maribor, Slovenia, avrelija.cencic@uni-mb.si

- Bruno **De Meulenaer**
  Ghent University, Department of Food Safety and Food Quality, Coupure Links 653, B-9000 Gent, België, Bruno.DeMeulenaer@UGent.be

- Frank **Devlieghere**
  Ghent University, Laboratory for Food Microbiology and Food Preservation, Department of Food Safety and Food Quality, Coupure Links 653, 9000 Gent, Belgium, Frank.Devlieghere@UGent.be

- Eva **Domenech**
  Universidad Politecnica de Valencia, Departamento de Tecnologia de Alimentos, Instituto de Ingenieria de Alimentos para el Desarrollo, Camino de Vera s/n, 46022 Valencia, Spain, evdoan@tal.upv.es

- Isabel **Escriche**
  Universidad Politecnica de Valencia, Departamento de Tecnologia de Alimentos, Instituto de Ingenieria de Alimentos para el Desarrollo, Camino de Vera s/n, 46022 Valencia, Spain, iescrich@tal.upv.es

- Kjell **Francois**
  Ghent University, Department of Food Safety and Food Quality, Coupure Links 653, 9000 Gent, Belgium, Kjell.Francois@UGent.be

- Lynn **Frewer**
  Wageningen University, Marketing and Consumer Behaviour Group, Hollandseweg 1, 6706 KN Wageningen, The Netherlands, Lynn.Frewer@wur.nl

- Kriton **Grigorakis**
  12 Souliou str., 15232 Halandri, Athens, Greece, kgrigo@ath.hcmr.gr

- Stefanie **Gymnich**
  University Bonn, Institute of Animal Science, Preventive Health Management Group, Katzenburgweg 7-9, 53115 Bonn, Germany, sgymnich@uni-bonn.de

- Mogens **Jakobsen**
  Food Microbiology, Department of Food Science, The Royal Veterinary and Agricultural University, Rolighedsvej 30, 4, 1958 Frederiksberg C, Denmark, moj@kvl.dk

- Susanne **Knura**
  Institute of Animal Science, Preventive Health Management Group, Katzenburgweg 7-9, 53115 Bonn, Germany, s.knura@uni-bonn.de

- Krystof **Krygier**
  Agricultural University of Warsaw, Department of Food Technology, Nowoursynkowska 159c, 02-787 Warszawa, Poland, krygier@alpha.sggw.waw.pl

- Pieternel **Luning**
  Wageningen University, Product Design and Quality Management, Biotechnion building, P.O. Box 8129, 6700 EV Wageningen, The Netherlands, Pieternel.Luning@wur.nl

- Adriane **Mack**
  University Bonn, Institute of Animal Science, Preventive Health Management Group, Katzenburgweg 7-9, 53115 Bonn, Germany, a.mack@uni-bonn.de

- Anna **Maraz**
  Dept. of Microbiology and Biotechnology, Faculty of Food Science, Corvinus University of Budapest, Somloi ut 14-16, 1118 Budapest, Hungary, anna.maraz@uni-corvinus.hu

- Willem **Marcelis**
  Wageningen University, Management studies, Leeuwenborch building, P.O. Box 8130, 6700 EW Wageningen, The Netherlands, Willem.Marcelis@wur.nl

- Fulgencio **Marin**
  Universidad de Murcia, Facultad de Veterinaria, Tecnología de los Alimentos, Nutrición y Bromatología, Campus Universitario de Espinardo-Murcia, 30071 Murcia, Spain, fmarin@um.es

- Brigitte **Petersen**
  University Bonn, Institute of Animal Science, Preventive Health Management Group, Katzenburgweg 7-9, 53115 Bonn, Germany, b-petersen@uni-bonn.de

- Andreja **Rajkovic**
  Ghent University, Department of Food Safety and Food Quality, Coupure Links 653, 9000 Gent, Belgium, Andreja.Rajkovic@UGent.be

- Ewa **Rembialkowska**
  Organic Foodstuffs Division, Faculty of Human Nutrition and Consumer Sciences, Nowoursynowska 159c, 02-787 Warszawa, Poland, rembialk@alpha.sggw.waw.pl

- Jordi **Rovira**
  University of Burgos, Department of Biotechnology and Food Science, Plaza Misael Bañuelos s/n, 09001 Burgos, Spain, jrovira@ubu.es

- Eva **Santos**
  Centro de Investigaciones Químicas, Universidad Autónoma del Estado de Hidalgo, Pachuca. Hidalgo 42076, Mexico, emsantos@uaeh.reduaeh.mx

- Thomas **Schmitz**
  Plato Ag, Breite Str. 6-8, 23552 Lübeck, Germany, tschmitz@plato-ag.com

- Joachim **Scholderer**
  MAPP, Aarhus School of Business, Haslegaardsvej 10, 8210 Aarhus V, Denmark, sch@asb.dk

- Gereon **Schulze Althoff**
  GIQS e.V., c/o University Bonn, Institute of Animal Science, Katzenburgweg 7-9, 53115 Bonn, Germany, GAlthoff@uni-bonn.de

- Carmen **Socaciu**
  Department of Chemistry and Biochemistry, University of Agricultural Sciences and Veterinary Medicine, str.Manastur 3-5, 400372 Cluj-Napoca, Romania, csocaciu@usamvcluj.ro

- Jacques **Trienekens**
  Wageningen University, Management studies group, Leeuwenborch, P.O. Box 8130, 6700 EW Wageningen, the Netherlands, Jacques.Trienekens@wur.nl

- Bernd **van der Meulen**
  Wageningen University, Law and Governance, P.O. Box 8130, 6700 EW Wageningen, the Netherlands, Bernd.vanderMeulen@wur.nl

- Marjolein **van der Spiegel**
  Wageningen University, Product Design and Quality Management, Biotechnion building, P.O. Box 8129, 6700 EV Wageningen, The Netherlands, Marjolein.VanderSpiegel@wur.nl

- Menno **van der Velde**
  Wageningen University, Law and Governance, P.O. Box 8130, 6700 EW Wageningen, the Netherlands, menno.vandervelde@wur.nl

- Jack **van der Vorst**
  Wageningen University, Operations Research and Logistics group, Leeuwenborch, P.O. Box 8130, 6700 EW Wageningen, the Netherlands, Jack.vanderVorst@wur.nl

- Wim **Verbeke**
  Ghent University, Deparment of Agricultural Economics, Coupure links 653, 9000 Gent, Belgium, wim.verbeke@UGent.be

- Roland **Verhé**
  Ghent University, Department of Organic Chemistry, Coupure links 653, 9000 Gent, Belgium, Roland.Verhe@UGent.be